Theoretical and Mathematical Physics

The series founded in 1975 and formerly (until 2005) entitled *Texts and Monographs in Physics* (TMP) publishes high-level monographs in theoretical and mathematical physics. The change of title to *Theoretical and Mathematical Physics* (TMP) signals that the series is a suitable publication platform for both the mathematical and the theoretical physicist. The wider scope of the series is reflected by the composition of the editorial board, comprising both physicists and mathematicians.

The books, written in a didactic style and containing a certain amount of elementary background material, bridge the gap between advanced textbooks and research monographs. They can thus serve as basis for advanced studies, not only for lectures and seminars at graduate level, but also for scientists entering a field of research.

T0207205

Roger Balian

From Microphysics
to Macrophysics

Methods and Applications
of Statistical Physics

Volume II

Translated by D. ter Haar

With 90 Figures

 Springer

Roger Balian

CEA, Direction des Sciences de la Matière, Service de Physique Théorique de Saclay
F-91191 Gif-sur-Yvette, France and
Ecole Polytechnique, F-91128 Palaiseau, France

Translator:

Dirk ter Haar
P.O. Box 10, Petworth, West Sussex, GU28 0RY, England

Title of the original French edition: *Du microscopique au macroscopique,*
Cours de l'Ecole Polytechnique
©1982 Edition Marketing S.à.r.l., Paris; published by Ellipses

2nd Printing of the Hardcover Edition with ISBN 3-540-53599-3

Library of Congress Control Number: 2006933534

ISBN-10 3-540-45478-0 Springer Berlin Heidelberg New York
ISBN-13 978-3-540-45478-6 Springer Berlin Heidelberg New York

Springer is a part of Springer Science+Business Media
springer.com
© Springer-Verlag Berlin Heidelberg 1991, 2007

Typesetting: Data conversion by Springer-Verlag
Cover design: eStudio Calamar, Girona, Spain
Production: LE-TEX Jelonek, Schmidt & Vöckler GbR, Leipzig

Printed on acid-free paper   SPIN: 11870883   55/3100/YL   5 4 3 2 1 0

Preface

Although it has changed considerably in both coverage and length, this book originated from lecture courses at the Ecole Polytechnique. It is useful to remind non-French readers of the special place this institution occupies in our education system, as it has few features in common with institutes with a similar name in other parts of the world. In fact, its programme corresponds to the intermediate years at a university, while the level of the students is particularly high owing to their strict selection through entrance examinations. The courses put a stress on giving foundations with a balance between the various natural and mathematical sciences, without neglecting general cultural aspects; specialization and technological instruction follow after the students have left the Ecole. The students form a very mixed population, not yet having made their choice of career. Many of them become high-level engineers, covering all branches of industry, some devote themselves to pure or applied research, others become managers or civil servants, and one can find former students of the Ecole amongst generals, the clergy, teachers, and even artists and Presidents of France.

Several features of the present volume, and in particular its contents, correspond to this variety and to the needs of such an audience. Statistical physics, in the broadest meaning of the term, with its many related disciplines, is an essential element of modern scientific culture. We have given a comprehensive presentation of such topics at the advanced undergraduate or beginning graduate level. The book, however, has to a large extent moved away from the original lecture courses; it is not only intended for students, but should also be of interest to a wider public, including research workers and engineers, both beginning and experienced. A prerequisite for its use is an elementary knowledge of quantum mechanics and general physics, but otherwise it is completely self-contained.

Rather than giving a systematic account of useful facts for specialists in some field or other, we have aimed at assisting the reader to acquire a broad and solid scientific background knowledge. We have therefore chosen to discuss amongst the applications of statistical physics those of the greatest educational interest and to show especially how rich and varied these applications are. This is the reason why, going far beyond the traditional topics of statistical mechanics – thermal effects, kinetic theory, phase transitions, radiation laws – we have dwelt on microscopic explanations of the mechanical, magnetic, electrostatic, electrodynamic, ... properties of the various states of matter. Examples from other disciplines, such as astrophysics, cosmology, chemistry, nuclear physics, the quantum theory of measurement, or even biology, enable us to illustrate the broad scope of statistical physics and to show its universal nature. Out of a concern for culture, and also in trying to keep engineers and scientists away from too narrow a specialization, we have also included introductions to various physical problems arising in important technological fields, ranging from the nuclear industry to lighting by incandescent lamps, or from solar energy to the use of semiconductors for electronic devices.

Throughout this abundance we have constantly tried to retain a unity of thought. We have therefore stressed the underlying concepts rather than the technical aspects of the various methods of statistical physics. Indeed, one can see everywhere in the book under various guises two main guiding principles: on the one hand, the interpretation of entropy as a measure of disorder or of lack of information and, on the other hand, a stress on symmetry and conservation laws. At a time when excessive specialization tends to hide the unity of science, we have deemed it instructive to present unifying points of view, showing, for instance, that the laws of electrodynamics, of fluid dynamics, and of chemical kinetics all go back to the same underlying, basic ideas.

The French tradition, both in secondary education and in the entrance examinations to the Ecole Polytechnique, has to some extent given pride of place to mathematics. We have tried to benefit from this training by putting our treatment on a strict logical basis and giving our arguments a structured, often deductive, character. Mathematical rigour has, however, been tempered by a wish to present and to explain many facts at an introductory level, to avoid formalistic stiffness, and to discuss the validity of models. We have inserted special sections to present the less elementary mathematical tools used.

A first edition of this book was published in French in 1982. When the idea of publishing an English translation started to take shape, it seemed desirable to adapt the text to a broader, more international audience. The first changes in this direction brought about others, which in turn suggested a large number of improvements, both simplifications and more thorough discussions. Meanwhile it took some time to find a translator. Further lecture courses, especially one given at Yale in 1986, led to further modifications. One way or another, one thing led to another and finally there was little left of the original text, and a manuscript which is for more than eighty per cent new was finally translated; the present book has, in fact, only the general spirit and skeleton in common with its predecessor.

The actual presentation of this book aims at making it easier to use by readers ranging from beginners to experienced researchers. Apart from the main text, many applications are incorporated as exercises at the end of each chapter and as problems in the last chapter of the second volume; these are accompanied by more or less detailed solutions, depending on the difficulty.

At the end of each of the two volumes we give tables with useful data and formulae. Parts of the text are printed in small type; these contain proofs, mathematical sections, or discussions of subjects which are important but lie outwith the scope of the book. For cultural purposes we have also included historical and even philosophical notes: the most fundamental concepts are, in fact, difficult to become familiar with and it is helpful to see how they have progressively developed. Finally, other passages in small type discuss subtle, but important, points which are often skipped in the literature. Many chapters are fairly independent. We have also tried clearly to distinguish those topics which are treated with full rigour and detail from those which are only introduced to whet the curiosity. The contents and organization of the book are described in the introduction.

I am indebted to John Gregg and Dirk ter Haar for the translation. The former could only start on this labour, and I am particularly grateful to the main translator, Dirk ter Haar, for his patience (often sorely tried by me) and for the care bestowed on trying to present my ideas faithfully. I have come to appreciate how difficult it is to find exact equivalents for the subtleties of the French language, and to discover some of the subtleties of the English language. He has also accomplished the immense task of producing a text, including all the mathematical formulae, which could be used directly to produce the book, and which, as far as I can see, should contain hardly any misprints.

The Service de Physique Théorique de Saclay, which is part of the Commissariat à l'Energie Atomique, Direction des Sciences de la Matière, and in which I have spent the major part of my scientific research career, has always been like a family to me and has been a constant source of inspiration. I am grateful to all my colleagues who through many discussions have helped me to elaborate many of the ideas presented here in final form. They are too numerous to be thanked individually. I wish to express my gratitude to Jules Horowitz for his suggestions about the teaching of thermodynamics. As indicated in the preface to the first edition, I am indebted to the teaching staff who worked with me at the Ecole Polytechnique for various contributions brought in during a pleasant collaboration; to those mentioned there, I should add Laurent Baulieu, Jean-Paul Blaizot, Marie-Noëlle Bussac, Dominique Grésillon, Jean-François Minster, Patrick Mora, Richard Schaeffer, Heinz Schulz, Dominique Vautherin, Michel Voos, and Libero Zuppiroli, who to various degrees have helped to improve this book. I also express my thanks to Marie-Noëlle Bussac, Annie Gervois, Albert Lumbroso, Madeleine Porneuf, and Marcel Vénéroni, who helped me in the tedious task of reading the proofs and made useful comments, to Anne Désalos, Valérie Lambert, and Sylvie Zaffanella, who efficiently typed urgent matter, and to Dominique Bouly, who drew the figures. Finally, Lauris and the other members of my family should be praised for having patiently endured the innumerable evenings and weekends at home that I devoted to this book.

Roger Balian

Preface to the Original French Edition

The teaching of statistical mechanics at the Ecole Polytechnique used for a long time to be confined to some basic facts of kinetic theory. It was only around 1969 that Ionel Solomon started to develop it. Nowadays it is the second of the three physics "modules", courses aimed at all students and lasting one term. The first module is an introduction to quantum mechanics, while the last one uses the ideas and methods of the first two for treating more specific problems in solid state physics or the interaction of matter with radiation. The students then make their own choice of optional courses in which they may again meet with statistical mechanics in one form or another.

There are many reasons for this development in the teaching of physics. Enormous progress has been made in statistical physics research in the last hundred years and it is now the moment not only to reflect this in the teaching of future generations of physicists, but also to acquaint a larger audience, such as students at the Ecole Polytechnique, with the most useful and interesting concepts, methods, and results of statistical physics. The spectacular success of microscopic physics should not conceal from the students the importance of macroscopic physics, a field which remains very much alive and kicking. In that it enables us to relate the one to the other, statistical physics has become an essential part of our understanding of Nature; hence the desirability of teaching it at as basic a level as possible. It alone helps to unravel the meaning of thermodynamic concepts, thanks to the light it sheds on the nature of irreversibility, on the connections between information and entropy, and on the origin of the qualitative differences between microscopic and macroscopic phenomena. Despite being a many-faceted and expanding discipline with ill-defined boundaries, statistical physics in its modern form has an irreplaceable position in the teaching of physics; it unifies traditionally separate sciences such as thermodynamics, electromagnetism, chemistry, and mechanics. Last and not least its numerous applications cover a wide range of macroscopic phenomena and, with continuous improvements in the mathematical methods available, its quantitative predictions become increasingly accurate. The growth of micro-electronics and of physical metallurgy indicates that in future one may hope to "design" materials with specific properties starting from first principles. Statistical physics is thus on the way to becoming one of the most useful of the engineering sciences, sufficient justification for the growth of its study at the Ecole Polytechnique.

This book has evolved from courses given between 1973 and 1982 in the above spirit. The contents and teaching methods have developed considerably during that period; some subjects were occasionally omitted or were introduced as optional extras, intended only for a section of the class. Most of the major threads of statistical mechanics were reviewed, either in the course itself, or in associated problems. Nevertheless, on account of their difficulty, it has been possible to treat some important topics, such as irreversible processes or phase transitions, only partially, and to mention some of them, like superconductivity, only in passing. The published text contains all the material covered, suitably supplemented and arranged. It has been organized as a basic text, explaining first the principles and methods of statistical mechanics and then using them to explain the properties of various systems and states of matter. The work is systematic in its design, but tutorial in its approach; it is intended both as an introductory text to statistical physics and thermodynamics and as a reference book to be used for further applications.

Even though it goes far beyond the actual lecture programme, this is the text circulated to the students. Its style being half way between a didactic manual and a reference book, it is intended to lead the student progressively away from course work to more individual study on chosen topics, involving a degree of literature research. Typographically, it is designed to ease this transition and to help the first-time reader by highlighting important parts through italics, by framing the most important formulae, by numbering and marking sections to enable selective study, by putting items supplementary to the main course and historical notes in small type, and by giving summaries at the end of each chapter so that the students can check whether they have assimilated the basic ideas. However, the very structure of the book departs from the order followed in the lecture course, which, in fact, has changed from year to year; this is the reason why some exercises involve concepts introduced in later chapters.

Classes at the Ecole Polytechnique tend to be mixed, different students having different goals, and some compromises have been necessary. It is useful to take advantage of the mathematical leanings of the students, as they like an approach proceeding from the general to the particular, but it is equally essential that they are taught the opposite approach, the only one leading to scientific progress. The first chapter echoes this sentiment in using a specific example in order to introduce inductively some general ideas; it is studied at the Ecole as course work in parallel with the ensuing chapters, which provide a solid deductive presentation of the basis of equilibrium statistical mechanics. Courses at the Ecole Polytechnique are intended to be complemented later on by specialized further studies. When we discuss applications we have therefore laid emphasis on the more fundamental aspects and we have primarily selected problems which can be completely solved by students. However, we have also sought to satisfy the curiosity of those interested in more difficult questions with major scientific or technological implications, which are only qualitatively discussed. Conscious of the coherence of the book as a whole, we have tried to maintain a balance between rigour and simplicity, theory and fact, general methods and specific techniques. Finally, we have tried to keep the introductory approach of the book in line with modern ideas. These are based upon quantum statistical mechanics, richer in applications and conceptually simpler than its classical counterpart, which is commonly the first topic taught, upon the entropy as a measure of information missing because of the probabilistic nature of our description, and upon conservation laws.

Capable of being read at various different levels, and answering to a variety of needs, this course should be useful also outside the Ecole Polytechnique. Given its introductory nature and its many different purposes, it is not intended as a substitute for the more advanced and comprehensive established texts. Nevertheless, the latter are usually not easy reading for beginners on account of their complexity or because they are aimed at particular applications and techniques, or because they are aimed at an English-speaking audience. The ever increasing part played by statistical physics in the scientific background essential for engineers, researchers, and teachers necessitates its dissemination among as large an audience as possible. It is hoped that the present book will contribute to this end. It could be used as early as the end of undergraduate studies at a university, although parts are at the graduate level. It is equally well geared to the needs of engineering students who require a scientific foundation course as a passport to more specialized studies. It should also help all potential users of statistical physics to learn the ideas and skills involved. Finally, it is hoped that it will interest readers who wish to explore an insufficiently known field in which immense scientific advances have been made, and to become aware of the modern understanding of properties of matter at the macroscopic level.

Physics teaching at the Ecole Polytechnique is a team effort. This book owes much to those who year after year worked with me on the statistical mechanics module: Henri Alloul, Jean Badier, Louis Behr, Maurice Bernard, Michel Bloch, Edouard Brézin, Jean-Noël Chazalviel, Henri Doucet, Georges Durand, Bernard Equer, Edouard Fabre, Vincent Gillet, Claudine Hermann, Jean Iliopoulos, Claude Itzykson, Daniel Kaplan, Michel Lafon, Georges Lampel, Jean Lascoux, Pierre Laurès, Guy Laval, Roland Omnès, René Pellat, Yves Pomeau, Yves Quéré, Pierre Rivet, Bernard Sapoval, Jacques Schmitt, Roland Sénéor, Ionel Solomon, Jean-Claude Tolédano, and Gérard Toulouse, as well as our colleagues Marcel Fétizon, Henri-Pierre Gervais, and Jean-Claude Guy from the Chemistry Department. I have had the greatest pleasure in working with them in a warm and friendly environment, and I think they will excuse me if I do not describe their individual contributions down the years. Their enthusiasm has certainly rubbed off onto the students with whom they have been in contact. Several of them have given excellent lectures on special topics for which there has regrettably not been room in this book; others have raised the curiosity of students with the help of ingenious and instructive experiments demonstrated in the lecture theatre or classroom. This book has profited from the attention of numerous members of the teaching staff who have corrected mistakes, simplified the presentation, and thought up many of the exercises to be found at the end of the chapters. Some have had the thankless task of redrafting and correcting examination problems; the most recent of these have been incorporated in the second volume. To all of them I express my heartfelt thanks. I am especially indebted to Ionel Solomon: it is thanks to his energy and dedication that the form and content of the course managed to evolve sufficiently rapidly to keep the students in contact with live issues. On the practical side, the typing was done by Mmes Blanchard, Bouly, Briant, Distinguin, Grognet, and Lécuyer from the Ecole's printing workshop, efficiently managed by M. Deyme. I am indebted to them for their competent and patient handling of a job which was hampered by the complexity of the manuscript and by numerous alterations in the text. Indeed, it is their typescript which, with some adjustments by the publisher, was reproduced for the finished work. The demanding and essential task of proofreading was performed by Madeleine Porneuf from our group at Saclay. I also thank the staffs of the Commissariat à l'Energie Atomique and of the Ecole Polytechnique, in particular, MM. Grison, Giraud, Servières, and Teillac for having facilitated publication. Finally, I must not forget the many students who have helped to improve my lectures by their criticism, questions, and careful reading, and from whose interest I have derived much encouragement.

Roger Balian

Contents of Volume II

Contents of Volume I

10. Quantum Gases Without Interactions

"La preuve de fait de Leibnitz était que, se promenant un jour dans le jardin de l'évêque de Hanovre, on ne put jamais trouver deux feuilles d'arbre indiscernables."

Voltaire

"L'un dit une chose, l'autre allait justement dire la même chose et répète cette même chose. Il semble qu'il était impossible de parler autrement. On est strictement jumeaux. Se distinguer, on n'y songe plus. Identité! Identité!"

Henri Michaux, La nuit remue

"Quand nous voulons nous mêler, nos élans l'un vers l'autre ne font que nous heurter l'un à l'autre."

Maupassant, Solitude

"Soubdain, Panurge, sans aultre chose dire, jette en pleine mer son mouton criant et bellant. Tous les aultres moutons, crians et bellans en pareille intonation, commencerent soy jecter et saulter en mer après, à la file. La foulle estoit à qui premier y saulteroit après leur compaignon. Possible n'estoit les en guarder, comme vous sçavez estre du mouton le naturel, tous jours suyvre le premier, quelque part qu'il aille."

Rabelais, Quart Livre

So far we have studied gases and liquids; in those states of matter quantum mechanics plays a hidden rôle, mainly through the internal structure of the molecules. The classical approximation was sufficient to treat the translational degrees of freedom of the molecules, both for gases (Chap.7), and for liquids (Chap.9). However, we saw in Chap.8 that the progressive freezing in of the internal degrees of freedom made it necessary for us to treat them quantum mechanically when the gas cools off. We expect that the same must also occur for the translational degrees of freedom of molecules, since the entropy of a classical gas would become negative if the temperature were lowered sufficiently, and this is impossible.

The *freezing in of the translations* gives rise to a large number of macroscopic effects where quantum mechanics plays a much more direct rôle than for gases and liquids. For most substances it manifests itself in the *crystalliza-*

tion of the sample; the characteristic temperature associated with translations is the solidification temperature. Therefore, solids (Chap.11) are *macroscopic quantum systems*. In Chaps. 12 and 13, we shall study other macroscopic quantum systems: liquid helium at low temperatures and electromagnetic radiation at equilibrium in a container. Further examples, such as white dwarfs in astrophysics and magnetism, will be treated in the Problems section.

For most of these applications we use simplified models of *quantum gases of non-interacting indistinguishable particles*. The present chapter is devoted to the formalism suitable for such gases and to its simplest applications. We shall start reminding ourselves of what quantum mechanics says about the *Pauli principle* (§ 10.1), and shall mention several physical phenomena where the indistinguishability of the constituent particles leads to remarkable effects, either on a microscopic or on a macroscopic scale (§ 10.1.4).

The formalism involving wavefunctions is not suitable for a description of systems consisting of indistinguishable particles, because it requires the labelling of these particles. We present in § 10.2 an alternative, more adequate, representation, due to Fock, in which the particles remain anonymous, as should be the case. This representation is essential in quantum field theory, as it accounts easily for the creation or annihilation of particles. It is also particularly suited for our purpose, since it is convenient for dealing with large numbers of non-interacting particles.

We shall thus be in a position to study quantum gases without interactions in grand canonical equilibrium, first in general terms (§ 10.3), and then by treating examples of gases consisting of *fermions* (§ 10.4) or of *bosons* (§ 10.5). We shall stress the differences between these two kinds of quantum gases, and we shall discuss both the classical limit (§ 10.3.4) and the large volume limit (§ 10.3.3).

Note that the "particles" which we are considering here are not necessarily elementary particles such as electrons, nor even composite particles such as atomic nuclei. We shall see, when constructing models to describe some material or another – for instance, crystals in Chap.11 – that we may be dealing more generally with elementary entities which behave practically as non-interacting objects, and which may have lost the features of the constitutive elementary particles.

10.1 The Indistinguishability of Quantum Particles

10.1.1 Exchange of Two Identical Particles

When two particles have the same characteristics, such as mass, charge, and so on, there is no way one can distinguish them; this is true both in classical and in quantum mechanics. Nevertheless, indistinguishablity has a more subtle character and more far-reaching consequences in quantum mechanics, as we can see from the following example. Consider a collision between two identical point particles which interact through a finite-range potential. In classical dynamics we could think of following their trajectories; if we label them 1 and 2 before the collision, it is then possible to find out after the collision which of the emerging particles was previously labelled 1. In quantum mechanics, if before the collision the two incident particles are represented by wavepackets which are separated in space, one can still label them 1 and 2; however, as soon as the particles approach at distances smaller than the spatial extension of the wavepackets, their individuality is lost, and it becomes impossible to know finally what was the original label of a particular particle which emerges from the region where the collision occurred. Hence, labelling identical particles is in quantum mechanics not only arbitrary, but also inadequate, as the particles can exchange their labels. This is the reason why in § 10.2 we shall construct a formalism which allows us to preserve the anonymity of indistinguishable particles.

Let us retain for the time being the more usual formalism of wavefunctions; we write $\psi(1,2)$ for the wavefunction of a pair of particles, denoting by 1 (or 2) the space and, possibly, also the spin coordinates of the first (or the second) particle. The exchange operator $\widehat{E}_{12}$ transforms the wavefunctions as follows:

$$\widehat{E}_{12}\,\psi(1,2) \;\equiv\; \psi(2,1), \tag{10.1a}$$

interchanging the labels of the particles. A physical observable $\widehat{A}$ cannot distinguish them, which means that the expectation value of $\widehat{A}$ over any function $\psi(1,2)$ is the same as its expectation value over $\psi(2,1)$; we can express this by the equation $\widehat{E}_{12}^{\dagger}\widehat{A}\widehat{E}_{12} = \widehat{A}$. As $\widehat{E}_{12}$ is Hermitean and its square is the unit operator, it follows that *each physical observable must commute with the exchange operator*. The momentum of one particle, its position, the operator $\widehat{E}_{12}$ itself, are *not* observables associated with physical quantities – even though they are Hermitean operators – whereas the total momentum or the position of the centre of mass of the pair are such observables.

The Hamiltonian $\widehat{H}$ which generates the evolution with time of the wavefunction $\psi(1,2)$ cannot distinguish the particles, and thus commutes with $\widehat{E}_{12}$. We can therefore diagonalize these two operators at the same time, so that amongst the quantum numbers which characterize the eigenstates of $\widehat{H}$ we have the eigenvalue η of $\widehat{E}_{12}$. As $\left(\widehat{E}_{12}\right)^{2} = \widehat{I}$, we have $\eta^{2} = 1$. The eigenfunctions of $\widehat{H}$, which also satisfy the equation

$$\widehat{E}_{12}\,\psi(1,2) \;=\; \eta\,\psi(1,2) \tag{10.1b}$$

with $\eta = \pm 1$, fall therefore into two classes: those which are symmetric ($\eta = +1$), and those which are antisymmetric ($\eta = -1$) under the exchange of the labels 1 and 2 of the two particles.

One would expect *a priori* to find the two kinds of levels with $\eta = \pm 1$ by analyzing spectra of systems which consist of two identical particles. Let us therefore consider a nitrogen molecule, N_2. In the far infra-red, neither the electronic nor the vibrational motions take part in the corresponding transitions (§ 8.4.1); one expects thus that the only relevant energy levels of $\widehat{H}$ are the lowest rotational levels, given by

$$\varepsilon_l \;=\; \varepsilon_0 + l(l+1)\,\frac{\hbar^2}{2\mathcal{I}}, \tag{10.2}$$

with a multiplicity equal to $2l+1$. The corresponding eigenfunctions have the form $R(\rho)Y_{lm}(\theta,\varphi)$, where ρ, θ, φ are the relative spherical coordinates of the two nuclei (§ 8.4.4); they are symmetric with respect to the exchange $1 \leftrightarrow 2$ when l is even, and they are antisymmetric when l is odd ($\eta = (-)^l$). However, even though experiments show transitions between the levels (10.2) for even l, the odd l levels are never observed, at least not for natural nitrogen which consists of the ^{14}N isotope. In contrast, if one considers $^{14}N^{15}N$ molecules which consist of two different isotopes, the whole of the spectrum (10.2) is seen. The indistinguishability of the nuclei of ^{14}N thus has the effect of forbidding the states with $\eta = -1$. (We shall see in § 10.1.3 why we may here neglect the electrons.)

On the other hand, let us consider the spectrum of a helium atom consisting of two electrons in the field of the He nucleus which we assume to be fixed. The Hamiltonian depends little on the spin of the electrons so that we can classify its 4 lowest eigenstates, which are practically degenerate, as forming one spin singlet and one spin triplet. The orbital wavefunction is the same for these four states; it is isotropic and symmetric under the exchange of the coordinates of the two electrons, so that we have $\eta = -1$ for the singlet state, which is antisymmetric under the exchange of the two spins, and $\eta = +1$ in the triplet states. Experimentally we find in this case that only the $\eta = -1$ state exists; this shows up, for instance, in the absence of magnetism – an external magnetic field has practically no effect on a singlet, but it would split the components of a triplet which have a total spin 1.

In these two cases one notices a remarkable property of quantum systems of indistinguishable particles: only $\eta = +1$ states are allowed for the ^{14}N nuclei, and only $\eta = -1$ states for electrons. Such a *restriction on the possible wavefunctions has no equivalent in classical mechanics*; it is one of the fundamental principles of quantum physics and we shall now state it in a general form.

10.1.2 The Pauli Principle

A considerable amount of experimental evidence, of which we just gave two examples and which we shall survey in § 10.1.4, supports the *symmetrization principle*. As we mentioned in § 2.1.5, this principle completes the principles of quantum mechanics given in § 2.2.7. We can state it as follows.

> The particles which make up any physical system can be classified either as *bosons* or as *fermions*. If one *exchanges the* (spatial and, if applicable, spin) *coordinates of two identical particles*, any wavefunction of the system remains *invariant, if we are dealing with bosons, or changes its sign, if we are dealing with fermions*. The Hilbert space $\mathcal{E}_H$ associated with the system therefore does not contain all possible wavefunctions, but only functions which are *symmetric* (bosons) or *antisymmetric* (fermions) under any exchange of the labels of two indistinguishable particles. Moreover, *bosons* have *integral spin* (usually 0 or 1) and *fermions* have *half-odd-integral spin* (usually $\frac{1}{2}$).

Thus, every wavefunction of a system of N indistinguishable bosons must remain invariant under any permutation of the labels of the N particles. For a system of fermions, it must be multiplied by $(-)^P$, where P is the parity of the permutation, that is, the number *modulo* 2 of exchanges of particles which are necessary to produce that permutation (note that this number, though itself not uniquely defined, has a well defined parity). This property is consistent with the identity $\widehat{E}_{23} = \widehat{E}_{12}\widehat{E}_{13}\widehat{E}_{12}$ which shows that a function which is odd under the exchanges $1 \leftrightarrow 2$ and $1 \leftrightarrow 3$ is also odd under the exchange $2 \leftrightarrow 3$. The signature η defined by (10.1), $\eta = +1$ for bosons, $\eta = -1$ for fermions, is clearly conserved during the time evolution; this is another consistency requirement.

Let q denote the set of quantum numbers characterizing a single-particle wavefunction (for examples see § 10.2.1), and let $q(1)$ denote this wavefunction in the single-particle Hilbert space $\mathcal{E}_H^{(1)}$; we use in $q(1)$ the single symbol 1 to denote the set of space and possibly spin coordinates of a particle with label 1. The product $q_1(1)q_2(2)\ldots q_N(N)$, where $q_1, q_2, \ldots, q_N$ are N (different or not) single-particle states, is an allowed wavefunction for a system of N *distinguishable* particles. However, this product which is an element of the product Hilbert space $\left[\otimes\mathcal{E}_H^{(1)}\right]^N$ (§ 2.1.1) does, in general, possess no simple symmetry properties under the exchange of particle labels, and it can therefore not be accepted as a wavefunction for N *indistinguishable* particles. Nonetheless, we can derive from $q_1(1)q_2(2)\ldots q_N(N)$ a family of $N!$ wavefunctions – which are all different, if $q_1, q_2, \ldots, q_N$ are different – by taking all possible permutations of the particle labels. Their (normalized) sum describes a state which is symmetric under any exchange, and it can thus be accepted for N bosons. When the quantum numbers $q_1, q_2, \ldots, q_N$ are all different, one can similarly construct, by adding all possible $N!$ permutations, but in this case each multiplied by the appropriate factor $(-)^P$, a state

which is antisymmetric under any exchange and which is thus acceptable for
N fermions; this state can alternatively be written in the form of a *Slater determinant*:

$$\frac{1}{\sqrt{N!}} \begin{vmatrix} q_1(1) & q_1(2) & \cdots & q_1(N) \\ q_2(1) & q_2(2) & \cdots & q_2(N) \\ \vdots & \vdots & \ddots & \vdots \\ q_N(1) & q_N(2) & \cdots & q_N(N) \end{vmatrix}. \tag{10.3}$$

This *symmetrization* or *antisymmetrization* procedure generates a *complete base* in the Hilbert space associated with either N bosons or N fermions, provided the base q of single-particle states, which was our starting point, itself is a complete base in $\mathcal{E}_H^{(1)}$. We shall use this method in § 10.2.2 to construct the states of systems with arbitrary numbers of indistinguishable particles, starting from the single-particle states.

The principle just given, which expresses in mathematical language in the wavefunction formalism the properties connected with the indistinguishability of the particles, is usually called the Pauli principle. Actually, Pauli himself stated only the *exclusion principle* for fermions. Less abstract than the principle of antisymmetric wave functions, but equivalent, the exclusion principle *forbids two fermions to have the same quantum numbers q*: in Eq.(10.3) the single-particle states $q_1, q_2, \ldots, q_N$ must all be different in order that one can construct an antisymmetric function. The indistinguishability thus forces identical fermions to be *correlated by mutual exclusions*.

For bosons, the symmetry of the wave functions also imposes correlations, but in a more subtle manner. Let us consider two orthogonal single-particle states, q and q'. Two *distinguishable* particles can occupy them independently so that the two-particle states are linear combinations of the 4 orthogonal states

$$q(1)q(2), \qquad q(1)q'(2), \qquad q'(1)q(2), \qquad q'(1)q'(2). \tag{10.4}$$

Two *fermions* exclude one another and thus can only be in the state

$$\frac{1}{\sqrt{2}}\left[q(1)q'(2) - q'(1)q(2)\right]. \tag{10.5}$$

For two *bosons*, the symmetrization procedure produces, starting from (10.4), on the one hand, the two states

$$q(1)q(2), \qquad q'(1)q'(2), \tag{10.6a}$$

and, on the other hand, the state

$$\frac{1}{\sqrt{2}}\left[q(1)q'(2) + q'(1)q(2)\right]. \tag{10.6b}$$

The numbering of the micro-states is therefore different for bosons, for fermions, or for distinguishable particles. In particular, we should note that, corresponding to the single state (10.6b) for bosons, there are two states, $q'(1)q(2)$ and $q(1)q'(2)$ for distinguishable particles. As a result, if the micro-states considered above in (10.6), (10.5), or (10.4) are equiprobable, two bosons have 2 chances out of 3 of having the same quantum numbers, either q or q', whereas this chance is 0 for two fermions, and 1 out of 2 for two distinguishable particles. The bosons are thus correlated in a way which is the opposite of that for fermions: they *behave gregariously*, as the presence of one boson in a single-particle state q produces a preference for another boson to occupy the same state.

For an arbitrary number of particles, one calls the different ways of numbering the micro-states "statistics", respectively, *Bose-Einstein statistics* ($\eta = +1$), *Fermi-Dirac statistics* ($\eta = -1$), and *Boltzmann statistics* (distinguishable particles). In order to study the thermal equilibrium of a gas of indistinguishable particles we must thus calculate Z as a sum over symmetric (boson) or antisymmetric (fermion) states, rather than include all eigenstates of the Hamiltonian; the numbering of the symmetric or antisymmetric states will thus be crucial.

Wolfgang Pauli (Vienna 1900–Zürich 1958) stated the exclusion principle in 1925 for systems of electrons and this was the discovery[1] for which he was awarded the Nobel Prize in 1945. The principal experimental fact on which he relied was the periodic classification of the chemical elements, which had been known since the work by Dmitri Ivanovich Mendeleev (Tobolsk 1834–St Petersburg 1907) in 1869, but had remained unexplained. On the other hand, the indistinguishability of particles had equally been for a long time a problem in the field of statistical mechanics. Gibbs's paradox (see §§ 3.4.3 and 8.2.1) had shown up a difficulty in classical physics: as Gibbs discussed in his book "Elementary Principles in Statistical Mechanics", published in 1902, in classical mechanics it is not clear whether or not one must consider two configurations which go over into one another through a permutation of identical particles as being the same (§ 10.1.1). The answer to that question was only given by the Pauli principle; as we shall see in § 10.3.4 for the case of a gas, it implies that the common classical limit of the Bose-Einstein and Fermi-Dirac statistics yields the classical measure (2.55), including the factor $1/N!$, a factor which is essential to resolve the Gibbs paradox and which the old Boltzmann statistics could not justify.

The first quarter of the twentieth century has seen heavy traffic between the emerging quantum mechanics (Bohr atom 1913, L. de Broglie's thesis 1925, Schrödinger equation 1926) and the statistical mechanics of systems consisting of a large number of indistinguishable particles. The first quantum theory which gave an explanation for the form of the black-body radiation spectrum (Planck 1900, Einstein 1905; see §§ 3.4.4, 13.2.2 and 13.3.1) led after a while to the concept of a photon, a particle which does not obey Boltzmann statistics. Important stages were the interpretation of the photoelectric effect[2] and the analysis of exchange of energy between atoms and radiation[3] by Einstein, and then the proof by Satyendra Nâth

---

[1] Zs. Physik **31** (1925) 765–783.

[2] Ann. Physik (Leipzig) **17** (1905) 132.

[3] Zs. Physik **19** (1917) 301.

Bose (Calcutta 1894–1974) that a quantized mode of electromagnetic oscillations in a cavity is equivalent to a system of photons[4]. However, notwithstanding Einstein's papers in 1924–5 on the new Bose-Einstein statistics which is obeyed by a gas of photons, the *raison d'être* of these statistics was still obscure in 1925 (as Einstein said, "von vorläufig ganz rätselhafter Art"). At about the same time, Pauli was led by the study of atoms and molecules to recognize the exclusion principle for electrons. Enrico Fermi (Rome 1901–Chicago 1954) discovered soon after that[5] the important consequences of this principle for the electron gas in metals (§ 10.4.1); we shall come across his name many times in Chaps.10 and 11. It was left to Paul Adrien Maurice Dirac (Bristol 1902–Tallahassee 1984) to show[6] how one can understand these results by connecting the indistinguishability of particles with their wave nature and with the symmetry character of the wavefunctions.

Finally, the relation between spin and statistics, which up to that time was regarded as an empirical fact, was *proved* in 1939–40 by Pauli and Belinfante. The (difficult) proof is based upon relativistic field theory and, in particular, uses the fact that measurements on two particles in space-time points which are lying at a space-like distance from one another cannot perturb one another. Bearing in mind the simplicity of the result, it is remarkable that so far it has been impossible to find a simple proof.

10.1.3 Bosons and Fermions; Elementary and Composite Particles

Amongst the particles which one encounters most often, the *photon* is a *boson* with zero mass and spin 1 (with a particular property which we shall discuss in § 13.1.4); the *electron*, the *proton*, and the *neutron* are *fermions* with spin $\frac{1}{2}$.

One often is dealing with *composite particles* whose internal structure plays no rôle at the scale considered. For instance, the ^{14}N nucleus in the nitrogen molecule (§ 10.1.1) behaves like a single particle, consisting of 7 protons and 7 neutrons bound together by nuclear forces. Exchanging two such nuclei amounts to exchanging at the same time the 7 protons and 7 neutrons which make them up. This does not change the wavefunction, as the number of fermions which is exchanged is even; in this way one is led to treat the ^{14}N nucleus as a boson in atomic, molecular, or solid state physics, and to regard it then as a single point particle, which behaves as if it were elementary. Of course, its composite nature shows up at the nuclear scale of fermis. More generally, if a particle consists of an *odd number of fermions*, possibly together with any number of bosons, it behaves as a *fermion*; if it consists of an *even number of fermions*, it behaves as a *boson*. This rule is compatible with the relation between spin and statistics, as one obtains the total spin of a composite particle by taking the vector sum of the

---

[4] Zs. Physik **26** (1924) 178.
[5] Zs. Physik **36** (1926) 902.
[6] Proc. Roy. Soc. (London) **A112** (1926) 661.

spins of the constituents, which is integral for bosons and half-odd-integral for fermions, together with the orbital angular momentum, which is integral.

In particular, a *nucleus* is a boson or a fermion according to the parity of the number of its nucleons, the protons and neutrons. A (neutral) *atom* is a boson or a fermion according to the parity of the number of neutrons in its nucleus, since the number of protons is equal to the number of electrons. Thus, ^{4}He atoms are bosons, whereas atoms of the ^{3}He isotope are fermions; this has important consequences which we shall examine in Chap.12.

The situation is slightly less simple in the case of *molecules* since these can be built up from several kinds of indistinguishable particles, atomic nuclei and electrons. For instance, any wavefunction of a nitrogen molecule $^{14}N_2$ must be symmetric under the exchange of its two nuclei and antisymmetric with respect to its 14 electrons. In the Born-Oppenheimer approximation (§§ 8.4.1 and 8.4.4) an eigenfunction ψ of $\widehat{H}$ can be factorized into a product (i) of an electronic part ψ_e which depends directly on the electron coordinates and involves indirectly the coordinates R_1 and R_2 of the nuclei as parameters and (ii) a purely nuclear part, $\psi_n(R_1, R_2)$, which was studied in § 8.4.4. For the lowest lying states, which are sufficient to determine the infrared spectrum and the gas properties at thermal equilibrium, the electron cloud is frozen in into its ground state ψ_e which is completely antisymmetric with respect to the electrons and invariant under the exchange of the two attractive centres R_1 and R_2. As a result, in the study of the symmetry character of ψ it is legitimate, on the one hand, to forget about the electrons and, on the other hand, to forget about the R_1- and R_2-dependence of ψ_e; it is therefore sufficient to consider ψ_n, requiring it to be symmetric under the exchange of the two ^{14}N nuclei, and antisymmetric in the case of a $^{15}N_2$ molecule. This argument justifies the considerations of § 10.1.1 where we restricted ourselves implicitly to the nuclear part ψ_n of the wavefunction, the symmetry of which is governed in final reckoning by $Y_l^m(\theta, \varphi)$. The Pauli principle then implies that l must be even for $^{14}N_2$; in the case of $^{15}N_2$ the antisymmetry also involves the spin-$\frac{1}{2}$ of the nuclei, and l takes on odd values for the spin triplet and even values for the singlet. Note that the exchange operation refers here to the ^{14}N *nuclei* which are bosons, and not to the *atoms* which are fermions. A simplistic argument where the nitrogen molecule is thought to be a bound state of two composite fermions, the ^{14}N atoms, and where one neglects the antisymmetry with respect to the constituent electrons and the symmetry with respect to the nuclei, would lead to a wrong result: requiring the wavefunction to be antisymmetric with respect to the atoms one would find l to be odd, rather than even, for all levels of $^{14}N_2$. Indeed, one cannot regard a nitrogen molecule as just the combination of two ^{14}N fermions, because the tight binding of the two atoms has radically changed the electron wavefunction; the valence electrons no longer belong to one atom or the other, but to both.

As illustrated by this example, we must be careful when defining a composite particle and assigning to it the character of a boson or of a fermion, since the Pauli principle only applies to the global exchange of two particles, each of which must behave *as a whole*. In principle, the analysis of a complex system, such as a piece of matter or even a molecule, should start from its most elementary constituents, which are unambiguously fermions or bosons. Nevertheless, in order to make the structure and the properties of this system intelligible, it is necessary to bring into the discussion composite entities which behave like weakly interacting particles. The composite particles introduced in this way are always the result of *approximations*: such objects have an individual nature only if their size is small as compared to the characteristic dimensions of the system which we try to describe through them, and if their binding energy is large as compared to the characteristic energies of the system. For instance, it is legitimate to describe liquid helium as an assembly of bosons, the He atoms, which are small and strongly bound, forgetting about the electrons and nuclei which are their constituents. On the other hand, we have just seen that it is incorrect to describe a nitrogen molecule as being formed from two fermions, the N atoms, and that we must go back to the nitrogen nuclei and the electrons, at least the valence electrons: even though a nitrogen atom can exist on its own and consists of an odd number of fermions, it does not behave like a fermion in the nitrogen molecule.

In *solid state physics* one introduces various kinds of "quasi-particles", which are elementary excitations of the material. One is dealing with composite entities in which often a large number of elementary particles, electrons and atomic nuclei, are involved, and this gives them a pronounced collective nature. For instance, we shall discuss in § 11.3.2 conduction electrons and holes, which are quasi-particles used to describe an insulator or a semiconductor; they are fermions, consisting of an extra or a missing electron, together with the perturbation which it produces in its vicinity. Other quasi-particles, formed by binding an electron and a hole together, then behave like a boson, provided their size is suffiently small and their binding energy sufficiently large. For instance, the "Cooper pairs" which are pairs of bound electrons responsible for superconductivity in many metals at low temperatures (§ 12.3.3) behave only very approximately as bosons and retain some memory of the exclusion principle of the constituent electrons, due to their small binding energy, which is typically 10^{-3} eV, and their large size, which is typically 1 μm.

On subnuclear scales, *elementary particle theory* helps us to classify these particles by letting bosons and fermions play different rôles. One distinguishes two kinds of elementary fermions: on the one hand, the *quarks* and, on the other hand, the *leptons* – electron, μ, and τ particles and their associated neutrinos – with each particle having its own antiparticle. The elementary bosons are responsible for the interactions between the fermions, namely, the *photon* for the electromagnetic interactions, the *weak bosons* Z^0, $W^\pm$ for the weak interactions, and the *gluons* for the strong interactions. The attraction between quarks due to the exchange of gluons is so strong that one never observes free quarks or free gluons, but only states with several quarks which are bound together by the gluons and which are called *hadrons*. Amongst those, the *baryons* appear above the fm scale as composite particles, consisting of 3 quarks – plus an undetermined number of gluons and quark-antiquark pairs; they therefore are fermions. In particular, the proton and the neutron have, respectively, the structures uud and udd, where u and d denote the so-called up and down quarks. The *mesons*, such as, for example, π and K consist of a quark-antiquark pair and are thus bosons. The prediction by Pauli and Fermi in 1933 of the existence of the neutrino was connected, on the one hand, with the necessity to explain the energy balance in β-radioactivity and, on the other hand, with a consequence of the Pauli principle, the conservation of the parity of the number of fermions in weak interactions. In fact, the angular momentum is integral for an even number of fermions and half-odd-integral for an odd number; hence, angular momentum could not be conserved if, for instance, the decay of the neutron followed the scheme $n \rightarrow p + e$, rather than producing a neutrino according to the scheme $n \rightarrow p + e + \bar{\nu}_e$. In terms of more elementary particles, this reaction is nowadays interpreted as $d \rightarrow u + W^- \rightarrow u + e + \bar{\nu}_e$, where the W^- intermediate boson can be transformed into the fermion-antifermion pairs d, $\bar{u}$ or e, $\bar{\nu}_e$.

10.1.4 Effects Connected with Indistinguishability

Very many phenomena, both at the microscopic and at the macroscopic scale, are explained by the Pauli principle. To demonstrate the extent of the field of applications of the latter we shall review several examples, giving a quick explanation. Other examples will be treated in more detail in later chapters and in the problem section.

The Periodic Table of the Chemical Elements. Rydberg noted in 1913 that the different periods in Mendeleev's table had the length $2n^2$ (where $n = 1$, 2, 3, or 4), a formula considered "cabalistic" by Sommerfeld. Pauli explained in a simple way the existence of the periodicity and the length of the periods by noting that the eigenstates of an electron in the central potential of an atom are grouped in shells separated by wide energy gaps of order 10 eV, with a multiplicity $2n^2$, where the factor 2 comes from the spin and the $n^2 = \sum_{l=0}^{n-1} (2l + 1)$ from the orbital angular momentum; the electrons fill successively each shell and elements with the same number of electrons in the last shell behave similarly. The situation is somewhat similar in the case of *the structure of the nuclei*; in that case one must, however, take into account the presence of two kinds of spin $\frac{1}{2}$ nucleons, the protons and neutrons, rather than one kind for the atoms, the electrons. This produces "nuclear magic numbers", which are different from the atomic ones.

Instability of Nuclei with a Large Neutron Excess. The ^{14}N and ^{15}N isotopes are stable, whereas ^{16}N, ^{17}N, ^{18}N, ^{19}N are unstable, with lifetimes decreasing from 7 s to 0.4 s, and there is no nitrogen nucleus with more than 19 nucleons. This is a quite general situation. Nevertheless, nuclear forces are attractive for neutrons, which should favour an increasing stability when the number of neutrons increases. To solve this paradox, we note that the number of single-neutron bound states in a nucleus is finite, and the exclusion principle limits the number of neutrons which can be accommodated in these states.

β-radioactivity of Nuclei. An isolated neutron is β-unstable, according to $n \to p + e + \bar{\nu}_e$, with a lifetime of 12 s. However, the neutrons in a nucleus such as ^{4}He do not show β-decay. This is due to the fact that the protons are fermions. In fact, the lowest ($s\frac{1}{2}$) shell of ^{4}He is completely occupied by two protons with opposite spin and two neutrons with opposite spin. In order that one of the latter decays, it must emit a proton which must go to an unoccupied level, that is, to an excited shell. However, the energy difference between the occupied and the first free proton shells is larger than the β-decay energy of the neutron; the decay is therefore forbidden.

Chemical Binding. Let us consider the example of H_2 and let us begin by neglecting the interaction between the electrons. Each electron sees the effective potential of the two nuclei so that its lowest bound state, which is two-fold degenerate because of the spin, has a lower energy than that of the electron in an H atom. One can thus, by putting the two electrons in this pair of bound states, form a molecule with an energy which is lower than that of the two separated atoms, provided the two nuclei are not too close; in this case the repulsion between them and the repulsion between the electrons would lose the energy that we have just gained through the effect discussed here. The Pauli principle prevents the same effect to produce HeH, as there are then 3 electrons available for only one pair of single-electron lowest bound states.

Ferromagnetism. As Exerc.9a shows, ferromagnetism can be explained by starting from the idea that matter consists of interacting spins with the contribution to the energy from a pair of spins $\boldsymbol{\sigma}_1$ and $\boldsymbol{\sigma}_2$ having the form $-J(\boldsymbol{\sigma}_1 \cdot \boldsymbol{\sigma}_2)$. The Curie transition temperature, below which the material is ferromagnetic, is thus of order of magnitude J/k. Nevertheless, the direct magnetic interaction,

$$\frac{\mu_0}{4\pi r^3}\left[3(\boldsymbol{\mu}_1 \cdot \boldsymbol{u})(\boldsymbol{\mu}_2 \cdot \boldsymbol{u}) - (\boldsymbol{\mu}_1 \cdot \boldsymbol{\mu}_2)\right],$$

between the two moments $\boldsymbol{\mu} = \mu_B \boldsymbol{\sigma}$ of the spins, where $\boldsymbol{u}$ denotes the unit vector in the direction connecting the two spins, r their distance apart, and where $\mu_B = e\hbar/2m = 0.927.10^{-23}$ J T^{-1}, does not have the necessary features. Not only does it change sign depending on the relative position of the spins, but especially its numerical value is too small, as $\mu_0 \mu_B^2 / 4\pi r^3 k$ in temperature units reaches only 0.6 K for a distance of 1 Å. However, there are many ferromagnetic materials at room temperature. In fact, it is an indirect combined effect of the *Pauli principle* and the *Coulomb repulsion* between electrons which produces the "exchange force", an effective force between spins which is much stronger than their magnetic interaction. The antisymmetry of the wavefunction of two electrons can be realized in two different ways. Either the spatial wavefunction is antisymmetric and the spins of the electrons are in one of the symmetric triplet states (total spin 1, that is, $(\boldsymbol{\sigma}_1 \cdot \boldsymbol{\sigma}_2) = +1$), or the spatial wavefunction is symmetric and the spins are in the antisymmetric singlet state ($\boldsymbol{\sigma}_1 + \boldsymbol{\sigma}_2 = 0$, that is, $(\boldsymbol{\sigma}_1 \cdot \boldsymbol{\sigma}_2) = -3$). In the first case, the electrons are on average further apart than in the second case, as the antisymmetry forces the wavefunction to vanish at small distances apart. The Coulomb repulsion is thus less effective, so that the energy is lower, by an amount which we shall denote by $4J$. The energy thus contains a contribution $-J(\boldsymbol{\sigma}_1 \cdot \boldsymbol{\sigma}_2)$ which looks like a strong interaction between the spins, tending to align them.

Extensivity of the Thermodynamic Quantities. This property expresses the experimental fact that the volume and the internal energy of a substance are, at fixed pressure and temperature, proportional to its mass. To explain this microscopically is far from easy and is a major problem in mathematical physics, depending crucially on the interactions between the constituent particles (§ 5.5.2). In particular, if these attract one another at short distances, they tend to agglomerate the more strongly, the larger their number, in order to minimize the energy, so that the density and the energy per particle, E/N, cannot have a limit for large N. Even for a non-interacting gas of bosons, extensivity raises questions (Exerc.12c). In the most fundamental description, a substance consists of N electrons and a number of atomic nuclei such that electrical neutrality is ensured. Let us strike an energy balance between the repulsive and attractive Coulomb forces. If both the electrons and nuclei were bosons, there would be configurations where the attraction would dominate at short distances; for instance, the state described by a constant wavefunction would be acceptable, and one sees easily that the average value of the energy in that state is proportional to $-N\Omega^{-1/3}$, if the substance occupies a volume Ω. The volume would thus tend to decrease and the substance would collapse. The Pauli principle for the electrons guarantees the *stability of the matter*, by forbidding such low energy wavefunctions where large numbers of particles become stuck together. More intuitively, the exclusion principle does not permit more than two (two because of the spin) electrons to be at the same point. It thus produces a short-range repulsion which ensures the extensivity of the thermodynamic quantities.

Stellar Stability. The gravitational equilibrium of a star (Exerc.6e) is realized only if its weight is counterbalanced by pressure forces which prevent the collapse of the star. In a rather young star which is hot and not very dense, like the sun, the matter behaves like a perfect gas where the pressure, which is proportional to the temperature, counterbalances the gravitational forces. During its evolution, the star emits radiation, using up nuclear energy. As the nuclear "fuel" gets exhausted, the temperature decreases, the pressure is no longer sufficient to counterbalance the gravitation, and the star collapses. In certain cases one reaches in this way a new type of very dense star, a *white dwarf* which consists of a plasma of electrons and nuclei (Probs.9 and 10). The internal pressure which ensures the stability of the star is then dominated by the repulsion (10.61) between the electrons due to the Pauli principle. We study this effect in Prob.9, where we show that equilibrium remains possible, provided the stellar mass is not too large. For more massive stars, collapse produces a violent implosion – a supernova – and may lead to an even denser residue, a *neutron star*, which can be modelled as a system of non-interacting neutrons. Here also, gravitational equilibrium is guaranteed by the fact that neutrons are fermions which effectively repel one another due to the exclusion principle, even when their interactions are neglected (Exerc.10e).

Lasers. As they are bosons, photons are created in a given mode more easily if there are already some photons in that mode. This effect, discovered in 1917 by Einstein, is called *stimulated emission* (Exerc.15e). It plays an important rôle in the possibility to construct lasers, devices which allow one to accumulate a large number of photons in the same mode.

Superfluidity of Helium. The ^{4}He isotope of helium, which is the most abundant one, is a boson. At low temperatures it possesses the remarkable property of superfluidity: it can move without viscosity. This phenomenon is connected with the fact that at low temperatures the atoms accumulate in the single-particle state with the lowest energy because they are bosons (§ 12.3 and Prob.14).

Superconductivity. Many substances become superconducting at sufficiently low temperatures when, in particular, they can let an electrical current pass through them without resistance (§ 12.3.3). This effect is related to superfluidity, but here the bosons which condense in the same state and thus easily transport the current are *electron pairs* (with opposite spins). The so-called Cooper mechanism which permits a pair of electrons to get bound together and to behave like a boson cannot be understood intuitively. It again is based upon the Pauli principle, but for fermions. The Fermi-Dirac statistics obeyed by the electrons forbid a pair to occupy states which are already occupied by other electrons. This modifies the dynamics of that pair, so that an effective attraction, however weak, is sufficient to produce a bound state of two electrons which resembles a boson[7].

10.2 Fock Bases

In this section we shall construct a formalism which will allow us to take the Pauli principle automatically into account *without having to label* the indistinguishable particles. This formalism is particularly adapted to deal with the classification of the eigenvalues and eigenfunctions of a Hamiltonian

---

[7] J. Bardeen, L.N. Cooper, and J.P. Schrieffer, Phys.Rev. **108** (1957) 1175.

of non-interacting particles. Such a Hamiltonian has, in the N-particle Hilbert space $\mathcal{E}_H^{(N)}$, the form,

$$\widehat{H}_N = \sum_{i=1}^{N} \widehat{h}_i, \tag{10.7}$$

of a sum of identical operators referring, respectively, to the particles $i = 1, 2, \ldots, N$. If we want to avoid the complications due to the symmetrization or antisymmetrization of the wavefunctions (Exerc.10a), it will be useful to leave N arbitrary and work in the *Fock space* already introduced in §§ 2.3.6 and 4.3.2. Let us remind ourselves that we are dealing with the space $\mathcal{E}_H = \bigoplus_{N=0}^{\infty} \mathcal{E}_H^{(N)}$ which is the direct sum of spaces of N indistinguishable particles which can be bosons or fermions. Note that we shall use the same notation for those two kinds of particles, even though the corresponding spaces are not equivalent as soon as $N \geq 2$. We refer to §§ 2.1.1 and 2.1.2 for the algebraic concepts used below when we shall be dealing with Hilbert spaces and operators.

10.2.1 Single-Particle States

We start from the Hilbert space $\mathcal{E}_H^{(1)}$ of the single-particle states. In contrast to the complete Fock space, this space has the same structure for fermions as for bosons: the nature of the particles does not manifest itself until there are several particles present. We choose for the base $\{q\}$ of the space $\mathcal{E}_H^{(1)}$ the set of eigenvectors of the single-particle Hamiltonian $\widehat{h}$. We indicate the set of quantum numbers which characterize each of these vectors by q, and the value of the corresponding energy eigenvalue by ε_q. Sometimes ε_q depends only on some of the quantum numbers q and one must take care in such cases of degeneracy to sum over the other quantum numbers when one evaluates traces.

To fix the ideas, we shall give a few examples. For a particle, such as a helium atom, enclosed in a parallelepiped shaped box of edge lengths L_x, L_y, L_z, the single-particle Hamiltonian $\widehat{h}$ reduces to the kinetic energy, plus a potential which is zero inside and infinite outside the box. The eigen-kets which span the space $\mathcal{E}_H^{(1)}$ are stationary plane waves which vanish at the walls of the box and which are characterized by three quantum numbers $m_x, m_y, m_z, = 1, 2, \ldots$. The absolute values of the momentum components are given by

$$p_x = m_x \frac{\hbar\pi}{L_x}, \qquad p_y = m_y \frac{\hbar\pi}{L_y}, \qquad p_z = m_z \frac{\hbar\pi}{L_z}. \tag{10.8}$$

The corresponding (single-particle) eigenfunction is

$$\left(\frac{8}{L_x L_y L_z}\right)^{1/2} \sin\frac{p_x x}{\hbar} \sin\frac{p_y y}{\hbar} \sin\frac{p_z z}{\hbar},$$

where the origin of the coordinate system is taken at one of the corners of
the box. If the particle also has a spin s, its ket will be characterized by an
extra quantum number s_z which can take on $2s + 1$ values. The energy ε_q
corresponding to the set of quantum numbers $q = (m_x, m_y, m_z, s_z)$ equals

$$\varepsilon_q = \frac{p^2}{2m} = \frac{\hbar^2 \pi^2}{2m} \left(\frac{m_x^2}{L_x^2} + \frac{m_y^2}{L_y^2} + \frac{m_z^2}{L_z^2} \right), \tag{10.9}$$

where m is the particle mass.

It is sometimes convenient to get rid of the surface effects which are due to
the presence of the walls of the box and to use periodic boundary conditions
– also called the Born-von Kármàn conditions – as a mathematical artifice:
rather than asking the wavefunctions to vanish at the walls, one requires
that the walls are pairwise identical; this means that the wavefunctions must
satisfy periodicity conditions such as

$$\psi(0, y, z) = \psi(L_x, y, z), \qquad \psi_x'(0, y, z) = \psi_x'(L_x, y, z).$$

The eigenfunctions of the single-particle Hamiltonian $\widehat{h}$ then become simply
travelling waves,

$$\left(\frac{1}{L_x L_y L_z} \right)^{1/2} \exp \mathrm{i}(p_x x + p_y y + p_z z)/\hbar;$$

the momenta are still determined by (10.8) and the energies ε_q by (10.9), but
the quantum numbers m_x, m_y, m_z now take the values $0, \pm 2, \pm 4, \dots$.

In the case of the electrons in a solid, the single-particle Hamiltonian $\widehat{h}$
in (10.7) contains, apart from the kinetic energy, also an effective periodic
potential due to the nuclei and the other electrons. The base states q and the
corresponding eigenvalues ε_q are obtained by solving a Schrödinger equation
in the three variables x, y, z (§ 11.2.2).

If one represents an atomic nucleus as a system of protons and neutrons
in an effective central potential (which models the interactions) q denotes,
both for the protons and for the neutrons, the angular quantum numbers l, m,
the radial quantum number n, and the spin quantum number s_z. The single-
particle energies ε_q depend only on l and n, and the levels are $2(2l + 1)$-fold
degenerate.

Finally, we shall in what follows also consider massless particles such
as photons enclosed in a container. These represent the vibrations of the
quantized electromagnetic field inside the container; in that case the photon
momentum $\boldsymbol{p} = \hbar \boldsymbol{k}$ is again given by (10.8), but now the energy equals

$$\varepsilon_q = h\nu = \hbar\omega = cp. \tag{10.10}$$

Each *vibrational mode*, characterized by the quantum numbers q, can be
identified as a *single-particle state*.

Just as we numbered the paramagnetic ion sites in § 1.1.2, we assume here that the single-particle states q are arranged in some standard order which is fixed once and for all. Note that these *states* are *distinguishable* even though the particles which occupy them are not.

10.2.2 Occupation Numbers

We shall associate with the base $\{q\}$ of the single-particle states defined above, a base in the space $\mathcal{E}_H^{(N)}$ of states of N indistinguishable particles, bosons or fermions. The method used has already been sketched in § 10.1.2. It consists of symmetrizing (for bosons) or antisymmetrizing (for fermions) the product wavefunctions $q_1(1)q_2(2)\ldots q_N(N)$. The symbol q_1 denotes both a set of quantum numbers and the associated eigenfunction of $\hat{h}$, while (1), (2), ..., (N) represent the (space and spin) coordinates of the particles numbered $1, 2, \ldots, N$. We must be careful about the terminology, as the word "state" has different meanings: according to the general principles, we shall later be interested in a thermal equilibrium *macro-state* which is defined through assigning probabilities to the *micro-states*; the latter, which are symmetric or antisymmetric eigenstates of (10.7), are *N-particle states* constructed from the *single-particle states* (or modes) q through the symmetrization or antisymmetrization of a product. We shall not always indicate which type of state we are dealing with, but that should be clear from the context.

In order to carry out this programme explicitly, it is convenient to have at our disposal a simple and adequate notation for specifying the different N-boson or N-fermion eigenstates. The first stage of the symmetrization procedure is to specify the states $q_1, q_2, \ldots, q_N$ which the particles $1, 2, \ldots, N$ occupy, respectively. However, a permutation of these single-particle states does not change the symmetric or antisymmetric N-particle state which we have produced; moreover, we must specify at the start that for fermions $q_1, q_2, \ldots, q_N$ are different. To avoid these complications of multiple counting and of exclusion, we shall *change our point of view*: instead of specifying which state is occupied by each of the particles, we *shall look at the set of single-particle states* q and specify for each of them *whether it is empty or whether it is occupied* by one or more of the N particles. This will define for each single-particle state q an *occupation number* n_q, which may be zero.

We are thus led to characterize the *vacuum*, which is the only state in the Hilbert space $\mathcal{E}_H^{(0)}$ of no particles, by stating that every one of the single-particle states q is empty, or equivalently, that *all occupation numbers* n_q *are zero*.

Let us return to the Hilbert space $\mathcal{E}_H^{(1)}$ of the single-particle states q. To express that the particle is in the state q, we use the following language: we say that *the occupation number* n_q *of the state* q *equals 1* and that the occupation numbers $n_{q'}$ of the other states q' are all zero. Each single-particle state of the base $\{q\}$ will thus be denoted by a set of occupation numbers $\{n_q\}$ which are all equal to 0, except one which equals 1. We shall write the ket $|q\rangle$ representing a particle in the state q as follows:

$$|0,\cdots,0,n_q = 1,0,\cdots\rangle,\qquad\qquad\qquad (10.11)$$

where we have arranged the single-particle states q in a *standard order* and where the occupation numbers are written down in that standard order – *all zero except* n_q.

We now turn to the two-particle states which span the Hilbert space $\mathcal{E}_{\mathrm{H}}^{(2)}$. For non-interacting particles, the Hamiltonian has the form

$$\widehat{H}^{(2)} \;=\; \widehat{h}_1 + \widehat{h}_2.$$

If one disregards indistinguishability, the eigenfunctions of $\widehat{H}^{(2)}$ are simply the product of two functions, $q(1)$ for particle 1 and $q'(2)$ for particle 2, while their energy is $\varepsilon_q + \varepsilon_{q'}$; the quantum numbers q and q' may or may not be different. In the first case, $q'(1)q(2)$ differs from $q(1)q'(2)$, but has the same energy. However, for *fermions* the Hilbert space $\mathcal{E}_{\mathrm{H}}^{(2)}$ is the Hilbert space of *antisymmetric* functions. We associate with each pair $q < q'$ of states of $\mathcal{E}_{\mathrm{H}}^{(1)}$ the wavefunction (10.5) of $\mathcal{E}_{\mathrm{H}}^{(2)}$, which is a suitably antisymmetrized eigenstate of $\widehat{H}^{(2)}$ with energy $\varepsilon_q + \varepsilon_{q'}$. We denote this state by letting *two occupation numbers* n_q *and* $n_{q'}$ *be equal to* 1 $(q < q')$, all other occupation numbers being equal to zero, and we write it as

$$|0,\cdots,0,n_q = 1,0,\cdots,0,n_{q'} = 1,0,\cdots\rangle,\qquad (10.12)$$

rather than as (10.5).

For two *bosons* the wavefunctions of $\mathcal{E}_{\mathrm{H}}^{(2)}$ must be *symmetric*. A first class of functions corresponds as before to each pair $q < q'$ for which we form the symmetric combination (10.6b) with energy $\varepsilon_q + \varepsilon_{q'}$. We characterize this state by *two occupation numbers* n_q *and* $n_{q'}$ *equalling* 1 $(q < q')$, the other occupation numbers being zero, and we write it again as (10.12). The use of the same notation (10.12) for the antisymmetric state (10.5) and for the symmetric state (10.6b) should not lead to any confusion as long as we make clear from the start whether we are dealing with fermions or with bosons. However, we can also put two bosons in the same state q according to (10.6a). The corresponding symmetric wavefunction, which in the usual notation is $q(1)q(2)$, has an energy $2\varepsilon_q$. We characterize this state by a *single non-vanishing occupation number* $n_q = 2$, and we write its wavefunction as

$$|0,\cdots,0,n_q = 2,0,\cdots\rangle.\qquad\qquad\qquad (10.13)$$

One can easily extend these arguments to an arbitrary particle number N. For N *fermions* each micro-state (10.3) was constructed by choosing N different single-particle states. One assigns to those states, arranged in the order $q_1 < q_2 < \cdots < q_N$, occupation numbers $n_{q_1} = \ldots = n_{q_N} = 1$, and to the other states q occupation numbers $n_q = 0$. An N-fermion micro-state is thus characterized by

$$|n_1, \cdots, n_q, \cdots\rangle, \tag{10.14}$$

where each n_q equals either 0 or 1, and where the sum of the n_q over all single-particle states equals

$$N = \sum_q n_q. \tag{10.15}$$

The energy associated with the micro-state (10.3) was $\varepsilon_{q_1} + \varepsilon_{q_2} + \cdots + \varepsilon_{q_N}$, an eigenvalue of (10.7). In the occupation number language it can also be written as

$$E\{n_q\} = \sum_q \varepsilon_q n_q, \tag{10.16}$$

as each term equals either ε_q or 0.

For N *bosons* the single-particle states $q_1, q_2, \ldots, q_N$, needed to construct the N-particle micro-states by symmetrizing the product $q_1(1)\ q_2(2)$ $\cdots q_N(N)$, are not necessarily different from one another; a permutation of them does not modify the N-particle state. Each N-boson micro-state is thus characterized by specifying, for each q, the number n_q of times that q appears in the sequence $q_1 \leq q_2 \leq \cdots \leq q_N$; the occupation number set $\{n_q\}$ is a one-to-one representation for each of the different micro-states. We still write it as (10.14), but each number n_q is now a positive integer or zero. The particle number N and the energy are again given by (10.15) and (10.16).

We have altogether, starting from the base of *single*-particle states of the $\mathcal{E}_{\mathrm{H}}^{(1)}$ Hilbert space, constructed a complete base $\{n_q\}$ for the Fock space of states with an *arbitrary number* of particles,

$$\mathcal{E}_{\mathrm{H}} = \mathcal{E}_{\mathrm{H}}^{(0)} \oplus \mathcal{E}_{\mathrm{H}}^{(1)} \oplus \mathcal{E}_{\mathrm{H}}^{(2)} \oplus \cdots,$$

in the two cases where the wavefunctions must be antisymmetric (fermions) or symmetric (bosons). Each state of this base, written in the form (10.14), is characterized by giving the quantum numbers n_q, *each of which can take on the values* $0, 1$ *for fermions, or* $0, 1, 2, \ldots$ *for bosons*. The base (10.14) is called the *Fock base* associated with the base $\{q\}$ of the single-particle states. Although the notation (10.11), (10.12), (10.13) seems unnecessarily complicated to denote states consisting of a small number of particles, the Fock base is very useful for systems consisting of a *large number of indistinguishable particles*. In contrast to the representation in terms of wavefunctions, this one *does not need labelled coordinates to describe the particles*: the *symmetry* or *antisymmetry* of the wavefunctions is *automatically* taken into account by the definition itself of the ket (10.14), and by the set of values which the occupation numbers can take on.

The above construction of the Fock base, starting from the space of the single-particle states, is called *second quantization*. This name is justified by the existence of two levels of quantum numbers, which are different in character and in the rôle they play. The vectors of the base $\{q\}$ of the $\mathcal{E}_{\mathrm{H}}^{(1)}$ Hilbert space of the single-particle states are characterized by the quantum numbers q. When one switches to the $\mathcal{E}_{\mathrm{H}}$ Fock space for an arbitrary number of bosons or fermions, one has available a new base (10.14), the subset (10.11) of which is isomorphic with the base $\{q\}$ of the subspace $\mathcal{E}_{\mathrm{H}}^{(1)}$. However, the new quantum numbers classifying the vectors of the Fock base are the occupation numbers $\{n_q\}$. In the Fock representation one must no longer consider the numbers q as quantum numbers; they henceforth play the rôle of *indices* of the occupation numbers n_q.

10.2.3 Operators in the Fock Representation

The introduction of the Fock base allows us to express all physical observables in a simple way. First of all, we define the *occupation number operator* $\hat{n}_q$, the eigenstates of which are the states (10.14) of the Fock base, with eigenvalues $n_q = 0$ or 1 for fermions, $n_q = 0, 1, 2, \ldots$ for bosons. The operators $\hat{n}_q$ form a complete set of commuting observables, associated with the base (10.14).

The *particle number operator* is their sum

$$\hat{N} = \sum_q \hat{n}_q, \tag{10.17}$$

and its eigenvalues are given by (10.15). Similarly, one can write down the *Hamiltonian for non-interacting particles* as

$$\hat{H} = \sum_q \varepsilon_q \hat{n}_q \tag{10.18}$$

in the complete Fock space. Each of its eigenvalues, the energy (10.16) of an N-particle state, is obtained simply by adding the energies ε_q of the occupied states, weighted by the number n_q of particles occupying the state q. One should note that the representation (10.18) of the operator $\hat{H}$ is valid for all its components $\hat{H}_1, \hat{H}_2, \ldots$ in the $N = 1, 2, \ldots$-particle Hilbert spaces, whereas the usual representation (10.7), in terms of the $\hat{r}_i$ and $\hat{p}_i$ operators associated with the positions and momenta of the labelled particles, changes when the number of particles changes and, moreover, does not automatically take their nature (fermions or bosons) into account.

Apart from being better adapted to the physical reality – identical particles need no longer be labelled, and their bosonic or fermionic nature is directly taken into account through the possible values of the occupation numbers – Fock space brings about a major simplification for the study of quantum gases at equilibrium. At first sight, the Hamiltonian (10.18) in Fock space hardly looks simpler than the Hamiltonian (10.7) in the N-particle Hilbert space. However, the evaluation of the canonical partition function

$$\mathrm{Tr}_N \exp\left[-\beta \sum_{i=1}^{N} \hat{h}_i\right]$$

is impracticable, notwithstanding the fact that the density operator seems to factorize into contributions associated with *separate particles*; indeed, the wavefunctions of the base have a complicated, unfactorized structure which involves the $N!$ permutations necessary to account for the symmetry or antisymmetry. Adding another particle thus necessarily modifies the state of the other particles through the symmetry structure of the wavefunctions; this introduces correlations of a quantum nature between particles, even when there are no interactions. The indistinguishability thus prevents the factorization of the canonical partition function into a product of factors referring to separate particles. On the other hand, second quantization with its change in point of view – the accent is now on states and their filling rather than on the particles themselves – and the use of the grand canonical ensemble make another kind of factorization possible. The Hamiltonian (10.18) in Fock space is a sum of terms which are now associated *each with a single-particle state q*. The latter are distinguishable, in contrast to the particles, and they *fill up independently* of one another: each quantum number n_q takes on all permissible values without interfering with the filling up of the states different from q. The price paid in order to benefit from this independence is that one needs to work in the grand canonical ensemble.

From a more mathematical point of view, Fock space has the *simple structure of a direct product of spaces relating to different single-particle states q*, in contrast to $\mathcal{E}_{\mathrm{H}}^{(N)}$ which cannot be factorized into simpler spaces:

$$\mathcal{E}_{\mathrm{H}} \;=\; \overset{\infty}{\underset{N=0}{\oplus}} \; \mathcal{E}_{\mathrm{H}}^{(N)} \;=\; \underset{q}{\otimes}\, \mathcal{E}_{\mathrm{H}q}.$$

This shows up clearly in the base (10.14); each subspace q is spanned by the vectors $n_q = 0, 1, 2, \ldots$ for bosons, and by the vectors $n_q = 0$ or 1 for fermions. This property allows the factorization (4.23) of the Boltzmann-Gibbs grand canonical density operator in terms of elementary operators,

$$\widehat{D}_q \;=\; \frac{1}{Z_q}\, \exp\left(-\beta\varepsilon_q \widehat{n}_q + \alpha \widehat{n}_q\right) \;\equiv\; \frac{1}{Z_q}\,\left(X_q\right)^{\widehat{n}_q}, \tag{10.19}$$

each acting in a q subspace of Fock space, and the factorization (4.22) of the grand canonical partition function. The *single-particle states q*, which are more abstract entities than the particles, play here the rôle of the *statistically independent subsystems* of § 4.2.5.

The introduction of the Fock base also has other advantages. When an observable commutes with $\widehat{N}$ one can represent it, as we did with the Hamiltonian (10.7), in terms of its components in the $\mathcal{E}_{\mathrm{H}}^{(N)}$ spaces; this was the way we characterized the grand canonical ensemble in § 4.2.3 through its components for each value of N. Nonetheless, some physical systems involve observables which do not commute with the particle number. For instance, the interaction of quantized electromagnetic radiation with charged particles induces changes in the state of the field through changes in the number of photons; a transition between different energy levels involves the emission or absorption of a photon. Similarly, in particle physics, in an electron-positron collision these two fermions can annihilate one another producing two photons in the process. Even when the particle numbers are conserved, it may be useful to describe, for instance, an elastic collision process as the annihilation of one pair, followed immediately by the creation of a pair with different momenta. The methods of the so-called many-body problem take advantage of this idea by using second quantization. In particular, theories of superconductivity or of the superfluidity of helium (Exerc.12d) describe the systems of the electrons or of the He atoms by *breaking the invariance* (end of § 9.3.3) which is associated with the particle number conservation: even though the Hamiltonian and the exact grand canonical density operator $\widehat{D}$ commute with $\widehat{N}$, one can construct a fair approximation to $\widehat{D}$ by replacing it by an operator $\widehat{\mathcal{D}}$ which is sufficiently simple to make calculations possible, but which does not commute with $\widehat{N}$; this operator can be determined by the variational method, as in § 9.3.1. Superconductivity and superfluidity are closely connected with this symmetry breaking which occurs when T lies below the transition temperature. To describe such phenomena we need operators which do not conserve the number of particles and which therefore cannot be written in terms of wavefunctions. One introduces them as follows.

We first of all introduce operators $\widehat{c}_q$ whose effect is to reduce the occupation number n_q by unity, without changing the other occupation numbers. It is convenient to include in their definition a factor given by

$$\widehat{c}_q \, |n_1, \ldots, n_q, \ldots\rangle \; = \; (\eta)^{\sum_{q' < q} n_{q'}} \sqrt{n_q} \, |n_1, \ldots, n_q - 1, \ldots\rangle, \qquad (10.20a)$$

where η is 1 for bosons and -1 for fermions, and where the indices q are arranged in the standard order. The $\widehat{c}_q$ operators are called *annihilation operators* and their Hermitean conjugates, $\widehat{c}_q^\dagger$, which add a particle in the state q, according to the rule

$$\widehat{c}_q^\dagger \, |n_1, \ldots, n_q, \ldots\rangle \; = \; (\eta)^{\sum_{q' < q} n_{q'}} \sqrt{1 + \eta n_q} \, |n_1, \ldots, n_q + 1, \ldots\rangle, \qquad (10.20b)$$

are called *creation operators*. The factors introduced in (10.20) guarantee, in particular, that $\widehat{c}_q$ applied to the vacuum gives 0 and that for fermions $\widehat{c}_q^\dagger$ applied to a state where q is already occupied also gives 0. Moreover, they lead to the following simple algebraic relations:

$$\widehat{c}_q \widehat{c}_{q'}^\dagger \, - \, \eta \, \widehat{c}_{q'}^\dagger \widehat{c}_q \; = \; \delta_{qq'}, \qquad \widehat{c}_q \widehat{c}_{q'} \, - \, \eta \, \widehat{c}_{q'} \widehat{c}_q \; = \; 0. \qquad (10.21)$$

The definition (10.20) itself of the creation and annihilation operators implies that *any operator* in Fock space can be expressed as an element of the algebra generated by the $\widehat{c}_q$ and $\widehat{c}_q^\dagger$ through multiplications and linear combinations. In particular, one can check that the occupation number observables are given by the equation

$$\widehat{n}_q \; = \; \widehat{c}_q^\dagger \widehat{c}_q, \qquad (10.22)$$

so that we get for the Hamiltonian (10.18):

$$\widehat{H} = \sum_q \widehat{c}_q^\dagger \varepsilon_q \widehat{c}_q.$$

More generally, let us consider a single-particle observable $\widehat{F}$, for instance, a component of the angular momentum; its components in $\mathcal{E}_{\mathrm{H}}^{(N)}$ have, in terms of wavefunctions, the form $\sum_{i=1}^{N} \widehat{f}_i$, which is similar to (10.7), where $\widehat{f}$ is characterized by its matrix elements $\langle q|f|q'\rangle$ with respect to the single-particle states. In second quantization we find for $\widehat{F}$:

$$\widehat{F} = \sum_{q,q'} \widehat{c}_q^\dagger \langle q|f|q'\rangle \widehat{c}_{q'}.$$

The creation and annihilation operators are also helpful to find expressions for observables which do not commute with $\widehat{N}$. In particular, we shall see in § 13.1.3 that the observables which in quantum mechanics describe the magnetic and electrical field at one point can be expressed as linear combinations of $\widehat{c}_q + \widehat{c}_q^\dagger$ and $i(\widehat{c}_q - \widehat{c}_q^\dagger)$, respectively, where $\widehat{c}_q$ absorbs and $\widehat{c}_q^\dagger$ creates a photon. The quantization of the electromagnetic field thus presupposes that one uses the second quantization formalism. Let us also note that any – symmetric or antisymmetric – N-particle wave function is, according to (10.20b) the result of operating with a product of N creation operators on the vacuum $|\phi\rangle$.

We have assumed in the foregoing that the base $\{q\}$ of the single-particle states was the base of the eigenfunctions of the Hamiltonian $\widehat{h}$. However, second quantization can start from any single-particle base. *Changing the base* in the space $\mathcal{E}_{\mathrm{H}}^{(1)}$ thus induces a change of base in Fock space and a linear transformation of the annihilation operators. In particular, starting from the base $|r\rangle$ of states which represent a particle localized at the point r, second quantization produces annihilation operators $\widehat{\psi}(r)$ and creation operators $\widehat{\psi}^\dagger(r)$, called *field operators* – which may have a spin index, omitted here for the sake of simplicity. When operating on the vacuum $|\phi\rangle$, $\widehat{\psi}^\dagger(r)$ produces the state $\widehat{\psi}^\dagger(r)|\phi\rangle = |r\rangle$, while $\widehat{c}_q^\dagger$ produces $\widehat{c}_q^\dagger|\phi\rangle = |q\rangle = \int d^3r\, |r\rangle\langle r|q\rangle$. Hence, we have

$$\widehat{c}_q^\dagger = \int d^3r\, \widehat{\psi}^\dagger(r)\langle r|q\rangle \quad , \quad \widehat{\psi}(r) = \sum_q \langle r|q\rangle \widehat{c}_q. \tag{10.23}$$

The operator associated with the number of particles per unit volume at the point r, the average of which is the *particle density*, is therefore given by

$$\widehat{\psi}^\dagger(r)\widehat{\psi}(r) = \sum_{q,q'} \widehat{c}_q^\dagger \langle q|r\rangle\langle r|q\rangle \widehat{c}_q. \tag{10.24}$$

The second quantization procedure, through eliminating the wavefunctions which do not possess the necessary symmetry, has at the same time eliminated all operators which do not commute with the exchange operator, such as, for instance, the momentum of *one* of the particles. In fact, such operators cannot be written in terms of the elementary operators $\widehat{c}_q, \widehat{c}_q^\dagger$; acting upon the base (10.14) of symmetric or antisymmetric wavefunctions, they would induce a departure from

Fock space. They do therefore have no meaning in the occupation number formalism. The questions raised in § 10.1.1 which appeared natural as long as the particles were labelled do no longer show up.

The Pauli principle itself has become implicit in the new formalism. It is expressed simply by the structure of Fock space. By requiring that the N-particle states are all symmetric for bosons, all antisymmetic for fermions, we have identified any exchange operator in $\mathcal{E}_{\mathrm{H}}^{(N)}$ with just the unit operator $\widehat{I}$ for bosons, and with $-\widehat{I}$ for fermions. Accordingly, *any* Hermitean operator in $\mathcal{E}_{\mathrm{H}}^{(N)}$ commutes with the exchange operator and is thus invariant under a relabelling of the particles. Conversely, one can show that, *if any Hermitean operator* in some N-particle Hilbert space *represents a physical observable* which does not distinguish between the particles, the wavefunctions of this space must be all symmetric, or all antisymmetric. This is an alternative formulation of the Pauli principle, to which we alluded in §§ 2.1.3 and 2.2.7. The distinction between bosons and fermions is mathematically expressed in second quantization through the algebraic relations (10.21). The exchange of the labels of two particles, which has no longer a meaning, is replaced by the exchange of two single-particle states in the commutation or anticommutation relation $\widehat{c}_q^\dagger \widehat{c}_{q'}^\dagger = \eta \widehat{c}_{q'}^\dagger \widehat{c}_q^\dagger$: when applied to the vacuum, the operator $\widehat{c}_q^\dagger \widehat{c}_{q'}^\dagger$, which first fills the state q' according to (10.20b) and then the state q, leads to the ket (10.12) if $q < q'$, whereas $\widehat{c}_{q'}^\dagger \widehat{c}_q^\dagger$ leads to this same ket, multiplied by $+1$ for bosons, by -1 for fermions. When $q = q'$, the anticommutation relation of $\widehat{c}_q^\dagger$ and $\widehat{c}_{q'}^\dagger$ for fermions reduces to $(\widehat{c}_q^\dagger)^2 = 0$, which expresses the exclusion principle.

10.3 Equilibrium of Quantum Gases

We have just emphasized in § 10.2.3 that the occupation number formalism enables us to factorize the contributions from the single-particle *states* for non-interacting quantum gases, notwithstanding the correlations between the *particles* introduced by the indistinguishability. However, in order to be able to sum freely over each $n_{\dot{q}}$ it is necessary to get rid of the constraint (10.15) and to work thus in a *grand canonical ensemble* (§ 4.3.2). For a macroscopic system, the ensembles are equivalent (§ 5.5.3), and it is therefore legitimate to choose the one which is technically the most convenient.

10.3.1 Grand Canonical Partition Function

The Hamiltonian (10.18) of non-interacting particles and the particle number (10.17) are diagonal operators in the Fock base (10.14), with eigenvalues (10.16) and (10.15). Calculating a trace over Fock space amounts to summing over the permitted quantum numbers, that is, over $n_q = 0$ or 1 for each single-particle state q when the system consists of fermions, and over $n_q = 0, 1, 2, \ldots$ when it consists of bosons. The grand partition function can thus be expressed as

$$Z_{\mathrm{G}} = \mathrm{Tr}\, e^{-\beta\widehat{H}+\alpha\widehat{N}} = \sum_{\{n_q\}} \exp\left[-\beta\sum_{q'}\varepsilon_{q'}n_{q'} + \alpha\sum_{q'}n_{q'}\right]$$

$$= \sum_{\{n_q\}} \prod_{q'}(X_{q'})^{n_{q'}} = \sum_{n_1,n_2,\ldots} X_1^{n_1}X_2^{n_2}\cdots, \tag{10.25}$$

where we have written

$$X_q \equiv e^{-\beta\varepsilon_q+\alpha}, \tag{10.26}$$

and where we must sum independently over each quantum number n_q.

The calculation now falls into the general framework of § 4.2.5, as Fock space, the density operator (10.19), and expression (10.25) all *factorize* into contributions associated with each state q. We have thus

$$\ln Z_{\mathrm{G}} = \sum_q \ln Z_q, \tag{10.27}$$

where Z_q is a trace in the elementary Fock subspace associated with the single-particle state q. In the case of Bose-Einstein statistics ($\eta = 1$), n_q takes on all non-negative integral values and we get

$$Z_q = \sum_{n=0}^{\infty} X_q^n = \frac{1}{1-X_q}; \tag{10.28a}$$

in the case of Fermi-Dirac statistics ($\eta = -1$), n_q equals 0 or 1, and we have

$$Z_q = \sum_{n=0,1} X_q^n = 1 + X_q. \tag{10.28b}$$

Combining Eqs.(10.27), (10.28), and (10.26), we get finally for the grand partition function

$$\boxed{\ln Z_{\mathrm{G}} = -\eta\sum_q \ln\left(1 - \eta e^{-\beta\varepsilon_q+\alpha}\right)}, \tag{10.29}$$

which is valid both for a gas of bosons ($\eta = 1$) and for a gas of fermions ($\eta = -1$).

The average number of particles, the internal energy, and the entropy can be derived from (10.29) through differentiation, as was indicated in § 4.3.2, and the pressure is $\mathcal{P} = \ln Z_{\mathrm{G}}/\beta\Omega$ for an extensive system. Below we shall give these quantities in an explicit and more convenient form.

10.3.2 Occupation Factors

All thermodynamic quantities follow from (10.29), but other, more detailed quantities can be derived from the grand canonical equilibrium density operator (10.19). In particular, the *probability for the occupation number* n_q has an *exponential* Boltzmann-Gibbs form:

$$p(n_q) = \frac{1}{Z_q} (X_q)^{n_q} = \frac{1}{Z_q} e^{-(\beta \varepsilon_q - \alpha) n_q}. \tag{10.30}$$

Using it, we can derive the expectation value, $f_q \equiv \langle n_q \rangle$ of the occupation number for the state q, which is called the *Fermi factor* or the *Bose factor* depending on the case. Explicitly, we get for fermions

$$f_q = \frac{1}{Z_q} \sum_{n_q=0,1} n_q (X_q)^{n_q} = \frac{X_q}{Z_q} = \frac{X_q}{1 + X_q},$$

and for bosons

$$f_q = \frac{1}{Z_q} \sum_{n_q=0}^{\infty} n_q (X_q)^{n_q} = \frac{X_q}{Z_q(1 - X_q)^2} = \frac{X_q}{1 - X_q}.$$

More simply, one uses the standard procedure of § 4.2.6 and derives f_q from $\ln Z_G$, noting that, if $\ln Z_G$ is regarded as a function of β, α and the energies ε_q, its partial derivative with respect to ε_q is just $-\beta \langle n_q \rangle$. Thus we get

$$f_q = \frac{\eta}{\beta} \frac{\partial}{\partial \varepsilon_q} \ln \left(1 - \eta e^{-\beta \varepsilon_q + \alpha}\right).$$

However we calculate it, the final, important expression for the Fermi or Bose occupation factor is

$$\boxed{f_q = \frac{1}{e^{\beta \varepsilon_q - \alpha} - \eta}} . \tag{10.31}$$

We can use this expression, together with the exponential form of (10.30) and the statistical independence of the different q, to characterize the probability distribution of the occupation numbers $\hat{n}_q$. For instance, the fluctuations of $\hat{n}_q$ are given by

$$\left\langle (\hat{n}_q - f_q)^2 \right\rangle = f_q(1 + \eta f_q), \tag{10.32}$$

and (10.30) is equivalent to

$$p(n_q) = \frac{(f_q)^{n_q}}{(1 + \eta f_q)^{n_q + \eta}}.$$

We can also express all thermodynamic quantities in terms of the f_q, thanks to the relation

$$\beta\varepsilon_q - \alpha \ = \ \ln \frac{1+\eta f_q}{f_q}, \tag{10.33}$$

which follows from (10.31). In particular, the grand partition function (10.29) can be written as

$$\ln Z_G \ = \ \eta \sum_q \ln(1+\eta f_q). \tag{10.34}$$

The particle number and the internal energy, which are the averages of (10.17) and (10.18) are given by

$$N \ = \ \sum_q f_q, \tag{10.35}$$

$$U \ = \ \sum_q \varepsilon_q f_q. \tag{10.36}$$

Using Eqs.(10.33) to (10.36) and (4.36) we find the entropy:

$$\begin{aligned} S \ &= \ k(\ln Z_G + \beta U - \alpha N) \\ &= \ k \sum_q [\eta(1+\eta f_q)\ln(1+\eta f_q) \ - \ f_q \ln f_q], \end{aligned} \tag{10.37}$$

which alternatively can be expressed in terms of ε_q, if we use (10.31) and the relation

$$1 + \eta f_q \ = \ \left[1 - \eta e^{-\beta\varepsilon_q + \alpha}\right]^{-1}.$$

10.3.3 Large Volume Limit; Density of States

We have expressed the various thermodynamic functions (10.34–37) as sums, over the single-particle states q, of functions $\varphi(\varepsilon_q)$ of the eigenenergies ε_q. One often has to consider the limit where the extent of the system becomes infinite; the ε_q spectrum then becomes a continuum, and expressions (10.34–37) integrals.

Let us, as an example, consider a gas of non-interacting bosons or fermions enclosed in a box, which we assume to be a parellepiped. If we forget about the spin of these particles, the states q are, according to (10.8), characterized by three quantum numbers $m_x, m_y, m_z = 1, 2, \ldots$. The summation over m_x becomes, as $L_x \to \infty$, an integration over the momentum component p_x, with a weight which follows from (10.8):

$$\sum_{m_x} \rightarrow \int_0^\infty dm_x \ = \ \frac{L_x}{\pi\hbar} \int_0^\infty dp_x \ = \ \frac{L_x}{h} \int_{-\infty}^{+\infty} dp_x$$

(the integrand is independent of the sign of p_x). Altogether, for a large size box the sum over q becomes a triple integral which is proportional to the volume:

$$\boxed{\sum_{m_x, m_y, m_z} \xrightarrow[\Omega \to \infty]{} \frac{\Omega}{h^3} \int d^3 p}. \tag{10.38}$$

Of course, we should not forget to sum over spin indices, if they occur.

In the case of a thin layer – liquid helium film or metallic layer; see Exerc.10d, Prob.11, and Prob.18 – we obtain in the same way a double integral, $(S/h^2) \int d^2 p$, with a weight which is proportional to the area S, but the quantum number corresponding to the transverse momentum remains discrete.

For a box with periodic boundary conditions (§ 10.2.1) the values of m_x, m_y, and m_z in (10.8) become positive and negative, but restricted to even values. In the limit as $\Omega \to \infty$ these two effects cancel one another and one gets again the same result (10.38) as for particles enclosed in a box with rigid walls. More generally, one can prove that (10.38) remains valid for a large box of arbitrary shape with arbitrary boundary conditions. This result ensures the existence of the thermodynamic limit (§ 5.5.2) for quantum gases without interactions and the *extensivity* of the various quantities (10.34–37): in the large volume limit the thermodynamic quantities are independent of the shape and of the surface conditions of the container, and are either proportional to the volume or constant. A notable exception is the gas of non-interacting bosons (§ 12.3) which becomes pathological when its temperature is sufficiently low or its density sufficiently high. In fact, α then tends to zero, the integrand becomes singular at $p = 0$, and the replacement (10.38) is no longer justified. We shall see that this leads to a phase transition (§ 12.3.2) and that the extensitivity of the Bose gas poses some problems (Exerc.12c).

The result (10.38) bears a similarity to the classical limit (2.69) of the trace for a single-particle Hilbert space,

$$\text{Tr} \xrightarrow[\hbar \to 0]{} \int \frac{d^3 r \, d^3 p}{h^3}, \tag{10.39}$$

which after integration over r reduces to (10.38). One should, however, note the differences between these limits. On the one hand, going over to the infinite volume limit (10.38) is, as we shall see in later sections and chapters, applicable to systems where quantum phenomena play an important rôle, even on the macroscopic scale. On the other hand, this limit (10.38) is only relevant to single-particle states in sums such as (10.34–37) for a large box with a constant potential, whereas the classical limit (10.39) can be generalized, as indicated in § 2.3.4, to any number of particles in any potential.

As the thermodynamic quantities (10.34–37) depend on the indices q only through ε_q, it is often useful in the case of three-dimensional systems to write them as single integrals over the variable ε_q rather than as the triple integrals

(10.38). To do this, we introduce the *single-particle density of states* $\mathcal{D}(\varepsilon)$, defined through

$$\mathcal{D}(\varepsilon) = \sum_q \delta(\varepsilon - \varepsilon_q), \tag{10.40}$$

which means that $\mathcal{D}(\varepsilon)\,d\varepsilon$ is the *number of states* q *with energies* ε_q *between* ε *and* $\varepsilon + d\varepsilon$. The sums (10.34–37) can then be written in the form

$$\boxed{\sum_q \varphi(\varepsilon_q) = \int d\varepsilon \mathcal{D}(\varepsilon)\,\varphi(\varepsilon)} , \tag{10.41}$$

and the study of a quantum gas at equilibrium is directly connected with finding the *distribution of the eigenvalues* ε_q, (10.40), of the single-particle Hamiltonian $\hat{h}$. In particular, the particle number and internal energy,

$$N = \int d\varepsilon\, \mathcal{D}(\varepsilon)\, f(\varepsilon) \quad , \quad U = \int d\varepsilon\, \mathcal{D}(\varepsilon)\, \varepsilon\, f(\varepsilon), \tag{10.42}$$

involve solely the product of the density of states $\mathcal{D}(\varepsilon)$, which characterizes the single-particle properties, and the occupation factor

$$f(\varepsilon) \equiv 1/(e^{\beta\varepsilon - \alpha} - \eta),$$

through which the temperature, the chemical potential, and the statistics obeyed by the particles are introduced. Equations such as (10.42) thus clearly *separate the dynamics* of the particular system under consideration, which enter through $\mathcal{D}(\varepsilon)$, from the *statistical parameters* β, α, and $\eta = \pm 1$.

For a finite system, $\mathcal{D}(\varepsilon)$ is a distribution and (10.41) is useless. Nonetheless, in the large volume limit, $\mathcal{D}(\varepsilon)/\Omega$ tends to a function which should be evaluated before we can use (10.42) directly. As an example, for a nonrelativistic spin-s particle, ε_q is given by (10.9) if there is no magnetic field, and using (10.38) in the definition (10.40) gives us (the Heaviside step function $\theta(\varepsilon)$ equals 0 when $\varepsilon < 0$ and equals 1 when $\varepsilon > 0$)

$$\begin{aligned}
\mathcal{D}(\varepsilon) &= (2s+1)\frac{\Omega}{h^3}\int d^3p\, \delta\left(\varepsilon - \frac{p^2}{2m}\right) \\
&= (2s+1)\frac{\Omega}{h^3} 4\pi \int_0^\infty p^2\, dp\, \frac{m}{p} \delta(\sqrt{2m\varepsilon} - p) \\
&= \frac{(2s+1)(2m)^{3/2}}{4\pi^2\hbar^3}\, \Omega \sqrt{\varepsilon}\, \theta(\varepsilon);
\end{aligned} \tag{10.43}$$

we have integrated over the angles and afterwards used the properties of the δ distribution which are summarized at the end of this volume. In a more elementary way we find the same result by rewriting (10.38) in spherical coordinates, integrating over the direction of p, and finally changing from the radial coordinate p to $\varepsilon = p^2/2m$.

On account of the practical interest of $\mathcal{D}(\varepsilon)$, we give yet another method of calculating it. We start by looking for the *number* $\mathcal{N}(\varepsilon)$ *of single-particle states with energies lower than* ε, the derivative of which is simply $\mathcal{D}(\varepsilon)$:

$$\mathcal{N}(\varepsilon) = \sum_q \theta(\varepsilon - \varepsilon_q) = \int_{-\infty}^{\varepsilon} d\varepsilon' \, \mathcal{D}(\varepsilon'). \tag{10.44}$$

For the non-relativistic particle considered, $\mathcal{N}(\varepsilon)$ is the number of points in $\boldsymbol{p}$-space belonging to the lattice (10.8) and situated inside the sphere of radius $p = \sqrt{2m\varepsilon}$. For a box with rigid walls, they have coordinates (10.8) with integer values of m_x, m_y, m_z and only one octant of the sphere should be retained; for periodic boundary conditions, the mesh is doubled and the sum extends over the whole of the sphere. In the limit as $\Omega \to \infty$ the points lie densely, and asymptotically their number is in both cases

$$\mathcal{N}(\varepsilon) = (2s+1)\frac{\Omega}{h^3}\frac{4\pi}{3}(2m\varepsilon)^{3/2}\,\theta(\varepsilon), \tag{10.45}$$

which leads again to (10.43).

The density of states depends crucially on the form of the energy $\varepsilon(p)$ and on the dimensionality of the system. For three-dimensional relativistic particles (Exerc.13c, Prob.9) we must replace in the above calculations $\varepsilon = p^2/2m$ by $\varepsilon = \sqrt{m^2c^4 + p^2c^2}$. In particular, for massless particles, with energies (10.10), we find

$$\mathcal{D}(\varepsilon) = \frac{1}{2\pi^2\hbar^3c^3}\,\Omega\,\varepsilon^2\,\theta(\varepsilon) \tag{10.46}$$

for each possible spin value (for photons we must multiply (10.46) by 2). Other examples of density of states which show up the rôle of an external potential and of the dimensionality are given in Exerc.10c, Chap.11, and Probs.11, 12, and 18.

Writing the thermodynamic quantities as integrals with the weight function $\mathcal{D}(\varepsilon)$ enables us to perform useful integrations by parts. In particular, we can use Eqs.(10.29), (10.41), and (10.44) to rewrite the grand potential and the pressure $\mathcal{P}$ in the form

$$-\mathcal{P}\Omega = A = -\frac{1}{\beta}\ln Z_G = \frac{\eta}{\beta}\int d\varepsilon\,\mathcal{D}(\varepsilon)\ln\left(1 - \eta e^{-\beta\varepsilon + \alpha}\right)$$

$$= -\int d\varepsilon\,\mathcal{N}(\varepsilon)\,f(\varepsilon), \tag{10.47}$$

and similarly we get for the particle number (10.42)

$$N = -\int d\varepsilon\,\mathcal{N}(\varepsilon)\frac{\partial f}{\partial \varepsilon}. \tag{10.48}$$

Expression (10.47) is particularly useful to calculate the *low-temperature pressure* of a gas of fermions, in contrast to the original expression (10.29) for the grand potential, for which it is more difficult to take the limit as $\beta \to \infty$.

10.3.4 Classical Limit; Low Densities

We indicated in § 7.1.2 that the classical perfect gas model was justified for the low density or for the high temperature limit. One should therefore expect that in those limits the results obtained above reduce to the properties of a perfect gas. In order to discuss the validity of the classical approximation, we rewrite the relation (10.42), which connects the density with the temperature and the chemical potential, in terms of dimensionless quantities, for the case of a non-relativistic gas in a box with the density of states given by (10.43). Putting $x \equiv \beta \varepsilon$, we have

$$\frac{N}{\Omega} \frac{h^3}{(2\pi m k T)^{3/2}} \equiv \frac{N}{\Omega} \lambda_T^3 = (2s+1) \frac{2}{\sqrt{\pi}} \int_0^\infty \frac{\sqrt{x}\, dx}{e^{x-\alpha} - \eta}. \qquad (10.49)$$

We have introduced here the thermal length λ_T which is proportional to $T^{-1/2}$ and was already defined in Chap.7. The relation (10.49) determines α as an increasing function of the quantity $N\lambda_T^3/\Omega$. In order that the latter be small compared to unity, we need to have $e^{-\alpha} \gg 1$, and the term in η is then negligible in the integrand on the right-hand side of (10.49). After integrating over x we find in this classical limit

$$\frac{N}{\Omega} \lambda_T^3 \frac{1}{2s+1} \sim e^\alpha \ll 1. \qquad (10.50)$$

This condition defines the domain of *validity of the perfect gas approximation, in terms of the density and the temperature* or in terms of the chemical potential $\mu = kT\alpha$. It is satisfied, as we saw in § 7.3.1, when the density is sufficiently low or the temperature sufficiently high for the thermal length to be short compared to the distance between the particles; this is the case for all atomic and molecular gases – bar helium below a few K. In that situation we have, both for bosons and for fermions,

$$f_q \sim e^{-\beta \varepsilon_q + \alpha} \ll 1 \qquad (10.51)$$

in all single-particle states. If we use (10.51) and (10.38) we find for the grand partition function (10.34)

$$\ln Z_G \sim \sum_q f_q \sim (2s+1) \frac{\Omega}{h^3} e^\alpha \int d^3 p\, e^{-\beta p^2/2m}$$

$$= \frac{(2s+1)\, e^\alpha\, \Omega}{\lambda_T^3}, \qquad (10.52)$$

the same expression as equation (7.28) which was calculated directly in the framework of classical statistical mechanics; there is no classical equivalent for the factor $(2s+1)$ and it can be interpreted as an internal partition function of the type discussed in § 8.3.1. Thus, in the limit (10.50) or (10.51), we recover all thermodynamic properties of the perfect gas. In particular, we find again from the quantum entropy (10.37) the classical expression

$$S = k \sum_q f_q \left(1 - \ln f_q\right) = k \sum_q e^{-\beta \varepsilon_q + \alpha} \left(1 + \beta \varepsilon_q - \alpha\right).$$

The number of particles with momenta within the volume element $d^3 p$ is, in the limit (10.50) and taking the spin into account, equal to

$$(2s+1) \frac{\Omega}{h^3} d^3 p \, f_p \sim (2s+1) \frac{\Omega}{h^3} e^{\alpha - \beta p^2 / 2m} d^3 p. \tag{10.53}$$

We thus recover the *Maxwell distribution*. Moreover, the normalization of (10.53) is in agreement with Eqs.(7.31) or (8.13) for the density of a perfect gas as function of the chemical potential.

In order to understand better why the Bose-Einstein and Fermi-Dirac statistics have as common limit the perfect gas when (10.50) is satisfied, we note that, if $e^\alpha \ll 1$, very few single-particle states are occupied. The probabilities (10.30) are practically equal to 1 for $n_q = 0$, they are small and close to (10.51) for $n_q = 1$, and, for the case of bosons, are completely negligible for $n_q > 1$; thus the distinction between the statistics no longer plays a rôle. Note, nevertheless, that the lowest-order corrections to (10.51), and hence to the thermodynamic quantities (10.37), (10.42), or (10.47) have opposite signs for fermions and for bosons: the perfect gas thus appears to lie half-way between the non-interacting fermion or boson quantum gases (Exerc.10f, 10g).

The perfect gas limit which we have just discussed only deals with particles in a box. In the general case where the single-particle Hamiltonian $\hat{h}$ is arbitrary, quantum mechanics appears in Eqs.(10.29) or (10.34) for the grand potential in two different ways which must be distinguished. On the one hand, we need to solve a *single-particle quantum problem*, that is, find the eigenvalues ε_q of $\hat{h}$. On the other hand, the *Pauli principle* gives us the form of the occupation factor which depend on the statistics and differs from the classical factor (10.51). As a consequence, depending on the circumstances, two different "classical" simplifications can occur even though classical statistical mechanics, in the strict sense of the word (§2.3), is not valid.

(a) *Low Densities.* When the chemical potential is very large and negative – to be more precise, when $e^{\alpha - \beta \varepsilon_0} \ll 1$, where ε_0 is the ground state energy of $\hat{h}$ – the approximation (10.51) is justified. As a result, (10.34) reduces to

$$\ln Z_{\mathrm{G}} \underset{\alpha \to -\infty}{\sim} \sum_q e^{-\beta \varepsilon_q + \alpha} = e^\alpha \, \mathrm{tr} \, e^{-\beta \hat{h}}, \tag{10.54}$$

where the trace is taken over the single-particle Fock space. In this limit Bose-Einstein or Fermi-Dirac statistics no longer enter the discussion, but the quantum nature of the single-particle problem is, in general, still there through the diagonalisation of $\hat{h}$. We give in §11.3 an important example: the conduction electrons in an insulator are few so that (10.54) is justified, but they are subject to a potential which varies rapidly on the scale of the lattice distances of the crystal and that makes it impossible to calculate the trace in (10.54) using semiclassical methods. A similar situation occurred in § 8.4.4 for the molecular internal degrees of freedom which had to be treated quantum mechanically. Note that in the low-density limit (10.54) implies that the canonical partition function equals $\left(\mathrm{tr}\, e^{-\beta \widehat{h}}\right)^N /N!$; we recover the factor $1/N!$ of Eq.(2.55) which follows from the Pauli principle and which persists even though the Bose or Fermi statistics no longer appear.

(b) *Slowly Varying Potential.* We saw in § 10.3.3 that, if the potential is constant, (10.38) is the same as the single-particle classical limit (10.39). More generally, let us assume that $\widehat{h}$ contains, apart from the kinetic energy of the particle, only a potential which varies slowly in space. We shall meet with this situation in § 11.3 when studying the electrostatic equilibrium of matter, the electrons of which are subject to a slowly varying electric field; similarly, the gravitational equilibrium of a star (Exerc.6e,10e and Prob.9) involves a slowly varying gravitational potential. In such cases the considerations of § 2.3.4 justify us to replace the sum over q by the integral (10.39) in the evaluation of the various thermodynamic functions (10.29), (10.34–37). Everything proceeds as if the potential were constant in the various volume elements which make up the sample so that expression (10.39) is obtained by regrouping the contributions (10.38) from each volume element. Using dimensional arguments one sees that this approximation is valid when the *potential varies sufficiently slowly* and the *temperature is sufficiently high* so that $\lambda_T |\nabla V| \ll kT$. One could use the Wigner representation to calculate the corrections (§§ 2.1.2 and 2.3.4). However, it can happen that the above condition is satisfied without the low density condition $e^{\alpha - \beta \varepsilon_0} \ll 1$ being satisfied. In that case the classical limit is valid only with regard to the summations over q, but quantum mechanics must be used for the form (10.31) of the occupation factors; the Pauli principle remains essential due to the high density.

10.4 Fermi-Dirac Statistics

10.4.1 Examples of Fermion Gases

A variety of systems can be described by the model of non-interacting fermions in a box.

A *metal* is a solid in which a number of *electrons* move freely within the sample, producing a high electrical and thermal conductivity. The other electrons remain strongly bound to the atomic nuclei. The resulting ions are fixed on a lattice and we assume that the conduction electrons move freely through that lattice. In fact, each of those electrons is subject to the combined potential of the other electrons and of the ion lattice. The model consists in replacing this potential to a first approximation by its average value over the lattice, taken to be constant inside the solid and becoming large outside it, since the Coulomb attraction from all other particles prevents an electron from escaping (see Fig.11.7). The difference between the potential outside and inside the metal is of the order of several eV, that is, very much larger than the temperature. We can therefore assume that the potential is infinite outside, except if we want to study phenomena when the electrons are pulled out of the metal (Exerc.10c); we choose the potential inside the metal as the origin of the energy, and we are thus back at the model of a box introduced in § 10.2.1. We may hope that this model will give us at least qualitative ideas about how the conduction electrons in a metal behave. We shall justify and improve it in Chap.11.

The number of free electrons depends on the valence of the metal. It is equal to 1 per atom for Cu, and 3 for Al. We can easily evaluate N/Ω: for Cu the atomic mass is 63.6 g mol^{-1} and the density 8.59 g cm^{-3}, which gives $N/\Omega = 0.85 \times 10^{29}$ m^{-3}, and hence distances between the electrons are of the order of 2 Å. On the other hand, the electron mass is small so that λ_T is much larger than for a gas of atoms or molecules: $\lambda_T = 43$ Å at room temperatures. Combining these results we find that $N\lambda_T^3/\Omega$ is very large (0.67×10^4 for the electron gas in Cu) and we have the opposite situation of the classical limit (10.50). In such a case where Fermi-Dirac statistics are essential, one often says that the fermion gas is "*degenerate*".

The first electron theory of metals[8] was worked out by Paul Drude (Brunswick 1863–Berlin 1906). It was based on the work of Hendrik Antoon Lorentz (Arnhem 1853–Haarlem 1928) who laid the foundations of a theory of the interaction between electromagnetic radiation and charged particles in matter in 1895, and that of Joseph John Thomson (Manchester 1856–Cambridge 1940) who showed that the "cathode rays" emitted by metals (Exerc.10c) consisted of light particles, the electrons, whose mass and charge he measured in 1897 and 1898, respectively. Drude succeeded in explaining metallic brightness and the fact that the electrical and thermal conductivities are proportional to one another (Wiedemann-Franz law, 1853) using the above model of free electrons in a box. However, he treated the electrons as particles of a classical perfect gas, obeying Boltzmann statistics, so that certain properties of metals such as their paramagnetism remained unexplained for another quarter of a century. One had to wait until 1926 before the new statistics following from the Pauli principle were applied by Fermi and by Dirac to the gas of electrons in a metal (§ 10.1.2). Pauli soon showed[9] that the paramagnetism of metals was due to the exclusion principle and the electron spin (Exerc.10b). It was mainly Arnold Sommerfeld (Königsberg 1868–Munich 1951) who developed the quantum electron theory of metals[10]. As we have already stressed, the classical approximation is completely unjustified so that Drude's explanation of the Wiedemann-Franz law appears to be a happy accident (§ 15.2.3).

---

[8] Ann. Physik (Leipzig) **1** (1900) 566 and **3** (1900) 369.
[9] Zs. Physik **41** (1927) 81.
[10] Zs. Physik **47** (1928) 1.

The inert gases with nuclei having an odd number of nucleons are another example of fermion gases since their atoms interact weakly with one another. Nonetheless, even for the lightest of them, the ^{3}He isotope of helium, condition (10.50) is well satisfied at ordinary pressures and temperatures: the ^{3}He mass is 5000 times that of the electron and the number density is of the order of 1000 times smaller than that of the electrons in a metal. Hence the perfect gas approximation is valid, and Fermi statistics bring in only small, though measurable, corrections (Exerc.10f). Only for *liquid helium 3* at low temperatures ($T < 3$ K) is the "degenerate" fermion gas model useful (§ 12.2). All other substances are found either in the solid state or in the classical gas or liquid states, depending on the temperature, as their atoms either are heavier or interact more strongly with one another.

The *atomic nuclei* consist of protons and neutrons which are fermions. They can to a first approximation be described as gases of non-interacting *nucleons* enclosed in a box, the size of which is that of the nucleus. This model is qualitatively correct, even though the actual interactions between nucleons are strong. The short-range part of these interactions is inhibited by the exclusion principle, whereas the average effect of their longer-range part is taken into account through the box potential which confines the nucleons to the nucleus. The density of nuclear matter is so large that Fermi-Dirac statistics play their full rôle and a nucleus is practically at zero temperature. However, the number of nucleons is such that the infinite volume approximation is, in general, insufficient.

The matter of a *neutron star* (Exerc.10e) consists to a first approximation of non-interacting neutrons. Notwithstanding the high temperature, the density is so huge that the matter behaves as a "degenerate" fermion gas. Similarly, in a *white dwarf* (Probs.9 and 10), the completely ionized matter consists of light nuclei and electrons; the latter produce a "degenerate" gas as in a metal, whereas the – heavier – nuclei satisfy condition (10.50) and behave like a classical gas.

10.4.2 Fermion Gas at Zero Temperature; Fermi Temperature

At zero temperature the canonical and grand canonical Gibbs distributions are the same, even if the system is finite. The N-fermion gas is in its ground state, obtained by putting these N fermions in the N single-particle states with lowest energies ε_q. (For the sake of simplicity we assume that the Nth and the $N + 1$-st level have different energies, so that the N-particle ground state is not degenerate.) The maximum energy ε_q attained in this filling up of states is called the *Fermi energy*, or *Fermi level*, ε_F. The number of particles in the gas is connected with ε_F through

$$N = \int_{-\infty}^{\varepsilon_F} d\varepsilon\, \mathcal{D}(\varepsilon) = \mathcal{N}(\varepsilon_F), \tag{10.55}$$

where we have used the definitions (10.44) of the function $\mathcal{N}(\varepsilon)$ and (10.40) of the distribution $\mathcal{D}(\varepsilon)$, while the energy of the gas equals

$$U = \int_{-\infty}^{\varepsilon_F} d\varepsilon \, \varepsilon \, \mathcal{D}(\varepsilon). \tag{10.56}$$

In the grand canonical formalism the Fermi factor, $f(\varepsilon) = \left[e^{\beta(\varepsilon-\mu)} + 1\right]^{-1}$, tends to the step function

$$f(\varepsilon) \underset{\beta \to \infty}{\longrightarrow} \theta(\mu - \varepsilon) = \begin{cases} 0, & \text{if } \varepsilon > \mu; \\ 1, & \text{if } \varepsilon < \mu, \end{cases} \tag{10.57}$$

which expresses the fact that the occupation numbers n_q are 1 when $\varepsilon < \varepsilon_F$ and 0 when $\varepsilon > \varepsilon_F$. (Using (10.32) and (10.57) we can check that there are no fluctuations.) The *Fermi energy* ε_F is the same as the *chemical potential at zero temperature*; in agreement with general principles, we can identify it with the energy needed to add a particle to the system, at constant volume and entropy.

In the case of a non-relativistic gas in a box, filling the single-particle states up to the Fermi level defines in momentum space the *Fermi surface*, a sphere, the radius of which, $p_F = \sqrt{2m\varepsilon_F}$, is called the Fermi momentum. It is useful to measure the Fermi energy in kelvins, in this way defining the *Fermi temperature* $\Theta_F = \varepsilon_F/k$. Using (10.45) and (10.55) we find for spin-$\frac{1}{2}$ particles

$$\Theta_F = \frac{p_F^2}{2mk} = \frac{\hbar^2}{2mk}\left(3\pi^2\frac{N}{\Omega}\right)^{2/3} = \frac{\mathcal{D}(\varepsilon_F)}{2k\mathcal{D}'(\varepsilon_F)}. \tag{10.58}$$

Numerically, for the free electrons in copper Θ_F is equal to 80 000 K, or, in energy units, 7 eV. (We could have expected this order of magnitude as ε_F is of the order of $\hbar^2/2md^2$, where $d \simeq 2$ Å is the average distance between electrons; this can be compared with the binding energy of 13.6 eV of a hydrogen atom, $\hbar^2/2ma_0^2$, where $a_0 = 0.53$ Å is the Bohr radius.) The Fermi temperature (10.58) is the "characteristic temperature" of the problem: it is the only quantity with the dimensions of a temperature which one can construct from the Fermi gas density or from the density of states. This is the temperature with which we must compare the actual temperature T of the metal. The very large value of Θ_F ensures that at all temperatures, even up to melting, we have $T \ll \Theta_F$. Hence the gas of the electrons in a metal is *always at very low temperatures* as compared to its characteristic temperature Θ_F, which means that the approximation $T = 0$ is a good one. Nonetheless, the velocity v_F of the electrons at the Fermi level is huge, as it equals

$$v_F = \sqrt{2k\Theta_F/m} \simeq 1500 \text{ km s}^{-1}. \tag{10.59}$$

Pauli's exclusion principle therefore results in giving the electrons very large characteristic velocities, even at absolute zero. The characteristic velocities of a classical gas of particles with the same mass would be much smaller at room temperatures, by a factor of $\sqrt{3T/2\Theta_F} \simeq 1/10$. Note that the quantity $N\lambda_T^3/\Omega$, defined in (10.49), behaves as $4(s+1)\alpha^{3/2}/3\sqrt{\pi}$ for $e^\alpha \gg 1$; it is thus the same as $(\Theta_F/T)^{3/2}$ when $T \ll \Theta_F$, apart from a factor $8/3\sqrt{\pi}$. The condition (10.50) for the validity of the classical approximation is thus exactly the opposite of the condition $T/\Theta_F \ll 1$ which expresses that the Fermi gas is "degenerate".

In the case of the non-relativistic Fermi gas, to which we are restricting ourselves here, the ground state *energy* (10.56), (10.43) is

$$U = \tfrac{3}{5} N \varepsilon_F, \tag{10.60}$$

where we have used (10.58). Due to the Pauli principle the mean energy per particle, $\tfrac{3}{5}\varepsilon_F$, is large, whereas it would have been zero for the Bose-Einstein and Boltzmann statistics; in the latter case it vanishes as $\tfrac{3}{2}kT$.

The *pressure*, given either by (10.47) or by the Gibbs-Duhem relation, at zero temperature equals

$$\mathcal{P} = \frac{1}{\Omega} \int_{-\infty}^{\varepsilon_F} d\varepsilon\, \mathcal{N}(\varepsilon) = \frac{1}{\Omega} (\varepsilon_F N - U). \tag{10.61}$$

Using (10.60) we get for a non-relativistic Fermi gas $\mathcal{P} = 2\varepsilon_F N/5\Omega$, whereas the pressure $\mathcal{P} = kTN/\Omega$ of a perfect gas tends to zero at zero temperature. It is just the large value (10.61) of the pressure of the neutrons in a neutron star (Exerc.10e) or of the electrons in a white dwarf (Prob.9) which enables the matter of those stars, for which $kT \ll \varepsilon_F$, to resist gravitational collapse.

To summarize, the exclusion principle, by forcing the fermions to occupy *distinct* single-particle states, considerably raises their typical velocities. Hence, at temperatures $T \ll \Theta_F$, the internal energy and the pressure take on values (10.60) and (10.61) which are much larger than the classical perfect gas values, by a factor $2\Theta_F/5T$, and which remain practically constant in that temperature range.

10.4.3 Equilibrium Properties at Low Temperatures

The properties of a fermion gas at finite temperatures are governed by the shape of the Fermi factor $f_q \equiv f(\varepsilon_q)$, which appeared in (10.34–37) or in (10.42), (10.47), and which equals

$$f(\varepsilon) = \frac{1}{e^{(\varepsilon-\mu)/kT} + 1} = \frac{1}{2} - \frac{1}{2}\tanh\frac{\varepsilon-\mu}{2kT}. \tag{10.62}$$

It is essential to master a feeling for the shape of this function.

Given that the occupation number n_q of a state q can only take on the two values 0 and 1, the Fermi factor, which is its expectation value, must necessarily lie between 0 and 1. It is a decreasing function of ε_q, as should be

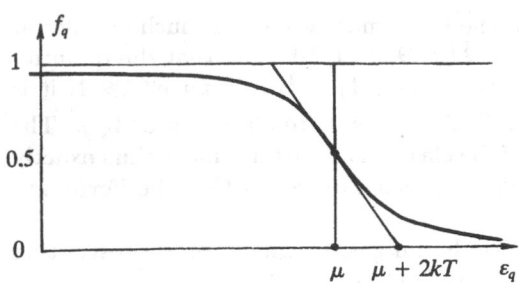

Fig. 10.1. The Fermi factor

the case: states q with a high energy are less populated than those with a low energy. Figure 10.1 shows that the Fermi factor f_q, as function of the energy ε_q, has a sudden drop around the value $\varepsilon_q = \mu = \alpha/\beta$, where the energy of the single-particle state q equals the chemical potential μ. At sufficiently low temperatures μ is close to ε_F as defined by (10.55), and one usually still calls μ the Fermi level. The lower the temperature, the faster the variation of f_q around the Fermi level; the width of the region where f_q decreases rapidly is of the order of $4kT$. As a result, at very low temperatures the states situated below the Fermi level ($\varepsilon_q < \mu$) are practically *all occupied* by a fermion and the states situated above it are practically *all empty*. When the system is heated up, particles are excited from states q below the Fermi level to states above it, in a range $|\varepsilon_q - \mu|$ of order kT. Thus, the occupied states become partially depopulated: to express that particles have been taken away, one says that "*holes have been created*" in the states $\varepsilon_q < \mu$. On the other hand, the empty states $\varepsilon_q > \mu$ become partially filled.

Note the *symmetry of the $f(\varepsilon)$-curve around the Fermi level μ*. In fact, we see from (10.62) that if we change the sign of the energy $\varepsilon_q - \mu$ of the state q, measured with respect to the Fermi level, f_q transforms to $1 - f_q$. This symmetry reflects a *symmetry between holes and particles*: $n_q = 1$ means that the state q is occupied by one particle, or equivalently, that there is no hole present, whereas $n_q = 0$ means a state q without a particle, or equivalently, the presence of a hole. Thus, whereas $\widehat{n}_q$ is the number operator for the particles occupying the state q, $\widehat{I} - \widehat{n}_q$ is the number operator for holes in the state q; if f_q is the average number of particles in the state q, $1 - f_q$ can be interpreted as the average number of holes in that state. When one creates a hole, the energy and the particle number decrease, which explains the change in sign of $\varepsilon_q - \mu$ in the particle-hole symmetry. This symmetry is also very evident in expression (10.37) for the entropy of a gas of fermions ($\eta = -1$), which is invariant under an exchange of f_q and $1 - f_q$.

The symmetry of the Fermi factor becomes clear, if we note the formal analogy between the calculation of (10.28b) and that of the partition function (1.13') of the spins in a paramagnetic crystal. The sites i are here replaced by the single-particle states q; the quantum number σ_i, which took on the two values $+1$ and -1, is replaced by the occupation number n_q, which takes on the two values 0 and 1 so that $2n_q - 1$ plays the rôle of σ_i. Expression (10.62) must be compared formally with the average magnetic moment (1.37) of a paramagnetic ion, and the $f_q \leftrightarrow 1 - f_q$ symmetry corresponds to the obvious $\langle \sigma_i \rangle \leftrightarrow \langle -\sigma_i \rangle$ symmetry.

When one is far from the Fermi level on either side, f tends exponentially to 0 or 1. More precisely, we have

$$f(\varepsilon) \sim e^{-(\varepsilon-\mu)/kT} \quad \text{when} \quad \varepsilon - \mu \gg kT, \tag{10.63a}$$

$$1 - f(\varepsilon) \sim e^{-(\mu-\varepsilon)/kT} \quad \text{when} \quad \mu - \varepsilon \gg kT. \tag{10.63b}$$

Compared with (10.51), expression (10.63a) means that for sufficiently high energy levels ε_q the particles behave classically. In fact, when the probability for the occupation of a single-particle state q is small, the Fermi-Dirac statistics reduce to the Boltzmann statistics. In particular, the classical limit of § 10.3.4 corresponded to a situation where the condition $\varepsilon_q - \mu \gg kT$ was implied by (10.50) for all single-particle states q. The two limits (10.63a and b) are related to each other by the hole-particle symmetry, and Eq.(10.63b) means that the holes behave classically when they are sufficiently far below the Fermi level. This remark will be very important when we study insulators and semiconductors (§ 11.3).

The behaviour of the Fermi factor $f(\varepsilon)$ enables us to understand qualitatively a number of properties of metals and other degenerate fermion gases. When T/Θ_F is small, $f(\varepsilon)$ does not differ significantly from its limit (10.57) except in a narrow range with a width of the order of kT around the Fermi level. When the material is subjected to a weak external action, such as an electric or magnetic field, or heating, the only transitions allowed are the ones for which the particles gain or lose little energy. That, however, is possible only for particles whose energies lie close to μ, as the states which lie far inside the Fermi surface are practically completely occupied and hence the particles remain frozen to them. This is the reason why the gas is called "degenerate": the electrons which take part in the infinitesimal changes of state have practically the same energy $\varepsilon_q \simeq \mu$. Their proportion is very small, even though they are responsible for important effects.

For example, the electrons in a metal move rapidly – the Fermi velocity (10.59) is of the order of $10^6\,\mathrm{m\,s^{-1}}$ – but when averaged their motions produce a zero current when there is no field. The electrical conduction, a non-equilibrium phenomenon which we shall study in § 15.2, is due to a differential effect in the vicinity of the Fermi velocity: the only single-electron states of which the average occupation can change under the action of an applied field are those whose energy is close to the Fermi energy, as an electron cannot be scattered into a state which is already occupied. Therefore, amongst the electrons with average velocities which are in absolute magnitude close to v_F there are more which move in one direction than in the other one, and this produces a net non-vanishing current.

The explanation of most of the other metallic properties is equally based upon merely considering the single-electron states close to the Fermi surface.

The *Pauli paramagnetism* associated with the electron spin (Exerc.10b) is, for instance, calculated using Sommerfeld's expansion (10.64) like the specific heat, which we evaluate below; the Fermi-Dirac statistics of the electrons lead to characteristics which are very different from those of the paramagnetism of localized spins (Chap.1). In particular, the magnetic susceptibility tends to a constant value at low temperatures, rather than following a $1/T$ Curie law. Moreover, there is a diamagnetic contribution associated with the orbital motion of the electrons: the *Landau diamagnetism* (Prob.11).

The *Volta effect*, that is, the appearance of an electrostatic potential difference between two different metals which are brought into contact, can be explained by noting that their chemical potentials μ, which are *a priori* unequal when the samples are separated, must become equal when it is possible to exchange electrons. The transfer of electrons from the metal which initially had the higher chemical potential towards the other metal creates then near the boundary a double layer of charges, positive on the side of the first metal and negative on the side of the second metal; this itself produces a potential difference equal to the initial difference between the chemical potentials (§ 11.3.5). The *thermionic effect* (Exerc.10c) is the emission of electrons by a heated metal. In order to understand it, we must improve the model and take into account the fact that the potential outside the box which confines the electrons is finite, equal to V. Thermally excited electrons with kinetic energies larger than V inside the metal may escape when they hit the wall; the number of electrons which is thus emitted is controlled by the Fermi factor (10.63a) in the range of energies higher than V. The emission of an electron current by a heated cathode leads to numerous practical applications, such as the electron gun in a television tube, or high-power rectifying and amplifying lamps. The applications of the *photoelectric effect* (cells) are just as well known. Here, the absorption of a photon with a sufficiently high frequency allows an electron to acquire sufficient energy to leave the sample, and ultimately to be captured by the anode. The abrupt shape of the Fermi factor explains the existence of a threshold for the photon frequency, which must be higher than $(V - \mu)/h$. Above this threshold, the luminous energy is transformed into electrical energy.

To obtain quantitative results we must for $T \ll \Theta_F$ expand the Fermi factor $f(\varepsilon)$ in the vicinity of its zero temperature limit (10.57). The terms of lowest order in T/Θ_F will, generally speaking, be sufficient for our purposes. Nevertheless, the limiting form $-\delta(\varepsilon - \mu)$ of $\partial f/\partial \varepsilon$ suggests that we must look for an expansion of $f(\varepsilon)$ in the sense of distributions. We therefore shall consider the following integral, which is of the general type of the thermodynamic quantities we want to evaluate:

$$\int d\varepsilon \, f(\varepsilon) \, \varphi(\varepsilon) \, - \, \int_{-\infty}^{\mu} d\varepsilon \, \varphi(\varepsilon),$$

where φ is a regular test function. We distinguish in this expression the regions $\varepsilon > \mu$ and $\varepsilon < \mu$, and use as variable $x = \beta(\varepsilon - \mu)$ in order to stretch the dominant region $\varepsilon - \mu \lesssim kT$. We find

$$\frac{1}{\beta} \int_0^\infty dx \, \frac{1}{e^x + 1} \varphi\left(\mu + \frac{x}{\beta}\right) - \frac{1}{\beta} \int_{-\infty}^0 dx \left[1 - \frac{1}{e^x + 1}\right] \varphi\left(\mu + \frac{x}{\beta}\right)$$

$$= \frac{1}{\beta} \int_0^\infty dx \, \frac{1}{e^x + 1} \left[\varphi\left(\mu + \frac{x}{\beta}\right) - \varphi\left(\mu - \frac{x}{\beta}\right)\right] .$$

Changing x to $-x$ in the second term we have used the hole-particle symmetry. We now only have to expand φ as $\beta \to \infty$, and we then get the asymptotic expansion (see formulas at the end of this volume)

$$\frac{2}{\beta} \int_0^\infty dx \, \frac{1}{e^x + 1} \sum_{n=1}^\infty \left(\frac{x}{\beta}\right)^{2n-1} \frac{1}{(2n-1)!} \left(\frac{d}{d\mu}\right)^{2n-1} \varphi(\mu)$$

$$= \sum_{n=1}^\infty 2\left(1 - 2^{-2n+1}\right) \zeta(2n) \, (kT)^{2n} \left(\frac{d}{d\mu}\right)^{2n-1} \varphi(\mu).$$

As a consequence we can write down the *low-temperature expansion of the Fermi factor*, called the *Sommerfeld expansion*:

$$\boxed{f(\varepsilon) = \theta(\mu - \varepsilon) - \frac{\pi^2}{6}(kT)^2 \delta'(\varepsilon - \mu) - \cdots} , \qquad (10.64)$$

which means that, for any regular function $\varphi(\varepsilon)$,

$$\int d\varepsilon \, f(\varepsilon) \, \varphi(\varepsilon) \underset{T \to 0}{\approx} \int_{-\infty}^\mu d\varepsilon \, \varphi(\varepsilon) + \frac{\pi^2}{6}(kT)^2 \, \varphi'(\mu). \qquad (10.64')$$

The thermodynamic properties of fermion gases at low temperatures are obtained by replacing $f(\varepsilon)$ by (10.64) in Eqs.(10.37), (10.42), (10.47). Using (10.44) we thus get for the expansion of the grand potential (10.47):

$$A \approx -\int_{-\infty}^\mu d\varepsilon \, \mathcal{N}(\varepsilon) - \frac{\pi^2}{6}(kT)^2 \, \mathcal{D}(\mu). \qquad (10.65)$$

Similarly we get from (10.37), (10.42), and (10.64) the particle number, the internal energy, and the entropy as functions of T and μ:

$$N \approx \mathcal{N}(\mu) + \frac{\pi^2}{6}(kT)^2 \, \mathcal{D}'(\mu), \qquad (10.66)$$

$$U \approx \int_{-\infty}^\mu d\varepsilon \, \varepsilon \, \mathcal{D}(\varepsilon) + \frac{\pi^2}{6}(kT)^2 \left[\mathcal{D}(\mu) + \mu \mathcal{D}'(\mu)\right], \qquad (10.67)$$

$$S \approx \frac{\pi^2}{3} k^2 T \, \mathcal{D}(\mu). \tag{10.68}$$

These expressions can also be derived from (10.65). For the non-relativistic gas model describing the electrons in a metal we can use (10.43) and (10.45) to write them in the form

$$N \approx \frac{\Omega}{3\pi^2 \hbar^2} (2m\mu)^{3/2} \left[1 + \frac{\pi^2}{8} \left(\frac{kT}{\mu} \right)^2 \right], \tag{10.69}$$

$$U \approx \frac{\Omega}{5\pi^2 \hbar^2} (2m\mu)^{3/2} \mu \left[1 + \frac{5\pi^2}{8} \left(\frac{kT}{\mu} \right)^2 \right], \tag{10.70}$$

$$S \approx \frac{\Omega}{6\hbar^2} (2m\mu)^{3/2} \frac{k^2 T}{\mu}. \tag{10.71}$$

Comparing (10.42) with (10.47) we see moreover that when $\mathcal{D}(\varepsilon) \propto \varepsilon^{1/2}$ the pressure is related to the energy through

$$P = \frac{2U}{3\Omega}, \tag{10.72}$$

a relation which is valid at all temperatures. The origin of this relation, which is also satisfied by the perfect gas, is discussed in Exerc.13a.

In order to use (10.69–71) to find the thermal properties of a sample with a fixed number N of particles we must eliminate μ, which varies with the temperature. Comparing (10.69) with its limit (10.55) at zero temperature, which defines Θ_F as function of the density N/Ω, we get

$$\mu \approx \varepsilon_F \left[1 - \frac{\pi^2}{12} \left(\frac{T}{\Theta_F} \right)^2 \right]. \tag{10.73}$$

As the temperature rises, the low-energy states empty to the advantage of the high-energy states; however, the latter lie more densely and it is necessary to *decrease the chemical potential* in order to keep the number of particles constant. Substituting (10.73) into (10.70) we find the correction to (10.60):

$$U \approx \frac{3}{5} N \varepsilon_F \left[1 + \frac{5\pi^2}{12} \left(\frac{T}{\Theta_F} \right)^2 \right]. \tag{10.74}$$

Hence we find the *specific heat at constant volume* at temperatures $T \ll \Theta_F$:

$$C = \frac{dU}{dT} \approx \frac{\pi^2}{2} Nk \frac{T}{\Theta_F}. \tag{10.75}$$

Note that the variation of μ with temperature has played a part in this calculation; if we had taken the derivative of (10.70) with respect to T with μ constant, we would have obtained a wrong result. We could also have derived Eq.(10.75) from (10.69) and (10.71) which give us $S = \pi^2 NkT/2\Theta_F$. The fact that S/N vanishes with the temperature expresses that the *Nernst principle* is satisfied.

The specific heat of metals thus contains a *linear contribution* (10.75) which comes from the conduction electrons. In Chap.11 we shall show that this contribution, which is swamped at room temperature by those from the other degrees of freedom of the crystal, mainly the vibrations of the nuclei, dominates at low temperatures (see the curve of the specific heat of potassium in Fig.11.22). A comparison of (10.75) with the specific heat of a classical perfect gas, $\frac{3}{2}Nk$, shows that the Pauli principle has the effect of introducing a small factor $\pi^2 T/3\Theta_F$. This corresponds to the fact that, in a metal, one can thermally excite only a small fraction, of the order of T/Θ_F, of the electrons, namely those which are situated in a layer with a thickness of the order of kT around the Fermi surface, while the other, deeper, electrons remain frozen in. The effective number of degrees of freedom which contribute to the specific heat is thus reduced by a factor of the order of kT/ε_F.

10.5 Bose-Einstein Statistics

10.5.1 Examples of Boson Gases

The formalism developed in § 10.3 applies to quantum gases for which the energy and the number of particles are constants of the motion. There are few examples of physical systems of bosons for which this model is relevant; the most notable one is *helium* in the form of its most common ^{4}He isotope. We discussed in § 10.4.1 that the temperature must be very low, of the order of a few K, in order that this fluid is not a classsical gas or fluid. The model where one neglects the interactions between the helium atoms is a very coarse one, but it gives a qualitative explanation of several phenomena (Chap.12) and it can be improved (Prob.14).

In Nature there are also boson systems for which the *number of particles is not conserved* and for which these particles can be *created or absorbed by their interactions* with other particles. For instance, a *photon* gas in an enclosure (Chap.13) represents in quantum mechanics the electromagnetic field in that enclosure. The exchange of energy between the field and the charged particles within the walls corresponds to a change in the state of the field, that is, to a change in the number of photons. The typical elementary process is the one where an electron in the wall changes its quantum state, emitting or absorbing a photon (§§ 10.2.3, 13.1.3, and 13.1.4).

Another example, studied in § 11.4, is that of *phonons* which are particles representing the quantized mechanical oscillations in a solid. Here also, exchange of energy with the other degrees of freedom is accompanied by the creation or annihilation of phonons.

10.5.2 Chemical Potentials of Boson Gases

When the *number of particles is conserved* the thermodynamics of a boson gas follows from Eqs.(10.34–37) with $\eta = +1$, and the Bose factor (10.31) is equal to

$$f_q = \left[1 - e^{(\varepsilon_q - \mu)/kT}\right]^{-1} = e^{-(\varepsilon_q - \mu)/kT}\left(1 + f_q\right);\qquad (10.76)$$

it is shown in Fig.10.2. The summation of the series (10.28a) for the grand partition function, the necessity for the occupation probability (10.30) to be bounded as $n_q \to \infty$, or the fact that f_q must be positive, imply that we have

$$\varepsilon_q - \mu > 0 \quad \text{for all} \quad q. \qquad (10.77)$$

As a result, whereas for a fermion gas μ can have any value, for a boson gas it is necessary that the chemical potential is lower than the lowest energy level ε_q. In particular, in the case of a gas of bosons of mass m enclosed in a box with energies (10.9), the *chemical potential must be negative* at all temperatures. This means that, as in the case of a classical gas (§ 8.1.4), the energy of the system decreases when one adds a particle to it: whereas a Fermi gas at low temperatures with a positive chemical potential has a tendency to *yield* particles, a Bose gas has the tendency to *absorb* them. Even though the particles do not interact, the symmetry or antisymmetry of the wavefunctions thus produces an effect which is similar to that of a force. Fermions seem to repel one another, which is easy to understand as the presence of a fermion in a state q saturates that state and forces the other fermions to go to the other states q'. In contrast, bosons seem to attract one another, since the free energy decreases more than for the case of a classical gas when one adds particles (Exerc.10g).

The limit $\alpha \equiv \mu/kT \to -\infty$ corresponds, as in the case of fermions, to the classical low-density limit, and (10.76) then reduces to the Boltzmann exponential. The high-density limit is found here as $\alpha \to 0$, in which case each low-energy state $\varepsilon_q \sim |\mu|$ is occupied by a very large number of bosons. We shall study this effect, the so-called *Bose condensation*, in § 12.3.1. It is a remarkable demonstration of the gregarious nature of bosons, resulting from

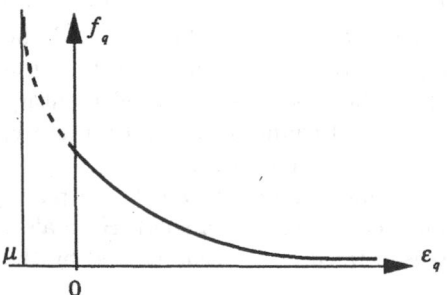

Fig. 10.2. The Bose factor

the fact that the Bose factor (10.76) has no upper bound as $\varepsilon_q - \mu \to 0$, in contrast to the Fermi and Maxwell factors.

For a boson gas in which the *number is not conserved* such as the photon gas, there is only one constraint, on the energy, instead of two independent constraints, on the energy and on the number of particles. The particle number is no longer a constant of the motion; the thermal equilibrium state is found by looking for the maximum of the entropy, given the total energy, but *leaving N unspecified*. One needs introduce solely a single Lagrangian parameter, β, associated with the energy, so that we find for the density operator in Fock space

$$\widehat{D} = \frac{1}{Z} e^{-\beta\widehat{H}}. \tag{10.78}$$

Comparing this with the grand canonical density operator we see that not introducing a constraint on $\langle N \rangle$ amounts to *putting the chemical potential $\mu = \alpha/\beta$ equal to zero*. The evaluation of the various thermodynamic quantities is thus the same as in § 10.3 but with a Bose factor equal to

$$f(\varepsilon) = \frac{1}{e^{\beta\varepsilon} - 1}. \tag{10.79}$$

The free energy F associated with the canonical distribution (10.78) is the same as the grand potential (10.47) with μ put equal to 0. The average number of particles,

$$N = \sum_q \frac{1}{e^{\beta\varepsilon_q} - 1}, \tag{10.80}$$

is no longer an independent variable which one can use to determine μ, as in the case of systems with a conserved number of particles; when the temperature is given, (10.80) determines automatically the average particle number.

Another way to understand why the chemical potential of a photon gas is zero in equilibrium consists of using the properties of equilibria of the kind occurring in chemistry (§ 6.6.3). Here the photon can, when in contact with a wall, undergo annihilation or creation reactions, while the number of the constituent particles of the wall remains unchanged. Using the language of chemical reactions, this is, if we disregard the particles which are only spectators, equivalent to $\gamma \leftrightarrows 0$, where γ denotes the photon. There is thus one allowed reaction of the type (6.74), with $\nu = 1$ for the photon. The general relation (6.78) characterizing the equilibrium then simply gives $\mu = 0$.

The formal analogy between (10.79) and the expressions found when we studied the equilibrium of the quantum harmonic oscillator (Exerc.4f) is not by chance. We shall, in fact, show (§§ 11.4 and 13.1) that a set of quantized – mechanical or electromagnetic – oscillation modes is the same as a boson gas with non-conserved particle number: these are two equivalent descriptions of the same physical situation.

Summary

The Pauli principle allows us to classify particles as fermions (electrons, protons, neutrons, ...) or bosons (photons, nuclei with an even number of nucleons, ...). The N-particle micro-states are constructed from the single-particle states q by giving the occupation numbers n_q of each of them. For fermions n_q can take on the values 0 and 1, and for bosons it can take on the values, $0, 1, 2, \ldots, \infty$.

When there are no interactions between the particles, the grand canonical partition function (10.29) can be factorized into contributions associated with each of the eigenstates of the single-particle Hamiltonian. The thermodynamic quantities can be expressed in terms of the Fermi or Bose factors (10.31) which are the expectation values of the occupation numbers. In the case of particles in a macroscopic box, these quantities become integrals, either over the momentum, (10.38), of one particle or over its energy, (10.41), with the single-particle density of states as weight function. The Fermi or Bose factor then accounts for the nature of the particles, the temperature, and the density of the gas, while the density of states accounts for the single-particle energies. When the density is sufficiently low and the temperature sufficiently high (molecular gases) we come back to the perfect gas.

The opposite conditions are satisfied for the gas of electrons in a metal, since room temperatures are very low as compared to the characteristic Fermi temperature (10.58). Most properties of the system are then governed by the neigbourhood of the Fermi surface. They can be evaluated by using the expansion (10.64) of the Fermi factor. Far above or far below the Fermi level, the Fermi factor tends to 0 or 1 according to (10.63). Particles and holes play symmetric rôles in Fermi-Dirac statistics.

The Bose factor diverges as $\varepsilon - \mu \to 0$. The chemical potential is negative for bosons if the particle number is conserved (^{4}He) and zero if the particle number is not conserved (photons).

Exercises

10a Indistinguishable Particles in a Harmonic Potential

1. We consider a harmonic potential well in which there may be present one or more non-interacting particles. What is the spectrum of the single-particle states? What are the changes in the canonical and in the grand canonical partition functions, in the internal energy, and the chemical potential – for fixed $\langle N \rangle$ – when one changes the zero of the energies, replacing ε_q by $\varepsilon_q' \equiv \varepsilon_q + \delta$? Take afterwards the ground state energy as the origin of the energy and write $e^{-\beta\hbar\omega} = y$.

2. Evaluate the canonical partition function

    – for a single particle;
    – for two distinguishable particles;
    – for two fermions, assuming they have no spin;
    – for two spin-zero bosons;
    – for two spin-$\frac{1}{2}$ fermions.

3. Compare the internal energies and the entropies in these cases. Study the limits as $T \to 0$, as $T \to \infty$, and as $\hbar \to 0$. Explain the results.

4. Write the grand partition functions for spin-zero fermions and bosons and for spin-$\frac{1}{2}$ fermions in the form of series. Write down the relations between μ and $\langle N \rangle$ for fixed $\langle N \rangle$ in the case of high and low temperatures.

5. Check the results of 2, starting from the form of the grand partition functions.

Solution:

1. The energy levels of the harmonic oscillator are given by $\varepsilon_q = (q + \frac{1}{2})\hbar\omega$, where the single quantum number q takes on the values 0, 1, 2, ... When ε_q is replaced by $\varepsilon'_q = \varepsilon_q + \delta$, Z_C becomes

$$Z'_C \equiv \sum e^{-\beta E'} = e^{-N\beta\delta} Z_C,$$

where $E' \equiv \sum_q \varepsilon'_q n_q$, and where the n_q must satisfy the constraint (10.15). The internal energy becomes $U' = -\partial \ln Z'_C / \partial \beta = U + N\delta$, and the change in the free energy is the same, whereas the entropy remains unchanged. The grand partition function becomes

$$Z'_G(\beta, \alpha') = \sum e^{-\beta E' + \alpha' N} = \sum e^{-\beta E + (\alpha' - \beta\delta)N}$$
$$= Z_G(\beta, \alpha' - \beta\delta).$$

For a given particle number $\langle N \rangle$, $\partial \ln Z'_G / \partial \alpha' = \partial \ln Z_G / \partial \alpha$ implies $\alpha' = \alpha + \beta\delta, \mu' = \mu + \delta$. The chemical potential changes along with the single-particle energies, but the value of the grand partition function remains unchanged.

2. The single-particle canonical partition function is

$$Z_1 = \sum_{q=0}^{\infty} y^q = \frac{1}{1-y}.$$

For N distinguishable particles we have $Z_N = Z_1^N$, and, in particular,

$$Z_2^D = \sum_{q,q'} y^{q+q'} = \frac{1}{(1-y)^2}.$$

For two spin-zero fermions the states, which must have antisymmetric wavefunctions, are characterized by stating which two different levels are occupied. Hence we have

$$Z_2^{F0} = \sum_{q'>q\geq0} y^{q+q'} = \frac{1}{2}Z_2^{D} - \frac{1}{2}\sum_{q=0}^{\infty} y^{2q} = \frac{y}{1+y}\frac{1}{(1-y)^2}.$$

For two bosons the two particles can occupy any two states, different or not, whence

$$Z_2^{B} = \sum_{q'\geq q\geq0} y^{q+q'} = Z_2^{F0} + \sum_{q=0}^{\infty} y^{2q} = \frac{1}{1+y}\frac{1}{(1-y)^2}.$$

For two spin-$\frac{1}{2}$ fermions the three triplet states, symmetric in the spins, must be associated with an antisymmetric orbital wavefunction and contribute $3Z^{F0}$; the singlet state, antisymmetric in the spins, must be associated with a symmetric orbital wavefunction, as for bosons. Combining the two, we get

$$Z_2^{F} = 3Z_2^{F0} + Z_2^{B} = \frac{1+3y}{1+y}\frac{1}{(1-y)^2}.$$

3. The internal energy is given by $U \equiv \hbar\omega u = -\partial \ln Z_C/\partial\beta = \hbar\omega y \partial \ln Z_C/\partial y$,

or

$$u^{D} = \frac{2y}{1-y}, \qquad u^{F0} = \frac{2y}{1-y} + \frac{1}{1+y},$$

$$u^{B} = \frac{2y}{1-y} - \frac{y}{1+y}, \qquad u^{F} = \frac{2y}{1-y} + \frac{2y}{(1+3y)(1+y)}.$$

In each case the entropy equals $S = k(\ln Z - u\ln y)$.
For a fixed temperature, we find

$$U^{B} < U^{D} < U^{F} < U^{F0}.$$

Even though the particles are non-interacting, the exclusion principle has the same effect as a repulsion, which is less pronounced if the fermions have spin $\frac{1}{2}$ since in that case two fermions can have the same orbital quantum number q. In contrast, bosons have a lower energy than distinguishable particles, as if they attracted one another.

We also find that for a fixed temperature the entropies and the free energies show the following ordering:

$$S^{F0} = S^{B} < S^{D} < S^{F}, \qquad F^{F0} > F^{B} > F^{D} > F^{F},$$

whereas for a fixed energy

$$S^{F0} < S^{B} < S^{D} < S^{F}.$$

One can understand these inequalities for the entropies by interpreting $e^{S/k}$ as an average number of configurations which are available at the given temperature or energy: this number is larger for distinguishable particles because of the possibility of exchanging particles, and even larger when the spin is non-zero, as there then exists an additional freedom.

At low temperatures, $y \to 0$, and

$$u^D \sim 2y, \quad u^{F0} \sim 1+y, \quad u^B \sim y, \quad u^F \sim 4y.$$

At zero temperature the exclusion principle forces the two spin-zero fermions to occupy the two levels $q = 0$ and $q = 1$, whence $u = 1$, whereas in the other cases it is possible for both particles to be in the $q = 0$ level, and hence $u = 0$. All entropies tend to zero, as

$$S^{F0} = S^B \sim \tfrac{1}{2} S^D \sim \tfrac{1}{4} S^F \sim ky(1 - \ln y).$$

At high temperatures and also in the classical limit we find for all cases, through expanding around $y \approx 1 - \hbar\omega/kT$,

$$U \sim 2kT,$$

in agreement with the equipartition theorem, and $S \sim 2k \ln(kT/\hbar\omega)$. In the expressions for U and S there also occur constants which depend on the system considered.

The results are gathered in Figs.10.3 and 10.4 which show $U(T)$, $S(T)$, and $S(U)$, with $dS/dU = 1/T$.

4. By applying (10.29) we find readily

$$\ln Z_G^{F0} = \sum_{q=0}^{\infty} \ln\left(1 + e^\alpha y^q\right),$$

$$\ln Z_G^B = -\sum_{q=0}^{\infty} \ln\left(1 - e^\alpha y^q\right),$$

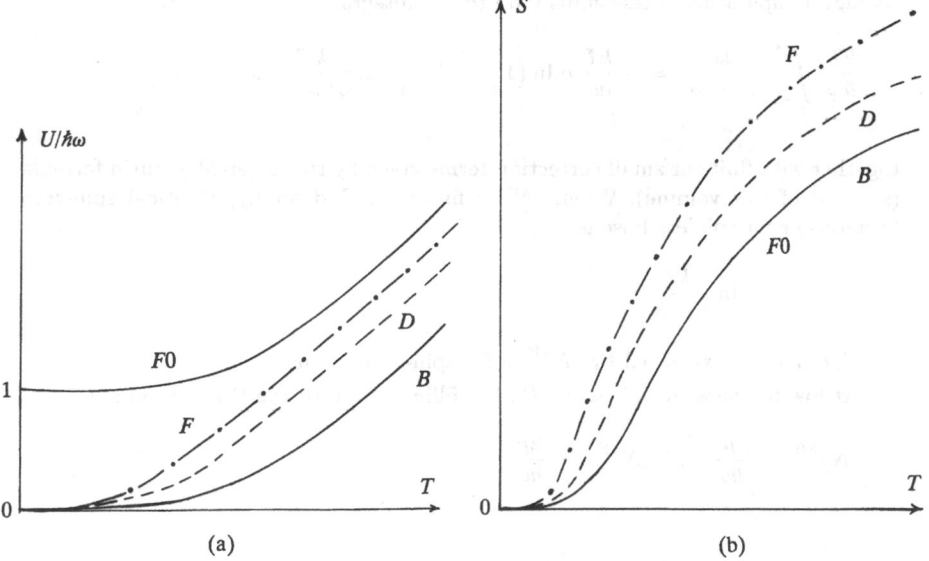

Fig. 10.3. The internal energy (**a**) and the entropy (**b**) as functions of the temperature

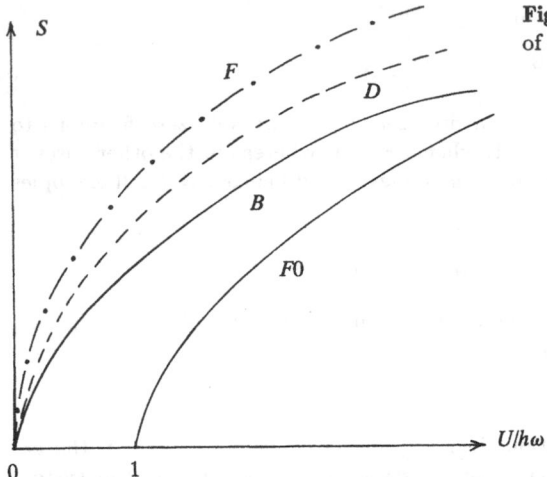

Fig. 10.4. The entropy as function of the internal energy

$$\ln Z_{\mathrm{G}}^{\mathrm{F}} = 2 \ln Z_{\mathrm{G}}^{\mathrm{F0}}.$$

The number of particles is given by

$$\langle N \rangle^{\mathrm{F0}} = \frac{1}{2} \langle N \rangle^{\mathrm{F}} = \sum_q \frac{1}{e^{\beta \hbar \omega q - \alpha} + 1},$$

$$\langle N \rangle^{\mathrm{B}} = \sum_q \frac{1}{e^{\beta \hbar \omega q - \alpha} - 1}.$$

At high temperatures these sums tend to the integrals

$$\frac{kT}{\hbar \omega} \int_{-\alpha}^{\infty} \frac{dx}{e^x - \eta} = -\frac{kT}{\hbar \omega} \eta \ln \left(1 - \eta e^{\alpha}\right) \underset{\alpha \to -\infty}{\widetilde{}} \frac{kT}{\hbar \omega} e^{\alpha},$$

together with finite or small correction terms given by the Euler-Maclaurin formula (see end of this volume). When $\langle N \rangle$ is finite, we find for hypothetical spin-zero fermions or for spinless bosons

$$\mu \sim -kT \ln \frac{kT}{\hbar \omega \langle N \rangle},$$

and $\langle N \rangle$ must be replaced by $\langle N \rangle^{\mathrm{F}}/2$ for spin-$\frac{1}{2}$ fermions.

At low temperatures, $T \ll \hbar \omega / k$, the filling-up up to the Fermi level gives

$$\langle N \rangle^{\mathrm{F0}} \sim \frac{\mu}{\hbar \omega} \quad , \quad \langle N \rangle^{\mathrm{F}} \sim \frac{2\mu}{\hbar \omega}.$$

For bosons, only the $q = 0$ state is occupied, and we find

$$\langle N \rangle^{\mathrm{B}} \sim \frac{1}{e^{-\alpha} - 1} \quad , \quad \mu \sim -kT \ln\left(1 + \frac{1}{\langle N \rangle}\right).$$

5. By looking at the definition of the grand partition function Z_{G} we see that the canonical partition function for two particles is the term with $e^{2\alpha}$ in Z_{G}. Hence, expanding the expressions sub 4 up to second order in e^{α} enables us to get the expressions sub 2. If we wanted to evaluate directly the canonical partition function for $N > 2$, the counting would become cumbersome. As in Exerc.4f, using the grand canonical ensemble solves this combinatorial problem.

10b Pauli Paramagnetism

We model the conduction electrons in a metal as a gas of free non-interacting fermions in a box. We apply a uniform magnetic field B parallel to the z-axis and neglect its interaction with the currents, but not with the electron spins. The magnetic moment $\widehat{M}$ equals $-\mu_{\mathrm{B}} \sum_i \widehat{\sigma}_i$, where $\sigma_i = 2S_i^z/\hbar$ takes on the values ± 1. Calculate the grand potential. Use it to find the spin susceptibility χ and its limits at room temperatures and at high temperatures.

We generalize the model by taking, for $B = 0$, an arbitrary density of states $\mathcal{D}(\varepsilon)$. Show that χ is proportional to $\partial N/\partial \mu$. Find χ and its limits in terms of $\mathcal{D}(\varepsilon)$.

Hints. The single-particle states characterized by their momentum p and their spin σ have energies $\varepsilon_q = p^2/2m + \mu_{\mathrm{B}} B \sigma$. We get the grand potential by using (10.29) and (10.38), and its derivative with respect to B is $-\langle M \rangle$, whence, using the fact that $\partial \mu/\partial B|_N = 0$ when $B = 0$, we get

$$\chi = \frac{1}{\Omega} \frac{\partial M}{\partial B}\bigg|_{N,B=0} = \frac{1}{\Omega} \frac{\partial M}{\partial B}\bigg|_{\mu,B=0} = \frac{1}{\beta\Omega} \frac{\partial^2 \ln Z_{\mathrm{G}}}{\partial B^2}$$

$$= \frac{\beta\mu_{\mathrm{B}}^2}{2h^3} \int \frac{d^3 p}{\cosh^2\left[\frac{1}{2}\beta\left(p^2/2m - \mu\right)\right]}.$$

At room temperatures, $T \ll \Theta_{\mathrm{F}}$, only the Fermi surface contributes and we find the finite positive (paramagnetic) Pauli susceptibility $\chi = 3\mu_{\mathrm{B}}^2 N/2\varepsilon_{\mathrm{F}}\Omega$. At high temperatures, $T \gg \Theta_{\mathrm{F}}$, we would have $\chi \sim \mu_{\mathrm{B}}^2 N/kT\Omega$ which is the same Curie law as in Chap.1: the exclusion principle does no longer operate and everything behaves as if we had N independent spins. Note, however, that the limit $T \gg \Theta_{\mathrm{F}}$ is unrealistic for metals.

In the general case, we note that a shift in the energies ε_q is equivalent to the opposite shift in μ so that

$$A(T, \mu, B) = \frac{1}{2}\left[A(T, \mu - \mu_{\mathrm{B}} B, 0) + A(T, \mu + \mu_{\mathrm{B}} B, 0)\right]$$

$$= A(T, \mu, 0) + \frac{(\mu_{\mathrm{B}} B)^2}{2} \frac{\partial^2 A}{\partial \mu^2} + \mathcal{O}(B^4).$$

Hence we find

$$\chi = \frac{\mu_B^2}{\Omega} \frac{\partial N}{\partial \mu} = -\frac{\mu_B^2}{\Omega} \int d\varepsilon \, \mathcal{D}(\varepsilon) \frac{\partial f}{\partial \varepsilon}.$$

Sommerfeld's formula gives for $T \ll \Theta_F$

$$\chi \approx \frac{\mu_B^2}{\Omega} \left\{ \mathcal{D}(\varepsilon_F) + \frac{\pi^2}{6} (kT)^2 \left(\mathcal{D}'' - \frac{\mathcal{D}'^2}{\mathcal{D}} \right) \right\},$$

where the last terms accounts for the shift in μ with temperature. In the classical limit, we would have

$$\chi \sim \frac{\mu_B^2 \beta}{\Omega} \int d\varepsilon \, \mathcal{D}(\varepsilon) \, e^{-\beta\varepsilon + \alpha} = \frac{N}{\Omega} \frac{\mu_B^2}{kT}.$$

10c Thermionic Effect

In order to study the emission of electrons by a heated metallic cathode we model this metal by a box. The potential, assumed to be zero inside the metal, is not infinite outside it, as in § 10.2.1, but constant and equal to V, the potential needed to extract an electron with zero momentum. Few electrons can escape, as $V - \mu \gg kT$. We assume that all electrons which reach the wall with a sufficient velocity to escape do so, following the laws of classical dynamics. These electrons are collected by an anode and are replaced by other electrons thanks to the negative potential of the cathode, so that the metal remains neutral and lets the electrons escape. (Otherwise the bulk of the metal would gain a positive charge and would attract a thin shell of electrons in electrostatic equilibrium around the metal.) Calculate the electrical current emitted as a function of temperature.

Solution. Since the energy of an electron is much higher than the Fermi energy μ when $p^2 > 2mV$, the Fermi factor reduces to the Maxwell factor and the number of electrons per unit volume with momenta within d^3p is given by the expression

$$2 \frac{d^3 p}{h^3} e^{(\mu - p^2/2m)/kT}.$$

The number of electrons which hit a unit surface element during a time interval dt can be evaluated as in (7.50), the only difference being the value of μ, and equals

$$\frac{p_x}{m} dt \, 2 \frac{d^3 p}{h^3} e^{(\mu - p^2/2m)/kT}.$$

Classical dynamics indicates that the particles which leave the metal are those for which $p_x > \sqrt{2mV}$. The emitted current density is thus

$$i = e \int_{\sqrt{2mV}}^{\infty} dp_x \int\!\!\int_{-\infty}^{\infty} dp_y\, dp_z\, \frac{p_x}{m} \frac{2}{h^3}\, e^{(\mu - p^2/2m)/kT}.$$

Evaluating this expression we find

$$i = \frac{4\pi m e k^2}{h^3}\, T^2\, e^{-(V-\mu)/kT},$$

in agreement with *Richardson's empirical law* (1912):

$$i = A T^2 e^{-B/T}.$$

The coefficient B equals $(V - \varepsilon_F)/k$ when $T \ll \Theta_F$. If the electrons had formed a Maxwell gas, μ would have had for the same density the value, following from (10.49) and (10.50),

$$kT \ln \left[\frac{4}{3\sqrt{\pi}} \left(\frac{\Theta_F}{T} \right)^{3/2} \right],$$

which is of the order of 0.1 eV, whereas ε_F and $V - \varepsilon_F$ are of the order of a few eV; hence, B would have been close to V/k instead of $(V - \varepsilon_F)/k$ and the thermionic effect would have been imperceptible. If, for instance, $\varepsilon_F = V - \varepsilon_F \simeq 1$ eV, we find $e^{-V/kT} \simeq 2 \times 10^{-35}$ at room temperature. Although $B/T = (V - \varepsilon_F)/kT$ is large, it is not as large as V/kT, and $e^{-B/T}$ may be sufficiently large (4×10^{-18} in our example) so that one obtains significant currents from a heated cathode.

10d Metallic Film

We model a thin metallic layer as a gas of electrons in a box in the shape of a parallepiped with macroscopic dimensions L_x and L_y but with a thickness L_z which is small. Show that the p_z degree of freedom remains frozen in, when the film is so thin that $kT L_z^2 \ll 3\pi^2 \hbar^2$. We assume henceforth that this condition is realized.

We can by a suitable choice of the zero of the single-particle energies represent the free electrons in the film as a two-dimensional fermion gas. Calculate the density of states $\mathcal{D}(\varepsilon)$, the Fermi energy, the chemical potential as function of the temperature and of the density N/S per unit area of the electrons, the specific heat for $T \ll \Theta_F$ and for $T \gg \Theta_F$, and the spin susceptibility per unit area.

Answers:

$$\mathcal{D}(\varepsilon) = \frac{Sm}{\pi\hbar^2}\, \theta(\varepsilon),$$

$$\varepsilon_F = \frac{N}{S} \frac{\pi\hbar^2}{m},$$

$$\mu = kT \ln\left[\exp\left(\frac{N}{S}\frac{\pi\hbar^2}{mkT}\right) - 1\right],$$

$$C \sim \frac{Sm\pi k^2}{3\hbar^2}T\left[1 - \mathcal{O}\left(e^{-\Theta_F/T}\frac{\Theta_F^2}{T^2}\right)\right],$$

$$C \sim Nk\left[1 - \mathcal{O}\left(\frac{\Theta_F^2}{T^2}\right)\right],$$

$$\chi = \frac{\mu_B^2 m}{\pi\hbar^2}\left[1 - \exp\left(-\frac{N}{S}\frac{\pi\hbar^2}{mkT}\right)\right].$$

10e Neutron Stars

1. The life of the most massive stars ends, in general, in an explosion (supernova). When the stellar mass is of the order of 10 solar masses, the centre of the star implodes and gives birth to a residue, a neutron star. The latter consists almost completely of neutrons, the interactions between which we shall neglect. Knowing that the density is comparable with that of nuclear matter, 0.17 nucleons per fm^3, and that the temperature in the interior of such a star is typically 10^8 K, show that one may treat the matter in the neutron star as a non-relativistic fermion gas at zero temperature.

2. Compare the thermodynamic pressure with the kinetic pressure.

3. Assuming the density to be uniform, determine the radius R of the star as function of its mass M, taking into account the internal energy and the gravitational energy, which is $-3GM^2/5R$ for a homogeneous sphere. Calculate that radius R_0 as well as the particle density $n_0 = N/\Omega$ for a neutron star of 1 solar mass, $M = 2 \times 10^{30}$ kg.

4. In actual fact, the density $n(r)$ varies from the centre to the surface of the star, due to the variation of the gravitational potential $V(r)$. Write down the equation for $n(r)$. Justify the result obtained sub 3.

5. Neutron stars rotate with very large angular velocities due to the conservation of angular momentum during the stages when they are formed, when they contract – the period may be well below 1 s. They are detected just because of their periodically varying luminosity, and they are for that reason called *pulsars*. How can we microscopically take this rotation into account when we write down the equation for the equilibrium distribution?

Hints and Results:

1. Assuming that the conditions which we are trying to find are satisfied, we get $\varepsilon_F \simeq 60$ MeV, which is much larger than $kT \simeq 10$ keV, and much less than $mc^2 \simeq 900$ MeV. This justifies our hypotheses.

2. We find $\mathcal{P} = 2\varepsilon_F N/5\Omega$ either using (10.61), or using the kinetic approach of § 7.4.2 in which we replace the Maxwell distribution by $\theta(\varepsilon_F - p^2/2m)$. This value is 2000 times larger than for a classical perfect gas at the same temperature.

3. The equilibrium of the star (Exerc.6e) corresponds to the minimum of the total energy, since the temperature is negligible, whence we find

$$R = \frac{\hbar^2}{G} \left(\frac{9\pi}{4}\right)^{2/3} \frac{1}{M^{1/3}m^{8/3}}.$$

When $M = M_\odot$, we find $R_0 \simeq 10$ km, $n_0 \simeq 0.2$ fm^{-3}, in agreement with the assumptions made in 1.

4. Even though the gas is degenerate, one can treat the single-particle problem classically (§ 10.3.4b). The Fermi factor is thus locally equal to

$$\theta\left(\mu - \frac{p^2}{2m} - V(r)\right),$$

where $V(r)$ is the gravitational potential, at a distance r from the centre of the star. In the relation

$$\mu - V(r) = \frac{\hbar^2}{2m} \left(3\pi^2 n(r)\right)^{2/3},$$

we must at equilibrium have a uniform chemical potential, while $V(r)$ is related to $n(r)$ through Newton's law, that is, $\nabla^2 V = 4\pi m^2 G n(r)$, or

$$V(r) = -4\pi m^2 G \left[\frac{1}{r}\int_0^r dr'\, r'^2\, n(r') + \int_r^\infty dr'\, r'\, n(r')\right].$$

In terms of the dimensionless variables r/R_0 and $\varphi \equiv n/n_0$ we find

$$R_0^2 \nabla^2 \varphi^{2/3} + 6\varphi = 0,$$

which shows that the values obtained sub 3 were qualitatively correct.

We would have obtained the same equation, if we had expressed the fact that at each point the pressure of the Fermi gas is in equilibrium with the weight.

5. We can introduce a Lagrangian multiplier connected with the angular momentum, as in Exerc.7b.

Notes. We discuss the equation obtained sub 4 in Prob.9. We show there that the density varies slowly for small r and falls to 0 for a certain value of r/R_0 of order unity beyond which it stays zero. The approximation sub 3 is thus qualitatively justified.

The way neutron stars are formed implies that they have masses which are a little larger than the solar mass $M_\odot$. If the mass were significantly larger, the density at the centre would become so large as to violate the condition $v_F \ll c$. One should then take relativistic effects into account and replace $\varepsilon_q = p^2/2m$ by $\varepsilon_q = \sqrt{m^2 c^4 + p^2 c^2} - mc^2$. Calculations, similar to those in Prob.9, show that there would then be a maximum mass of the order of $6M_\odot$ beyond which the star would be gravitationally unstable.

A neutron star behaves like a huge nucleus, and we could ask, as in § 10.1.4, whether it is β-stable. In fact, through the reaction n → p+e+v̄, a certain number of protons and electrons will be produced. However, a kind of chemical equilibrium will be established; the relation (6.78) between the chemical potentials then implies, if we take the kinetic, gravitational, and reaction contributions into account, that the fraction of electrons and protons remains small for stars with a mass of the order of $M_\odot$.

10f Quantum Corrections to the Perfect Gas

Calculate the lowest-order corrections due to the Pauli principle to the chemical potential, the equation of state, the internal energy, and the specific heat of perfect gases. Such corrections are, in general, swamped by the interactions, but they can be seen by performing accurate measurements on the lightest rare gases and their isotopes, where the interactions between the atoms are weak.

Results. Starting from the expansion of (10.29) for small e^α we get

$$\ln Z_G \approx \frac{2\Omega}{\lambda_T^3} e^\alpha \left(1 + \frac{\eta}{2^{5/2}} e^\alpha\right),$$

$$e^\alpha \approx \frac{\lambda_T^3 N}{2\Omega} \left(1 - \frac{\eta}{2^{3/2}} \frac{\lambda_T^3 N}{2\Omega}\right),$$

$$\mathcal{P}\Omega \approx NkT \left(1 - \frac{\eta}{2^{5/2}} \frac{\lambda_T^3 N}{2\Omega}\right),$$

$$U \approx \frac{3}{2} NkT \left(1 - \frac{\eta}{2^{5/2}} \frac{\lambda_T^3 N}{2\Omega}\right),$$

$$C_v \approx \frac{3}{2} Nk \left(1 + \frac{\eta}{2^{7/2}} \frac{\lambda_T^3 N}{2\Omega}\right),$$

where λ_T is defined in (10.49). For fixed density and temperature Fermi-Dirac statistics increase the pressure, the chemical potential, and internal energy, whereas Bose-Einstein statistics decrease them. This agrees with the interpretation of these statistics as effective repulsion for fermions, effective attraction for bosons. However, the specific heat at constant volume changes in the opposite way.

We can also obtain these results by using the Wigner representation (§§ 2.1.2 and 2.3.4) to calculate the lowest-order correction term coming from the symmetrization (Eqs. (2.74–77)).

10g Spatial Correlations Due to Indistinguishability

1. Use the formalism of § 10.2.3 to calculate in terms of the Fermi or the Bose factor the expectation value $X = \langle \hat{c}_{q_1}^\dagger \hat{c}_{q_2} \hat{c}_{q_3}^\dagger \hat{c}_{q_4} \rangle$ over a grand canonical ensemble.

2. Use this result to evaluate the correlation function, defined by

$$C(r - r') = \langle \hat{n}(r)\hat{n}(r') \rangle - \langle \hat{n}(r) \rangle \langle \hat{n}(r') \rangle,$$

where $\widehat{n}(r) \equiv \widehat{\psi}^\dagger(r)\widehat{\psi}(r)$ is the operator (10.24) representing the number of particles per unit volume at the point r. First disregard the spin, and then take it into account. Discuss the results; find limiting expressions for short distances apart, for low densities, and for zero temperature for fermions.

Solution:

1. As the density operator is diagonal in the base (10.14), and as $\widehat{c}_q$ and $\widehat{c}_q^\dagger$ change the states of that base according to (10.20), the two creation operators in X must refer to the same states q as the two annihilation operators. There are three possible cases:

(i) $q_1 = q_2 \neq q_3 = q_4$: in that case X is, because of (10.22), the average of $\widehat{n}_{q_1}\widehat{n}_{q_3}$ for which, if we use the factorization of § 10.3.1, we find $f_{q_1} f_{q_3}$.

(ii) $q_1 = q_4 \neq q_2 = q_3$: from (10.21) and (10.22) it then follows that X is the average of $\widehat{n}_{q_1}(1 + \eta \widehat{n}_{q_2})$, which is equal to $f_{q_1}(1 + \eta f_{q_2})$.

(iii) $q_1 = q_2 = q_3 = q_4$: in that case X is the average of $(\widehat{n}_{q_1})^2$, which is given by (10.32).

Combining these results we can write

$$X = \delta_{q_1 q_2}\delta_{q_3 q_4} f_{q_1} f_{q_3} + \delta_{q_1 q_4}\delta_{q_2 q_3} f_{q_1}(1 + \eta f_{q_2})$$

$$= \langle \widehat{c}_{q_1}^\dagger \widehat{c}_{q_2} \rangle \, \langle \widehat{c}_{q_3}^\dagger \widehat{c}_{q_4} \rangle + \langle \widehat{c}_{q_1}^\dagger \widehat{c}_{q_4} \rangle \, \langle \widehat{c}_{q_2} \widehat{c}_{q_3}^\dagger \rangle.$$

The last equation is a special case of *Wick's theorem*, which expresses the expectation value of any number of creation and annihilation operators, for a density operator of the form $\widehat{D} \propto \exp\left[-\sum \widehat{c}_q^\dagger f_{qq'} \widehat{c}_{q'}\right]$, as a sum of terms, each of which is constructed by pairing the creation and annihilation operators in all possible ways.

2. If we forget about the spin, q represents the three components of the momentum and the wavefunction $\langle r|q\rangle$ in (10.24) equals $\exp(ip \cdot r/\hbar)/\sqrt{\Omega}$. If we substitute (10.24) into the definition of $C(r - r')$, X appears. The first term in X is cancelled by the last term in C and if we use (10.38), there remains

$$C(r - r') = \frac{1}{h^6} \int d^3p \, d^3p' \, e^{i(p'-p)\cdot(r-r')/\hbar} \, f_p(1 + \eta f_{p'})$$

$$= \frac{1}{h^3} \int d^3p \, e^{-ip\cdot(r-r')/\hbar} \, f_p \, \delta^3(r - r') + \eta \left| \int \frac{d^3p}{h^3} e^{ip\cdot(r-r')/\hbar} f_p \right|^2$$

$$= n \, \delta^3(r - r') + \eta \left| \int \frac{d^3p}{h^3} \frac{e^{ip\cdot(r-r')/\hbar}}{e^{\beta p^2/2m - \alpha} - \eta} \right|^2.$$

The first term describes simply the correlation of a particle with itself. It would be present even for a set of points distributed randomly in the volume Ω; therefore, the only interesting term is the second one. It has the expected sign.

One must, in order to take the spin into account, include in the definition of the operator $\widehat{n}(r)$ a summation over the spin index σ by writing $\widehat{n}(r) = \sum_{\sigma} \widehat{\psi}_{\sigma}^{\dagger}(r)\widehat{\psi}_{\sigma}(r)$. This gives us for $r \neq 0$

$$C(r) = \eta(2s+1)\left[\frac{1}{4\pi^2\hbar^2 ir}\int_{-\infty}^{+\infty}\frac{p\,dp\,e^{ipr/\hbar}}{e^{\beta p^2/2m-\alpha}-\eta}\right]^2.$$

At short distances, $C(r)$ reaches its largest value, and behaves as

$$C(r) \approx \frac{\eta n^2}{2s+1}\left[1 - \frac{\langle p^2\rangle r^2}{3\hbar^2}\right].$$

The conditional probability of finding a particle in a volume element d^3r around $r \neq 0$ when there is already a particle present at the origin equals $[n + C(r)/n]\,d^3r$. For Boltzmann statistics it would be equal to $n\,d^2r$, as expected. Near the origin the correlations $C(r) \to -n^2$ reduce that probability to zero for fermions with the same spin, the exclusion principle prohibiting two particles to be at the same point; the probability is divided by 2 for spin-$\frac{1}{2}$ fermions, as two fermions with opposite spins are uncorrelated. In contrast, for spin-zero bosons the presence of one particle at a point doubles the probability of finding another particle in the neighbourhood.

We can evaluate the integral in the perfect gas limit and we get

$$C(r) \sim \frac{\eta n^2}{2s+1}e^{-mkTr^2/\hbar^2} = \frac{\eta n^2}{2s+1}e^{-2\pi r^2/\lambda_T^2}, \qquad T \gg \Theta_F.$$

The correlations due to the indistinguishability vanish at distances much longer than the thermal length λ_T, but the Pauli principle continues to play a rôle for shorter distances. In fact, bar for helium at a few K (Exerc.12d), λ_T is for ordinary gases shorter than the range of the interatomic forces and the quantum correlations are drowned by the correlations due to the interactions.

For a degenerate fermion gas we get through evaluating the integral in $C(r)$

$$C(r) \sim -\frac{n^2}{2s+1}\left[\frac{3}{x^3}(\sin x - x\cos x)\right]^2, \qquad T \ll \Theta_F,$$

where $x \equiv p_F r/\hbar$. The correlations decrease as $1/r^4$ at large distances, but show oscillations on the scale $\hbar/2p_F$ which is of the order of a few Å for electrons in a metal. These so-called Friedel oscillations are indirectly the source of magnetic effects in alloys.

11. Elements of Solid State Theory

"La cristallisation d'un sel toujours assujetti à prendre une
même forme n'est elle pas aussi admirable que la génération
constante des animaux?"

A. de Condorcet, Haller

"Sans ce mot (cristallisation), qui, suivant moi, exprime le
principal phénomène de cette folie nommée amour, ... la
description que je donne de ce qui se passe dans la tête et
dans le cœur de l'homme amoureux devenait obscure, lourde,
ennuyeuse, ..."

Stendhal, De l'amour, III

"Ils allaient conquérir le fabuleux métal
Que Cipango mûrit dans ses mers lointaines."

J.-M. de Heredia, Les conquérants

The existence of the solid state is one of the most remarkable manifestations
of *quantum mechanics on a macroscopic scale*: if microphysics were subject
to the laws of classical mechanics, systems consisting of nuclei and electrons
could not, not even at high densities and low temperatures, produce solids
having the mechanical, electric, and optical properties with which we are familiar. Since we are dealing with relatively high densities, the interactions
between the particles play an important rôle, even more so than in a liquid.
Moreover, at sufficiently low temperatures the translational degrees of freedom become frozen in, after all the other degrees of freedom: the constituents
can no longer move freely, and their motion must be treated quantum mechanically. Matter acquires a great degree of coherence and the resulting solid
phase (or phases) has (have), in general, an ordered crystalline structure: the
atomic nuclei are arranged in a *regular lattice*, with a three-dimensional periodicity which is reflected in all the properties of the crystal. Increasing the
temperature decreases the order, without making the periodic structure disappear; this persists up to the melting point, but vacancies are created and
the nuclei oscillate around their equilibrium positions. We must consider the
whole of the crystal as one huge molecule and our theoretical starting point
is the same as in Chap.8, namely the Born-Oppenheimer method (§ 11.1.1).
However, due to the macroscopic size of a crystal, statistical mechanics is

needed to explain the extremely diverse properties that are found experimentally, such as elasticity, magnetism, thermal, optical, or electric properties, or conductivity.

We shall in this chapter survey the simplest properties of crystalline solids in thermal equilibrium. In Chaps. 14 and 15 we shall come back to the transport properties of electrons in metals and semiconductors. Our guideline is to analyze the system in terms of *independent* degrees of freedom, at least approximately, in order to study those separately along the lines of § 4.2.5. The elementary entities, or *"quasi-particles"*, which we treat as non-interacting objects, are not the particles making up the solid, that is, the nuclei and electrons, but represent composite objects to be constructed by the theory. In any case, once we have identified the independent degrees of freedom, we can use the methods of Chap.10 to find the thermodynamic properties of the material.

We shall in this way see that some properties are connected with the *lattice* structure (§ 11.1.2), others with the *electron* motion (§§ 11.2 and 11.3), and yet others with lattice *vibrations* (§ 11.4). The first set reflects at the macroscopic scale the regular geometric arrangement of the atomic nuclei. It also includes properties resulting from *defects* in this arrangement. In the latter case the nearly independent entities serving to describe the micro-states simply are the defects themselves.

The study of the electronic properties encounters from the start a serious difficulty: the Coulomb interactions of the electrons with one another and with the nuclei are strong and should not be neglected. We attack this problem, using the Hartree approximation (§ 11.2.1) where one assumes that each electron moves in an effective average potential created by all the other particles, which depends on the macro-state considered. The Coulomb interactions between the particles are thus modelled by a *mean field* which varies strongly on the microscopic scale and which, in principle, should be determined by a self-consistent variational calculation. In fact, this field acts upon the electrons, but is itself determined by the electronic charge density, while we treat the nuclei as fixed point charges. In this approximation the electron cloud is represented by a set of non-interacting electrons embedded in an external potential. The presence of the latter, on the other hand, prevents us from identifying the quasi-particles used to describe the material with the original "bare" electrons; the quasi-particles are here "dressed" electrons, modified by the medium surrounding them which, for instance, gives them an effective mass which differs from the electron mass.

We have stressed that the periodicity of the crystalline lattice affects all the properties of the solid. As regards the electrons, described as non-interacting fermions moving in a potential, the latter's periodicity has important consequences which are the subject of *band theory*. We give its essential features (§§ 11.2.2 to 11.2.5) by using various approaches to discuss the shape of the spectra of the single-electron states in crystalline media. The combination of the results thus obtained with the Pauli principle explains

the existence of materials which are as different as *metals*, on the one hand
(§ 11.3.1), and *insulators* (§ 11.3.2) or *semiconductors* (§ 11.3.4), on the other
hand. In the first case, we shall in this way justify the model of fermions in a
box which we used in the elementary theory of metals in § 10.4, at the same
time finding its limitations. In the second case, we shall see emerging a new
kind of quasi-particles, the *holes* which carry a positive charge.

The microscopic theory sketched in the present chapter enables us not
only to explain and even to predict the properties of various substances, but
also to recover the concepts and to establish the general laws of *macroscopic
electrostatics*. Just as in Chap.5 we used microscopic physics and statistical
mechanics to derive the laws of thermodynamics, we carry out a similar deduction for the laws of electrostatic equilibrium, at least in crystalline solids
(§ 11.3.3). In particular, we explain the microscopic significance of the screening effect, of the polarization, of the dielectric constant, and of macroscopic
fields.

Finally we shall show that the *vibrational* micro-states of the crystal
lattice can also conveniently be represented as sets of quasi-particles, the
phonons. They are bosons, the number of which is not conserved and the
relation of which with the constituent particles of the solid is rather indirect;
their effective interactions with one another and with the other degrees of
freedom are weak. They contribute both to the propagation of sound and
to the specific heat of the solid. Here again, we invoke the periodic lattice
structure to find their properties and to establish, for instance, the relation
between the energy of a phonon and its momentum.

Solid state physics has since the start of this century been developed
enormously and continuously. In the present book we shall only discuss a
few topics, and if one wants to delve deeper into this vast topic, we suggest
consulting the classic books by C. Kittel (*Introduction to Solid State Physics*,
Wiley, New York, 1986) and by N.W. Ashcroft and D.N. Mermin (*Solid State
Physics*, Holt, Rinehart, and Winston, Philadelphia, 1979). In particular, we
shall hardly consider at all the many phenomena for the explanation of which
we need to take into account the residual interactions between the quasi-
particles. These interactions are, generally speaking, treated by complicated
perturbation theory methods, in the framework of a branch of statistical
physics which started towards the end of the fifties and which is called the
Many-Body Problem. We shall only briefly discuss superconductivity at the
end of Chap.12, and treat ferromagnetism in metals as a problem (Exerc.11f).

Solid state physics is also of major importance in technology and engineering. If ferromagnetism did not exist, it would be very difficult to transform
electrical and mechanical energies into one another by means of motors or
alternators. Semiconductors have entered our every-day life and their use has
revolutionized electronics; even though this broad subject goes beyond the
framework of the present book, its importance has led us to give in §§ 11.3.4
and 11.3.5 a brief account of the physical effects on which the manifold applications of semiconductors are based.

11.1 Crystal Order

11.1.1 The Born-Oppenheimer Method

We have stressed at the beginning of this chapter that a crystal resembles a very huge molecule with its atomic nuclei arranged in a regular lattice. This is a profound analogy, as solid state theory is based upon the same Born-Oppenheimer method which we have used in § 8.4.1 to describe the structure of a molecule in a gas. Its essential idea is approximately to separate the motions of the nuclei, which are heavy and hardly move, and of the electron cloud, in order to describe the solid in terms of *independent* entities.

The aim of the theory is to find an approximation for the eigenenergies and eigenstates of the Hamiltonian

$$\widehat{H} \equiv \widehat{T}_{\mathrm{n}} + \widehat{T}_{\mathrm{e}} + \widehat{V}, \tag{11.1}$$

which contains the kinetic energies of the nuclei and of the electrons,

$$\widehat{T}_{\mathrm{n}} \equiv \sum_n \frac{\widehat{\boldsymbol{P}}_n^2}{2M_n}, \qquad \widehat{T}_e \equiv \sum_i \frac{\widehat{\boldsymbol{p}}_i^2}{2m}, \tag{11.2}$$

and their potential interaction energy, $\widehat{V}$. The index i refers to electrons and the index n to nuclei, whose masses M_n and charges $Z_n e$ may have varying values if the solid contains different kinds of atoms. The magnetic interactions involving the spins are, in general, weak so that the potential energy,

$$\widehat{V} \equiv \widehat{V}_{\mathrm{nn}} + \widehat{V}_{\mathrm{en}} + \widehat{V}_{\mathrm{ee}}. \tag{11.3}$$

only contains the Coulomb interactions between nuclei and nuclei, between nuclei and electrons, and between electrons and electrons,

$$\widehat{V}_{\mathrm{nn}} = \frac{e^2}{4\pi\varepsilon_0} \sum_{n>n'} \frac{Z_n Z_{n'}}{|\widehat{\boldsymbol{R}}_n - \widehat{\boldsymbol{R}}_{n'}|}, \qquad \widehat{V}_{\mathrm{en}} = -\frac{e^2}{4\pi\varepsilon_0} \sum_{in} \frac{Z_n}{|\widehat{\boldsymbol{r}}_i - \widehat{\boldsymbol{R}}_n|},$$

$$\widehat{V}_{\mathrm{ee}} = \frac{e^2}{4\pi\varepsilon_0} \sum_{i>j} \frac{1}{|\widehat{\boldsymbol{r}}_i - \widehat{\boldsymbol{r}}_j|}. \tag{11.4}$$

The Born-Oppenheimer approximation consists in constructing a model which avoids the impossible diagonalization of $\widehat{H}$. It is based upon the fact that the nuclear masses M_n are so much larger than the electron mass m. This allows us to treat, to begin with, the positions $\boldsymbol{R}_n$ of the nuclei as parameters, while their momenta $\boldsymbol{P}_n$ vanish (§ 8.4.1).

We then assume that the *electron Schrödinger equation* (8.39) has been solved. The *eigenvalue* $W(\{\boldsymbol{R}_n\}, \lambda)$ of this equation, which depends on the positions $\{\boldsymbol{R}_n\}$ of the system of nuclei and on the state λ of the electrons, can be interpreted as an *effective potential* for the nuclei. We finally must solve the *nuclear Schrödinger equation* (8.40) with $\widehat{W}$ as potential. Both those tasks, already hard for small molecules, seem here completely hopeless, as we should, in principle, solve equations with 10^{23} variables. However, there is one major simplification, which makes it possible to produce solid state theory: solids are periodic structures. Each lattice unit contains only a few atoms so that the final problem is hardly more complicated than that of a *small* molecule with the size of a single cell, although the complete crystal behaves like a gigantic molecule of some 10^{23} atoms.

In order to explain the periodic structure of a solid, we shall consider its ground state, that is, its state at zero temperature and zero pressure. In the Born-Oppenheimer approximation the nuclei then see the effective potential $W(\{\widehat{\boldsymbol{R}}_n\}, \lambda = 0)$, corresponding to the lowest electronic state, $\lambda = 0$. In order to simplify the discussion, we shall assume that the nuclear masses are so large that we can neglect their kinetic energy $\widehat{T}_n$, as compared to W. Just as for the diatomic molecule of Chap.8 the minimum of the effective potential $W(\varrho)$ between the two nuclei determined their mean distance apart, $\bar{\varrho}$, here the *minimum of the effective internuclear potential* $W(\{\widehat{\boldsymbol{R}}_n\}, \lambda = 0)$ must determine the average positions, $\{\overline{\boldsymbol{R}_n}\}$, of these nuclei and, hence, the crystal structure, at zero temperature and zero pressure. Because of the large number of variables characterizing the relative positions of the nuclei, the problem is not as simple as it was for the diatomic molecule. Nevertheless, one can show, and we shall assume, that the effective potential energy W has, under rather general conditions, its minimum when the nuclei are *regularly placed upon a lattice*; the dimensions and shape of the lattice corresponding to this minimum depend on the effective forces between the nuclei which, in turn, are determined by the nature of the material.

A one-dimensional model will help us to understand why the stable equilibrium configuration is a regular lattice. Consider heavy particles interacting with one another pairwise through a potential $W(\varrho)$ which varies with the distance apart ϱ in the same way as for a diatomic molecule. As in §8.4.1, $W(\varrho)$ has a minimum at a distance $\bar{\varrho}$. In the ground state each particle tries to place itself at a distance, close to $\bar{\varrho}$, from its two neighbours, in order to make the total potential,

$$\sum_{n>n'} W(|R_n - R_{n'}|), \tag{11.5}$$

a minimum when the positions R_n are arbitrarily varied. More precisely, each particle sees the total potential created by all other particles; in Fig.11.1 particle 1 feels the repulsion of particles 0 and 2, the attraction of 3 and -1, an even weaker attraction of 4 and -2, and so on. The minimum of (11.5) is reached when the particles are equidistant from one another, at a distance apart ϱ_0, slightly smaller than $\bar{\varrho}$, which is determined by the relation

$$W'(\varrho_0) + 2W'(2\varrho_0) + 3W'(3\varrho_0) + \ldots = 0. \tag{11.6}$$

This model fits rather well for organic molecules with long chains such as alkanes or polymers, where the carbon atoms are placed equidistantly along

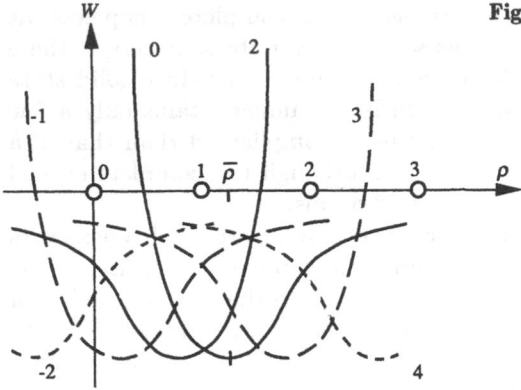

Fig. 11.1. Potentials felt by particle 1

the molecular axis. In a real, three-dimensional, solid the nuclei are situated similarly on a regular lattice, with mesh size close to $\bar{\varrho}$, the minimum of the interparticle potential, which is of the order of a few Å. In several cases, the structure of a solid reminds one of that of molecules which contain the same constituents: in the diamond lattice each C atom is surrounded by 4 neighbours, placed at the vertices of a tetrahedron, as in saturated organic compounds; graphite has a lamellar structure, and in each plane the C atoms form a hexagonal lattice as in the polycyclic compounds with fused benzene rings. In all cases, the strong repulsion at short distances apart, which is the combined effect of the Coulomb forces and the Pauli principle for electrons, ensures a lower bound for the interatomic distances, whereas the attraction at larger distances apart tends to pack the atoms like billiard balls, the more compactly, the stronger the attraction. Nevertheless, although one can produce a two-dimensional hexagonal pattern which repeats the elementary compact shape consisting of three circles touching one another, it is impossible to completely fill three-dimensional space regularly by spheres arranged compactly like tetrahedra. There are in three dimensions two kinds of regular packing, which are as compact as possible, and where each sphere has 12 nearest neighbours. Moreover, the ground state energies associated with different lattices may be close to one another. This explains the fact that the same material can occur in different crystal forms, and also that certain materials, such as glasses, solidify in an irregular lattice. The absence of periodicity makes the study of these so-called *amorphous* solids difficult. Recently intermediate materials, the so-called *quasi-crystals*, have been discovered where the energy minimum is reached for a lattice which is not periodic, but which has an orientational order at large distances.

The crystal structure of a solid is thus determined, at least at zero temperature and zero pressure, by the preliminary stage of the Born-Oppenheimer method, which consists in looking for the minimum of W. Thanks to its three-dimensional structure, a crystal lattice is much more *rigid* than a polymer chain molecule: in fact, this lattice *remains ordered* when the nuclei *vibrate* due to the thermal agitations and when the solid is *deformed* by constraints; one therefore continues to see the crystal order up to the melting point and when there are external forces present. Further on we shall return to the next stages in the Born-Oppenheimer method, the study of the state of the electron cloud (§ 11.2.1), and the vibrations of the atomic nuclei around their equilibrium positions (§ 11.4.1).

11.1.2 Properties Related to the Crystal Structure

General observations indicate readily how the crystal lattice shows up almost directly at the macroscopic scale: *optical or mechanical anisotropy* of a crystal reflects that of its lattice, and its *plane faces* which make *well defined angles* with one another clearly exhibit the lattice planes.

At the microscopic scale, a detailed experimental study of the lattice is carried out through wave diffraction by the regularly arranged nuclei; the wavelengths must be of the same order of magnitude as the lattice distances, that is, a few Å. *Diffraction of X-rays*, or of synchrotron radiation photons, is useful for lattices with atoms, the atomic numbers of which are not too different, as the X-ray scattering cross-section is an increasing function of the nuclear charge. One also currently uses *neutron diffraction*; the neutrons, which are produced in a nuclear reactor, are "thermal" neutrons, as they are a gas in equilibrium with the matter of the reactor. At room temperature, where the mean neutron energy ε equals $\frac{3}{2} \cdot \frac{1}{40}$ eV, the neutron de Broglie wavelength is $h/p = h/\sqrt{2m\varepsilon} \simeq 1.5$ Å, which is comparable with the interatomic distances. The scattering of neutrons may be important for light nuclei and it depends on the nuclear spin. This makes it possible to analyze lattices which contain both heavy and light nuclei, and also magnetic crystal structures.

The *long-range order* which occurs in a crystal is truly spectacular: between two parallel faces of a quartz crystal, at a distance of 5 cm, there are 10^8 planes which regularly repeat the same pattern! The microscopic structure of the elementary cell, with a size of a few Å, is thus directly shown at our scales. This macroscopic manifestation of quantum mechanics must be emphasized: as in the liquid-gas transition in § 9.3, the simplicity of the observations risks hiding the remarkable features of the phenomena.

In § 11.1.1 we restricted ourselves to giving reasons for the existence of a regular crystal lattice for a solid at zero temperature and zero pressure. The crystal structure is characterized by the group of space transformations, including translations, rotations, symmetries, and their products, which leave the crystal invariant. *Crystallography* is concerned with the study of the various possible groups and with their classification; one must note that each of the 230 groups predicted by geometry is represented in Nature. A theoretical solid state physicist therefore wants to understand *why a given material crystallizes in a given group* for given temperature and pressure. More generally, he wants to explain the *phase diagrams* by determining the curves which, in the temperature-pressure plane, separate the various crystal structures exhibited by a particular substance, and also the *melting and sublimation curves* which separate the regions of crystal order from the liquid and gas regions (Prob.8).

To do this we use the same kind of variational method (§4.2.2) as for the gas-liquid transition, a method which is very suitable to explain a *qualitative change* of the state of the material (§ 9.3). The rough argument which we used a moment ago, relying on looking for the minimum of the potential energy W, is only useful for finding the ground state of the solid. In order to study its state as function of the temperature and the pressure – or, what amounts to the same but is, in general, simpler, as function of the temperature and the chemical potentials of the electrons and the nuclei – we must look for the minimum of a test grand potential which depends on a number of adjustable parameters. We start therefore by assuming that the nuclei are placed on a test lattice. This test lattice may have the same invariance group as the one corresponding to the minimum of W, but it may also correspond to other crystal structures which could, for instance, become more stable at high pressures. On the other hand, the lattice distances, of the order of a few Å, are also adjustable. We then construct, using the approximation methods of §§ 11.2 and 11.4, for each test lattice the corresponding grand potential. We find thus a *test grand potential $\mathcal{A}$ which depends on our choice of crystal structure, and on the shape and size of the lattice cell.* To determine the various parameters in thermal equilibrium, as functions of the temperature and of the chemical potentials, we must thus look for the minimum of $\mathcal{A}$, for given T and μ. Often one finds *several relative minima*, each corresponding to a given crystal structure, which depend on the thermodynamic variables. The lowest of them determines the equilibrium structure, and we obtain the separatrices of the phase diagram by looking for what values of the thermodynamic variables the absolute minimum of $\mathcal{A}$ is reached simultaneously for two different crystal structures. As to the melting and sublimation lines, we must similarly compare the test grand potential $\mathcal{A}$ determined in this way with the grand potential of the fluid phases found in § 9.3: the stable phase in thermal equilibrium is the one corresponding to the lowest test grand potential (Probs.8 and 10).

An equivalent method consists in comparing the test free enthalpies $\mathcal{G}(N, \mathcal{P}, T)$ calculated for the geometries corresponding to the various crystal or fluid phases considered. Although the evaluation of the free enthalpy is often less simple than that of the grand potential, its more traditional use has the advantage of providing us directly with the usual phase diagram in the $T, \mathcal{P}$ plane, since the separating curves are obtained by equating the two trial free enthalpies.

If two minima of $\mathcal{A}$, or of $\mathcal{G}$, cross when a thermodynamic parameter, such as T, is varied, the lowest one corresponds to the equilibrium phase; the other is often associated with a phase which has become *metastable*. Its lifetime can be more or less short, depending on whether there exists a set of many intermediate configurations which allow the system to evolve gradually from one phase to the other. If such configurations are rare, the metastable state corresponds to a Boltzmann-Gibbs equilibrium in the restricted set of accessible micro-states (§ 4.1.6, Exerc.9d and Prob.3).

The method we have just sketched determines in principle not only the phase diagram, but also the shape and size of the elementary cell as function of the temperature, the constraints, and the chemical potentials. We can use it to determine the *equation of state* of the crystal, its *elasticity constants*, and its *expansion coefficients* in various directions (Prob.19).

There are still other properties of solids which are related to the existence of crystal order. In particular, we have so far assumed that the order is perfect. In fact, there are always crystallization *defects*, either point defects (*vacancies* or *interstitial* atoms), or linear defects (*dislocations*), or planar defects (*stacking faults*). The higher the temperature, the more numerous the defects at thermal equilibrium (Exerc.11a and Prob.8). They do not always disappear when we cool the crystal, as they should do if the crystal remained in thermal equilibrium, since the migration of atoms is, in general, a slow process in a solid (Prob.19). They play an essential rôle for the *mechanical properties* of solids, such as elasticity, plasticity, hardness, and so on, which are of great practical importance (Prob.19). In particular, the study of dislocations is essential in physical metallurgy: a crystal lattice is too rigid to deform significantly in bulk under external constraints; those act, in fact, through the intermediary of the displacement of dislocations, which results in a change in shape without a large cost in energy. The point defects behave as weakly interacting localized quasi-particles. They play a rôle in the *optical properties* of transparent crystals, and often produce their colours.

The arrangement of atoms in the crystal lattice determines the properties of *alloys* or composite solids. In particular, depending on the conditions, two kinds of atoms may alternate regularly, they may be arranged randomly on an ordered lattice, or they may segregate, thus giving rise to phases with different characteristics (Probs.4, 5, and 19). A convenient approximate approach is provided by the Ising model (Exerc.9a) for binary alloys, or by other magnetic models (Exerc.9d); the two values $\sigma_i = \pm 1$ of the spin at each site should here be understood as denoting the presence of the one or of the other of the two species. Moreover, a host of phenomena related to the crystal geometry, or to the order and disorder of the atomic positions, concern *crystal surfaces* (Prob.3), which may have a rounded equilibrium shape. (The plane faces that we observe are generally the result of a dynamic growth process which only produces a metastable, but long-lived, state.)

Let us note that the Born-Oppenheimer method described earlier, although elementary in principle, is hardly practicable except for the simplest solids, notwithstanding the large simplification brought about by the existence of a periodic crystal lattice. For most theoretical studies of realistic solids we are led to use more schematic models, such as we used in Chap.1 for paramagnetism, and such as we shall use in the remainder of this chapter, or in the exercises and problems mentioned above. Nevertheless, some phenomena depend on the interplay of the three characteristics of solids that we distinguished, namely, lattice structure, electrons, and lattice vibrations. In particular, we have just indicated that the temperature affects the lattice shape and the equation of state of the crystal, through the grand potential determined by the vibrational energy. There exist also materials (Prob.12) where the electron state affects the crystal structure.

11.2 Single-Electron Levels in a Periodic Potential

In the application of the Born-Oppenheimer method to diatomic molecules
(§ 8.4.1) the study of the electron cloud was simple, as we could assume that
it was frozen in into its ground state for all temperatures of interest. However,
in a solid, the energy levels of the electron states λ lie very closely, so that we
need to take excited electron states into account, even at low temperatures.
In the present section we study methods leading to a simple description of
the electron cloud in a crystal.

11.2.1 Independent Electron Approximation

In the Born-Oppenheimer approximation the energy levels W of the electron
cloud states λ are, for fixed positions $\{R_n\}$ of the nuclei, determined by
Eq.(8.39) which has the form

$$\left[\widehat{T}_e + \widehat{V}(\{\widehat{r}_i\}, \{R_n\})\right]|\psi_{\lambda e}\rangle = W(\{R_n\}, \lambda)|\psi_{\lambda e}\rangle, \tag{11.7}$$

where $\widehat{V}$ is given by (11.3) and (11.4).

In thermal equilibrium the nuclei vibrate around their average positions
$\{\overline{R_n}\}$ which are arranged, as we have seen, in a lattice. They do *not stray
far from their average positions*, not even when the temperature is close to
the melting point (end of § 11.4.1), and we can, when we study the electron
motion, to a first approximation replace $\{R_n\}$ by $\{\overline{R_n}\}$. To simplify our
formulae, we shall in §§ 11.2 and 11.3 denote the positions of the nuclei by R
rather than by $\overline{R}$.

The potential $\widehat{V}(\{\widehat{r}_i\}, \{R_n\})$ which occurs in the electron Schrödinger
equation (11.7) consists of three terms. The first one, $\widehat{V}_{nn}$, describes the
Coulomb repulsion *between the nuclei* which are assumed to be fixed to the
lattice sites; it is a *constant* and occurs directly in each of the eigenenergies $W(\{R_n\}, \lambda)$. The second is the Coulomb attraction *between nuclei and
electrons*,

$$\widehat{V}_{en} = -\frac{e^2}{4\pi\varepsilon_0}\sum_i\left(\sum_n\frac{Z_n}{|\widehat{r}_i - R_n|}\right); \tag{11.8}$$

it has the form of a periodic *external potential* produced by point charges, in
which the electrons move independently of one another. If these two terms
were the only ones, we would need only solve a one-electron Schrödinger
equation; the eigenenergies of the N-electron states would then simply be
found by using the occupation number formalism of § 10.2.2. Unfortunately,
the third term, the Coulomb repulsion *between electrons*,

$$\widehat{V}_{ee} = \frac{e^2}{4\pi\varepsilon_0}\sum_{i<j}\frac{1}{|\widehat{r}_i - \widehat{r}_j|}, \tag{11.9}$$

is an interaction term and it makes the Schrödinger equation (11.7) too complicated to be soluble. The difficulty is exactly the same as when one tries to find the energy levels of an atom, and we are similarly led to use the *Hartree approximation*. The idea is to *replace the interaction potential* between the electrons by an *effective average potential*, acting upon the electrons as if they were independent and simulating the interaction (11.9) as well as possible, so that $\widehat{V}_{en} + \widehat{V}_{ee}$ would be replaced by a one-electron potential of the form

$$\widehat{\mathcal{V}} = \sum_i \mathcal{V}(\widehat{\boldsymbol{r}}_i). \tag{11.10}$$

The contribution from the nuclei to $\mathcal{V}(\widehat{\boldsymbol{r}}_i)$ is clearly the ith term in (11.8). As to (11.9), the ith electron sees all the other electrons, which are treated as a classical charged fluid of density $n(\boldsymbol{r})$. This gives us the Hartree effective potential,

$$\mathcal{V}(\widehat{\boldsymbol{r}}) = \frac{e^2}{4\pi\varepsilon_0} \left[-\sum_n \frac{Z_n}{|\widehat{\boldsymbol{r}} - \boldsymbol{R}_n|} + \int d^3r' \, \frac{n(\boldsymbol{r}')}{|\widehat{\boldsymbol{r}} - \boldsymbol{r}'|} \right], \tag{11.11}$$

which, when acting upon the ith electron, only depends on the other electrons through the expectation value $n(\boldsymbol{r})$ of the density. In this approximation, each electron moves in the *mean field* created by its surroundings, and we neglect the fluctuations and the interparticle correlations.

We have thus been led to study a *gas of independent fermions*, subject to an *external potential* $\mathcal{V}$ produced by the nuclei and the electron cloud. Instead of the N-electron Schrödinger equation (11.7) we are left with an approximate *one-electron* problem with the Hamiltonian

$$\widehat{h} \equiv \frac{\widehat{\boldsymbol{p}}^2}{2m} + \mathcal{V}(\widehat{\boldsymbol{r}}). \tag{11.12}$$

As in § 10.2, we have to find the solution by diagonalizing $\widehat{h}$ as follows:

$$\widehat{h}\,|q\rangle = \varepsilon_q\,|q\rangle. \tag{11.13}$$

In the next few subsections we shall discuss the general characteristics of the single-electron energy spectrum ε_q. The micro-states λ of the electron cloud are in the approximation used here characterized by an ensemble n_q of occupation numbers; if we assume that the energy of such a micro-state can be evaluated as if the electrons behaved as non-interacting quasi-particles with energies ε_q, we expect the eigenvalue W of (11.7) to be given by

$$W(\{\boldsymbol{R}_n\}, \lambda) \simeq c + \sum_q \varepsilon_q n_q, \tag{11.14}$$

where c is a constant. The thermodynamics of the electron cloud in the solid is then finally reduced to that of a non-interacting *Fermi-Dirac gas* with single-particle energies which are the eigenvalues of the Hamiltonian (11.12). In actual fact, we shall see in the remainder of this sub-section that it is not altogether consistent to treat the electron cloud in the Hartree approximation as a gas of non-interacting quasi-particles with energies ε_q which add up; however, it will also be seen that this point has hardly any consequences.

Our solution scheme is as yet incomplete since Eq.(11.11) contains the density $n(\boldsymbol{r})$ of the electron cloud at the position $\boldsymbol{r}$, which still has to be determined. In our independent electron approximation, each single-particle state q, with wavefunction $\langle \boldsymbol{r}s|q\rangle$, where $s = \pm 1$ denotes the spin coordinate, contributes $|\langle \boldsymbol{r}s|q\rangle|^2$ to the density of electrons with spin s. At equilibrium the mean occupation number of the state q is the Fermi factor f_q which, as function of ε_q, is given by (10.62), so that we have

$$n(\boldsymbol{r}) = \sum_{qs} f_q \, |\langle \boldsymbol{r}s|q\rangle|^2. \tag{11.15}$$

The Hartree equations which we have found form a *self-consistent* system, with the density $n(\boldsymbol{r})$ being expressed in terms of the effective potential $\mathcal{V}(\boldsymbol{r})$ through (11.12), (11.13), and (11.15), while $\mathcal{V}(\boldsymbol{r})$ itself is connected with $n(\boldsymbol{r})$ through (11.11). An efficient solution involves an iterative interplay: starting from an assumed form of $\mathcal{V}(\boldsymbol{r})$, we solve the three-dimensional Schrödinger equation (11.13), we then determine $n(\boldsymbol{r})$ which we substitute into (11.11) to find an improved form of $\mathcal{V}(\boldsymbol{r})$, and so on. This is a numerically heavy programme, and we shall assume in what follows that it has been solved.

In fact, among the equations which we have just written down, Eq.(11.14) for the energy of the electron gas is incorrect. We have used an extrapolation of (10.16) which was valid for a non-interacting fermion gas in an *external* potential assumed to be given. However, the *self-consistency* of the potential (11.11), which depends on the electron cloud state through $n(\boldsymbol{r})$, implies that the total energy is not the sum of the individual energies, even if the independent electron approximation is a good one. Actually, the energy ε_q of an electron takes the interactions with the other electrons into account through the second term in (11.11). Thus, the interaction between each i,j pair is counted twice in (11.14), once for the ith and once for the jth electron.

In order to rectify this mistake, to better understand the nature of the Hartree approximation, and possibly to apply corrections to it, we shall give a *variational justification* based on the methods of §4.2.2. Our aim is to simulate as well as possible the electron grand canonical density operator,

$$\widehat{D} \propto \mathrm{e}^{-\beta \widehat{H} + \alpha \widehat{N}}, \qquad \widehat{H} \equiv \sum_i \widehat{K}_i + \widehat{V}_{\mathrm{ee}}, \tag{11.16}$$

where $\widehat{K}_i$ contains the kinetic energy of the ith electron, the potential (11.8) produced by the nuclei, and, possibly, an external potential applied to the material. We introduce an approximate density operator,

$$\widehat{\mathcal{D}} \propto \mathrm{e}^{-\beta \sum_i \widehat{h}_i + \alpha \widehat{N}}, \tag{11.17}$$

where $\widehat{h}$ is a trial single-electron Hamiltonian which we want to optimize. To do this, folowing the general scheme, we look for the minimum of the trial grand potential,

$$\mathcal{A} \equiv \operatorname{Tr} \widehat{\mathcal{D}} \left(\widehat{H} - \mu \widehat{N} \right) - T S(\widehat{\mathcal{D}}) = \sum_q f_q \langle q | \widehat{K} | q \rangle + \operatorname{Tr} \widehat{\mathcal{D}} \, \widehat{V}_{\mathrm{ee}}$$

$$- \mu \sum_q \left[f_q + kT \sum_q f_q \ln f_q + (1 - f_q) \ln (1 - f_q) \right]. \tag{11.18}$$

In writing down (11.18) we have used the diagonalization (11.13) of the as yet unknown trial single-particle Hamiltonian $\widehat{h}$ and second quantization (§ 10.2). The first term in (11.18) follows from the fact that the single-particle part, $\sum \widehat{K}_i$, of $\widehat{H}$ gives a contribution $\langle q | \widehat{K} | q \rangle$ for each state $|q\rangle$ occupied by an electron; the last term is the entropy (10.37) of the macro-state (11.17). Using the methods of §10.2.3 we can express the second term in (11.18) also in terms of the wavefunctions $|q\rangle$ and their occupation factors f_q; we thus get the *Hartree-Fock expression*

$$\operatorname{Tr} \widehat{\mathcal{D}} \, \widehat{V}_{\mathrm{ee}} = \frac{1}{2} \frac{e^2}{4\pi\varepsilon_0} \int d^3 r \, d^3 r' \frac{1}{|r - r'|} \sum_{qq'ss'} f_q f_{q'}$$

$$\times \left[\langle rs|q \rangle \langle q|(|rs\rangle \langle r's'| - |r's'\rangle \langle rs|)|q'\rangle \langle q'|r's'\rangle \right]. \tag{11.19}$$

Earlier we had restricted ourselves to the Hartree method, which involves an extra approximation. Expression (11.19) is complicated, as it takes into account the spatial correlations between the electrons introduced by the Pauli principle, even though (11.17) describes a gas without interactions (Exerc.10g). Let us neglect these correlations, which have a short range, of order $\hbar/p_F$. The average number of pairs of electrons within volume elements $d^3 r$ and $d^3 r'$ around the points r and r' is then given by the expression $\frac{1}{2} n(r) n(r') d^3 r \, d^3 r'$, where $n(r)$ is given by (11.15); the factor $\frac{1}{2}$ has been introduced in order not to count twice configurations where two electrons are interchanged – this is the counterpart to the restriction $i > j$ in (11.11). In this Hartree approximation we thus have for the electron-electron interaction energy

$$\operatorname{Tr} \widehat{\mathcal{D}} \, \widehat{V}_{\mathrm{ee}} \simeq \frac{1}{2} \frac{e^2}{4\pi\varepsilon_0} \int d^3 r \, d^3 r' \frac{1}{|r - r'|} n(r) \, n(r'), \tag{11.20}$$

which amounts to neglecting the second, so-called "exchange", term from (11.19).

We must find the minimum of (11.18), (11.20) in terms of the trial single-particle operator $\widehat{h}$, or equivalently in terms of the eigenfunctions $|q\rangle$ and eigenvalues ε_q of $\widehat{h}$. Alternatively, the trial operator

$$\widehat{f} \equiv \sum_q |q\rangle f_q \langle q| = \frac{1}{e^{\beta \widehat{h} - \alpha} + 1} \tag{11.21}$$

is in one-to-one correspondence with $\widehat{h}$, and we shall find it convenient to take in our calculations the matrix elements of (11.21) as independent variables; in those variables we get for (11.18), (11.20)

$$\mathcal{A} = \text{tr}\,\widehat{f}\widehat{K} + \frac{1}{2}\frac{e^2}{4\pi\varepsilon_0}\int d^3r\,d^3r'\,\frac{1}{|r-r'|}\,n(r)\,n(r')$$

$$- \mu\,\text{tr}\,\widehat{f} + kT\,\text{tr}\big[\widehat{f}\ln\widehat{f} + (1-\widehat{f})\ln(1-\widehat{f})\big] \tag{11.22}$$

with $n(r) = \sum_s \langle rs|\widehat{f}|rs\rangle$. After a short calculation using (11.21), we get for the variation of $\mathcal{A}$ with $\widehat{h}$ or $\widehat{f}$,

$$\delta\mathcal{A} = \text{tr}\left\{\delta\widehat{f}\left[\widehat{K} + \frac{e^2}{4\pi\varepsilon_0}\int d^3r'\,\frac{1}{|\widehat{r}-r'|}\,n(r') - \widehat{h}\right]\right\}. \tag{11.23}$$

Putting this variation equal to zero for any $\delta\widehat{f}$ gives us the self-consistent form (11.11), (11.12) of the effective Hartree Hamiltonian $\widehat{h}$, coupled with the same expression (11.21) of the Fermi factor as for independent electrons. We thus recover equations (11.11), (11.12), (11.13), and (11.15).

Note that, in the present variational approximation, the equilibrium internal energy of the electrons, which follows from (11.20) or (11.22), is given by

$$\langle W\rangle - V_{\text{nn}}(\{R_n\}) = \left\langle \sum_i \widehat{K}_i + \widehat{V}_{\text{ee}} \right\rangle$$

$$= \text{tr}\,\widehat{f}\widehat{K} + \frac{1}{2}\frac{e^2}{4\pi\varepsilon_0}\int d^3r\,d^3r'\,\frac{1}{|r-r'|}\,n(r)\,n(r'),$$

whereas the energy of a set of *independent* quasi-particles would be

$$\sum_q \varepsilon_q\,\langle n_q\rangle = \sum_q \varepsilon_q\,f_q = \text{tr}\,\widehat{f}\widehat{h} = \left\langle \sum_i \widehat{K}_i + 2\widehat{V}_{\text{ee}} \right\rangle.$$

Expression (11.14), which would lead to an average energy $\langle W\rangle = c + \sum_q \varepsilon_q f_q$ is thus *incorrect*. Nevertheless, consider a *shift in equilibrium* associated with an infinitesimal change in the temperature, the electron density, or the field applied to the electrons from the outside or by the nuclei. This shift is characterized by variations δT, $\delta\mu$, $\delta\widehat{K}$ of the parameters which through the various self-consistent equations give rise to variations in $\mathcal{V}(r)$, $\widehat{h}$, $|q\rangle$, ε_q, and f_q. The corresponding variation in the internal energy of the electrons equals

$$\delta\langle W\rangle = \text{tr}\,\widehat{f}\,\delta\widehat{K} + \text{tr}\,\delta\widehat{f}\,\widehat{K} + \frac{e^2}{4\pi\varepsilon_0}\,\text{tr}\,\delta\widehat{f}\int d^3r'\,\frac{1}{|\widehat{r}-r'|}\,n(r'),$$

that is, if we bear in mind the variational equation $\delta\mathcal{A} = 0$,

$$\delta\langle W\rangle = \text{tr}\,\widehat{f}\,\delta\widehat{K} + \text{tr}\,\delta\widehat{f}\,\widehat{h}$$

$$= \sum_q \big(f_q\,\langle q|\delta\widehat{K}|q\rangle + \varepsilon_q\,\delta f_q\big). \tag{11.24}$$

The two terms in (11.24) are *the same as if the energy of the quasi-particles were additive*: the first is associated with the change in the Hamiltonian and the second with the change in the average occupation of the single-electron states.

In order to understand this simplification, note that the variational expression (11.22) differs from the grand potential of a gas of non-interacting fermions only in the interaction energy (11.20); the entropy and the particle number have a trivial form. Moreover, the grand potential, which is the minimum of (11.22) with respect to $\widehat{f}$, depends on the equilibrium parameters partly explicitly and partly indirectly through the occupation matrix $\widehat{f}$. In an infinitesimal shift of the equilibrium, the second relation has no effect, since (11.22) is stationary with respect to *any* variation in $\widehat{f}$, and (11.23) vanishes identically. Due to the variational nature of the Hartree approximation, δA thus reduces to the explicit variation

$$\delta A \;=\; \operatorname{tr}\widehat{f}\,\delta\widehat{K} - \delta\mu\operatorname{tr}\widehat{f} + k\delta T\operatorname{tr}\left[\widehat{f}\ln\widehat{f} + (1-\widehat{f})\ln(1-\widehat{f})\right]; \qquad (11.24')$$

this expression is independent of the term $\langle\widehat{V}_{ee}\rangle$ and has just the same form as for a gas without interactions. This result also holds for $\delta U = \delta A + \delta(\mu N + TS)$.

The variational derivation of the Hartree approximation therefore justifies all equations we had above, *except* (11.14) which differs from the correct approximate form for the internal energy through the presence of a *factor* $\frac{1}{2}$ *in the interaction term* (11.20). The electrons are not really independent in this approximation; correlations due to their Coulomb repulsion are partially taken into account through the self-consistency of the potential $\mathcal{V}$, and that part of the energy which comes from the interaction between a pair of electrons should be attributed simultaneously to each of them and counted twice, but should be shared between them. Moreover, the interactions between electrons result in $\mathcal{V}$ depending on the equilibrium parameters such as the temperature, the chemical potential, or external fields.

Nevertheless, even though the internal energy at equilibrium is not equal to $c + \sum_q \varepsilon_q f_q$, its first-order *variations* when the electron cloud is changed through perturbations, such as the application of an external field, heating, or changes in the electron density, are well represented by the simplistic expression (11.24). Everything follows as if $\mathcal{V}$ were not an effective self-consistent potential depending on the equilibrium parameters, but a fixed external potential. The two errors – on the one hand, the omission of the factor $\frac{1}{2}$ in the interaction energy and, on the other hand, the neglect of the variation of ε_q in $\sum_q \varepsilon_q f_q$, when f_q is varied – cancel one another. It is permissible here to forget the self-consistency of $\mathcal{V}$ and the changes it implies in the energies ε_q when the occupation numbers f_q vary. This point will be useful for us in what follows: when we calculate in §§ 11.3.1 and 11.3.3 first derivatives of the grand potential or of the energy of some substance, say, as function of its temperature or its charge density, we shall rely on (11.24) or (11.24') without reservations.

11.2.2 Introduction to Band Theory

We shall use the remainder of this section to study the general characteristics of the solution of the single-electron Schrödinger equation (11.11–13), where we assume that $n(r)$ is fixed at its equilibrium value for the substance considered. The potential $\mathcal{V}(r)$, which is strongly attractive near the nuclei, reaches maximum values between them. It has a remarkable feature which will entail a number of electronic properties of solids, namely, it has the *symmetries of the crystal lattice*. We shall not consider symmetries which are specific for particular lattices, such as invariance under some rotations over $\frac{1}{2}\pi$ in a cubic crystal, or over $\frac{1}{3}\pi$ in a hexagonal crystal; we shall merely consider the generic symmetry which is common to all crystals, namely, the *periodicity in three spatial directions* with periods equal to the dimensions of the crystal cell. This periodicity of $\mathcal{V}(r)$ is clear for the first term of (11.11), which is governed by the positions R_n of the nuclei. For the second term, it results from the fact that the electron density $n(r)$ has, in general, the same symmetry properties as the lattice: unless these are broken spontaneously (Prob.12), they are induced by the nuclei through the self-consistency of the Hartree equations. In order that the periodicity be a strict one, it is nevertheless necessary that the crystal lattice can be regarded as perfect and infinite. For the time being, we shall not discuss surface effects or the effect of defects on the electrons. We shall later on return to these problems when we study how an excess or shortage of electrons is taken up in a semiconductor near an impurity or in a charged substance near the surface.

A rough approximation, which is qualitatively correct for *metals*, consists in *neglecting the spatial variations* of the potential $\mathcal{V}$. The electrons in a metal then behave like independent particles enclosed in a box, and their energy levels are given by (10.9). They lie extremely densely, in contrast to the electronic levels in a gas (§ 8.4.1), where only the lowest electron state contributes to the thermodynamic properties. In a metal, the electron degrees of freedom are not frozen in, the electrons themselves contituting a gas, and it is essential to take them into account as in § 10.4.3, when we study the properties of the solid. Nevertheless, in general, the approximation, which consists in replacing the potential $\mathcal{V}$ by its mean value, is inadequate. It cannot explain why solids can be classified as metals or insulators. Moreover, even for a metal, it is clear that not *all* electrons can be treated as being free in a box; whereas the conduction electrons move freely, those from the low-lying shells of the atoms which form the solid must remain bound. If we want to construct a less gross theory, it is thus necessary to take into account the spatial variation of the effective periodic potential $\mathcal{V}$ which acts on the electrons. This is the subject of *band theory*, the main results of which we shall summarize.

The most striking consequence of the periodicity of the potential is the nature of the energy spectrum ε_q of the single-electron states $|q\rangle$. As for all macroscopic systems (§ 10.3.3), this spectrum is continuous, with a *density of states $\mathcal{D}(\varepsilon)$ which is proportional to the volume*. Nevertheless, instead of covering a complete range $\varepsilon = (\varepsilon_{\min}, +\infty)$, it has gaps, or *forbidden bands*, energy ranges of varying width in which there is not a single one-electron level. The levels are grouped into *allowed bands*, the structure of which depends on the crystal geometry. We see thus the coexistence of *two complementary aspects*, which we shall meet with again in various forms: the allowed bands, separated by forbidden bands, remind us of the *discrete nature* of the *bound states* of an electron in an atom or molecule, whereas the *continuous nature* of the spectrum inside an allowed band reminds us of the spectrum of the *plane waves* inside a macroscopic box.

In order to understand these results, to determine the form of the eigenfunctions $|q\rangle$ of $\widehat{h}$, and to find the quantum numbers which characterize each state $|q\rangle$, we shall, to simplify the discussion, consider a one-dimensional lattice, with N cells of size a. Retaining the notation r for the only coordinate, we write down the periodicity condition for the potential: $\mathcal{V}(r+a) = \mathcal{V}(r)$. In order to get rid of boundary effects and to obtain a perfect periodicity without having to deal with an infinite crystal, we introduce periodic boundary conditions, as in § 10.2.1. We thus identify point 0 with point L, where L is the size of the crystal, which is a large integral multiple of the cell size, $L = Na$. We shall also omit the electron spin, on which $\widehat{h}$ does not depend. In § 11.2.5 we shall come back to the electron band structure of real three-dimensional solids.

The mathematical formalism which takes into account the invariance under a translation by a is based upon the introduction of the unitary operator $\widehat{U}$ which produces that translation in the single-electron Hilbert space (§ 2.1.5). This operator can be expressed in terms of the momentum operator $\widehat{p}$, which is the infinitesimal generator of translations; it transforms the ket $|r\rangle$, which describes a particle localized at r, as follows:

$$\widehat{U}|r\rangle \equiv e^{-i\widehat{p}a/\hbar}|r\rangle = |r+a\rangle. \tag{11.25}$$

The existence of periodic boundary conditions means that any ket in the Hilbert space remains invariant under a translation by Na, or equivalently, that

$$\widehat{U}^N = \widehat{I}. \tag{11.26}$$

The invariance of $\widehat{h}$ under a translation by a implies that the operators $\widehat{h}$ and $\widehat{U}$ commute:

$$[\widehat{U},\widehat{h}] = 0, \tag{11.27}$$

so that we can diagonalize $\widehat{h}$ and $\widehat{U}$ simultaneously by a unitary transformation (§ 2.1.2). The *common eigenfunctions* $|q\rangle$ *of* $\widehat{h}$ *and* $\widehat{U}$ are called *Bloch waves* (Felix Bloch, Zürich 1905–1983).

As the operator $\widehat{U}$ is unitary, its eigenvalues are complex numbers of unit absolute magnitude, so that we can define its eigenvectors and eigenvalues as

$$\widehat{U}|q\rangle = e^{-ipa/\hbar}|q\rangle. \tag{11.28}$$

We can thus assign a quantum number p to each Bloch wave $|q\rangle$; this is its *quasi-momentum*, which characterizes how it transforms under a translation over a. More generally, under a translation over ja which is a multiple of the cell size a of the lattice, the Bloch function is simply *multiplied by a phase factor* with an argument which is proportional both to the quasi-momentum p and to the translation ja:

$$\widehat{U}^j \, |q\rangle \; = \; \mathrm{e}^{-\mathrm{i}p(ja)/\hbar} \, |q\rangle. \tag{11.29}$$

Since the periodic boundary conditions amount to (11.26), we must have

$$\mathrm{e}^{-\mathrm{i}pL/\hbar} \; = \; 1,$$

so that p must be an integral multiple of $2\pi\hbar/L$:

$$p \; = \; m \frac{2\pi\hbar}{L}, \qquad \text{where } m \text{ is an integer.} \tag{11.30}$$

The quasi-momentum is a quantum number which bears some resemblance to ordinary momentum, as suggested by comparison of (11.25) and (11.28). In fact, let us consider a plane wave $|P\rangle$ characterized by its momentum P. (For future convenience, we have denoted the quasi-momentum by p, and to avoid confusion we shall use in the present chapter a capital P to denote the ordinary momentum, which is the eigenvalue of the momentum operator $\widehat{p}$.) For a particle in a box of length L with periodic boundary conditions, the values of P are, like those of p, quantized by (11.30). Under a translation over some distance α, represented by the unitary operator $\widehat{U}_\alpha$, the plane wave transforms according to

$$\widehat{U}_\alpha \, |P\rangle \; = \; \mathrm{e}^{-\mathrm{i}P\alpha/\hbar} \, |P\rangle, \tag{11.31}$$

as one can see immediately in the $\langle r|$ representation. This equation resembles (11.29) which holds for a Bloch wave. In particular, like the Bloch waves, the plane waves are eigenfunctions of the operator $\widehat{U}$, with the same eigenvalues (11.28); hence for a plane wave the momentum and the quasi-momentum are the same. Nevertheless, one should not confuse these two concepts. Whereas the property (11.31) of a plane wave is valid for *any* translation, a Bloch wave is transformed according to the simple equation (11.29) solely for particular translations over $\alpha = ja$ which *leave the lattice invariant* and which form a discrete group. On the other hand, (11.28) *only defines p modulo $2\pi\hbar/a$*. Thus, whereas P and $P+2\pi\hbar/a$ are different momenta, the quantum number p must be taken to be the same as $p+2\pi\hbar/a$, so that all values $p, p\pm2\pi\hbar/a, p\pm 4\pi\hbar/a, \ldots$ are equivalent.

It is convenient to lift this ambiguity, which is inherent in the definition (11.28) of the quasi-momentum, through the following convention. We require that p is confined to the interval

$$-\frac{\pi\hbar}{a} < p \le \frac{\pi\hbar}{a}, \qquad (11.32)$$

which defines the *Brillouin zone* (also called first Brillouin zone). The quasi-momentum then takes on N non-equivalent values, given by (11.30), (11.32), which are uniformly distributed over the Brillouin zone. (Incidentally, the eigenvalues (11.28) of $\widehat{U}$ are the Nth power roots of unity, uniformly distributed over the unit circle.) In the limit of an infinite crystal, $N \to \infty$, $L \to \infty$, $a = L/N$ fixed, the possible values of the quantum number p become a continuum, and a sum over the quasi-momentum p becomes an integral,

$$\sum_p \to \frac{L}{h} \int_{-\pi\hbar/a}^{\pi\hbar/a} dp, \qquad (11.33)$$

in the Brillouin zone. The weight L/h is the same as for the ordinary momentum P, but the integration range is here finite.

The quantum number p is not sufficient to characterize a Bloch wave $|q\rangle$, not even for our simplified one-dimensional problem, in contrast to P which characterizes the plane wave $|P\rangle$. Equation (11.28) where the quasi-momentum p takes on a *fixed* value defines a subspace of the single-electron Hilbert space. The restriction $\widehat{h}_p$ of $\widehat{h}$ to that subspace is an operator which must still be diagonalized. One can prove that, in contrast to $\widehat{h}$, the eigenvalues of $\widehat{h}_p$ form a discrete spectrum, even in the limit as $L \to \infty$. Moreover, in one dimension they are not degenerate. Let us arrange them in increasing magnitude, and characterize them by an index b, which takes on the values $b = 1, 2, \ldots$, and which is called the *band index* (Fig.11.2). As a result, in the

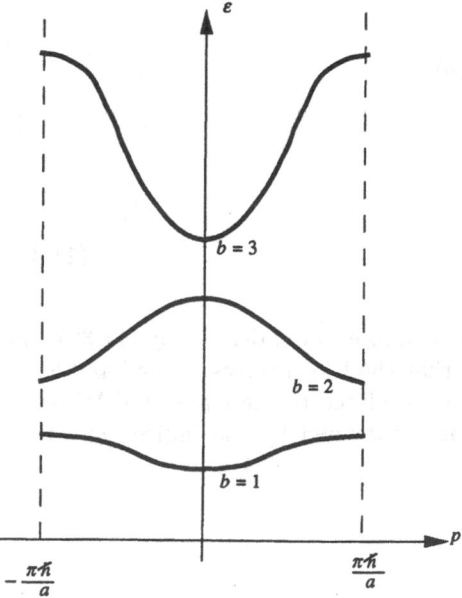

Fig. 11.2. The eigenenergies ε in the Brillouin zone for the lowest bands

case of one spatial dimension, each Bloch wave $|q\rangle \equiv |p, b\rangle$ is characterized by two quantum numbers, the quasi-momentum p which varies continuously in the Brillouin zone in the limit as $L \to \infty$, and the band index b which is discrete. If, for a given value of b, we vary the quasi-momentum p, the eigenenergy ε_q also varies continuously with p. Each (continuous) permitted band of the spectrum is thus characterized by its index b. One can show that in one dimension there is no overlap between the bands $b = 1, b = 2, \ldots$, which are thus separated by forbidden bands that do not contain any energy level, as we mentioned earlier. We shall verify this in §§ 11.2.3 and 11.2.4 for two extreme cases.

Let us finally define more precisely the structure of the Bloch waves by introducing for each band b its *Wannier wavefunction*, defined by

$$\chi_b(r) \equiv \frac{1}{\sqrt{N}} \sum_p \langle r | p, b \rangle \tag{11.34}$$

in terms of the Bloch functions of the given band. This definition assumes that we have made a definite choice of phase for each Bloch function, which so far had been defined by (11.13) apart from an overall phase. One can show, and we shall assume, that provided these phases are suitably chosen, the Wannier function (11.34) is *localized* in the vicinity of some characteristic point r_0 of the lattice. Moreover, in one dimension, it is either symmetric or antisymmetric with respect to r_0, and $2r_0/a$ is an integer. We shall write down explicitly examples of Wannier waves in §§ 11.2.3 and 11.2.4 and check these properties. For definiteness, we assume here that $r_0 = 0$. By carrying out each of the N lattice translations we can use (11.34) to construct the set of functions $\chi_b(r - ja)$, each of which is localized around the site ja. Using (11.25) and (11.29) we get

$$\chi_b(r - ja) = \frac{1}{\sqrt{N}} \sum_p \langle r - ja | p, b \rangle$$

$$= \frac{1}{\sqrt{N}} \sum_p \langle r | \widehat{U}^j | p, b \rangle$$

$$= \frac{1}{\sqrt{N}} \sum_p \langle r | p, b \rangle \, \mathrm{e}^{-ipja/\hbar}. \tag{11.35}$$

The summation over p is over N successive integers m defined by (11.30) and (11.32), and expression (11.35) shows that the Bloch waves of the band b are through a discrete Fourier transformation related to the translated Wannier waves. More precisely, the $N \times N$ matrix defined by the indices m and j through

$$\frac{1}{\sqrt{N}} \, \mathrm{e}^{-ipja/\hbar} = \frac{1}{\sqrt{N}} \, \mathrm{e}^{-2\pi i m j/N}$$

is unitary, so that relation (11.35) can be inverted to give

$$\langle r|p, b\rangle = \frac{1}{\sqrt{N}} \sum_j e^{ipja/\hbar} \chi_b(r - ja), \qquad (11.36)$$

where the summation is over all lattice sites. All Bloch waves of one band are thus produced, starting from a single Wannier function. They are constructed by subjecting this Wannier function to discrete translations and by adding the terms thus obtained, after having multiplied them by a phase factor which increases progressively with each translation. We can compare Eq.(11.36) with the equation

$$\langle r|P\rangle = \frac{1}{\sqrt{L}} \int d\alpha \, e^{iP\alpha/\hbar} \, \delta(r - \alpha),$$

which holds for a plane wave: the discrete translations over ja replace the continuous translations over α, and the Wannier wave plays the rôle of the wave $\delta(r)$, localized at the origin. Nevertheless, (11.36) applies inside each band b; moreover, the Wannier wave has a certain spatial extension which, one can prove, increases with the band index.

Expression (11.36) for the Bloch waves clearly demonstrates their *double nature* and separates the rôles of the two quantum numbers p and b. The *travelling wave nature* shows up directly through the exponential factor, which is the only place where the quasi-momentum p occurs. For small p (11.36) has the form of a plane wave $e^{ipr/\hbar}$ of momentum p, modulated by the factors χ. The *localized nature* shows up through the factors χ_b, which change from one band to another and which are centred around each lattice site. We shall see in §11.2.4 that these functions are a reminder of the atomic *orbitals* in which the electrons could be as long as the substance was gaseous. Note that the Wannier waves are not eigenfunctions of either $\hat{h}$ or $\hat{U}$, but that they form an orthonormalized base:

$$\int_0^L dr \, \chi_b^*(r - ja) \, \chi_{b'}(r - j'a) = \delta_{bb'}\delta_{jj'}. \qquad (11.37)$$

We shall summarize the various aspects of band theory in §11.2.5 in a form suitable for what follows. Subsections 11.2.3 and 11.2.4 are mainly for illustration and can be skipped in a first reading.

11.2.3 Weak Binding Limit

As an exercise we shall study the band structure of a one-dimensional model where the potential $\mathcal{V}(r)$ is small. When $\mathcal{V} = 0$, the problem is that of a particle in a box of length L with periodic boundary conditions, and the solutions of the Schrödinger equation (11.13) are the plane waves (11.31). The momentum P, which is quantized by (11.30), is a good quantum number when $\mathcal{V} = 0$, since the problem is invariant under translation over an arbitrary distance α, but it is no longer a good quantum number when $\mathcal{V} \neq 0$; however, we still have invariance under a translation over a, which implies the existence of the quasi-momentum quantum number p. It is therefore proper to find out how a *plane wave* $|P\rangle$ can be considered, in the limit as $\mathcal{V} \to 0$, as the *limit of a Bloch wave* $|p, b\rangle$. First of all, if we give the momentum P, we can deduce the quasi-momentum p. In fact, comparing (11.28) with (11.31) for $\alpha = a$, we get $p = P \bmod(2\pi\hbar/a)$, and condition (11.32) completely determines p. Conversely, with a given value of p in the Brillouin zone there is associated a family of plane waves, all with the same quasi-momentum p, which are characterized by their momentum

$$P = p + s\frac{2\pi\hbar}{a}, \qquad s = 0, \pm 1, \pm 2, \dots . \tag{11.38}$$

The band index is then determined by ordering, for given p, in increasing magnitude the eigenenergies of this family,

$$\varepsilon_{p,b}^{(0)} = \frac{P^2}{2m} = \frac{1}{2m}\left(p + s\frac{2\pi\hbar}{a}\right)^2, \tag{11.39}$$

where the index 0 indicates that $\mathcal{V} = 0$. When $0 < p < \pi\hbar/a$ these energies are ordered according to $s = 0, -1, +1, -2, \dots$; whereas, when $-\pi\hbar/a < p < 0$, the order is $s = 0, +1, -1, +2, \dots$. The band $b = 1$ thus corresponds to $s = 0$, the band $b = 2$ to $s = -\operatorname{sign} p$, the band $b = 3$ to $s = +\operatorname{sign} p$, the band $b = 4$ to $s = -2\operatorname{sign} p$, and so on (Fig.11.3). The bands 1 and 2 join each other when $\varepsilon^{(0)} = (\pi\hbar/a)^2/2m$, and, more generally, the bands b and $b + 1$ when $\varepsilon^{(0)} = (b\pi\hbar)^2/2m$.

In the case when there are no interactions, the Bloch wave $|p, b\rangle_0$ is identified with the plane wave

$$\langle r|p, b\rangle_0 = \frac{1}{\sqrt{L}}\,e^{iPr/\hbar} = \frac{1}{\sqrt{L}}\,e^{ipr/\hbar}\,e^{2\pi isr/a}, \tag{11.40}$$

where the index s is related to the band index b as we have just indicated. Using (11.34) and (11.33) we now find the Wannier waves corresponding to the potential $\mathcal{V} = 0$, for the choice (11.40) of the phases of the Bloch waves:

$$\chi_b^{(0)}(r) = \frac{\sqrt{a}}{\pi r}\left[\sin\frac{b\pi r}{a} - \sin\frac{(b-1)\pi r}{a}\right]. \tag{11.41}$$

These waves are relatively localized, notwithstanding a long oscillatory tail coming from the singularities at the points $p = 0, p = \pm\pi\hbar/a$, where the bands are joined together. The oscillations of the Bloch wave (11.36) or (11.40) can partly be attributed to the phase associated with the quasi-momentum, and partly to the Wannier wave, which only contributes to the wavelengths shorter than a.

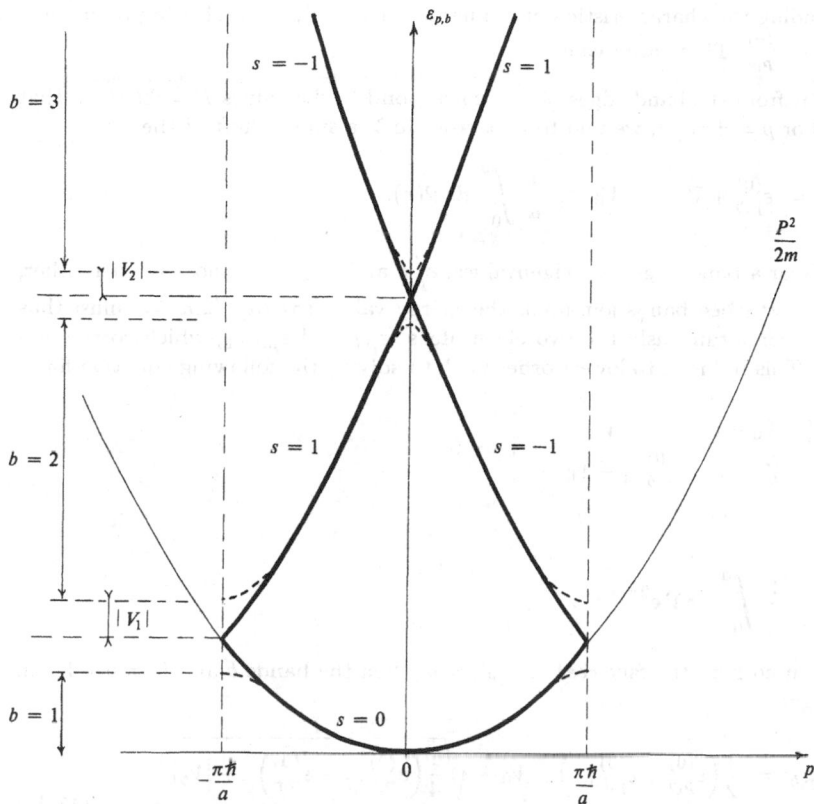

Fig. 11.3. The band spectrum in the weak binding limit: the thinly drawn line corresponds to a free particle; the thick line is the same spectrum, brought into the Brillouin zone, in terms of the quasi-momentum; the dashed curves are the perturbed spectrum, where we have assumed that $V_0 = 0$

Introducing the potential $\mathcal{V}$ modifies both the Bloch waves (11.40) and the spectrum (11.39). Nevertheless, the invariance of $\mathcal{V}$ under a translation over a, which is expressed by (11.27), implies that, if we write $\widehat{h}$ in the base (11.40) of plane waves, regarded as Bloch waves, its matrix elements between states with different quasi-momenta vanish. The non-vanishing matrix elements are

$$_0\langle p, b|\widehat{h}|p, b'\rangle_0 \;=\; \varepsilon^{(0)}_{p,b}\, \delta_{bb'} + \frac{1}{a}\int_0^a dr\, \mathcal{V}(r)\, e^{2\pi i(s'-s)r/a}. \qquad (11.42)$$

The determination of the Bloch waves $|p, b\rangle$ and their energies $\varepsilon_{p,b}$ in the presence of the potential thus reduces to diagonalizing the matrix (11.42) in b, b' for a given p, a task which we now shall achieve to first order in $\mathcal{V}$.

In the weak binding limit ($\mathcal{V} \to 0$) the off-diagonal elements of the matrix (11.42) are small and we can determine its eigenvalues $\varepsilon_{p,b}$, which are close to $\varepsilon^{(0)}_{p,b}$, by expanding the characteristic determinant, $\det(\hat{h} - \varepsilon)$, of (11.42) in powers of $\mathcal{V}$ and of $\varepsilon - \varepsilon_{p,b}^{(0)}$. There are two cases:

(i) Far from the band edges, which correspond to the values $P = \pm b\pi\hbar/a$, that is, $p = 0$ or $p = \pm\pi\hbar/a$, we find to first order in $\mathcal{V}$ a simple shift of the set:

$$\varepsilon_{p,b} = \varepsilon_{p,b}^{(0)} + V_0, \qquad V_0 \equiv \frac{1}{a} \int_0^a dr\, \mathcal{V}(r).$$

(ii) Near a band edge, two eigenvalues, $\varepsilon_{p,b}^{(0)}$ and $\varepsilon_{p,b+1}^{(0)}$, lie close to each other, as the unperturbed bands join up at the energy value $(b\pi\hbar/a)^2/2m$. We must thus determine simultaneously the two eigenvalues, $\varepsilon_{p,b}$ and $\varepsilon_{p,b+1}$, which correspond to them. This reduces, to lowest order in $\mathcal{V}$, to solving the following equation for ε:

$$\begin{vmatrix} \varepsilon_{p,b}^{(0)} + V_0 - \varepsilon & V_b \\ V_b^* & \varepsilon_{p,b+1}^{(0)} + V_0 - \varepsilon \end{vmatrix} = 0,$$

with

$$V_b \equiv \frac{1}{a} \int_0^a dr\, \mathcal{V}\, e^{2\pi i b r/a}.$$

We have used here the fact that $|s - s'| = b$ when the bands b and $b' = b + 1$ join up. We get

$$\varepsilon_{p,b} = \frac{1}{2}\left(\varepsilon_{p,b}^{(0)} + \varepsilon_{p,b+1}^{(0)}\right) + V_0 - \sqrt{\frac{1}{4}\left(\varepsilon_{p,b+1}^{(0)} - \varepsilon_{p,b}^{(0)}\right)^2 + |V_b|^2},$$

$$\varepsilon_{p,b+1} = \frac{1}{2}\left(\varepsilon_{p,b}^{(0)} + \varepsilon_{p,b+1}^{(0)}\right) + V_0 + \sqrt{\frac{1}{4}\left(\varepsilon_{p,b+1}^{(0)} - \varepsilon_{p,b}^{(0)}\right)^2 + |V_b|^2};$$

(11.43)

this also includes the first case, which we find when $\varepsilon_{p,b+1}^{(0)} - \varepsilon_{p,b}^{(0)} \gg |V_b|$. If we denote by $\delta p \ll \pi\hbar/a$ the distance of p from the band edge $p = 0$ or $p = \pm\pi\hbar/a$, we find from (11.43) for the two bands b and $b + 1$:

$$\varepsilon = \frac{1}{2m}\left(\frac{b\pi\hbar}{a}\right)^2 + \frac{\delta p^2}{2m} + V_0 \pm \sqrt{\left(\frac{b\pi\hbar}{ma}\right)^2 \delta p^2 + |V_b|^2},$$

(11.44)

which is shown by the dashed curve in Fig.11.3.

The most noticeable feature of the spectrum is the *appearance of forbidden bands* of width $2|V_b|$ at the points, $\varepsilon = (b\pi\hbar/a)^2/2m$, where the two unperturbed bands b and $b + 1$ join up. However weak the periodic potential $\mathcal{V}$, it is sufficient to change the spectrum, which is continuous when there is no potential, into a spectrum with alternately allowed and forbidden bands. This fact shows the general nature of the band structure. Nevertheless, in general, $|V_b|$ decreases as b increases, so that the forbidden bands which are relatively wide at low energies, become narrower at high energies, while the width $(2b - 1)(\pi\hbar/a)^2$ of the allowed bands increases.

The fact that a weak periodic potential $\mathcal{V}(r)$ manages to produce a significant, qualitative, change in the spectrum can be understood as a resonance effect. This potential, which has a period a in space, induces transitions between plane waves with wavenumbers $P/2\pi\hbar$ which differ by multiples of $1/a$. These transitions are particularly effective between waves of the same energy $P^2/2m$, that is, between waves $|P\rangle$ and $|-P\rangle$ with opposite wavenumbers; the effect is thus important if $2P/2\pi\hbar$ is a multiple of $1/a$. These values $P = \pm b(\pi\hbar/a)$ are just the momentum values corresponding to the joining of the two bands b and $b + 1$, that is, the points where the forbidden bands crop up.

The determination of the Bloch waves to lowest order in $\mathcal{V}$ is similar, but less simple. One can show as an exercise that, with a suitable choice of phase, $\langle r|p, b\rangle$ is for each b an analytical function of p of period $2\pi\hbar/a$, and that the Wannier functions decrease for large $|r|$ as $\exp\left[-ma|V_b r|/b\pi\hbar^2\right]$. One can also show that, depending on the band index and on the sign of V_b, these Wannier functions are either symmetric or antisymmetric with respect to $r = 0$ or to $r = \frac{1}{2}a$.

11.2.4 Tight Binding Limit

In the preceding subsection we have shown how a weak periodic potential creates narrow forbidden bands in the continuous spectrum of a free electron. We shall now study as a further exercise the opposite problem, and start from bound electron states in order to show how an isolated, degenerate level spreads out into a narrow allowed band. To do this, we introduce a one-dimensional, attractive, short-range potential $v(r)$, with its centre at the origin of the coordinates; this potential models the action of an ion to which the electron may be bound. The electron Hamiltonian, $\widehat{p}^2/2m + v(\widehat{r})$, produces a set of bound states $b = 1, 2, \ldots$, with eigenfunctions $\varphi_b(r)$, and with increasing energies $\varepsilon_b^{(0)} \equiv -\hbar^2\kappa_b^2/2m$. Outside the range of the potential v the orbital $\varphi_b(r)$ decreases as $e^{-\kappa_b|r|}$. We shall not be concerned with the unbound part of the spectrum, as we shall consider solely the lowest $\varepsilon_b^{(0)}$ levels.

Our model describes a hypothetical, very dilute crystal with a large cell size a, which we may gradually let decrease. Neglecting self-consistency effects, we have for the single-electron Hamiltonian, when there are N attractive, equidistant centres present, with periodic boundary conditions,

$$\widehat{h} = \frac{\widehat{p}^2}{2m} + \mathcal{V}(r) = \frac{\widehat{p}^2}{2m} + \sum_j v(\widehat{r} - ja). \tag{11.45}$$

We shall look for its spectrum and its Bloch waves. As $a \to \infty$ the spectrum $\varepsilon_b^{(0)}$, which is discrete, is the same as for a single attractive centre, but each level now has the large multiplicity N; letting a decrease, we shall see how *each discrete level blows up* and produces an *allowed band*.

We start from the limit where the distance a between the attractive sites is much larger than the range $1/\kappa_b$ of the orbital $\varphi_b(r)$. The latter remains an eigenfunction of $\hat{h}$, as it is practically equal to zero in the regions where the potentials associated with the other centres ja $(j \neq 0)$ act. Similarly, for a given j $(j = 0, 1, \ldots)$, the function $\varphi_b(r - ja)$ only feels the potential $v(r - ja)$ of that particular j, and it is also an eigenfunction of $\hat{h}$ with eigenvalue $\varepsilon_b^{(0)}$. The spectrum of $\hat{h}$ is thus discrete, and each of its eigenvalues $\varepsilon_b^{(0)}$ has a multiplicity N, with eigenfunctions $\varphi_b(r - ja)$. All intervals $\varepsilon_b^{(0)}, \varepsilon_{b+1}^{(0)}$ play the rôle of forbidden bands, with the allowed bands being infinitely narrow. In this limit as $a \to \infty$, the Wannier waves can be identified with the orbitals: $\chi_b^{(0)} = \varphi_b(r)$, and the Bloch waves, which are eigenkets common to $\hat{h}$ and $\hat{U}$, are thus according to (11.36) given by

$$\langle r|p, b\rangle_0 = \frac{1}{\sqrt{N}} \sum_j e^{ipja/\hbar} \varphi_b(r - ja). \tag{11.46}$$

In a realistic crystal structure, the cell size a is of the same order as the range of the potential, a few Å. We shall treat this problem here by starting from the above very dilute crystal limit, and *reducing gradually the interatomic distances* a which, nevertheless, will remain large as compared to the range $1/\kappa_b$ of the orbitals. When a decreases, the Bloch waves are modified, because we can no longer neglect the overlap of a φ function corresponding to a particular site with the potential v corresponding to another site. The energies $\varepsilon_{p,b}$ are therefore no longer degenerate. It is convenient to use the base (11.46), which was the eigenbase of $\hat{h}$ in the limit as $a \to \infty$. This base has the disadvantage of not being orthonormalized when a is finite. However, the fact that it has the quasi-momentum p as a quantum number simplifies the solution, as it allows us to work with a given p: in fact, thanks to the periodicity of the potential $\mathcal{V}(r)$, the Hamiltonian $\hat{h}$, written in the base (11.46), has non-vanishing matrix elements only between states with the same quasimomentum p, and that is also true for the unit operator. The energies $\varepsilon = \varepsilon_{p,b}$ are thus, for given p, obtained by solving the eigenvalue equation

$$\det{}_{bb'} \left[{}_0\langle p, b|\hat{h} - \varepsilon|p, b'\rangle_0 \right] = 0. \tag{11.47}$$

In the tight binding limit which we are considering here, the orbitals φ_b of interest are sufficiently tightly bound that their value at a neighbouring site, which is of the order of $\exp(-\kappa_b a)$, remains small. Under those conditions, the matrix elements of (11.47) contain, apart from the unperturbed term $\delta_{bb'}(\varepsilon_b^{(0)} - \varepsilon)$, only exponentially small contributions. Nevertheless, it is difficult to treat those by a simplistic perturbation expansion, as the orders of magnitude of the factors $\exp(-\kappa_b a)$ vary greatly when b varies. In particular, the overlap matrix

$$R_{bb'} \equiv {}_0\langle p, b | p, b' \rangle_0 = \delta_{bb'} + \sum_{j \neq 0} e^{-ipja/\hbar} \int dr\, \varphi_b(r - ja)\, \varphi_{b'}(r)$$

is dominated by the terms $j = \pm 1$, which behave as the largest of the two factors $\exp(-\kappa_b a)$ and $\exp(-\kappa_{b'} a)$, the one for which b or b' is the largest.

We can similarly write the Hamiltonian matrix in the form

$$_0\langle p, b | \widehat{h} | p, b' \rangle_0 = \varepsilon_b^{(0)} R_{bb'} + W_{bb'},$$

where

$$W_{bb'} = \sum_{j \neq j'} e^{-ipja/\hbar} \int dr\, \varphi_b(r - ja)\, v(r - j'a)\, \varphi_{b'}(r).$$

Comparing the range of the potential with a, κ_b, and $\kappa_{b'}$ enables us to retain, in the dominant order, only the terms $j' = 0$, $j = \pm 1$ in $W_{bb'}$, which behave as $\exp(-\kappa_b a)$; here we do not have a contribution behaving as $\exp(-\kappa_{b'} a)$, even when $b' > b$. The eigenvalue equation (11.47) therefore becomes

$$\det{}_{bb'} \left[\left(\varepsilon_b^{(0)} - \varepsilon \right) \delta_{bb'} + \left(W R^{-1} \right)_{bb'} \right] = 0,$$

where each row of the matrix $(W R^{-1})_{bb'}$ is small like $\exp(-\kappa_b a)$. In evaluating the determinant when ε is close to $\varepsilon_b^{(0)}$, it therefore suffices in the dominant order to retain in each row b', different from b, merely the diagonal element, of order $\varepsilon_{b'}^{(0)} - \varepsilon_b^{(0)}$. The eigenvalue $\varepsilon_{p,b}$ close to $\varepsilon_b^{(0)}$ is thus found to order $\exp(-\kappa_b a)$ by writing down the vanishing of the diagonal b, b element. Since the matrix R is close to unity, the result is

$$\varepsilon_{p,b} = \varepsilon_b^{(0)} + 2 \cos \frac{pa}{\hbar} \int dr\, \varphi_b(r - a)\, v(r)\, \varphi_b(r). \tag{11.48}$$

The degeneracy of the band is thus lifted through the action of neighbouring sites. When $\kappa_b a$ is large, the band remains narrow, its width, given by (11.48), being proportional to $e^{-\kappa_b a}$. The shape of the band as function of p, which is a section of a sinusoid, looks like what we did find for the weak binding limit, but now the allowed, rather than the forbidden, bands are narrow. In practice, the tight binding approximation is justified for deep lying bands for which κ_b is large. When b increases, κ_b decreases, so that the allowed bands widen more and more; this conclusion is the same as in the opposite limit of § 11.2.3.

To the order we have used, the Wannier and Bloch functions are, respectively, given by

$$\chi_b(r) = \varphi_b(r) - \tfrac{1}{2} \left[\varphi_b(r - a) + \varphi_b(r + a) \right]$$

$$\times \int dr'\, \varphi_b(r' - a)\, \varphi_b(r'), \tag{11.49}$$

and by (11.36), which are expressions consistent with the energy value (11.48). The corrections to the atomic orbitals $\varphi_b(r)$ given by (11.49) allow the functions $\chi_b(r)$ and $\chi_{b'}(r+ja)$, and hence the Bloch waves $|p, b\rangle$ and $|p', b'\rangle$, to be orthogonal. The Wannier waves can thus be interpreted as atomic orbitals, modified by the presence of neighbouring atoms in the solid, which are mutually orthogonal, and which can be used to construct, through the linear superposition (11.36), the eigenfunctions of the single-electron Hamiltonian.

11.2.5 Three-Dimensional Bands

The one-dimensional models considered in §§ 11.2.2–4 help us to understand the band structure of a realistic three-dimensional solid, which we shall only briefly describe. Each single-electron state q is characterized by five quantum numbers. The first three are the components of the *quasi-momentum* p, which is associated with the invariance of the solid under the discrete translation group of the crystal. Each translation, represented by a unitary operator generalizing $\widehat{U}^j$ to three dimensions, is characterized by a vector which generalizes ja and which we denote by R. The set of possible values of R form a lattice, and the quasi-momentum p enters solely in the combinations $e^{-i(p \cdot R)/\hbar}$, similar to (11.29); it is therefore defined only modulo certain translations forming another lattice in momentum space. We can, as we did through (11.32), lift this ambiguity, by requiring p to be confined to a certain region, the *Brillouin zone*, the shape of which in three dimensions depends on the crystal lattice. For instance, in the case of a simple cubic lattice of cell size a, the Brillouin zone is itself a cube in momentum space with its centre at the origin and edgelength $2\pi\hbar/a$. The components of p are quantized by (11.30) in each of the three spatial directions as in the case of plane waves in a box. The number of non-equivalent values of the quasi-momentum p is equal to the number N of cells in the crystal lattice. Whatever the crystal structure, the volume of the Brillouin zone is $(2\pi\hbar)^3/v$, where v is the cell volume, and the summation over p becomes in the case of a crystal with a large volume Ω the integral (10.38) in the Brillouin zone.

The fourth, discrete, quantum number is the *band index* b. Finally, the last quantum number is the z-component s_z of the electron *spin*; the electron energy does not depend on it. The *Bloch waves* $|p, b\rangle$, which are the common eigenstates of the single-particle Hamiltonian $\widehat{h}$ and of the translation operators, can be written, in a similar way as (11.36), in terms of localized *Wannier functions*, where now the translations ja have been replaced by the lattice vectors R.

In general, the spectrum $\varepsilon(p, b)$ for a given direction of p, as function of the length p, behaves similarly as in one dimension. Nevertheless, in some directions, two bands may join, cross, or even coincide. Moreover, it can happen that the maximum of the band b, reached in one direction of p, is higher than the minimum of the band $b + 1$, reached in another direction. Therefore, whereas in one dimension the number of electron states in each band was always twice (because of the spin) the number of lattice cells, the multiplicity and overlap effects imply that, in general, the allowed bands in three dimensions contain a number of single-electron states which is an *even multiple of the number of cells*. Between two successive gaps in the spectrum we may thus find a *multiple band* containing several subbands.

As in one dimension, the single-electron states show the features both of *free waves* and of *bound states*. This is reflected directly in the quantum numbers $q = (\boldsymbol{p}, b, s_z)$, in the eigenfunctions, the form of which is a generalization of (11.36) to three dimensions, and in the spectrum of $\widehat{h}$. Either the one, or the other of these two aspects was dominant in the approximations of §§ 11.2.3 and 11.2.4, and will be the origin of the difference between metals and insulators. The free electron aspect shows up in the fact that a Bloch wave extends through the whole of the crystal, that it resembles a plane wave in respect to discrete lattice translations, and that its quasi-momentum $\boldsymbol{p}$, like its energy, can vary continuously. The bound electron aspect is reflected in the localization of the Wannier functions around a crystal site, or around a bond between two sites, in the discrete nature of the band index b, and in the arrangement of the spectrum as bands separated by gaps.

More precisely, in the tight binding approximation (§ 11.2.4), we start from a gas state where the atomic nuclei are far from one another and we build up the solid by letting these nuclei gradually get closer until they reach the lattice positions which they actually occupy in the crystal considered. At the start, the eigenstates are the atomic orbitals $1s$, $2s$, $2p$, $3s$, $3p$, ..., which are localized at one of the N sites. At each site, the multiplicity of a level is $d = 2(2l+1)$, due to the electron spin and to the invariance of the effective atomic potential under rotations. The Nd states which are in this way associated with each *electron shell* $1s$, $2s$, ..., and which have all the same energy, produce in the solid a *continuous band* containing Nd single-electron states; this band is a multiple one for $l \geq 1$. The fact that an electron can pass from one site to another lifts the multiplicity N. During that process in which the electron shells are spread out, two neighbouring bands may *encroach* one onto the other, thus increasing the multiplicity of the allowed band. Conversely, a multiple band may split up. As an example we give in Fig.11.4 the single-electron energy levels ε_q of a hypothetical sodium crystal where we can change the cell size a arbitrarily. For the sodium gas ($a \to \infty$) the deepest shells, $1s$, $2s$ (which are not shown), and $2p$ are all occupied, by 10 electrons, while the eleventh electron is in the $3s$ shell. As a decreases, each atomic shell expands into a band with Nd levels, which in the case of the deepest shells is very narrow, but which is wider for the outer shells because they feel more strongly the perturbation from the presence of neighbouring atoms. When the physical value of a has been reached, the $3s$ and $3p$ shells have mixed, producing a multiple $3s$-$3p$ band with $2(1 + 3)N = 8N$ states. The Wannier waves, which in the deepest bands are practically the atomic orbitals, are in such a case constructed through hybridization (linear combination) of orbitals; here, for instance, of the four $3s$-$3p$ orbitals of each atom, with possible hybridization

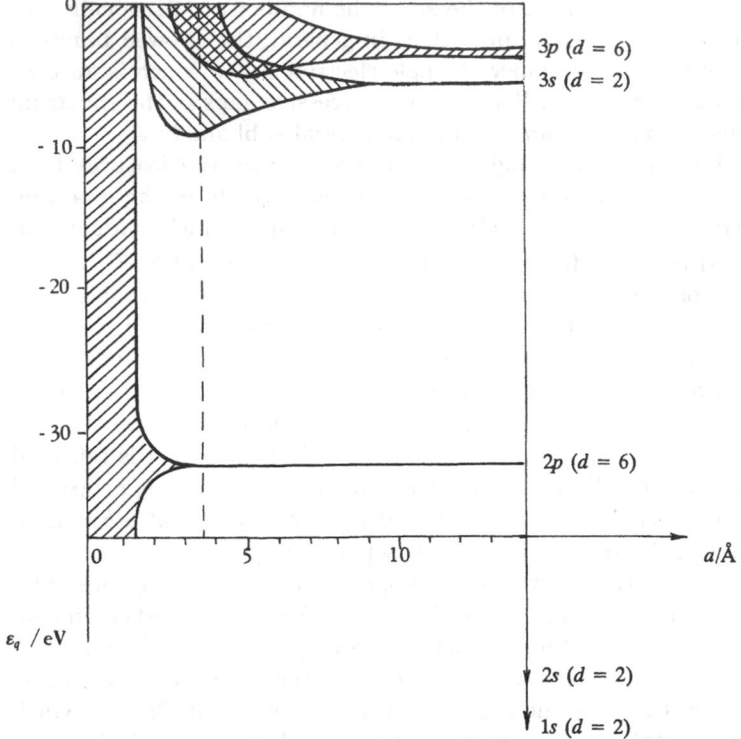

Fig. 11.4. The energy bands of a hypothetical sodium crystal with a variable mesh size a. The dotted vertical line represents the experimental value $a = 3.7$ Å

between neighbouring atoms. The examples of diamond, given below, and of graphite (§ 11.3.2) will show that the band structure obtained in this way for a solid may have little in common with the initial structure of the atomic shells.

Moreover, one should take into account the invariance of the potential, not only under the translation group of the crystal, but also under certain rotations or symmetries which characterize the crystal structure, especially for multiple bands. One can show that the band shape depends on this structure. (In one dimension this kind of property shows up already in a very simplified manner in that a symmetry of the potential $\mathcal{V}(r) = \mathcal{V}(-r)$ leads to a symmetry of the spectrum under the exchange $p \leftrightarrows -p$.) The Wannier waves can similarly be related to one other, not only through translations, but also through the rotations and symmetries of the crystal group, which either leave them invariant or associate them in non-equivalent multiplets. The corresponding band is in the latter case a multiple one, and we obtain a localized Wannier wave by a superposition of the Bloch waves of the subbands; conversely, the expression for a Bloch wave which generalizes (11.36) contains a summation over Wannier functions which are derived from one another not only through translations R, but also through rotations or symmetries. The

Wannier waves can be localized either around a nucleus, like the atomic orbitals, or between two nuclei, like the electrons responsible for the covalent bond in a molecule such as H_2.

On the other hand, the breaking of the rotational invariance of the atomic orbitals, due to the anisotropy of the crystal lattice, lifts the $2l + 1$ degeneracy of the atomic shells when a decreases. If that effect is larger than the broadening of the multiple band, a single atomic shell may produce several bands.

As an example, let us consider *diamond*. Without going into the details of its crystal structure, we note that each C atom is surrounded by 4 neighbours, placed at the vertices of a tetrahedron of which it is the centre. A crystal cell contains two C atoms, the tetrahedra of which have different orientations. The $1s$ atomic shell produces a deep, very narrow band containing $4N$ states which are all occupied by electrons; here N is the number of cells and $2N$ the number of C atoms. Each atom supplies four more electrons, in the $2s$ and $2p$ shells. These shells lie closely together and produce in the solid altogether $16N$ states, organized in 8 subbands which are expected more or less to overlap. (We must divide by 2 because of the spin, and by N because of the quantum number p.) On the other hand, the $8N$ corresponding Wannier waves are localized around the bonds between neighbouring atoms as in the covalent bond of tetrahedric carbon. The number of such bonds is $4N$. (Each of the $2N$ atoms has four bonds, but each bond is shared by two atoms; there are thus 4 non-equivalent bonds per cell, which can be derived the one from the other through symmetries and rotations.) Altogether, through hybridization of the atomic $2s$-$2p$ orbitals, one can thus construct at each bond 2 Wannier waves; one, the so-called bonding one, is a function which is symmetric with regard to the centre of the bond, and the other, called antibonding, is antisymmetric. The 8 subbands can thus be classified into two groups: the lower 4, which overlap, are associated with the symmetric Wannier function and with those derived from it through the operations of the crystal group, and the 4 higher ones are associated with the antisymmetric function. The two groups are separated by a forbidden band. At zero temperature, the $8N$ $2s$-$2p$ electrons occupy exactly the lower multiple band, and the gap is large enough so that they remain frozen in at room temperatures: diamond is an insulator.

To study the equilibrium properties we need to evaluate, as we mentioned in § 10.3, sums over the quantum numbers q, which can be reduced to the integral (10.38) over p in the Brillouin zone together with sums over the band index b and the spin. Another method consists, as in § 10.3.3, to introduce the *single-electron density of states* $\mathcal{D}(\varepsilon)$, defined by (10.40). In the model where these states are represented by plane waves in a box, $\mathcal{D}(\varepsilon)$ was given by (10.43), that is,

$$\mathcal{D}(\varepsilon) = \frac{(2m)^{3/2}\Omega}{2\pi^2\hbar^3} \sqrt{\varepsilon}\, \theta(\varepsilon).$$

For a more realistic band spectrum we have

$$\mathcal{D}(\varepsilon) = \frac{2\Omega}{(2\pi\hbar)^3} \sum_b \int d^3p\, \delta[\varepsilon - \varepsilon(p, b)], \qquad (11.50)$$

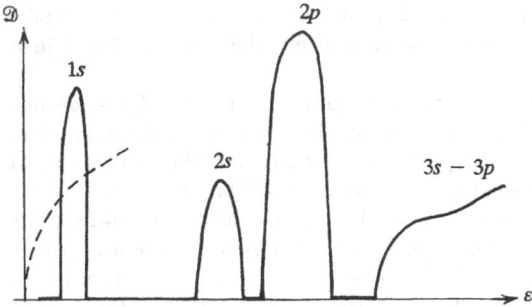

Fig. 11.5. Density of states for electrons in a crystal. The dashed lines represent the density of states in the box model

where the integral is over the Brillouin zone. The density of states, which is proportional to the volume and vanishes for ε values in the forbidden bands, makes it easier in three dimensions to visualize the spectrum than the functions $\varepsilon(\boldsymbol{p}, b)$. We have sketched it in Fig.11.5 strongly exaggerating the width of the lowest allowed bands. In fact, these bands are associated with very tightly bound electron orbitals, with a spatial extension which is small compared to the crystal cell size. The tight binding model (§ 11.2.4) shows that under those conditions the band width is very small, as a bound state at a site is practically unaffected by the potential produced by the other sites. The $1s$, $2s$, $2p$ bands should, in fact, have been represented by extremely sharp peaks. The area of each allowed band represents the number of levels in that band: $2N$ for $1s$ and $2s$, $6N$ for $2p$, and $8N$ for $3s$-$2p$.

11.3 Electrons in Metals, Insulators, and Semiconductors

The properties of the electron cloud in a solid depend directly on the band structure. We shall see several examples in what follows. The large mobility of the electrons, due to their small mass, explains that they determine all the *electromagnetic properties* of solids; on the other hand, energy exchanges between electrons correspond to frequencies of the order of those of visible light, one to three eV, so that the electrons also determine the *optical properties* of solids. The great variety of the electromagnetic and optical properties of crystals reflects the complexity of band structures.

Since we shall treat the electron cloud in the Hartree approximation as a non-interacting Fermi-Dirac gas, we shall restrict ourselves to explaining those properties of solids for which the interactions of electrons with electrons or with the lattice do not play a rôle. Phenomena, such as *magnetism* or *superconductivity*, which involve these interactions, can only be studied by using more refined models which go beyond the framework of the present book (see, however, Exerc.11f and § 12.3.3).

In Chapters 14 and 15 we shall also study *heat and electricity transport* properties which, in non-equilibrium situations, are connected with the existence of temperature or chemical potential gradients. Again, by virtue of their great mobility, it is the electrons which are, in general, responsible for conduction effects in solids. Let us note right now the importance of band theory for their explanation. Electric conduction, for instance, is the result of the combination of two effects: the acceleration of the electrons by the applied field, and their *collisions* with fixed obstacles which tend to isotropize their velocity distribution. However, these obstacles *are not the nuclei*. In fact, we have just taken the effect of nuclei into account for the case where they are fixed to a regular lattice; the Bloch wave functions (11.36) of the electrons in the presence of fixed nuclei are similar to the plane waves which would have been asssociated with the free electrons, and we shall see in § 11.3.2, by writing down the equations of motion for a Bloch wavepacket, that its dynamics resemble that of a free wavepacket, possibly with a change in the mass. As a result, *a regular lattice is transparent to electrons* in contrast to a randomly distributed system of nuclei. This is an important fact, as the nuclei are densely packed, and if we assumed that the electrons are scattered by them, we would be unable to explain the observed high conductivity of metals. In fact, the electrons are scattered by lattice *imperfections*: crystallization defects, impurities, and, above all, *displacements of the nuclei* from their equilibrium positions (or "phonons", see § 11.4.2). These are random displacements which are excited thermally and which produce a non-periodic perturbation of the potential seen by the electrons.

11.3.1 Metals

At zero temperature the system of electrons in a solid occupies the single-electron states with the lowest energies ε_q, up to the Fermi level ε_F, which is the $T = 0$ limit of the chemical potential μ. Let us assume that the *Fermi level lies inside an allowed band*, which we shall then call the *conduction band*. We shall see that in this case the solid is a *metal*; in contrast, for an insulator, the available electrons will occupy exactly a number of the lowest energy bands. For example, in a crystal of N sodium atoms, $10N$ electrons occupy the deep lying $1s$, $2s$, and $2p$ bands (we have greatly exaggerated the width of these bands and greatly decreased their height in Fig.11.6); there remain N electrons which occupy the bottom of the $3s$-$3p$ conduction band.

The spacing of the bands is, in general, several eV (about twenty for Na) whereas room temperature corresponds to $\frac{1}{40}$eV. Hence, raising the temperature has no effect whatever on the electrons in the deep lying bands, which remain *frozen in*. Therefore, only the conduction band is involved in the thermal excitation of electrons, as in all effects, such as electric conduction, where the relevant energies are small compared to an eV. In fact, we shall show below that we can describe metallic sodium either by the band scheme of Fig.11.6, where the $11N$ electrons are subject to a self-consistent potential

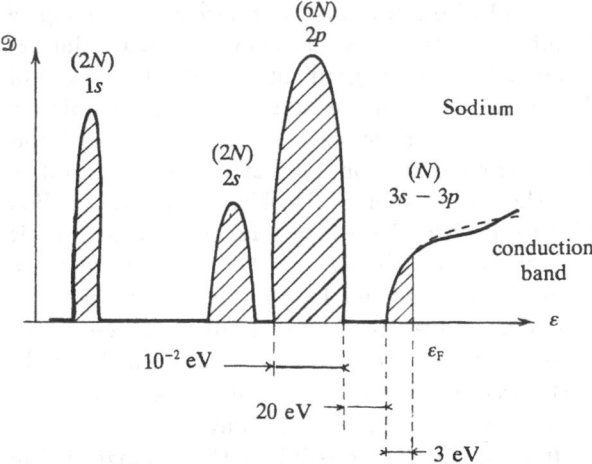

Fig. 11.6. The band structure and band occupation of sodium. The dashed lines refer to the box model

created by themselves and by the N sodium nuclei, or by the following model. We replace the N Na nuclei by N Na$^+$ ions placed on the same lattice, and we introduce just the N conduction electrons, neglecting the $10N$ electrons fixed to the ions. Each electron sees the effective potential of the Na$^+$ ions and a self-consistent potential due solely to the conduction electrons; in that case we need only take the conduction band into account, which is the lowest band for the new problem.

The fact that the level density in the conduction band, which determines the thermodynamic properties of the electron gas in a metal, is similar to the level density of an electron gas in a box (see the dashed line in Fig.11.6) justifies the latter model which we studied in §10.4.3. In particular, even though there are $11N$ electrons in the sodium crystal, one must model the crystal by solely N electrons enclosed in a box. Figure 11.7 allows us to clarify the relation between the box model for a metal and the more sophisticated band model: in contrast to the electrons of the deep lying bands which remain localized near the nuclei, those in the conduction band have sufficiently high energies so that we can reasonably replace the periodic potential by a constant average value, close to the energy at the bottom of the conduction band. Everything is happening in this approximation as if the system of Na$^+$ ions seen by the electrons were equivalent to a uniform positive charge density.

One can justify the replacement of the nuclei and the electrons in the deep lying bands by fixed ions as follows. First of all, we note that those bands can be treated in the tight binding approximation of § 11.2.4. Even for the $2p$ band of Na, the binding energy, of the order of 20 eV, leads to a range $1/\kappa$ of the electron atomic orbitals $(\varepsilon_b^{(0)} = -\kappa^2/2m)$ of about 0.5 Å, whereas the cell size a is 3.7 Å, so that the expansion parameter $e^{-\kappa a}$ is only 0.5×10^{-3}. Hence, the $1s$, $2s$, $2p$ bands are very narrow, with all energies ε_q equal to the atomic binding energies; their

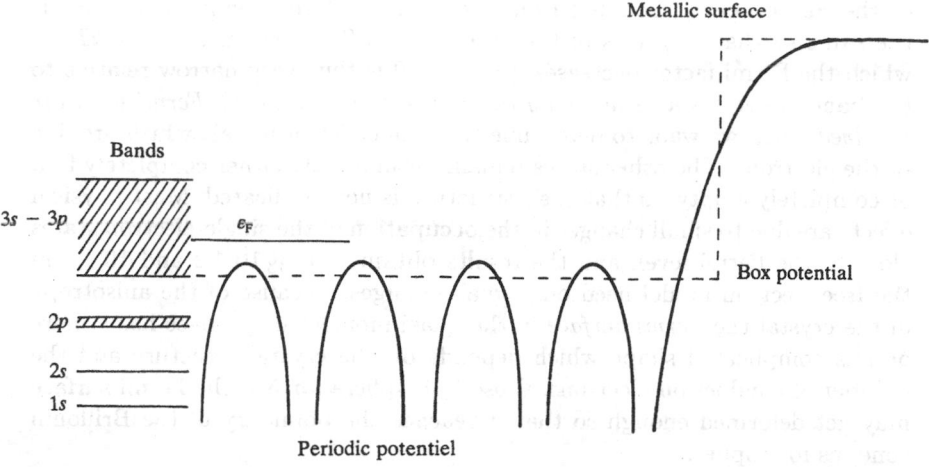

Fig. 11.7. The box potential for sodium

Wannier waves reduce to the atomic orbitals. On the other hand, the set of Bloch waves of a band can be derived from the corresponding set of Wannier waves, localized near each of the Na nuclei, through the unitary transformation (11.36). It is thus *equivalent* to saying that *all the Bloch states of a band are occupied* or that all Wannier states, that is, *all atomic orbitals of the corresponding shell are occupied.* Filling the $1s$, $2s$, $2p$ orbitals amounts to placing a Na^+ ion at each lattice site. Altogether, it is legitimate to neglect band theory for the set of deep lying shells and to treat the corresponding electron cloud, together with the nuclei, as a rigid charge distribution produced by N Na^+ ions arranged on the lattice. From this periodic distribution one can then construct the Bloch waves of the conduction band where one will find the other N electrons. Here one can use the weak binding limit (§ 11.2.3), as the potential produced by the Na^+ ions does not vary greatly in space. The conduction band thus appears as the lowest band of this problem, which is not greatly perturbed, so that the density of states $\mathcal{D}(\varepsilon)$ in this energy range is similar to that of the box model, apart from a change in origin.

The same idea also simplifies considerably the theory of (insulating) ionic crystals, such as NaCl. Here, each of the N cells contains one Na atom and one Cl atom. In the ground state the electrons occupy exactly the bands resulting from the $1s$, $2s$, $2p$ shells of Na and the $1s$, $2s$, $2p$, $3s$-$3p$ shells of Cl, which are all subject to the tight binding approximation. From the electron point of view, the state can be described in terms of bands, but equivalently and more simply by stating that all these *atomic orbitals are occupied* (the highest ones are slightly perturbed by the periodic structure), which justifies the intuitive picture of a crystal consisting of N Na^+ ions and N Cl^- ions. The effective potential W which determines the crystal structure (§ 11.1.1) reduces thus to the sum of the effective interactions between those ions, which contain not only the Coulomb forces associated with their total charge, but also a short range repulsion due to the Pauli principle, and other contributions associated with the spatial extension of the ions.

We have seen in § 10.4.2 that in the box model for metals the Fermi energy $\varepsilon_F = k\Theta_F$ equals a few eV. In band theory the order of magnitude of the Fermi energy, counted from the bottom of the conduction band, is the same so that kT/ε_F is of the order of 1%. The region, of order kT, in which the Fermi factor decreases from 1 to 0 is thus very narrow relative to the band shape: as a result, *only the states q closest to the Fermi level are involved* when we want to determine the properties of metals which are due to the electrons. The other states remain permanently either completely full, or completely empty, so that their structure is not implicated. Most physical effects are due to small changes in the occupation of the single-electron states close to the Fermi level, and the results obtained in §§ 10.4.2 and 10.4.3 in the free electron model need only small changes. Because of the anisotropy of the crystal the *Fermi surface* in the quasi-momentum p space has a more or less complicated shape which depends on the crystal structure and the number of conduction electrons. Close to a sphere for Na, the Fermi surface may get deformed enough so that it reaches the boundary of the Brillouin zone, as for copper.

The various thermodynamic expressions (10.65) to (10.68) remain unchanged, provided the electrons are treated as independent particles with energies ε_q. They are valid for $T \ll \Theta_F$, where the characteristic Fermi temperature Θ_F is here defined in relation to the shape of $\mathcal{D}(\varepsilon)$ in the neighbourhood of ε_F, for instance, through $k\Theta_F = \mathcal{N}_c(\varepsilon_F)/\mathcal{D}(\varepsilon_F)$, or through $k\Theta_F = \mathcal{D}(\varepsilon_F)/2\mathcal{D}'(\varepsilon_F)$, where $\mathcal{N}_c$ indicates the number of states counted from the bottom of the conduction band. Eliminating μ for given N and $T \ll \Theta_F$ gives

$$\mu - \varepsilon_F \sim -\frac{\pi^2}{6}(kT)^2 \frac{\mathcal{D}'(\varepsilon_F)}{\mathcal{D}(\varepsilon_F)}, \tag{11.51}$$

$$S \sim \frac{\pi^3}{3}k^2 T \mathcal{D}(\varepsilon_F) \sim C, \tag{11.52}$$

so that the electron contribution, $C = TdS/dT$ to the *specific heat is linear in T*, as in § 10.4.3. The chemical potential still remains close to ε_F, whatever the temperature.

Similarly, the magnetic susceptibity due to the electron spin, that is, the *Pauli susceptibility*, is nearly *constant* and equals

$$\chi = \mu_B^2 \frac{\mathcal{D}(\varepsilon_F)}{\Omega} = \frac{\mu_B^2}{\Omega} \frac{\partial N}{\partial \mu}, \tag{11.53}$$

as in the box model. In order to find this expression (Exerc.10b) one notes that the magnetic term $\mu_B B \hat{\sigma}$, which must be included in the Hamiltonian $\hat{h}$, is for each value $\sigma = \pm 1$ simply a constant which can be added to the energies $\varepsilon(p, b)$ of band theory without a magnetic field B. Hence, we get from Eq.(10.47)

$$A = -\frac{1}{2}\int d\varepsilon\, \mathcal{N}(\varepsilon) \sum_\sigma f(\varepsilon + \mu_B B\sigma), \tag{11.54}$$

where $\mathcal{N}(\varepsilon)$ is evaluated for the case without field. As a result we find, for small B and using the fact that $f(\varepsilon) \sim \theta(\varepsilon_{\mathrm{F}} - \varepsilon)$, the expression for the magnetization,

$$\frac{M}{\Omega} = -\frac{1}{\Omega}\frac{\partial A}{\partial B} = \frac{\mu_{\mathrm{B}}}{2\Omega}\int d\varepsilon\, \mathcal{N}(\varepsilon)\left[\frac{\partial f(\varepsilon + \mu_{\mathrm{B}}B)}{\partial\varepsilon} - \frac{\partial f(\varepsilon - \mu_{\mathrm{B}}B)}{\partial\varepsilon}\right]$$

$$\sim \frac{\mu_{\mathrm{B}}^2 B}{\Omega}\int d\varepsilon\, \mathcal{N}(\varepsilon)\frac{\partial^2 f}{\partial\varepsilon^2} \sim \frac{\mu_{\mathrm{B}}^2 B}{\Omega}\mathcal{D}(\varepsilon_{\mathrm{F}}), \qquad (11.55)$$

and thus χ.

We have used for the evaluation of the thermodynamics quantities in a metal the theory of a gas of non-interacting fermions, without taking into account the self-consistency of the potential $\mathcal{V}$ which depends on parameters such as T, μ, or B. Hence, expressions (10.47), (10.65), or (11.54) which we have written down for the grand potential, like Eq.(10.67) for the internal energy, are incorrect. In fact, as we saw in § 11.2.1, they overestimate the interaction energy (11.20) by a factor 2. In the Hartree approximation, the correct expression for the grand potential of the electrons is the minimum of (11.22), and it differs from (11.54). Nevertheless, the other equations (11.51–53) and (11.55) remain correct. One sees this by noting that N, S, and M all follow from the grand potential through differentiating with respect to μ, T, or B. According to (11.24') these quantities can therefore be calculated in the Hartree approximation as if the electrons were a gas of non-interacting fermions with energies ε_q.

The most remarkable property of metals is their high *electric conductivity*. We leave its study to the end of § 11.3.2 and to § 15.2.3.

11.3.2 Insulators; Dynamics of Electrons and Holes

One of the properties which can vary most widely from one solid to another is the electric conductivity. For instance, at room temperature the resistivity of copper is $1.7 \times 10^{-6}\ \Omega$ cm, whereas that of sulphur is $10^{17}\ \Omega$ cm. We shall see that the distinction between metals and insulators arises from an essential difference in the way the electron bands are filled.

Band Occupation

A metal was characterized by a conduction band which was partially filled at zero temperature, and hence at all temperatures. An *insulator* is a solid such that *the number of electrons is exactly equal to the number of states in the lowest bands*. At zero temperature the Fermi level thus lies between two bands: the band above the Fermi level, the *conduction band*, is therefore completely empty, while the one below it, the *valence band*, is completely occupied (Figs.11.8 and 11.9).

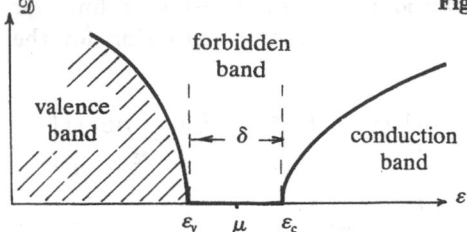

Fig. 11.8. Band occupation in an insulator

The interesting part of the ε_q spectrum, the only part which will play a rôle in what follows, is the bottom of the conduction band and the top of the valence band. In the simplest cases these parts correspond to small values of the quasi-momentum in both bands. We shall restrict ourselves in what follows to such a situation and we shall denote by ε_c and ε_v the limits of the forbidden band, which are reached for $\boldsymbol{p} = 0$.

One might think that this situation is an exception, as it implies the equality of two large numbers: the number of electrons and the number of states in the lowest bands. However, that is not the case, since we have seen in § 11.2.5 that the number of states in each band is a multiple of the number of cells in the crystal; the number of electrons is a multiple of the number of atoms, and this makes it possible, depending on the band structure, to fill a certain number of bands *exactly*. For instance, the crystals of the inert gases are insulators. The filled atomic shells produce, in fact, in this case filled bands, well separated from the empty bands which lie above them. On the other hand, most solids with an odd number of electrons per cell are metals: alkalis, Cu, Ag, Au, Al. In fact, due to the level degeneracy connected with the spin inside each band, the number of states in each band is an *even* multiple of the number of cells, which forbids the conduction band to be filled exactly.

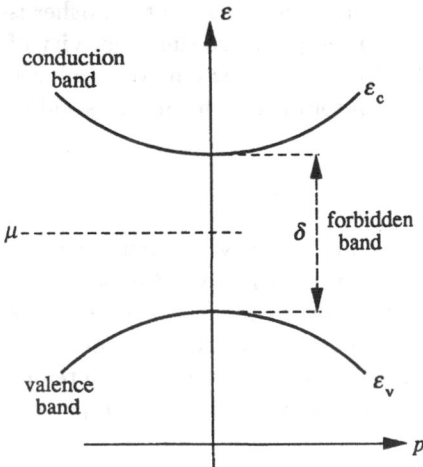

Fig. 11.9. The bands involved for an insulator, in quasi-momentum space

The situation is less simple for other substances; one can only decide about whether a solid is an insulator or a conductor through a detailed and difficult study of the band structure. This is particularly clear for substances such as carbon which may be either insulators or conductors, depending on their crystal structure: in § 11.2.5 we described the bands of *diamond* which is a typical insulator, whereas *graphite* is an anisotropic conductor. Its atoms are arranged in hexagonal lattices in parallel planes. Let the z-axis be taken at right angles to those planes. As in diamond, the $1s$ orbitals produce a deep lying, filled band. Hybridization similar to that in diamond, however, now occurs between the three $2s$, $2p_x$, and $2p_y$ orbitals; they produce two Wannier waves for each bond on the hexagonal lattice, one of which is bonding, and the other anti-bonding. (There are three such bonds per two C atoms.) Two sets of subbands correspond to them, which are separated by a forbidden band; the highest set is empty and the lowest set is filled with electrons – 3 per atom. The conducting nature of graphite shows up when one studies the remaining $2p_z$ orbitals and the remaining electrons – one per atom. The bands produced by those orbitals do not have a simple structure, because the unit cell contains more than one C atom; but they overlap and constitute a multiple allowed band, which contains $2M$ single-electron states, when we take spin into account, where M is the number of C atoms. We still need to place M electrons in those bands, and their partial filling explains that graphite is a conductor, but not quite as good as a metal.

We can easily understand why a complete filling of bands leads to an *insulating* crystal: it needs at least an energy equal to the gap $\delta = \varepsilon_c - \varepsilon_v$ between the valence band and the conduction band to let an electron move freely through exciting it and thus to produce a current, whereas in a metal an infinitesimally small energy is sufficient to excite electrons in the neighbourhood of the Fermi surface. An insulating crystal thus resembles a gas molecule: in both cases the electronic degrees of freedom are frozen in.

If the temperature increases, some electrons are thermally excited into the conduction band, leaving holes in the valence band. This is a very weak effect, as most electrons remain frozen in into the valence band: the temperature is, in fact, always small compared to the characteristic temperature, which here is δ/k, where δ is the width of the forbidden band. Nevertheless a number of physical phenomena are the consequence of this small variation in population, as we shall see later on. Let us evaluate the *average number of electrons excited* into the conduction band:

$$N_c = 2 \sum_{\boldsymbol{p}} f[\varepsilon_c(\boldsymbol{p})]. \tag{11.56}$$

The sum is over the quasi-momenta; $\varepsilon_c(\boldsymbol{p})$ denotes the energies ε_q of the conduction band, and the factor 2 comes from the spin. In the Fermi factor,

$$f(\varepsilon) = \frac{1}{e^{\beta(\varepsilon-\mu)} + 1},$$

the exponential dominates the denominator: we shall see, in fact, that the chemical potential μ is close to the middle of the forbidden band so that the exponential is large, of order $e^{\beta\delta/2}$. Therefore, the conditions of §10.3.4a hold. The Fermi-Dirac gas of the electrons in the conduction band – for the moment we forget about the other bands – behaves like a *classical gas* and we can replace f in that band by

$$e^{-\beta[\varepsilon_c(p)-\mu]}. \tag{11.57}$$

On the other hand, this factor decreases fast with increasing p so that only the *small quasi-momenta* play a rôle. We can therefore replace the energies in the conduction band by the start of their expansion in powers of p:

$$\varepsilon_c(p) \simeq \varepsilon_c + \frac{1}{2}\sum_{ij} K_{ij}p_ip_j, \qquad i,j, = x,y,x. \tag{11.58}$$

To simplify matters we assume that the tensor K is *isotropic*, which is the case for crystals with cubic symmetry. The conduction band spectrum then behaves like a *free particle spectrum*,

$$\varepsilon_c(p) \simeq \varepsilon_c + \frac{p^2}{2m_c}, \tag{11.59}$$

where the constant m_c can be interpreted as an *effective mass for the electrons in the conduction band*. Finally, one can neglect the distinction between quasi-momenta and momenta: the quasi-momentum values for a finite size crystal are the same as the momentum values in a box of the same shape, bar that the quasi-momenta are restricted within the Brillouin zone; however, since only the small values of p play a rôle, one can extend the summations to infinity. Finally, we thus find

$$N_c \simeq 2\sum_p e^{\beta(\mu-\varepsilon_c)-\beta p^2/2m_c}$$

$$\simeq 2\,\frac{\Omega}{h^3}\int d^3p\, e^{\beta(\mu-\varepsilon_c)-\beta p^2/2m_c}$$

$$= 2\,e^{\beta(\mu-\varepsilon_c)}\,\frac{\Omega}{h^3}\left(2m_c\pi kT\right)^{3/2}, \tag{11.60}$$

which is the same expression as that for the number of molecules with mass m_c of a *classical* perfect gas with $\mu - \varepsilon_c$ as its chemical potential (§7.2.2).

Let us now evaluate the *average number of holes, left in the valence band* at a temperature T:

$$N_v = 2\sum_p \left\{1 - f[\varepsilon_v(p)]\right\}. \tag{11.61}$$

We use therefore the *particle-hole symmetry* which we exhibited in §10.4.3. The energies $\varepsilon_v(p)$ of the valence band states have a form which is similar to

(11.58), with the matrix K_{ij} now being *negative* so that ε_v is the maximum of $\varepsilon_v(\boldsymbol{p})$. In the approximation where we neglect the anisotropy, we have thus

$$\varepsilon_v(\boldsymbol{p}) \;\simeq\; \varepsilon_v - \frac{p^2}{2m_v}. \tag{11.62}$$

We know that by changing the sign of ε_q and of μ we replace the expression for the average number of holes, $1 - f_q$, by that for the average number of particles, f_q. Equation (11.61) has thus exactly the same form as (11.56) if we replace $\mu - \varepsilon_c$ by $\varepsilon_v - \mu$ and m_c by m_v. As a result we find

$$N_v \;\simeq\; 2\,e^{\beta(\varepsilon_v - \mu)}\,\frac{\Omega}{h^3}\left(2m_v\pi kT\right)^{3/2}. \tag{11.63}$$

The thermal excitation of the electron cloud thus produces two kinds of objects: electrons in the conduction band and holes in the valence band, which behave like *independent particles*. These elementary excitations, called "*quasi-particles*" or "*charge carriers*", are a convenient method to describe how the electron cloud deviates from its ground state. The latter plays the rôle of a "vacuum" and each excited state can be interpreted as a collection of quasi-particles which move in this vacuum. An excess electron in the conduction band resembles through its charge a "bare" electron, that is, a free electron. However, the fact that it moves through the insulator rather than through the true vacuum modifies its energy spectrum to (11.58) or (11.59). The change in – and possible anisotropy of – its mass is the effect of the "dressing" of the electron by its interactions with the other electrons and the lattice. The holes are new objects which appear because in our "vacuum" the valence band is filled. Producing a hole of quasi-momentum $\boldsymbol{p}$ is equivalent to suppressing an electron of quasi-momentum $-\boldsymbol{p}$ in the valence band. This operation increases the energy by an amount $-\varepsilon_v(-\boldsymbol{p})$ so that, according to (11.62), the holes in the valence band have a *positive effective mass* m_v. Similarly, their *charge is positive*.

Dynamics of Quasi-Particles

The analogy between the conduction electrons (or the holes in the valence band) and a gas of particles with mass m_c (or m_v) and charge $-e$ (or $+e$), which we have just established for thermal equilibrium properties, also extends to the *dynamics* of the electron cloud. To see that, consider a single-electron wavepacket

$$\widehat{\varrho} \;\equiv\; \sum_{\boldsymbol{p},\,\boldsymbol{p}'} |\boldsymbol{p}\rangle\langle\boldsymbol{p}|\widehat{\varrho}|\boldsymbol{p}'\rangle\langle\boldsymbol{p}'|, \tag{11.64}$$

which we assume has been formed by a superposition of Bloch waves, only from the conduction band. The single-electron density operator (11.64) can represent just as well a pure state as a statistical mixture and it satisfies

$\operatorname{tr} \widehat{\varrho} = 1$. We denote the Bloch waves $|\boldsymbol{p}\rangle$ of energy $\varepsilon_c(\boldsymbol{p})$ solely by their quasi-momentum, omitting the conduction band index and the spin. We want to study the dynamics of the excess electron when it moves in the lattice under the influence of a force $\boldsymbol{f}$ which is added to the periodic potential $\mathcal{V}(\boldsymbol{r})$. The force field is assumed to be uniform over distances of the order of the extension Δr of the wavepacket; it includes the effect of the screening due to the self-consistency of the potential (§ 11.3.3) so that the effective Hamiltonian of the electron is $\widehat{h} - (\boldsymbol{f} \cdot \widehat{\boldsymbol{r}})$. We shall neglect interband transitions which might be induced by the field, so that $\widehat{\varrho}$ retains its form (11.64). If $\langle \boldsymbol{p}|\widehat{\varrho}|\boldsymbol{p}'\rangle$ is concentrated at values of $\boldsymbol{p} \simeq \boldsymbol{p}'$ which are small compared to $\hbar/a$, or, more generally, if the fluctuation Δp of the quasi-momentum around its average value $\boldsymbol{p}$ is small compared to $\hbar/a$, the spatial extension Δr of the packet will be large compared to the cell size a. The form (11.36) of the Bloch functions shows that the state $\widehat{\varrho}$ will then on a large scale resemble an ordinary wavepacket in the vacuum, where $|\boldsymbol{p}\rangle$ should be replaced by a plane wave: it is localized in a region Δr around $\langle \boldsymbol{r}\rangle$ and oscillates as $\exp[\mathrm{i}(\boldsymbol{p} \cdot \boldsymbol{r})/\hbar]$; however, on the scale of a cell, it involves a rapid supplementary spatial modulation coming from the Wannier wave χ of the conduction band. To simplify matters we assume that the latter is a simple band, produced by a single Wannier function $\chi(\boldsymbol{r})$ per cell.

We shall show that this modulation on the small scale and the presence of the potential $\mathcal{V}(\boldsymbol{r})$ in $\widehat{h}$ have as sole effect the substitution, in the classical equations of motion, of the kinetic energy of the bare electron by the band energy $\varepsilon_c(\boldsymbol{p})$. Let us, therefore, write down Ehrenfest's equations (2.29) which govern the evolution of the expectation values of the quasi-momentum and of the position.

Using (11.64) we find

$$\frac{d}{dt} \langle \boldsymbol{p}\rangle = \frac{1}{\mathrm{i}\hbar} \operatorname{tr} \widehat{\varrho} \big[\widehat{\boldsymbol{p}}, \widehat{h} - (\boldsymbol{f} \cdot \widehat{\boldsymbol{r}}) \big]$$

$$= \frac{1}{\mathrm{i}\hbar} \sum_{\boldsymbol{p}, \boldsymbol{p}'} \langle \boldsymbol{p}|\widehat{\varrho}|\boldsymbol{p}'\rangle \langle \boldsymbol{p}'|(\boldsymbol{f} \cdot \widehat{\boldsymbol{r}})|\boldsymbol{p}\rangle (\boldsymbol{p} - \boldsymbol{p}'),$$

$$\frac{d}{dt} \langle \boldsymbol{r}\rangle = \frac{1}{\mathrm{i}\hbar} \operatorname{tr} \widehat{\varrho} \big[\widehat{\boldsymbol{r}}, \widehat{h} - (\boldsymbol{f} \cdot \widehat{\boldsymbol{r}}) \big]$$

$$= \frac{1}{\mathrm{i}\hbar} \sum_{\boldsymbol{p}, \boldsymbol{p}'} \langle \boldsymbol{p}|\widehat{\varrho}|\boldsymbol{p}'\rangle \langle \boldsymbol{p}'|\widehat{\boldsymbol{r}}|\boldsymbol{p}\rangle \big[\varepsilon_c(\boldsymbol{p}) - \varepsilon_c(\boldsymbol{p}') \big].$$

Strictly speaking, the operator $\widehat{\boldsymbol{p}}$ is not well defined due to the ambiguity in the definition of the quasi-momentum $\boldsymbol{p}$. Nevertheless, if the wavepacket is localized in quasi-momentum space well away from the boundaries of the Brillouin zone, we can use as a definition $\widehat{\boldsymbol{p}} = \sum |\boldsymbol{p}\rangle \boldsymbol{p} \langle \boldsymbol{p}|$ where $\boldsymbol{p}$ is chosen inside that zone. The expression which we have just written down uses that convention; however, it should be changed, if the wavepacket passes through the boundary of the Brillouin zone. To calculate $\langle \boldsymbol{p}'|\widehat{\boldsymbol{r}}|\boldsymbol{p}\rangle$ we use the representation (11.36) of the Bloch waves in terms of Wannier waves. We assume that the origin is at the centre of the crystal and by $\boldsymbol{R}$ we denote the translations which leave the lattice invariant. We find

$$
\langle \boldsymbol{p'}|\widehat{\boldsymbol{r}}|\boldsymbol{p}\rangle = \frac{1}{N} \sum_{\boldsymbol{R},\boldsymbol{R'}} e^{i(\boldsymbol{p}\cdot\boldsymbol{R}-\boldsymbol{p'}\cdot\boldsymbol{R'})/\hbar} \int d^3r\,\chi^*(\boldsymbol{r}-\boldsymbol{R'})\,\boldsymbol{r}\,\chi(\boldsymbol{r}-\boldsymbol{R})
$$

$$
= \frac{1}{N} \sum_{\boldsymbol{R},\boldsymbol{R'}} e^{i(\boldsymbol{p}\cdot\boldsymbol{R}-\boldsymbol{p'}\cdot\boldsymbol{R'})/\hbar} \left[\boldsymbol{R}\,\delta_{\boldsymbol{R},\boldsymbol{R'}} + \int d^3r\,\chi^*(\boldsymbol{r})\,\boldsymbol{r}\,\chi(\boldsymbol{r}-\boldsymbol{R}+\boldsymbol{R'}) \right]
$$

$$
= i\hbar\,\nabla_{\boldsymbol{p'}}\,\delta_{\boldsymbol{p},\boldsymbol{p'}} + \delta_{\boldsymbol{p},\boldsymbol{p'}}\,\sqrt{N} \int d^3r\,\chi^*(\boldsymbol{r})\,\boldsymbol{r}\,\langle\boldsymbol{r}|\boldsymbol{p}\rangle. \tag{11.65}
$$

We have replaced $\boldsymbol{r}$ by $\boldsymbol{R'}+(\boldsymbol{r}-\boldsymbol{R'})$ and used the invariance under the translations $\boldsymbol{R'}$. The second term in (11.65) does not contribute to the equations of motion, as they contain factors $\boldsymbol{p}-\boldsymbol{p'}$ and $\varepsilon_c(\boldsymbol{p})-\varepsilon_c(\boldsymbol{p'})$ which vanish when $\boldsymbol{p}=\boldsymbol{p'}$. In the first term, we must treat $\boldsymbol{p}$ as a continuous variable, writing $\delta_{\boldsymbol{p},\boldsymbol{p'}} = (h^3/\Omega)\delta^3(\boldsymbol{p}-\boldsymbol{p'})$, whence we get

$$
\frac{d}{dt}\langle\boldsymbol{p}\rangle = \frac{\Omega}{h^3} \int d^3p\,\langle\boldsymbol{p}|\widehat{\varrho}|\boldsymbol{p}\rangle\,\boldsymbol{f} = \boldsymbol{f}, \tag{11.66}
$$

$$
\frac{d}{dt}\langle\boldsymbol{r}\rangle = \frac{\Omega}{h^3} \int d^3p\,\langle\boldsymbol{p}|\widehat{\varrho}|\boldsymbol{p}\rangle\,\nabla_{\boldsymbol{p}}\,\varepsilon_c(\boldsymbol{p}) = \langle\nabla_{\boldsymbol{p}}\,\varepsilon_c(\boldsymbol{p})\rangle. \tag{11.67}
$$

Hence, the motion of a quasi-particle in the conduction band resembles that of a bare particle in the vacuum. The variable which is conjugated to the position is no longer the momentum, but the quasi-momentum, and the kinetic energy is replaced by $\varepsilon_c(\boldsymbol{p})$. If the wavepacket is localized in a region Δr containing a large number of cells but still microscopic, and in a region Δp small compared to the size of the Brillouin zone, we can neglect the statistical fluctuations of $\boldsymbol{p}$ and $\boldsymbol{r}$, and Eqs.(11.66) and (11.67) describe the motion of a classical particle with Hamiltonian $\varepsilon_c(\boldsymbol{p}) - (\boldsymbol{f}\cdot\boldsymbol{r})$. Alternatively we can interpret the velocity $\nabla_{\boldsymbol{p}}\,\varepsilon_c$ of the conduction electrons, using the de Broglie relations $\boldsymbol{p} = \hbar\boldsymbol{k}$, $\varepsilon_c(\boldsymbol{p}) = \hbar\omega(\boldsymbol{k})$, as the group velocity of a wave characterized by the dispersion relation $\omega(\boldsymbol{k})$. When $\boldsymbol{p}$ is small, Eqs.(11.58) and (11.67) give as components of the velocity $v_i = \sum_j K_{ij}p_j$. If, moreover, the bottom of the band is isotropic, (11.59) gives $\boldsymbol{v} = \boldsymbol{p}/m_c$, so that the quasi-particle has the same dynamics as a particle of mass m_c in the vacuum.

Similarly, when we consider a valence band with one electron missing, we can equivalently describe its dynamics by considering the motion of a wavepacket which represents the missing particle; this motion is described by Eqs.(11.66) and (11.67), with ε_c replaced by ε_v. However, the total momentum $\boldsymbol{p}$ of the cloud, which can be identified as the quasi-momentum of the hole, is the opposite of the quasi-momentum associated with the wavepacket and occurring in (11.66) and (11.67). The hole dynamics thus obeys the equations

$$\frac{d}{dt} \langle \boldsymbol{p} \rangle = \boldsymbol{f}_{\mathrm{h}} \equiv -\boldsymbol{f}, \tag{11.66'}$$

$$\frac{d}{dt} \langle \boldsymbol{r} \rangle = -\langle \nabla_{\boldsymbol{p}} \, \varepsilon_{\mathrm{v}}(\boldsymbol{p}) \rangle \simeq \frac{\langle \boldsymbol{p} \rangle}{m_{\mathrm{v}}}. \tag{11.67'}$$

The relation (11.67') between the velocity and the momentum has the same form as for a particle with a *positive mass* m_{v}. On the other hand, from (11.66') it follows that the force $\boldsymbol{f}_{\mathrm{h}}$, felt by the hole, is the opposite of the force $\boldsymbol{f}$ applied to the electrons. This is consistent: since an electron in an electric field $\boldsymbol{E}$ is subjected to a force $-e\boldsymbol{E}$, a hole must be regarded as a classical particle subject to the force $e\boldsymbol{E}$, that is, a particle with a *positive charge e*.

The equations of motion (11.66) and (11.67) look classical, but their derivation has not needed any of the conditions from § 10.3.4, and in actual fact we are dealing with *quantum mechanical* equations. The potential $\mathcal{V}(\boldsymbol{r})$ is clearly not a slowly varying one, and the electron density is not necessarily small. These equations are therefore valid not only for an insulator, but also for a *metal*. They are compatible with the exclusion principle at equilibrium, since they conserve the single-particle energy; if there are no external forces, $\varepsilon_{\mathrm{c}}(\boldsymbol{p})$ *remains constant* during the motion of a wavepacket, and this is thus also true of the expectation value of the occupation number which depends solely on $\varepsilon_{\mathrm{c}}(\boldsymbol{p})$. Note, however, that the relevant part of the band spectrum for a metal is not the bottom or the top of a band, but the region where $\varepsilon_{\mathrm{c}}(\boldsymbol{p}) \simeq \varepsilon_{\mathrm{F}}$. The gradient which occurs in the expression (11.67) for the velocity must be calculated at the Fermi surface, where an approximation such as (11.58) is, in general, not justified. In the case of the quasi-particles which are the conduction electrons in a metal the relation (11.67) between the velocity and the quasi-momentum is therefore not just a proportionality.

Insulator as a Classical Gas of Carriers

At equilibrium, in an *electrically neutral* insulator the total number of electrons remains fixed when the temperature varies. The number of electrons excited into the conduction band is thus equal to the number of holes left behind in the valence band. This condition, $N_{\mathrm{c}} = N_{\mathrm{v}}$, determines the *chemical potential* of the electrons, if we use (11.60) and (11.63):

$$\mu = \frac{1}{2}\left(\varepsilon_{\mathrm{c}} + \varepsilon_{\mathrm{v}}\right) + \frac{3}{4} \, kT \, \ln \frac{m_{\mathrm{v}}}{m_{\mathrm{c}}}. \tag{11.68}$$

The value thus obtained lies close to the middle of the band, since $kT \ll \delta$, which justifies the approximations made about the low densities of the conduction electrons and of the holes. As $T \to 0$, μ has a well defined limit, $\frac{1}{2}(\varepsilon_{\mathrm{c}} + \varepsilon_{\mathrm{v}})$, whereas the chemical potential would be undetermined between ε_{v} and ε_{c} in the ground state itself. Eliminating μ from (11.60) and (11.68), we find

$$N_c = N_v = 2\Omega \left(\frac{\sqrt{m_c m_v} kT}{2\pi\hbar^2} \right)^{3/2} e^{-\delta/2kT}. \tag{11.69}$$

As expected, this is a very small number because of the exponential factor and the large value of the characteristic temperature δ/k.

The introduction of quasi-particles is thus seen to be much more than artificial semantics. It considerably simplifies the description of the electron system in the insulator and makes physical reasoning easier. In fact, one can forget the complications connected with the original electrons, that is, their mutual interactions and their interactions with the lattice, the band structure – apart from the bottom of the conduction band and the top of the valence band – and even the Fermi-Dirac statistics, however important it was for the construction of our quasi-particle model. This new description only involves two kinds of *non-interacting* quasi-particles or charge carriers, the conduction electrons with a charge $-e$ and the holes in the valence band with a charge $+e$. These quasi-particles *do not see the lattice*, which has dropped out of the description. At zero temperature there are no quasi-particles present and the electron state of the insulator can be interpreted as the *vacuum* for the quasi-particles. Supplying energy to the electron cloud, for instance, by heating it, is equivalent to *creating pairs* of opposite charge carriers. Their density remains always extremely small so that we can treat them as a *mixture of two very dilute classical perfect gases*.

The quasi-particle energies have the same form as the ordinary kinetic energy, with *effective masses* m_c and m_v which are different from the original electron mass; these effective masses occur both in the occupation numbers at thermal equilibrium and in the dynamics which is the same as that of classical particles. Finally, Eqs.(11.60) and (11.63), which relate the densities of the two gases of carriers to the chemical potential, have the same form (8.7), (8.5) as for a classical gas; the density of conduction electrons contains the factor $e^{\beta(\mu - \varepsilon_c)}$, replacing the factor $\zeta e^{\beta\mu}$ which usually occurs in the density of a classical gas, but the hole density contains a factor $e^{\beta(\varepsilon_v - \mu)}$. We are thus led to assign to the conduction electrons and the holes chemical potentials $\mu_c = \mu$ and $\mu_v = -\mu$, which satisfy the relation $\mu_c + \mu_v = 0$ when the material is at equilibrium and, moreover, the relation $N_c = N_v$ when it is electrically neutral. The equation $\mu_c + \mu_v = 0$ is nothing but a special case of the chemical equilibrium condition (6.78), applied to the reaction "electron + hole $\leftrightarrows$ 0", which describes the creation or annihilation of a pair of carriers. In contrast to the situation for a mixture of ordinary gases, *the number of carriers is not conserved*; it varies with temperature, following (11.69), changing in such a way that the material is electrically neutral and at equilibrium. It will be essential to rely on this model of two carrier gases, which we have justified above, when we want to understand the behaviour of an insulator or of a semiconductor.

The various properties of insulators are governed by the presence of a forbidden band and by the fact that the electrons in the valence band are nearly completely frozen in. They are easily evaluated in the description in terms of two classical gases. The grand potential is given by (11.22), where f_q is replaced by its classical approximations, (10.63a) for the conduction band and (10.63b) for the valence band, while $f_q = 1$ for deeper bands. Accordingly, the entropy is expressed by the Sackur-Tetrode formula (7.41) for each of the two gases of quasi-particles; it is extremely small, since the densities N_c/Ω and N_v/Ω are very low, and cannot be experimentally observed. In particular, the contribution from the electrons to the *heat capacity is negligible*. In fact, heating an insulator just amounts to exciting electrons from the valence band into the conduction band and the number of pairs $N_c = N_v$ produced in this way is given by (11.69). The energy absorbed for producing a pair is practically equal to δ so that the electron specific heat is equal to

$$C_{el} \sim \frac{dN_c}{dT}\delta \sim \frac{k}{2}\left(\frac{\delta}{kT}\right)^2 N_c(T). \tag{11.70}$$

The factor $e^{-\delta/2kT}$ in N_c makes this quantity negligible at all temperatures as compared to the contributions from other degrees of freedom.

The fact that many insulating crystals are *transparent* is a consequence of the fact that only photons with energies above δ can be absorbed, with the creation of a pair of carriers. The photons of the visible spectrum, with wavelengths of 0.4 to 0.8 μm, have energies of 1.5 to 3 eV, which is less than δ for many insulators, so that they cannot interact with these substances. This property is in contrast to the metallic brightness, the result of elastic scattering of photons by the electrons in the metal which are close to the Fermi surface and which can easily be excited, thus creating surface currents. We shall see in § 13.3.3 that the *electrostatic properties* of insulators also follow from the fact that the number of free charge carriers is small, which itself is a consequence of the wide forbidden band.

Conduction

The (low) electric conductivity of an insulator originates from the charge transport by the two gases of carriers, the density of which increases when the insulator is heated. The quasi-particles, which are equivalent to the electron cloud, are accelerated according to (11.66) and (11.67) by an electric field – the electrons in the opposite direction and the holes in the same direction. Nevertheless, when we studied the dynamics of the quasi-particles we assumed that the electrons interacted with a *regular* lattice of *fixed* nuclei; we neglected the interaction of those electrons with *variations* in the nuclear positions, or with crystallization *defects* in the lattice. If we subtract the periodic potential, which has already been taken into account in (11.66) and (11.67), a nucleus missing from a given lattice site behaves like a negative charge added at that point, and a displaced nucleus behaves like an electric dipole (Prob.5). As a result of those residual interactions, the conduction electrons and the holes which are accelerated by the electric field also

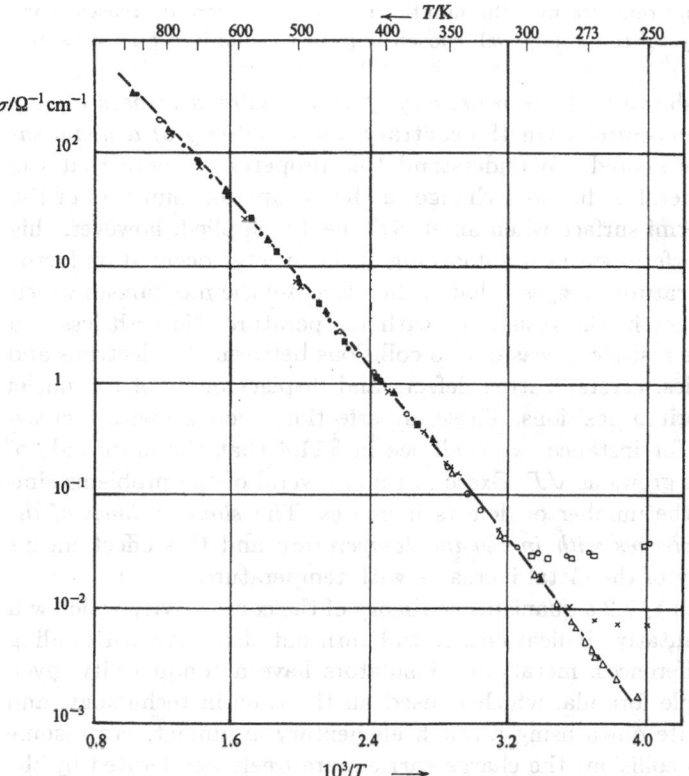

Fig. 11.10. The conductivity of germanium as function of the temperature

undergo collisions and can *transfer energy to the lattice* which will increase its vibrations and heat up. This is the origin of the *Joule effect.* A stationary, *non-equilibrium*, regime is produced for the conduction electrons and the holes, where they acquire a non-vanishing mean velocity. Below and in § 15.2 we shall give more or less elementary arguments which enable us to calculate the resistivity. Nevertheless, it is clear, without having to develop the theory in more detail, that the conductivity must be *proportional to the numbers of charge carriers*, N_c and N_v. This can be checked from Fig.11.10 which shows the conductivity of germanium as function of temperature. As we shall see in § 11.3.4, semiconductors, of which Ge is an example, are insulators with a small gap δ. The curve clearly is dominated by the factor $e^{-\delta/2kT}$ occurring in (11.69) and varying very fast: the logarithm of the conductivity is, in fact, nearly a linear function of the inverse temperature. One could use these experimental results to calculate, as an exercise, the value of the forbidden band gap δ in germanium; one finds $\delta = 0.7$ eV.

The various experimental points shown in Fig.11.10 correspond to germanium crystals with different impurity densities. We note that below a certain temperature the amount of impurities governs the conductivity. In fact, we shall see in § 11.3.4

that, with decreasing temperature, the number of charge carriers decreases much more slowly in a crystal containing well chosen impurities than in a pure crystal.

We have just shown that *the resistivity of an insulator decreases rapidly with increasing temperature.* On the contrary, *the resistivity of a metal increases* when it is heated. To understand this property we note that the conduction in a metal is due to a change in the occupation numbers of the states near the Fermi surface when an electric field is applied; however, this change hardly interferes with the smoothing of the average occupation factor f when the temperature changes. That is therefore *not* the mechanism which controls the changes in the resistivity with temperature. Nevertheless, we have seen that the resistivity was due to collisions between the electrons and lattice irregularities, crystallization defects and displacements of the nuclei from their equilibrium positions. These imperfections increase with increasing temperature; for instance, we shall see in § 11.4 that the amplitude of nuclear vibrations grows as $\sqrt{T}$; Exerc.11a and several of the problems similarly show that the number of defects increases. The *slowing down of the electrons thus increases with increasing temperature* and this effect means that the resistivity of the metal increases with temperature.

We shall give in § 15.2 a quantitative theory of the conductivity which will confirm these qualitative indications. It will turn out that, notwithstanding their essential differences, metals and insulators have a conductivity given by the same simple formula, which is used all the time in technology, and which we now write down using a rough elementary argument. We assume that between two collisions the charge carriers are freely accelerated by the electric field $\boldsymbol{E}$, like classical particles of mass m and charge q; their velocity at time t is thus given by the formula

$$\boldsymbol{v} \;=\; \boldsymbol{v}_0 + \frac{q\boldsymbol{E}}{m}\,(t - t_0), \tag{11.71}$$

where $\boldsymbol{v}_0$ is the velocity at the moment t_0 of the last collision. We also assume that as the result of a collision the charge carrier acquires a random velocity $\boldsymbol{v}_0$ with an isotropic distribution; the average $\langle \boldsymbol{v}_0 \rangle$, taken over the system of carriers, is thus equal to zero. The average drift velocity of the carriers is thus equal to

$$\langle \boldsymbol{v} \rangle \;=\; \frac{q\boldsymbol{E}}{m}\,\tau, \tag{11.72}$$

where τ is the average time elapsed since the last collision, of the order of the *average time between collisions.* We define the *electric mobility* μ_{el} by

$$\mu_{\text{el}} \;=\; \frac{q\tau}{m} \;=\; \frac{\langle v \rangle}{E}, \tag{11.73}$$

which characterizes the average velocity imparted to the charge carriers by a unit field. Traditionally the mobility is denoted by μ; this should not be confused with the chemical potential. We thus find for the *conductivity*, the ratio of the electric current density J_{el} to the field,

$$\sigma = \frac{J_{el}}{E} = \frac{nq\langle v \rangle}{E} = nq\,\mu_{el}, \qquad (11.74)$$

where n is the number of carriers of the type considered per unit volume.

This relation, which we shall derive in a more elaborate way in § 15.2.2, is useful not only for an insulator or a semiconductor but also for an *ionic solution*; in both cases the hypotheses which we have made are well justified. In the former case, the collisions are with lattice imperfections, especially with phonons; there are two kinds of carriers, with charges $q = \mp e$ and with masses m_c and m_v, which contribute additively to the conductivity; their mobilities (11.73) differ, not only because of their different effective masses, but also because the times between collisions are different – the mean free paths are about the same, but the average velocities, of the order of $\sqrt{kT/m}$, differ. The large change in carrier density with temperature dominates (11.74), masking the changes in mobility, as we can see in Fig.11.10. We shall show (§ 15.2.3) that Eq.(11.74) happens to remain valid for a *metal* where now n is of the order of the density of the *whole set* of conduction electrons, notwithstanding the fact that in the conduction process only those near the Fermi surface are involved. The number n is thus enormous and this explains the large conductivity of a metal. Moreover, contrary to what happens for insulators, it does not change with temperature so that the conductivity (11.74) varies as the mobility. Hence the resistivity is proportional to the density of crystal imperfections, and it increases with temperature.

11.3.3 Microscopic Foundations of Electrostatic Equilibrium

In §§ 11.3.1 and 11.3.2 we have considered neutral substances, where the total charge of the electrons was exactly the same as that of the nuclei. Various electrostatic perturbations can change that situation. If we *increase the chemical potential*, the number of electrons increases and we thus produce a negative charge density. Moreover, when we took the potential $\mathcal{V}$ to be periodic, we assumed implicitly that the substance was pure and without defects. Adding *impurities* changes the charge distribution: if in metallic sodium we replace a Na nucleus with $Z = 11$ by an Al nucleus with $Z = 13$, there will occur at that spot an excess charge $+2e$, while we have added two electrons; the situation resembles that of an atom with $Z = 2$ being embedded in a pure metal. Finally, we can operate on the material by applying an *external electric field*.

In all those cases the electrostatic perturbation is accompanied by a rearrangement of the electron cloud, whose equilibrium state is changed. We expect, in particular, that the electron density will increase where the electric potential is the highest. However, this density variation, $\delta n(\mathbf{r})$, itself will induce a field which counteracts the applied field and reduces its effects. This is called the *screening effect*. It is particularly pronounced in *conductors*, inside which *one cannot produce electric fields* on a scale larger than the interatomic distances. For instance, the potential due to the excess charge $+2e$ of an impurity Al nucleus in Na attracts electrons, which are free to rearrange themselves near the Fermi surface. At equilibrium, the average number of electrons increases by 2 in a region of a few Å round the impurity, of the order of the atomic size. The distance over which the perturbation of the electron cloud extends is called the *screening length*. As soon as we are further away from the impurity than one screening length, the potential produced by the two bound electrons balances exactly that of the excess charge $+2e$ of the nucleus: the electrons screen the perturbation introduced by this charge $+2e$, and the presence of the impurity has no effect beyond the screening length. Similarly, if the metal is globally charged, or if it is subject to an external field, the electron cloud is deformed in such a way that the total charge density differs from zero only at the surface of the sample, in a *thin layer* with a thickness of the order of the screening length, a few Å. This layer, which is negatively or positively charged, according to whether there is an excess or deficit of electrons as compared to their average equilibrium density in the neutral metal, produces a field which inside the metal compensates exactly any applied field.

The effect of screening is important in metals because near the Fermi surface it costs so little energy to move an electron from an occupied state to an empty state. This ease with which the electron cloud can be perturbed explains the small value of the screening length. On the other hand, in *insulators* the *screening effect is weak*. At zero temperature, there are *no free charge carriers* which can get bound to an impurity to neutralize it, or which accumulate at the surface to counterbalance an external field; each electron transition needs at least an energy δ; moreover, changing μ by less than $\delta/2$ does not introduce any change. When the temperature is finite, a few quasi-particles appear, which can move from one region to another to screen applied fields; however, because of their low density this process is not efficient in a good insulator. Furthermore, we shall see in § 11.3.4 that in a semiconductor the size of the quasi-particle orbitals which are bound to an impurity nucleus can be as large as a hundred Å.

Nevertheless, there exists another phenomenon which reduces the effect of an applied field, the *polarization* of the medium. In a neutral insulator, each nucleus is surrounded by an electron cloud, formed by filling all bands below the forbidden band, or, what amounts to the same, all associated Wannier orbitals. Apart from a crystal deformation, this cloud resembles that of the atomic orbitals in an ion or an atom. Under the influence of an external field, it becomes deformed, even at zero temperature. For example, in the neighbourhood of a positive excess charge, the centres of the orbitals are displaced in the direction of the impurity; in Fig.11.11 the dashed lines indicate their positions in the pure insulator and the asterisks the fixed nuclei. At each site we thus get a *dipole moment* due to the change in the electron density. The resulting polarization produces a field which reduces the external field. We

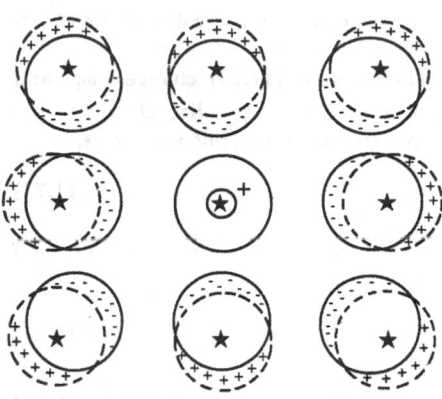

Fig. 11.11. Polarization of a medium by a positive excess charge

shall show in what follows that this perturbation of the electron cloud results in *dividing the applied field* by a factor $\varepsilon/\varepsilon_0$ which is characteristic for the medium and which is its relative *dielectric constant* or *permittivity*.

The microscopic theory of electric equilibrium in solids is based upon the following principles. Both the screening effect and the polarization, which are a result of a rearranging of the electron cloud, will be explained as consequences of the *self-consistency* of the potential $V(r)$ seen by each electron. This potential is produced not solely by the nuclei and by sources outside the medium, but also by the Coulomb interactions between the electrons. Its formula (11.11) shows that a change $\delta n(r)$ in the electron density $n(r)$, starting from its value in a neutral medium without impurities or defects, produces a change $\delta V_{el}(r)$ which is added to the applied potential. The solution of the equations which we get in this way will allow us to justify the laws of macroscopic electrostatics, to understand quantities such as the screening length, and to calculate their values in a given medium.

We start from the coupled equations (11.11–15) and treat all effects due to excess charges, applied fields, or deformations of the electron cloud, as perturbations. In our reference situation, the solid is neutral and there are no impurities, defects, or applied fields. The corresponding Hartree Hamiltonian $\widehat{h}_0$ is periodic, and it is thus associated with a band structure, $|q\rangle$, ε_q, and an unperturbed electron density $n(r)$, all of which we assume to be known. The perturbation,

$$\widehat{h} - \widehat{h}_0 \equiv \delta V(\widehat{r}) = \widehat{V}_{ext} + \widehat{V}_{imp} + \delta\widehat{V}_{el}, \qquad (11.75)$$

contains contributions from various sources. The term $\widehat{V}_{ext} = -e\Phi_{ext}$ comes from the external electric potential Φ_{ext} to which the medium may be subject. Each impurity nucleus or each defect produces an excess or deficit charge at a point in the solid, and $\widehat{V}_{imp}$ is the potential created by this change in the nuclear charges. Finally, $\delta\widehat{V}_{el}$ is the contribution to the Hartree potential (11.11) coming from the change $\delta n(r)$ in the self-consistent electron density as compared to its reference value $n(r)$. Altogether, the *perturbation added to the band Hamiltonian* $\widehat{h}_0$ is

$$\delta V(r) = V_{ext}(r) + V_{imp}(r) + \frac{e^2}{4\pi\varepsilon_0} \int \frac{d^3r'}{|r-r'|} \delta n(r'). \qquad (11.76)$$

The total electron charge at equilibrium is controlled by the value of the *chemical potential*. We denote by $\delta\mu$ its deviation from its reference value μ.

The perturbation $\delta\mathcal{V}$ induces through (11.12) and (11.13) changes $|\delta q\rangle$ and $\delta\varepsilon_q$ in the eigenfunctions and eigenenergies of $\widehat{h}$. The corresponding change in the electron density (11.15) can, in the linear approximation, be split as follows,

$$\delta n(r) = \delta n_1(r) + \delta n_2(r), \tag{11.77}$$

$$\delta n_1(r) = \sum_q \delta f_q \, |\langle r|q\rangle|^2, \tag{11.77'}$$

$$\delta n_2(r) = \sum_q f_q \, \delta|\langle r|q\rangle|^2. \tag{11.77''}$$

The first term comes from the change in the occupation of the states $|q\rangle$ associated with the changes $\delta\varepsilon_q$ and $\delta\mu$; it describes a local electrostatic change in the *filling* of these electron states, and we shall interpret it as representing the *screening* effect. The second term is associated with the *deformation* $|\delta q\rangle$ of the single-electron states, and describes the *polarization*.

To evaluate $\delta n(r)$ explicitly we rewrite (11.15) using (11.21), as we did for (11.22), and after that vary $\widehat{h}$ and μ in the operator $\widehat{f}$. This gives

$$\delta n(r) = \langle r|\delta\widehat{f}|r\rangle = \langle r|\delta\big[e^{\beta\widehat{h}-\alpha} + 1\big]^{-1}|r\rangle.$$

Using the identities

$$\delta\big(\widehat{X}^{-1}\big) = -\widehat{X}^{-1}\delta\widehat{X}\,\widehat{X}^{-1}, \quad \delta e^{\beta\widehat{Y}} = \int_0^\beta du\, e^{u\widehat{Y}}\delta\widehat{Y}\, e^{(\beta-u)\widehat{Y}}, \tag{11.78}$$

for the variations of operators, and using as base the unperturbed Bloch waves, we get

$$\delta n(r) = -\sum_{q,q'} \langle r|q'\rangle\langle q'|f_{q'}\left(\delta e^{\beta\widehat{h}-\alpha}\right) f_q \, |q\rangle\langle q|r\rangle$$

$$= -\sum_{q,q'} \langle r|q'\rangle\langle q'|\delta\widehat{h}-\delta\mu|q\rangle\langle q|r\rangle$$

$$\times \int_0^\beta du\, f_{q'}\, f_q\, \exp[u\varepsilon_{q'} + (\beta - u)\varepsilon_q - \alpha],$$

or, after integrating and using (10.33),

$$\delta n(r) = \sum_{q,q'} \langle r|q'\rangle\langle q'|\big[\delta\mathcal{V}(\widehat{r}) - \mu\big]|q\rangle\langle q|r\rangle \frac{f_{q'} - f_q}{\varepsilon_{q'} - \varepsilon_q}. \tag{11.79}$$

The coupled equations (11.76) and (11.79) determine the *electrostatic equilibrium on the microscopic scale*.

To change to a larger scale we must *smooth out* the rapid oscillations in $\delta n(r)$ with the crystal cell size as period. We also assume that $\delta\mathcal{V}(r)$ *varies slowly in space* as compared to the cell size. First of all, we shall even calculate (11.79), using the approximation procedure of § 10.3.4b: we imagine that the solid can be divided into small cells each of which is at the same time sufficiently small that in it we can replace the potential by a constant, and sufficiently large that one can use the large volume limit for it. Near the point r we therefore simply replace in (11.79) the operator $\delta\mathcal{V}(\hat{r})$ by the number $\delta\mathcal{V}(r)$. The sum (11.79) reduces to the terms with $q = q'$, and in those $(f_{q'} - f_q)/(\varepsilon_{q'} - \varepsilon_q)$ must be interpreted as a derivative, so that we get for (11.77′) the approximation

$$\delta n_1(r) \simeq \sum_q |\langle r|q\rangle|^2 \frac{\partial f(\varepsilon_q)}{\partial \varepsilon_q} [\delta\mathcal{V}(r) - \delta\mu], \tag{11.80}$$

while (11.77″) vanishes. We could have obtained (11.80) directly starting from (11.15), if we had assumed that $\delta\mathcal{V}$ were a constant. In fact, adding $\delta\mathcal{V}$ to the energies ε_q and $\delta\mu$ to the chemical potential only changes the Fermi factors, and thus changes $n(r)$ to

$$n(r) + \delta n(r) = \sum_q |\langle r|q\rangle|^2 f(\varepsilon_q + \delta\mathcal{V} - \delta\mu). \tag{11.81}$$

This expression, which is valid provided $\delta\mathcal{V}(r)$ varies sufficiently slowly, does not require that δn, $\delta\mathcal{V}$, and $\delta\mu$ are small. It reduces to (11.80) to first order in $\delta\mathcal{V}$ and $\delta\mu$.

The average of $-e\delta n(r)$, taken over a cell, can be interpreted as the *macroscopic electron charge density* $\varrho_{el}(r)$. If $\varrho_{ext}(r)$ denotes the external charge density which produces the applied field, and $\varrho_{imp}(r)$ the impurity charge density, the smoothing out of $-\delta\mathcal{V}(r)/e$ performed on the last term of (11.76) gives, in the approximation (11.81), for the *macroscopic electric potential* just the Coulomb potential:

$$\Phi(r) = \frac{1}{4\pi\varepsilon_0} \int \frac{d^3r'}{|r - r'|} [\varrho_{ext}(r') + \varrho_{imp}(r') + \varrho_{el}(r')], \tag{11.82}$$

or, equivalently, the Poisson equation

$$\varepsilon_0 \nabla^2 \Phi + \varrho_{ext} + \varrho_{imp} + \varrho_{el} = 0. \tag{11.83}$$

The density ϱ_{el} itself can be expressed as a function of Φ by taking the spatial average over a cell of (11.80) or (11.81). Noting that $|q\rangle$ is normalized in a volume Ω and that $|\langle r|q\rangle|^2$ is periodic, we find that the average of $|\langle r|q\rangle|^2$ is Ω^{-1} and we get the *Thomas-Fermi equation*:

$$\varrho_{el}(r) = -e\overline{\delta n_1(r)} \simeq -e \int d\varepsilon \frac{\mathcal{D}(\varepsilon)}{\Omega} \{f[\varepsilon - e\Phi(r) - \delta\mu] - f(\varepsilon)\}, \tag{11.84}$$

$$\sim e[e\Phi(r) + \delta\mu] \int d\varepsilon \frac{\mathcal{D}(\varepsilon)}{\Omega} \frac{\partial f}{\partial \varepsilon}. \tag{11.84′}$$

We can then explain the *screening effect* in a simple manner. Assuming that $e\Phi + \delta\mu$ is small, we eliminate Φ from (11.84′) and (11.83), and find inside the material

$$(1 - \lambda^2 \nabla^2) \varrho_{el}(r) = -\varrho_{imp}(r). \tag{11.85}$$

The *screening length* λ is here defined by

$$\lambda^{-2} = \frac{e^2}{\varepsilon_0} \int d\varepsilon \, \frac{\mathcal{D}(\varepsilon)}{\Omega} \left(-\frac{\partial f}{\partial \varepsilon}\right). \tag{11.86}$$

In the neighbourhood of an impurity with charge $+e$ at the origin, the solution of (11.85) with $\varrho_{imp} = e\delta^3(\boldsymbol{r})$ gives

$$\varrho_{el}(\boldsymbol{r}) = -\frac{e}{4\pi\lambda^2 r} e^{-r/\lambda}, \tag{11.87}$$

which represents a cloud with a total charge $-e$, localized within a region of size λ. Similarly, near the surface of a charged substance, (11.85) shows that the electron density decreases exponentially over a depth λ. Beyond that one finds that the matter is neutral, and (11.84') shows that the potential $\Phi(\boldsymbol{r})$ is *uniform and equal to* $-\delta\mu/e$ over distances from electrostatic irregularities larger than λ. This justifies the name *electrochemical potential* given to μ in this context. One should bear in mind, however, that in electrostatic equilibrium μ is *everywhere uniform* – which is, in fact, the equilibrium condition – in contrast to $\Phi(\boldsymbol{r})$ which satisfies the Poisson equation (11.83). Moreover, in non-equilibrium situations the changes in the local electrochemical potential $\mu(\boldsymbol{r}, t)$ and of the macroscopic electrical potential $\Phi(\boldsymbol{r}, t)$ are not directly related to one another (§ 15.2.2).

In the case of a *metal* the integral occurring in (11.86) reduces to $\mathcal{D}(\varepsilon_F)/\Omega$. In the box model, the screening length (11.86) can thus be expressed in terms of the Bohr radius $a_0 = 4\pi\varepsilon_0\hbar^2/me^2 \simeq 0.53$ Å and the electron density N/Ω, as

$$\lambda = \frac{1}{2} \left(\frac{\pi\Omega}{3N}\right)^{1/6} a_0^{1/2} \simeq 0.5 \text{ Å},$$

where the numerical value is for copper. As we mentioned earlier, we find a screening length of the order of interatomic distances. However, the small value we found makes our hypothesis, that $\Phi(\boldsymbol{r})$ varies slowly, invalid, and that was the basis of the Thomas-Fermi approximation. The result can thus only be considered to give us an order of magnitude, underestimating slightly the screening effect in a metal.

However, the approximation is justified for other substances where the screening length is longer. We shall use it in § 11.3.5 for *semiconductors*. Its extension to dilute solutions of *strong electrolytes* gives us the *Debye-Hückel* equation which governs the ion distribution when there is a field present. In that case, (several kinds of) ions play the rôle of electrons, their interaction with the solvent changes their mass to an effective mass, and the factor f occurring in (11.84) must be replaced by its classical limit $e^{-\beta\varepsilon+\alpha}$. The same idea applies to *plasma physics*, where the effect of a charge is screened by the classical electron gas with density $n \ll (mkT/\hbar^2)^{3/2}$, over a range given by (11.86), $\lambda = (ne^2/\varepsilon_0 kT)^{-1/2}$, called the *Debye length*.

For *insulators* the macroscopic charge density (11.84) represents a local lack of balance between the numbers of holes and conduction electrons. Its value does not become appreciable, except for insulators with a narrow forbidden band (semiconductors; see § 11.3.4), or when $|e\Phi + \delta\mu|$ becomes sufficiently large for the Fermi level to approach the edge of the conduction, or the valence, band. Nevertheless, there is another effect, even if the applied field is weak, namely, the *polarization*. It comes about on the microscopic scale from the deformation (11.77'') of the Bloch waves in an electrostatic field. So far we have neglected it. Let us come back to (11.79) to evaluate it in an arbitrary solid. By treating the potential $\delta\mathcal{V}(r)$ as nearly constant, we have neglected the effects of the electric field on the scale of a cell, and those we just need here. To calculate the polarization induced by the field $E = -\nabla\Phi$, we treat this field as a constant in the volume considered, and thus take $\delta\mathcal{V}$ to be linear, putting $\delta\mathcal{V}(r) = e(E \cdot r)$ in (11.79). We neglect the thermal excitation of quasi-particles, replacing $f(\varepsilon)$ by $\theta(\mu - \varepsilon)$ so that of the states $|q\rangle$ and $|q'\rangle$ in (11.79) one is occupied and the other one empty. The result,

$$\delta n_2(r) \simeq -eE \cdot \sum_{\varepsilon_q > \mu > \varepsilon_{q'}} \langle r|q'\rangle\langle q'|\widehat{r}|q\rangle\langle q|r\rangle \frac{1}{\varepsilon_q - \varepsilon_{q'}} + \text{c.c.}, \tag{11.88}$$

is the same as that of first-order perturbation theory for the wavefunction $|q\rangle$ in (11.77''), as expected. We can use (11.65) to write (11.88) in a more explicit form; the two Bloch waves $|q\rangle$ and $|q'\rangle$ have for an insulator different band indices so that only the last term in (11.65) contributes, and we can write (11.88) in the form

$$\delta n_2(r) \simeq -2eE \cdot \sum_{p,b,b'} \sqrt{N} \int d^3r'$$

$$\times \langle r|p,b'\rangle\chi_{b'}^*(r')r'\langle r'|p,b\rangle\langle p,b|r\rangle \frac{1}{\varepsilon_b(p) - \varepsilon_{b'}(p)} + \text{c.c.}, \tag{11.88'}$$

where b stands for the empty and b' for the full bands.

The electron charge distribution $-e\delta n_2(r)$, which is periodic, has a *zero average* in a cell, since the integration of (11.88') over r gives a factor $\langle p,b|p,b'\rangle = 0$. It therefore does not contribute to (11.84). However, it has a *dipole moment* so that each cell behaves, as predicted, as an electric dipole. The polarization, that is, the mean dipole moment per unit volume, follows from (11.88):

$$P = -\frac{e}{\Omega} \int d^3r\, r\, \delta n_2(r) = \chi E, \tag{11.89}$$

where the *electric susceptibility tensor* χ is given by

$$\chi_{ij} = \frac{e^2}{\Omega} \sum_{\varepsilon_q > \mu > \varepsilon_{q'}} \langle q|\widehat{r}_i|q'\rangle\langle q'|\widehat{r}_j|q\rangle \frac{1}{\varepsilon_q - \varepsilon_{q'}} + \text{c.c.}$$

$$= \frac{2e^2N}{\Omega} \sum_{p,b,b'} \int d^3r\, d^3r'\, \langle p,b|r\rangle r_i \chi_{b'}(r)$$

$$\times \chi_{b'}^*(r')r_j'\langle r'|p,b\rangle \frac{1}{\varepsilon_b(p) - \varepsilon_{b'}(p)} + \text{c.c.} \,. \tag{11.90}$$

The susceptibility thus depends on the overlap between the Wannier functions of the full bands b' and the Bloch functions of the bands b into which the electrons can be excited. The narrower the forbidden band, the larger it is. It is a symmetric tensor, and for substances with a simple crystal structure it reduces to a constant, $\chi_{ij} = \chi\delta_{ij}$.

We must still find the contribution from the charge density $-e\delta n_2(r)$ to the potential (11.76), when the field E and the polarization P vary in space. To do that we first write the integral over r' inside a cell centred at a point R of the lattice in the form

$$\int_R \frac{d^3r'}{|r-r'|}\,\delta n_2(r') \simeq \int_R d^3r'\left[\frac{1}{|r-R|} + (r'-R)\cdot\nabla_R\,\frac{1}{|r-R|}\right]\delta n_2(r')$$

$$= \frac{\Omega}{Ne}\,P\cdot\nabla_R\,\frac{1}{|r-R|},$$

where P is the polarization (11.89) at the point R. If P varies slowly, the remaining summation over the cells R again becomes an integral, and we must add to (11.82) the contribution

$$\Phi_{\text{pol}} = \Phi - \Phi_{\text{ext}} - \Phi_{\text{imp}} - \Phi_{\text{el}}$$

$$= -\text{div}\,\frac{1}{4\pi\varepsilon_0}\int\frac{d^3r'}{|r-r'|}\,P(r') = \frac{1}{4\pi\varepsilon_0}\int\frac{d^3r'}{|r-r'|}\,\varrho_{\text{pol}}(r'). \qquad (11.91)$$

The smoothing out of $-e\delta n_2(r)$ has thus on the macroscopic scale produced a charge density

$$-\overline{e\delta n_2(r)} = \varrho_{\text{pol}} = -\text{div}\,P. \qquad (11.92)$$

In the case of a uniformly polarized sample, ϱ_{pol} is localized at the surface. The polarization produces a field $E_{\text{pol}} = -\nabla\Phi_{\text{pol}} = -P/\varepsilon_0$, which is proportional to the applied field and directed against it, thus reducing the effects of that field.

Altogether we have used the self-consistency of the Hartree potential to derive Eqs.(11.76) and (11.79) which connect the microscopic potential $\mathcal{V} + \delta\mathcal{V}$ and the electron density $n + \delta n$ in electrostatic equilibrium. Getting the macroscopic laws was then achieved by smoothing, replacing the various quantities by their averages in each cell of the crystal. The average of $-\delta\mathcal{V}(r)/e$ can be interpreted as the macroscopic electric potential Φ which is associated with the *macroscopic electric field* $E = -\nabla\Phi$. The electrons make two contributions to the *macroscopic charge density*. The first one, ϱ_{el}, given by (11.84), can be interpreted as deriving from *"free" charges*; it is the dominant one in the behaviour of metals, of plasmas, and of electrolyte solutions, and it also contributes to semiconductor properties. The second one, ϱ_{pol}, given by (11.92), describes the effects of the *polarization* of the medium, especially in insulators and semiconductors. We can eliminate it from the equations by introducing the *electric displacement or induction*,

$$D \equiv \varepsilon_0 E + P = (\varepsilon_0 + \chi)E = \varepsilon E. \qquad (11.93)$$

We thus find, starting from (11.91) the equations of *macroscopic electrostatics*:

$$\text{curl}\,E = 0, \quad E = -\nabla\Phi, \quad \text{div}\,D = \varrho_{\text{ext}} + \varrho_{\text{imp}} + \varrho_{\text{el}}, \qquad (11.94)$$

where the elimination of P has the effect of replacing the constant $\varepsilon_0 = 1/(4\pi \times 10^{-7}c^2)$ in SI units, by the *permittivity* or *dielectric constant* $\varepsilon = \varepsilon_0 + \chi$. Our microscopic theory has, moreover, given us formula (11.90) for the latter. In an insulator or a semiconductor we can thus *forget about the self-consistency effect* associated with the deformation of the wavefunctions, by everywhere replacing ε_0 by ε. In particular, in the quasi-particle dynamics, the force which occurs in (11.66) is $e\nabla\Phi$; if it is produced by an impurity, we must calculate it by replacing the impurity charge by an *effective charge* which is reduced by the factor $\varepsilon/\varepsilon_0$. If there are no macroscopic localized charges in the substance, that is, if $\Phi_{\text{imp}} = \Phi_{\text{el}} = 0$, D/ε_0 can be interpreted as the external field E_{ext} which gives rise to the total field $E = E_{\text{ext}} + E_{\text{pol}}$. Let us, finally, remind ourselves that the free charges ϱ_{el} are responsible for the screening effect which is included in the macroscopic equations (11.94) and (11.84), and which is more explicitly described by (11.85) and (11.86).

11.3.4 Semiconductors

When the width δ of the forbidden band is sufficiently small, the conductivity of an insulating solid may become appreciable at room temperatures. Such solids are called *intrinsic semiconductors* when they are pure. For germanium, for instance, we have $\delta = 0.7$ eV, and the effective masses are of the order of one tenth of the electron mass so that at room temperatures (11.69) leads to 7×10^{11} excited electrons per cm^3. Relatively speaking the ratio is completely negligible, since a band contains typically 10^{23} states per cm^3; however, the electrons which are excited and the holes which are left behind in the valence band behave as *free charge carriers*. These particles which are easily excited and accelerated can transport significant electric currents: in the case of germanium the total charge density of the excited electrons and of the holes is about 0.2 C m^{-3} and their mobilities, given by Eq.(11.72), are of the order of 0.1 m^2 V^{-1} s^{-1}.

The semiconductor which is currently used most is *silicon* with $\delta = 1.12$ eV, but there are many others, for instance, the III-V compounds, such as GaAs or InAs. We shall illustrate our theory mainly by using germanium as an example. The section of the periodic table which concerns us most is shown in Fig.11.12 and contains parts of columns III, IV, V, and VI. The essential facts are the following. Introducing a very small proportion of certain

	B	C	N	O
	Al	Si	P	S
	Ga	Ge	As	Se
Cd	In	Sn	Sb	
		Pb		

Fig. 11.12. Part of the periodic table containing the elements of importance for semiconductors

impurities considerably changes the number of quasi-particles. Moreover, one can control these changes, for instance, by applying potentials, which enables one to play with the properties of the substance and as a result to realize different electronic devices. More precisely, an *impurity or extrinsic semiconductor* is an insulator with a narrow forbidden band, the density of states $\mathcal{D}(\varepsilon)$ of which has been changed by introducing impurities in the crystal. In practice, this addition is carried out by *"doping"* (Exerc.15a), that is, by a hot migration of impurities controlled in such a way that they will diffuse to the required regions. Depending on the nature of the impurities one distinguishes two kinds of extrinsic semiconductors, the n type in which most of the charge carriers are negative, and the p type in which most of them are positive.

In *n-type semiconductors* the impurities are elements which bring in electrons, such as As or Sb for Ge – in the next column of the periodic table. We shall see that doping Ge with a number, N_{I}, of As impurities produces N_{I} extra levels, or *donor levels*, with energies ε_{d} which lie in the forbidden band, at a very small distance (0.0127 eV) below ε_{c} (Fig.11.13). One can easily understand how these impurity levels arise: each impurity corresponds to the replacement in the lattice of a (tetravalent) Ge nucleus by a (pentavalent) As nucleus. This As nucleus has a charge which is larger by unity than the Ge nucleus that it replaces; the impurity also brings in an extra electron. Compared to the reference state without impurity, the situation resembles that of a hydrogen atom: the extra positive charge of the As nucleus tends to retain the extra electron; the donor level, associated with the As impurity, and at zero temperature occupied by an electron if the substance is neutral, can be described as a *bound state of that electron near the excess impurity charge*. We explain below why the binding energy $\varepsilon_{\mathrm{c}} - \varepsilon_{\mathrm{d}}$ is so weak, and accordingly why the spatial extension of the bound state is large, of the order of 100 Å.

To study the properties of a donor state and to calculate its binding energy $\varepsilon_{\mathrm{c}} - \varepsilon_{\mathrm{d}}$ we start from the reference situation of pure Ge, with a full valence band, and we assume that an As atom is substituted for the Ge atom at the origin. The effective single-electron Hamiltonian differs from the Hamiltonian for the Ge bands through the addition of the Coulomb potential $-e^2/4\pi\varepsilon_0 r$, associated with the extra charge of the As nucleus. Moreover, we saw in § 11.3.3 that the redistribution of the electron charges due to the presence of an impurity gives, through self-

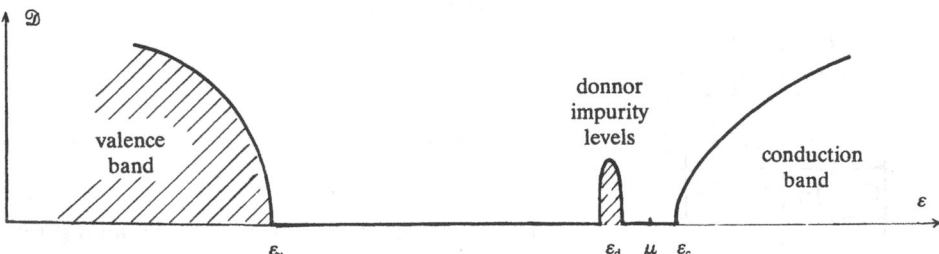

Fig. 11.13. Density of states for an n-type semiconductor

consistency, rise to an extra potential. We shall show in what follows that the effect of the impurity and of the electron cloud which screens it extends over a distance which is large compared to the crystal cell. Under those conditions, we can use the results found for an insulator in § 11.3.3, and simply represent the effect of the self-consistency of the potential by the dielectric constant. The latter is large ($\varepsilon = 15.8\varepsilon_0$ in Ge) because the forbidden band is narrow. Hence, the effective potential $-e^2/4\pi\varepsilon r$ due to the impurity is considerably weaker than that of a hydrogen atom.

If we only consider the lowest states of the conduction band, with Bloch waves $|p, c\rangle$ and approximate energies (11.59), we must diagonalize the Hamiltonian

$$\widehat{h} = \varepsilon_c + \sum_p |p, c\rangle \frac{p^2}{2m_c} \langle p, c| - \frac{e^2}{4\pi\varepsilon} \frac{1}{|\widehat{r}|}. \tag{11.95}$$

Once again we use the fact that we expect the wavefunction of the donor state to extend over a large number of cells. The $1/r$ potential can thus be considered to vary slowly. Its matrix elements between Bloch waves $|p, c\rangle$, calculated as in (11.65) by using the representation (11.36) in Wannier waves, are approximately

$$\left\langle p', c \left| \frac{1}{|\widehat{r}|} \right| p, c \right\rangle = \frac{1}{N} \sum_{R,R'} e^{i(p \cdot R - p' \cdot R')/\hbar} \int d^3 r \, \chi_c(r - R') \frac{1}{r} \chi_c(r - R)$$

$$\simeq \frac{1}{N} \sum_R e^{i(p-p') \cdot R/\hbar} \frac{1}{R} \simeq \frac{1}{\Omega} \int \frac{d^3 R}{R} e^{i(p-p') \cdot R/\hbar}, \tag{11.96}$$

where we have used the fact that the Wannier functions are localized to replace r by R, and that they are orthonormalized. Expression (11.96) is the same as between plane waves with momenta p and the Hamiltonian (11.95) is thus the same as that for a hydrogen atom in the p representation, apart from that m is replaced by m_c and ε_0 is replaced by ε; the spatial modulation due to the Wannier waves and, more generally, the band structure do not play a rôle at all. The binding energy, following from that of the hydrogen atom, is therefore

$$\varepsilon_c - \varepsilon_d = \left(\frac{e^2}{4\pi\varepsilon}\right)^2 \frac{m_c}{2\hbar^2} = \left(\frac{\varepsilon_0}{\varepsilon}\right)^2 \frac{m_c}{m} 13.6 \text{ eV} \simeq 5 \times 10^{-3} \text{ eV}.$$

Due to the screening and to the smallness of the effective mass, it is considerably reduced. The value we have found is comparable with the experimental one. One can obtain a better result by taking into account the large anisotropy of the conduction band of Ge: the eigenvalues of the matrix K in (11.58) have a ratio of 1 to 20.

A *large spatial range* is associated with the *weak binding energy* of the donor state. As compared to the hydrogen atom, we get a Bohr radius,

$$\frac{\hbar^2}{m_c} \frac{4\pi\varepsilon}{e^2} = \frac{m}{m_c} \frac{\varepsilon}{\varepsilon_0} 0.53 \text{ Å} \simeq 100 \text{ Å},$$

which is large compared to the crystal cell size; this a posteriori justifies the use of the dielectric constant and the approximation (11.96).

In principle, the impurity produces also other levels, but they do not play any rôle as their binding energy is much smaller than kT so that they do get mixed up with the bottom of the conduction band. Moreover, we have forgotten about the spin. If there were no interactions between the electrons, the N_I impurities would produce $2N_I$ levels with an energy ε_d. However, the Coulomb repulsion prevents two electrons with opposite spin to be bound to the same impurity site. This is the reason why we have said, not quite properly, that N_I impurities produce N_I donor levels. In what follows we shall write that at equilibrium those levels are occupied by $N_I/[e^{\beta(\varepsilon_d - \mu)} + 1]$ electrons. Being more realistic, and taking into account the fact that there are 3 rather than 2 possibilities for each level – no electron, electron with spin up, and electron with spin down – we should write $2N_I/[e^{\beta(\varepsilon_d - \mu)} + 2]$, as can easily be checked as an exercise.

We still must justify that in (11.95) we have neglected the valence band. The excess Coulomb potential has, in fact, matrix elements not only between the Bloch waves of the conduction band c, but also between those of the valence band v. In the Bloch wave base, this matrix contains, first of all, (c,c) elements which, as we saw in (11.96), are similar to those of the hydrogen potential in the plane wave base. Then there are the interband (c,v) and (v,c) elements; however, they are small, since the potential varies slowly on the scale of one cell so that a calculation similar to the one leading to (11.96) would involve the orthogonal Wannier functions χ_c^* and χ_v. Finally, it contains (v,v) elements which are again similar to those of the hydrogen potential in the plane wave base. The corresponding problem is decoupled from that of the conduction band. However, as the unperturbed term associated with the valence band is $\varepsilon_v - p^2/2m_v$, there is a change in relative sign between the potential and the kinetic energies. Everything thus happens as if the impurity is *repulsive* for the valence band, and there is no essential change in the spectrum of that band.

The second class of impurity semiconductors consists of the *p-type semiconductors*, such as Si or Ge, doped with B, Al, Ga, or In impurities; such trivalent impurities lead to *acceptor levels* with energies ε_a slightly above ε_v and empty at zero temperature. These semiconductors play a *symmetric* rôle as compared to the n-type semiconductors, with electrons being replaced by holes. The extra negative charge of the impurity repels the electrons, that is, attracts the holes which have a positive charge. A bound state for positive charge carriers is thus produced. For a substance which is neutral at zero temperature the acceptor level is empty. One can express this fact by saying that a hole is bound to that impurity.

The study of acceptor levels is similar to that of donor levels. The contribution to the potential due to the excess negative charge of the impurity nucleus is $e^2/4\pi\varepsilon r$ so that the single-electron Hamiltonian only differs from the effective Hamiltonian for a donor studied above by a change in sign and by interchanging the valence and the conduction bands. This implies that when there are N_I acceptor impurities, there are N_I single-electron states, of energy ε_a slightly larger than ε_v and such that $\varepsilon_a - \varepsilon_v$ is of the same order of magnitude as $\varepsilon_c - \varepsilon_d$. Just as the donor levels were produced starting from the conduction band, the acceptor levels are *produced starting from the valence band*. Hence, the total number of states of that band is reduced by N_I so that, if N_I electrons are missing, as is the case for a neutral substance, the acceptor levels remain empty at zero temperature. The symmetry between n- and p-type semiconductors is complete with respect to an exchange between the two types of quasi-particles: the acceptor level is a state of an excess hole, of mass m_v and charge e, bound to the impurity with excess charge $-e$.

In a homogeneous and neutral semiconductor at equilibrium the *numbers of conduction electrons and of holes* are determined by the temperature and the doping. They are related to the chemical potential through (11.56) and (11.61) while in an n-type semiconductor a number $N_d = N_I f(\varepsilon_d)$ of electrons remain fixed to the donor sites at non-zero temperatures. We get μ from the condition of electric neutrality, $N_c + N_d - N_v = N_I$. Solving these equations is essential for semiconductor technology; one can only do this numerically or using more or less rough approximations (Exerc.11b). One difficulty comes from the fact that the classical approximations (10.63) for the Fermi factors, which led to the simple formulæ (11.60) and (11.63) for insulators, cannot always be used for a semiconductor. Depending on the temperature and the doping, one can distinguish three limiting regimes.

At low temperatures or strong doping, we are in the *extrinsic regime* or the *impurity* or *"freeze-out"* regime. Let us focus on an n-type semiconductor. Noting that at zero temperature all donor levels are occupied by the extra electrons and that the conduction band is empty, we conclude that this regime resembles the state of an insulator with forbidden band $(\varepsilon_d, \varepsilon_c)$. The chemical potential is thus equal to $\frac{1}{2}(\varepsilon_c + \varepsilon_d)$ and it varies very little with temperature, if the number N_I of the donor centres is large (Exerc.11b). It therefore remains close to the bottom of the conduction band, since $\frac{1}{2}(\varepsilon_c - \varepsilon_d)$ is only $\frac{1}{160}$ eV for Ge doped with As, as compared to room temperature, corresponding to $\frac{1}{40}$ eV. The *number of conduction electrons* $N_c \simeq N_I(1 - f(\varepsilon_d))$ *is large* at not too low temperatures, being a fraction, of the order of half, of N_I. On the other hand, the holes, whose number was given by (11.69) for a pure semiconductor, have practically completely disappeared through the effect of n-type doping, as μ has increased by about $\frac{1}{2}\delta$ and (11.63) is here dominated by a factor $e^{-\beta\delta}$.

In the opposite case, the *intrinsic regime*, corresponding to weak doping, resembles the insulator described in § 11.3.2, as there are too few impurities to fix μ in the range $(\varepsilon_d, \varepsilon_c)$. The value of μ lies therefore near the centre of the forbidden band $(\varepsilon_v, \varepsilon_c)$. The numbers N_c and N_v given by (11.60) and (11.63) remain small, but now may differ significantly from one another, if we take into account that $N_d \simeq N_I e^{-\beta(\varepsilon_d - \mu)}$ electrons are bound to the donor sites. Eliminating μ between those expressions gives, for an n-type semiconductor at equilibrium, *mass action* type laws:

$$\frac{N_c}{\Omega} \frac{N_v}{\Omega} = 4 e^{-\beta\delta} \left(\frac{\sqrt{m_c m_v} kT}{2\pi\hbar^2} \right)^3, \qquad (11.97)$$

$$\frac{N_d^2}{N_I^2} \frac{N_v}{N_c} = e^{-\beta(2\varepsilon_d - \varepsilon_c - \varepsilon_v)} \left(\frac{m_v}{m_c} \right)^{3/2}, \qquad (11.98)$$

which, together with the relation $N_c + N_d - N_v = N_I$, determine the carrier numbers N_c and N_v as functions of N_I. We note, in particular, that N_c and N_v are inversely proportional to one another.

Finally, an intermediate doping, or temperature, leads to the *exhaustion or saturation regime* where $\mu - \frac{1}{2}(\varepsilon_c + \varepsilon_v) \gg kT$ and where $\varepsilon_d - \mu \gg kT$. Equations (11.97) and (11.98) remain valid. The first condition means that $N_v \ll N_c$, and the second that $N_d \ll N_I$ so that the donor levels are practically empty and there are practically no holes. As a result, we have $N_c \simeq N_I$: everything happens, as if *all the extra electrons* introduced by the donor impurities were *injected into the conduction band*.

A sufficient amount of doping therefore enables us significantly to increase the *conductivity* given by (11.73), which here takes the form $\sigma = e(n\mu_n + p\mu_p)$; the symbols $n = N_c/\Omega$ and $p = N_v/\Omega$ denote, respectively, the negative and positive charge carrier densities, and μ_n and μ_p their mobilities, which are of the order of 0.1 m^2 V^{-1} s^{-1}. The electrons fixed to the donor centres do not take part in the conductivity. Figure 11.10 showed the conductivity of Ge in the intrinsic regime, but one could already see at low temperatures plateaus depending on the amount of impurities. We can now explain these by noting that they correspond to the saturation regime where the conductivity, for the case of As impurities, reduces to $\sigma \simeq en\mu_n \simeq e(N_I/\Omega)\mu_n$ which *varies little with temperature, but strongly with doping*. Fractional amounts of As of the order of 10^{-6} are sufficient to dominate the conductivity completely. This strong sensitivity to a parameter which is difficult to control explains why for a long time semiconductivity was a badly understood phenomenon.

For a p-type semiconductor the situation is symmetric. In particular, in the extrinsic regime, for a sufficiently large impurity density and not too high a temperature, the Fermi level μ lies near ε_v, whereas it was lying near ε_c for extrinsic n-type semiconductors. The crystal then contains a number of holes of the order of N_I at room temperatures and the *conduction* is now ensured mainly through the *displacement* of these (positively charged) *holes*, whereas it is through the displacement of the conduction electrons in n-type semiconductors. One says that among the quasi-particles the positive charge carriers (holes) are the *majority* carriers in a p-type semiconductor, and they are the *minority* carriers in an n-type semiconductor. The *Hall effect* (1879) shows the sign of the majority carriers. If there is a current in the x-direction flowing through the semiconductor, applying a magnetic field in the z-direction tends to deflect the carriers, whatever their sign, into the $-y$ direction. A stationary regime is established, where an excess of carriers on the $-y$ boundary of the sample and a deficit of carriers on the $+y$ boundary create an electric field along $\pm y$, depending on the sign of the carriers, which enables the current to remain parallel to x. One can detect this effect by measuring the potential difference between the $-y$ and $+y$ boundaries: its sign is the same as that of the majority carriers (Exerc.15d). The frequent occurrence of positive carriers – in p-type semiconductors – was only explained in 1931, by A.H.Wilson, who clarified the rôle of bands in different substances and who introduced the hole concept.

The temperature dependence of the conductivity of semiconductors has been used to advantage to construct *resistance thermometers*, for example, between 0.1 and 100 K, by using Ge doped with As in the extrinsic regime. One obtains the temperature immediately by a simple resistivity measurement.

When a semiconductor is illuminated, a photon with an energy larger than the width δ of the forbidden band can be absorbed, by exciting an electron from the valence band into the conduction band, that is, by producing a pair of charge carriers of opposite sign. Continuous illumination produces a stationary, non-equilibrium state where there is an excess of carriers and where the conductivity is thus the larger, the stronger the light flux. This effect, *photoconductivity*, has many applications. For instance, in a *photoelectric cell*, measuring the resistivity of a semiconductor immediately gives us the illuminance. One uses, for example, CdS with a forbidden band of 2.4 eV (in the green part of the visible spectrum). *Infrared detectors* use semiconductors with a narrower band (0.2 to 0.04 eV). The sensitive part of a *video camera* is a semiconductor, such as PbO for which $\delta = 2.3$ eV, covered by a conducting transparent anode, and swept from behind by an electron beam. If the spot where this beam hits is not illuminated, the resistivity is large, the electrons are stopped in the semiconductor and cannot reach the anode; at points which receive light, a current circulates which is proportional to the light intensity and it is evacuated by the anode. Treating the signal thus obtained enables one to reconstruct the image afterwards. Finally, *photocopiers* use a similar principle. A layer of Se ($\delta = 2$ eV) deposited on the surface of a metal is charged positively, before being exposed to light. As it is a bad conductor, it retains in the dark that charge on its surface notwithstanding the proximity of the metal substrate. However, the points where it is illuminated become conducting, the charge which was there moves away through the metal so that an electric image is formed of those regions which were in the dark and remained charged. One then sprinkles a pigment powder on the Se layer, which is attracted to the charged zones. Finally, one applies a piece of paper to which the pigment is transferred and is fixed by heating.

Many other industrial applications involve semiconductors (§ 11.3.5). The interest in these substances arises from the fact that they are very *flexible in their use*, which results from the possibilities to bring into contact several types of semiconductors and metals, to vary the proportions of the impurities, and to operate easily on the electrons by applying electric potentials at various spots. Moreover, the *orders of magnitude* involved are particularly convenient: the binding energies of the donor and acceptor levels and the width δ of the forbidden band, vary from a small fraction of an eV to several eV, that is, they are comparable with room temperatures ($\frac{1}{40}$ eV) and with optical (1.5 to 3 eV) or infrared photons, and they correspond to electric potentials of the order of a volt. The various devices have many advantages: reliability, strength, longevity, and low energy consumption. Above all, in the last thirty years one has developed inexpensive mass production methods, which allow both *miniaturization* and arrangements of many elementary components by techniques related to *printing* or *engraving*. Selected substances can, for instance, be deposited from a plasma or by sputtering and removed by chemical attack. Starting from a Si layer which is as pure as possible, one introduces at chosen points the required amounts of chosen impurities, or one uses masks to cut the surface, controlling the migration of the impurities which is more or less fast, depending on the temperature. In this way one can get side by side, for instance, weakly n-doped Si and strongly p-doped Si; through oxidization one can produce insulating regions. One also knows how to make heterojunctions (between different intrinsic semiconductors) or metal-semiconductor interfaces by successive deposits. One can thus practically on demand produce composite devices which are extremely small and through which one is able to realize predetermined complicated operations.

11.3.5 p-n Junctions, Photocell, Diode, and Transistor

The essential element of many systems based on the use of semiconductors is the p-n *junction*, that is, the juxtaposition of two semiconductor pieces, one of which is doped with acceptors, and the other with donors. Let us start by studying the electrostatic equilibrium of such a junction, using the results of § 11.3.3. The inhomogeneity of the system has remarkable consequences. Before they are in contact, the n- and p-type semiconductors have different chemical potentials, situated near ε_c and ε_v, respectively. When they are brought into contact they must therefore exchange electrons, and the equilibrium state corresponds to an *equalization of the chemical potentials*. The electric neutrality, which is locally ensured in each lattice cell of the pure substance or within a Bohr radius (of the order of 100 Å) around each impurity, does no longer exist in a junction, because of this transfer of electrons from the n side where the chemical potential is the higher to the p side. A *spontaneous macroscopic charge density* appears, which is positive on the n side and negative on the p side. This charge produces, through the self-consistency of the microscopic potential $\mathcal{V}(r)$, a *macroscopic field* $E(r)$ and a *macroscopic electric potential* $\Phi(r)$. Here "macroscopic" refers to scales of 1000 Å = 0.1 μm. All devices using junctions are based upon the existence of these electrostatic effects, produced by the contact between different substances, like the Volta effect in metals (§ 10.4.3).

A quantitative study can be based on the equations which we found in § 11.3.3. The total macroscopic charge density associated at each point with the electrons and the impurities is here

$$\varrho = e\big(p - n + \nu_d - n_d - \nu_a + p_a\big), \tag{11.99}$$

where $n = N_c/\Omega$ and $p = N_v/\Omega$ are the standard notations for the local densities of the carriers which take part in the conduction, ν_d and ν_a are the donor and acceptor centre densities, and n_d and p_a the numbers of electrons and holes fixed to them per unit volume. We take into account the charge due to the polarisation by means of the dielectric constant, so that we have for the electric potential (11.82) and (11.91)

$$\Phi(r) = \frac{1}{4\pi\varepsilon} \int \frac{d^3r'}{|r - r'|}\, \varrho(r'). \tag{11.100}$$

In each point, one must use the Thomas-Fermi approximation (11.84), which adds $-e\Phi$ to the band energy, so that we have locally

$$\left.\begin{aligned}
n &= \int_{\varepsilon_c}^{\infty} d\varepsilon\, \frac{\mathcal{D}(\varepsilon)}{\Omega}\, \frac{1}{e^{\beta(\varepsilon-e\Phi-\mu)} + 1}, & n_d &= \frac{\nu_d}{e^{\beta(\varepsilon_d-e\Phi-\mu)} + 1}, \\
p &= \int_{-\infty}^{\varepsilon_v} d\varepsilon\, \frac{\mathcal{D}(\varepsilon)}{\Omega}\, \frac{1}{e^{\beta(-\varepsilon+e\Phi+\mu)} + 1}, & p_a &= \frac{\nu_a}{e^{\beta(-\varepsilon_a+e\Phi+\mu)} + 1}.
\end{aligned}\right\} \tag{11.101}$$

The coupled equations (11.99), (11.100), and (11.101) characterize the electrostatic equilibrium of a semiconductor when it is, for instance, globally charged or inhomogeneously doped; they determine the shapes of $\Phi(r)$ and $\varrho(r)$. A *graphic representation* much used in technology consists in taking one of the spatial coordinates along the abscissa axis, and drawing along the ordinate axis the energies of the band edges $\varepsilon_c - e\Phi$ and $\varepsilon_v - e\Phi$, and possibly, the donor and acceptor level energies $\varepsilon_d - e\Phi$ and $\varepsilon_a - e\Phi$. This shows the shift by $-e\Phi(r)$ of the energy levels, when the electrostatic potential is read from top to bottom. In thermal and electrostatic equilibrium the chemical potential is represented by a horizontal line. Its position represents the filling of the bands: where the conduction band is close to the chemical potential, the density n of negative carriers is larger; similarly, the density p of positive carriers increases when the chemical potential comes close to the valence band. Figure 11.14 shows the neighbourhood of the surface (on the left) of an n-type semiconductor which has been positively charged. Far from the surface, the situation is the same as in the case of equilibrium of the neutral semiconductor, apart from a shift by $-e\Phi(\infty)$. The potential Φ decreases close to the surface, in order to satisfy Eq.(11.100), which implies $\varepsilon\nabla^2\Phi + \varrho = 0$. Its

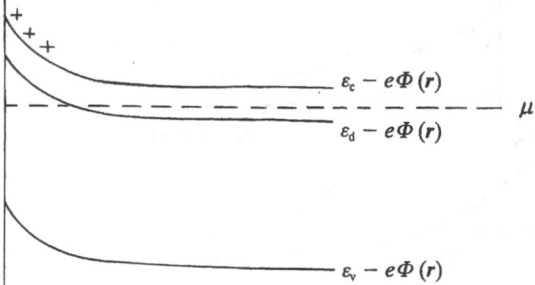

Fig. 11.14. Position of the bands and donor levels in a positively charged n-type semiconductor near its surface

variation produces the curving of the conduction band and of the donor levels, and so engenders a partial emptying of those, thus creating the necessary positive charge to satisfy the coupled equations (11.99) and (11.101), and the condition imposed upon the global charge of the sample. As in a conductor, this charge distributes itself over the surface according to Eq.(11.84), but here the approximation (11.84′) is no longer valid; the screening effect is very weak so that the thickness of the charged region is macroscopic, of the order of a fraction of a micron, whereas the screening length in a conductor is of the order of an Å. An order of magnitude of this thickness is evaluated in Exerc.11c.

The above description of the surface of a positively charged n-type semi-conductor can be applied to the n side of the p-n junction. Indeed, the latter can be modelled as a sudden discontinuity in doping, and at equilibrium it is equivalent to the juxtaposition of two n- and p-type semiconductors with opposite charges. In the p part, which is negatively charged, the potential has an inverted shape (Fig.11.15). The parameters which are characteristic for the junction, such as the charge transferred from the p side to the n side, the thickness of the charged zone, and the shapes of the potential and of the electric field, are obtained by writing down the electrostatic equations (11.99–101), together with the conditions for the matching of the electric and chemical potentials (Exerc.11c). The electric potential, $\Phi(r)$, must be continuous, as well as its first derivatives, when it passes through the p-n partition. The chemical potential, μ, must be constant; far from the junction it must have the same position in relation to the bands as for the neutral pure substances, apart from the global shifts $-e\Phi_n = -e\Phi(+\infty)$ and $-e\Phi_p = -e\Phi(-\infty)$. Figure 11.15 shows schematically the junction in equilibrium. In order that μ can be constant, *there occurs in the junction a spontaneous potential difference $\Phi_n - \Phi_p$ of the order of a volt*, slightly smaller than the width of the forbidden band. In fact, $e(\Phi_n - \Phi_p)$ is equal to the difference $\mu_n - \mu_p$ between the chemical potentials of the two separated neutral semiconductors. This

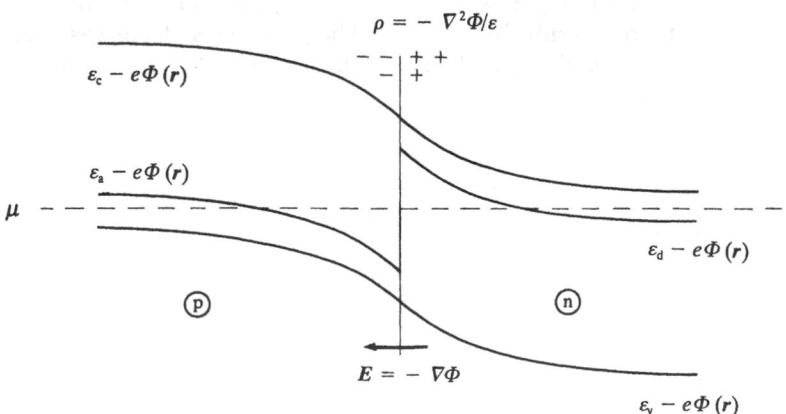

Fig. 11.15. Potential, field, and charges in a p-n junction at equilibrium

potential difference $\Phi_n - \Phi_p$ is produced, in accordance with Eq.(11.100), by an electrostatic *double layer of charges* situated astride the junction: the acceptor levels get partly filled by electrons on the p side and the donor levels get partly emptied on the n side; the populations of the two bands are also changed, which has as a result a *reduction of the conductivity*. The variation of Φ implies that there exists at equilibrium a *spontaneous electric field* $E = -\nabla\Phi$ at the level of the junction, which is directed from the n to the p side. The thickness of the zone where this field and the charges exist is of the order of a fraction of a micron.

Even though various semiconductor devices use *non-equilibrium* junctions, we are now in a position to understand their functioning. In § 15.2.2 we shall give complementary discussions of transport phenomena in semiconductors, and of non-equilibrium situations where the chemical potentials are not constant. That should make it easier to undertake the study of more specialized books such as S.M. Sze, *Physics of Semiconductor Devices*, Wiley, New York, 1981.

A *photocell* or a *solar cell* has as its essential element a junction between two thin p and n layers, which is used to catch light. As in the case of photoconductivity (§ 11.3.4), the absorption of photons produces an extra population of pairs of carriers of both signs. However, here the spontaneous field, acting in the junction, pushes the excess electrons to the n side and the holes to the p side before they have had time to recombine. In an open circuit one produces thus an electromotive force between the electrodes which are connected with the n and p faces, and a current when the circuit is closed (*photovoltaic effect*). The photocells are used to transform solar energy into electricity, and also as radiation detectors. Their efficiency, of the order of 15 %, the low density of the solar energy (1 kW m^{-2}), and the small e.m.f. (a fraction of a volt) restrict their use to the production of low powers.

The *electroluminescent diodes* used for the display and transmission of low power signals (remote control, signal indicators) are based on the inverse phenomenon: when a current flows through a junction from p to n, the majority carriers, holes on the p side and electrons on the n side, are forced to move towards one another and they recombine in the junction, producing photons with energies close to δ, in the visible or the infrared.

A semiconductor *diode* consists of a single p-n junction. It functions as a *rectifier* as it lets electric current from p to n pass more easily than from n to p. An explanation of this non-equilibrium effect, which is more elementary but less rigorous than transport theory (Chap.15), can be given by kinetic theory (Chap.7), using simply the expression for the flux of classical particles which hits a wall.

Let us first examine how the conduction electron density, at equilibrium, remains important at the n side and small at the p side: thermal agitation tends to equalize the populations, but the potential $\Phi(r)$ pushes the electrons from p to n. Quantitatively, the minority electrons at the p side carry, when they move to n, a current I_0, in the direction from n to p, equal to

$$I_0 \;=\; \frac{2e\Delta S}{h^3}\int_{p_x>0} d^3 p\,\frac{p_x}{m_c}\,\exp\left[-\beta\left(\varepsilon_c-\mu+\frac{p^2}{2m_c}-e\Phi_p\right)\right].\qquad(11.102)$$

This current is hardly at all sensitive to the field occurring in the junction, and it remains unchanged if we apply to the junction an extra potential difference, in a non-equilibrium situation. Nevertheless, the majority electrons arriving at the junction from the n side, which are much more numerous, are decelerated by the potential so that only the fastest ones manage to pass through; the others are reflected. At equilibrium, the corresponding current, which is directed from p to n, must be equal to I_0, so as to cancel it. However, in contrast to the current in the opposite direction, this current is changed when one applies a potential V to the junction: *only those electrons with a sufficiently high kinetic energy*, from among the electrons in n moving towards p, can overcome the total electric potential difference $\Phi_n-\Phi_p$ and pass through. The number of those fast electrons varies approximately as $\exp[-\beta e(\Phi_n-\Phi_p)]$, a factor which occurs outside the integration over p because of the lower bound $e(\Phi_n-\Phi_p)$ on the kinetic energy $p^2/2m_c$. If we denote by V the electric potential *applied* to the p semiconductor and added to the spontaneous potential, the current due to the electrons going from n to p will be equal to $I_0\,e^{\beta eV}$.

Altogether, the equilibrium is broken by the potential V applied to p: there occurs a current due to conduction electrons which, if measured as going from p to n, equals

$$I \;=\; I_0\left(e^{\beta eV}-1\right).\qquad(11.103)$$

We see easily that the same considerations when applied to holes do not change the conclusion, although the definition of I_0 must be modified. Equation (11.103) thus gives us the *junction characteristic*, that is, the relation between the current and the applied potential; its asymmetric form enables us easily to understand the rectifying effect of a p-n junction. Notwithstanding the approximations made to find that expression, its agreement with experiments is remarkable, as Fig.11.16 shows. The order of magnitude of I_0 following from (11.102) is, however, not correct, as our discussion assumed an electron mean free path longer than the thickness of the junction. A more rigorous theory must take into account the scattering of charge carriers in the junction and also the recombination of holes and conduction electrons; these effects reduce (11.103) by a factor 1000. Nevertheless, the expression for I_0 is dominated as in (11.102) by an exponential factor close to $e^{-\delta/kT}$, so that it is necessary to use semiconductors with a *narrow forbidden band* in order to obtain appreciable currents.

Another fundamental electronic device is the *transistor*, invented in 1948 by J. Bardeen and W.H. Brattain in the form of a point-contact transistor, and in 1949–51 by W. Shockley in the form of a *junction transistor* or bipolar transistor which we shall describe. Such a transistor is obtained by combining three alternating semiconductors, for instance, n-p-n. The first is the *emitter*, the second the *base*, and the third the *collector* (Fig.11.17). We bring the base to a potential V_b which is positive relative to the emitter and which can be

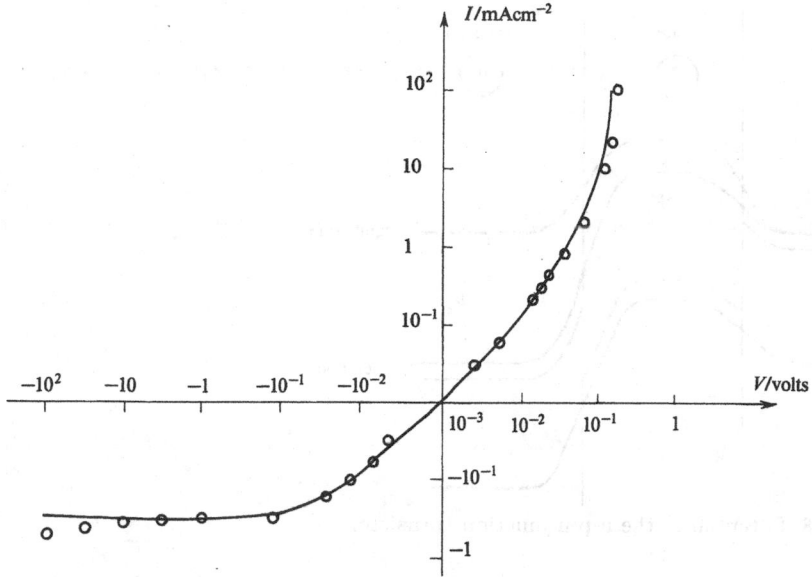

Fig. 11.16. Characteristic of a p-n junction

varied, and the collector to a large, fixed, positive potential V_c. As a result, currents I_e, I_b, and I_c flow in the three connections; we want to evaluate these currents. We sketch in Fig.11.18 the spatial variation of the band edges, which are the constants ε_c and ε_v plus $-e\Phi(r)$, before and after the application of the potentials.

Let us calculate the currents through each of the two junctions, first examining the dominant contribution from the conduction electrons, and analyzing the fluxes in both directions as we did for a single junction. Because V_c is large, practically no electron passes in the collector-base direction. On the other hand, the electrons moving from the emitter to the base produce, as in a diode, a current

$$I_e \simeq I_0\, e^{\beta e V_b}, \tag{11.104}$$

and we can neglect I_0 as compared to this current. The operation of the transistor is based upon an essential characteristic of its geometry: *the base is very thin,*

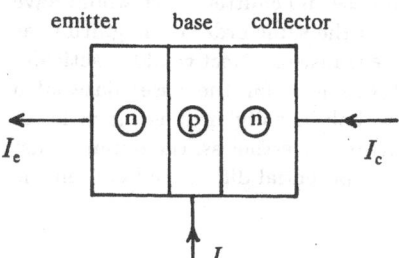

Fig. 11.17. The n-p-n junction transistor

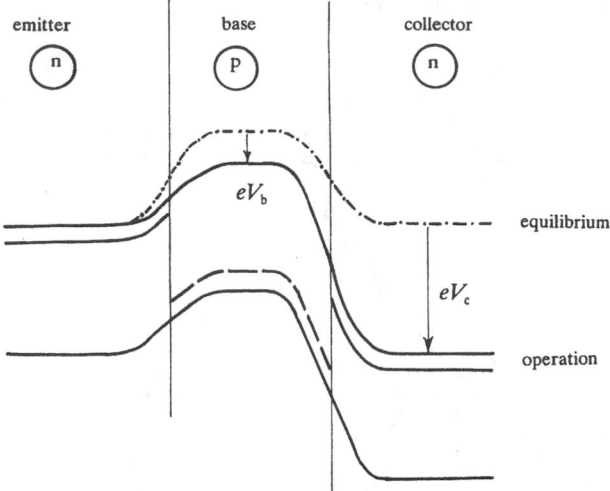

Fig. 11.18. Potential in the n-p-n junction transistor

with a thickness of the order of a micron. Most electrons coming from the emitter, with a mean free path of the same order of magnitude as this thickness, are thus accelerated by the strong field acting between the base and the collector, and are *captured* by the latter before having had the time to diffuse in the base. Let ξ be the ratio of the electrons diffused within the base and then channelled by the conductor connection which maintains its potential at V_b, to the number of electrons captured by the collector. This is a small number, of the order of 1 %, and it depends mainly on the shape and size of the various parts of the transistor. The currents I_b and I_c are then given by the relations

$$I_e \; = \; I_b + I_c, \qquad I_b \; = \; \xi I_c. \tag{11.105}$$

The thinness of the base and the fact that there the electrons are minority carriers imply that the conduction electrons do not make other significant contributions to the currents than the ones which we have considered.

In this discussion we have neglected the current carried by the *holes*. They are minority carriers in the collector and only make a small contribution to the current I_c. On the other hand, the large potential V_c prevents the holes to pass from the base to the collector. Nevertheless, they can easily pass from the base to the emitter, producing a current which must be added to I_b and thus decreases the amplification ratio I_c/I_b. For a junction between base and emitter which would have a hole-particle symmetry, this current would be of the same order of magnitude as the current I_e of the conduction electrons, and the transistor effect would practically disappear. To get rid of this unwanted effect one uses for the base material a *weakly doped* p-type semiconductor so that the hole density in it is much lower than the conduction electron density in the emitter. Nevertheless, the doping must be sufficient to maintain a significant spontaneous potential difference between the emitter and the base.

The currents and the potentials in the transistor are thus approximately connected with one another through Eqs.(11.104) and (11.105). The parameters I_0 and ξ, which is a small number, thus characterize its behaviour in a circuit. The remarkable phenomenon is that small changes in V_b or in I_b produce large changes in I_c or $I_e = (1 + \xi)I_c$: the transistor is an *amplifier*. Moreover, we get a *linear* amplification of the current, since I_c is proportional to I_b, with an amplification ratio $1/\xi$. Finally, it can also operate as a *gate*, since I_e varies, according to (11.104), abruptly with V_b, which enables us to control the current $I_e \simeq I_c$ by means of the base potential.

Many other devices based upon the p-n junction are used in electronics. Depending on the potential V applied between n and p, the thickness of the charged region varies. We have seen that the conductivity is weaker in that region where the free-carrier density is reduced. One can therefore operate upon a current flowing in one of the semiconductors along the junction by simply controlling V. This idea gives the principle of the *field effect transistor* which is a device where the resistance along the junction is governed by a transverse voltage between n and p (Prob.18). The most often used type is the MOSFET – from Metal Oxide Semiconductor Field Effect Transistor; a layer of oxide (dielectric) and a layer of metal, put on one of the semiconductors of the junction, together with it form a condenser which provides the controlling voltage.

We may also mention the *variable-capacitance diode*, or *varactor*, used, for instance, to control the frequency of a radio receiver. Here also the potential V applied to the p-n junction changes the way it is charged and thus controls its capacity.

Let us finally remark that the semiconductor properties described here are not limited to the perfect crystals which we have considered. It is true that the simplifications produced by the geometry, the existence of a quasi-momentum and Bloch waves, do not exist for amorphous substances; however, the analysis of the tight binding theory shows that the electron structure of an amorphous substance retains a certain memory of the discrete nature of the levels of its atoms. In particular, the single-electron density of states retains a shape which is similar to that of an insulator: there is no longer a forbidden band, but it is replaced by a region which contains few levels and this plays an analogous rôle. The manufacture of badly crystallized or amorphous semiconductors is relatively cheap: for this reason the use of amorphous silicon is likely to develop rapidly, for applications where a completely forbidden band is not required.

11.4 Phonons

So far we have firstly studied the properties of solids which are connected with the average positions of the atomic nuclei. After that we considered their electronic properties, and there now remains for us to investigate the rôle of the nuclear displacements. Motions with relatively large amplitudes do not occur until the solid melts (Probs.8 and 10). Otherwise, the nuclei in the solid state move, in general, only slightly from their equilibrium positions and the harmonic approximation is already sufficient for describing their vibrations. These vibrations are responsible for the heat capacity of the solid. However, it is essential to treat them *quantum mechanically*. This fact was realized in 1907 by A. Einstein (Exerc.11d) even though quantum mechanics of the simplest systems (Bohr's atom model of 1913, the de Broglie relations of 1926, or the Schrödinger equation of 1926) had not been worked out at that time. A major step in the quantum theory of solid vibrations was the work by Petrus Debye (Maastricht 1884–Ithaca, NY 1966) who in 1912 gave a quantitative explanation of the specific heats and in 1914 introduced the concept of phonons, applying it to heat conduction in insulators.

11.4.1 Lattice Vibrations

As we mentioned in § 11.1.1, the last stage in the Born-Oppenheimer method (§ 8.4.1) consists in treating the eigenvalue $W(\{R_n\})$ of Eq.(11.7) as an interaction potential for the nuclei in the effective Schrödinger equation (8.40). Eliminating the electron degrees of freedom leads thus to an effective Hamiltonian,

$$\widehat{H}_{\mathrm{n}} = \widehat{T}_{\mathrm{n}} + W(\{\widehat{R}_n\}), \tag{11.106}$$

where the coordinates R_n of each nucleus n, which so far had been considered to be parameters, are replaced by operators. It is difficult, if not impossible, to calculate the potential $W(\{R_n\})$ explicitly, as this would require that we had solved the electronic problem (11.7) not only when the nuclei are fixed to their equilibrium positions in the lattice as in § 11.3, but also when they are displaced from those positions. However, it will be sufficient to be aware of two essential properties of W. Firstly, $W(\{R_n\})$ is a *minimum* when the nuclei occupy their average positions $\overline{R_n}$ which form the lattice; as the nuclei do not move far from these positions, one can expand W around that minimum. Secondly, the periodicity of the lattice implies that W is *invariant under displacements* of the crystal group, such as translations, rotations, symmetries, and their combinations.

The energy W defined by Eq.(11.7) depends, in fact, not only on the nuclear coordinates, but also on the state λ of the electrons. Those remain practically frozen in into their ground state for an insulator, as for a molecule. We have seen that this is not the case for a metal. To be more rigorous, we must then take for the potential $W(\{R_n\})$ which occurs in (11.106) the average, at the temperature considered, over the electron microstates λ, of $W(\{R_n\}, \lambda)$ so that this effective potential depends on T and μ. This dependence is, however, weak and can be neglected. Moreover, the fact that W depends both on the coordinates $\{R_n\}$ and on the state λ generates an effective interaction between the nuclear and electron degrees of freedom, an interaction which we shall neglect (§ 11.4.2).

For molecules, the search for the eigenstates of the effective nuclear Hamiltonian (8.40) consisted of a study of global rotations and of internal motions, which for the simple molecules considered reduced to vibrations. Among the various possible motions of the nuclei in a crystal, the global rotations and translations, which are macroscopic displacements, are not thermally excited: in contrast to the molecules in a gas which can be rotated by thermal excitation, a crystal retains its fixed orientation and its fixed position in space when it is heated. Hence, we must only consider from amongst the nuclear motions those which correspond to *vibrations*, excluding the global rotations and translations. We remind ourselves that for a diatomic molecule there was a single vibrational mode, associated with changes in the interatomic distance, with an angular frequency ω which could be calculated by expanding the potential around its minimum reached at the equilibrium position. The *number of modes* of the lattice vibrations is $3N - 6$, where N is the number of nuclei: we have to subtract from the total number of coordinates the 3 translations and the 3 rotations of the whole system. As N is large, we shall in what follows replace $3N - 6$ by $3N$.

As in any *small vibrations* problem, we look for the *normal modes* of the vibrations of the crystal nuclei, replacing in (11.106) the potential $W(\{R_n\})$ by its *quadratic approximation* around the average values $\overline{R_n}$, and then *diagonalizing* the quadratic form in $\{\delta R_n\} = \{R_n\} - \{\overline{R_n}\}$ which we have thus obtained. To simplify the discussion, let us assume henceforth that we have only *one atom per cell* and let its mass be M. The expression

$$W(\{R_n\}) - W(\{\overline{R_n}\}) \tag{11.107}$$

is a quadratic form of the $3N$ variables $\{\delta R_n\}$, $1 \leq n \leq N$. We write the eigenvalues of the corresponding matrix as $\frac{1}{2}M\omega_q^2$; they are all positive as we have a stable equilibrium. Let ξ_q be the eigenvectors, which are linear functions of the displacements $\{\delta R_n\}$; their number is $3N$. Finally, the momenta π_q, conjugate to the variables ξ_q, are the corresponding linear combinations of the nuclear momenta P_n. The effective nuclear Hamiltonian (11.106) can thus be written in the small vibration approximation as

$$\widehat{H}_n = W(\{\overline{R_n}\}) + \sum_{n=1}^{N} \frac{\widehat{P}_n^2}{2M} + \left[\widehat{W}(\{\widehat{R}_n\}) - W(\{\overline{R_n}\})\right]$$

$$= W(\{\overline{R_n}\}) + \sum_{q} \left(\frac{\widehat{\pi}_q^2}{2M} + \frac{1}{2}M\omega_q^2 \widehat{\xi}_q^2 \right), \tag{11.108}$$

where there are $3N$ terms in the sum over q. In fact, (11.108) includes 6 terms associated with the translations and rotations of the crystal, for which $\omega_q = 0$ since W remains constant under those displacements.

In terms of the normal coordinates ξ_q and their conjugate momenta, the nuclear small vibration Schrödinger equation can thus be split into a set of $3N$ equations for one-dimensional *independent harmonic oscillators* with angular frequencies ω_q. (Strictly speaking the frequency is $\nu = \omega/2\pi$ but ω itself is often simply called frequency.) The lattice vibrations are *quantized*; the *energy levels* of $\widehat{H}_{\mathbf{n}}$ are given by

$$W(\{\overline{\boldsymbol{R}_n}\}) + \sum_q \hbar\omega_q \left(n_q + \frac{1}{2}\right) = b + \sum_q n_q \hbar\omega_q, \tag{11.109}$$

and characterized by $3N$ independent quantum numbers $n_q = 0, 1, 2, \ldots$. The thermodynamic properties will mainly depend on the frequency spectrum ω_q of the various vibrational modes for small ω_q and we must thus determine it by explicitly diagonalizing the quadratic form (11.107).

We solve this problem for a one-dimensional model of a crystal, containing only one kind of atoms which can oscillate around their equilibrium positions $\overline{R_j} = ja$. To simplify the calculations we assume that the effective potential W only includes interactions w between *nearest neighbour nuclei*. It is therefore expanded in powers of $\delta R_j \equiv R_j - ja$ as

$$\begin{aligned}
W &= \sum_{j=1}^{N-1} w(R_{j+1} - R_j) \\
&= \sum_{j=1}^{N-1} \Big[w(a) + w'(a)\big(\delta R_{j+1} - \delta R_j\big) \\
&\qquad + \tfrac{1}{2} w''(a)\big(\delta R_{j+1} - \delta R_j\big)^2 + \cdots \Big],
\end{aligned}$$

and the conditions that it is a minimum for $R_j = ja$ are $w'(a) = 0$ and $w''(a) \equiv C > 0$. In the harmonic approximation the Hamiltonian (11.106) thus reduces to

$$\widehat{H}_{\mathbf{n}} = \sum_{j=1}^{N} \left[\frac{\widehat{P}_j^2}{2M} + \frac{1}{2} C\big(\delta\widehat{R}_{j+1} - \delta\widehat{R}_j\big)^2 \right], \tag{11.110}$$

where we have dropped the additive constant $W(\{\overline{\boldsymbol{R}_n}\})$.

As in band theory (§ 11.2.2), we wish to take advantage of the invariance group associated with the periodicity of the crystal structure. Since the number of nuclei, N, is large, the boundary effects due to the fact that the crystal is finite must play a negligible rôle. We can therefore slightly change our model and introduce periodic boundary conditions (§ 10.2.1), that is, assume that the crystal ends are joined together. We therefore identify $N + 1$ with 1 in the last term in (11.110). An important simplification when we look for the normal modes is now introduced by the fact that (11.110) is *invariant under a lattice translation* over a distance a. We use that property by carrying out the same discrete Fourier transformation as the one which enabled us to connect the localized Wannier orbitals to the Bloch waves for the electrons through (11.35) and (11.36). Through this transformation we change from the lattice sites j to the wavenumbers k or the quasi-momenta $\hbar k$. One should take care to distinguish between these quasi-momenta, which are associated with the translational invariance of the lattice, and the nuclear momenta P_j. In this way we introduce the normal coordinates

$$\xi_k = \frac{1}{\sqrt{N}} \sum_j \delta R_j \, e^{-ik\overline{R_j}}, \qquad \delta R_j = \frac{1}{\sqrt{N}} \sum_k \xi_k \, e^{ik\overline{R_j}}, \qquad (11.111a)$$

and their conjugate momenta

$$\pi_k = \frac{1}{\sqrt{N}} \sum_j P_j \, e^{-ik\overline{R_j}}, \qquad P_j = \frac{1}{\sqrt{N}} \sum_k \pi_k \, e^{ik\overline{R_j}}, \qquad (11.111b)$$

where the indices k take on the N values

$$k = m\frac{2\pi}{L}, \qquad -\frac{\pi}{a} < k \leq \frac{\pi}{a}. \qquad (11.112)$$

In terms of these new variables the Hamiltonian (11.110) has the form

$$\widehat{H} = \sum_k \left(\frac{\widehat{\pi}_k^\dagger \widehat{\pi}_k}{2M} + C\big(1 - \cos ka\big)\widehat{\xi}_k^\dagger \widehat{\xi}_k \right). \qquad (11.113)$$

Although the operators $\widehat{\xi}$ and $\widehat{\pi}$ are not independent and non-Hermitean, since we have $\widehat{\xi}_k^\dagger = \widehat{\xi}_{-k}$, $\widehat{\pi}_k^\dagger = \widehat{\pi}_{-k}$, they satisfy the same commutation relations as pairs of conjugated variables, namely $[\widehat{\xi}_k, \widehat{\pi}_{k'}^\dagger] = i\hbar\delta_{kk'}$, a property which one can check by using their definitions (11.111). The Hamiltonian (11.112) thus looks, for each value of $k \neq 0$, like that of a harmonic oscillator of frequency

$$\omega_k = \sqrt{\frac{2C}{M}\big(1 - \cos ka\big)} = 2\sqrt{\frac{C}{M}} \left| \sin \frac{1}{2}ka \right|. \qquad (11.114)$$

To achieve its diagonalisation, we introduce, as for the ordinary harmonic oscillator, the operators

$$\widehat{c}_k = \frac{1}{\sqrt{2\hbar}} \left(\widehat{\xi}_k \sqrt{M\omega_k} + \frac{i\widehat{\pi}_k}{\sqrt{M\omega_k}} \right) \qquad (11.115)$$

and their conjugates, which satisfy the commutation relations

$$\big[\widehat{c}_k, \widehat{c}_{k'}\big] = 0, \qquad \big[\widehat{c}_k, \widehat{c}_{k'}^\dagger\big] = \delta_{kk'}. \qquad (11.116)$$

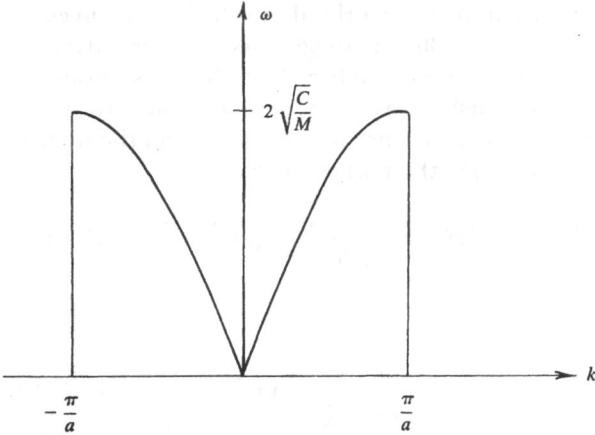

Fig. 11.19. The frequency spectrum of a one-dimensional lattice

We thus get for the Hamiltonian (11.113)

$$\widehat{H}_n = \sum \hbar\omega_k\left(\widehat{c}_k^\dagger \widehat{c}_k + \tfrac{1}{2}\right) = \sum \hbar\omega_k\left(\widehat{n}_k + \tfrac{1}{2}\right), \qquad (11.117)$$

and its eigenvalues follow from those of $\widehat{n}_k = \widehat{c}_k^\dagger \widehat{c}_k$, which are $n_k = 0, 1, \dots$. We show in Fig.11.19 the frequency spectrum ω_k given by (11.114); we bear in mind that the possible values of k are given by (11.112).

The term with $k = 0$ in (11.113) does not contain the operator $\widehat{\xi}_0$, which is associated with the global translations $\delta R_j = \xi_0/\sqrt{N}$. It does not describe a vibrational mode, but represents simply the global kinetic energy of the crystal. It needs only weak pinning forces to prevent the crystal from moving as a whole; if there were no such forces, the mean square thermal equilibrium velocity would be of order $\sqrt{kT/m}$, or 2×10^{-9} m s^{-1} for a 1 g crystal at room temperature.

In three dimensions, for example in the case of a cubic lattice with cell size a and one atom per cell, each *mode* occurs as a *vibrational wave* with the displacement δR_n of a nucleus situated at the site $\overline{R_n}$ being, as in (11.111a), proportional to the real or the imaginary part of

$$e^{i k \cdot \overline{R_n}}.$$

A mode q is characterized by the *wavenumbers*

$$k_x = m_x \frac{2\pi}{L}, \qquad k_y = m_y \frac{2\pi}{L}, \qquad k_z = m_z \frac{2\pi}{L}, \qquad (11.118)$$

for a crystal of size L, where the (integer) values of the m are *bounded* by

$$-\frac{L}{2a} < m \le \frac{L}{2a}. \qquad (11.119)$$

Moreover, for each direction of propagation there are 3 vibrational modes, associated with the displacements of atoms in particular directions, depending on the direction of the wavevector, and being mutually orthogonal. We thus recover the $3N$ modes which we expected from the counting of degrees of freedom. If the medium were isotropic – were an amorphous solid – one of the modes would be *longitudinal* with the atoms vibrating parallel to k and the other two modes would be *transverse* for symmetry reasons. The same property holds for a cubic cystal, in the special case when k is parallel either to the edges or to the diagonals of the cell; however, in general, the direction of the vibrations of the atoms in an eigenmode is not oriented in a simple way relative to the wavevector, due to the anisotropy of the crystal.

The mode frequencies are functions of the wavenumber. As the one-dimensional model we have just studied suggests, and as more detailed theories show, the frequencies ω_k of the three modes tend *linearly* to zero when k decreases. In the isotropic approximation ω/k is independent of the direction of k so that for the longitudinal and the two transverse modes we have

$$\omega \sim u_l k, \qquad \omega \sim u_t k. \tag{11.120}$$

Small values of k correspond to macroscopic wavelengths $\lambda = 2\pi/k$, and the lattice vibrations thus propagate the *acoustic waves* in the solid. The velocity $\nabla_k \omega$ of the displacement of a wavepacket is just the *sound velocity*. It is independent of the frequency when $ka \ll 1$, but it may take on three different values, depending on the direction of the oscillations of the atoms, and it may also be not parallel to k. A typical value is 5000 m s^{-1} in metals, with the longitudinal velocity u_l being larger than the transverse velocity u_t, as compared to the sound velocity of 300 m s^{-1} in air. The mechanisms for sound propagation in solids and in fluids are quite different. In the first case, we are dealing with coherent displacements of the atoms in the lattice over wavelengths which are large compared to the crystal cell size, a microscopic mechanical effect governed by the effective forces $W(\{R_n\})$; in the second case, we are dealing with the propagation of a pressure oscillation, a macroscopic thermodynamic effect governed by the hydrodynamic equations of § 14.4.6.

For a solid with an arbitrary crystal structure the wavevectors k of the eigenmodes are restricted to the Brillouin zone which generalizes (11.119), and they continue to take on N values, where N is the number of cells in the lattice. If there are s atoms per cell, the $3Ns$ vibrational modes split into $3s$ branches, which generalize the 3 branches of a simple cubic lattice. Nevertheless, in the limit as $k \to 0$, these $3s$ branches can be classified as 3 so-called "acoustic" branches for which ω decreases linearly with k, and $3s - 3$ so-called "optical" branches for which ω remains finite. The latter correspond to oscillations for which different kinds of atoms in the same cell oscillate with opposite phases. On the contrary, for the acoustic branches all atoms oscillate in phase; the lattice is thus locally very little distorted, and this is the reason why for those modes the energy $\hbar\omega_k$ tends to 0 with k.

The acoustic modes show up directly as mechanical vibrations of solids with long wavelengths and low frequencies. The optical modes, which hardly propagate, as their propagation velocity $d\omega/dk$ tends to 0 with k, play a rôle in the interaction between the crystal and electromagnetic radiation, because of the order of magnitude of their energy $\hbar\omega_k$. For instance, in NaCl, with 2 atoms per cell, apart from the three acoustic modes, there are three optical modes, two transverse and one longitudinal, with frequencies which are, respectively, equal to $3.09 \times 10^{13}\,\mathrm{s}^{-1}$ and $4.87 \times 10^{13}\,\mathrm{s}^{-1}$ in the $k = 0$ limit. Infrared waves which have frequencies of that order of magnitude interact strongly with these *optical modes* and this explains their name. A photon can be transformed into an optical phonon, that is, it can be absorbed, while exciting lattice vibrations; the conservation of energy $\hbar\omega$ and of momentum $\hbar k = \hbar\omega/c$ can here be satisfied, thanks to the large value of the light velocity c.

The calculation of the eigenfrequencies, starting from the effective interactions W between the atoms, can be done as in one dimension, using the transformation (11.111) which will for each k produce $3s$ pairs of ξ, π variables. One must thus eventually diagonalize a $3s \times 3s$ matrix (Exerc.11e).

The masses of the nuclei are sufficiently large that their average displacements in a solid remain practically always small compared to the cell size (Prob.10). Even at the melting temperature the mean square displacement rarely exceeds $\frac{1}{8}$ of the cell size (*Lindemann's criterion*). This justifies an approximation which we have made implicitly and which consists in neglecting the Pauli principle for the atomic nuclei. Their fermion or boson nature plays, in general, no rôle whatever in their vibrations – although those are quantized – as the wavefunctions remain localized round each lattice site. Under those conditions, the symmetrization or antisymmetrization of the wavefunction of the N indistinguishable nuclei will not change anything. The indistinguishability only plays a rôle in solid helium where the nuclei, bosons for ^{4}He and fermions for ^{3}He, are sufficiently light and sufficently weakly bound that the states of neighbouring nuclei overlap.

11.4.2 Interpretation of a Mode as a Boson State

Expression (11.109) for the eigenenergies of the Hamiltonian of the lattice vibrations shows a large similarity with the expression $\sum_q \varepsilon_q n_q$ of the *energy levels of a gas with an arbitrary number of bosons* (§ 10.5.2). It is therefore natural to talk about the system of *quantized lattice vibration modes* as being a gas of particles, the *"phonons"*, which satisfy Bose-Einstein statistics. We shall see in § 11.4.4 and in § 13.1, where we shall study the quantized oscillation modes of the electromagnetic field, that this is more than just a simple analogy, and that the photons and phonons show all the characteristics of elementary particles. We shall restrict ourselves here to pointing out the correspondence between the two languages describing the same reality, that of the quantized vibrations and that of the phonons.

Each *oscillation mode* q corresponds to a *single-phonon state*, that is, a *plane wave* as for a particle in a box (§ 10.2.1). The *wavevector* k corresponds to the *quasi-momentum* $p = \hbar k$ given by the de Broglie relation and, according to (11.118) and (11.119), taking on the same values (11.30) and (11.32) as the quasi-momentum of an electron in the same crystal. Moreover, the vibrational modes are characterized by an index which can take on three values for crystals with one atom per cell, and which describes the (in the simplest cases longitudinal or transverse) *polarization* of the vibrational wave; this index plays the same rôle as an *internal quantum number* – like the spin – of the phonon. The energy (11.109) of a *vibrational micro-state* corresponds, apart from an additive constant, to that of a *system of bosons*, (10.16), provided one associates the *frequency* ω_q with the *phonon energy* $\varepsilon_q = \hbar\omega_q$ and interprets the quantum number n_q of each harmonic oscillator as the *occupation number* of the state q. A vibrational micro-state corresponds to a micro-state (10.14) of Fock space, and a change in the vibrational state is described as the *creation or annihilation* of phonons. In the harmonic oscillation approximation considered here, the phonons *do not interact*, as is shown by Eq.(11.109) for the energy. Their total *number* is not a *constant of the motion*. As a result (§ 10.5.2), their *chemical potential is zero*.

The phonons, like the conduction electrons and the holes in an insulator, are a characteristic example of *quasi-particles*. Instead of describing the crystal lattice and its motion as a system of nuclei, interacting strongly with one another, we have been led to introduce the phonons, fictitious bosons with practically *no interactions*, which enables us to describe the same physical situation, but much more simply. Even though the quasi-particles have the same properties as real particles, they are only distantly related to the true particles which make up the crystal, the electrons and the nuclei. They represent, in fact, *collective aspects*. For instance, in an insulator, a hole or a conduction electron describes how the whole of the electron cloud is changed when one takes away or adds an electron. The collective nature of phonons is even more pronounced, as the creation of a phonon amounts to changing the vibrational state of the *system* of nuclei; the quasi-momentum $\hbar k$ of the phonon is related to the wavevector k of this vibrational state, but has nothing to do with the momenta of the nuclei.

The correspondence between the quantized vibrations of the lattice and bosons is completed by identifying the operators (11.115), which through (11.117) diagonalize the harmonic oscillators associated with each mode, with the annihilation and creation operators (10.20) of the phonons. The algebra (10.21) and (11.116) of these operators is, indeed, exactly the same. According to the general properties of § 10.2.3, each observable can be expressed as a function of the operators $\widehat{c}$ and $\widehat{c}^\dagger$. In particular, the *displacement of each nucleus* occurs as a *linear combination of phonon annihilation and creation operators*. For example, for the one-dimensional crystal model studied in § 11.4.1, we find from (11.111) and (11.115) that

$$\delta\widehat{R}_j = \sqrt{\frac{\hbar}{2N}} \sum_k e^{ik\overline{R}_j} \frac{\widehat{c}_k + \widehat{c}^\dagger_{-k}}{\sqrt{M\omega_k}}. \tag{11.121a}$$

Similarly, the momentum P_j of a nucleus is given by

$$\widehat{P}_j = \sqrt{\frac{\hbar}{2N}} \sum_k e^{ik\overline{R}_j} \left(\widehat{c}_k - \widehat{c}^\dagger_{-k}\right) \frac{\sqrt{M\omega_k}}{i}, \tag{11.121b}$$

which is another combination of the annihilation and creation operators. Equations (11.121) enable us to translate all physical quantities from one representation to another, for example, to find from the average number of phonons $\langle \widehat{c}_k^\dagger \widehat{c}_k \rangle = f_k$ in each mode the statistical fluctuations in the position of each nucleus in thermal equilibrium.

The existence of *anharmonicity* is in the phonon language translated into the addition of extra terms to the Hamiltonian (11.117). For instance, a term with δR^4 or with δR^3 in the potential W produces, if we use (11.121), terms like $\widehat{c}_{k_1}^\dagger \widehat{c}_{k_2}^\dagger \widehat{c}_{k_3} \widehat{c}_{k_4}$ describing the scattering of two phonons with quasi-momenta k_3 and k_4 into the modes k_1 and k_2 (with conservation of total quasi-momentum), like $\widehat{c}_{k_1} \widehat{c}_{k_2} \widehat{c}_{k_3}$ describing the annihilation of 3 phonons ($k_1 + k_2 + k_3 = 0$), or like $\widehat{c}_{k_1}^\dagger \widehat{c}_{k_2}^\dagger \widehat{c}_{k_3}$ describing the transformation of one phonon into two others. Even though they are small, those terms contribute to the establishment of equilibrium in the phonon gas, where the number of phonons is not conserved ($\mu = 0$). When they are significant, we treat them using perturbative expansions similar to those used in particle physics; the phonon language is then eminently suitable.

We have similarly treated the electrons in § 11.2 in the approximation where the nuclei were fixed at their average position. A displacement $\delta \boldsymbol{R}_n$ of the nuclei adds to (11.8) a perturbation

$$- \frac{e^2}{4\pi\varepsilon_0} \sum_{i,n} \delta \widehat{\boldsymbol{R}}_n \cdot \left(\widehat{\boldsymbol{r}}_i - \boldsymbol{R}_n \right) \frac{Z_n}{\left| \widehat{\boldsymbol{r}}_i - \boldsymbol{R}_n \right|^3} ,$$

which can be interpreted, if we take (11.121) into account, as an *electron-phonon interaction* describing the scattering of an electron, with the creation or annihilation of a phonon. This interaction is responsible for numerous physical phenomena, such as the Joule effect – the transfer of electron energy to the lattice, thus heating it. It also enables two electrons to exchange a phonon; this gives rise to an effective attraction between them. This attraction is enhanced by the sudden jump in $f(\varepsilon)$ at the Fermi surface, to such an extent that it is possible for two electrons with energy ε_F and with opposite momenta and spins to form a bound pair in spite of their Coulomb repulsion. The resulting pairs, the so-called Cooper pairs, resemble bosons and can condense; this is the mechanism for superconductivity in metals at low temperatures (§ 12.3.3).

The phonons are a prototype of "*Goldstone bosons*", which are quasi-particles, or particles, associated with a continuous invariance property of the system. Here we are concerned with the *translational invariance* which the Hamiltonian satisfies, but which is *spontaneously broken* (§ 9.3.3) at thermal equilibrium where the atoms occupy well-defined equilibrium positions. An arbitrary translation transforms this equilibrium state into *another* equivalent equilibrium state, with the same energy. Let us assume that we excite a vibrational mode, in the long wavelength limit. Locally this (non-equilibrium) state is obtained by a translation with an amplitude which varies slowly in space. We understand thus why the energy of this mode tends to 0 as k decreases, since for $k = 0$ we would have just a global translation. This property holds generally (Goldstone's theorem) for any long wavelength excitation of a system, occurring because a continuous invariance is broken. The linear behaviour of ω_k for small k is a specific property of the phonons, but the existence of 3 acoustic branches, for which the energy vanishes with k, is itself a mere consequence of the breaking of the translational invariance in the crystal.

Another example of a Goldstone boson is provided by the elementary excitations, called *"magnons"*, of a *ferromagnetic* solid (Exerc.9a). In the Heisenberg model where the spins $\boldsymbol{\sigma}_i$ interact through an effective potential $-J(\widehat{\boldsymbol{\sigma}}_i \cdot \widehat{\boldsymbol{\sigma}}_j)$, the Hamiltonian is invariant under a *rotation* of the spins. This continuous invariance is spontaneously broken in the ground state and, more generally, in any equilibrium state at a temperature below the Curie temperature (Exerc.9a), as the spins are in that case all oriented along a privileged direction. A global rotation of the spins does not change the energy. A magnon, or spin wave, describes an oscillation of the spins around their equilibrium orientation which propagates from one spin to another. It is the Goldstone mode associated with the spin rotations, and its energy vanishes with k. The same concept exists in *particle physics* where the ground (or equilibrium) state is replaced by the *vacuum* and where an elementary excitation with momentum p and energy $\varepsilon = \sqrt{m^2 c^4 + p^2 c^2}$ represents a particle with a rest mass m. The Goldstone bosons have an energy which vanishes with k, so that their *mass is zero*. An example is the *photon*; one can show that this is the Goldstone boson associated with *gauge invariance*, which is broken because one must choose a particular gauge to write down the potentials occurring in the Hamiltonian of charged particles.

11.4.3 Specific Heats of Solids

The statistical mechanics of quantized lattice vibrations can be studied in either one of the two equivalent descriptions: the canonical partition function (11.117) for the independent oscillation modes, calculated by using its factorization (§ 4.2.5), is the same as the grand canonical partition function for non-interacting phonons with $\mu = 0$, written down in §§ 10.3.1 and 10.5.2. Thus, the thermodynamic functions associated with the vibrational modes of the lattice, that is, with the phonon gas in the crystal, are given by the formulæ of § 10.3 with $\mu = 0$. In particular, apart from an additive constant, the *internal energy* (10.42) of the phonons, which is the average of (11.109), equals

$$U_{\text{ph}} = \sum_q \varepsilon_q f_q = \int d\varepsilon \, \mathcal{D}(\varepsilon) \frac{\varepsilon}{e^{\beta \varepsilon} - 1}, \tag{11.122}$$

where $\mathcal{D}(\varepsilon)$ is the density of the modes,

$$\mathcal{D}(\varepsilon) = \sum_q \delta(\hbar \omega_q - \varepsilon). \tag{11.123}$$

At low temperatures the important modes in the integral in (11.122) are those with *low energies*. They have a linear spectrum so that in three dimensions the mode density (11.123) behaves as ε^2 for small ε, since $\sum_q$ introduces an integral $\int d^3 k$. As a result, if we take $\beta \varepsilon$ as the variable in (11.122), we find an internal energy which is proportional to T^4 as $T \to 0$, which means a specific heat proportional to T^3. We shall determine its coefficient in what follows, in (11.129).

At high temperatures (11.122) reduces to

$$U_{\mathrm{ph}} \sim kT \int d\varepsilon\, \mathcal{D}(\varepsilon) = 3NkT, \tag{11.124}$$

where $3N$ is the total number of modes. More exactly, in a solid with several atoms per cell and N cells, the $3N$ acoustic modes contribute to (11.124) while the optical modes do not contribute, provided kT lies between the maximum energy of the acoustic modes and the minimum energy of the optical modes. The specific heat is therefore nearly constant and equal to $3Nk$.

The *Debye model* bridges these results. It is based upon the use of the isotropic approximation (11.120) for the phonon spectrum, which in the large volume limit (§ 10.3.3) and for small ε gives

$$\begin{aligned}
\mathcal{D}(\varepsilon) &= \sum_{p} \left[\delta(u_1 p - \varepsilon) + 2\delta(u_t p - \varepsilon) \right] \\
&= \frac{\Omega}{h^3} \int d^3 p \left[\delta(u_1 p - \varepsilon) + 2\delta(u_t p - \varepsilon) \right] \\
&= \frac{\Omega}{2\pi^2 \hbar^3} \left(\frac{1}{u_1^3} + \frac{2}{u_t^3} \right) \varepsilon^2.
\end{aligned} \tag{11.125}$$

Nevertheless, $\mathcal{D}(\varepsilon)$ must vanish when ε becomes large as the total number of modes, $3N$, is finite; it must satisfy the *normalization* condition (11.124). The Debye approximation, which enables us to avoid the detailed determination of $\mathcal{D}(\varepsilon)$, consists in extrapolating (11.125) up to a certain maximum value after which one takes $\mathcal{D}$ to vanish. This value $k\Theta_{\mathrm{D}}$ is determined by the normalization condition, which determines the only parameter of the theory, the *Debye temperature* of the crystal,

$$\Theta_{\mathrm{D}} = \frac{\hbar}{k} \left[\frac{18\pi^2 N}{\Omega \left(u_1^{-3} + 2u_t^{-3} \right)} \right]^{1/3}. \tag{11.126}$$

One can determine it from measurements of the longitudinal and transverse sound velocities. Typical values are 90 K for Pb, 400 K for Al, and 3000 K for diamond. One can therefore rewrite the mode density (11.125) in the approximate form

$$\mathcal{D}(\varepsilon) = \frac{9N\varepsilon^2}{(k\Theta_{\mathrm{D}})^3} \, \theta(k\Theta_{\mathrm{D}} - \varepsilon). \tag{11.125'}$$

The resulting internal energy (11.122) equals

$$U_{\mathrm{ph}} = \frac{9NkT^4}{\Theta_{\mathrm{D}}^3} \int_0^{\Theta_{\mathrm{D}}/T} \frac{x^3\, dx}{e^x - 1}, \tag{11.127}$$

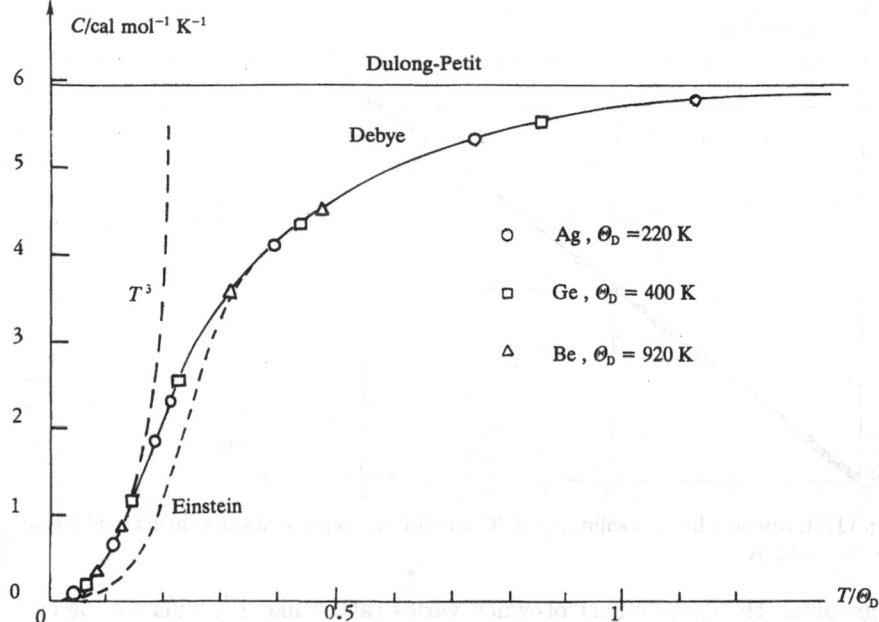

Fig. 11.20. The specific heat of some solids

where we have taken $x \equiv \beta\varepsilon$ as variable. Hence we get for the *phonon contribution to the specific heat of the crystal*

$$C_{\mathrm{ph}} = \frac{dU_{\mathrm{ph}}}{dT} = \frac{9NkT^3}{\Theta_{\mathrm{D}}^3} \int_0^{\Theta_{\mathrm{D}}/T} \frac{x^4 e^x \, dx}{(e^x - 1)^2}, \qquad (11.128)$$

which we show in Fig.11.20. The dashed curve corresponds to the Einstein model (Exerc.10d).

The experimental results are also shown in Fig.11.20 for a few substances which have a sufficiently simple crystal structure for the phonon spectrum to be well represented by the Debye approximation. The agreement with the theoretical curve is remarkable: the Debye temperatures which produce the best agreement between theory and experiment are, up to a few %, equal to those which one evaluates from (11.126), using the sound velocities.

We find, as predicted, the T^3 *dependence at low temperatures* if in (11.128) we replace the upper integration limit by ∞, which for $T \ll \Theta_{\mathrm{D}}$ gives (see formulae section at the end of the book)

$$C_{\mathrm{ph}} \simeq \frac{9NkT^3}{\Theta_{\mathrm{D}}^3} \int_0^\infty \frac{x^4 e^x \, dx}{(e^x - 1)^2} = \frac{12\pi^4}{5} Nk \frac{T^3}{\Theta_{\mathrm{D}}^3}. \qquad (11.129)$$

Figures 11.20 and 11.21 illustrate that experiments check the T^3 law very satisfactorily for most solids, and even with a remarkable accuracy for *insulators* at low temperatures (solid argon for $T < 2$ K). A notable exception

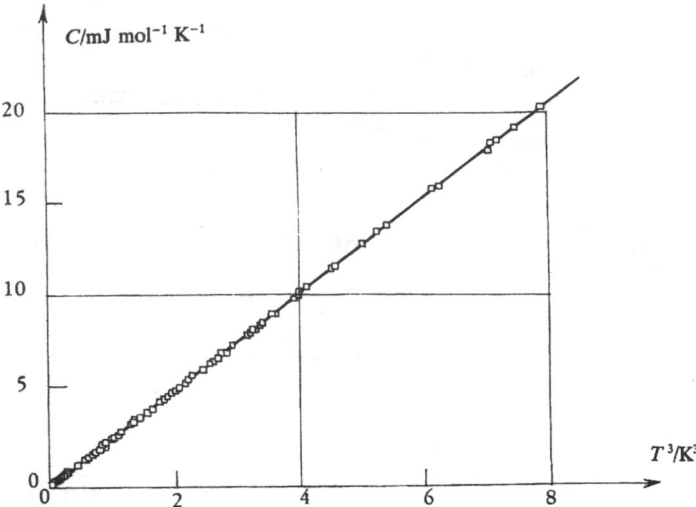

Fig. 11.21. Specific heat of solid argon. The solid line represents the theoretical curve for $\Theta_D = 92$ K

is graphite, the specific heat of which varies rather like T^2. This can be explained by the practically two-dimensional structure of this substance, which changes the mode density (11.125) and produces a behaviour $\mathcal{D}(\varepsilon) \propto \varepsilon$ instead of $\propto \varepsilon^2$ in a sufficiently large region. Nevertheless for *metals* at very low temperatures the specific heat is dominated by the *electron contribution* (11.52) which is *linear in T*, and which tends to zero less rapidly than the contribution (11.129) from the lattice vibrations. One can distinguish these two contributions easily by plotting C/T as function of T^2, as the experimental results for potassium below 0.5 K show (Fig. 11.22).

When the temperature increases, the crystal vibrations become more and more unfrozen and C_{ph} increases. When $T \gg \Theta_D$, we expect to find the result of *classical* statistical mechanics, predicted by the equipartition theorem (§ 8.4.2): the number of degrees of freedom is $6N$ so that C_{ph} should tend to $3Nk$. One can check this result by expanding the integrand in (11.128) in the vicinity of $x = 0$, which yields

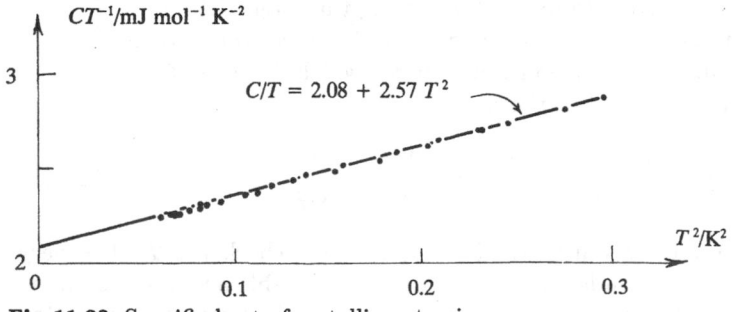

Fig. 11.22. Specific heat of metallic potassium

$$C_{\text{ph}} \sim 3Nk \left(1 - \frac{\Theta_D^2}{20T^2} \cdots \right). \tag{11.130}$$

The value $3Nk \simeq 25 \, \text{J mole}^{-1} \text{K}^{-1}$ for the specific heat of solids had been observed experimentally for number of them at the beginning of the nineteenth century (*Dulong and Petit law*). We see that this concerns solids with a Debye temperature which is rather low, below room temperature. We also note that the electron contribution is negligible compared to $3Nk$: for potassium, an extrapolation to $T = 300 \, \text{K}$ of the linear law only gives $C_{\text{el}} \simeq 0.6 \, \text{J mole}^{-1} K^{-1}$.

The Debye approximation thus enables us to connect the *specific heats* of solids with a simple structure to their *sound velocities*. For crystals with a less simple structure, or in order to obtain a greater accuracy, we must use (11.122) and (11.123) which connect the *thermodynamic* properties with the properties of the *vibrations*: another example of the unifying power of statistical physics. For many substances one has checked the agreement between the measured specific heat and its value calculated starting from the $\omega(\boldsymbol{k})$ spectrum, which itself is determined in experiments, for instance, on inelastic scattering of neutrons or photons by phonons.

11.4.4 Thermal Equilibrium of a Vibrating String

As an introduction to quantum field theory and as an exercise we shall study a model describing the vibrations of a *continuous* medium, rather than of a discrete system of atoms as in § 11.4.1. This will help us to deepen the equivalence which we have established between quantized oscillators and quasiparticles such as phonons (§ 11.4.2); the discrete nature of a substance should not play any role when $ka \ll 1$. This will help us also to introduce in § 13.1 the photon concept. In fact, the dynamical variables describing the deformations of a continuous medium constitute a *field* and the quantization of the electromagnetic field will follow the same stages as the present quantization of the deformation field, the eigenmodes of which are mechanical oscillations.

To simplify matters we shall restrict ourselves to a one-dimensional system, that is, to an elastic string. We assume that this string, which is fixed at its ends, $x = 0$ and $x = L$, can only be deformed in one, transverse, direction. Moreover, we limit ourselves to small displacements and we assume that the tension τ and the linear mass density ϱ are given. In *classical mechanics* the displacement $\varphi(x,t)$ of the string at the point x, which is the solution of the equations of motion (11.134) and which depends on the initial conditions $\varphi(x,0)$ and $\partial\varphi(x,0)/\partial t$, must be considered to be a *classical field*, that is, a continuum of dynamic variables φ each of which is associated with a point x of the string. The oscillation *eigenmodes* are particular solutions, forming a base. They are characterized by their wavenumber

$$k = \frac{m\pi}{L}, \qquad m = 1, 2, \ldots, \tag{11.131}$$

which takes on discrete positive values, and by their angular frequency ω_k (or frequency ν_k),

$$\omega_k = 2\pi\nu_k = \sqrt{\frac{\tau k^2}{\varrho}} = ck, \tag{11.132}$$

where $c = \sqrt{\tau/\varrho}$ is the velocity of propagation of sound along the string. The general solution $\varphi(x,t)$ can be expanded as a superposition of eigenmodes:

$$\varphi(x,t) = \sqrt{\frac{2}{L}} \sum_k \xi_k \sin kx. \tag{11.133}$$

The amplitude ξ_k of each mode in $\varphi(x)$, which satisfies the equation $\ddot{\xi}_k + \omega_k^2 \xi_k = 0$, is a sinusoidal function of the time with frequency $ck/2\pi$. *Each mode thus appears as a classical harmonic oscillator* ξ_k *and the string itself as a set of an infinite number of harmonic oscillators* characterized by the index k. In the linear chain of atoms of § 11.4.1 the number of degrees of freedom was finite, which, according to (11.112), restricted the number of modes to N. The decomposition (11.133) is the continuous version of (11.111), with stationary rather than travelling waves; here x replaces $\overline{R_j}$ and φ replaces δR_j. The eigenfrequency spectrum (11.132) replaces (11.114); here it remains linear for all wavenumbers, and is not bounded.

One should take care not to confuse the equation of motion of the string,

$$\frac{\partial^2 \varphi}{\partial x^2} - \frac{1}{c^2} \frac{\partial^2 \varphi}{\partial t^2} = 0, \tag{11.134}$$

with boundary conditions $\varphi(0,t) = \varphi(L,t) = 0$, which is a *classical* wave equation, with a wave equation from quantum mechanics. It is true that there exists a certain *formal* analogy between $\varphi(x,t)$ and a wavefunction $\psi(x,t)$ which would describe the motion of a quantum particle enclosed in a one-dimensional box of length L. However, it is essential not to be confused either by this analogy or by the notation. Whereas in the case of a classical or quantum particle moving along an axis, the dynamic variable is x, together with its conjugate momentum p, with the wavefunction ψ being a means of calculating the expectation values of observables, there are here an *infinite number of dynamic variables*, the values of φ at each point x; the coordinate x is a continuous *index* which enables us to characterize each of the dynamic variables. The transformation (11.133) is just a change of variables from the dynamic variables $\varphi(x)$ to the equivalent dynamic variables ξ_k; these are characterized by the wavenumber k, a discrete index which takes on an infinite number of values (11.131).

The equation of motion (11.134), which expresses the acceleration $\partial^2 \varphi/\partial t^2$ of a point of the string in terms of the restoring force, formulates the dynamics in *Newtonian* form. In order to quantize the system, we need to know the conjugate momenta of the dynamic variables $\varphi(x)$ or ξ_k, that is, to know the *Hamiltonian* formulation of the classical problem (§ 2.3.3). To do this, we pass through the intermediate stage of the *Lagrangian* formulation in terms of the variables ξ_k and $\dot{\xi}_k$. The equations of motion $\ddot{\xi}_k + \omega_k^2 \xi_k = 0$ must be identified as the Lagrangian equations (2.61) so that we can take for the Lagrangian which produces these equations:

$$L\{\xi_k, \dot{\xi}_k\} = \frac{\varrho}{2} \sum_k \left(\dot{\xi}_k^2 - \omega_k^2 \xi_k^2\right).$$ (11.135)

As a Lagrangian remains the same under a change of variables, we can use (11.133) to rewrite (11.135) in terms of the original variables:

$$L\{\varphi(x), \dot{\varphi}(x)\} = \frac{1}{2} \int_0^L dx \left\{ \varrho \left[\dot{\varphi}(x)\right]^2 - \tau \left(\frac{\partial \varphi}{\partial x}\right)^2 \right\}.$$ (11.136)

We find the difference between the kinetic and the potential energy of the string as we might have expected. The equations of motion (11.134) are found as the Lagrangian equations following from (11.136), provided we treat the derivatives $\partial L/\partial \varphi$ and $\partial L/\partial \dot{\varphi}$ as functional derivatives. By introducing the conjugate momenta $\pi_k = \partial L/\partial \dot{\xi}_k$ of the variables ξ_k, the general relation (2.63), which defines a classical Hamiltonian, here gives, if we use (11.135),

$$H\{\xi_k, \pi_k\} = \sum_k \left(\frac{1}{2\varrho} \pi_k^2 + \frac{1}{2} \varrho \omega_k^2 \xi_k^2 \right).$$ (11.137)

To treat this problem *quantum mechanically* we must now replace the classical, conjugate variables ξ_k, π_k, the time-development of which is governed by the Hamiltonian equations, by *operators* $\widehat{\xi}_k, \widehat{\pi}_k$ which satisfy the commutation relations

$$\left[\widehat{\xi}_k, \widehat{\pi}_{k'}\right] = i\hbar \delta_{kk'}, \qquad \left[\widehat{\xi}_k, \widehat{\xi}_{k'}\right] = \left[\widehat{\pi}_k, \widehat{\pi}_{k'}\right] = 0.$$ (11.138)

The Hamiltonian (11.137) of the string thus takes the form of a sum of Hamiltonians of one-dimensional *quantum mechanical harmonic oscillators* for each mode. The variables $\widehat{\xi}_k$ play the rôle of "positions", and their conjugates $\widehat{\pi}_k$ that of momenta, ϱ that of mass, and $-\tau k^2 \widehat{\xi}_k$ that of the restoring force. One should note again that the quantum mechanical dynamic variables are $\widehat{\xi}_k$ and $\widehat{\pi}_k$ *and not* the abscissa x and the wavenumber k, or the variable $p = \hbar k$ which will be interpreted as the phonon momentum; those in quantum field theory remain indices and commute. The *field* $\widehat{\varphi}(x)$ itself, which is related to $\widehat{\xi}_k$ through (11.133), forms a *continuous set of operators* characterized by x. In the Heisenberg picture (§ 2.1.5) it depends on the time as follows:

$$\widehat{\varphi}(x, t) = e^{i\widehat{H}t/\hbar} \, \widehat{\varphi}(x, 0) \, e^{-i\widehat{H}t/\hbar},$$

but in the Schrödinger picture which we are using here, the state and not the observables $\widehat{\varphi}(x)$ depends on the time.

One can diagonalize $\widehat{H}$ by introducing, as in (11.115), the phonon operators

$$\widehat{c}_k = \frac{1}{\sqrt{2\hbar}} \left(\widehat{\xi}_k \sqrt{\varrho \omega_k} + \frac{i \widehat{\pi}_k}{\sqrt{\varrho \omega_k}} \right),$$ (11.139)

and their Hermitean conjugates. The Hamiltonian becomes

$$\widehat{H} = \sum_k \hbar \omega_k \left(\widehat{c}_k^\dagger \widehat{c}_k + \frac{1}{2} \right),$$ (11.140)

and the commutation relations (1.138) are equivalent to the commutation relations (10.21), that is,

$$\left[\widehat{c}_k, \widehat{c}_{k'}^\dagger\right] = \delta_{kk'}, \qquad \left[\widehat{c}_k, \widehat{c}_{k'}\right] = 0,$$

which define boson *annihilation and creation operators*. The operator $\widehat{n}_k = \widehat{c}_k^\dagger \widehat{c}_k$ is, as we saw from (10.22), the *occupation number observable* of the single-boson state k, and its eigenvalues $n_k = 0, 1, \ldots$ give us the eigenenergies,

$$\sum_k \hbar \omega_k \left(n_k + \frac{1}{2}\right) = \sum_k n_k \hbar \omega_k + E_0, \qquad (11.141)$$

of the Hamiltonian (11.140). The latter can thus be identified with the Hamiltonian of a *gas of bosons* in Fock space, with E_0 as the vacuum energy.

A complete isomorphism appears in this way between the Hilbert space of the vibrations of a quantum string and the Fock space of a system of bosons enclosed in a one-dimensional box of length L. The classical vibrational modes k have become the single-phonon states. Using (11.133) and (11.139) we can write the *field operator* at a point x as

$$\widehat{\varphi}(x) = \sqrt{\frac{\hbar}{L\varrho}} \sum_k \frac{1}{\sqrt{\omega_k}} \sin kx \left(\widehat{c}_k + \widehat{c}_k^\dagger\right). \qquad (11.142)$$

The form $\sin kx$ of the classical wave associated with the mode k appears here as the amplitude, related to this mode, of the operator describing the deformation of the string at the point x. It is proportional to the matrix elements of this operator between the micro-states with n_k and $n_k + 1$ phonons in the state k. Hence we can interpret $p = \hbar k$ as the momentum of a phonon in that state. (More precisely, it is the absolute magnitude of that momentum; to find its sign we should introduce periodic boundary conditions which would give classical oscillations with $e^{ikx} = e^{ipx/\hbar}$.) The whole story of § 11.4.2 of quantum oscillators in terms of bosons follows. For instance, (11.140) shows that the energy of a phonon in the state k is $\hbar \omega_k = \hbar c k = cp$. Equation (11.142) implies that deforming the string is quantum mechanically equivalent to creating or annihilating phonons. In particular, this is what happens when the string interacts with a thermostat, which is coupled to $\widehat{\varphi}(x)$.

One obtains the *thermodynamics* of the string in equilibrium at a temperature T by writing down the free energy associated with the spectrum (11.141), or, what amounts to the same thing, the grand potential for the non-conserved phonons with $\mu = 0$. We thus find the free energy,

$$F(T) = kT \sum_{m=1}^{\infty} \ln\left(1 - e^{-m\pi \hbar c/LkT}\right) + E_0. \qquad (11.143)$$

One can simplify this function in the limit when the string is sufficiently long so that $LT/c \gg \pi \hbar / k \simeq 2.4 \times 10^{-11}$ K s. The sum over m then becomes an integral and gives (see formulae at the end of this volume)

$$F - E_0 \sim \frac{Lk^2T^2}{\pi\hbar c} \int_0^\infty dx \, \ln\left(1 - e^{-x}\right)$$

$$= -\frac{Lk^2T^2}{\pi\hbar c} \int_0^\infty dx \, \frac{x}{e^x - 1}$$

$$= -\frac{\pi Lk^2T^2}{6\hbar c}. \tag{11.144}$$

From F we derive the entropy $S = -\partial F/\partial T$, the internal energy $U = F+TS \sim -F$, and the specific heat

$$C = \frac{\partial U}{\partial T} \sim S \sim \frac{\pi Lk^2T}{3\hbar c}, \tag{11.145}$$

which is proportional to the temperature when $LkT \gg \hbar c$.

The use of *quantum* mechanics is *essential* to calculate the equilibrium properties of the string. In fact, one would obtain for the free energy in classical statistical mechanics:

$$F = -\frac{1}{\beta} \sum_k \ln \int \frac{d\xi_k \, d\pi_k}{h} \exp\left[-\frac{1}{2\varrho kT}\left(\pi_k^2 + \omega_k^2\xi_k^2\right)\right]$$

$$= -kT \sum_{m=1}^\infty \ln \frac{LkT}{m\pi\hbar c}, \tag{11.146}$$

which is a divergent series. The terms in (11.146) for which $m \ll LkT/\pi\hbar c$ are good approximations for the corresponding terms in (11.143), but this is not the case for the higher-order terms. In fact, the equipartition theorem (§ 8.4.2) implies that the classical specific heat equals the product of k with the number of modes, which is infinite.

One can easily check that (11.145) is the same as the expression, similar to (11.129), which would have been obtained at low temperatures for the one-dimensional chain of atoms with the spectrum (11.114). At temperatures which are low as compared to the Debye temperature the dominant vibrational modes are those with wavelengths long as compared to the cell size a, and then one can replace the linear chain (11.110) by a continuous string with a linear density $\varrho = M/a$ and a tension $\tau = C/a$, where $\delta R \sim \varphi$. The sound velocities in those two models are the same. We can thus treat the crystal as a continuous medium not only to account for sound propagation, but also to find the specific heat, provided we quantize, as we have just done, the deformation field $\varphi(x)$, and provided we restrict ourselves to low temperatures, $T \ll \Theta_{\mathrm{D}}$.

From the Boltzmann-Gibbs distribution we can find the *probability law for the deformations* of the string at thermal equilibrium. In *classical* statistical physics we know from (11.146) that this law cannot be normalized. Nevertheless, since the divergence arises from the summation over the modes, nothing prevents us from studying the distribution of a finite number of variables π_k and ξ_k. These are uncorrelated, Gaussian, random variables, as the phase density is proportional to $e^{-\beta H}$, where H is the quadratic form (11.137). The fluctuations in ξ_k are given by the equipartition theorem:

$$\frac{1}{2}\varrho\omega_k^2\,\langle\xi_k^2\rangle \;=\; \frac{\pi^2\tau m^2}{2L^2}\,\langle\xi_k^2\rangle \;=\; \frac{1}{2}kT. \tag{11.147}$$

The classical field at each point, which is a sum (11.133) of Gaussian random variables ξ_k, is also a Gaussian variable, characterized by the fluctuations and correlations,

$$\langle\varphi(x)\varphi(x')\rangle \;=\; \frac{2}{L}\sum_{m=1}^{\infty}\sin\frac{\pi m x}{L}\sin\frac{\pi m x'}{L}\frac{kT}{\tau(\pi m/L)^2}.$$

Using the Fourier series

$$\sum_{m=1}^{\infty}\frac{\cos\pi m y}{m^2} \;=\; \pi^2\left(\frac{1}{6}-\frac{y}{2}+\frac{y^2}{4}\right),\qquad 0\le y\le 2,$$

we find, for $x\le x'$,

$$\langle\varphi(x)\varphi(x')\rangle \;=\; \frac{kT}{L\tau}\,x(L-x'). \tag{11.148}$$

Figure 11.23 shows, with a large magnification of the ordinates, the string at one particular time. Its shape fluctuates all the time: this is a new example of *thermal noise* due to coupling to a thermostat (§ 5.7.3). The amplitude of the fluctuations, which is obviously zero at the ends of the string, varies as $x(L-x)$, and its maximum, at the mid-point of the string, equals $\sqrt{LkT/4\tau}$. At room temperatures, a 1 m string under a tension of 1 g is in this way randomly displaced from its equilibrium position over a distance of 3 Å. Such small displacements seem to be difficult to measure. However, one has been able to measure experimentally a completely analogous effect: thermal fluctuations of the free surface of a liquid around the horizontal plane.

Extending these results to *quantum* statistical mechanics is based on evaluating the probability distribution at equilibrium of the *"position" variable ξ_k for the harmonic oscillator* with a Hamiltonian (11.140). The factorization of the various modes enables us to consider just a single one. As the spectrum of $\widehat{\xi}_k$ is continu-

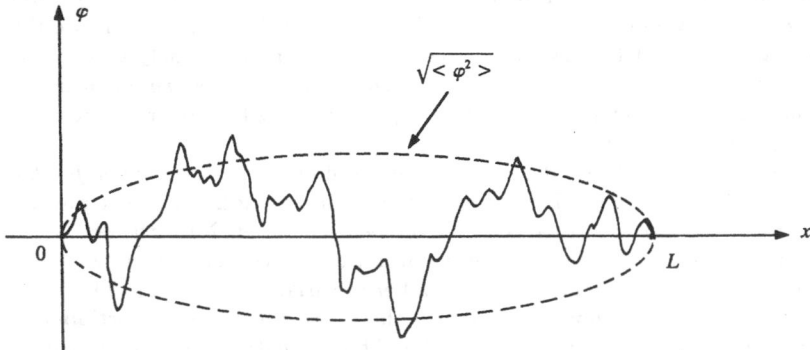

Fig. 11.23. Fluctuating shape of a classical string at thermal equilibrium

ous, the probability distribution for its eigenvalues can be found directly from the characteristic function

$$\left\langle e^{i\lambda\widehat{\xi}_k} \right\rangle = \frac{\operatorname{tr} e^{-\beta\hbar\omega_k\widehat{n}_k} e^{i\lambda\widehat{\xi}_k}}{\operatorname{tr} e^{-\beta\hbar\omega_k\widehat{n}_k}}, \tag{11.149}$$

which we shall evaluate. To simplify the notation we write

$$a \equiv \beta\hbar\omega_k, \qquad b \equiv \lambda\sqrt{\frac{\hbar}{2\varrho\omega_k}},$$

so that the required function (11.149) becomes $g(b)/g(0)$, with

$$g(b) \equiv \operatorname{tr} e^{-a\widehat{n}} e^{ib(\widehat{c}+\widehat{c}^\dagger)}.$$

The derivative $dg/db \equiv i(g_1 + g_2)$ involves functions which we shall express in terms of g:

$$g_1 = \operatorname{tr} e^{-a\widehat{n}} e^{ib(\widehat{c}+\widehat{c}^\dagger)} \widehat{c}, \qquad g_2 = \operatorname{tr} e^{-a\widehat{n}} e^{ib(\widehat{c}+\widehat{c}^\dagger)} \widehat{c}^\dagger.$$

In order to evaluate g_1 we note that

$$e^{ib(\widehat{c}+\widehat{c}^\dagger)} \widehat{c}\, e^{-ib(\widehat{c}+\widehat{c}^\dagger)} = \widehat{c} - ib, \tag{11.150a}$$

which can be checked by differentiating both sides of the equation with respect to b, and that

$$e^{-a\widehat{n}} \widehat{c}\, e^{a\widehat{n}} = e^a \widehat{c}, \tag{11.150b}$$

which can be checked by letting both sides operate on the ket $|n\rangle$. By letting $\widehat{c}$ pass through $e^{ib(\widehat{c}+\widehat{c}^\dagger)}$ and after that through $e^{-a\widehat{n}}$, and using the cyclic invariance of the trace, we find

$$\begin{aligned} g_1 &= \operatorname{tr} e^{-a\widehat{n}} (\widehat{c} - ib)\, e^{ib(\widehat{c}+\widehat{c}^\dagger)} = \operatorname{tr}\left(e^a\widehat{c} - ib\right) e^{-a\widehat{n}} e^{ib(\widehat{c}+\widehat{c}^\dagger)} \\ &= e^a g_1 - ib\, g. \end{aligned}$$

We can similarly express $g_2 = e^{-a}g_2 + ibg$ as a function of g, and we finally find

$$\frac{1}{g}\frac{dg}{db} = -b \coth\frac{a}{2}.$$

The solution of this equation gives us the characteristic function we were looking for

$$\left\langle e^{i\lambda\widehat{\xi}_k} \right\rangle = \exp\left[-\lambda^2 \frac{\hbar}{4\varrho\omega_k} \coth\frac{\hbar\omega_k}{2kT}\right].$$

Hence, the quantum mechanical amplitudes ξ_k of the oscillator are still uncorrelated *Gaussian random variables* but their statistical fluctuations (11.147) have been changed to

$$\left\langle \widehat{\xi}_k^2 \right\rangle \;=\; \frac{\hbar}{2\varrho\omega_k}\,\coth\frac{\hbar\omega_k}{2kT}. \tag{11.151}$$

Quantum mechanics *enhances the fluctuations* of a harmonic oscillator, especially for high frequencies $\hbar\omega_k > kT$.

The probability distribution for the field $\varphi(x)$ follows from (11.133) and (11.151): it again describes a *Gaussian noise*, characterized at any time by the *correlation function*

$$\langle\widehat{\varphi}(x)\widehat{\varphi}(x')\rangle \;=\; \frac{\hbar c}{\pi\tau}\,\sum_{m=1}^{\infty}\sin\frac{\pi m x}{L}\,\sin\frac{\pi m x'}{L}\,\frac{1}{m}\,\coth\frac{\pi m \hbar c}{2LkT}. \tag{11.152}$$

We have found a new effect: when $x = x'$ the series (1.152), which gives the fluctuation $\langle\widehat{\varphi}^2(x)\rangle$ at a given point, diverges, whereas the classical approximation (11.148), which is valid only at the beginning, $m \ll LkT/\hbar c$, of the series, remains finite. The quantization of the field has introduced *short range singularities*.

The existence of such divergences which are called *"ultraviolet" divergences*, because they involve large values of the wavenumber k, is one of the difficulties of quantum field theory. We have earlier skipped another ultraviolet divergence: the constant E_0 which occurs in the quantum Hamiltonian (11.140), and which represents the *ground state energy* of the string, or the energy of the *vacuum* for the equivalent gas of bosons, is a divergent series, $\sum \frac{1}{2}\hbar\omega_k$. This divergence has, however, no physical consequences, as it is not involved if we restrict ourselves to energy differences. This simplification is due to the fact that the classical field equation (11.134) is linear or, equivalently, that there are no interactions between the phonons; other divergences of the same kind appear for interacting particles and their elimination poses huge problems which are the subject of the so-called renormalization theory.

Actually, (11.146) has shown us that even in classical statistical mechanics there occur ultraviolet divergences. We encounter those again if we try to calculate directly, starting from (11.148), the potential energy $\frac{1}{2}\tau\int_0^L dx\,\langle(\partial\varphi/\partial x)^2\rangle$ at equilibrium, since

$$\tau\left\langle\left[\frac{\varphi(x)-\varphi(x')}{x-x'}\right]^2\right\rangle \;=\; kT\left(\frac{1}{|x-x'|}-\frac{1}{L}\right)$$

diverges as $x' \to x$. The curvature of the classical string is thus on average infinite at each point at non-zero temperatures.

However, the situation is even more pathological in quantum theory since (11.152) itself diverges as $x' \to x$, in contrast to (11.148). Moreover, this occurs even as $T \to 0$ when the string tends towards its ground state, which in phonon language is the vacuum. In fact, in quantum field theory *the vacuum has a nontrivial structure*. Even though the mean value of the field $\widehat{\varphi}(x)$ vanishes, (11.152) shows that in the quantum vacuum there remain correlations and fluctuations which are characterized by

$$\langle\widehat{\varphi}(x)\widehat{\varphi}(x')\rangle \;=\; \frac{\hbar c}{2\pi\tau}\,\ln\frac{\sin\pi(x+x')/2L}{\sin\pi|x-x'|/2L}, \tag{11.153}$$

whereas (11.148) tends to 0 as $T \to 0$. *No phonons does therefore not mean no field.* Conversely, according to (11.142), asking the field to vanish amounts to looking for the eigenstate of $\widehat{c}_k + \widehat{c}_k^\dagger$ for all values of k; however, that eigenstate, which is proportional to

$$\exp\left[-\frac{1}{2} \sum_k \left(\widehat{c}_k^\dagger\right)^2\right] |0\rangle,$$

cannot be normalized, and gives a divergent value for the expectation value $\langle \widehat{n}_k \rangle$ of the number of phonons in each mode. *Decreasing the fluctuations of the field makes the number of phonons diverge.* We shall return in what follows to this complementarity between field and phonons.

The ultraviolet divergence of (11.152) as $x \to x'$ has the following consequences: as the statistical fluctuation, in a point x, of the field $\varphi(x)$ is infinite at any temperature, one cannot measure this quantity. This important and surprising effect is connected with the *continuity of the string.* If, in fact, we introduce, instead of the field variables $\widehat{\varphi}(x)$, the M average displacements

$$\widehat{\varphi}_j = \frac{M}{L} \int_{jL/M}^{(j+1)L/M} dx\, \widehat{\varphi}(x), \qquad (11.154)$$

which correspond to cutting the string into M small sections, their probability distribution will be completely reasonable: $\langle \widehat{\varphi}_j \widehat{\varphi}_{j'} \rangle$ will remain finite, and it will be approximated by (11.148) if $1 \ll M \ll LkT/\hbar c$. In quantum theory one *cannot measure a field at a perfectly localized point.* Even in non-relativistic physics it is necessary that the measuring apparatus has a certain spatial size. One can also check that, if one tries to make the measurement more and more localized, one must for the reduction of the wavepacket supply an energy which becomes infinite.

The vibrating string model also illustrates another difficulty of quantum field theory, the existence of *"infrared" divergences.* As the size L of the system becomes infinite, the free energy per unit length (11.144) has a finite limit, but the *phonon density,*

$$\frac{\langle N \rangle}{L} = \frac{1}{L} \sum_{m=1}^{\infty} \frac{1}{e^{\pi m \hbar c / LkT} - 1} > \frac{1}{L} \int_{1/2}^{\infty} \frac{dx}{e^{\pi x \hbar c / LkT} - 1}$$

$$> \frac{kT}{\pi \hbar c} \ln \frac{2LkT}{\pi \hbar c},$$

diverges logarithmically with increasing L. This divergence would also occur for a discrete linear chain. It is connected with the small values of the wavenumber or with large distances, and corresponds to the existence of an *infinite number of bosons with a very small energy* in the system.

Let us, finally, note that, if we want a state to describe the string in the *classical limit* (§ 2.3.4), the expectation value $\langle \widehat{\varphi}(x) \rangle$ of the field $\widehat{\varphi}(x)$ in this state should be identified with the classical displacement of the string, and its fluctuations should be relatively small, that is, should satisfy

$$\langle \widehat{\varphi}(x) \widehat{\varphi}(x') \rangle \sim \langle \widehat{\varphi}(x) \rangle \langle \widehat{\varphi}(x') \rangle,$$

since the vanishing of fluctuations implies the vanishing of correlations. The fluctuations of the velocity or of the momentum conjugate to $\varphi(x)$, and, as a result, those of the ξ_k and π_k variables, must also be small. By virtue of Eqs.(11.139) or (11.142) we find that in the classical limit the phonon creation and annihilation operators themselves must have *non-vanishing expectation values*, related to those of the field $\widehat{\varphi}(x)$ and its velocity, and relatively *small fluctuations*:

$$\langle \widehat{c}_k \widehat{c}_{k'} \rangle \sim \langle \widehat{c}_k \rangle \langle \widehat{c}_{k'} \rangle, \qquad \langle \widehat{c}_k^\dagger \widehat{c}_{k'} \rangle \sim \langle \widehat{c}_k^\dagger \rangle \langle \widehat{c}_{k'} \rangle \sim \langle \widehat{c}_{k'} \widehat{c}_k^\dagger \rangle. \tag{11.155}$$

If we take the commutation relations into account, we see that (11.155) implies

$$|\langle \widehat{c}_k \rangle| = |\langle \widehat{c}_k^\dagger \rangle| \gg 1, \tag{11.156}$$

at least for the low-frequency modes; the high-frequency modes are in the classical limit not excited if the string is not too irregular in shape. These relations imply $\langle \widehat{n}_k \rangle \geq |\langle \widehat{c}_k \rangle|^2 \gg 1$, or, that the number of phonons in the excited modes is large. Moreover, it follows from (11.156) that a density operator which can describe the string in the *classical limit* must necessarily have *off-diagonal elements* between states with n_k and with n_k+1 phonons, which is *not the case in thermal equilibrium*, when the density operator $\widehat{D} \propto e^{-\beta \widehat{H}}$ gives $\langle \widehat{c}_k \rangle = 0$. In a state which can be interpreted classically, not only is the number of phonons large and ill defined, but also the eigenstates of the density operator themselves are *coherent sums* of kets in Fock space, associated with *variable phonon numbers*. For a single mode, an example of such a state is the projection operator on the so-called "coherent" state vector

$$e^{\gamma(\widehat{c}+\widehat{c}^\dagger)} |0\rangle e^{-\gamma(\gamma+\gamma^*)/2} \;=\; e^{\gamma \widehat{c}^\dagger} |0\rangle e^{-\gamma\gamma^*/2} \;=\; \sum_n |n\rangle \frac{\gamma^n}{\sqrt{n!}} e^{-\gamma\gamma^*/2}, \tag{11.157}$$

which is the (normalized) eigenket of $\widehat{c}$ with eigenvalue γ, as can be seen from (11.150a) and (10.20b). The off-diagonal elements $|n\rangle\langle n'|$ of the corresponding density operator are essential for getting the relations (11.155); the latter reduce here to $\langle \widehat{c} \rangle = \gamma$, $\langle \widehat{c}^2 \rangle = \gamma^2$, $\langle \widehat{c}^\dagger \widehat{c} \rangle = \gamma\gamma^*$, and $\langle \widehat{c} \widehat{c}^\dagger \rangle = \gamma\gamma^*+1$, which is the same as $\langle \widehat{c} \rangle \langle \widehat{c}^\dagger \rangle$ if $|\gamma| \gg 1$. The fluctuation of n in the state (11.157) is $\Delta n = |\gamma|$.

These remarks illustrate the *complementarity* between the descriptions in terms of *fields* and in terms of *particles*. As the field operator (11.142) does not commute with the phonon number operator, one cannot simultaneously assign precise values for those two quantities. To speak of a *well-defined field*, we must place ourselves in situations where $\widehat{\varphi}(x)$ or $\widehat{c}$ does not fluctuate much; this implies that the density operator is not diagonal in n and that n *is large and fluctuates*. Conversely, if n is well defined, or, more generally, in states like the canonical equilibrium state which are *diagonal in n*, the field vanishes on average and fluctuates; such states contain off-diagonal elements when one writes them down in the ξ-representation in terms of the "position" variable for each oscillator or of the fields.

Summary

At sufficiently low temperatures substances usually show crystal order, with their nuclei arranged on a regular lattice. The geometric and mechanical properties of solids depend on this structure and its defects; the electromagnetic and optical properties depend mostly on the electrons, and the acoustic and thermal properties on the lattice vibrations. The microscopic theory associates these three elements with each of the three stages of the Born-Oppenheimer method. It explains the crystal structures by connecting each phase with a minimum of the effective internuclear potential or, more generally, of a variational grand potential. The theoretical analyses can be simplified by replacing the constituent particles of the solid by entities which have a collective nature and interact weakly, the quasi-particles.

The Hartree approximation allows us to treat the electrons like a gas of fermions with each of them subject to a potential produced by the other electrons and the nuclei. Because of the periodicity of this potential, the spectrum of the single-electron states consists of continuous bands, inside which each state is characterized by its quasi-momentum, and which are separated by forbidden bands. The band structure of a crystal reminds one of that of the atomic spectrum of its constituents, each shell giving rise to one or more bands, the width of which increases with energy. The single-electron wavefunctions, or Bloch waves, have a double nature: itinerant, due to their resemblance to plane waves, and bound, as they are superpositions of localized Wannier orbitals.

The incomplete or complete filling of the allowed bands exhibits at the macroscopic scale the itinerant or bound nature of the electrons, and determines whether a solid is a metal or an insulator. The characteristic electron temperatures, the Fermi temperature in a metal or the width of the forbidden band in an insulator, are sufficiently high so that the equilibrium state at room temperatures hardly differs from the ground state. The elementary excitations, which behave as independent quasi-particles for both equilibrium and dynamic properties, are the electrons at the Fermi surface for metals, the conduction electrons and the holes for insulators. An insulator is modelled as a classical gas of positive and negative charge-carrier quasi-particles. Its conductivity increases with increasing temperature, as the number of quasi-particles; that of a metal decreases.

The macroscopic electrostatic properties, such as the screening effect, the surface charge, the spontaneous potential difference between different substances, or the polarization in a dielectric, follow from the self-consistency of the Hartree potential. Adding impurities to an insulator with a narrow forbidden band makes it into a semiconductor, by increasing either the number of conduction electrons (n-type) or the number of holes (p-type). The sensitivity of semicondutors to outside influences, such as an electric potential, illumination, or heating, leads to many effects having industrial applications. In particular, the properties of a junction between p- and n-type semiconductors make it possible to manufacture photocells, rectifiers, and transistors.

The vibrational modes of the nuclei in a crystal are quantum waves; they are equivalently described as a gas of bosons, the phonons, the energy of which is proportional to their quasi-momentum. The number of phonons is not conserved. Their contribution to the specific heat dominates over that from the electrons, except for metals at low temperatures. At high temperatures, it satisfies the Dulong-Petit law.

Exercises

11a Point Defects in Crystals

At absolute zero the N atoms in a crystal are arranged on a regular lattice. The effective potential which is produced by the system of atoms and is acting on one of them has a minimum at the site occupied by that atom at equilibrium, but it can also have other secondary minima, that are less deep, at interstitial sites which at zero temperature are unoccupied. By providing an energy ε one can let one of the atoms migrate from one of the N lattice sites to one of the N' interstitial sites: in this way one produces a pair of defects, a so-called Frenkel pair. More generally, by heating the crystal one can produce a fairly large, although small compared to N, number of Frenkel defects, and the disorder increases together with the internal energy. A simplified model assumes that the energy of any configuration where n atoms are displaced is equal to $n\varepsilon$. Determine the average number $\langle n \rangle$ of defects at a temperature T; for a numerical application, take $\varepsilon = 1$ eV, and for T room temperature. Calculate the contribution from these defects to the specific heat of the crystal.

Another kind of defect, the so-called Schottky defect or vacancy, corresponds to the migration of an atom from a lattice site to the crystal surface, with an increase of energy equal to ε. Evaluate the contribution of Schottky defects to the temperature dilatation coefficient of the solid and to the specific heat. One may assume that the number of different configurations formed by placing n atoms at the surface is small compared to the number of ways of creating n defects. One may also assume (in the grand canonical ensemble) that the crystal is in equilibrium with its vapour and that the chemical potential in the vapour is small compared to ε.

Hints:

One can treat this exercise (like Exerc.5a) by various methods using one or another of the canonical ensembles. We shall, as an example, use the microcanonical ensemble, but one could as an exercise try to find the same results in the canonical or grand canonical formalisms, bearing in mind that all ensembles are equivalent in the limit where the sample is large (§ 5.6.3).

The micro-states are characterized by specifying whether each of the N lattice sites and each of N' interstitial sites is or is not occupied by an atom; the sites are distinguishable, but the atoms are not. In microcanonical equilibrium a number of $n = U/\varepsilon$ atoms are displaced from n among the N lattice sites to n among the N' interstitial sites. There are therefore

$$W = \frac{N!}{n!(N-n)!} \frac{N'!}{n!(N'-n)!}$$

equiprobable micro-states, and the entropy is

$$
\begin{aligned}
S &= k \ln W \\
&\sim k \big[N \ln N + N' \ln N' - 2n \ln n \\
&\quad - (N-n)\ln(N-n) - (N'-n)\ln(N'-n) \big],
\end{aligned}
$$

where we have used Stirling's approximation for the factorials.

Expressing S as function of U, N, and N', we get the temperature

$$\frac{1}{T} = \frac{\partial S}{\partial U} = \frac{k}{\varepsilon} \ln \frac{(N-n)(N'-n)}{n^2},$$

and hence the equation for n:

$$\frac{n^2}{(N-n)(N'-n)} = e^{-\varepsilon/kT},$$

or, when the number of defects is small:

$$n \sim \sqrt{NN'}\, e^{-\varepsilon/2kT}.$$

We calculate the specific heat, using either the formula $C = T\, dS/dT$ or the formula $C = dU/dT = \varepsilon\, dn/dT$.

In the canonical ensemble one calculates approximately

$$F = -\frac{1}{\beta} \ln \sum_n W e^{-\beta n \varepsilon},$$

noting that only the largest term in the sum contributes in the case of a macroscopic system (Exerc.5b). In the grand canonical ensemble one factorizes the contributions from the two kinds of site, and writes down the condition that their chemical potentials are equal.

Similarly, whatever method one uses, one finds that there are $n = N/\left(e^{\beta\varepsilon}+1\right) \sim N e^{-\beta\varepsilon}$ Schottky defects. A less simplistic theory for these kinds of defects is presented in Prob.8.

11b Semiconductors in the Extrinsic and Saturation Regimes

Show by studying the extrinsic and saturation regimes (§ 11.3.4) and making the effective mass approximation that the number of conduction electrons in a neutral n-type semiconductor at equilibrium is a monotonically increasing function of the temperature, but that the chemical potential has a maximum. Discuss the limiting cases.

Hints:

Since the valence band remains empty, the condition for electric neutrality can be written as $N_c = N_I(1 - f(\varepsilon_d))$, or in terms of the dimensionless variables $\tau \equiv kT/(\varepsilon_c - \varepsilon_d)$, $\lambda \equiv (\mu - \varepsilon_d)/(\varepsilon_c - \varepsilon_d)$, and $\nu \equiv 4\hbar^3 [2\pi m_c(\varepsilon_c - \varepsilon_d)]^{-3/2} N_I/\Omega$,

$$\left(1 + e^{\lambda/\tau}\right) \frac{2}{\sqrt{\pi}} \int_0^\infty \frac{\sqrt{x}\,dx}{e^{(x+1-\lambda)/\tau} + 1} = \nu.$$

This equation determines $\lambda(\tau)$, that is, μ and $N_c = N_I(1 + e^{\lambda/\tau})$, as function of T for a given doping ν. Using λ/τ and τ as variables and differentiating, we find that λ/τ decreases with τ, therefore, that N_c increases with temperature from 0 to N_I. In the saturation regime, reached when $\eta \equiv \nu\tau^{-3/2}e^{1/\tau} \ll 1$, we have $N_I - N_c \sim \eta N_I$, and λ decreases as $-\tau \ln(1/\eta)$ so that μ decreases with T. At low temperatures, for $e^{2/\tau} \gg \eta \gg 1$, we have $N_c \sim N_I\eta^{-1/2}$ and 2λ varies as $\tau \ln \eta$, so that μ first *increases*, starting from $(\varepsilon_c + \varepsilon_d)/2$, and then *decreases*. It passes again through its original value when $\lambda = \frac{1}{2}$, that is, when τ is a solution of

$$\frac{2}{\sqrt{\pi}} \int_0^\infty \sqrt{y}\,dy \frac{1 + e^{-1/2\tau}}{e^y + e^{-1/2\tau}} = \nu\tau^{-3/2},$$

or $\tau = A\nu^{2/3}$, where A varies between 1 when $\nu \ll 1$ and 0.75 when $\nu \gg 1$. The value $\mu = \varepsilon_d$, or $\lambda = 0$, corresponds to a half filling, $N_c = N_I/2$. When we have strong doping, $\nu > 2.92$, the maximum of μ lies in the conduction band, and μ crosses ε_c for the two solutions for τ of

$$\frac{\nu\tau^{-3/2}}{1 + e^{1/\tau}} = \frac{2}{\sqrt{\pi}} \int_0^\infty \frac{\sqrt{y}\,dy}{e^y + 1} = \left(1 - \frac{1}{\sqrt{2}}\right)\zeta\left(\frac{3}{2}\right) \simeq 0.77.$$

When $\nu \gg 3$, there is even a temperature range, satisfying $\eta \gg e^{2/\tau}$, where the behaviour is metallic, as we have $\mu > \varepsilon_c$ and $kT \ll \mu - \varepsilon_c$. In that region where $\lambda - 1 \gg \tau$, the chemical potential, given by $e^{\lambda/\tau}(\lambda - 1)^{3/2} = 3\sqrt{\pi}\nu/4$, increases, and N_c/N_I is small as $e^{-\lambda/\tau}$.

11c p-n Junction at Zero Temperature

Solve the coupled equations which determine at $T = 0$ the charge density, field, and electrostatic potential near the surface $x = 0$ of an n-type semiconductor, which is positively charged and occupies the $x > 0$ region; assume that external charges in the region $x < 0$ force the potential to be a constant as $x \to +\infty$. What is the thickness l_n of the charged zone and the potential difference between the interior and the surface of the semiconductor, for a donor density $\nu_d = 3 \times 10^{21}$ m^{-3}, a dielectric constant $\varepsilon = 15\varepsilon_0$, and a surface charge density $\varrho_s = 3 \times 10^{-4}$ C m^{-2}? What is the chemical potential?

Hence derive the charge distribution and the shape of the field in a p-n junction which globally is uncharged, at $T = 0$, and find an expression for the maximum field strength as function of the doping densities. How do the thicknesses l_n and l_p vary with doping?

Hints:

At $T = 0$, a donor site is full or empty, according to whether its local energy is lower or higher than μ. Thus, the electrons which are missing from the donor sites produce between the surface $x = 0$ and the plane $x = l_n$, that is, in a layer where $\varepsilon_d - e\Phi(x) > \mu$, a constant volume charge density $\varrho = e\nu_d = \varrho_s/l_n$. When $x > l_n$, the charge density (11.99) vanishes, as $n_d = \nu_d$, $n = p = 0$. Hence we get, because $\varepsilon\nabla^2\Phi = -\varrho$,

$$\Phi(x) = \Phi_n - \frac{\varrho_s}{2\varepsilon l_n}(l_n - x)^2, \qquad 0 \le x \le l_n.$$

In the interior of the sample, $x \ge l_n$, $\Phi = \Phi_n$ is constant (see Figs.11.14 and 11.15). Outside it, $x \le 0$, Φ varies linearly. The field,

$$E(x) = -\frac{d\Phi}{dx} = -\frac{\varrho_s}{\varepsilon l_n}(l_n - x), \qquad 0 \le x \le l_n,$$

is continuous and thus equal to $-\varrho_s/\varepsilon$ for $x \le 0$, and vanishing for $x \ge l_n$. Because there are surface charges present, the value of μ at $T = 0$ inside the sample is $\varepsilon_d - e\Phi_n + 0$, immediately above the donor levels, whereas it is $\frac{1}{2}(\varepsilon_c + \varepsilon_d)$ when there are no charges. This change, $\frac{1}{2}(\varepsilon_c - \varepsilon_d)$ in the chemical potential is due to its great sensitivity to external perturbations for $T = 0$ when μ lies inside a forbidden band.

For a p-n junction one joins this solution up with the inverted solution on the p side, taking the values ϱ_s on the n side and $-\varrho_s$ on the p side, and requiring that the chemical potentials are the same, $\mu = \varepsilon_d - e\Phi_n = \varepsilon_a - e\Phi_p$, and also that $\Phi(x)$ and $E(x)$ are continuous. Using the fact that $l_n = \varrho_s/e\nu_d$, $l_p = \varrho_s/e\nu_a$, we find

$$\Phi_n - \Phi_p = \frac{\varepsilon_d - \varepsilon_a}{e} = \frac{\varrho_s}{2\varepsilon}(l_n + l_p) = \frac{\varrho_s^2}{2\varepsilon e}\left(\frac{1}{\nu_d} + \frac{1}{\nu_a}\right),$$

and hence ϱ_s, l_n, and l_p as functions of $\varepsilon_d - \varepsilon_a$ and the doping densities. The field $E(x)$ varies linearly in the regions $(-l_p, 0)$ and $(0, l_n)$. The maximum of its absolute magnitude is reached at $x = 0$, where

$$E(0) = -\frac{\varrho_s}{\varepsilon} = -\left[\frac{2(\varepsilon_d - \varepsilon_a)\nu_d\nu_a}{\varepsilon(\nu_d + \nu_a)}\right]^{1/2}.$$

It increases with doping, and so does the displaced charge ϱ_s, whereas the total thickness of the barrier decreases; but l_n increases with ν_a for fixed ν_d.

11d Einstein Model

We model the quantum vibrations of a crystal by assuming that the restoring force which pushes each nucleus back to its equilibrium position is independent of the position of the other nuclei. Evaluate the specific heat of the crystal as function of temperature. Compare it with the model of classical vibrations and with the results of the Debye model.

Answer. Introducing the Einstein temperature

$$\Theta = \frac{h\nu}{k},$$

where ν denotes the oscillation frequency of each nucleus, we get

$$C = \frac{3Nk(\Theta/T)^2 e^{\Theta/T}}{\left(e^{\Theta/T} - 1\right)^2}.$$

The Einstein model gives the general behaviour of $C(T)$ and the Dulong-Petit law, but it deviates from experimental data at low temperatures (Fig.11.20), as it does not reproduce the mode density (11.125) at low frequencies. In fact, Einstein's hypothesis is unrealistic for long wavelengths and it violates Goldstone's theorem (see the end of § 11.4.2); this is the result of the fact that the potential W in the Einstein model does not remain constant under a global translation of the crystal.

11e Acoustic and Optical Phonons

Find the vibrational eigenfrequencies and the phonon creation and annihilation operators for a one-dimensional model with two atoms per cell, with masses M' and M'', $M' > M''$, which oscillate, respectively, around the sites $R'_j = ja$ and $R''_j = (j + \frac{1}{2})a$, and which have a nearest neighbour interaction governed by an effective potential $\frac{1}{2}C(\delta\widehat{R}' - \delta\widehat{R}'')^2$. To do this, let $M'M'' \equiv M^2$ and $M'/M'' \equiv e^{2u}$, carry out the transformation (11.111) for each type of atoms, and end up with the diagonalization of a 2×2 matrix, after having changed the normalization of the conjugate variables π'_k, ξ'_k and π''_k, ξ''_k by factors $e^{\pm u/2}$. Study the shape of the spectrum and the motions of the atoms in the acoustic and optical modes, at the centre and at the edge of the Brillouin zone.

Answer:

Using the notation $\pm$ to distinguish the two modes that we obtain for each value of k, with the upper sign corresponding to the acoustic phonons, and the lower sign to the optical ones, we have

$$\omega_{k\pm} = \sqrt{\frac{C}{M}} \left(\sqrt{\cosh u + \sin|ka/2|} \mp \sqrt{\cosh u - \sin|ka/2|}\right).$$

The acoustic branch has the same shape as in the case of a monatomic chain, with a sound velocity $\omega_{k+}/k \sim \sqrt{Ca^2/2(M'+M'')}$ and a maximum $\omega_{k+} = \sqrt{2C/M'}$ at the edge $k = \pm\pi/a$ of the Brillouin zone. The optical branch has a maximum $\omega_{0-} = \sqrt{2C(M'+M'')/M'M''}$ for $k = 0$ and a minimum $\omega_{k-} = \sqrt{2C/M''}$ for $k = \pm\pi/a$; the group velocity of a wavepacket vanishes in these extrema.

The phonon annihilation operators are

$$\widehat{c}_{k\pm} = \frac{1}{2\sqrt{\hbar}} \left[\sqrt{1 \pm \frac{\sinh u}{A}} \left(\widehat{\xi}'_k \sqrt{M'\omega} + \frac{i\widehat{\pi}'_k}{\sqrt{M'\omega}} \right) \right.$$

$$\left. \pm \sqrt{1 \mp \frac{\sinh u}{A}} \left(\widehat{\xi}''_k \sqrt{M''\omega} + \frac{i\widehat{\pi}''_k}{\sqrt{M''\omega}} \right) \right],$$

where ω stands for $\omega_{k\pm}$, and where $A \equiv \left(\sinh^2 u + \cos^2 \tfrac{1}{2}ka \right)^{1/2}$. Inversely, the displacement operators of the atoms are given by

$$\delta\widehat{R}'_j = \sqrt{\frac{\hbar}{NM'}} \sum_k \left[\sqrt{1 + \frac{\sinh u}{A}}\, \widehat{\gamma}_{k+} + \sqrt{1 - \frac{\sinh u}{A}}\, \widehat{\gamma}_{k-} \right] e^{ik\overline{R}'_j},$$

$$\delta\widehat{R}''_j = \sqrt{\frac{\hbar}{NM''}} \sum_k \left[\sqrt{1 - \frac{\sinh u}{A}}\, \widehat{\gamma}_{k+} - \sqrt{1 + \frac{\sinh u}{A}}\, \widehat{\gamma}_{k-} \right] e^{ik\overline{R}''_j},$$

with

$$\widehat{\gamma}_{k\pm} \equiv \frac{\widehat{c}_{k\pm} + \widehat{c}^{\dagger}_{-k\pm}}{2\sqrt{\omega_{k\pm}}}.$$

For an acoustic phonon ($k \to 0$, whence $A \to \cosh u$, and index $+$), the displacements of all the atoms are the same, $\delta R' \sim \delta R''$, as $(1 + \sinh u/A)e^{-u} \sim (1 - \sinh u/A)e^u$. For an optical phonon of zero momentum (index $-$), the two kinds of particle oscillate in opposite directions with a larger amplitude for the lighter species, $M'\delta R' \sim -M''\delta R''$. At the edge of the zone ($k = \pi/a$, whence $A = \sinh u$), only the heavier atoms oscillate in the acoustic mode; their displacements alternate in directions, $\delta R'_j \sim -\delta R'_{j+1}$. In the optical mode with $k = \pi/a$ only the light atoms vibrate, $\delta R''_j \sim -\delta R''_{j+1}$, again with opposite phases.

11f Ferromagnetism of Metals

The Ising model (Exerc.9a), where the localized spins tend to orient themselves parallel to each other under the action of their short-range interactions, gives a simple explanation of the ferromagnetism of certain substances. However, this explanation is not suitable for the ferromagnetic metals, such as iron or nickel. In fact, the elementary magnetic moments are in that case attached to the conduction electrons which are *not localized*; moreover, the *direct* magnetic interaction between the spins is negligible. We shall show how the Coulomb repulsion between the electrons in the metal, combined with the exclusion principle, provides us with a mechanism which makes it indirectly possible to orient their spins, that is, to give rise to a ferromagnetic phase.

We shall be interested in elements with $24 \leq Z \leq 30$ electrons per atom, and we shall make the following simplifying assumptions. The $\mathcal{N}$ atoms of the crystal of volume Ω are positioned on the sites of a simple cubic lattice with mesh size a, which are denoted by $\boldsymbol{R}_j$ $(j = 1, \ldots, \mathcal{N})$. We assume that $24\mathcal{N}$ of the $Z\mathcal{N}$ electrons are frozen in into the low-lying bands, and that the $N = (Z - 24)\mathcal{N}$ remaining electrons are in a conduction band $-\Delta \leq \varepsilon \leq \Delta$, the centre of which we choose as energy zero. We assume that this band contains three subbands $b = 1, 2, 3$ so that altogether it contains $6\mathcal{N}$ single-electron states, denoted by $q \equiv (\boldsymbol{p}, b, \sigma)$; $\boldsymbol{p}$ denotes the quasi-momentum and $\sigma = \pm 1$ the z-component of the spin. We also assume that its density of states equals

$$
\mathcal{D}(\varepsilon) = \frac{9\mathcal{N}}{2\Delta} \left(1 - \frac{\varepsilon^2}{\Delta^2} \right) \theta(\Delta^2 - \varepsilon^2),
$$

and that the single-electron energies $\varepsilon_q \equiv \varepsilon_{\boldsymbol{p}}$ are independent of b. The other bands will play no rôle, since $kT \ll \Delta$.

The N-electron effective Hamiltonian $\widehat{H} \equiv \widehat{H}_0 - B\widehat{M} + \widehat{V}$ contains: (i) the conduction-band Hamiltonian $\widehat{H}_0$, with eigenvalues $\sum_q n_q \varepsilon_{\boldsymbol{p}}$; (ii) the magnetic energy of the spins in a possible field B, applied along the z-axis, where $\widehat{M} = -\mu_B \sum_{i=1}^N \widehat{\sigma}_i$ denotes the z-component of the magnetic moment; and (iii) the Coulomb repulsion between the conduction electrons, reduced by screening (§ 11.3.3). (This repulsion was not included in the self-consistent potential of the band Hamiltonian $\widehat{H}_0$.) The screening length is short, of the order of the lattice mesh size a, so that the only significant matrix elements of the effective potential $\widehat{V}$ are, within the conduction band, between Wannier orbitals $\chi_b(\boldsymbol{r} - \boldsymbol{R}_j)$ localized on the same site. To simplify matters, we assume also that this matrix is proportional to the unit matrix in b, b' space, so that $\widehat{V}$ has the form

$$
\widehat{V} = u \sum_{j=1}^N \sum_b \sum_{\sigma,\sigma'} \widehat{n}(\boldsymbol{R}_j, b, \sigma)\, \widehat{n}(\boldsymbol{R}_j, b, \sigma').
$$

We have denoted by $\widehat{n}(\boldsymbol{R}_j, b, \sigma)$ the occupation number operator in the Wannier orbital $\chi_b(\boldsymbol{r} - \boldsymbol{R}_j)$ for spin σ; the positive constant u, which is the diagonal matrix element of the Coulomb interaction between Wannier waves of extension a, is of the order of magnitude of $e^2/4\pi\varepsilon_0 a$, that is, of eV, and the summation over σ and σ' indicates the fact that $\widehat{V}$ is spin-independent.

The problem will be solved by a mean-field variational method, similar to the one of § 11.2.1. Taking into consideration that it is practically impossible to work with the exact grand canonical density operator $\widehat{D} = Z^{-1} \exp[-\beta\widehat{H} + \alpha\widehat{N}]$, the latter will be approximated by a simpler trial density operator, $\widehat{D} =$

$Y^{-1} \exp\left[-\sum_q y_{p,\sigma} \widehat{n}_q\right]$, where $\widehat{n}_q$ denotes the occupation number operator for the single-electron state $q \equiv (p, b, \sigma)$. The y_q, which are independent of the band index b, are adjustable parameters which will be best determined by looking for the minimum of the trial grand potential $\mathcal{A} \equiv \langle\widehat{H}\rangle - \mu\langle\widehat{N}\rangle - TS(\widehat{\mathcal{D}})$, where $\langle\ \rangle$ denotes an average value with respect to $\widehat{\mathcal{D}}$.

1. Find expressions for Y, $S(\widehat{\mathcal{D}})$, $\langle\widehat{H}_0\rangle$, $N_\pm$, $M \equiv \langle\widehat{M}\rangle$, and $\langle\widehat{V}\rangle$ in terms of the parameters $f_q \equiv \langle n_q\rangle$, which are equivalent to the y_q; $N_\pm \equiv \langle\widehat{N}_\pm\rangle$ denotes the number of electrons with spins $\sigma = \pm 1$. Derive, without solving them, the equations determining the y_q and the f_q. Show, in particular, that $y_{p,\sigma}$ has the form $\beta(\varepsilon_p - x_\sigma)$ where the remaining two parameters x_σ should be determined by coupled equations to be written down in terms of $\mathcal{D}(\varepsilon)$. Interpret the result by representing the electron system as a mixture of two Fermi gases with $\sigma = \pm 1$, each of which is subject to a uniform mean potential produced by the other.

2. Show that for $Z = 27$ the parameters x_+ and x_- are each other's opposite and find μ. Write down the equations determining $x \equiv \frac{1}{2}(x_+ - x_-)$ and M for temperatures $T \ll \Delta/k$. Solve these equations for the case when there is no field. Show especially that at zero temperature the metal is ferromagnetic, that is, has a non-vanishing spontaneous magnetization M_s/Ω, provided u exceeds a certain magnitude u_0 to be determined as function of Δ. Show also that for a fixed u exceeding u_0 there exists a critical temperature, the Curie temperature T_c, above which there is no ferromagnetism. Determine the spontaneous magnetization M_s/Ω and the Curie temperature in terms of u/u_0 and sketch the $M_s(T)$-curve. Are the solutions the same for $B = 0$ and for $|B| \to 0$?

3. Show, still with $N = 3\mathcal{N}$, that in the regions where there is no ferromagnetism the metal is paramagnetic and determine its susceptibility χ. Sketch $\chi(T)$ for $u > u_0$ and for $u < u_0$. Compare the result with Pauli paramagnetism ($u = 0$). For $u > u_0$, determine the behaviour in the vicinity of the Curie temperature: (i) of the magnetization as function of the field B for $T = T_c$ and $B \to 0$; (ii) of the susceptibility in the paramagnetic phase as function of $T - T_c \to 0$ and in the ferromagnetic phase as function of $T_c - T \to 0$ (compare the results); (iii) of the specific heat in zero field; sketch the $C(T)$-curve. Compare the results with the Landau theory (Exerc.6d).

4. Cobalt is a ferromagnetic metal below 1404 K. Its mass number is 59, its density 8.71 g cm^{-3}, and its atomic number $Z = 27$. Assume that it can be described by the present model, with a density of states at the Fermi surface equal to $\mathcal{D}(\varepsilon_F)/\Omega = 0.47$ eV^{-1}Å^{-3}. Calculate the mesh size a, the band width 2Δ, and the interaction strength u which follow from these data.

5. Assume now that the number N of electrons in the conduction band is arbitrary, but fixed as T varies. Write down the equations determining the Curie temperature. Take as parameters $x = \frac{1}{2}(x_+ - x_-)$, which is small, and $z = \frac{1}{2}(x_+ + x_-)$, and do not specify $\mathcal{D}(\varepsilon)$ until the end of the calculation. Show

that for given values of Δ and u the temperature T_c has a maximum for a half-filled band and that the substance is not ferromagnetic at any temperature if $\overset{\circ}{N}$ exceeds limits which should be determined. Experiments show that from among the successive elements in the atomic table, chromium ($Z = 24$), manganese ($Z = 25$), iron ($Z = 26$), cobalt ($Z = 27$), nickel ($Z = 28$), copper ($Z = 29$), and zinc ($Z = 30$), only Fe, Co, and Ni are ferromagnetic with Curie temperatures equal to 1043, 1404, and 631 K, respectively. Discuss these facts, assuming that all these elements have the same model Hamiltonian, the same density of states $\mathcal{D}(\varepsilon)$, the same interaction u, and the same number of atoms per unit volume, so that they differ solely in the number of electrons in their conduction bands.

Solution:

1. As in §10.3 we find

$$\ln Y = \sum_q \ln\left(1 + e^{-y_q}\right), \qquad f_q = -\frac{\partial}{\partial y_q} \ln Y = \frac{1}{e^{y_q} + 1},$$

$$S = k \operatorname{Tr} \widehat{\mathcal{D}} \left[\ln Y + \sum_q y_q \widehat{n}_q\right] = k \sum_q \left[\ln\left(1 + e^{-y_q}\right) + \frac{y_q}{e^{y_q} + 1}\right],$$

$$\langle \widehat{H}_0 \rangle = \sum_q \varepsilon_q \langle n_q \rangle = 3 \sum_{p,\sigma} \varepsilon_p f_{p,\sigma},$$

$$N_\sigma = 3 \sum_p f_{p,\sigma}, \qquad M = -\mu_B \left(N_+ - N_-\right).$$

By separating the terms with $\sigma = \sigma'$ from those with $\sigma = -\sigma'$ we can rewrite the potential $\widehat{V}$ as follows:

$$\widehat{V} = u\widehat{N} + 2u \sum_{j,b} \widehat{n}(\boldsymbol{R}_j, b, +)\widehat{n}(\boldsymbol{R}_j, b, -).$$

The factorized form of $\widehat{\mathcal{D}}$ does not give rise to any correlations between electrons with opposite spins so that

$$\langle \widehat{V} \rangle = uN + 2u \sum_{j,b} \langle \widehat{n}(\boldsymbol{R}_j, b, +) \rangle \langle \widehat{n}(\boldsymbol{R}_j, b, -) \rangle.$$

The state $\widehat{\mathcal{D}}$ treats the sites $\boldsymbol{R}_j$ and the subbands b in the same way; one should thus expect that $\langle \widehat{n}(\boldsymbol{R}_j, b, \sigma) \rangle$ is independent of $\boldsymbol{R}_j$ and of b, and therefore equals $N_\sigma/3\mathcal{N}$; hence

$$\langle \widehat{V} \rangle = uN + \frac{2u}{3\mathcal{N}} N_+ N_-.$$

To prove that result, we use (10.22) and the change in base (10.36), which can now be written as

$$\langle r | p, b \rangle \;=\; \frac{1}{\sqrt{\mathcal{N}}} \sum_j e^{i p \cdot R_j / \hbar} \chi_b(r - R_j),$$

leading to

$$
\begin{aligned}
\widehat{n}(R_j, b, \sigma) &= \widehat{c}^\dagger(R_j, b, \sigma) \widehat{c}(R_j, b, \sigma) \\
&= \sum_{p,p'} \widehat{c}^\dagger_{p,b,\sigma} \langle p, b | R_j, b \rangle \langle R_j, b | p', b \rangle \widehat{c}_{p',b,\sigma} \\
&= \frac{1}{\mathcal{N}} \sum_{p,p'} e^{i(p'-p) \cdot R_j / \hbar} \widehat{c}^\dagger_{p,b,\sigma} \widehat{c}_{p',b,\sigma}.
\end{aligned}
$$

The average over $\widehat{D}$ then gives, as expected,

$$\langle \widehat{n}(R_j, b, \sigma) \rangle \;=\; \frac{1}{\mathcal{N}} \sum_p f_{p,\sigma} \;=\; \frac{N_\sigma}{3\mathcal{N}}.$$

Looking for the minimum of $\mathcal{A}$ with respect to the f_q leads to

$$kT y_{p,\sigma} \;=\; \varepsilon_p + \mu_B B \sigma + u + \frac{2u}{3\mathcal{N}} N_{-\sigma} - \mu,$$

an equation which is coupled with the expression for $N_\pm$, or, equivalently,

$$x_\sigma \;=\; \mu - \mu_B B \sigma - u - \frac{2u}{3\mathcal{N}} N_{-\sigma},$$

$$N_\sigma \;=\; \sum_{p,b} f_{p,\sigma} \;=\; \frac{1}{2} \int_{-\Delta}^{+\Delta} d\varepsilon \, \mathcal{D}(\varepsilon) \, \frac{1}{e^{\beta(\varepsilon - x_\sigma)} + 1}.$$

One can easily check that the equilibrium of a gas without interactions is found for $u = 0$.

If one treats the electrons with spin $-\sigma$ as a fixed continuum of density $N_{-\sigma}/\Omega$, as we did in (11.11) for the whole set of electrons, the expectation value of $\widehat{n}(R_j, b, -\sigma)$ is equal to $N_{-\sigma} a^3 / 3\Omega = N_{-\sigma}/3\mathcal{N}$. The form of $\widehat{V}$ implies that in this approximation the electrons with spin σ have single-particle energies $\varepsilon_p + \mu_B B \sigma + u + 2u N_{-\sigma}/3\mathcal{N}$. Writing down the associated Fermi factor with the same chemical potential for the two gases we find again the coupled equations for N_σ and x_σ.

2. For $Z = 27$ we have $N = 3\mathcal{N}$, and the conduction band is half filled. The equation $2(N_+ + N_-) - 6\mathcal{N} = 0$ leads to

$$
\begin{aligned}
0 &= \int_{-\Delta}^{\Delta} d\varepsilon \, \mathcal{D}(\varepsilon) \left[\frac{1}{e^{\beta(\varepsilon - x_+)} + 1} + \frac{1}{e^{\beta(\varepsilon - x_-)} - 1} - 1 \right] \\
&= \int_{-\Delta}^{\Delta} d\varepsilon \, \mathcal{D}(\varepsilon) \left[\frac{1}{e^{\beta(\varepsilon - x_+)} + 1} - \frac{1}{1 + e^{-\beta(\varepsilon - x_-)}} \right] \\
&= \int_{-\Delta}^{\Delta} d\varepsilon \, \mathcal{D}(\varepsilon) \left[\frac{1}{e^{\beta(\varepsilon - x_+)} + 1} - \frac{1}{e^{\beta(\varepsilon + x_-)} + 1} \right],
\end{aligned}
$$

where we have successively used the hole-particle symmetry and the relation $\mathcal{D}(\varepsilon) = \mathcal{D}(-\varepsilon)$. The square bracket has the same sign as $x_+ + x_-$; it vanishes only when $x_+ + x_- = 0$. Hence, it follows that $\mu = 2u$, giving us a chemical potential which is independent of T and B. The increase of μ with u reflects the fact that the repulsion between electrons tends to make it easier for them to be released to an external system.

The coupled equations for x_σ and N_σ lead, after subtraction, to

$$x = -\mu_B B - \frac{u}{N\mu_B} M,$$

$$M = -\mu_B(N_+ - N_-)$$

$$= \frac{1}{2} \mu_B \int_{-\Delta}^{\Delta} d\varepsilon \, \mathcal{D}(\varepsilon) \left[\frac{1}{e^{\beta(\varepsilon+x)} + 1} - \frac{1}{e^{\beta(\varepsilon-x)} + 1} \right],$$

or, if we use the Sommerfeld expansion, which assumes that $|x| < \Delta$, and integrate over ε,

$$M = -\frac{3N\mu_B x}{2\Delta} \left[1 - \frac{x^2}{3\Delta^2} - \frac{\pi^2}{3} \left(\frac{kT}{\Delta} \right)^2 \right].$$

If there is no field, the equation for x becomes

$$x = \frac{3u}{2\Delta} x \left[1 - \frac{x^2}{3\Delta^2} - \frac{\pi^2}{3} \left(\frac{kT}{\Delta} \right)^2 \right].$$

If $u < u_0$, where $u_0 \equiv \frac{2}{3}\Delta$, the only solution is $x = 0$, and hence $M = 0$, whatever the temperature: ferromagnetism is not possible. If $u > u_0$, and, more generally, if

$$\frac{u_0}{u} + \frac{\pi^2}{3} \left(\frac{kT}{\Delta} \right)^2 < 1,$$

there exists, apart from the $x = 0$ solution, two opposite solutions, given by

$$x = \pm \Delta \sqrt{3} \sqrt{1 - \frac{u_0}{u} - \frac{\pi^2}{3} \left(\frac{kT}{\Delta} \right)^2},$$

leading to a non-vanishing spontaneous magnetization, even though $B = 0$:

$$\frac{M_s}{\Omega} = \mp \mu_B \frac{N}{\Omega} \Delta \sqrt{3 - \frac{2\Delta}{u} - \pi^2 \left(\frac{kT}{\Delta} \right)^2}.$$

When we have only one solution, $x = 0$, $\mathcal{A}$ has its minimum for $x = 0$. When there are three solutions, the two non-trivial solutions appear, as one lets u increase, or T decrease, starting from the point $x = 0$. Moreover, $\mathcal{A}(x)$ is symmetric in x when $B = 0$. Hence, it follows from continuity considerations that $\mathcal{A}(x)$ can only have a maximum at $x = 0$, and two minima of the same depth. As we are looking for the absolute minimum of $\mathcal{A}(x)$, we must eliminate the solution $x = 0$, and the other two give us solutions with a non-vanishing spontaneous magnetization. The substance is therefore ferromagnetic. When $u > u_0 = \frac{2}{3}\Delta$, this behaviour occurs only if $T < T_c$, where the critical temperature is defined by

$$T_c = \frac{\Delta}{\pi k} \sqrt{3} \sqrt{1 - \frac{u_0}{u}};$$

if $T > T_c$, the only solution is the trivial $x = 0$ solution. The curve $M_s(T) = \pi k \mu_B N \sqrt{T_c^2 - T^2}$ is one quarter of an ellipse, and behaves as $\sqrt{T_c - T}$ near the Curie point.

For $B = 0$ the Hamiltonian is invariant under rotations, and for $T < T_c$ the spontaneous magnetization can be in any arbitrary direction: the invariance is broken (§ 9.3.3). However, our approximation has given us only two particular solutions where M_s is directed along $\pm z$, since $\widehat{\mathcal{D}}$ introduced a selected direction which prevented us from finding the general solution. When $B \neq 0$, one of the two minima of $\mathcal{A}(x)$ becomes deeper than the other one due to the term $-BM$. The presence of a field, even an infinitesimal one, thus selects that solution M which is oriented in the same direction as B.

Note. If we wish to avoid the continuity argument and show directly that, when we find three solutions, the point $x = 0$ must be eliminated, we can use Sommerfeld's formula, for $kT \ll \Delta - |x_\sigma|$, to calculate

$$\mathcal{A}(x_+, x_-) = \frac{9\mathcal{N}}{4\Delta} \sum_\sigma \left[\frac{1}{2} x_\sigma^2 - \frac{x_\sigma^4}{4\Delta^2} - \frac{1}{4} \Delta^2 - \frac{\pi^2}{6} (kT)^2 \left(1 + \frac{x_\sigma^2}{\Delta^2} \right) \right]$$

$$+ \sum_\sigma \left(\mu_B B \sigma + u - \mu \right) N_\sigma + \frac{2}{3} \frac{u}{\mathcal{N}} N_+ N_-,$$

$$N_\sigma = \frac{9\mathcal{N}}{4\Delta} \left[x_\sigma - \frac{x_\sigma^3}{3\Delta^2} + \frac{2}{3}\Delta - \frac{\pi^2}{3} \left(\frac{kT}{\Delta} \right)^2 x_\sigma \right].$$

For $B = 0$ and with $x_+ = -x_- = x$, $\mu = 2u$, this expression becomes

$$\mathcal{A}(x) = -\mathcal{N}\Delta \left[\frac{9}{8} + \frac{u}{u_0} + \frac{3\pi^2}{4} \left(\frac{kT}{\Delta} \right)^2 \right] - \frac{3\mathcal{N}}{8\Delta^3} x^4$$

$$+ \frac{9\mathcal{N}}{4\Delta} \frac{u}{u_0} \left[1 - \frac{x^2}{3\Delta^2} - \frac{\pi^2}{3} \left(\frac{kT}{\Delta} \right)^2 \right] \left[\frac{x^2}{3\Delta^2} - 1 + \frac{u_0}{u} + \frac{\pi^2}{3} \left(\frac{kT}{\Delta} \right)^2 \right] x^2.$$

The last term vanishes in each stationary point; hence, $\mathcal{A}(x) < \mathcal{A}(0)$ for a nontrivial $x \neq 0$ solution. Our calculations assume that $|x| < \Delta$, but one can check that the minimum of $\mathcal{A}(x)$ is always reached for $|x| \leq \Delta$; if $u > \frac{3}{2}u_0 = \Delta$, we have $x = \pm\Delta$ at $T = 0$, and in that case the spins of all the conduction electrons are pointing in the same direction: the spontaneous magnetization, $\mu_B N/\Omega$, is a maximum. We shall restrict ourselves to $u < \Delta$, a condition which is realized for all elements studied.

3. When $B \neq 0$, we find, after eliminating M, that x is determined by

$$x \left[1 - \frac{u_0}{u} - \frac{x^2}{3\Delta^2} - \frac{\pi^2}{3} \left(\frac{kT}{\Delta} \right)^2 \right] = \frac{u_0}{u} \mu_B B.$$

If $u < u_0$, or if $u > u_0$, $T > T_c$, x tends to 0 proportionally to B, as $B \to 0$; the same is true for M, and we find

$$\frac{M}{\Omega} = -\frac{N\mu_B}{\Omega u}\left[x + \mu_B B\right] \sim -\frac{N}{\Omega u}\mu_B^2 B\left[\frac{u_0}{u - u_0 - \frac{1}{3}\pi^2 u(kT/\Delta)^2} + 1\right],$$

$$\chi = \frac{N}{\Omega}\mu_B^2\frac{1 - \frac{1}{3}\pi^2(kT/\Delta)^2}{u_0 - u + \frac{1}{3}\pi^2 u(kT/\Delta)^2}.$$

The susceptibility is positive: the metal is paramagnetic. It decreases with increasing temperature, which reflects the increase in spin disorder with temperature when B is small, and fixed. When $u < u_0$, this decrease occurs starting from a finite value, and χ is larger than the Pauli susceptibility, which we get for $u = 0$. When $u > u_0$, $T > T_c$, the susceptibility,

$$\chi = \frac{N}{\Omega(u - u_0)}\mu_B^2\left[\frac{u_0}{u}\frac{T_c^2}{T^2 - T_c^2} - \left(1 - \frac{u_0}{u}\right)\right],$$

diverges as $T \to T_c + 0$.

At $T = T_c$ we find

$$-\frac{x^3}{3\Delta^2} = \frac{u_0}{u}\mu_B B,$$

and as a result the magnetization,

$$\frac{M}{\Omega} \sim \frac{N}{\Omega}\Delta\left(\frac{\mu_B}{u}\right)^{4/3}(2B)^{1/3},$$

varies more slowly than B, as $B^{\frac{1}{3}}$: the susceptibility diverges.

If x_0 is the value of x for $B = 0$, which was calculated sub 2, we find that $\delta x \equiv x - x_0$ is for $T < T_c$ given by the equation

$$-(x_0 + \delta x)\frac{1}{3\Delta^2}\left(2x_0\delta x + \delta x^2\right) = \frac{u_0}{u}\mu_B B,$$

whence, as $B \to 0$ and replacing x_0 by its value, we obtain

$$\frac{M - M_s}{\Omega} = -\frac{N\mu_B}{\Omega u}\left[\delta x + \mu_B B\right] \sim -\frac{N}{\Omega u}\mu_B^2 B\left[-\frac{3u_0\Delta^2}{2ux_0^2} + 1\right],$$

$$\chi = \frac{N}{\Omega(u - u_0)}\mu_B^2\left[\frac{u_0}{2u}\frac{T_c^2}{T_c^2 - T^2} - \left(1 - \frac{u_0}{u}\right)\right].$$

The susceptibility is thus again positive as in the paramagnetic phase. It increases with the temperature while M_s decreases, and tends to infinity as $T \to T_c - 0$. In both phases χ behaves in the vicinity of T_c as $|T - T_c|^{-1}$, but its coefficient on the ferromagnetic side is smaller by a factor 2.

The specific heat at constant N, which is here equivalent to at constant μ, equals

$$C = T\frac{dS}{dT} = -\beta\frac{dS}{d\beta}.$$

The entropy depends on the temperature directly through $y_q = \beta(\varepsilon_p - x_\sigma)$ and indirectly through the x_σ; this gives us

$$C = -\beta \frac{dS}{d\beta} = -k\beta \sum_q y_q \frac{\partial f_q}{\partial y_q} \left[(\varepsilon_p - x_\sigma) - \beta \frac{dx_\sigma}{d\beta} \right]$$

$$= -\frac{1}{2} k\beta \sum_\sigma \int_{-\Delta}^{\Delta} d\varepsilon\, \mathcal{D}(\varepsilon) \left(\varepsilon - x_\sigma \right)$$

$$\times \frac{\partial}{\partial \varepsilon} \frac{1}{e^{\beta(\varepsilon - x_\sigma)} + 1} \left[\varepsilon - x_\sigma - \beta \frac{dx_\sigma}{d\beta} \right].$$

In the paramagnetic phase $T > T_c$ we have $x_\sigma = 0$ and this expression reduces to

$$C_{\rm pm} = -\frac{1}{T} \int_{-\Delta}^{\Delta} d\varepsilon\, \mathcal{D}(\varepsilon) \varepsilon^2 f'(\varepsilon)$$

$$\simeq -\frac{9\mathcal{N}}{2T\Delta} \int_{-\Delta}^{\Delta} d\varepsilon \left(1 - \frac{\varepsilon^2}{\Delta^2} \right) \varepsilon^2 \left[-\delta(\varepsilon) - \tfrac{1}{6}\pi^2 (kT)^2 \delta''(\varepsilon) \right]$$

$$= \frac{\pi^2}{2} Nk \frac{kT}{\Delta}.$$

This result is the same as for a gas of non-interacting electrons. In the ferromagnetic phase ($u > u_0$, $T < T_c$) we have $x_+ = -x_- = x$ and, using the $\pm\varepsilon$ symmetry,

$$C_{\rm fm} = -\frac{1}{T} \int_{-\Delta}^{\Delta} d\varepsilon\, \mathcal{D}(\varepsilon) \left[(\varepsilon - x)^2 + T(\varepsilon - x) \frac{dx}{dT} \right] f'(\varepsilon - x).$$

Differentiating the equation which determines x as function of T we find

$$\frac{2x\,dx}{3\Delta^2} + \frac{2\pi^2}{3} \frac{k^2 T\,dT}{\Delta^2} = 0,$$

and hence, replacing x by its value,

$$C_{\rm fm} \simeq \frac{9\mathcal{N}}{2T\Delta} \int_{-\Delta}^{\Delta} d\varepsilon \left(1 - \frac{\varepsilon^2}{\Delta^2} \right) \left[(\varepsilon - x)^2 - (\varepsilon - x) \frac{\pi^2}{x} (kT)^2 \right]$$

$$\times \left[\delta(\varepsilon - x) + \frac{\pi^2}{6} (kT)^2 \delta''(\varepsilon - x) \right]$$

$$= \frac{\pi^2 N}{4\Delta} k^2 T \frac{d^2}{d\varepsilon^2} \left\{ \left(1 - \frac{\varepsilon^2}{\Delta^2} \right) \left[(\varepsilon - x)^2 - (\varepsilon - x) \frac{\pi^2}{x} (kT)^2 \right] \right\} \bigg|_{\varepsilon = x}$$

$$= \frac{\pi^2 N}{2\Delta} k^2 T \left[\left(1 - \frac{x^2}{\Delta^2} \right) + 2\pi^2 \left(\frac{kT}{\Delta} \right)^2 \right]$$

$$= \frac{\pi^2}{2} Nk \frac{kT}{\Delta} \left[1 + \pi^2 \left(\frac{kT_c}{\Delta} \right)^2 \left(3\frac{T^2}{T_c^2} - 1 \right) \right].$$

Apart from the linear contribution we get for $T < T_c$ an extra term which is associated with the spin order. It is negative when $T < T_c/\sqrt{3}$ and slightly reduces the slope at the origin. It is positive for $T_c/\sqrt{3} < T < T_c$ and gives rise to a discontinuity at $T = T_c$, corresponding to

$$C_{\text{fm}} - C_{\text{pm}} = \pi^4 N k \left(\frac{k T_c}{\Delta}\right)^3.$$

The specific heat decreases abruptly when T crosses T_c. It also has a slope discontinuity:

$$\frac{d C_{\text{fm}}}{dT} - \frac{d C_{\text{pm}}}{dT} = \frac{9\pi^4}{2} \frac{N k^2}{\Delta} \left(\frac{k T_c}{\Delta}\right)^2.$$

The entropy is continuous at T_c; its derivative is positive and decreases abruptly. Experiments confirm these predictions (Fig.9.10); we must remember to add the phonon contribution to the electron specific heat. The shape of the predicted anomaly differs here slightly from the one provided by the Ising model (Exerc.9a): especially, the value of the discontinuity in C is less simple than in the case of localized spins.

All results obtained here in the vicinity of T_c are in agreement with Landau's theory.

4. The number of conduction electrons per unit volume is

$$\frac{3\mathcal{N}}{\Omega} = \frac{N}{\Omega} = 3 \times 6 \times 10^{23} \times \frac{8.71 \times 10^3}{59 \times 10^{-3}} = 2.7 \times 10^{29} \text{ m}^{-3},$$

so that $a = (\Omega/\mathcal{N})^{1/3} = 2.2$ Å. The bandwidth follows from $\mathcal{D}(\varepsilon_F)/\Omega = 9\mathcal{N}/2\Omega\Delta$, whence we find $2\Delta = 1.7$ eV, and for the limiting interaction $u_0 = \frac{2}{3}\Delta \simeq 0.57$ eV. The band being half filled we precisely satisfy the conditions of the questions sub 2 and 3 so that

$$1 - \frac{u_0}{u} = \frac{\pi^2}{3} \left(\frac{k T_c}{\Delta}\right)^2 = \frac{\pi^2}{3} \left(\frac{1.38 \times 10^{-23} \times 1404}{0.5 \times 1.7 \times 1.6 \times 10^{-19}}\right)^2 = 6.7 \times 10^{-2},$$

and hence

$$u \simeq 1.07 \, u_0 \simeq 0.6 \text{ eV}.$$

5. In terms of the variables z, x, N, and M, the coupled equations for x_σ and N_σ can, in the low-temperature expansion, be written as

$$z = \mu - u - \frac{uN}{3\mathcal{N}}, \qquad x = -\mu_B B - \frac{uM}{3\mathcal{N}\mu_B},$$

$$N_\sigma = \frac{1}{2}\left(N - \frac{M\sigma}{\mu_B}\right) = \frac{1}{2} \int_{-\Delta}^{z+\sigma x} d\varepsilon \, \mathcal{D}(\varepsilon) + \frac{\pi^2}{12} (kT)^2 \, \mathcal{D}'(z + \sigma x).$$

In the vicinity of the Curie point x is small, for $B = 0$, and we find

$$N = \int_{-\Delta}^{z} d\varepsilon \, \mathcal{D}(\varepsilon) + \frac{1}{6}\pi^2 (kT)^2 \mathcal{D}'(z) = \frac{3N}{u}(\mu - u - z),$$

$$-\frac{M_s}{\mu_B} = \frac{3N}{u} x = x\mathcal{D}(z) + \frac{1}{6}x^3 \mathcal{D}''(z) + \frac{1}{6}\pi^2 (kT)^2 x\mathcal{D}''(z).$$

The two equations which connect N and z determine μ as function of T and N. The equation for x has either the trivial, paramagnetic, $x = 0$ solution, or the ferromagnetic solutions given by

$$-\frac{1}{6}x^2 \mathcal{D}''(z) = \mathcal{D}(z) - \frac{3N}{u} + \frac{1}{6}\pi^2 (kT)^2 \mathcal{D}''(z).$$

Since $\mathcal{D}'' < 0$, those solutions exist, provided the right-hand side be positive, which implies that

$$u\mathcal{D}(z) > 3N, \qquad T \leq T_c, \qquad \frac{1}{6}\pi^2 (kT_c)^2 = \left.\frac{3N - u\mathcal{D}(z)}{u\mathcal{D}''(z)}\right|_{z=z(T_c,N)}.$$

If it exists, T_c is determined by coupled equations for z, T_c, and N.

Since $kT_c \ll \Delta$ we can use Sommerfeld's expansion to determine z. Let us put $z_0 \equiv \varepsilon_F - u - uN/3N$ for the value of z at zero temperature, which is given as a function of the band filling by

$$N = \int_{-\Delta}^{z_0} d\varepsilon \, \mathcal{D}(\varepsilon) = 3N \left[1 + \frac{3z_0}{2\Delta} - \frac{1}{2}\left(\frac{z_0}{\Delta}\right)^3\right].$$

Hence we find

$$z = z_0 - \frac{\pi^2}{6}(kT)^2 \frac{\mathcal{D}'(z_0)}{\mathcal{D}(z_0)},$$

so that the Curie temperature is determined by

$$\frac{\pi^2}{6}(kT_c)^2 \left[\frac{\mathcal{D}'(z_0)^2}{\mathcal{D}(z_0)} - \mathcal{D}''(z_0)\right] = \mathcal{D}(z_0) - \frac{3N}{u},$$

or, finally,

$$\frac{\pi^2}{3}\left(\frac{kT_c}{\Delta}\right)^2 = \frac{\Delta^2 - z_0^2}{\Delta^2 + z_0^2}\left(1 - \frac{z_0^2}{\Delta^2} - \frac{u_0}{u}\right).$$

If $u < u_0$, the metal is never ferromagnetic. If $u > u_0$, the maximum value of T_c is reached when $z_0 = 0$, that is, for the half-filled band which we studied above. If $|z_0|$ increases, T_c decreases, until it vanishes when

$$z_0 = \pm \Delta \sqrt{1 - \frac{u_0}{u}}.$$

These values correspond to

$$N = 3\mathcal{N} \left[1 \pm \left(1 + \frac{u_0}{2u} \right) \sqrt{1 - \frac{u_0}{u}} \right].$$

The metal can thus never be ferromagnetic, whatever the temperature, if N lies outside that range, the centre of which corresponds to half filling the conduction band.

Under the conditions in the statement of the problem, the region where the metal can be ferromagnetic at low temperatures is given by

$$|Z - 27| < 3 \left(1 + \frac{u_0}{2u} \right) \sqrt{1 - \frac{u_0}{u}} \simeq 1.1.$$

This excludes all elements bar those with $Z = 26$, 27, or 28, in agreement with the experimental evidence. For $Z = 28$ we find for z_0 the equation

$$\frac{N - 3}{\mathcal{N}} = Z - 27 = \frac{9 z_0}{2\Delta} - \frac{3}{2} \left(\frac{z_0}{\Delta} \right)^2,$$

whence $z_0 = 0.23\Delta$. From this we get for the Curie temperature

$$T_{\mathrm{c}} = \frac{\Delta}{k} \frac{\sqrt{3}}{\pi} \sqrt{\frac{\Delta^2 - z_0^2}{\Delta^2 + z_0^2} \left(1 - \frac{z_0^2}{\Delta^2} - \frac{u_0}{u} \right)} = 640 \text{ K},$$

in agreement with the experimental value for Ni. For $Z = 26$ we find $z_0 = -0.23\Delta$ and the same Curie temperature, whereas the experimental data for Fe give a higher value, albeit lower than for Co. Even though the model is coarse, it reproduces the actual facts qualitatively correctly, namely, the maximum of T_{c} in the middle of the band for Co, and the presence of ferromagnetism only for the elements next to it.

Note. The mean field theory used here neglects the effects of fluctuations which tend to hinder the establishment of ferromagnetism. To obtain realistic values for Co and Ni we have deliberately underestimated the band width 2Δ, which is in actual fact of the order of 10 eV. Moreover, the shape of the conduction band is more complex. The elements with $24 \leq Z \leq 30$, considered here, are in the periodic table of elements situated in the middle of the fourth row, $18 \leq Z \leq 36$, which corresponds to filling the 4s, 3d, and 4p shells. These shells give rise to many bands which more or less overlap, and which change their shape when they are being filled. As a result, the density of states is not symmetric, and this is shown by the difference between Fe and Ni; it varies, like u, from one element to another. Finally, our model would give an exactly empty band for Cr, and an exactly filled band for Zn, which would therefore be insulators; in reality, the density of states at the Fermi level, although smaller than for the elements in the cobalt region, does not vanish, and Cr and Zn are metals.

The fluctuations do not only reduce the Curie temperature, they also change the critical behaviour: the results of the question sub 3 give curves which are roughly correct, but they do not agree with experiment as $T \to T_{\mathrm{c}}$.

12. Liquid Helium

"Les soleils mouillés ...

Ch.Baudelaire, Les Fleurs du Mal

"Il est de forts parfums pour qui toute matière
Est poreuse. On dirait qu'ils pénètrent le verre."

ibidem

"La Science, la nouvelle noblesse! Le progrès.
Le monde marche!"

A.Rimbaud, Une Saison en Enfer

In the present short chapter we shall give a sketch of the remarkable properties of helium at temperatures below a few K. The specific effects which occur then have been, and are still, the subject of many investigations. Their thorough study goes beyond the framework of the present book, but their scentific interest justifies at least a short account. Helium in its two isotopic forms ^{3}He and ^{4}He has the peculiar property of remaining liquid down to zero temperature at ordinary pressures, whereas all other simple substances crystallize. Nevertheless, because of the small mass of the atoms and the low temperatures, the translational degrees of freedom are frozen in and cannot be treated classically: we have *quantum fluids* which are different for the two isotopes, as ^{3}He is a fermion and ^{4}He is a boson (§ 12.1).

Helium 3 is a Fermi liquid at low temperatures: its equation of state (§ 12.2.1) and its specific heat (§ 12.2.2) are governed by Pauli's exclusion principle. We shall also see that the disorder due to the two spin states of each of the atomic nuclei in the solid phase makes this solid phase within a certain range of pressures less stable than the liquid when one lowers the temperature (§ 12.2.3).

Helium 4 resembles a Bose-Einstein gas, even though the interactions between the atoms play an important rôle. The gregarious nature of the bosons shows up in their condensation at low temperatures which gives rise to a new ordered state of matter (§§ 12.3.1 and 12.3.2). As a result, helium 4 shows around 2 K a phase transition, from the ordinary liquid phase to a new super-fluid liquid phase which can flow without viscosity and which in some respects looks like a wave (§ 12.3.3). This is not just a laboratory curiosity, as the phenomenon is related to the superconductivity of certain substances, which lose their resistivity below a certain temperature, and which have started to become the subject of industrial applications. We shall restrict ourselves to qualitative discussions, as the theory of superconductivity and of super-fluidity requires that one takes into account interactions between particles or quasi-particles by using the techniques of the many-body problem, which are too complicated to be included in the present book. However, as an exercise (Prob.14) we shall show how the Landau model gives a rudimentary explanation of the superfluidity of helium. A useful introduction to this vast subject can be found in the book by D.R. and J.Tilley (*Superfluidity and Superconductivity*, Adam Hilger, Bristol, 1986).

12.1 Peculiar Properties of Helium

12.1.1 Phase Diagrams

Among all substances helium has one remarkable special property. At not too high pressures it *does not solidify at low temperatures.*

Natural helium consists practically 100 % of the ^{4}He isotope. Its phase diagram (Fig.12.1) shows that one can only obtain solid helium for pressures above 25 atm. One also sees that helium has *two liquid phases* which are separated by a transition curve around 2 K.

There is also a lighter stable isotope, ^{3}He, which is present in natural helium as a trace (1.4×10^{-6}), and the phases of which have also been studied, after it has been isolated and purified. Roughly speaking its phase diagram (Fig.12.2) resembles that of ^{4}He: the stability region of the solid phase is again bounded on the low pressure side, in contrast to the usual phase dia-

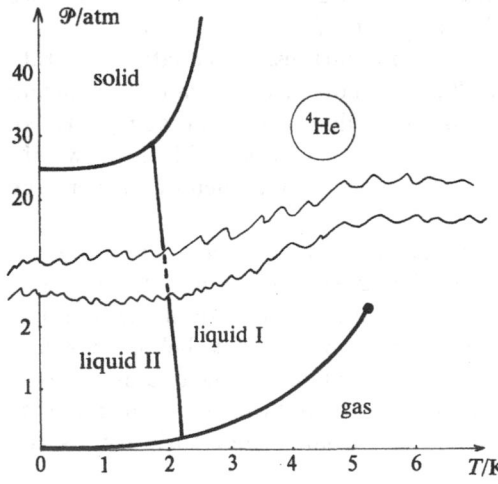

Fig. 12.1. Phase diagram of ^{4}He

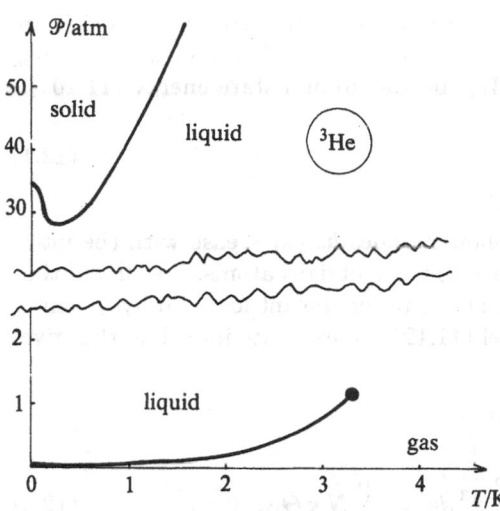

Fig. 12.2. Phase diagram of ^{3}He

grams where we have a solid-gas sublimation line passing through the origin. However, the melting curve shows a *minimum* and, in the range of temperatures of the order of kelvins, there exists only one liquid phase, in contrast to the case of ^{4}He. The difference in the atomic mass is insufficient to explain these qualitative differences which show directly the need to appeal to the Pauli principle to understand the properties of helium in the kelvin range of temperatures.

New effects appear in ^{3}He when one lowers the temperature to the millikelvin region: one has found two other liquid phases, predicted by theory, which show unusual properties (end of § 12.3.3). The scale of Fig.12.2 is too small to show these phases.

12.1.2 Quantum Liquids

The reason why the low pressure and low temperature phase is not a solid for helium is a *quantum phenomenon* connected with the *small mass* of its atoms. Let us remind ourselves of the discussion in § 11.1.1: we find the state of the crystal for the case of nuclei with an infinite mass, at low pressures and low temperatures, by looking for the minimum W_m of the total effective interaction potential $W(\{\boldsymbol{R}_n\})$ between the nuclei, which is reached when they occupy the sites $\{\overline{\boldsymbol{R}_n}\}$ of a certain lattice. The order of magnitude of W_m/N is that of the minimum of the effective potential W between two atoms. For the inert gases this is very small (see Fig.9.1 for the potential W for argon) and especially for helium where the electrons are very strongly bound to the nucleus; we get in this case for W_m/N an estimate of -2.6×10^{-3} eV per atom.

However, in order to take into account the lattice vibrations which are quantized as a system of harmonic oscillators (§ 11.4.1), we must add to W_m the *zero-point energy* and replace W_m by the ground state energy (11.109):

$$W_m + \frac{1}{2} \sum_k \hbar\omega_k. \tag{12.1}$$

We see from (11.114) that the phonon energies $\hbar\omega_k$ decrease with the interactions, but increase when the lattice consists of light atoms. We should thus expect that the second term in (12.1) can be significant for solid hydrogen or helium. Let us use the Debye model (11.125) to estimate its value; this gives

$$\frac{1}{2} \sum_k \hbar\omega_k = \frac{1}{2} \int_0^\infty d\varepsilon\, \mathcal{D}(\varepsilon)\, \varepsilon$$

$$\simeq \frac{9N}{2(k\Theta_D)^3} \int_0^{k\Theta_D} \varepsilon^3\, d\varepsilon = \frac{9}{8} N k \Theta_D. \tag{12.2}$$

For He Θ_D is of the order of 20 K; this gives a zero-point energy of 2×10^{-3} eV per atom, which practically completely cancels the weak binding energy of the lattice (-2.6×10^{-3} eV per atom). For H the binding energy of two atoms is large (4.7 eV) so that we have to deal with H_2 molecules; those are lighter than He and have a fairly large zero-point energy, but the attraction W_m between H_2 molecules is much larger than for He; as a result, at ordinary pressures and low temperatures, hydrogen is a molecular crystal in spite of the zero-point fluctuations.

The combination of two effects, the small value of the minimum of the potential between atoms (inert gas) and the large zero-point energy (small mass), thus makes helium an exceptional substance in which the *binding energy in the solid state is very small*. One can thus understand that this binding energy may be smaller than that of the liquid state. In fact, even though in the liquid state the average potential energy is higher, the zero-point energy is smaller, as the amplitudes of the displacements are larger. This therefore explains why *at low pressures, the liquid state is the stable one*. Naturally, at higher pressures, the solid state is the stable one: looking for the minimum of the grand potential gives us the phase diagram, as in §§ 9.3.2 and 11.1.2, and denser states are preferred when the pressure (and hence μ) increases, because of the presence of the term $-\mu N$ in $A = U - TS - \mu N$. The model studied in Prob.13 confirms these qualitative indications.

Nevertheless, in ^{3}He and ^{4}He in the liquid state below a few K, *quantum effects* play an essential rôle. Let us, first of all, note that ^{3}He, which contains an odd number of fermions, behaves like a fermion, as the exchange of two atoms changes the sign of the wavefunction, whereas a ^{4}He atom, which contains an even number of fermions, behaves like a boson. To understand why helium at low temperatures is a quantum liquid, and not an ordinary liquid, let us remind ourselves of the condition (10.49) and (10.50) for the classical approximation,

$$\frac{N}{\Omega} \left(\frac{2\pi\hbar^2}{mkT} \right)^{3/2} \ll 1. \tag{12.3}$$

At the density of liquid helium, about 0.1 g cm^{-3}, the left-hand side is of the order of 1 at a temperature of about 5 K. Use of *quantum statistics*, Fermi-Dirac for ^{3}He and Bose-Einstein for ^{4}He, thus becomes necessary at low temperatures. In what follows we shall neglect the *interactions* between the atoms, which is hardly justified at liquid densities. Nevertheless, the principal characteristics of ^{3}He and ^{4}He will be rather well explained in this approximation.

The Pauli principle for the nuclei also plays a rôle at higher pressures in the dynamics of the lattice of solid helium. In fact, because they interact so weakly and their masses are so small, the atoms oscillate with amplitudes which are significant compared to their distances apart. The theory of phonons (§ 11.4.1) must thus be changed, as now the wavefunctions of the oscillators centred on neighbouring lattice sites overlap.

12.2 Helium Three

12.2.1 Equation of State

We describe ^{3}He at low temperatures as a Fermi-Dirac gas, neglecting the interactions between the atoms. The grand potential is then given by (10.47), where the number of single-particle states $\mathcal{N}(\varepsilon)$ with energies below ε is given by (10.45), so that

$$A = -\mathcal{P}\Omega = -\frac{\Omega(2m)^{3/2}}{3\pi^2\hbar^3} \int_0^\infty d\varepsilon \, f(\varepsilon) \, \varepsilon^{3/2}. \tag{12.4}$$

Hence, we get, through differentiation, the number of particles,

$$N = \frac{\Omega(2m)^{3/2}}{2\pi^2\hbar^3} \int_0^\infty d\varepsilon \, f(\varepsilon) \, \varepsilon^{1/2}, \tag{12.5}$$

the internal energy,

$$U = -\tfrac{3}{2} A = \tfrac{3}{2} \mathcal{P}\Omega, \tag{12.6}$$

and the entropy,

$$S = \frac{1}{T}(U - A - \mu N) = \frac{1}{T}\left(-\frac{5}{2} A - \mu N \right). \tag{12.7}$$

At high temperatures and low densities these expressions are the same as those for a classical perfect gas, but they are different in the quantum region. In particular, at sufficiently low temperatures we get, by eliminating μ from (10.65) and (10.73) and using (10.43), (10.45), and (10.58),

$$A = -\frac{2\Omega(2m)^{3/2}}{15\pi^2\hbar^3}\,(k\Theta_F)^{5/2}\left[1 + \frac{5\pi^2}{12}\left(\frac{T}{\Theta_F}\right)^2 + \ldots\right]. \tag{12.8}$$

If we take the mass density of helium to be 0.082 g cm^{-3}, the experimental value at ordinary pressures and very low temperatures, we find that the Fermi temperature Θ_F, defined by (10.58), is 5 K, which is 10 000 times lower than the electron Fermi temperatures in metals. Therefore, whereas we can treat the electron cloud in a metal as a Fermi gas at all temperatures, for ^{3}He the expansion (12.8) becomes valid only *well below the Fermi temperature* $\Theta_F = 5$ K.

In that low-temperature region the Fermi fluid shows a behaviour which is very different from that of a classical gas. In particular, the *equation of state* obtained by eliminating μ from (12.4) and (12.5) becomes, in the (12.8) and (10.58) low-temperature approximation,

$$\mathcal{P}^3\left(\frac{\Omega}{N}\right)^5 = \frac{9\pi^4\hbar^6}{125m^3}\left[1 + \frac{5\pi^2m^2k^2T^2}{\hbar^4}\left(\frac{\Omega}{3\pi^2N}\right)^{4/3}\right]. \tag{12.9}$$

As we are neglecting the interactions, we can expect only qualitative agreement of this equation of state with experimental data for ^{3}He below 5 K. Nevertheless, the order of magnitude of the density following from (12.9) is correct since at atmospheric pressure (12.9) gives a density of 0.033 g cm^{-3} as against the measured density of 0.082 g cm^{-3}. The shape of the isotherms is also predicted by (12.9) with a reasonable accuracy. In § 12.3.2 we shall see that such, at least qualitative, successes are due to Pauli's exclusion principle which reduces the effects of interactions.

The *Landau theory* enables one to obtain a quantitative agreement with most experimental data for ^{3}He, by taking interactions into account in a simple manner. The idea is the same as in § 11.2.1, where we replaced the interactions between the electrons in a solid by an effective self-consistent potential. Here ^{3}He is considered as a gas of non-interacting quasi-particles, each of which represents a "dressed" atom in the average potential of those surrounding it. Since most properties depend only on what happens in the vicinity of the Fermi surface, as shown by Eqs.(10.65)–(10.68), we need only find out how the density of states $\mathcal{D}(\varepsilon)$ is changed by this effective potential near $\varepsilon \sim \varepsilon_F$. As for the electrons in a solid, this change can be represented by a simple change in the mass m of the atoms to an *effective mass* m^*. The ratio m^*/m varies with T and $\mathcal{P}$, typically between 3 and 5. This increase in the mass is a consequence of the fact that an atom partially drags its neighbours along. It significantly reduces the Fermi temperature, and makes (12.9) a quantitative relation.

12.2.2 Thermal Properties

The relation (12.6) between the internal energy and the pressure is the same as for a perfect classical gas. Nevertheless, because the equation of state has a different form, the internal energy per atom is no longer only a function of the temperature, but it depends also on the density: the Joule law does not hold for the Fermi-Dirac gas.

The *specific heat* at constant volume is the derivative of the internal energy (12.6) with respect to T, for constant N and Ω. It is shown in Fig.12.3: at temperatures well above the Fermi temperature, in practice above a few K, we recover the value $\frac{3}{2}Nk$ of a perfect classical gas. At low temperatures the specific heat decreases, and tends to zero, as it should, at zero temperature (§ 6.5.2); its behaviour, given by (10.75),

$$C_{\mathrm{v}} \sim \frac{1}{2}\pi^2 Nk \frac{T}{\Theta_{\mathrm{F}}}, \tag{12.10}$$

is *linear*. Experiments on liquid ^{3}He confirm these predictions about the specific heat, which is again remarkable in view of the fact that we have neglected the interatomic interactions. Nevertheless, we should calculate Θ_{F} using the effective mass of the atoms, which is increased through the interactions.

Finally, the *entropy* (12.7), which at high temperatures is the same as expression (8.15) for a classical perfect gas, also tends linearly to zero at low temperatures, as

$$S \sim \frac{1}{2}\pi^2 Nk \frac{T}{\Theta_{\mathrm{F}}}. \tag{12.11}$$

We thus check the *Nernst principle* for liquid ^{3}He: the vanishing of S at zero temperature comes about because the ground state, obtained by populating all single-particle states up to the Fermi level, is non-degenerate.

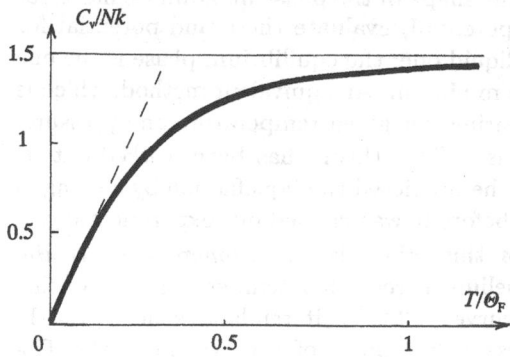

Fig. 12.3. Specific heat of ^{3}He

12.2.3 Solid-Liquid Transition

One consequence of the form (12.11) of the entropy of liquid ^{3}He at low temperatures is a rather surprising feature of the phase diagram given in Fig.12.2. If one cools liquid ^{3}He down at a pressure of around 30 atmospheres, it starts by solidifying, which is normal. However, if one continues the cooling, it *again becomes liquid*! As cooling a system means that we bring it into states which are more and more ordered, this means that at low temperatures, in the region of the pressures considered, *liquid* ^{3}He is more ordered than solid ^{3}He. This looks paradoxical, as the atoms are regularly arranged in a solid, but not in a liquid.

To understand this effect, let us consider solid ^{3}He: its Debye temperature being 20 K, the lattice vibrations are frozen in below 1 K, and the corresponding entropy is zero. However, the existence of a spin $\frac{1}{2}$ for the nuclei is an extra disorder factor, as each of them can independently be in *two spin states*. The entropy of the crystal thus remains constant and equal to

$$S_{\mathrm{cr}} = Nk \ln 2 \tag{12.12}$$

at temperatures in the kelvin range. The nuclear spins start to align themselves and the crystal entropy starts to decrease only at temperatures below 10^{-9} K: the energy of the interaction between the magnetic moments μ of the nuclei is, indeed, of the order of $(\mu_0/4\pi)(\mu^2/r^3) \simeq 10^{-13}$ eV $\simeq 10^{-9}$ K, because these magnetic moments are so small.

We bear in mind (§ 8.3.1) that at low densities and high temperatures, when the classical approximation is justified, the entropy also contains a spin contribution, equal to (12.12). However, at low temperatures, when the effects of the Fermi statistics become important, the entropy of the fluid, given by (12.11), tends to zero, even though we are taking the nuclear spin into account. *The lack of spatial order in the liquid is compensated by an ordering of the nuclear spins*, which is much larger than in the solid phase, so that at zero temperature the liquid is more ordered than the solid. To produce a less qualitative theory and to explain the shape of the phase diagram we must, for given temperature and chemical potential, evaluate the grand potential for the two phases, the solid and the liquid one; the equilibrium phase is the one which makes the grand potential a minimum. An equivalent method, which is used in Prob.13, consists in comparing, for given temperature and pressure, the free enthalpies in the two phases. This theory has been worked out by Pomeranchuk in 1950; remarkably he predicted the liquefaction by cooling of ^{3}He around 30 atmospheres long before it was carried out experimentally.

An interesting application of this effect is the *Pomeranchuk cooling* method. If we compress liquid helium three, at a temperature below that of the minimum of the melting curve (0.32 K), it tends to solidify. As the solidification proceeds, the representative point of the system in the $T, \mathcal{P}$ plane moves along the melting curve (Fig.12.4); the pressure increases as the proportion of solid increases, while the temperature decreases. To better un-

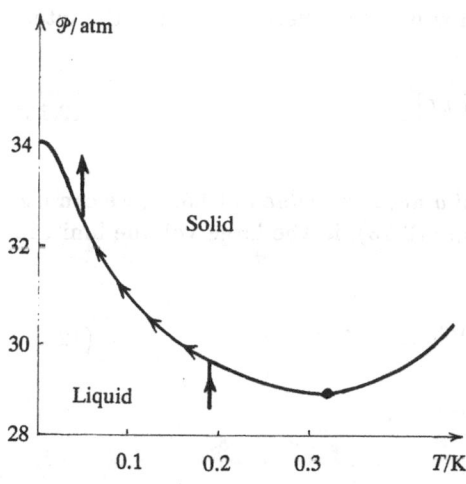

Fig. 12.4. Pomeranchuk cooling

derstand this phenomenon we note that the transformation takes place while the entropy remains practically constant, as the system does not exchange heat during the compression. Since the compression increases the proportion of the (denser) solid phase, it tends to increase the spin disorder, which is larger in the solid than in the liquid phase; this increase must therefore be compensated by a decrease in the disorder of the translational degrees of freedom – the lattice vibrations in the solid and the translations of atoms in the liquid – hence by a cooling. *Order is transferred* from the spins in the liquid to the crystal vibrations, the amplitude of which thus decreases.

The Pomeranchuk cooling method is very efficient. It is used in laboratories, like adiabatic demagnetization (§ 1.4.4), to reach very low temperatures, of the order of mK, in macroscopic samples.

The coupling between the nuclear moments connected with the spin of ^{3}He and an external field produces interesting *magnetic* effects. In particular, if all spins are polarized, everything happens as if the atoms had no spin. The density of states is divided by 2 so that the Fermi level increases for a given density of atoms; the Pomeranchuk effect disappears (Prob.13).

12.3 Helium Four and Bose Condensation

12.3.1 Bose Condensation

The ^{4}He isotope of helium, which is by far the more common one in nature, differs from ^{3}He by its nuclear mass and the absence of spin. However, the most important difference at temperatures of a few K, where the translational degrees of freedom must be treated quantum mechanically, is that the ^{4}He atoms have a *boson* nature.

The *grand potential* is, in the approximation where we neglect the inter-actions, given by (10.29), that is

$$A = kT \sum_{p} \ln \left[1 - e^{-(p^2/2m - \mu)/kT} \right].$$
(12.13)

When the *chemical potential is fixed at a negative value* (§ 10.5.2), we can use (10.47) and (10.45) to write expression (12.13), in the large volume limit, in the form

$$A = -\frac{\Omega(2m)^{3/2}}{6\pi^2\hbar^3} \int_0^\infty d\varepsilon \, f(\varepsilon) \, \varepsilon^{3/2} = -\mathcal{P}\Omega,$$
(12.14)

where $f(\varepsilon)$ is the Bose factor,

$$f(\varepsilon) = \frac{1}{e^{(\varepsilon - \mu)/kT} - 1}.$$
(12.15)

Hence we get the internal energy

$$U = \tfrac{3}{2}\mathcal{P}\,\Omega,$$
(12.16)

and the relation between the number of particles and the chemical potential

$$N = \sum_p f(\frac{p^2}{2m}) \implies \frac{\Omega(2m)^{3/2}}{4\pi^2\hbar^3} \int_0^\infty d\varepsilon \, f(\varepsilon) \, \varepsilon^{1/2}, \quad \mu < 0.$$
(12.17)

The high temperature limit is the same as for a perfect gas or for helium three. At temperatures of a few K, where one can no longer neglect the 1 in the denominator in (12.15), the equation of state and the thermodynamic properties are considerably changed. In particular, a new effect occurs, when one *cools down helium at constant density*.

Equation (12.17) defines the chemical potential μ as function of density and temperature for $\mu < 0$. Figure 12.5 shows this relation between N and μ at various temperatures. For a given value of T, N remains bounded as μ increases up to its bound $\mu = -0$. When the temperature decreases at constant density, μ increases. For a certain temperature T_λ, which is related to the density N/Ω through (see formulae at the end of this volume)

$$\frac{N}{\Omega} = \frac{(2m)^{3/2}}{4\pi^2\hbar^3} \int_0^\infty \frac{\sqrt{\varepsilon}\,d\varepsilon}{e^{\varepsilon/kT_\lambda} - 1} = 2.61 \left(\frac{mkT_\lambda}{2\pi\hbar^2} \right)^{3/2},$$
(12.18)

μ becomes equal to zero. One can thus find μ from (12.17), for given N/Ω and T, only if $T > T_\lambda(N/\Omega)$. The above equations (12.14) and (12.17), which have been derived for $\mu < 0$, are therefore no longer valid at temperatures $T \leq T_\lambda$

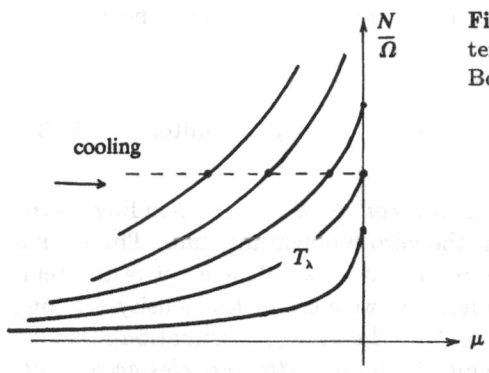

Fig. 12.5. Relation between chemical potential, density, and temperature for a Bose gas

for the density considered. We must, in particular, reconsider the solution of (12.17) for μ values close to zero. We have assumed that, as $\Omega \to \infty$, we could replace the discrete sum over all plane waves $\boldsymbol{p}$ by an integral; this is justified as long as the integrand is continuous, since we are then dealing with the definition itself of a Riemann integral, but not as $\mu \to 0$, when the integrand is no longer bounded. In that case we must analyze the contribution from the levels with the lowest (discrete) energies, corresponding to small values of $\boldsymbol{p}$, before letting the volume tend to infinity. If we use periodic boundary conditions, the plane wave $\boldsymbol{p} = 0$, with zero energy, has an average occupation number equal to

$$N_0 \;=\; f(0) \;=\; \frac{1}{e^{-\mu/kT} - 1} \;\sim\; \frac{kT}{|\mu|}. \tag{12.19}$$

By itself this single-particle state can make a significant contribution to the density N/Ω of the fluid: it is sufficient for μ to go to zero in such a way that $|\mu|\Omega$ remains finite as Ω tends to infinity. Whereas normally each single-particle state is on average occupied by a finite number $f(\varepsilon)$ of particles, a *macroscopic number of bosons is concentrated into a single state* $\boldsymbol{p} = 0$, that of the lowest energy. This spectacular consequence of the gregarious nature of bosons is called *Bose condensation* (F. London, 1938).

The occupation factors of the lowest excited states with energies $\varepsilon_{\boldsymbol{p}} = p^2/2m \propto \Omega^{-2/3}$ are also large, but only of order $kT/\varepsilon_{\boldsymbol{p}} \propto \Omega^{2/3}$. Let us show that their contribution to the total density is not pathological. If we split off from the sum (12.17) the term $\boldsymbol{p} = 0$, there are no problems in taking the limit as $\Omega \to \infty, \mu \to -0$, and we can easily prove that

$$\lim_{\substack{\Omega \to \infty \\ \mu \to -0}} \frac{1}{\Omega} \sum{}' f(\varepsilon_{\boldsymbol{p}}) \;=\; \frac{1}{h^3} \int \frac{d^3\boldsymbol{p}}{e^{\beta p^2/2m} - 1}.$$

In fact, the integrand is a decreasing function of the components of $\boldsymbol{p}$; this enables us to bound the sum by larger and smaller values through two integrals which tend to each other and to the integral on the right-hand side.

Altogether, equation (12.17) which connects the density with the chemical potential is for $T < T_\lambda$ replaced by

$$\frac{N}{\Omega} = \frac{kT}{|\mu|\Omega} + 2.61 \left(\frac{mkT}{2\pi\hbar^2}\right)^{3/2}, \qquad \mu \to -0, \quad \mu\Omega \quad \text{finite.} \qquad (12.20)$$

The first term is the density N_0/Ω of the *condensate*, corresponding to the particles which have accumulated in the zero momentum state. The second term, the density of particles in the other states, is independent of the temperature: therefore, if at a given temperature we increase the density, starting from the critical value $2.61(mkT/2\pi\hbar^2)^{3/2}$, the average occupation numbers of all states with $p \neq 0$ remain unchanged, and *all extra particles accumulate into the $p = 0$ state.*

The Bose condensation is associated with the convergence of the integral (12.17) as $\mu \to 0$, that is, of $\int d^n p/\varepsilon(p)$ around the origin of p. It would not occur for $n \leq 2$ dimensions with a spectrum $\varepsilon(p) \propto p^2$, nor for three dimensions with a spectrum $\varepsilon(p) \propto p^m$ with $m \geq 3$. Moreover, we see that the conservation of the number of bosons plays an essential rôle. For a gas of photons or phonons, where the number of quasi-particles is not conserved, μ is always equal to zero. The number of bosons in the state with the lowest energy $\varepsilon_0(> 0)$ is then large as $1/\beta\varepsilon_0$ at all temperatures, without any qualitative changes as happens here with N_0/Ω.

The boundary conditions also play an important rôle. In the above case of periodic boundary conditions, the Bose condensation occurs in the zero momentum state which has a constant wavefunction, so that the spatial density N_0/Ω of the condensate is uniform. If non-interacting bosons are enclosed in a cubic box of edge length L with rigid walls, the single-particle ground state has a wavefunction

$$\psi_0(r) = \left(\frac{2}{L}\right)^{3/2} \sin\frac{\pi x}{L} \sin\frac{\pi y}{L} \sin\frac{\pi z}{L},$$

and an energy $\varepsilon_0 = \pi^2\hbar^2/2mL^2$. The Bose condensation occurs when $\varepsilon_0 - \mu \propto \Omega^{-1}$ in this state ψ_0 which will then acquire a macroscopic average occupation number N_0. As a result, the condensate has a density $N_0|\psi_0(r)|^2$ which varies widely in space, over macroscopic distances of order L. This great sensitivity of the equilibrium state of the gas of bosons to external influences, such as the surface conditions, also exists in liquid helium (§ 12.3.3), but it is less pronounced, because of the interactions between the atoms. In particular, the condensate there is uniform, but not necessarily with zero momentum: the wavefunction $\psi_p(r) = \exp[i(p \cdot r)/\hbar]/\sqrt{\Omega}$ of the macroscopically occupied state varies, depending on the circumstances, even though $|\psi_p(r)|^2$ always remains constant in space.

The fact that at equilibrium the density depends on the boundary conditions, at macroscopic distances from the walls so that the system is *not extensive*, is a specific pathological property of a gas of non-interacting bosons at temperatures below T_λ. For most systems, on the contrary, the density is not affected by the presence of a wall, except over finite distances from it, which are small compared to L. As an exercise, one could check, using (11.15), that in a box with rigid walls the density vanishes at the walls, but becomes practically constant at a distance from the wall which is independent of L, both for a gas of fermions at all temperatures, and for a gas of bosons at temperatures $T > T_\lambda$.

Note that the single-particle state $p = 0$ makes a contribution $kT \ln(1 - e^{\mu/kT})$ to the grand potential which must be added to the integral (12.14). This term looks negligible, since it is not extensive; even in the limit as $\mu \rightarrow -0$ with $\mu\Omega$ finite it behaves as $-kT \ln \Omega$ which is small as compared to Ω. However, its derivative with respect to μ will not be negligible, as it represents the number $N_0 = kT/|\mu|$ of particles in the condensate which is extensive when $|\mu| \propto \Omega^{-1}$. Exercise 12c shows that notwithstanding the problems we have with the extensivity of the Bose gas in phase II, the thermodynamic relations derived here in the grand canonical ensemble remain valid for the canonical ensemble.

12.3.2 Bose-Einstein Phase Transition

The Bose gas thus shows a *phase transition*: there are two regions where the thermodynamic functions have different analytical forms. In region I, where

$$T > T_\lambda = \frac{2\pi\hbar^2}{mk} \left(\frac{N}{2.61\Omega} \right)^{2/3}, \tag{12.21}$$

we have $\mu < 0$ and the density is given by (12.17). In the low-temperature or high-density region II where $T < T_\lambda$, the chemical potential is practically equal to zero and is connected with the density through (12.20). The transition temperature (12.21) of the non-interacting Bose gas equals $T_\lambda = 3.1$ K for a density of 0.14 g cm^{-3} which is that of liquid helium four. We saw in § 12.1.1 that helium four also has two distinct liquid phases I and II, in contrast to helium three, with a transition temperature of the same order of magnitude, $T_\lambda = 2.18$ K. The existence of this transition is thus an effect not of the attractions between the atoms, as for the gas-liquid or liquid-solid transitions, but of the *indistinguishability* of the atoms and of their *boson* nature.

The properties of helium four in the two phases I and II and near the transition are more or less well described by the model where one neglects the interactions. In fact, the *rôle of the interactions* is here more important than in helium three. In a fermion liquid the antisymmetry of the wavefunctions implies that they vanish when two particles with the same spin are at the same point; the probability for two fermions to be close together is thus reduced by the Pauli principle. Even in the absence of any repulsion between them (Exerc.10g), fermions have little chance to lie near one another. We thus understand why the presence of short-range interactions is less effective than for distinguishable particles, while longer range forces can rather well be taken into account by an average effective potential. This is true both for ^{3}He as for the electrons in a solid: in those two cases the distances between particles are sufficiently small that one expects major qualitative effects to be produced by the interactions, like in classical liquids; nevertheless, the Fermi gas model and the Hartree approximation have been sufficient for us to obtain theoretical results in good agreement with experiments. In nuclear physics we similarly understand the success of the *shell model* where the protons and neutrons in a nucleus are, in spite of their strong interactions, described as a gas of fermions in a self-consistent potential which replaces their mutual interactions. On the contrary, the interparticle interactions profoundly change the properties of Bose liquids, even more than of classical liquids, because of the positive correlations due to the Bose statistics (Exerc.10g). Those reinforce the effect of the short-range forces which we cannot legitimately neglect here.

In particular, for the non-interacting Bose gas model, the *equation of state* in the phase I is given by (12.14) and (12.17); in the phase II it can be written as (see formulae at the end of this volume)

$$\mathcal{P} = -\frac{A}{\Omega} = \frac{(2m)^{3/2}}{6\pi^2\hbar^3} \int_0^\infty \frac{\varepsilon^{3/2}\,d\varepsilon}{e^{\varepsilon/kT}-1} = 1.34kT \left(\frac{mkT}{2\pi\hbar^2}\right)^{3/2}. \quad (12.22)$$

The pressure thus remains constant for a given temperature T, when the density increases beyond the value $2.61(mkT/2\pi\hbar^2)^{3/2}$. The compressibility is infinite and the isotherms are horizontal lines in that region; in fact, the condensate, consisting of zero momentum particles, does not contribute to the pressure nor to the energy. On the other hand, in helium II the interactions prevent the condensate to be independent of the non-zero momentum atoms (Exerc.12d) and the equation of state depends on the density. In the phase diagram of Fig.12.1 helium II occupies a region in the $T, \mathcal{P}$ plane whereas for a non-interacting Bose gas, the phase II would be represented solely by the line (12.22); the region to the right of it would correspond to the phase I and the region to the left of it would be forbidden. Moreover, the slope of this line is positive, in contrast to that of the line separating He I and He II. Note also that there exists a triple point between He I, He II, and vapour, and a coexistence line between He I and vapour in the phase diagram of Fig.12.1, whereas these two phases cannot be distinguished for the Bose gas.

The *specific heat* of the phases I and II can be calculated by differentiation of the grand potential, given, respectively, by Eqs.(12.14) and (12.22). In particular, in the phase II the entropy and the internal energy, equal to

$$S = 1.34k\,\Omega\,\frac{5}{2}\left(\frac{mkT}{2\pi\hbar^2}\right)^{3/2} = 1.28\,k(N-N_0) = \frac{5}{3}\frac{U}{T}, \quad (12.23)$$

are just proportional to the number of non-condensed particles. The specific heat at constant volume follows, if we use (12.18):

$$C_v = \frac{15}{4} \times \frac{1.34}{2.61}\,k(N-N_0) = 1.93Nk\left(\frac{T}{T_\lambda}\right)^{3/2}. \quad (12.24)$$

The calculation is less simple in the phase I (Exerc.12a) and it gives a curve $C_v(T)$ which has a discontinuous slope at the transition temperature. The

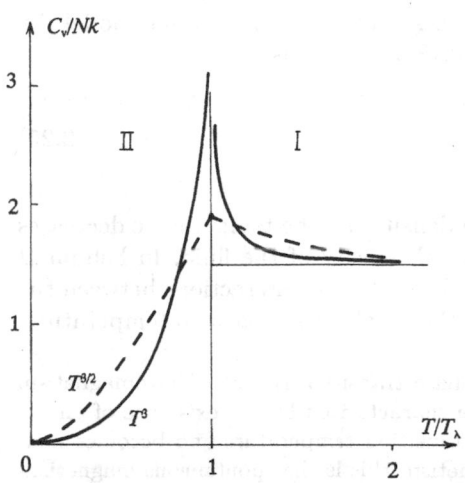

general shape is similar to the experimental curve (Fig.12.6). It was the characteristic shape of this curve which suggested the name λ *point* for the liquid helium transition. As one might expect, the specific heat of helium tends to zero at low temperatures in agreement with the Nernst principle, and to the classical value $\frac{3}{2}kT$ at high temperatures. On the other hand, it becomes infinite at the λ point; moreover, it behaves in helium II not as $T^{3/2}$ like the specific heat (12.24) of the Bose gas, but as T^3 like that of the phonons in a solid, (11.129). The theory of liquid helium, which needs the interactions between the atoms to be taken into account (Exerc.12d), is difficult. A simple model, due to Landau, consists in treating the excited states of the fluid as a gas of quasi-particles, non-interacting bosons which have a semi-empirical spectrum looking like that of the phonons in a solid. This model is studied as a problem (Prob.14). It gives a specific heat in excellent agreement with experimental data in the phase II, except near $T = T_\lambda$.

The phase transition of the Bose gas and the λ point of helium show characteristics which are very different from those of gas-liquid or liquid-solid transitions. In the latter cases, which are *first-order transitions* (§ 6.4.6), the density is discontinuous and there is a latent heat (an evaporation or melting latent heat), except at the critical point of the gas-vapour transition (§ 9.3.2) which is a second-order point. For the Bose gas or for helium, where we have *so-called second-order transitions* (§ 6.4.5), the *density and the internal energy change continuously* through the λ point. Only the higher-order derivatives of the thermodynamic functions show discontinuities. Moreover, the line separating He I and He II in the phase diagram is not a coexistence line, as for first-order transitions: *at no time does a sample of liquid helium separate into two phases* which occupy different parts of space. The change from He I to He II corresponds to the appearance, in the bulk of the sample, of a condensate with a density which is zero at the λ point and which increases progressively as the temperature decreases. In the non-interacting model the density of the condensate, given by (12.18–20), is thus

$$\frac{N_0}{\Omega} = \frac{N}{\Omega} \left[1 - \left(\frac{T}{T_\lambda} \right)^{3/2} \right],$$

(12.25)

where T_λ is the function (12.21) of the density. As the temperature decreases from T_λ to 0, it rises from 0 to the total density of the fluid. In helium II the condensate density is drastically reduced by the interactions between the atoms: it is only of the order of 0.1 of the total density at zero temperature.

Another example of a second-order phase transition is that of ferromagnetism (Exercs.9a and 11f). Such transitions are characterized by the existence of an *order parameter* which is zero above the transition temperature and becomes non-vanishing below it (§ 9.3.3). In ferromagnetism this is the spontaneous magnetization when there is no field. For the Bose condensation this rôle is played by N_0/Ω, given by (12.25). In general, the appearance of an order parameter is accompanied by the spontaneous breaking of some invariance: the continuous translations in a crystal (§ 11.4.2), the change in the spin orientation for ferromagnetism. This effect is not obvious in the λ transition of helium. One can show that the invariance involved here is the conservation $[\widehat{H}, \widehat{N}] = 0$ of the number of particles, which implies that in the equilibrium state the mean value $\langle \widehat{c}_p \rangle$ of the annihilation operator $\widehat{c}_p$ of a helium atom with momentum p must vanish, if the invariance is unbroken. We have identified this operator (§ 11.4.2) with a field operator, in the representation where the gas of bosons is interpreted as a quantized field. One can show (Exerc.12d) that when $T < T_\lambda$ everything happens as if that invariance has been broken, with $\langle \widehat{c}_p \rangle$ as the order parameter. The phase II of helium thus shows an analogy with the *classical limit of a field* (§ 11.4.4). This classical field at the point r is

$$\varphi(r) = \sum_p \psi_p(r) \langle \widehat{c}_p \rangle,$$

(12.26)

and it is proportional to the wavefunction $\psi_p(r)$, if the Bose condensation occurs in a single state p. In particular, if $p = 0$, we can according to (11.155) identify $|\varphi(r)|^2$ with $\langle \widehat{c}_0^\dagger \rangle \langle \widehat{c}_0 \rangle / \Omega \sim N_0/\Omega$, that is, the density of the condensate (Exerc.12d).

In general, the statistical *fluctuations* at thermal equilibrium become important in a second-order transition point, or critical point. This effect is illustrated in Exerc.12b and Exerc.12c, where we estimate the fluctuations in the density N_0/Ω of the condensate in grand canonical and canonical equilibria.

12.3.3 Superfluidity and Superconductivity

Helium II has a number of extraordinary properties: it is *superfluid*, that is, it flows with an *apparently zero viscosity*. For instance, a dewar filled with liquid helium below 2 K empties by itself: the film of helium which is formed by capillarity along the wall acts as a siphon, and the helium easily flows through it, notwithstanding its very small thickness (Fig.12.7).

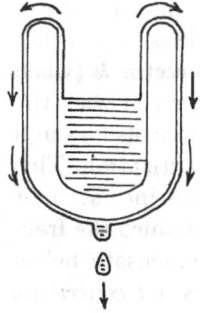

Fig. 12.7. Spontaneous leakage of helium from a dewar

It shows a *very large thermal conductivity* – of the order of 1000 times that of copper at room temperatures. A visible consequence is that it *does not boil* since vapour bubbles cannot be formed inside a liquid, the temperature of which becomes so easily uniform. One can thus directly observe the λ transition: when liquid helium is cooled down, it stops boiling when it passes from phase I to phase II; the latter vaporizes slowly through surface evaporation rather than fast through boiling like phase I.

Helium also shows remarkable *thermomechanical* effects. When it flows out of a vessel though a porous medium, the remaining liquid heats up. Conversely, a local heating within helium II near a tube filled with porous matter produces a helium jet which may reach a height of several tens of cm. This is called the *fountain effect* (Fig.12.8).

Finally, several experiments have exhibited *macroscopic wave properties* of helium II. For instance, the rotation of helium in a cylindrical vessel is quantized. Such a flow does not resemble that of a normal fluid, but rather the propagation of a cylindrically symmetric quantum wave behaving as $e^{im\varphi}$, where $m = 0, \pm 1, \pm 2, \cdots$ is associated with the discrete values $m\hbar$ of the angular momentum. One even observes interference effects. Such phenomena clearly exclude the possibility that helium II might obey the normal laws of hydrodynamics.

One can understand all these properties, if one uses the model of a noninteracting Bose gas with a finite fraction (12.25) of the atoms condensed in one single-particle state. This fraction tends to 1 at zero temperature, the ground state being the one where all particles are piled up in the lowest

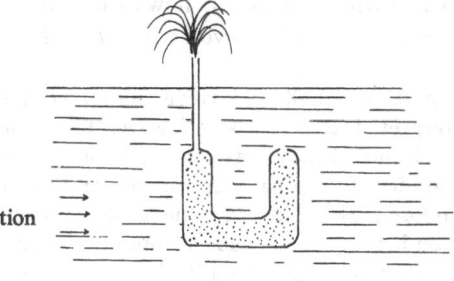

illumination

Fig. 12.8. The fountain effect

single-particle quantum state. Amongst the excited, non-equilibrium states of the Bose gas, some are obtained by placing all N particles in another single-particle state, for example, in a plane wave with wavevector k (when the fluid has a total momentum $N\hbar k$) or in a cylindrical wave m (when the total angular momentum is $Nm\hbar$). Such states resemble non-viscous flow and the second one exhibits moreover an angular momentum quantization. This picture persists to some extent when one takes the interactions into account and when one goes to finite temperatures, two effects which reduce the fraction of condensed atoms. In particular, there remains a condensate below T_λ, characterized by a function (12.26). *This condensate does not contribute either to the entropy or to the internal energy.* Moreover, it is not spatially separated from the remainder of the fluid. All particles of the condensate are in the same quantum state, with zero wavevector at equilibrium, and thus behave like a *wave which occupies the whole of the available space.* The state of the condensate, for instance, its wavevector, is reversibly, that is, without dissipation, changed when one acts upon the liquid.

This picture gives us an intuitive explanation of the various properties of helium II. Its absence of viscosity and its large thermal conductivity are a first result: the condensate wave (12.26) directly puts the various regions of space into contact, and tends to equalize their velocities and temperatures. *The non-viscous flow is, in fact, the propagation of this wave.* The thermomechanical effects come about because the condensate has zero entropy: that part of the fluid which flows across a porous channel mainly contains the condensate, and therefore *carries order with it* so that the remaining liquid becomes more disordered and heats up. In final reckoning, the wave properties of He II are a direct macroscopic manifestation of quantum mechanics: the number N_0 of particles in the condensate is large so that we *see on a macroscopic scale the form of the wavefunction* of the single-particle state $\psi(r)$ into which the N_0 particles have been condensed: the density, the momentum, or the angular momentum of $\psi(r)$ are multiplied by N_0.

A less qualitative explanation of the superfluid properties of helium II is given by Landau's model, where the interactions between the atoms are taken into account through describing the elementary excitations as quasiparticles. Problem 14 shows that this model leads to a satisfactory theory of flow in liquid helium, with a hydrodynamics which is very different from that of other fluids; it explains the thermomechanical effects, and also the existence of a *limiting velocity* below which helium II flows without viscosity, but above which the fluid *again becomes normal* even when $T < T_\lambda$.

The *macroscopic quantum* properties of helium involve the quantity (12.26), which can at each moment be interpreted as a "classical" wave. The equivalence which we have established (§§ 11.4.2 and 11.4.4) between a gas of bosons and a quantized field, which enabled us to identify the particle creation and annihilation operators as field operators, is therefore particularly suitable for describing helium. In the gas state and the liquid phase I, the particle aspect dominates. However, in phase II it is better to regard the system as a quantized field with small statistical fluctuations; the mean value (12.26) of this field obeys a classical field dynamics which we observe as macroscopic waves.

Unusual properties, similar to the superfluidity of helium II, are found in a large number of solids which become *superconductors* at sufficiently low temperatures. The most spectacular aspect of superconductors is the apparent absence of electric *resistivity*, an effect which looks like the absence of viscosity in helium II. Superconductivity was discovered experimentally in 1911 by Heike Kamerlingh Onnes (Groningen 1853–Leiden 1926), three years after he had succeeded to liquefy helium for the first time. Kamerlingh Onnes also showed in 1913 that superconductivity *disappears* when the material is put in a *magnetic field* larger than a certain critical value which depends on the temperature, an effect which is similar to the disappearance of superfluidity in helium when the flow velocity becomes large. For a long time superconductivity was observed only in metals at temperatures below about 10 K. The necessity to use liquid helium to maintain such low temperatures has so far limited practical applications of superconductivity. The most noteworthy application is the construction of coils carrying large currents to produce strong magnetic fields, for instance for the Tore Supra apparatus which is used in fusion studies, or for high-energy particle accelerators. However, the recent discovery by K.A. Müller and J.G. Bednortz (1986) that certain ceramic oxides can be superconducting at much higher temperatures, reaching 100 K, has been the revival of a large research effort in that direction.

Apart from their lack of resistivity, superconductors have many other properties reminding us of superfluidity. These effects appear suddenly below a well-defined critical temperature, which is characteristic of the substance. The *specific heat* has at the transition point an anomaly similar to the one at the λ point of helium. Another feature of superconductivity is the *Meissner effect*: an applied magnetic field does not penetrate into the interior of a superconductor. This expulsion of the magnetic field explains why a superconducting sample can remain suspended in air above a magnet (*levitation*). Because of the absence of resistivity a current induced in a superconducting ring remains fixed for ever and ever; it flows in the skin of the wire. Moreover, the *magnetic flux* across the ring is *quantized*, an effect comparable with the quantization of the angular momentum of rotating helium. The unit of flux, $h/2e$, is found by noting that, because of the Meissner effect, $\mu_0 j = \operatorname{curl} B$ is zero in the bulk of the ring; as j is proportional to the velocity $(p + 2eA)/2m$ of the condensed electron pairs, we find that the circulation $\oint (dr \cdot A)$ along the ring, equal to the flux, is proportional to the change in the phase of the wavefunction over one loop, which is a multiple of 2π. Another macroscopic wave effect is the *Josephson effect*; it shows up, for example, by the appearance of alternating currents when a continuous voltage is applied to a ring consisting of superconducting sections separated by junctions, through which the electrons can pass because of the tunnel effect.

Theoretically, superconductivity remained for a long time a mystery: its existence in metals was only explained in 1957 by John Bardeen, Leon N. Cooper, and J. Robert Schrieffer (BCS theory), and at the moment this is written, there is as yet no satisfactory theory for the high-temperature superconductivity. The analogy with the Bose condensation is based on the fact that bound *electron pairs* can *condense*. Such pairs effectively obey Bose-Einstein statistics, but the main difficulty lies in understanding why two electrons in a metal can form a bound state notwithstanding their Coulomb repulsion. The mechanism is the following. Because of interactions between conduction electrons and lattice vibrations, which were derived in § 11.4.2 but neglected in Chap.11, a phonon can be emitted by an electron and be absorbed by another. Such exchanges give rise to an effective attraction between a pair of electrons. By itself this attraction would not be sufficient to bind the pairs, but it is reinforced by the *Cooper effect* which is an unexpected consequence of the Pauli principle: the sudden jump of the occupation factor at the Fermi surface changes the dynamics in such a way that even a very small attraction is sufficient to bind a pair of electrons, with momenta p' and p'' which have absolute magnitudes close to p_F and which are practically oppositely directed, and with spins coupled into a singlet state. The bound pairs behave as a field $\widehat{c}_{p'\sigma'}\widehat{c}_{p''\sigma''}$ which here plays the same rôle as the field $\widehat{c}_{p'+p''}$ of a boson. The condensation manifests itself by the appearance of a non-vanishing average for this composite two-electron field operator, just as the superfluidity of helium was associated with a finite value of (12.26).

The electrons in superconducting metals are not the only fermions which can form boson-like pairs near the Fermi surface. This is an important feature in nuclear physics where both the protons and the neutrons tend to be bound in pairs in the interior of *atomic nuclei*. It also occurs in *helium three* at extremely low temperatures, below a few mK. The existence of a small attractive part in the interatomic interactions, reinforced by the Cooper effect, results in a pairing of ^{3}He atoms and their condensation. This entails the superfluidity of ^{3}He, even though we are dealing with a Fermi and not a Bose liquid. Moreover, the pairing occurs in a p, rather than an s wave, and the spins of the atoms are coupled in the triplet state; this entails anisotropy. In fact, there exist two superfluid phases: the A phase (Anderson-Brinkman-Morel) and the B phase (Balian-Werthamer) which are qualitatively distinguished from one another through a different anisotropy.

Summary

The ^{3}He and ^{4}He isotopes of helium are quantum fluids at a few K. Helium three, which is well described by the Fermi gas model, has a linear specific heat, and it returns from the solid to the liquid state when it is cooled at suitably chosen pressures. Helium four, which can roughly be described by the Bose gas model, has a phase transition leading from the normal liquid to a new low-temperature liquid phase, called helium II, where a macroscopic fraction of the atoms are condensed into the zero-momentum state. Its specific heat is infinite at the critical point. Helium II is superfluid, and has macroscopic quantum properties, similar to those of superconducting materials.

Exercises

12a Specific Heat of the Bose Gas

Express the specific heat at constant volume and its derivative with respect to T in the form of integrals, for the two phases of the non-interacting Bose gas. Hence, show that there is a discontinuity in the slope of the $C(T)$ curve at the point $T = T_\lambda$. Find the correction to the perfect gas expressions for $T \gg T_\lambda$.

Answers:

We write $B \equiv \Omega(2mk)^{3/2}/4\pi^2 \hbar^3$ and introduce the integrals

$$I_n(\alpha) \equiv \int_0^\infty \frac{x^n\, dx}{e^{x-\alpha} - 1}, \qquad \frac{dI_n}{d\alpha} = nI_{n-1}$$

(when $n < -1$ we replace the lower limit on the integral by ε and subtract the divergent part, a polynomial in $1/\varepsilon$, before letting $\varepsilon \to 0$, or, equivalently, define I_n through analytic continuation in the complex n-plane). We find for $T > T_\lambda$

$$N = BT^{3/2}I_{1/2}, \qquad U = BkT^{5/2}I_{3/2},$$

whence we get by taking the derivative of U with N constant

$$\frac{C}{Nk} = \frac{5}{2}\frac{I_{3/2}}{I_{1/2}} + \frac{3}{2}\frac{d\alpha}{dT} = \frac{5}{2}\frac{I_{3/2}}{I_{1/2}} - \frac{9}{2}\frac{I_{1/2}}{I_{-1/2}},$$

$$\frac{T}{Nk}\frac{dC}{dT} - \frac{3C}{2NK} = \frac{9}{4}\frac{I_{1/2}}{I_{-1/2}} + \frac{27}{4}\frac{\left(I_{1/2}\right)^2 I_{-3/2}}{\left(I_{-1/2}\right)^3}.$$

As $T \to T_\lambda + 0$, $\alpha \to -0$, the formulae from the end of the volume give

$$I_{1/2} \to \Gamma(\tfrac{3}{2})\zeta(\tfrac{3}{2}) \simeq 2.31, \qquad I_{3/2} \to \Gamma(\tfrac{5}{2})\zeta(\tfrac{5}{2}) \simeq 1.78,$$

$$I_{-1/2} \sim \pi|\alpha|^{-1/2}, \qquad I_{-3/2} = -2\frac{dI_{-1/2}}{d\alpha} = \sim -\pi|\alpha|^{-3/2},$$

whence we get, using the fact that $N = BT_\lambda^{3/2}I_{1/2}(0)$,

$$C \to 1.93Nk, \qquad T_\lambda\frac{dC}{dT_\lambda} \to -0.78Nk.$$

When $T < T_\lambda$, (12.24) gives in the limit as $T \to T_\lambda - 0$

$$C \to 1.93Nk, \qquad T_\lambda\frac{dC}{dT_\lambda} \to 2.89Nk.$$

As $T \to \infty$, we use the relation

$$I_n \approx \Gamma(n+1) e^\alpha \left[1 + 2^{-n-1} e^\alpha\right],$$

which gives

$$C - \frac{3}{2} Nk \sim \frac{3N^2 k}{8\sqrt{2\pi} BT^{3/2}} \sim 0.35 \, Nk \left(\frac{T_\lambda}{T}\right)^{3/2}.$$

We show the function $C(T)$ in Fig.12.6.

12b Critical Fluctuations

1. Evaluate the statistical fluctuation ΔM of the total magnetic moment of a ferromagnetic sample in zero field in the approximation of Exerc.9a, and show that it diverges at the Curie temperature.

2. Show that the fluctuations Δn_0 and ΔN in a non-interacting Bose gas in grand canonical equilibrium diverge as $T \to T_\lambda + 0$, and remain macroscopic for $T < T_\lambda$.

Hints:

1. According to Exerc.4a we have $\Delta M^2 = \Omega k T \chi$. Hence, $\Delta M^2 \sim N\mu_B^2 T_c/(T - T_c)$ for $T > T_c$ and $\Delta M^2 \sim 2N\mu_B^2 T_c/(T_c - T)$ for $T < T_c$.

2. From (10.32) we have $\Delta n_0^2 = N_0(1 + N_0) \sim |\alpha|^{-2}$ when $\alpha \to -0$. Alternatively, we can start from the probability law $p(n_0) \propto e^{\alpha n_0}$ for the occupation number n_0 and derive from it the moments N_0 and Δn_0^2. In the condensed phase n_0 is badly defined since $\Delta n_0 \sim N_0$. From (4.38) and using the notation of Exerc.12a we have

$$\Delta N^2 = \frac{\partial N}{\partial \alpha} = \frac{1}{2} BT^{3/2} I_{-1/2} + \Delta n_0^2$$

$$\sim 0.68 \, N \left(\frac{T}{T_\lambda}\right)^{3/2} |\alpha|^{-1/2} + |\alpha|^{-2}.$$

12c Bose Condensation in a Canonical Ensemble

1. Express the canonical partition function and the canonical averages $\langle n_0 \rangle$ and $\langle n_0^2 \rangle$ in the form of integrals, using the grand potential.

2. Show that the canonical and the grand canonical ensembles are equivalent, as far as the thermodynamic quantities are concerned.

3. Estimate the fluctuation Δn_0 in the number of particles in the condensate in canonical equilibrium for $T < T_\lambda$ and for $T > T_\lambda$. Discuss the result.

Solution:

1. In order to extract the term in N from a sum such as $\mathrm{Tr}\, e^{-\beta\widehat{H}+\alpha\widehat{N}}\widehat{n}_0^s$ with $s = 0, 1$ or 2, we note that

$$\frac{1}{2\pi i} \int_{\alpha-i\pi}^{\alpha+i\pi} dz\, e^{z(N'-N)} = \delta_{N,N'},$$

where the choice of $\alpha(< 0)$ is immaterial. The result is, for $s = 0$, the canonical partition function

$$Z_N = \frac{1}{2\pi i} \int_{\alpha-i\pi}^{\alpha+i\pi} dz\, e^{-zN}\, Z_{\mathrm{G}}(z, \beta),$$

whence we get, if we use the expression for the grand potential found in § 12.3.1 and the notation of Exerc.12a,

$$Z_N = \frac{1}{2\pi i} \int_{\alpha-i\pi}^{\alpha+i\pi} dz\, \exp\left[-zN + \frac{2}{3} BT^{3/2} I_{3/2}(z) - \ln\left(1 - e^z\right)\right].$$

The last term is the contribution from the zero-momentum state which is needed when $\alpha \to 0$. To find $Z_N\langle n_0\rangle$ and $Z_N\langle n_0^2\rangle$, which correspond to $s = 1$ and $s = 2$, we multiply the integrand by $1/(e^{-z} - 1)$ and $(e^{-z} + 1)/(e^{-z} - 1)^2$, respectively; these factors are for $z = \alpha$ the averages $\langle n_0\rangle$ and $\langle n_0^2\rangle$ in the grand canonical ensemble.

2. The exponent in Z_N is stationary when $z = \alpha$, where α is defined as the solution of

$$-N + BT^{3/2} I_{1/2}(\alpha) + \frac{1}{e^{-\alpha} - 1} = 0.$$

This equation is the same as (12.17) or as (12.20), depending on whether N is smaller or larger than $BT^{3/2} I_{1/2}(0)$. The saddle-point method (Exerc.5b) then indicates that the integrand is concentrated near the point $z = \alpha$ with $|z - \alpha| = \mathcal{O}(1/\sqrt{N})$ for $T > T_\lambda$ ($-\alpha$ finite), or near the point $z = 0$ with $z = \mathcal{O}(1/N)$ for $T < T_\lambda$, that is, $|\alpha| = \mathcal{O}(1/N)$. The result is that, if we restrict ourselves to the extensive parts, $F - N\alpha/\beta$ is the same as the grand potential A in the two cases.

By itself this identity is not sufficient to guarantee the equivalence between the canonical and grand canonical ensembles, since the Bose gas is not extensive in its phase II. The contribution $kT \ln(1 - e^\alpha)$ to A, even though negligible in the thermodynamic limit, is sufficiently singular at $\alpha = 0$ for its derivative with respect to μ to give $\langle n_0\rangle = N_0$, a finite part of N (see end of § 12.3.1). On the other hand, Exerc.12b has shown that $\Delta n_0/N_0$ equals 1 when $T < T_\lambda$ in the grand canonical ensemble, in contrast to the usual behaviour like $1/\sqrt{N}$ of all relative fluctuations in extensive systems (§ 5.7.1). Nevertheless, these difficulties only involve the condensate. The question sub 3 will show that, if we give the total number of particles, exactly in the canonical and on average in the grand canonical ensemble, those particles will split up between the condensate and the states with non-vanishing momenta, *on average* in the same way in the two ensembles, even though the fluctuations around $\langle n_0\rangle$ are large in the second ensemble. Moreover, the condensate contributes neither to the pressure, nor to the energy or the entropy. Hence, in the thermodynamic limit all macroscopic properties, such as the equation of state or the thermal properties, are independent of the ensemble, notwithstanding the violation of the extensivity.

3. In order to be able to subtract $\langle n_0 \rangle^2$ from $\langle n_0^2 \rangle$ for $T < T_\lambda$ we must calculate the first two terms in the expansions of Z_N, $Z_N \langle n_0 \rangle$, and $Z_N \langle n_0^2 \rangle$ for large N, since the dominant terms cancel one another. To do that, we put $z = x/N$, as the integrals are concentrated in the region $z = \mathcal{O}(1/N)$ for $T < T_\lambda$, and we expand in powers of $1/N$, using the expansions (see Exerc.12a):

$$ I_{3/2}(z) \approx I_{3/2}(0) + \frac{3}{2} I_{1/2}(0) \frac{x}{N} + \pi \frac{(-x)^{3/2}}{N^{3/2}} + \mathcal{O}\left(\frac{x^2}{N^2}\right), $$

$$ \frac{1}{1 - e^z} \approx -\frac{N}{x} + \mathcal{O}(1), \qquad \frac{1}{1 - e^z} \frac{1}{e^{-z} - 1} \approx -\frac{N^2}{x^2} + \mathcal{O}(1), $$

$$ \frac{1}{1 - e^z} \frac{e^{-z} + 1}{(e^{-z} - 1)^2} \approx -\frac{2N^3}{x^3} + \mathcal{O}\left(\frac{x}{N}\right). $$

Writing

$$ \zeta \equiv \exp\left[\frac{2}{3} BT^{3/2} I_{3/2}(0)\right], \qquad \eta \equiv \frac{2\pi}{3} \frac{BT^{3/2}}{N^{3/2}}, $$

$$ a \equiv 1 - \frac{B}{N} T^{3/2} I_{1/2}(0) = 1 - \left(\frac{T}{T_\lambda}\right)^{3/2}, $$

and restricting ourselves to the first two orders in $N^{-1/2}$, we find

$$ Z_N \approx -\frac{\zeta}{2\pi i} \int \frac{dx}{x} e^{-ax} \left[1 + \eta(-x)^{3/2}\right], $$

$$ Z_N \langle n_0 \rangle \approx \frac{\zeta N}{2\pi i} \int \frac{dx}{x^2} e^{-ax} \left[1 + \eta(-x)^{3/2}\right], $$

$$ Z_N \langle n_0^2 \rangle \approx -\frac{2\zeta N^2}{2\pi i} \int \frac{dx}{x^3} e^{-ax} \left[1 + \eta(-x)^{3/2}\right]. $$

The integration path over x can be deformed into a loop, clockwise encircling the real semi-axis $x > 0$, and we get

$$ Z_N \approx \zeta\left[1 - \frac{\eta a^{-3/2}}{\pi} \Gamma\left(\frac{3}{2}\right)\right], $$

$$ Z_N \langle n_0 \rangle \approx N\zeta\left[a + \frac{\eta a^{-1/2}}{\pi} \Gamma\left(\frac{1}{2}\right)\right], $$

$$ Z_N \langle n_0^2 \rangle \approx N^2 \zeta\left[a^2 - \frac{2\eta a^{1/2}}{\pi} \Gamma\left(\frac{-1}{2}\right)\right], $$

which finally leads to

$$ \langle n_0 \rangle \sim Na = N\left[1 - \left(\frac{T}{T_\lambda}\right)^{3/2}\right] = N_0, $$

$$\Delta n_0^2 \sim \frac{3N^2 \eta a^{1/2}}{2\sqrt{\pi}} \simeq 0.77 \, (N - N_0) \, N_0^{1/2}.$$

We find for $\langle n_0 \rangle$ the same expression (12.25) as was found in grand canonical equilibrium. However, the relative fluctuation $\Delta n_0 / N_0$ is small as $N^{-1/4}$ whereas in the grand canonical ensemble we had $\Delta n_0 / N_0 \sim 1$ so that N_0 / Ω was badly defined. Here $\Delta n_0 \to 0$ as $T \to 0$, as should be the case, since then N_0 tends to N which does not fluctuate. At the λ point, $\Delta n_0 / N_0 \sim 0.88 N^{1/2} N_0^{-3/4}$ diverges, since N_0 vanishes as $3N(T_\lambda - T)/2T_\lambda$.

For $T > T_\lambda$ the weight of the integrals which determine Z_N, $\langle n_0 \rangle$, and $\langle n_0^2 \rangle$ is concentrated near the point $z = \alpha$ within a range of order $N^{-1/2}$, where α is determined by $N = BT^{3/2} I_{1/2}(\alpha)$. Hence, we have

$$\langle n_0 \rangle = \frac{1}{e^{-\alpha} - 1}, \qquad \langle n_0^2 \rangle = \frac{e^{-\alpha} + 1}{(e^{-\alpha} - 1)^2},$$

whence $\Delta n_0^2 = N_0(N_0 + 1)$, as in the grand canonical ensemble. The fluctuations in N_0 which are finite as $N \to \infty$, diverge as N_0^{-1} as $T \to T_\lambda + 0$. When $T > T_\lambda$ the two ensembles are equivalent, not only for thermodynamic quantities but even for the fluctuation Δn_0, in contrast to the case $T < T_\lambda$.

12d Rôle of the Interactions in the Bose Transition

1. Use the formalism of § 10.2.3 to prove, by writing the matrix elements between states with 1, 2, ..., N particles in the Fock base, that, in terms of the creation and annihilation operators of plane waves with momentum $\hbar \boldsymbol{k}$, the Hamiltonian $\widehat{H} = \widehat{H}_0 + \widehat{V}$ of a system of spinless indistinguishable particles of mass m, contained in a box with periodic boundary conditions and with binary interactions governed by a potential $V(|\boldsymbol{r} - \boldsymbol{r}'|)$, is given by

$$\widehat{H}_0 = \sum_{\boldsymbol{k}} \frac{\hbar^2 k^2}{2m} \, \widehat{c}_{\boldsymbol{k}}^\dagger \widehat{c}_{\boldsymbol{k}},$$

$$\widehat{V} = \frac{1}{2} \int d^3 r \, d^3 r' \, V(|\boldsymbol{r} - \boldsymbol{r}'|) \, \widehat{\psi}^\dagger(\boldsymbol{r}) \widehat{\psi}^\dagger(\boldsymbol{r}') \widehat{\psi}(\boldsymbol{r}') \widehat{\psi}(\boldsymbol{r})$$

$$= \frac{1}{2} \sum_{\boldsymbol{k}_1 \boldsymbol{k}_2 \boldsymbol{k}_3 \boldsymbol{k}_4} \langle \boldsymbol{k}_1 \boldsymbol{k}_2 | V | \boldsymbol{k}_3 \boldsymbol{k}_4 \rangle \, \widehat{c}_{\boldsymbol{k}_1}^\dagger \widehat{c}_{\boldsymbol{k}_2}^\dagger \widehat{c}_{\boldsymbol{k}_3} \widehat{c}_{\boldsymbol{k}_4},$$

$$\langle \boldsymbol{k}_1 \boldsymbol{k}_2 | V | \boldsymbol{k}_3 \boldsymbol{k}_4 \rangle \equiv \frac{1}{\Omega} \delta_{\boldsymbol{k}_1 + \boldsymbol{k}_2, \boldsymbol{k}_3 + \boldsymbol{k}_4} \, \widetilde{V}(|\boldsymbol{k}_4 - \boldsymbol{k}_1|),$$

$$\widetilde{V}(\boldsymbol{k}) = \int d^3 r \, V(r) \, \cos(\boldsymbol{k} \cdot \boldsymbol{r}).$$

2. With the aim of determining approximately the grand potential of a gas of weakly interacting bosons, use the variational method (§ 4.2.2) and take a trial density operator of the form

$$\widehat{D} = \exp\left[-\sum_k x_k \widehat{c}_k^\dagger \widehat{c}_k + y\widehat{c}_0^\dagger + y^*\widehat{c}_0 + z\right].$$

If we get a value $y \neq 0$, it will indicate the appearance of a spontaneous breaking of the invariance $[\widehat{H}, \widehat{N}] = 0$ (§ 12.3.2). Determine z as function of the undetermined parameters x_k and y such that $\widehat{D}$ is normalized. Express the averages of $\widehat{c}_0$, $\widehat{c}_0^\dagger$, $\widehat{c}_k^\dagger \widehat{c}_k$ (for $k \neq 0$), and $\widehat{c}_0^\dagger \widehat{c}_0$ over $\widehat{D}$, as functions of x_k and y; these averages will be denoted by φ, φ^*, f_k, and $f_0 + |\varphi|^2$, respectively.

3. Express the trial energy, $\mathrm{Tr}\widehat{D}\widehat{H}$, the trial entropy, $-k\mathrm{Tr}\widehat{D}\ln\widehat{D}$, and the trial grand potential $\mathcal{A}$ in terms of f_k and φ. Write down the equations in f_k and φ which, for $\widetilde{V}(0) > 0$, determine the minimum of $\mathcal{A}$ and show that, depending on the values of T and μ, they take on two forms which are different according to whether $\langle\widehat{\psi}(\boldsymbol{r})\rangle = \varphi/\sqrt{\Omega}$ (or $N_0/\Omega = |\varphi|^2/\Omega$) vanishes or is non-vanishing. Compare this with the transition of a non-interacting Bose gas.

Results:

Noting that when the operator $\widehat{D}$ is expressed in terms of the boson operators $\widetilde{c}_k = \widehat{c}_k - \delta_{k,0}y/x_0$, it describes independent particles, one finds that

$$z + \frac{|y|^2}{x_0} = \sum_k \ln\left(1 - e^{-x_k}\right), \qquad \varphi = \frac{y}{x_0}, \qquad f_k = \frac{1}{e^{x_k} - 1}.$$

$$\mathrm{Tr}\,\widehat{D}\widehat{V} = \frac{1}{2\Omega}\sum_{kk'}\left[\widetilde{V}(0) + \widetilde{V}(k - k')\right]f_k f_{k'}$$

$$+ \frac{1}{\Omega}\sum_k\left[\widetilde{V}(0) + \widetilde{V}(k)\right]f_k|\varphi|^2 + \frac{1}{2\Omega}\widetilde{V}(0)|\varphi|^4,$$

$$\mathcal{A} = \sum_k\left(\frac{\hbar^2 k^2}{2m} - \mu\right)f_k - \mu|\varphi|^2 + \langle\widehat{V}\rangle$$

$$- kT\sum_k\left[(1 + f_k)\ln(1 + f_k) - f_k\ln f_k\right].$$

The equations $\partial\mathcal{A}/\partial f_k = \partial\mathcal{A}/\partial\varphi = 0$ become either

$$\varphi = 0, \quad kT\,x_k = \frac{\hbar^2 k^2}{2m} - \mu + \frac{1}{\Omega}\sum_{k'}\left[\widetilde{V}(0) + \widetilde{V}(k' - k)\right]f_{k'},$$

which characterize phase I, or

$$\widetilde{V}(0) \frac{|\varphi|^2}{\Omega} = \mu - \frac{1}{\Omega} \sum_{k'} \left[\widetilde{V}(0) + \widetilde{V}(k') \right] f_{k'},$$

$$kT x_k = \frac{\hbar^2 k^2}{2m} + \widetilde{V}(k) \frac{|\varphi|^2}{\Omega} + \frac{1}{\Omega} \sum_{k'} \left[\widetilde{V}(k' - k) - \widetilde{V}(k') \right] f_{k'},$$

which characterize phase II. These self-consistent equations can be solved by iteration, if the potential is sufficiently weak. If μ is negative, or slightly positive, only the solution with $\varphi = 0$ exists, as $\widetilde{V}$ is positive. The λ transition occurs at the point where

$$\mu = \frac{1}{\Omega} \sum_{k'} \left[\widetilde{V}(0) + \widetilde{V}(k') \right] f_{k'},$$

$$kT x_k = \frac{\hbar^2 k^2}{2m} + \frac{1}{\Omega} \sum_{k'} \left[\widetilde{V}(k' - k) - \widetilde{V}(k') \right] f_{k'}.$$

For larger values of μ, that is, for higher densities or lower temperatures, the solution with $\varphi \neq 0$ gives the minimum of $\mathcal{A}$, and the fluid is in phase II. In contrast to what happens for the non-interacting gas, the Bose factor f_k remains bounded for $k = 0$ in phase II, with the condensate density N_0/Ω being given by $|\varphi|^2/\Omega$ rather than by f_0/Ω. The chemical potential is then finite and equal to

$$\mu = \frac{1}{\Omega} \sum_{k'} \left[\widetilde{V}(0) + \widetilde{V}(k') \right] f_{k'} + \widetilde{V}(0) \frac{N_0}{\Omega},$$

instead of being nearly zero as $-kT/N_0$. Like the pressure

$$\mathcal{P} = -\frac{\mathcal{A}}{\Omega} = kT \sum_k \ln\left(1 + f_k\right) + \langle \widehat{V} \rangle,$$

it varies in phase II with the density, and the occupation factors f_k depend on the condensate density.

Notes:

The existence of the thermodynamic limit (§ 5.5.2) requires that $\mathcal{A}/\Omega$ be bounded from below. For this to be the case, it is necessary that the coefficient of the term with $|\varphi|^4$, $\widetilde{V}(0)$, be positive. It is also necessary that the term with $f_k f_{k'}$ be positive for any $f_k > 0$. If these conditions were not satisfied, the system would collapse onto itself, and would not be extensive.

Even if the interactions are weak, they couple the condensate to the rest of the system and this changes the properties of the fluid qualitatively. In particular, they cure the pathologies of the Bose gas, such as the absence of extensivity, the singularity of its phase diagram, or its infinite compressibility below T_λ.

The variational method used above introduces $\varphi = \langle \widehat{c_0} \rangle$ as the order parameter for the λ transition. The invariance which is spontaneously broken in phase II is characterized by the group of the unitary transformations $e^{i\gamma \widehat{N}}$ which leave $\widehat{H}$

invariant; this is a group which here plays the same rôle as the rotations for the ferromagnetic transition. In particular, in phase II where this invariance is broken there exists a continuous set of solutions, the order parameters of which follow one from the other through the transformation $e^{-i\gamma\widehat{N}}\widehat{c_0}e^{i\gamma\widehat{N}} = e^{i\gamma}\widehat{c_0}$. However, in the present case experimentally one cannot observe a change in the phase of $\varphi = \langle\widehat{c_0}\rangle$ directly, in contrast to a change in the orientation of the spontaneous magnetization in the case of ferromagnetism. A more detailed discussion of the two problems will help us to understand the origin of this difference. In any realistic situation, the Hamiltonian of a magnetic system is not exactly invariant under rotation. It contains weak perturbations from the walls of the sample, from defects or from external fields, which do not commute with the generators of the rotation group, that is, with the components of the angular momentum operator $\widehat{\boldsymbol{J}}$. Even if such perturbations are too small to produce any sizeable effects in the paramagnetic phase, they play a decisive rôle in the breaking of rotational invariance below the Curie temperature T_C (§ 9.3.3) and are essential to validate effective field approximations of the type of Exerc. 9a. Indeed, when they are present, such an approximation provides a satisfactory value for the magnitude M of the spontaneous magnetization $M = \text{Tr } \widehat{D}\widehat{M} \simeq \text{Tr } \widehat{D} \,\widehat{M}$, but it does not create any preferred direction; the actual orientation of $\boldsymbol{M}$ is selected by the weak perturbations. Likewise, the phase separation in the liquid-vapour equilibrium is governed by the gravity field (§ 9.3.3).

The situation is different if we study a model of ferromagnetism in which the Hamiltonian $\widehat{H}$ is *exactly* invariant under rotation. Since the exact canonical density operator $\widehat{D} \propto e^{-\beta\widehat{H}}$ commutes with the components of $\widehat{\boldsymbol{J}}$, the rotation characterized by the vector $\boldsymbol{\omega}$ transforms the exact magnetization $\boldsymbol{M}_{\text{ex}}$ into

$$\text{Tr } \widehat{D} \, e^{-i(\boldsymbol{\omega}\cdot\widehat{\boldsymbol{J}})/\hbar} \, \widehat{\boldsymbol{M}} \, e^{i(\boldsymbol{\omega}\cdot\widehat{\boldsymbol{J}})/\hbar} = \text{Tr } e^{i(\boldsymbol{\omega}\cdot\widehat{\boldsymbol{J}})/\hbar} \, \widehat{D} \, e^{-i(\boldsymbol{\omega}\cdot\widehat{\boldsymbol{J}})/\hbar} \, \widehat{\boldsymbol{M}}$$
$$= \text{Tr } \widehat{D}\widehat{\boldsymbol{M}} = \boldsymbol{M}_{\text{ex}}.$$

As the exact spontaneous magnetization $\boldsymbol{M}_{\text{ex}}$ is thus invariant under rotation, it must vanish at any temperature, in contrast to its Weiss field approximation $\boldsymbol{M} = \text{Tr } \widehat{D}\widehat{\boldsymbol{M}}$. Strictly speaking, the invariance is not broken, and $\boldsymbol{M}$ cannot be observed directly. The suppression of the weak perturbations had a dramatic effect, bringing the magnitude of the magnetization down from M to 0. Nevertheless, the order parameter M still governs, but indirectly, other quantities. In order to exhibit qualitative differences between the paramagnetic and ferromagnetic phases, we must resort here to rotationally invariant quantities, such as the two-point correlation $K(\boldsymbol{r},\boldsymbol{r}') \equiv \langle\varrho_M(\boldsymbol{r}) \cdot \varrho_M(\boldsymbol{r}')\rangle$ between the local densities of magnetization at two points $\boldsymbol{r}$ and $\boldsymbol{r}'$ which are distant from each other. One can show that such quantities are not affected by the presence of weak perturbations, in contrast to $\langle\varrho_M(\boldsymbol{r})\rangle$ which below T_C is equal either to M/Ω or to 0. Moreover, they are approximately given by the effective field method. In the model of Exerc. 9b, we can even check that the exact expression for the correlations is given by this method, whereas $\langle\varrho_M\rangle$ vanishes when there is no magnetic field. More generally, in any model with a rotationally invariant Hamiltonian, the two-point correlation function $K(\boldsymbol{r},\boldsymbol{r}')$ tends to 0 above T_C and to $(M/\Omega)^2$ below T_C, if $|\boldsymbol{r}-\boldsymbol{r}'|$ tends to infinity. The direction of $\boldsymbol{M}$ remains irrelevant but its magnitude comes out as a measure of the long range correlations, while the spontaneous magnetization $\boldsymbol{M}_{\text{ex}}$ vanishes. Thus, the *long range order* exhibited by the correlations below T_C characterizes the ferromagnetic ordering in a better way than does the spontaneous magnetization: it occurs even in the absence of perturbations which explicitly break the invariance under rotation.

Exactly the same situation prevails for the condensation of Bose liquids (or for superconductivity), if we replace the rotations by the transformations $e^{i\gamma \widehat{N}}$, $\widehat{\boldsymbol{J}}$ by $\widehat{N}$, M by $|\varphi|$, and the orientation of M by the phase of φ. Here, the Hamiltonian *always* commutes with $\widehat{N}$, even if realistic perturbations of various kinds are included. Hence, the order parameter cannot be observed directly, as the *exact* equilibrium value of $\langle \widehat{c_0} \rangle$ is zero both above *and below* T_λ. (For superconductivity, the order parameter is the expectation value of a product of two annihilation operators for fermions, as indicated near the end of § 12.3.3.) The above effective field theory yields in contrast $\langle \widehat{c_0} \rangle \simeq \varphi$; it would be valid as a first approximation for unphysical models involving perturbations that do not commute with $\widehat{N}$. However, even in the absence of such perturbations, it remains a convenient approach, providing sensible approximations for quantities such as $K(\boldsymbol{r}, \boldsymbol{r}') \equiv \langle \widehat{\psi}^\dagger(\boldsymbol{r}) \widehat{\psi}(\boldsymbol{r}') \rangle$, where $\widehat{\psi}^\dagger(\boldsymbol{r}) \widehat{\psi}(\boldsymbol{r}')$ commutes with $\widehat{N}$. Using (10.23) and the results of question 2, we thus find

$$K(\boldsymbol{r}, \boldsymbol{r}') \simeq \frac{|\varphi|^2}{\Omega} + \frac{1}{(2\pi)^3} \int d^3k \; e^{i\boldsymbol{k} \cdot (\boldsymbol{r}' - \boldsymbol{r})} f_k,$$

and $|\varphi|^2/\Omega$ comes out as the limit of $K(\boldsymbol{r}, \boldsymbol{r}')$ as $|\boldsymbol{r} - \boldsymbol{r}'| \longrightarrow \infty$. The order parameter $\varphi/\sqrt{\Omega}$ is not directly meaningful; its phase is unphysical, but its modulus characterizes the so-called *off-diagonal long range order* (ODLRO) exhibited by $K(\boldsymbol{r}, \boldsymbol{r}')$ in a Bose liquid below T_λ (or in a superconductor). The expression "off-diagonal" refers to the fact that the operator $\widehat{\psi}(\boldsymbol{r})$ relates states with different particle numbers. Higher order correlations involving the same number of creation and annihilation operators, such as

$$\langle \widehat{\psi}^\dagger(\boldsymbol{r}_1) \widehat{\psi}^\dagger(\boldsymbol{r}_2) \widehat{\psi}(\boldsymbol{r}_3) \widehat{\psi}(\boldsymbol{r}_4) \rangle \simeq K(\boldsymbol{r}_1, \boldsymbol{r}_3) K(\boldsymbol{r}_2, \boldsymbol{r}_4) + K(\boldsymbol{r}_1, \boldsymbol{r}_4) K(\boldsymbol{r}_2, \boldsymbol{r}_3) - \frac{|\varphi|^4}{\Omega^2}$$

$$\longrightarrow \frac{|\varphi|^4}{\Omega^2},$$

are evaluated in the above approximation by means of Wick's theorem (Exerc. 10g). Below T_λ, they do not tend to zero when all points get far apart from one another, but to a limit obtained by replacing each operator $\widehat{\psi}(\boldsymbol{r})$ by the order parameter $\varphi/\sqrt{\Omega}$ (and $\widehat{\psi}^\dagger(\boldsymbol{r})$ by $\varphi^*/\sqrt{\Omega}$). When some points remain at finite distances apart, the limit is less simple. For instance, if we let $\boldsymbol{r}_1 = \boldsymbol{r}_3 = \boldsymbol{r}$ and $\boldsymbol{r}_2 = \boldsymbol{r}_4 = \boldsymbol{r}'$ and substract $n(\boldsymbol{r}) n(\boldsymbol{r}') - n(\boldsymbol{r}) \delta^3(\boldsymbol{r} - \boldsymbol{r}')$, we get the "diagonal" correlation $C(\boldsymbol{r} - \boldsymbol{r}')$ between the particle densities at $\boldsymbol{r}$ and $\boldsymbol{r}'$, defined in Exerc. 10g, which tends to 0 as $|\boldsymbol{r} - \boldsymbol{r}'| \longrightarrow \infty$, both above and below T_λ.

A slightly different long range order occurs in the condensation of a non-interacting Bose gas (§ 12.3.2). Just as in the calculation leading to (12.20), we find in that case below T_λ

$$K(\boldsymbol{r}, \boldsymbol{r}') = \sum_k \langle k|\boldsymbol{r} \rangle \langle \boldsymbol{r}'|k \rangle f_k = \frac{N_0}{\Omega} + \frac{1}{(2\pi)^3} \int \frac{d^3k \; e^{i\boldsymbol{k} \cdot (\boldsymbol{r}' - \boldsymbol{r})}}{e^{\beta \hbar^2 k^2 / 2m} - 1},$$

and again $\langle \widehat{\psi}^\dagger(r)\widehat{\psi}(r')\rangle$ tends to N_0/Ω as $|r - r'| \longrightarrow \infty$. However, higher order correlations behave at large distances in a pathological manner: for instance,

$$\langle \widehat{\psi}^\dagger(r_1)\widehat{\psi}^\dagger(r_2)\widehat{\psi}(r_3)\widehat{\psi}(r_4)\rangle = K(r_1, r_3)K(r_2, r_4) + K(r_1, r_4)K(r_2, r_3)$$

tends to $2\,N_0^2/\Omega^2$, not to N_0^2/Ω^2; the density correlation $C(r - r')$ tends to N_0^2/Ω^2, not to 0.

This exercise is a first step towards a theory of helium II starting from the interactions $\widehat{V}$ between the atoms. The next step consists in extending the variational space so that the trial density operator has the form

$$\widehat{D} = \exp\left(- \sum_k x_k \widetilde{c}_k'^\dagger \widetilde{c}_k' + z \right),$$

where the $\widetilde{c}_k'$ are defined by the *Bogolyubov transformation*

$$\widetilde{c}_k' = \widehat{c}_k \sqrt{1 + |v_k|^2} + v_k \widehat{c}_{-k}^\dagger + w\,\delta_{k,0}.$$

The variational parameters to be determined by looking for the minimum of $\mathcal{A}$ are here the x_k, v_k, and w, which generalize the earlier parameters x_k, $v_k \equiv 0$, and $y = -w x_0$. One can easily check that the operators $\widetilde{c}_k'$, $\widetilde{c}_k'^\dagger$ satisfy boson operator commutation relations; hence $\widehat{D}$ again describes a set of *independent quasi-particles* with an energy spectrum x_k/β. These quasi-particles are related to the original helium atoms through the Bogolyubov transformation, which in the present exercise breaks the particle number invariance associated with $e^{i\gamma\widehat{N}}$, but not the translational invariance. The effective quasi-bosons which describe He II in this approach can be identified with the phonons in Landau's empirical model (Prob.14), which thus receives a certain justification. Quantitative fits with experiments have been obtained by more refined theories, which rely on series expansions of the difference between $\widehat{D}$ and the exact state in powers of the residual interaction between the quasi-particles.

13. Equilibrium and Transport of Radiation

"Dans cette édition, j'ai tâché d'insérer toutes les acquisitions nouvelles dont la Physique s'était enrichie; on pourra remarquer avec satisfaction que les nouvelles richesses acquises par la science ont toutes trouvé leurs places dans les grandes divisions déjà établies. Tel est le caractère d'une science faite. La progression rapide avec laquelle la physique se complète tous les jours peut faire regarder l'époque de sa stabilité entière comme peu éloignée de nous."

J.-B. Biot, Précis élémentaire de physique expérimentale, 1823

"Une circonstance éminemment propre à établir cette inégalité d'échanges, et en rendre les conséquences évidentes, c'est d'exposer un corps, la nuit, à l'aspect libre d'un ciel serein, en l'isolant d'ailleurs, aussi bien que possible, de toute cause de réchauffement. Car alors, tout ce que ce corps rayonnera de chaleur vers les espaces célestes sera perdu pour lui; et, si ce qu'il reçoit du contact de l'air et des corps environnans ne suffit pas pour compenser cette perte, sa température devra s'abaisser. ... On fait ainsi, en grand, de la glace au Bengale, depuis un temps immémorial."

J.-B. Biot, ibid.

In statistical physics as well as in thermodynamics, the study of electromagnetic radiation plays a prime rôle. It is of direct scientific importance, as this radiation is the simplest example of a *macroscopic quantum* system, side by side with solids. It is of indirect scientific importance because of its many applications in both observational and theoretical *astrophysics*. It is of technical, and even day-to-day, importance, as radiation is one of the most common forms of *energy transfer*. Finally, it has been of *historical* importance. Around 1900 physics seemed to be a completed science, and thermodynamics of radiation appeared one of the few open problems, an irritation, but a minor one. However, it eventually turned out to be the source of both quantum mechanics and modern statistical mechanics – at that time only kinetic gas theory existed. In the present chapter we shall indicate a few chronological landmarks showing how, historically, the various ideas developed in inverse order of the logical exposition which we follow. Moreover, we have already (§ 3.4.4) underlined the, *a priori* surprising, contribution of the study of radiation to the birth of the concept of entropy.

Even if there is no matter, that is, nuclei or electrons, present, an enclosure contains energy. In fact, thermal fluctuations in the walls produce in them random currents which in turn produce a random electromagnetic field inside; the energy density is proportional to the mean square of the field. We thus expect thermal phenomena to occur even in a vacuum due to the presence of electromagnetic fields. In § 13.1 we shall show that after quantization these fields can be described as systems of a new type of particles, the *photons*. Formally, the question is similar to the quantization of the vibrations in a solid (§ 11.4), and here again we find two equivalent descriptions, in terms either of oscillating fields or of particles (§ 13.1.4). The statistical mechanics study of the energy distribution of the radiation in an enclosed space thus reduces to that of the thermal equilibrium of a gas of non-interacting photons moving within that space (§ 13.2). It produces results which have been confirmed experimentally with a remarkable accuracy. We conclude the present chapter by giving the main empirical laws governing the exchange of radiation between different bodies (§ 13.3). This is a subject in non-equilibrium thermodynamics; as is usual in such a case, the use of balance equations provides us in a seemingly simplistic way with important results, and it helps us to deal with numerous practical applications (§ 13.3.4).

13.1 Quantizing the Electromagnetic Field

We want to study the properties of electromagnetic radiation contained in an enclosed space in thermal equilibrium. To do this we must first learn to treat this radiation quantum mechanically. As in the study of a quantum mechanical vibrating string (§ 11.4.4), a model which shall be our guideline at the beginning of the present chapter, we first analyze the solutions of the classical wave equations which here are the Maxwell equations. Then we quantize the system by replacing the physical observables by operators. Reinterpreting the results then leads us to the introduction of the photon, an elementary particle which describes the quantized electromagnetic radiation.

13.1.1 Classical Modes in a Cavity

We start with constructing the *classical modes* for the oscillations of the electromagnetic field within a cavity. We are dealing here with those solutions of the Maxwell equations (1864)

$$\operatorname{curl} \boldsymbol{E} + \frac{\partial \boldsymbol{B}}{\partial t} = 0, \qquad \operatorname{div} \boldsymbol{B} = 0, \tag{13.1}$$

$$\varepsilon_0 \operatorname{div} \boldsymbol{E} = \varrho, \qquad \frac{1}{\mu_0} \operatorname{curl} \boldsymbol{B} - \varepsilon_0 \frac{\partial \boldsymbol{E}}{\partial t} = \boldsymbol{j}, \tag{13.2}$$

which, as functions of the time, behave sinusoidally. As usual, such solutions can be shown to constitute a complete base, so that any arbitrary time-dependent solution of (13.1) and (13.2) is a linear superposition of modes. In SI units we have $\mu_0 = 4\pi \times 10^{-7}$ and $\varepsilon_0 \mu_0 c^2 = 1$, where c is the velocity of light.

We shall consider an enclosure with the shape of a parallelepiped, with edgelengths L_x, L_y, and L_z, and we assume that its walls are perfectly conducting. The right-hand sides of equations (13.2), the sources producing the fields, are therefore zero inside the enclosure. In the conductor, the charges satisfy the equations $\operatorname{div} \boldsymbol{j} + \partial \varrho / \partial t = 0$ and $\boldsymbol{j} = \gamma \boldsymbol{E}$, with $\gamma \to \infty$, which entails that $\boldsymbol{E} = 0$, and hence $\boldsymbol{B} = 0$, since $\boldsymbol{B}$ is a sinusoidal function of time satisfying $\partial \boldsymbol{B} / \partial t = 0$. The fields inside the enclosure are produced by surface charges and currents along the interior walls of the enclosure. According to equations (13.1) the tangential components of $\boldsymbol{E}$ and the normal component of $\boldsymbol{B}$ are continuous across such a wall; they must therefore vanish when we approach the interior surface of the enclosure. We must thus in the parallelepiped solve the Maxwell equations (13.1) and (13.2) when there are no sources and when we have the boundary conditions $\boldsymbol{E}_{\mathrm{t}} = 0$, $\boldsymbol{B}_{\mathrm{n}} = 0$.

We can satisfy the two equations (13.1) by expressing the *fields* $\boldsymbol{E}$ and $\boldsymbol{B}$ in terms of electromagnetic *potentials* through

$$\boldsymbol{E} = -\nabla \Phi - \frac{\partial \boldsymbol{A}}{\partial t}, \qquad \boldsymbol{B} = \operatorname{curl} \boldsymbol{A}. \tag{13.3}$$

This choice is not unique, as the fields remain unchanged when we make the substitution

$$\boldsymbol{A}, \quad \Phi \quad \mapsto \quad \boldsymbol{A} + \nabla \Lambda, \quad \Phi - \frac{\partial \Lambda}{\partial t}, \tag{13.4}$$

where Λ is an arbitrary function of the space-time coordinates (*gauge invariance*). We take advantage of this fact and choose Λ such that $\partial \Lambda / \partial t = \Phi$ so that in the new gauge we have $\Phi = 0$. Equations (13.2) then reduce for $0 < x < L_x, 0 < y < L_y, 0 < z < L_z$ to

$$\operatorname{div} \boldsymbol{A} = 0, \qquad c^2 \operatorname{curl} \operatorname{curl} \boldsymbol{A} + \frac{\partial^2 \boldsymbol{A}}{\partial t^2} = 0, \tag{13.5}$$

with at the walls the boundary conditions

$$\boldsymbol{A}_{\mathrm{t}} = 0, \qquad \operatorname{curl} \boldsymbol{A}\big|_{\mathrm{n}} = 0. \tag{13.6}$$

We thus obtain the *general solution of the Maxwell equations* in a cavity with perfectly conducting, that is, perfectly reflecting walls, in the form

$$E = -\sum_q \frac{\partial \xi_q(t)}{\partial t} A_q(r), \qquad B = \sum_q \xi_q(t) \operatorname{curl} A_q(r), \qquad (13.7)$$

where $\xi_q(t)$ denotes a function of the time which is proportional to $\cos(\omega_q t + \varphi_q)$, and where, for each mode characterized by the index q, $A_q(r)$ denotes a *stationary wave* which is a solution of

$$\operatorname{div} A_q = 0, \qquad c^2 \nabla^2 A_q + \omega_q^2 A_q = 0, \qquad (13.8)$$

and of (13.6). The explicit solution of (13.8), (13.6) is obtained for a parallelepipedal box by separating the variables and Fourier transforming. We introduce the *wavevector*

$$k_x = \frac{\pi m_x}{L_x}, \quad k_y = \frac{\pi m_y}{L_y}, \quad k_z = \frac{\pi m_z}{L_z},$$

$$m_x, m_y, m_z = 0, 1, \ldots . \qquad (13.9)$$

For a given wavevector (13.9) there are, in general, two modes, characterized by their *polarization* vectors ε_1 and ε_2, which are unit vectors perpendicular to k and to one another. Each mode q, which is thus characterized by the indices $q = (m_x, m_y, m_z, \sigma)$ has a frequency

$$\nu_q = \frac{\omega_q}{2\pi} = \frac{ck}{2\pi}, \qquad (13.10)$$

where k is the length of the vector (13.9). The index $\sigma = \pm 1$ denotes the two possible polarizations. The components of A_q have the form

$$A_q^x = \sqrt{\frac{8}{\Omega}} \, \varepsilon_x \cos k_x x \, \sin k_y y \, \sin k_z z, \qquad (13.11)$$

in the general case where none of the indices m_x, m_y, m_z equals zero. The two modes ε_1 and ε_2 then describe stationary waves with a wavevector which is directed obliquely in the parallelepiped. The electric and magnetic fields, given by (13.7) and (13.11), are at each point at right angles to each other for a given mode.

There are also modes for which $m_x = 0$, with m_y and m_z non-vanishing. In that case only one polarization ε is possible, in the x-direction, and we have

$$A_q^x = \frac{2}{\sqrt{\Omega}} \sin k_y y \, \sin k_z z, \qquad A_q^y = A_q^z = 0. \qquad (13.11')$$

There are no modes with more than one vanishing wavevector component. The various solutions (13.11) and (13.11') are orthonormal, that is,

$$\int d^3 r \, A_q(r) \cdot A_{q'}(r) = \delta_{qq'}. \qquad (13.12)$$

However, their closure relation is complicated and non local, because they form a complete base only for vector fields E satisfying the constraint $\operatorname{div} E = 0$.

To study the thermodynamics of radiation in the limit of large volumes, we need the mode density,

$$\mathcal{D}(\nu) = \sum_q \delta(\nu - \nu_q),$$

where ν_q is given by (13.9) and (13.10): the number of modes with frequencies between ν and $\nu + d\nu$ is equal to $\mathcal{D}(\nu)d\nu$. In evaluating $\mathcal{D}(\nu)$ as $\Omega \to \infty$, the summation over the integers m becomes an integration, which leads to

$$\mathcal{D}(\nu) \sim 2 \frac{\Omega}{\pi^3} \int\limits_{k_x, k_y, k_z > 0} d^3k \, \delta\left(\nu - \frac{ck}{2\pi}\right) = \frac{8\pi\Omega\nu^2}{c^3}. \qquad (13.13)$$

The special features occurring when one of the indices m is zero do not play any rôle in this limit. More generally, one can show that (13.13) remains valid for any shape of the enclosure and any boundary conditions, in the limit of large volumes. The conclusions which we shall reach in § 13.2 about the thermodynamic properties of radiation in a cavity are thus of a very general nature, independent of the nature and shape of the enclosure.

13.1.2 Quantized Radiation Spectrum

In classical electromagnetism, the physical system consisting of the interior of the cavity is characterized at any time by the values of the electric and magnetic fields at each point, with constraints imposed by the two Maxwell equations div $B = 0$, div $E = 0$, and by the boundary conditions. It follows from (13.7) that this is the same as taking the coefficients ξ_q as *dynamic variables*, since the vector potentials $A_q(r)$ of all the modes $q = (k, \sigma)$ are given once and for all and since they form a base for expanding any field. The time-dependence of the fields is governed by the other two Maxwell equations, which reduce to the equations of motion $d^2\xi_q/dt^2 + \omega_q^2\xi_q = 0$ (see also (13.17), (13.18) below). The need to describe the system by an infinity of dynamic variables ξ_q is the consequence of the fact that the number of degrees of freedom of the field, that is, the components of E and B at each point r, is infinite. We must bear in mind that although the electromagnetic field is a wave, in § 13.1.1 we were still talking about a *classical wave*: the components of E and B commute, and the energy can change continuously. The analogy between (13.7) and the expansion of a quantum wavefunction in terms of a base of stationary waves is purely *formal*.

   To change to a *quantum field theory* we note that *each* variable ξ_q behaves classically as the position variable of an *harmonic oscillator* with frequency ν_q. Although this amplitude ξ_q of the electromagnetic field (13.7) in the mode q is a more abstract dynamic variable than the position of an ordinary mechanical harmonic oscillator, the quantization proceeds in the same way, as we shall see in full detail in §§ 13.1.3 and 13.1.4. However, in order to describe the quantized electromagnetic field in thermal equilibrium, we need, in fact, only its energy spectrum. We shall be satisfied here with writing the latter down, proceeding by analogy.

Let us consider the oscillator q with angular frequency ω_q. In quantum mechanics, its energy takes on the discrete values $\left(n_q + \frac{1}{2}\right)\hbar\omega_q$, characterized by the quantum number n_q which can be equal to $0, 1, 2, \ldots$. By changing the energy origin we drop the constant $\frac{1}{2}\hbar\omega_q$, and we attribute in this way to the mode q a set of eigenenergies $n_q\hbar\omega_q$. The modes q can be separately excited, and they are independent degrees of freedom. The complete system, that is, the electromagnetic field in the enclosure, is thus a set of an infinite number of independent harmonic oscillators labelled by q. As a result, the eigenstates of the Hamiltonian of the field are characterized by the set $\{n_q\}$ of quantum numbers. Suitably choosing the origin of the energy, we find that the *eigenenergy of the quantum micro-state* $\{n_q\}$ is equal to

$$E(\{n_q\}) = \sum n_q\hbar\omega_q, \qquad q = (\boldsymbol{k}, \sigma), \tag{13.14}$$

where the summation is over all the modes q, which are characterized by the wavevector (13.9) and by one or other of the two polarization values, while ω_q is given by (13.10).

13.1.3 Field Operators

For the complete quantization of the electromagnetic field we need to construct an *algebra of observables*, that is, of operators which represent the various physical quantities such as $\boldsymbol{E}$, $\boldsymbol{B}$, $\boldsymbol{A}$, or ξ_q, and to write the *Hamiltonian* in terms of these operators in order to produce the dynamics. For completeness, we work out this programme below, to help improve the understanding of the photon concept (§ 13.1.4), and as a first step towards quantum field theory or towards problems involving the interaction between matter and radiation. For an introduction to the latter we refer to the two books by C.Cohen-Tannoudji, J.Dupont-Roc, and G.Grynberg, *Photons and Atoms*, Wiley, New York, 1989 and 1991, and for the former to C.Itzykson and J.-B.Zuber, *Quantum Field Theory*, McGraw-Hill, New York, 1980 and to J.Zinn-Justin, *Quantum Field Theory and Critical Phenomena*, Clarendon Press, Oxford, 1989.

Let us start by constructing, as we did with (11.137) for the vibrating string model, a *classical Hamiltonian* for the electromagnetic field in the enclosure. We are therefore looking for a function H of a set of conjugated variables ξ_q, π_q whose equations of motion, which are equivalent to the Maxwell equations, can be recognized as the Hamiltonian equations (2.64), namely,

$$\frac{d\xi_q}{dt} = \frac{\partial H}{\partial \pi_q}, \qquad \frac{d\pi_q}{dt} = -\frac{\partial H}{\partial \xi_q}.$$

Moreover, the value of $\dot{H}$, which is constant with time, should be the energy of the field. We shall therefore rely on the fact that the *energy density* of an arbitrary electromagnetic field inside the enclosure is equal to

$$\frac{1}{2}\left(\varepsilon_0\, E^2 + \frac{1}{\mu_0}\, B^2\right),\tag{13.15}$$

whereas its *momentum density* is equal to

$$\varepsilon_0\left[E \times B\right] = \frac{1}{c^2}\, N,\tag{13.16}$$

where N is the *Poynting vector*. The flux of N through a closed surface represents the *radiated power*. Using the equations $\operatorname{div} E = \operatorname{div} B = 0$ in the enclosure and the boundary conditions on E and B, we can account for the spatial dependence of E and B thanks to an expansion over the modes, defined by solving (13.8) and (13.6). At a given time we thus parametrize the electromagnetic field by

$$E = -\frac{1}{\varepsilon_0}\sum_q \pi_q A_q, \qquad B = \sum_q \xi_q \operatorname{curl} A_q,\tag{13.17}$$

where ξ_q and π_q represent the instantaneous *amplitudes of the magnetic and the electric fields over the various modes*. We have seen in § 13.1.2 that the Maxwell equations $\partial E/\partial t = c^2 \operatorname{curl} B$, $\partial B/\partial t = -\operatorname{curl} E$ are equivalent to Eqs.(13.7) where the evolution of $\xi_q(t)$ is generated by $d^2\xi_q/dt^2 + \omega_q^2\xi_q = 0$. In terms of the variables ξ_q, π_q, the dynamics of the field (13.17) can thus be expressed by the equations of motion

$$\frac{d\xi_q}{dt} = \frac{1}{\varepsilon_0}\,\pi_q, \qquad \frac{d\pi_q}{dt} = -\varepsilon_0\omega_q^2\xi_q.\tag{13.18}$$

On the other hand, let us write down the total energy of a field in terms of the variables ξ_q, π_q which parametrize it. We get this by integrating (13.15) over r. We use (13.12) for the term in E^2. For the term in B^2 we get from equations (13.6), (13.8) and an integration by parts,

$$\int d^3r\left(\operatorname{curl} A_q \cdot \operatorname{curl} A_{q'}\right) = -\int d^3r\left(A_q \cdot \nabla^2 A_{q'}\right) = \frac{\omega_q^2}{c^2}\,\delta_{qq'},$$

so that the total energy of the field is equal to

$$H = \sum_q\left(\frac{1}{2\varepsilon_0}\,\pi_q^2 + \frac{1}{2}\,\varepsilon_0\omega_q^2\xi_q^2\right).\tag{13.19}$$

We check immediately that equations (13.18) are the Hamiltonian equations corresponding to H, provided we regard ξ_q, π_q as pairs of *conjugate* variables. Thus, expression (13.19) is the classical Hamiltonian we were looking for, which governs the dynamics of the variables ξ_q, π_q, and hence of the field (13.17). We also see that H describes a set of independent classical harmonic oscillators q and that ε_0 plays the rôle of the mass and $\varepsilon_0\omega_q^2 = \varepsilon_0 c^2 k^2$ that of the restoring constant.

The *quantization* now proceeds as in § 11.4.4, by treating $\widehat{\xi}_q$ and $\widehat{\pi}_q$ as *operators* satisfying the canonical commutation relations (11.138). Expression (13.19) becomes the Hamiltonian of a set of quantum mechanical harmonic oscillators, with eigenenergies given by (13.14), apart from an additive constant.

More precisely, we find from (11.138) that the operators

$$\widehat{c}_q = \frac{1}{\sqrt{2k}} \left(\widehat{\xi}_q \sqrt{\varepsilon_0 \omega_q} + \frac{i\widehat{\pi}_q}{\sqrt{\varepsilon_0 \omega_q}} \right) \tag{13.20}$$

and $\widehat{c}_q^\dagger$ satisfy the same commutation relations (10.21) as boson annihilation and creation operators. The eigenvalues of $\widehat{c}_q^\dagger \widehat{c}_q$, which can be identified with the operator $\widehat{n}_q$ describing the number of photons in the mode q, are thus the integers $n_q = 0, 1, \dots$. If we use (13.20), we can write the quantum Hamiltonian derived from (13.19) in the form

$$\begin{aligned}
\widehat{H} &= \left(\frac{1}{2\varepsilon_0} \widehat{\pi}_q^2 + \frac{1}{2} \varepsilon_0 \omega_q^2 \widehat{\xi}_q^2 + \frac{i}{2} \omega_q \left[\widehat{\xi}_q, \widehat{\pi}_q \right] \right) \\
&= \sum_q \hbar \omega_q \, \widehat{c}_q^\dagger \widehat{c}_q,
\end{aligned} \tag{13.21}$$

which shows that, indeed, its eigenvalues are (13.14).

As we have already seen in § 11.4.4, when changing from classical to quantum mechanics, we had to change the Hamiltonian by adding a constant to it. This constant is, for each mode, proportional to the commutator $\left[\widehat{\xi}_q, \widehat{\pi}_q \right] = i\hbar$, and it vanishes when we go to the classical limit. Moreover, this constant is irrelevant, as only energy differences can be measured. However, its subtraction was necessary, as the sum $\sum_q \frac{1}{2} \hbar \omega_q$ over all modes would produce an ultraviolet divergence.

The *field observables* $\widehat{E}(r)$ and $\widehat{B}(r)$, which at each point are given by (13.17), and also the vector potential $\widehat{A}(r)$, become operators in quantum electromagnetism. Their algebra can be conveniently formulated by expressing them in terms of the operators (13.20), as follows

$$\widehat{E}(r) = -\frac{1}{\varepsilon_0} \sum_q \widehat{\pi}_q \, A_q(r) = i \sum_q \sqrt{\frac{\hbar \omega_q}{2\varepsilon_0}} \left(\widehat{c}_q - \widehat{c}_q^\dagger \right) A_q(r), \tag{13.22}$$

$$\widehat{B}(r) = \operatorname{curl} \widehat{A}(r), \tag{13.23}$$

$$\widehat{A}(r) = \sum_q \widehat{\xi}_q \, A_q(r) = \sum_q \sqrt{\frac{\hbar}{2\varepsilon_0 \omega_q}} \left(\widehat{c}_q + \widehat{c}_q^\dagger \right) A_q(r). \tag{13.24}$$

In particular, the components of $\widehat{E}(r)$ do not commute with those of $\widehat{A}(r')$, so that we cannot specify exactly the electric and magnetic fields at the same time, since their commutator $\varepsilon_0 \left[\widehat{E}^\alpha(r), \widehat{B}^\beta(r') \right] = i\hbar \varepsilon_{\alpha\beta\gamma} \partial \delta^3(r-r')/\partial r'_\gamma$ does not vanish. One should note that the classical wavefunctions $A_q(r)$ associated with each mode *do not* become either operators or kets in the quantization process, but their coefficients ξ_q in (13.7) become the operators $\widehat{\xi}_q$ in the expansion (13.24).

As an exercise, one could repeat the above considerations, imposing *periodic boundary conditions*, as we did in § 11.4.1 for the case of crystal vibrations. The main changes are the following. The components of k ($\neq 0$) take on the values $k_x = 2\pi m_x/L, \ldots$, with $m_x = 0, \pm 1, \pm 2, \ldots$, instead of (13.9). Each mode is described by a complex vector potential

$$A_q(r) \;=\; \frac{1}{\sqrt{\Omega}}\,\varepsilon\,e^{ik\cdot r}, \qquad q \;=\; (k, \varepsilon), \qquad\qquad (13.26)$$

with two possible polarization vectors for each k, which satisfy the relations

$$\varepsilon_1 \cdot k \;=\; \varepsilon_2 \cdot k \;=\; \varepsilon_1^* \cdot \varepsilon_2 \;=\; 0, \qquad \varepsilon_1^* \cdot \varepsilon_1 \;=\; \varepsilon_2^* \cdot \varepsilon_2 \;=\; 1. \qquad (13.27)$$

By letting ε be complex we are able to describe modes with a circularly or an elliptically polarized wave. The operators $\widehat{\xi}_q$ and $\widehat{\pi}_q$ are no longer either Hermitean or independent: if $-q = (-k, \varepsilon^*)$ denotes the mode $A_{-q}(r) = A_q^*(r)$, we have $\widehat{\xi}_q^\dagger = \widehat{\xi}_{-q}$, $\widehat{\pi}_q^\dagger = \widehat{\pi}_{-q}$, and the commutation relations become $\left[\widehat{\xi}_q, \widehat{\pi}_{q'}^\dagger\right] = \delta_{qq'}$. In the term with $\widehat{c}_q^\dagger A_q$ in (13.22) and (13.24) we must replace $A_q(r)$ by $A_q^*(r)$. The energy retains its form (13.14). However, whereas there was no simple expression for the *total momentum of the field* for the case of perfectly conducting walls, since each mode (13.11) was a superposition of plane waves with wavevectors $\pm k_x$, $\pm k_y$, $\pm k_z$, we get here an interesting expression. We start from the momentum density (13.16), which can be split into two contributions

$$\varepsilon_0 \left[E \times B\right] \;=\; \varepsilon_0 \left[E \times \text{curl}\, A\right] \;=\; \varepsilon_0\, E^\alpha \nabla A^\alpha - \varepsilon_0 (E \cdot \nabla) A. \qquad (13.28)$$

The integral of the second term vanishes, as one can see through integrating by parts; if we use successively (13.22), (13.24), (13.26), and (13.27) we get from the first term after integration and quantization

$$\widehat{P} \;=\; -\sum_{q,q'} \widehat{\pi}_q^\dagger \widehat{\xi}_{q'} \int d^3 r\, A_q^{\alpha*} \nabla A_{q'}^\alpha \;=\; -i \sum_{q,q'} \widehat{\pi}_q^\dagger \widehat{\xi}_{q'}\, k\, \delta_{kk'} \varepsilon_q^* \cdot \varepsilon_{q'}$$

$$= \; -i \sum_q \widehat{\pi}_q^\dagger \widehat{\xi}_q k,$$

or, finally, using (13.20) and the fact that $-q = (-k, \varepsilon^*)$,

$$\widehat{P} \;=\; \sum_{k,\sigma} \hbar k\, \widehat{c}_q^\dagger \widehat{c}_q. \qquad\qquad (13.29)$$

In agreement with § 2.1.5, the *momentum operator* (13.29) is the *generator of the translations* of the field.

Similarly, when we evaluate the *total angular momentum of the field*,

$$J \;=\; \varepsilon_0 \int d^3 r \left[r \times \left[E \times B\right]\right], \qquad\qquad (13.30)$$

using (13.28) produces two terms. The first one, according to the above calculation, is changed by subtracting $[\delta r \times P]$ when we shift the origin of the coordinates by δr; it can thus be interpreted as an "orbital angular momentum". The second one, which through integrating by parts can be written as

$$S = -\varepsilon_0 \int d^3r \left[r \times (E \cdot \nabla)A \right] = \varepsilon_0 \int d^3r \left[E \times A \right], \qquad (13.31)$$

is independent of the origin of the coordinates and is thus *intrinsic* in nature. Let us remind ourselves of the interpretation of angular momentum as the *generator of rotations* (§2.1.5). The splitting of J into two terms thus simply shows up the fact that a rotation of the field has two effects: rotating the system of spatial coordinates and rotating the components of the *vector* field. Following a calculation similar to the one which led to (13.29) gives the explicit form of (13.31) after quantization:

$$\widehat{S} = -\sum_{q,q'} \widehat{\pi}_q^\dagger \widehat{\xi}_{q'} \delta_{kk'} \left[\varepsilon_q^* \times \varepsilon_{q'} \right]$$

$$= -i\hbar \sum_{q,q'} \widehat{c}_q^\dagger \widehat{c}_{q'} \delta_{kk'} \left[\varepsilon_q^* \times \varepsilon_{q'} \right].$$

Let w be a unit vector in the direction of the propagation of the wave (13.26), and let u and v be two unit vectors which with w form a direct trihedral, and let us choose the two complex polarization vectors associated with k such that

$$\varepsilon_1 = \varepsilon_2^* = \frac{1}{\sqrt{2}} \left(u + iv \right).$$

We characterize the mode $q = (k, \sigma)$ by the number σ which takes on the value $+1$ for ε_1 and -1 for ε_2. By looking at the classical fields $E(r)$ and $B(r)$ associated with a single mode q, we see easily that $\sigma = +1$ describes a left-handedly *circularly polarized plane wave* and $\sigma = -1$ a right-handedly polarized wave, and that $\left[\varepsilon_1^* \times \varepsilon_1 \right] = iw$. We can therefore simplify the expression for the intrinsic angular momentum of the field as follows:

$$\widehat{S} = \sum_q \hbar \sigma w \, \widehat{c}_q^\dagger \widehat{c}_q, \qquad w = \frac{k}{k}. \qquad (13.32)$$

We have now quantized the electromagnetic field in vacuo. The problems of emission and absorption of radiation by matter involve the dynamics of the field and its coupling with charged particles. We know (see Eq.(2.65)) how to write down the dynamics of spinless non-relativistic particles in a *classical* field which is assumed to have been *given*, but the treatment of emission and absorption makes it necessary to quantize *at the same time* the particles and the field. For the particles it is sufficient to replace in the Hamiltonian (2.65) the variables r_i and p_i by conjugated operators. For the electromagnetic field we need to start by choosing a gauge, as the Hamiltonian is expressed in terms of the potentials rather than of the fields E and B. The most convenient gauge here is the *Coulomb gauge* which is characterized by $\operatorname{div} A = 0$. The scalar potential, which does not vanish when there are charges, must then satisfy the equation $\varepsilon_0 \nabla^2 \Phi = -\varrho$. The solution of this equation determines Φ as a function of the particle coordinates r_i and makes, after quantization, a contribution $-\nabla \widehat{\Phi}$ to the field $\widehat{E}(r)$, with

$$\widehat{\Phi}(r) = \sum_i \frac{e_i}{4\pi\varepsilon_0 |r - \widehat{r}_i|},$$

which must be added to (13.22). The field operators are, as before, the $\widehat{\xi}_q$, $\widehat{\pi}_q$, or the $\widehat{c}_q$, $\widehat{c}_q^\dagger$. In contrast to the vector potential (13.24), the scalar potential $\widehat{\Phi}$ is independent of the field operators; it is an operator only through the coordinates $\widehat{r}_i$ of the particles with charges e_i. The Hamiltonian $\widehat{\mathcal{H}}$ of the system of particles and field contains four contributions: (i) the contribution (13.21) of the *free field*; (ii) the *Coulomb potential*

$$\widehat{W} = \frac{1}{4\pi\varepsilon_0} \sum_{i>j} \frac{e_i e_j}{|\widehat{r}_i - \widehat{r}_j|}; \tag{13.33}$$

(iii) the *kinetic energy*, changed by replacing p_i by $p_i - e_i A(r_i)$,

$$\sum_i \frac{\left(\widehat{p}_i - e_i \widehat{A}(\widehat{r}_i)\right)^2}{2m_i}; \tag{13.34}$$

(iv) when the particles have a *spin* s_i, the coupling between the associated *magnetic moment* and the field, which is proportional to $\left(\widehat{B}(\widehat{r}_i) \cdot \widehat{s}_i\right)$. The *coupling between the field and the matter* enters (13.34) through the vector potential $\widehat{A}(\widehat{r}_i)$, which is obtained from (13.24) by replacing the coordinate r by the position operator $\widehat{r}_i$ of the ith particle, and which thus contains simultaneously the field operators $\widehat{c}_q$ and $\widehat{c}_q^\dagger$ and the particle operators $A_q(\widehat{r}_i)$. One can check that the equations $d\widehat{X}/dt = \left[\widehat{X}, \widehat{\mathcal{H}}\right]/i\hbar$ which describe the evolution of the observable $\widehat{X}$ in the Heisenberg picture (§ 2.1.5) are for $\widehat{X} = \widehat{E}(r)$ or $\widehat{B}(r)$ the same as the quantum version of the Maxwell equations (13.1) and (13.2) where the particle observables occur on the right-hand sides, for $\widehat{X} = \widehat{r}_i$ the same as the formula $\widehat{v}_i = \left[\widehat{p}_i - e_i \widehat{A}(\widehat{r}_i)\right]/m_i$ for the velocity, and for $\widehat{X} = m_i \widehat{v}_i$ the same as the quantum version of the Lorentz force,

$$m_i \frac{d\widehat{v}_i}{dt} = e_i \left\{ \widehat{E}_i(\widehat{r}_i) + \tfrac{1}{2} \left[\widehat{v}_i \times \widehat{B}(\widehat{r}_i)\right] - \tfrac{1}{2} \left[\widehat{B}(\widehat{r}_i) \times \widehat{v}_i\right] \right\}.$$

13.1.4 Photons

Although expression (13.14) for the levels of the quantized field would suffice to study its thermodynamics, it is useful to change the language for describing these levels. As we did in § 11.4.2 for lattice vibrations or in § 11.4.4 for vibrations of a string, we base ourselves upon the analogy between (13.14) and the energy of a system of bosons, and also upon the equivalence between the formal structure of the algebra of the operators (13.20) and that of the boson annihilation and creation operators. Just as quantized mechanical vibrations can conveniently be described as *phonons* we shall interpret the oscillations of the quantized electromagnetic field in terms of a *new type of bosons*, the *photons*. Historically (§§ 13.3.1 and 11.4), photons preceded phonons: after Planck's break-through in 1900, the photon idea was introduced by Einstein in 1905 in his interpretation of Wien's radiation law (13.48) as being produced by a set of particles; it was then used by him to explain the photoelectric effect. Proceeding by analogy, Einstein assumed in 1907 that the energy levels of any harmonic oscillator should be quantized, which led him to an explanation of the specific heats of solids; the theory of phonons was worked out by Debye in 1914, when the old quantum theory was still in its infancy.

Let us establish the correspondence between the two equivalent and complementary languages for describing an electromagnetic field in quantum physics, either in terms of waves, or in terms of photons. We rely on the simple arguments of § 13.1.2, and shall below make our statements more rigorous by using the results of § 13.1.3. Let us start from the oscillation eigenmodes which we constructed in § 13.1.1. Each *mode* $q = (\boldsymbol{k}, \sigma)$ can be interpreted as one of the *single-particle states* of § 10.2.1 (Table 13.1). In fact, the values (13.9) of the *wavevectors* are the same as for a particle in a box with rigid walls (Eq.(10.8)) provided we identify the photon *momentum* $\boldsymbol{p}$ with $\hbar\boldsymbol{k}$ in agreement with de Broglie's relation. The *polarization*, which takes on two values $\sigma = \pm 1$, plays the rôle of an internal quantum number of the photon; in what follows we shall show that it is equivalent to a *spin*. The *vector potential* (13.11) or (13.26) of the mode q plays the rôle of a *photon wavefunction* in the ket q, and the *classical* wave equation (13.8), which determines it, is the analogue of the single-photon Schrödinger equation. Equations (13.10) and (13.14) indicate that we should assign to a photon in the mode q an *energy* proportional to the wave *frequency*,

$$\varepsilon_q = \hbar\omega_q = \hbar ck = cp, \tag{13.35}$$

again in agreement with the Planck-de Broglie relation $\varepsilon = h\nu$. The relation $\varepsilon = cp$ between energy and momentum shows by the way that a photon must be considered as a relativistic particle with zero rest mass. In particle language, this property is the counterpart of the fact that electromagnetic waves propagate with a *velocity* c independent of the frequency, a velocity which can be identified with that of the photons.

Table 13.1. Equivalence between a single-mode $q = (\boldsymbol{k}, \sigma)$ of an electromagnetic wave and a single-photon state

Classical electromagnetic mode	Single-photon state
Vector potential $\boldsymbol{A}_q(\boldsymbol{r})$ of a mode	Photon wavefunction
Wavevector $\boldsymbol{k}$	Momentum $\boldsymbol{p} = \hbar\boldsymbol{k}$
Circular polarization	Spin helicity $\sigma = \pm 1$
Frequency $\omega = 2\pi\nu$	Photon energy $\varepsilon = \hbar\omega$
Wave velocity $c = \omega/k$	Photon velocity $c = \varepsilon/p$

Table 13.2. Equivalence between electromagnetic waves as quantum oscillators and many-photon states

Micro-state of the quantized field	Set of photons
Hilbert space of vibration states	Fock space of photons
Oscillator quantum numbers	Occupation numbers $\{n_q\}$
Energy of independent harmonic oscillators $$\sum_q n_q \hbar\omega_q$$	Energy of non-interacting photons $$\sum_q n_q \varepsilon_q$$
Field operators	Photon creation and annihilation operators
Canonical equilibrium	Grand canonical equilibrium with $\mu = 0$

We must stress that the equivalence between modes and single-photon states associates a *classical* concept with a *quantum* one. In wave language, the $A_q(r)$ are classical stationary waves, obtained as solutions of the classical Maxwell equations; nevertheless, the same quantities play for the photons the rôle of a base in the Hilbert space $\mathcal{E}_H^{(1)}$ of the single-particle states. In the first case, a linear superposition such as $\sum_q \xi_q A_q(r)$ or as (13.7) can be interpreted as a classical electromagnetic wave; in the second case, it plays the rôle of a *single*-photon wavefunction (provided it is normalized). The indices $q = (k, \sigma)$ which characterize a mode in the classical or in the quantum picture become for a photon its quantum numbers.

Let us now turn to a quantized electromagnetic field, which is the equivalent of a set of an arbitrary number of photons (Table 13.2). The Hilbert space of the field can be interpreted as the Fock space of the photons. A quantum micro-state describing oscillations of the electromagnetic field, characterized by the set $\{n_q\}$ of the *quantum numbers of the harmonic oscillators*, can be viewed as a state constructed by putting a *number n_q of photons* in each single-particle state q. The possible values of the *occupation numbers* $n_q = 0, 1, 2, \ldots$ correspond to *Bose-Einstein* statistics. The expression (13.14) for the energy of the set of harmonic oscillators can be understood as the energy of a set of photons, which *do not interact* with one another, since their energies are just simply added together.

Expression (13.21) for the Hamiltonian confirms this result, since the operator $\hat{c}_q^\dagger \hat{c}_q$ has as eigenvalue the number n_q of photons in the mode q so that each q photon contributes $\hbar\omega_q$ to the energy. Nevertheless, in contrast to the case of lattice vibrations (§ 11.4.2) where the anharmonicity of the interatomic forces produces terms in the Hamiltonian which describe interactions between the phonons and processes where several phonons are created or annihilated, expression (13.21) here is *exact*. This property is connected with the *linearity of the Maxwell equations* in vacuo, which implies that the oscillations of the free electromagnetic field are *purely harmonic* and that the modes are independent. After quantization, the independence of the photons occupying the various modes corresponds to the superposition of the classical electromagnetic waves. For the examples considered in the earlier chapters – conduction electrons, holes, phonons, helium atoms – treating the particles or quasi-particles as being independent was a more or less justified approximation. Here, there is rigorously no interaction between the photons.

The correspondence $p = \hbar k$ between momentum and wavevector has been introduced above only through a simple analogy. Its rigorous justification is based upon expression (13.29) for the operator $\widehat{P}$, the total momentum of the electromagnetic field which we derived from the momentum density (13.16). This equation, in fact, expresses simply that each photon in the mode $q = (k, \sigma)$ with *wavevector k* contributes $\hbar k$ to the *total momentum P*. The ratio c^2 between the radiated power and the momentum density (13.16) has a direct interpretation in the language of a kinetic theory of photons: a first factor c arises from the energy $\varepsilon = cp$, and a second factor c from the velocity of the photons crossing a surface element.

Similarly, expression (13.32) for the *intrinsic angular momentum of the field* can be understood by assigning to each photon $q = (k, \sigma)$ an intrinsic angular momentum with component along k equal to $\hbar \sigma = \pm \hbar$. As a result the photon has a *spin equal to* 1, with the peculiar feature that from amongst the three permitted values $m = 0, +1, -1$ for the component of the spin 1 along k, called the *helicity*, only two, $m = \pm 1$, can physically be realized. This feature may come about because the photon is a relativistic particle with zero rest mass so that a change of frame does not affect its helicity. The *two possible helicities* $m = +1$ and $m = -1$ correspond, respectively, to *left-handedly and right-handedly circularly polarized* waves.

The *field operators* $\widehat{c}_q$ and $\widehat{c}_q^\dagger$, linearly related through expressions (13.22), (13.23), and (13.24) to the quantized fields $\widehat{E}(r)$ and $\widehat{B}(r)$ and the potential $\widehat{A}(r)$, are the *photon annihilation and creation operators*. Let us now assume that the density operator describing a state of the field is diagonal in the occupation numbers n_q in the photon Fock space. This will be the case for the thermal equilibrium state, studied in § 13.2. This is also the case for states describing a given number of *photons in the limit of classical particles* $\langle n_q \rangle \ll 1$, when their Bose statistics becomes irrelevant. Under those conditions we have $\langle \widehat{c}_q \rangle = \langle \widehat{c}_q^\dagger \rangle = 0$, and hence the mean values $\langle \widehat{E}(r) \rangle$ and $\langle \widehat{B}(r) \rangle$ of the electric and the magnetic fields vanish. Inversely, the state describing a *classical electromagnetic field* in terms of photons must include coherences such that $\langle \widehat{c}_q \rangle$, $\langle \widehat{c}_q^\dagger \rangle$, and hence the $\langle n_q \rangle \geq |\langle c_q \rangle|^2$ are large, since the average values of $\widehat{E}(r)$ and of $\widehat{B}(r)$ must be large and their relative fluctuations small. We discussed this point in § 11.4.4 for the vibrating string problem, and noted that the *classical wave* limit and the *classical particle* limit are *not the same*; indeed, these two limits are even the opposite of each other.

The interaction of the electromagnetic field with matter, in particular with the walls of the enclosure, produces energy exchanges which, according to (13.14) involve changes in the quantum numbers n_q of the oscillators. Such changes in the quantum state of the field are in the photon language simply expressed as the *creation or annihilation of photons by the walls*. This is clear in expression (13.34) for the interaction between the electromagnetic field and a system of charged particles: the vector potential (13.24) is the sum of a term which creates a photon and a term which annihilates a photon, and hence $\widehat{\mathcal{H}}$ does not commute with the occupation number operators $\widehat{n}_q$.

Quantum theory has made it possible to combine and transcend the two old ideas of optics which seemed to be irreconcilable: Newton's corpuscular concept (1675) and Fresnel's wave concept (1819). We sketched the history of this synthesis in § 10.1.2. However, we have had to pay for this: the photon only resembles a particle when it takes the form of a more or less – but never completely – localized wavepacket, whereas the electromagnetic waves $\widehat{E}(r)$ and $\widehat{B}(r)$ which are represented by operators always show statistical fluctuations. The arguments of the present section have shown that the two languages are equally valid for describing the same reality. They also made it possible for us to understand through the example of the photon why elementary particle physics and quantum field theory are one and the same discipline.

13.2 Equilibrium of Radiation in an Enclosure

For reasons which will become clear in § 13.3.2 electromagnetic radiation in thermal equilibrium is traditionally called *black-body radiation*. We shall here use the methods of statistical physics to study it, in either the quantum-field or the photon language.

13.2.1 Extensive Quantities

Let us assume that the walls of the enclosure which may contain an electromagnetic field, the micro-states and eigenenergies (13.14) of which we have determined, are maintained at a temperature T. The currents produced by the thermal motion of the charged particles which make up the walls produce random fields inside the enclosure. Conversely, electromagnetic fields inside the enclosure produce currents in the walls and thus can exchange energy with the charged particles making up these walls. There is thus a coupling – relatively weak, if the enclosure is large – between the interior electromagnetic field, that is, the photon gas, and the matter of the walls: the latter can exchange energy with the field by changing the quantum numbers n_q, or, what amounts to the same, by creating or absorbing photons. This will ultimately lead to a thermal equilibrium macro-state of the electromagnetic field.

This state is characterized by two extensive variables, the volume and the internal energy, which is the expectation value of (13.14). This is obvious in the wave description of the field. In the particle description, we might also wonder about the number of photons $\sum n_q$. However, that is not a constant of the motion; for the system of the enclosure together with the walls, the total energy is conserved, but not the number of photons which does not commute with the coupling term (13.34) of the total Hamiltonian. We are therefore in the second situation of § 10.5.2 where equilibrium is described either as *canonical* or as *grand canonical with a zero chemical potential* for the photons.

The partition function is thus given by (10.29) with $\varepsilon_q = \hbar\omega_q = cp$ and $\mu = 0$, which in the large volume limit gives us the *free energy*

$$F = kT \sum_q \ln\left(1 - e^{-\beta\varepsilon_q}\right)$$

$$= 2kT \frac{\Omega}{h^3} \int d^3p \, \ln\left(1 - e^{-\beta cp}\right),$$

where the factor 2 comes from the summation over the polarizations σ. Integrating by parts we get (see formulæ at the end of this volume)

$$F(\Omega, T) = \frac{\Omega(kT)^4}{\pi^2(\hbar c)^3} \int_0^\infty x^2 \, dx \, \ln\left(1 - e^{-x}\right)$$

$$= -\frac{\Omega(kT)^4}{3\pi^2(\hbar c)^3} \int_0^\infty \frac{x^3 \, dx}{e^x - 1}$$

$$= -\frac{\Omega\pi^2(kT)^4}{45(\hbar c)^3}. \tag{13.36}$$

We could also have obtained the same result directly by using equation (10.47) and expression (13.13) for the mode density, which implies $\mathcal{N}(\varepsilon) = \Omega\varepsilon^3/3\pi^2\hbar^3c^3$.

From (13.36) we get the *entropy*

$$S = -\frac{\partial F}{\partial T} = \frac{4\pi^2 \Omega k^4 T^3}{45(\hbar c)^3}, \tag{13.37}$$

and the *internal energy* per unit volume

$$\frac{U}{\Omega} = \frac{F + TS}{\Omega} = \frac{\pi^2(kT)^4}{15(\hbar c)^3}. \tag{13.38}$$

The latter is very small at room temperatures (6×10^{-6} J m^{-3}) but it increases rapidly with temperature due to the power law T^4. The radiation energy becomes appreciable in astrophysics in stellar interiors and even considerable in cosmology in those stages where the Universe still had a very high temperature: when $T = 10^{11}$ K, the photon energy density in mass units reached $U/\Omega c^2 = 10^{12}$ kg m^{-3}, and it was still 10^4 kg m^{-3} three minutes after the Big Bang, when T was 10^9 K. Then, during 10^6 years, as long as the Universe was a homogeneous magma with a temperature above 3000 K, it was "radiation dominated", that is, the major part of the available energy was in the form of radiation, whereas it is now "matter dominated" (§ 13.2.2).

The *pressure* due to the radiation,

$$\mathcal{P} = -\frac{\partial F}{\partial \Omega} = \frac{U}{3\Omega}, \tag{13.39}$$

has the same characteristics. It contributes to the stellar equilibrium by opposing – but, in general, less strongly than the kinetic pressure due to the matter – stellar collapse under the effect of gravitational forces (Exerc.6e). Expression (13.39) differs by a factor $\frac{1}{2}$ from (10.72), which is the equation valid for gases of non-interacting particles with kinetic energies $p^2/2m$. This difference is due to the relativistic nature of the photon, which has an energy cp (Exerc.13a).

We can obtain (13.39) also by using the kinetic theory methods of §7.4 and interpreting the pressure as the consequence of the *collisions of the photons* with a wall. Striking a balance as in §7.4.1 or as in the coming §13.3.1 we see, first of all, that the volume occupied by photons with momentum p which will reach a surface area ΔS during a time interval Δt is equal to $\Delta S\, c\, \cos\theta\, \Delta t$, where θ is the angle between p and the normal to the surface area (Fig.7.3). The average number of such photons at thermal equilibrium is thus equal to $2\,h^{-3}\Delta S\, c\, \cos\theta\, \Delta t\, d^3p\, f_p$, where the factor $2h^{-3}\, d^3p$ is the number of modes per unit volume with momenta within d^3p. Let us assume to fix ideas that the wall is perfectly absorbing. The photons which hit it will per unit time and per unit area give off to the wall an amount of normal momentum equal to

$$\frac{2}{h^3}\, c \int p\cos^2\theta\, \frac{d^3p}{e^{\beta cp}-1} = \frac{U}{6\Omega}. \tag{13.40}$$

(The integration over θ from 0 to $\frac{1}{2}\pi$ gives a factor $\frac{1}{3}$ whereas in the calculation of the energy this integration gives a factor 2.) These photons are at equilibrium replaced by other photons which are emitted by the wall and which have on average an opposite momentum. The pressure force (13.39) means that an amount equal to $U/3\Omega$ of momentum is transferred per unit time to unit area of the wall.

In wave language the pressure $\mathcal{P}$ can be understood, like in classical electromagnetism, as the average of the *radiation pressure* exerted on the walls by the electromagnetic waves in the enclosure. We know, in fact, that an electromagnetic wave exerts a radiation pressure on material bodies; for instance, the solar radiation pressure orientates the comet tails in a direction away from the sun. The fact that the result is the same as the earlier one comes about because the pressure of a wave is connected with its momentum (13.16) which in turn is connected through (13.29) with that of the photons.

Radiation in equilibrium within a large enclosure is *homogeneous, isotropic, and unpolarized*. One can also show that it is independent of the *shape* of the enclosure and of the *conditions at the walls* (§§ 5.5.2 and 10.3.3).

The absence of polarization is the direct consequence of the fact that the two polarization modes have the same energy and thus are equally densely populated. The homogeneity, which can easily be ascertained from (13.26) for the case of periodic boundary conditions, is less obvious inside the enclosure with perfectly reflecting walls that we considered in § 13.1.1. In fact, for a given mode (13.11) the energy density (13.15) shows oscillations of the form $\sin^2 k_x x$ in each of the three directions. The energy density at equilibrium is obtained by adding together the contributions from all modes with a weight proportional to the Bose factor, and it has certainly spatial variations. Nevertheless, provided $L \gg \hbar c/kT$ the modes (13.9) are so closely bunched that changing m_x, m_y, m_z by one or a few units practically changes neither the wavevector nor the occupation factor of the mode. For calculating the energy density we can therefore take an average over such neighbouring modes. Provided we are well away from the boundaries of the enclosure, this averaging has the effect of replacing the oscillating factor $\sin^2 k_x x$ by $\frac{1}{2}$. To see this, let us consider a position with an abscissa ξ which is at a finite distance from the centre of the enclosure ($\xi = |x - \frac{1}{2}L| \ll L$); for a momentum k_x the energy density of the mode oscillates either as $\sin^2 k_x \xi$ or as $\cos^2 k_x \xi$, depending on whether $m_x = k_x L/\pi$ is even or odd, so that an average over two neighbouring modes is sufficient to give a constant contribution in ξ. This argument, which can be generalized to any point away from the wall, shows that the energy density is practically independent of the coordinates. It is valid even when we restrict ourselves to a given direction of k and this also proves the isotropy.

13.2.2 Planck's Law

To describe the characteristics of equilibrium radiation in more detail we introduce the *black-body radiation density* $u(\nu)$ which for photons plays the same rôle as the Maxwell distribution for the molecules of a classical perfect gas. It is defined as follows: $u(\nu)\,d\nu$ is the average energy per unit volume at equilibrium of the photons with frequencies between ν and $\nu + d\nu$. Because the radiation is homogeneous, we only need divide the energy which the enclosure contains in the frequency range $\nu, \nu + d\nu$ by its volume. To do this we multiply the Bose factor

$$\frac{1}{e^{h\nu/kT} - 1},$$

which is the average number of photons occupying a mode with frequency $\nu = \omega/2\pi$, by the energy $\hbar\omega = h\nu$ of that mode, and by the number of modes in the range $\nu, \nu + d\nu$. The latter is given by expression (13.13). We get the same result by noting that according to (10.38) the number of modes with their momentum inside a volume d^3p equals $2\Omega\, d^3p/h^3$, where we have taken into account that there are two polarizations, whence, integrating over the directions and replacing p by $h\nu/c$, we find

$$\mathcal{D}(\nu)\,d\nu \;=\; \frac{2\Omega}{h^3}\,4\pi p^2\,dp \;=\; \frac{8\pi\Omega}{c^3}\,\nu^2\,d\nu. \tag{13.41}$$

We thus get altogether

$$u(\nu)\,d\nu \;=\; \frac{8\pi h\nu^3\,d\nu}{c^3\left(e^{h\nu/kT}-1\right)}\;.\qquad\qquad (13.42)$$

This is *Planck's radiation law*, which characterizes the way the energy of the electromagnetic field is distributed over the wave frequencies for given T. The result *depends only on the temperature* and can be represented (Fig.13.1) by curves $u(\nu)$ which are obtained one from the other by simple magnifications, proportionally to T for the abscissa ν and to T^3 for the ordinate n. The radiation energy density increases rapidly with temperature for any fixed frequency. At the same time, its distribution as function of the frequency changes, with the maximum shifting to higher frequencies when the temperature increases. This maximum, which is determined by the equation

$$3 \;=\; \frac{h\nu_m}{kT}\,\frac{1}{1-e^{-h\nu_m/kT}}\,,$$

that is, by the relation

$$\nu_m \;=\; 2.82\,\frac{kT}{h}\,,\qquad\qquad (13.43)$$

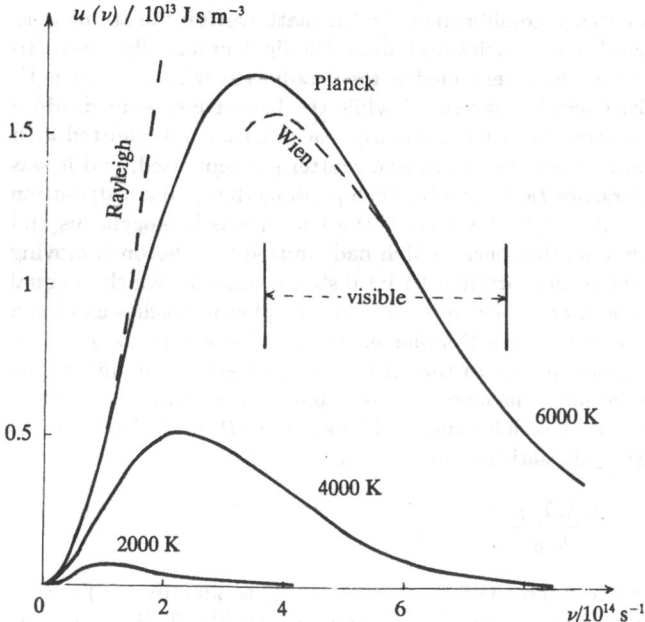

Fig. 13.1. Planck's radiation law

varies *linearly* with the temperature. This property, which was discovered empirically, is *Wien's displacement law*. At a temperature of 8500 K, the maximum is in the middle of the visible spectrum ($\lambda \simeq 6000$ Å). The area under each curve represents the total energy density (13.38), and increases with the temperature as T^4.

We shall see in §§ 13.3.1 and 13.3.4 that experimental checks of Planck's law are very accurate, but indirect. It would be extremely difficult to carry out calorimetric measurements inside an empty enclosure.

Observations in the microwave, centimeter or decimeter wavelength, range have enabled us to establish that we receive from the interstellar medium an *isotropic radiation*, the *so-called cosmic background radiation*. This discovery by Arno A. Penzias and Robert W. Wilson in 1965 is of great importance in cosmology. Recent satellite measurements have confirmed that the strength of the cosmic background radiation at the various frequencies follows with a remarkable precision Planck's law with a temperature of 2.74 K; however, one still continues to speak of the "3 K radiation". Everything takes place, as if, apart from the radiation emitted by the stars, the Universe were a cavity containing electromagnetic radiation at equilibrium at that temperature.

The cosmic background can be explained as follows. During the formation of the Universe, that took 10^6 years, its constituents were in thermal equilibrium with a temperature which during the expansion of the Universe decreased from 10^{11} to 3000 K. At the end of this period the matter had not yet been condensed into galaxies and stars; the atoms were ionized, as the characteristic ionization temperature is 3000 K, so that there was a permanent absorption and emission of photons and the radiation was in equilibrium with the matter. After the atoms were formed and they condensed into galaxies and stars, the light practically ceased to interact with the atoms and there remained a fossil radiation with a wavelength which due to the Doppler effect has increased while the Universe expanded. More precisely, let us consider a photon which is presently reaching us. It was emitted with a frequency ν_0 at the epoch where radiation and matter got separated, and it was at equilibrium at a temperature $T_0 \simeq 3000$ K. Such photons thus had a distribution which was Planck's law $u_0(\nu_0, T_0)$. However, if the Universe is homogeneous and its expansion is a uniform one, the source which had emitted the photon is moving away from us with a velocity proportional to its distance from us, which is equal to about the present radius R of the Universe. Hence, the photon reaches us with a frequency $\nu = \nu_0 R_0 / R$ because of the Doppler effect. In other words, its observed wavelength has grown proportionally to the size of the Universe. In addition, the number of photons with initial frequencies ν_0 has remained constant, but they are now distributed over a volume which is expanded by a factor $(R/R_0)^3$. The actually observed distribution $u(\nu)$ thus satisfies the relation

$$\frac{u(\nu)\, d\nu}{h\nu} = \left(\frac{R}{R_0}\right)^3 \frac{u_0(\nu_0, T_0)\, d\nu_0}{h\nu_0},$$

which enables us, bearing in mind the form (13.42) of u_0, to identify $u(\nu)$ with a Planck distribution characterized by an effective temperature $T = T_0 R_0 / R$. Observations confirm this theoretical result, already predicted at the end of the forties.

The value $T \simeq 3$ K indicates that the radiation was decoupled from the matter when the Universe was 1000 times smaller than now. An estimate of the expansion velocity, which is characterized by the Hubble constant, that is, the ratio of the recession velocity of an object to its distance, gives an age of the Universe of the order of 20×10^9 years.

Clearly the above arguments apply only to the cosmic background radiation and are not germane to the radiation emitted by the stars and the galaxies at a more recent epoch. The latter is also distributed according to a law which is close to Planck's law, but with a temperature equal to that of the emitting object. We also note that the effective temperature of 3 K of the cosmic background radiation is not a genuine temperature, as the photons of this radiation no longer interact either with one another or with charged particles; there is no mechanism to bring them into equilibrium, and their distribution only is a memory of their initial state. However, it turns out that one obtains the same result if one assumes that the expansion of the Universe is adiabatic (Exerc.13b).

13.3 Exchanges of Radiative Energy

Emission or absorption of radiation by matter covers a large variety of phenomena which are specific for the substance; using non-equilibrium statistical physics to study them lies outside the framework of the present text. Here we shall simply review the general thermodynamic properties which follow from the First and the Second law of Thermodynamics. Like all other non-equilibrium effects considered in this book our guiding principle will be *balance* methods.

13.3.1 Black-Body Radiation

In § 7.2.3 we saw that one can observe the equilibrium Maxwell distribution of the molecules of a perfect gas by letting molecules escape from their enclosure through a hole in the wall. In exactly the same way, the most direct method for experimentally checking Planck's formula is to analyze through a balance method the *radiation escaping from a hole* in the wall of the enclosure inside which the photons are in thermal equilibrium. The hole must be sufficiently small not to disturb the equilibrium state inside the enclosure, but sufficiently large to enable us to describe the escaping field in terms of photons: the wavepackets which represent the photons must be sufficiently localized so that their wavelength is small compared to the size of the hole. Let us calculate the energy carried away by photons with frequencies between ν and $\nu + d\nu$, which during a time interval dt escape through the hole of area dS, inside a solid angle $d^2\omega$ around a direction which makes an angle θ with the normal to the wall (Fig.13.2). Planck's formula (13.42) gives us the energy per unit volume $u(\nu)\, d\nu$, for the frequencies considered, of the photons propagating inside the enclosure in arbitrary directions with velocity c. The photons which leave the

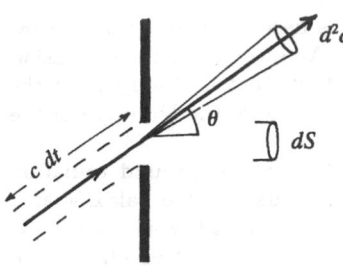

Fig. 13.2. Emission of radiation through a hole

enclosure in this way are contained in an oblique cylinder of base dS, slant height $c\,dt$, and height $c\,dt\cos\theta$, that is, of volume $c\,dt\cos\theta\,dS$; moreover, they must propagate in directions within the solid angle $d^2\omega$, which gives us, because of the isotropy of the radiation, a factor $d^2\omega/4\pi$. Altogether, the *power radiated* through the area dS within the solid angle $d^2\omega$ and within the frequency range $\nu,\nu+d\nu$ is thus given by the relation

$$\delta W = u(\nu)\,c\,\cos\theta\,dS\,\frac{d^2\omega}{4\pi}\,d\nu. \tag{13.44}$$

Hence we can measure $u(\nu)$ directly by spectrometry. Such experiments have made it possible to check Planck's law with a large accuracy, at least in those temperature and frequency ranges where $u(\nu)$ is not too small.

Experiments also check the *angular distribution*, with $\cos\theta$, of the emitted radiation, which simply reflects the fact that the power emitted through the hole in a given direction is proportional to the *apparent area* $\cos\theta\,dS$ rather than the area dS of the hole itself. This is known as *Lambert's law*.

Finally, the *total power emitted* into the exterior half-space per unit area of the hole, called the *emittance*, equals

$$R = \int \delta W = \frac{c}{4\pi}\int_0^\infty u(\nu)\,d\nu \int_0^{\pi/2}\cos\theta\,d^2\omega,$$

or, after integrating and using (13.38) for the integral of $u(\nu)$,

$$\boxed{R = \sigma T^4}\ . \tag{13.45}$$

This expression is *Stefan-Boltzmann's law* (1879) which was found originally in experiments and by thermodynamic arguments. Stefan's constant,

$$\sigma = \frac{\pi^2 k^4}{60\hbar^3 c^2} = 5.67\times10^{-8}\ \mathrm{W\,m^{-2}\,K^{-4}}, \tag{13.46}$$

is small, but the T^4 dependence implies that the emitted power soon becomes large when the temperature increases: at 3000 K, R equals 460 W/cm^2. Note that the energy per unit volume (13.38) of a photon gas in equilibrium can also be expressed as

$$\frac{U}{\Omega} = \frac{4}{c}\sigma T^4 = \frac{4}{c}R. \tag{13.38'}$$

On the other hand, measuring the radiation emitted through an opening in a cavity makes it possible to *determine the temperature* inside the cavity. One measures in this way the temperature of Martin-Siemens furnaces in steelmaking by collecting the light which passes through a small hole in the door. Such a measurement can involve either the total emitted intensity, or, more simply, the mean frequency of the emitted radiation (optical pyrometers). We have already seen that Wien's displacement law (13.43) states that the frequency of the maximum in the emission varies linearly with temperature.

We have found Planck's law as a direct consequence of statistical mechanics applied to the Bose gas of the photons. In fact, *historically* one first measured the radiation curves as function of frequency and temperature and their theoretical interpretation posed serious problems during the last decades of the nineteenth century, at a time when physics seemed to be a completed discipline. Treating the electromagnetic vibrational modes by classical statistical mechanics, the only one available at that time – Planck's formula dates from the end of 1900 – had led to the *Rayleigh-Jeans formula* (1900),

$$u(\nu)\,d\nu \sim kT\,\frac{8\pi\nu^2\,d\nu}{c^3}. \tag{13.47}$$

Lord Rayleigh (Langford Grove 1842–Witham, Essex 1919) derived this formula by using the equipartition theorem (§ 8.4.2) to attribute an energy $\frac{1}{2}kT$ to each degree of freedom; he found (13.13) for the number of modes in the range $\nu, \nu + d\nu$, and noted that each mode has two degrees of freedom (conjugated variables ξ and π). By comparing it with the exact formula (13.42), one sees (Fig.13.1) that expression (13.47) is correct at *low frequencies*: classical statistical mechanics is, in fact, applicable to harmonic oscillators in the limit of large quantum numbers n_q, and the occupation numbers $\langle \hat{c}_q^\dagger \hat{c}_q \rangle = n_q$ are large for the low frequency modes, those for which $h\nu \ll kT$. This makes it possible for $\langle \hat{c}_q \rangle$ to be large, which is a necessary condition for being able to treat the electromagnetic field as a *classical field* (§§ 13.1.4 and 11.4.4).

On the other hand, at *high frequencies* (Fig.13.1) we can replace the Bose factor in (13.42) by a Boltzmann factor and we get *Wien's formula* (1893),

$$u(\nu)\,d\nu \sim \frac{8\pi h\nu^3}{c^3}\,e^{-h\nu/kT}\,d\nu. \tag{13.48}$$

This can be interpreted as another classical limit, where one treats the photons as *classical particles* which obey Boltzmann statistics, as their density is low (§§ 10.3.4 and 13.1.4). Wien had derived the formula using arguments involving the Doppler effect and thermodynamics. It explained the displacement law, $\nu_m \propto T$, since (13.48) gives $\nu_m = 3kT/h$, and the difference between that relation and (13.43) could not be detected at a time when the constants k and h had not yet been determined.

The crisis of 1900 was provoked by the contradiction between formulae (13.47) and (13.48) for black-body radiation. They agreed with experiments in different frequency ranges, and were both based upon theoretical arguments which looked reasonable. Rayleigh's theory seemed to have the better theoretical foundation, but formula (13.47) was certainly wrong, since it led to a total radiative energy density, U/Ω, which is infinite!

Planck wrote down Eq.(13.42) in 1900, guided by the need to interpolate (13.47) and (13.48), and found excellent agreement with experimental data. However, the theoretical justification for his theory (see § 3.4.4) had hardly any basis. It fell to Einstein (1905) to open the door to modern theories by showing that Planck's law was the result of quantizing as in (13.14) the energy levels of the oscillators. The final identification of the energy quantum with a particle, the photon (§ 10.1.2), thus played an important rôle in the germination of the ideas which would lead to quantum mechanics. Rather paradoxically, the black body, historically the first example of quantization, is far from being the simplest one, as it also involves field theory and statistical physics.

13.3.2 Absorption and Emission

A class of phenomena which are of great practical importance is connected with the emission and absorption of radiation by material bodies, that is, with the interaction between photons and electrons or nuclei. We have already seen the rôle played by the walls in the establishment of equilibrium radiation inside an enclosure. However, the situations of interest are, in most cases, *nonequilibrium*: illumination by an electrical lamp, heating by infrared radiation, or solar radiation. Electromagnetic radiation is even one of the most common mechanisms for *energy transfer* between material bodies. Microscopic studies are based upon the coupling (13.34) between the quantized field and the particles which constitute the system under study, molecules in a gas or condensed matter; this coupling leads to emission or absorption of photons while at the same time inducing a transition between two energy eigenstates of the system of particles. Here we shall restrict ourselves to a macroscopic approach which will not enable us quantitatively to explain all effects, but which gives us relations between them and which shows how one can analyze them in terms of a few empirical coefficients.

Historically, this approach has preceded the study of black-body radiation, the importance of which it has helped to show. Already in 1801, half a century before the energy concept itself was worked out, the astronomer Sir William Herschel (Hanover 1738–Slough 1822) started a quantitative exploration of what we would now call the transformation of the solar radiation into heat and the energy distribution of the former as function of frequency. To do that he placed a small thermometer at various positions in a spectrum obtained when analyzing the light from the Sun; by extending his measurements beyond the visible he discovered by the way infrared radiation. The observations and experiments carried out during the whole of the

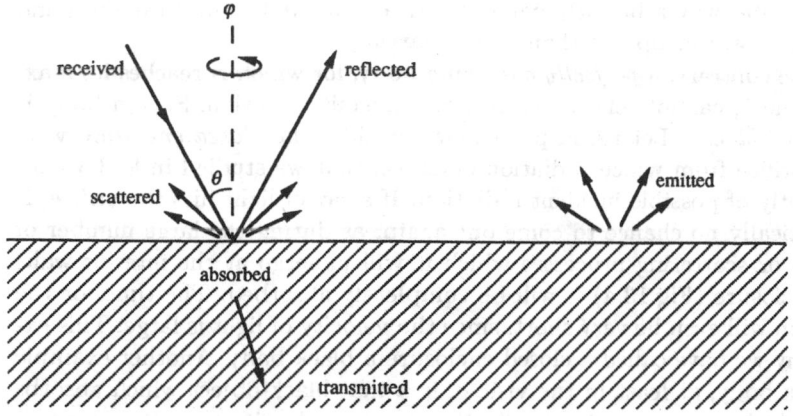

Fig. 13.3. Exchanges of radiation with a body

nineteenth century were accompanied by technical discoveries such as Edison's lamp (1878) and the Dewar vessel (1892) and resulted in posing the problem of explaining theoretically the thermal equilibrium of radiation.

Let us start by enumerating the various processes involved (Fig.13.3). When a body receives radiation, the latter can be *absorbed* in the matter in regions close to its surface. Depending on the nature of the material, it can be *reflected* following Snell's (Descartes's) laws, or *scattered* in any direction, or *transmitted* across the substance, if the latter is transparent. In addition, a body can spontaneously *emit* radiation with a more or less high frequency. The energy *flux* corresponding to each of these various effects, that is, the power crossing unit area, is characterized by a coefficient to be defined; we shall regard the existence and the magnitude of this coefficient as an experimental fact, although one can determine it theoretically, by a rather difficult analysis of the interaction between the radiation and the substance considered.

The *absorptivity* $A(\nu, \theta, \varphi, T)$ of a body at a temperature T, subject to parallel monochromatic radiation of frequency ν, incident at a direction (θ, φ) which makes an angle θ with the normal to the body (Fig.13.3), is defined as the fraction of the power of that radiation which is absorbed by the body. It depends on the nature of the body and its temperature.

Similarly one can define the *reflectivity*, the *scattering coefficient* into any direction different from $(\theta, \varphi + \pi)$, and the *transmittivity* as functions of the direction of incidence and of the frequency of the incoming radiation. An obvious balance, based upon the First Law, shows that $1 - A$ represents the sum of the reflected, scattered, and transmitted fractions of the incident radiation. Hence, a good reflector such as polished metal, a good scatterer such as snow, or a transparent material such as glass in the visible, have a very small absorptivity. This is the reason why clean snow melts very slowly in the sun and why a brightly painted carriage, as well as its chromium and its windows, warms up less than a dark carriage.

On the contrary, a *perfectly absorbing* body, for which A reaches its maximum value 1, cannot reflect, scatter, or transmit radiation. Such a body is said to be "*black*". Let us, in particular, consider the *closed enclosure* with a small orifice from which radiation emerges, that we studied in § 13.3.1 independently of possible incident radiation. If a ray of light hits the orifice, it has practically no chance to come out again, as during the large number of reflection or scattering processes which it will undergo at the interior walls of the enclosure (Fig.13.4) it will be completely absorbed. The time for the absorption will even be very short, since the velocity of light is large. This explains why we have called a closed enclosure a black body. Another example of a black body, at least in the visible, is a mat black object which absorbs all the light it receives without reflecting or scattering it.

The *emission* of radiation by a body in thermal equilibrium is characterized by its *luminance* or *radiance* $L(\nu, \theta, \varphi, T)$ which, like its absorptivity, depends on its nature and on its temperature. The definition of L is such that the power emitted by a surface element dS, into a solid angle $d^2\omega$ around the direction (θ, φ), between the frequencies ν and $\nu + d\nu$, is equal to

$$\delta W_{\text{emitted}} = L(\nu, \theta, \varphi, T) \cos\theta \, dS \, d^2\omega \, d\nu. \tag{13.49}$$

One can measure L by photometry using filters to select the wanted frequency. The factor $\cos\theta$ was introduced into (13.49) so that $\cos\theta \, dS$ is the apparent surface area of the emitting surface. Comparing this expression with (13.44) we see that the *luminance of a black body* simply equals

$$L_0(\nu, \theta, \varphi, T) = u(\nu, T)\frac{c}{4\pi}. \tag{13.50}$$

If a body satisfies Lambert's law, its luminance is independent of θ and φ.

By themselves, the concepts of absorptivity and luminance are useful only when the phenomena which are thus described are *independent*. This is often the case in practical situations. In particular, absorption is a linear effect when the powers involved are not too large: when one adds radiation of different frequencies or different directions of incidence, the powers absorbed by the substance can be added. Similarly, in many cases, the radiative energy emitted by a body is not changed by the radiation received by it. Neverthe-

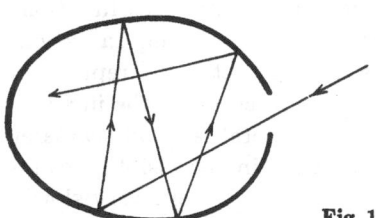

Fig. 13.4. The orifice of an enclosure as a black body

less, there are exceptions where the luminance, as defined by (13.49), would depend on the received radiation and thus lose its interest. For example, in a *laser* or in a *fluorescent lamp* primary radiation stimulates new radiation, with the same or a longer wavelength, which would be absent, if the active material were left to itself. In what follows we shall restrict ourselves to processes where the emission is not affected by absorption.

13.3.3 Kirchhoff's Law

It is easy to calculate the energy exchanges through radiation between two bodies, maintained at different temperatures T_1 and T_2, when one knows their luminances and the coefficients characterizing their behaviour when they receive radiation. The *Second Law* puts restrictions on the possible values of these quantities, as it states that the global received energy balance must always be positive for the colder body. We shall apply it here in the limiting case where the two bodies have the same temperature; this will give us a remarkable relation between the luminance and the absorptivity.

Let us assume that the body under consideration, with a temperature T, luminance L, and absorptivity A, is enclosed within a large volume at the same temperature. There is therefore around that body equilibrium radiation at the same temperature T, and *no energy exchange* can take place between the body and its surroundings. In fact, in the opposite case, one could produce work starting from a source at a single temperature, for instance, by using the radiation pressure of the extra radiation in one direction or the opposite. This argument shows, moreover, that we must have *detailed balance*: the total power passing through *any surface*, for *any direction of the radiation*, and for *any frequency* must be zero. Let us work out this balance across a surface element dS placed just at the exterior of the material under consideration. The only radiation passing through dS in the incident direction is that which comes from the walls of the enclosure. The power that it transports, which is the same as if the material were replaced by a hole, is given by (13.44), where we use (13.50),

$$\delta W_{\text{rec}} = L_0(\nu, T) \cos\theta \, dS \, d^2\omega \, d\nu. \tag{13.51}$$

In the opposite direction dS is crossed by the emitted radiation (13.49) and also by the radiation which is reflected, scattered, and possibly transmitted through the material, if it is transparent.

Let us first restrict ourselves to a material which *neither scatters, nor is transparent*. The power reflected into the solid angle $d^2\omega$ around (θ, φ) is equal to the product of the power (13.51) received in the direction $(\theta, \varphi + \pi)$ and the reflectivity $R(\nu, \theta, T)$, whence

$$
\begin{aligned}
\delta W_{\text{refl}} &= R(\nu, \theta, T) \, \delta W_{\text{rec}} \\
&= [1 - A(\nu, \theta, T)] \delta W_{\text{rec}}.
\end{aligned}
\tag{13.52}
$$

We have assumed for the sake of simplicity that the absorptivity, and thus the reflectivity, does not depend on the azimuthal angle φ. Writing down the balance $\delta W_{\text{rec}} = \delta W_{\text{emitted}} + \delta W_{\text{refl}}$ we get for any direction (θ, φ)

$$\delta W_{\text{abs}} = A \, \delta W_{\text{rec}} = \delta W_{\text{emitted}}. \tag{13.53}$$

This proves *Kirchhoff's law* (1859),

$$\boxed{\frac{L(\nu, \theta, \varphi, T)}{A(\nu, \theta, \varphi, T)} = L_0(\nu, T) = \frac{c}{4\pi} u(\nu, T)} . \tag{13.54}$$

It expresses that *the ratio of the luminance to the absorptivity of a body in thermal equilibrium at a temperature T is equal to the black-body luminance at the same temperature, provided the body neither scatters, nor is transparent or fluorescent.*

If the body considered *scatters* light, the energy balance can no longer be drawn up for each direction (θ, φ), or, more precisely, for each pair of directions (θ, φ) and $(\theta, \varphi + \pi)$, since part of the radiation which is leaving comes from scattering. However, one can still write down for any surface element dS and for any frequency range $d\nu$ that the absorbed energy equals the emitted energy, as for (13.53), but after integrating now over the solid angle $d^2\omega = \sin\theta \, d\theta d\varphi$. This gives a less detailed law than (13.54):

$$\int d^2\omega \left[L(\nu, \theta, \varphi, T) - A(\nu, \theta, \varphi, T) L_0(\nu, T) \right] \cos\theta = 0. \tag{13.55}$$

Nevertheless, more elaborate arguments show that Kirchhoff's law (13.54) remains valid for a very large class of substances, *even scatterers* which satisfy a symmetry condition for scattering.

To see this, let us write down the thermal equilibrium balance,

$$\delta W_{\text{rec}} = \delta W_{\text{emitted}} + \delta W_{\text{refl}} + \delta W_{\text{sc}}, \tag{13.56}$$

across the surface dS. The last term is added to those which we considered earlier to prove (13.54) for a non-scattering substance; it involves a scattering coefficient $D(\nu, \theta, \varphi, \theta', \varphi', T)$ which we simply write as $D(\omega \to \omega')$, and which is defined as follows. Consider a beam incident from a direction $\omega = (\theta, \varphi)$ with unit flux and frequency ν. The energy scattered by the surface dS into the solid angle $d^2\omega'$ around $\omega' = (\theta', \varphi')$ equals $D(\omega \to \omega') \, dS \, d^2\omega'$. For that unit flux, the power received by the surface dS, which makes an angle θ with the beam, equals $dS \cos\theta$ so that energy conservation gives us the relation

$$\cos\theta \left[1 - A(\omega) - R(\omega) \right] = \int d^2\omega' \, D(\omega \to \omega') \tag{13.57}$$

between the absorptivity, the reflectivity, and the scattering coefficient. Let us restrict ourselves to a substance where the reflecting and scattering properties are *invariant under a reversal of the direction of propagation* of the radiation. This means the symmetries $D(\omega \to \omega') = D(\omega' \to \omega)$ and $R(\theta, \varphi) = R(\theta, \varphi + \pi)$.

Under those conditions we obtain the power scattered into the direction ω by evaluating the integral over all incident directions ω'' of the received radiation, the power of which equals $L_0(\nu, T) \, d^2\omega'' \, d\nu$ in the incident solid angle $d^2\omega''$ and the chosen range of frequencies. We have therefore

$$\delta W_{sc} = dS \, d^2\omega \, d\nu \int d^2\omega'' \, D(\omega'' \to \omega) \, L_0(\nu, T). \tag{13.58}$$

Using (13.51), (13.49), and (13.58) we get from the balance (13.56)

$$L_0 \cos\theta = L \cos\theta + R(\theta, \varphi + \pi) L_0 \cos\theta + L_0 \int d^2\omega'' \, D(\omega'' \to \omega). \tag{13.59}$$

If we now use the symmetry of D and R and Eq.(13.57), we have proved Kirchhoff's law (13.54) for any direction.

If we now consider a *transparent* substance, we must add to the right-hand side of (13.56) part of the light received at other points of the body, which has passed through it. An extra complication comes from the fact that absorption occurs all through the body. An argument similar to the one which we have just used, nevertheless, shows that Kirchhoff's law remains valid, not only after integration over all directions and over the whole of the surface of the body, but even in detail. Finally, if the body is *fluorescent* its luminance is not sufficient to characterize its radiation, which now depends also on the received radiation; the energy balance, which is global, does not give much information, and Kirchhoff's law is violated.

13.3.4 Applications

Together with *conduction*, to be studied in Chaps.14 and 15, and *convection*, which is produced by fluid motion, *radiation* is one of the three simple forms of energy transport. Many problems of energy exchange by radiation can be solved by striking *balances* (Exerc.13d, e, f, 15f, Probs. 15, 16) but we must take care to state which is the system we are analyzing. If we are dealing with a balance of energies exchanged *by a substance*, we write down that its change in temperature per unit time is the ratio of the total power it receives and its heat capacity. If we are dealing with a balance *through a surface*, we write down that the total flux which passes through it vanishes in a stationary state, a special case of the conservation laws to be studied in Chap.14.

Kirchhoff's law has many important practical applications, as it helps us to determine the radiation emitted by an arbitrary body. In fact, one can relatively easily measure the absorption coefficient A, or calculate it theoretically, and hence, using Kirchhoff's law, one can derive the characteristic features of the emission.

Let us assume, for instance, that the body is *perfectly absorbing*, that is, "black". From (13.54) it follows that its luminance equals L_0: all black objects emit the same radiation as the one which is found inside a closed enclosure at the same temperature. This property justifies to call this radiation the radiation of *the* black body. At very low temperatures, for instance, the black-body radiation at 3 K in interstellar space, black bodies emit at radar (cm) wavelengths according to (13.54). At room temperature, a "black" body still looks black, as it mainly emits in the far infrared; when we heat it up, it emits in the nearer infrared, and it is still "black as in a black hole". However, it starts to emit in the visible towards 1000 K, where it looks dark red, for instance, a heated iron bar.

The stars absorb practically all the radiation they receive and are thus to a good approximation black bodies (Exerc. 15f); their colour, red, yellow, or white, gives us immediate information about their surface temperature. More precisely, accurate measurements of the temperature of the surface layers of a star are made by an analysis of the observed spectrum and by comparison with a Planck curve. The Sun emits as a black body at 6000 K; this is the reason why it looks like a *flat disk*, since according to Lambert's law the emission from a surface element on the Sun is proportional to its *apparent* area. We can thus understand why the solar radiation has been a prototype for the study of the black body. Deviations from Planck's law, which are small, are only due to absorption by the corona and the terrestrial atmosphere. The emission at the level of the solar surface is strictly according to Planck's law, as the photons do not interact – the Maxwell equations are linear – and the solar surface is a perfect absorber.

More generally, estimates of the absorptivity of an object enable us to determine its temperature through measuring the radiation it emits. For instance, near room temperature, one can evaluate *from a distance* the temperature within about half a degree by measuring the intensity of the millimeter wavelength radiation: infrared detectors.

Let us now consider the emission of radiation by an arbitrary, not black, body. Rather paradoxically, Kirchhoff's law implies that such an *emission is always lower than the black-body emission* at the same temperature. A simple observation shows that a substance must absorb light in order to be able to emit in the visible: if one heats transparent glass, a material which absorbs badly in the visible, to 1000 K, one does not see it emit, whereas charcoal or iron brought to the same temperature emit red light. Similarly, the flame of a gas stove, of matter which does not absorb in the visible, is weakly luminous, notwithstanding its temperature. It becomes luminous and yellow, if one introduces a bit of salt into it, as sodium has two strong absorbing lines in the yellow.

Therefore, if we want to *augment the radiation* of a body in a certain range of frequencies or a given direction, we should *increase its absorptivity* at those frequencies and in that direction. We could, of course, also increase its temperature to increase the factor L_0 in (13.54), but that has the drawback of increasing the emitted radiation globally, and thus decreasing the efficiency, if we are interested in mainly emitting in a given range of frequencies.

These arguments play a rôle in the design of *radiators*; one must increase not only their temperature, but also their absorbing power in order that they can efficiently heat through infrared radiation. For instance, for that purpose one should paint them with substances which are mat and absorb in the infrared.

Similarly, *lighting* by *incandescent lamps* requires high temperatures (2500 to 3000 K) which one can reach by using tungsten filaments (Coolidge, 1906); but the efficiency in luminous energy remains small, as an important fraction of the radiation is emitted in the form of heat, in the infrared. Using Planck's law shows (Prob.16) that only 8.5 % of the energy is emitted in the visible at 2700 K. To improve the yield one has tried to make the filament blacker, and to use substances which have a large absorptivity in the visible and a small absorptivity in the infrared. This is not a very efficient procedure, however, and the main technical advances have been related to increasing the temperature. In fact, the fraction emitted around 2700 K by the black body in the visible increases relatively by 15 % per 100 K. Moreover, it is not only that fraction which characterizes the optical efficiency, since the eye is not uniformly sensitive over the whole of the visible range. One should weight the emitted energy by the physiological sensitivity curve of the eye which has a maximum in the yellow and vanishes at the red (and blue) limit. At the temperatures of interest, the light emitted by the black body is still dominated by the red. It becomes progressively whiter as the temperature increases, which helps to improve the optical efficiency. One can characterize this efficiency by using *photometry*, through measurements where the energy is weighted by the sensitivity curve. The main unit, the *lumen*, is by definition a luminous flux equivalent to 1/683 W for yellow light at 5550 Å; the unit of illuminance, the lux, is 1 lm m^{-2}, and the unit of luminous intensity, the candela, is 1 lm sterad^{-1}. An ordinary electric bulb consuming 75 W emits 900 lm, which corresponds to a yield of only 1.8 %. Nevertheless, increasing the temperature is counteracted by the evaporation of the filament. To reduce the latter one fills the bulb with an inert gas (Langmuir, 1909). In iodine lamps one manages to have the evaporated tungsten being condensed again on the filament (Prob.16). Much higher yields are obtained using the *luminescence* of a vapour, for instance, sodium vapour, through which an electrical discharge passes, or by using *fluorescent lamps*: an electrical discharge in mercury vapour enclosed in a tube produces, without thermal equilibrium being established, ultraviolet radiation; that is absorbed by a fluorescent powder covering the interior of the tube which re-emits in the visible. The wall of the tube remains cold; Kirchhoff's law (13.54) does not apply to it, as the emitting substance is fluorescent and it manages to emit more strongly in the visible than a black body at the same temperature. Of course, Kirchhoff's law integrated over all frequencies does apply so that the total emission of the fluorescent tube is less than that of a black body: the large power emitted in the visible is compensated by a nearly complete absence of infrared radiation and by ultraviolet absorption; the luminous yield is thus excellent.

If, on the contrary, one tries to *avoid losses* of heat by the radiation, one gains an advantage by using badly absorbing, clear, or reflecting materials. This is the reason why the walls of *dewar flasks*, more commonly known as thermos flasks, which are used to maintain a liquid (for instance, a liquefied gas or hot coffee) at a low or at a high temperature are silvered. These thin metallic layers are strongly reflecting and thus radiate very little, which greatly reduces the heat exchanges (Exerc.13e). Similarly, chromium-plating kettles has the effect of keeping liquids inside them hot for a longer time.

The problem of the *absorption of solar energy* is subject to the same kind of considerations. This absorption will be small for strongly reflecting or scattering substances, such as white clothes or snow. If, on the other hand, one tries to increase it, one should use materials with a selective absorbing power, which as a consequence do not radiate strongly. For instance, glass in greenhouses (Exerc.13f) lets the visible light pass through, but not infrared; as a result, the energy emitted by the Sun in the visible reaches the soil, but the energy emitted in turn by the soil, which would have the effect of cooling it if it were not for the greenhouse, is emitted in the far infrared and remains trapped in the greenhouse. The improvement of solar collectors would similarly benefit from the use of specially chosen materials to take care of the limitations imposed by Kirchhoff's law.

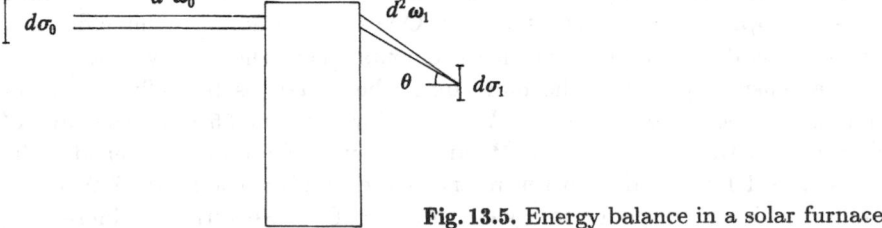

Fig. 13.5. Energy balance in a solar furnace

In this context one might ask the question of what temperatures one could reach in a *solar furnace* (Fig.13.5) by concentrating the received radiation as much as possible. Let $d\sigma_0$ be a surface element of the sun, which is sending to the optical system a beam $d^2\omega_0$, which is absorbed by the surface element $d\sigma_1$ of the receiver at an incident direction (θ, φ) and within a solid angle $d^2\omega_1$. The power transported by the beam is $L_0(\nu, T_0)\, d\sigma_0\, d^2\omega_0\, d\nu$, where T_0 is the solar temperature (6000 K). For the sake of simplicity, let us assume that the radiation propagates according to the rules of geometric optics, or, what amounts to the same, that it is described by the motion of photons according to classical (relativistic) mechanics. It follows from Liouville's theorem (§ 2.3.3) that $d^3r\, d^3p$ is conserved in this motion; as the velocity remains equal to c and the longitudinal momentum p to $h\nu/c$ we find that $d\sigma_0\, d^2\omega_0 = d\sigma_1 \cos\theta\, d^2\omega_1$. The power absorbed by the receiver, at a temperature T, is thus at most, if we bear in mind that there may be losses, equal to

$$\int L_0(\nu, T_0)\, A(\nu, \theta, \varphi, T)\, \cos\theta\, d\sigma_1\, d^2\omega_1\, d\nu.$$

The receiver heats up until a permanent regime is established, where its losses through radiation balance the absorption of the above energy:

$$\int L_0(\nu, T_0)\, A(\nu, \theta, \varphi, T)\, \cos\theta\, d\sigma_1\, d^2\omega_1\, d\nu$$

$$= \int L_0(\nu, T)\, A(\nu, \theta, \varphi, T)\, \cos\theta\, d\sigma\, d^2\omega\, d\nu, \tag{13.60}$$

an equation which determines T. In the second integral the integration domain over σ and ω is larger than in the first integral, so that $L_0(\nu, T)$ is necessarily smaller than $L_0(\nu, T_0)$. The efficiency of a solar furnace is maximal when the receiver is not larger than the solar image (σ minimum) and when it is illuminated at the largest possible solid angle (ω_1 maximum), but, of course, one can expect to reach only temperatures below 6000 K.

Summary

The quantized vibrations of the electromagnetic field are equivalent to a gas of bosons, the photons, non-interacting relativistic particles with zero rest mass, which can be created or annihilated by matter (§ 13.1.4). The thermal equilibrium of the radiation in an enclosure is characterised by the Planck distribution (13.42) of the energy as function of the frequency. The total energy and the pressure of the radiation vary as T^4.

A nearly closed enclosure emits as a perfectly absorbing black body. Black bodies are those which radiate the most strongly, according to (13.44) and (13.45). Kirchhoff's law (13.54) which connects the luminance with the absorptivity helps to determine the radiation from other bodies, and has many practical applications.

Exercises

13a Pressure and Internal Energy

1. Consider a gas of non-interacting particles whose energies are related to their momenta through $\varepsilon = a p^r$; r equals 2, if $\varepsilon = p^2/2m$ and it equals 1 for zero rest mass particles such as phonons, photons, or neutrinos. The particles may be fermions or bosons, with arbitrary spin, and possibly in the classical limit. The spatial dimension d may be 1, 2, or 3. Show that in grand canonical equilibrium the pressure is proportional to the internal energy per unit volume.

2. Prove this result also, using kinetic theory and leaning on §§ 7.4.1 and 13.2.1.

Solution:

1. The number of states $\mathcal{N}(\varepsilon)$ with energies below $\varepsilon = ap^r$, which is given by (10.40), (10.44), equals

$$\mathcal{N} = (2s+1)\frac{\Omega}{h^d} \int d^d p\, \theta(\varepsilon - ap^r)$$

$$= K\,\Omega\,\varepsilon^{d/r},$$

where $K = (2s+1)\pi^{d/2}/h^d a^{d/r}\Gamma(d/2+1)$ is a constant. For one or two dimensions, Ω is the length or the area of the box. The factor f in (10.47) depends on ε only through $x \equiv \beta\varepsilon$; it also depends on α. The grand partition function thus has the form

$$\ln Z_{\mathrm{G}}(\beta,\alpha) = K\,\Omega\,\beta^{-d/r} \int_0^\infty dx\, x^{d/r}\, f(x,\alpha).$$

Using the relations $U = -\partial \ln Z_{\mathrm{G}}/\partial\beta$ and $\mathcal{P} = \ln Z_{\mathrm{G}}/\beta\Omega$ we find

$$\mathcal{P} = \frac{r}{d}\frac{U}{\Omega},$$

which has (10.72) and (13.39) as special cases.

2. The speed of the particles is $d\varepsilon/dp$ so that the normal component of the velocity at the wall is $arp^{r-1}\cos\theta$. The momentum transferred to the wall equals $|p\cos\theta|$, both for an incident particle and for a particle leaving the wall, independent of whether the collision is elastic or inelastic or whether the particle was reflected, absorbed, or emitted. Denoting by $\varphi(p)\,d^d r\, d^d p$ the mean number of particles per unit single-particle phase space volume, with $\varphi(p) = (2s+1)f/h^d$, the total momentum received by an element of area ΔS of the wall during a time interval Δt equals

$$2\,\Delta S\,\Delta t \int_{\theta<\frac{1}{2}\pi} d^d p\, arp^{r-1}\cos\theta\,\varphi(p)\,p\cos\theta,$$

where the factor 2 comes from the distinction between particles moving towards or away from the wall, which all have the same equilibrium distribution $\varphi(p)$. Moreover, the energy per unit volume equals

$$\frac{U}{\Omega} = \int d^d p\, ap^r \varphi(p).$$

As the angular average of $\cos^2\theta$ equals $1/d$, we find again $\mathcal{P} = (r/d)(U/\Omega)$.

13b Density of Photons in the 3 K Radiation

1. Calculate the average density of photons in equilibrium radiation at an arbitrary temperature. Compare the value obtained for the 3 K radiation with the baryon (neutron and proton) density, estimated to be 1 per m^3. Compare also the energy densities of radiation and of matter. Are there temperatures where we can treat the photon gas as a classical gas?

2. How do the pressure and the photon number change under an isothermal expansion? And how under an adiabatic expansion?

3. Assuming that the photons have undergone an adiabatic expansion and have remained in equilibrium since they were decoupled from the matter, when they had a temperature of 3000 K, find the expansion of the Universe since that time. Are our hypotheses correct? What would have happened, if the photons had had a rest mass?

Answers:

1. The photon density is

$$\frac{\langle N \rangle}{\Omega} = \frac{2}{h^3} \int \frac{d^3 p}{e^{cp/kT} - 1} = \frac{2.404}{\pi^2} \left(\frac{kT}{\hbar c} \right)^3$$

(see formulae at the end of this volume). For $T = 3$ K, we find 0.5×10^9 photons per m^3, much larger than the baryon density. However, the energy density of the photons, 6×10^{-14} J m^{-3}, is much smaller than the energy density of the baryons, $mc^2/\Omega \simeq 1.5 \times 10^{-10}$ J m^{-3}. The average momentum $\langle p \rangle$ of a photon is $U/c\langle N \rangle = 2.7kT/c$, whereas the mean distance between the photons $\langle d \rangle$ is of the order of $\hbar c/kT$; the product $\langle p \rangle \langle d \rangle$ is thus of the order of $\hbar$, that is, never much larger than $\hbar$ so that the gas is never classical.

2. $\mathcal{P}$ does not change, but N increases as Ω in an isothermal expansion, whereas $\mathcal{P}$ decreases as $T^4 \propto \Omega^{-4/3}$ while N remains constant in an adiabatic expansion.

3. In the adiabatic expansion, $\Omega \propto T^{-3}$ so that the radius of the Universe increases by a factor 10^3, the same result as in § 13.2.2. The hypothesis about thermal equilibrium is incorrect as the photons do no longer interact. The fact that the expansion of the Universe produces simply a change in the temperature in the Planck law is due to the fact that the occupation factor only depends on the combination ν/T and to the conservation of the photon number. If the photons had had a finite rest mass, m, the expression $\sqrt{c^2 p_0^2 + m^2 c^4}/kT_0$ occurring in the Bose factor at the initial time would have been transformed into $\sqrt{c^2 p^2 R^2/R_0^2 + m^2 c^4}/kT_0$, which is not of the form $\sqrt{c^2 p^2 + m^2 c^4}/kT$. The present distribution of the cosmic background radiation would not have looked like an equilibrium one.

13c Neutrinos in Cosmology

According to present-day cosmological theories the Universe contains neutrinos with a distribution which is the same as if they were in equilibrium at 2 K. Neutrinos are fermions with a zero rest mass and spin $\frac{1}{2}$, but their helicity, the component of the spin along the direction of propagation, takes only one value. Associated with the electron and the positron we have a neutrino-antineutrino pair, produced, for instance, in the reactions $p^+ + e^- \leftrightarrows n + \nu_e$, $n \leftrightarrows p^+ + e^- + \bar{\nu}_e$; there are also two other neutrino pairs which are, respectively, associated with the μ, $\bar{\mu}$ and τ, $\bar{\tau}$ leptons. Up to a certain time t_1, about 1 s after the Big Bang, the matter of the Universe was so dense that the neutrinos were in equilibrium with the baryons and the leptons – through reactions such as the above ones – which themselves were in equilibrium with the photons, for instance, through the reaction $e^- + \bar{e}^+ \leftrightarrows \gamma$. The interaction between the neutrinos and matter is much weaker than that between photons and matter so that the time t_1 was much earlier than the time t_0, of the order of 10^6 years, when the photons were decoupled from the matter; the density had, in fact, to be much higher, and one can estimate the ratio R_0/R_1 between the corresponding dimensions of the Universe to be 5×10^6.

1. What was the energy distribution at the time t_1 as function of the frequency for each type of neutrino? What was the relation between the chemical potentials $\mu_1(\nu_e)$ and $\mu_1(\bar{\nu}_e)$?

2. Since the time t_1 the neutrinos interact no longer with anything, but their distribution changes due to the expansion of the Universe and the Doppler effect, as happened for the photons after t_0 (see § 13.2.2). Show that the present energy distribution is the same as if the neutrinos were in equilibrium at a time t_1 at a temperature T and a chemical potential μ, to be expressed as a function of the values T_1 and μ_1 at time t_1. Does the expansion of the Universe change the total entropy of the neutrinos?

3. Assuming that the numbers of neutrinos and antineutrinos are the same and that $T \simeq 2$ K, estimate their present density and compare their energy density with that of the photons of the 3 K radiation. What were the temperature T_1' of the neutrino gas at t_1, and its effective temperature T_0' at the moment t_0 of the decoupling of the photons? Compare that with the temperature T_0 of the photon gas at t_0 (see § 13.2.2).

Answers:

1. By multiplying the density of states and the Fermi factor, we get

$$u(\nu)\, d\nu = \frac{4\pi h \nu^3\, d\nu}{c^3 \left[e^{(h\nu - \mu_1)/kT_1} + 1 \right]}.$$

The equilibrium $\nu_e + \bar{\nu}_e \leftrightarrows \gamma$ implies that $\mu(\nu_e) + \mu(\bar{\nu}_e) = \mu(\gamma) = 0$.

2. A neutrino created with frequency ν_1 is observed with the frequency $\nu = \nu_1 R_1/R$; the number of neutrinos per unit volume, $u\, d\nu/h\nu$ must be divided by $(R/R_1)^3$. These changes retain the shape of $u(\nu)$, provided we replace T_1 by $T = T_1 R_1/R$ and μ_1 by $\mu = \mu_1 R_1/R$. The total entropy (10.37) remains unchanged under the transformation $T_1 |\rightarrow T$, $\mu_1 |\rightarrow \mu$, $\Omega_1 |\rightarrow \Omega = \Omega_1 (R/R_1)^3$, $p_1 |\rightarrow p = p_1 R_1/R$.

3. The symmetry between neutrinos and antineutrinos implies that their chemical potentials are equal; since they are also the opposite of one another, they must vanish.

$$\frac{\langle N \rangle}{\Omega} = \frac{6}{h^3} \int \frac{d^3p}{e^{cp/kT} + 1} = \frac{5.41}{\pi^2} \left(\frac{kT}{\hbar c} \right)^3 \simeq 0.4 \times 10^9 \ \mathrm{m}^{-3}.$$

$$\frac{U}{\Omega} = \frac{6}{h^3} \int \frac{cp \, d^3p}{e^{cp/kT} + 1} = \frac{7\pi^2}{40} \frac{(kT)^4}{(\hbar c)^3} \simeq 3 \times 10^{-14} \ \mathrm{J \ m}^{-3},$$

which is comparable to the value 6×10^{-14} J m^{-3} for the photons.

$$T_1 = \frac{TR}{R_1} \simeq 10^{10} \ \mathrm{K} \simeq 1 \ \mathrm{Mev}, \qquad T'_0 = \frac{TR}{T_0} \simeq 2000 \ \mathrm{K}.$$

The higher temperature of the photons, $T_0 \simeq 3000$ K at time t_0 is due to the reheating of the radiation and the creation of photons produced by pair annihilations, $e^- + \bar{e}^+ \rightarrow \gamma$, between the times t_1 and t_0.

13d Planetary Temperatures

Determine the temperature $T_\odot$ of the surface of the Sun, assuming that it radiates as a black body and knowing that the power we receive outside our atmosphere is 1.4 kW m^{-2}; the Sun-Earth distance is 8 light-minutes and the solar radius is $R_\odot = 109$ Earth radii.

Assuming that the planets behave as black bodies exposed to the solar radiation, calculate their temperature. Compare the results with the data from Table 13.3.

Table 13.3. Measured planetary temperatures

	Mercury	Venus	Earth	Moon	Mars	Jupiter	Saturn	Uranus	Neptune
Temperature (illuminated side)	600	740	295	400	250	120	90	65	50
Relative distance to the Sun	0.4	0.7	1	1	1.5	5.2	9.5	19.2	30

Hints:

The total power emitted by the Sun is $4\pi R_\odot^2 \sigma T_\odot^4$, or, at a distance D and per unit area,

$$\frac{\sigma T_\odot^4 R_\odot^2}{D^2}.$$

Hence we find that $T_\odot = 5700$ K. The power reaching the Earth's surface does not exceed 1 kW m^{-2} due to the absorption of some of the ultraviolet and infrared solar radiation by the atmosphere; 18 % of the power radiated by the Sun is in the UV and 35 % in the IR.

By expressing that the power received by a surface at right angles to the solar rays equals the power emitted by that surface at a temperature T we get for Mercury and the Moon

$$T = T_\odot \sqrt{\frac{R_\odot}{D}}.$$

On the other hand, if the planet rotates so fast that its temperature is practically uniform, like the Earth, a factor 4 appears since the energy received is proportional to the apparent surface πR^2, whereas the emitted energy is proportional to the Earth's surface $4\pi R^2$, whence

$$T = T_\odot \sqrt{\frac{R_\odot}{2D}}.$$

The discrepancies between the results obtained and the observed data are due to the fact that the planets are not black bodies. In particular, their atmospheres can reflect or absorb, and reemit, part of the received radiation, selectively depending on the frequency. The very high temperature at the surface of Venus, for instance, is due to an important greenhouse effect: the atmosphere is opaque to the IR. A more precise calculation can be carried out considering balances at different altitudes and using Kirchhoff's law.

13e Dewar Flasks

1. For industrial purposes and also for many physics experiments it is necessary to keep liquefied gases, such as N_2, O_2, H_2, or He, as long as possible at low temperatures. Similarly, domestically, one wants to keep drinks cold or hot for several hours. In an ordinary receptacle most of the heat losses are due to conduction through the wall. Write down the heat flux Φ_1 through a wall of thickness l and heat conductivity λ, assuming that its interior side is at the temperature T_i of the liquid, and its exterior side at the temperature T_e of the atmosphere (see Chap.14). Calculate this flux for $T_e = 300$ K, $T_i = 77$ K (liquid nitrogen), $l = 2$ cm, $\lambda = 1$ W m^{-1} K^{-1} (glass or porcelain). How long will it take 1 kg of liquid nitrogen with a vaporization heat of 5.6 kJ mol^{-1} to evaporate, if the area of the wall is 0.1 m^2?

2. An interesting idea for improving the insulation is to use a double wall. The air separating the two walls has a low conductivity, but it still transports heat through convection. One can suppress this convection by filling the gap, for instance, with expanded polystyrene, with $\lambda \simeq 0.01$ W m^{-1} K^{-1}. More efficiently, Dewar had the idea to pump the air out; to maintain a good vacuum he put charcoal inside the double wall which adsorbs gases strongly at low temperatures (Exerc. 4b). The only energy flux passing the double

wall is then due to radiation from the two sides. Calculate the flux Φ_2 in the case where the sides behave as black bodies.

3. Considerable progress was made by silver-plating the surfaces facing one another. Calculate the flux Φ_3, assuming a reflectivity R equal to 98 %.

Results:

1. The heat loss is

$$\Phi_1 \;=\; \frac{\lambda\left(T_e - T_i\right)}{l} \;\simeq\; 10^4 \text{ W m}^{-2}.$$

It will take 3 min to evaporate 1 kg of nitrogen.

2. By writing down the energy balance across a surface situated between the two walls we get

$$\Phi_2 \;=\; \sigma\!\left(T_e^4 - T_i^4\right) \;\simeq\; 500 \text{ W m}^{-2}.$$

3. The flux emitted by each wall, which is given by Kirchhoff's law, equals $A\sigma T^4$, where $A = 1 - R$. We must add to this the fluxes in both directions associated with successive reflections, namely

$$\Phi_3 \;=\; A\sigma T_e^4\left(1 - R + R^2 - \ldots\right) - A\sigma T_i^4\left(1 - R + R^2 - \ldots\right)$$

$$\;=\; \sigma\!\left(T_e^4 - T_i^4\right)\frac{1 - R}{1 + R} \;\simeq\; 5 \text{ W m}^{-2},$$

which decreases the rate of nitrogen evaporation to 1 kg per 100 hours.

Another way of calculating this is to write down the relations between the two fluxes Φ' and Φ'' which are, respectively, directed from the exterior towards the interior, and conversely:

$$\Phi' \;=\; A\sigma T_e^4 + R\Phi'',$$

$$\Phi'' \;=\; A\sigma T_i^4 + R\Phi',$$

$$\Phi_3 \;=\; \Phi' - \Phi'' \;=\; A\sigma\!\left(T_e^4 - T_i^4\right) - R\Phi_3.$$

Silver-plating helps us to gain a factor $\Phi_2/\Phi_3 \simeq 100$. In practice, one can then no longer neglect the losses at the orifice of the flask, due to conduction along the wall from the exterior to the interior, to convection in the neck, or to conduction across the stopper – which is the weakest point of a thermos flask.

13f Greenhouse Effect

1. Estimate the temperature reached in a stationary regime by the soil, assumed to be a black body, under the influence of solar radiation. Take the flux of that radiation to be equal to $\Phi = 0.8 \text{ kW m}^{-2}$ and assume that the Sun is 30° above the horizon. Neglect heat losses of the soil due to conduction and to convection in the air. What happens when the Sun is in the zenith? What, when the sky is cloudy; assume that the light is scattered isotropically

without absorption by the atmospheric water droplets. What happens if the soil has an absorptivity A?

2. One puts glass or plastic sheets over the soil. These are assumed to be perfectly transparent in the visible, which is 60 % of the solar energy reaching the soil, and perfectly absorbing in the infrared, which is 40 % of the solar energy. Estimate the temperature T_1 of the glass and the temperature T of the soil.

Answers:

1. The balance between the absorption and the emission of the soil gives $\Phi \cos \theta = \sigma T^4$, whence $T = 290$ K. When the Sun is in the zenith, $T = 345$ K; when the sky is cloudy, half of the radiation from the Sun goes towards the soil and the other half upwards, so that $T = 245$ K. The results are independent of A. In practice the values are lower, as there is heat exchange with the air and cooling during the night.

2. The glass emits as a black body at room temperatures, since $u(\nu)$ is significant only in the infrared, where glass is absorbing. Putting the flux just below it equal to zero in a stationary regime gives $\sigma T_1^4 = \Phi \cos \theta$, or $T_1 = 290$ K. The soil receives the radiation from the glass together with the visible radiation from the Sun. Hence, $\sigma T^4 = \sigma T_1^4 + 0.6\Phi \cos \theta = 1.6\Phi \cos \theta$, or $T = 325$ K instead of 290 K. Moreover, the greenhouse reduces heat losses from the soil due to air convection (Prob.15).

14. Non-Equilibrium Thermodynamics

"Notre nature est dans le mouvement; le repos entier est la mort."

Pascal, Pensées

"Un déséquilibre atroce, d'autant plus douloureux que j'ai connu le calme, la foi sereine ... ...: l'inerte n'existe pas."

Roger Martin du Gard, Jean Barois

"On peut dire qu'on n'a fait, depuis Descartes, que changer ce qui ne change pas: conservation de la quantité de mouvement, conservation de la force vive, conservation de la masse et celle de l'énergie; il faut convenir que les transformations de la conservation sont assez rapides."

Paul Valéry, Variété V

"La symétrie plait à l'âme par la facilité qu'elle donne d'embrasser d'abord tout l'objet."

Montesquieu, Essai sur le goût

So far we have studied mainly stable or metastable equilibrium situations. Dynamics have only incidentally been involved, when we studied the approach to equilibrium (§ 4.1.5), simple applications to a perfect gas (§ 7.4), and energy transport by radiation (§ 13.3). Nevertheless, there exists a great variety of non-equilibrium phenomena and many are of considerable practical importance: heat transfer, diffusion of charge carriers in a conductor, of neutrons in a nuclear reactor, or of a solute in a solvent, mechanical dissipation, the dynamics of chemical reactions, and so on. Such effects, in particular, determine the efficiency of all industrial techniques as well as the way electronic devices operate. Their theoretical study is, however, difficult. Whereas for the description of matter at equilibrium we have at our disposal a systematic and unified approach based upon the formalism of the canonical Boltzmann-Gibbs distributions, the methods used to study the dynamics of non-equilibrium processes on the microscopic scale are manifold and very varied, depending on the context. We shall hardly enter this field of physics which is so enormous. Nevertheless, on the macroscopic scale there are several useful guidelines which help us in analyzing the most diverse problems.

We shall review those in the present chapter (§§ 14.1 and 14.2), afterwards connect them with the underlying microscopic physics (§ 14.3), and finally show how they can be applied in practical cases (§ 14.4). However, we shall not be able to justify them in as complete a manner as we were able to do in Chap. 5 and 6 for the principles of thermostatics – the term we use, in order to avoid misunderstandings, for *equilibrium* thermodynamics.

Thermodynamics proper, that is, the study of the *evolution* with time of the processes which in an isolated system would lead to equilibrium is based upon a few principles that are more or less related to those of thermostatics, but that are sufficiently different for it to be a separate discipline. We shall here once again meet with the two basic ingredients of statistical mechanics. On the one hand, the microscopic *conservation, invariance, and symmetry* laws should be rigorously taken into account (§§ 14.3.1 and 14.3.2) and they give rise to several laws of thermodynamics (§§ 14.1.3 and 14.2.4). It is true that when we go over to the macroscopic scale, certain microscopic symmetry laws will be broken. For instance, the *irreversibility* of most processes on our scale is in contrast to the invariance under time reversal of the microscopic equations of motion. However, even in such cases the microscopic symmetries have a counterpart in the macroscopic dynamics (§§ 14.2.4 and 14.3.5). Notwithstanding their simplicity the conservation and symmetry laws are an effective aid for solving many concrete problems. In particular, in §§ 7.4 and 13.3 we have seen the power of simple counting and *balance* techniques which were sufficient to understand and analyze phenomena such as the effusion or viscosity of gases or the exchange of energy by radiation.

The second aspect of statistical mechanics, its probabilistic nature, shows up at the macroscopic scale through the *entropy* which is a measure of disorder. In situations close to equilibrium, to which we shall restrict ourselves, the entropy can be evaluated at each moment as if the system were at equilibrium. It plays a central rôle in writing down the equations of thermodynamics close to equilibrium (§ 14.2.1) and its increase with time has important consequences for the macroscopic equations of motion (§ 14.2.5).

Understanding the *microscopic* significance of the entropy and the reasons why it increases has been a major problem ever since statistical mechanics was born (§ 3.4.3). We have seen (§§ 5.3 and 6.1) how in thermostatics the entropy of a macroscopic equilibrium state can be identified with the statistical entropy $S(\widehat{D}) = -k \operatorname{Tr} \widehat{D} \ln \widehat{D}$, that is, with the lack of information associated with the macro-state which describes this equilibrium on the microscopic scale. In non-equilibrium thermodynamics information theory will again provide us with a solution for this conceptual problem, but the microscopic definition of the entropy will now be more subtle. In fact, the macro-state $\widehat{D}$ evolves for an isolated system according to the Liouville-von Neumann equation $i\hbar d\widehat{D}/dt = [\widehat{H}, \widehat{D}]$ which is governed by the Hamiltonian $\widehat{H}$; however, the associated statistical entropy $S(\widehat{D})$ remains constant with time. It can therefore not be identified with the thermodynamic entropy. This difficulty which is the *irreversibility paradox* will be elucidated in §§ 14.3.3 to 14.3.5. To do that we shall associate with $\widehat{D}$ a macro-state $\widehat{D}_0$ which at all times contains *only the information relating to the macroscopic variables*. This will enable us to define a *relevant statistical entropy* $S(\widehat{D}_0)$ which we shall identify with the macroscopic entropy. This quantity measures the *disorder* existing on the microscopic scale when only the *thermodynamic variables are given*. The analysis of non-equilibrium processes in statistical mechanics thus involves a so-called *mesoscopic description* in terms of the density operator $\widehat{D}_0$, which is intermediate between the macroscopic description of thermodynamics and the most detailed microscopic description which uses the density operator $\widehat{D}$.

Even though we shall restrict ourselves to processes close to equilibrium, our field of studies remains broad and it covers the most diverse objects and phenomena (§ 14.1.2). We shall show that, notwithstanding this variety of problems, their statement in mathematical terms obeys common general principles. In our analysis we first of all split the system into subsystems with macroscopic states which at all times are characterized by the same parameters as in equilibrium (§ 14.1.1). We define in this way intensive variables, such as the local temperature, the local chemical potential, or the hydrodynamic velocity (§ 14.2.1). The system reacts to the non-uniformity of such variables, which tend to return to their equilibrium values; this gives rise to exchanges of energy, matter, or momentum between subsystems. We consider regimes where the resultant fluxes are sufficiently small to be *linearly* related to the differences between intensive variables (§ 14.2.2). The *response coefficients* introduced in this way govern the macroscopic dynamics, and the principles of thermodynamics provide relations (§ 14.2.4) and inequalities (§ 14.2.5) between them. The scheme given in § 14.2.6 summarizes this *unified approach* and we shall illustrate its power in § 14.4 by examples of characteristic *applications*, ranging from heat propagation to diffusion, and from electrodynamic phenomena to the derivation of the hydrodynamic equations.

14.1 Conservation Laws

We start by giving a brief survey of various effects from the domain of thermodynamics, and then we shall introduce *fluxes*, which are quantities characterizing the speed with which exchanges between subsystems take place. These fluxes obey conservation laws, valid under all circumstances, even far from equilibrium; we thus obtain a first set of equations for thermodynamics.

14.1.1 The Problematics of Thermodynamics

As we started to do in Chap.6, we shall distinguish henceforth *thermostatics*, which enables us to determine *which equilibrium state* an isolated system reaches at the end of its evolution, from *thermodynamics*, which governs the *evolution* with time itself. This terminology still remains, however, inadequate as the two disciplines cover not only thermal, but also mechanical, chemical, or electromagnetic phenomena. From the beginning of the nineteenth century thermostatics and thermodynamics developed in parallel. The study of heat flow (Fourier, 1811) marks the start of the latter discipline which is earlier than that of the former discipline (Carnot's law, 1824; energy conservation, 1842–47). However, the working out of the general principles of thermodynamics proper only dates from the first half of our century. In both cases the aim is to explore from a general point of view the macroscopic consequences of the conservation and symmetry laws, on the one hand, and of the existence of disorder on the microscopic scale, on the other hand. In particular, this enables us to obtain relations between various properties of systems, either at, or off equilibrium.

In order to be able to place all problems of thermodynamics within a single framework we proceed as in thermostatics (§ 6.1.2). We start by dividing the system under consideration into macroscopic and homogeneous subsystems a, b, c, ... , and we characterize at any time the macroscopic state of the system by *independent variables* $A_i(a)$, where the index i denotes the nature of the quantity A_i, such as the energy E or the number N of particles of a given type. The aim we have here in mind is to study how the $A_i(a)$ vary as functions of t, whereas thermostatics uses the maximum entropy principle to determine the value of $A_i(a)$ at equilibrium, when all possible exchanges between the subsystems have taken place.

We restrict ourselves to the evolution of *quasi-equilibrium* systems; we have already met with an example (§ 7.4.5) in the framework of the physics of gases: even though globally the gas may not be in equilibrium, we considered regimes in which it was *locally* at equilibrium, as we can subdivide it into volume elements each of which is at all times close to equilibrium. More generally, we are interested in macroscopic systems consisting of *weakly coupled* parts a, b, c, ... If there were no coupling, each part would get to equilibrium, independently of the others, and its macroscopic state would then be characterized as in thermostatics (§§ 4.1.4 and 6.1.2) by giving the *extensive variables*. We assume that the same $A_i(a)$ variables also suffice to characterize at all times the macroscopic state of the system when it evolves close to equilibrium.

When we were dealing with true equilibria in thermostatics, we took the $A_i(a)$ to be conservative variables. We could, however, extend the results to metastable equilibria, provided we included among the $A_i(a)$ *nearly conservative* quantities. We shall see (§ 14.3.5) that in thermodynamics proper the existence of conservation laws allows the $A_i(a)$ to vary over *macroscopic times scales*, and our formalism will again apply to *nearly conservative* variables which themselves evolve slowly. For instance, we can include among the $A_i(a)$ an order parameter such as the magnetization in a ferromagnet, or a number of unstable particles such as neutrons. Nevertheless, we shall for the sake of simplicity usually refer to the $A_i(a)$ variables which at any time characterize the macroscopic state as the *"conservative" variables*.

We shall transpose the analysis of § 6.1.3 to dynamics; we distinguished there two stages when looking for the maximum of the statistical entropy which determined the equilibrium macro-state. On the one hand, there exist *within* each subsystem mechanisms (§ 4.1.5) for approaching equilibrium and we assume these to be much more efficient than those which, on the other hand, tend to move the whole system towards *global* equilibrium. Under such conditions each subsystem can during its evolution adjust itself near an equilibrium macrostate characterized by the $A_i(a)$ variables. However, due to the coupling between the subsystems a, b, ... , the $A_i(a)$ are no longer constants of the motion, in contrast to what would be the case if each part were isolated. If the *coupling between the subsystems is weak*, the characteristic time associated with bringing about global equilibrium is much longer than the time corresponding to bringing about internal equilibria. This introduces two time scales where the one is considered to be macroscopic and the other to be microscopic. In such a regime, called either a quasi-equilibrium or a local equilibrium regime, we can, if we observe phenomena on the macroscopic time scale, assume that each of the subsystems a, b, ... is at all times nearly in equilibrium. However, the parameters characterizing its macroscopic quasi-equilibrium state change – slowly – because there is a tendency to move towards global equilibrium; thermodynamics is interested in just this evolution.

As in thermostatics (§ 6.1.2) and even more often here, the slow macroscopic variables $A_i(a)$ cannot always easily be identified, for instance, when we are dealing with hysteresis effects. Only comparisons with experiments enable us to know whether the macroscopic description which we have adopted is a proper one and whether there are additional hidden variables which we need introduce.

14.1.2 Different Kinds of Processes

The abstract formulation which we have just given covers, in fact, a large variety of situations and actual phenomena.

(a) This variety is, to begin with, connected with the *nature of the quantities* involved. Thermal or mechanical phenomena are associated with transfers of *energy*, chemical or diffusion phenomena with transfers of neutral *particles*, and electromagnetic phenomena with transfers of *charges*. All this is the dynamic counterpart of properties studied in thermostatics where, however, the substances almost always were fixed in space (Exerc.4e and 7b are exceptions). For instance, for a fluid in equilibrium the conserved extensive A_i variables were the energy, the number of particles, and the volume. In thermodynamics we must in addition include among the state variables $A_i(a)$ the *components of the momentum*, which are conservative quantities on the same footing as the energy, as soon as we are interested in a fluid, or a solid, in motion. Transfer of momentum between one part and another of the system then plays an essential rôle.

(b) A second classification of thermodynamic processes is based upon the *spatial organization* of the parts of the system. In most cases of interest the interactions which couple the subsystems are short-range ones, so that transport takes place gradually between subsystems which are *close to one another* in space. The simplest problems involve two or just *a few homogeneous subsystems* which are in contact with one another and each of which is practically at equilibrium, or even a single homogeneous system interacting with an external reservoir. The prototype of this situation is *thermal contact* (§§ 5.1.2 and 6.3.3) between two homogeneous substances which can exchange heat through their common boundary and the evolution of which with time one now tries to find. (Note that the dynamics of heat exchange through radiation which we studied in § 13.3 do not enter the general macroscopic framework which we give here, as one of the subsystems – the photon gas – remains far from equilibrium due to the absence of interactions between photons.) Similarly, in a sufficiently slow *osmosis* process the two solutions which exchange particles through a semi-permeable partition can be considered as subsystems a and b which are practically at equilibrium; an analogous situation occurs in *electrochemistry*, or when a liquid evaporates into its undersaturated vapour, or again when we are dealing with the *adsorption* or desorption of a gas by the walls of the container which encloses it. The *friction* of two solids, each of which is treated as being at equilibrium, produces transfer of momentum from the one to the other, if their velocities are different. It can similarly be useful to consider, as is often done in thermostatics, that two fluids separated by a mobile piston can exchange *volume* when their pressures are different, a transformation which occurs in a quasi-equilibrium regime if it is sufficiently slow.

A situation which is slightly less simple but which often occurs is the one where the subsystems $a, b, \ldots$ form a *continuum* in space. That was the case in § 7.4.6 for a gas the temperature or the velocity of which changed from point to point. As we discussed in § 7.4.5 we are dealing with local equilibrium if we can divide the system into volume elements ω which are (i) sufficiently large for each of them to remain practically at equilibrium over macroscopic times under the action of collisions between the molecules, but (ii) sufficiently small so that the intensive variables which characterize this local equilibrium, such as temperature, chemical potential, velocity, pressure, are practically constant inside each of them. For this reason the establishing of a *hydrodynamic regime*, that is, a regime with transport in local equilibrium, requires not only that typical *times*, but also the typical *lengths* for the establishing of equilibrium are small as compared to the transfer times and to the distances over which the state of the matter changes. Here again there are many and important examples: *heat flow* in a solid or a fluid with a non-uniform temperature (Fourier's law), *diffusion* of neutrons in a nuclear reactor, *doping*, that is, migration of impurities in a heated solid (Exerc.15a). The electrical *current* is a transport of charged particles – ions in an electrolyte, electrons in a metal, conduction electrons or holes in a semi-conductor – under the action of a spatial variation in the chemical potential; this variation is itself proportional to the macroscopic electromotive force. In the mechanics of continuous media one can similarly divide the solid or fluid matter into macroscopic homogeneous volume elements. For instance, in the flow of a fluid in the laminar regime, the various layers move at different velocities, which causes transport, from one layer to another, of momentum parallel to the flow; Newton's law defining the viscosity and the hydrodynamical Navier-Stokes equations pertain to this situation.

Continuous media can be described in mechanics in two different ways. In the so-called *Lagrangian* description, the subsystems a, b, ... are volume elements of the *matter* which follow the latter during its motion and change its shape with it; each subsystem is closed, that is, it does not exchange matter with its neighbours. This description is not very adequate at the microscopic scale or when diffusion takes place. We shall in all what follows systematically adopt the so-called *Eulerian* description, where the subsystems a, b, ... are volume elements ω which are *fixed* once and for all. Each subsystem is *open* since the particles in it change with time, entering or leaving ω. It has thus a more abstract nature than the Lagrangian description since we are dealing with a *region in space* and not with a material object.

Finally, we have already met with quasi-equilibrium phenomena where the weakly coupled subsystems a, b, ... were *superimposed upon each other in space*. We saw, in particular, that it is useful to analyze in this way the chemical reactions in gases or dilute solutions (§§ 6.6.3 and 8.2.2): the subsystems a, b, ... in that case represent the sets of molecules of the various kinds, and the A_i variables include the numbers of elementary constituents – atoms, ions, or radicals – which can be transferred from one set of molecules to another. This *"chemical" situation* is found also in all cases where a set of particles of a given kind – a subsystem – can be transformed into another one, as in the problem of ortho- and parahydrogen (§ 8.4.5); the reactions which produce these transformations should, however, be less efficient than the thermalization phenomena so that each population remains close to thermal equilibrium. An important example is that of *semi-conductors* (§§ 11.3.4 and 11.3.5). Thermal equilibrium can rapidly be established both amongst the conduction electrons and amongst the holes in the valence band; however, the processes which annihilate or create a pair of positive and negative charge carriers are less efficient. A semi-conductor sample can thus present a quasi-equilibrium state for each of the populations a and b of the electrons and of the holes, with independent chemical potentials, although this sample is not at equilibrium as far as the numbers of charge carriers of the two species are concerned; if they are, for example, in excess, the ensuing *recombination* process enters the general framework of "chemical" reactions in a homogeneous phase.

(c) A last classification of the processes is related to their *temporal organization*. We have already noted (§§ 7.4.5 and 7.4.6) the existence of stationary phenomena where the macroscopic state of the system *does not change* with time, although remaining permanently *in a non-equilibrium state*: viscous flow, heat transport, or electricity transport through a conducting medium. This can occur only if the system is not isolated but weakly coupled to several *exterior reservoirs* – for example, two thermostats at different temperatures, or particle or momentum sources – which prevents the establishment of global equilibrium. In this case, *transport* of energy, particles, and so on, takes place across the system from one reservoir to another, in a *local equilibrium stationary regime*, without any possibility of global equilibrium being reached.

It is clear that thermodynamics also applies to more general *forced regimes*, where the system *responds to a varying external perturbation* which prevents it from remaining at equilibrium; this perturbation should change sufficiently slowly so that the local equilibrium of the subsystems a, b, ... be practically reached at all times. For instance, *the propagation of sound or of electromagnetic waves* in a medium falls into this category; it, moreover, presents us with a *wave* aspect. Amongst the phenomena produced by the response of a system to an external perturbation, *resonances* are the most striking ones which show up best in a *periodic* regime (Exerc.14b). All the same it is important to note that here we try only to describe processes which are sufficiently slow for the substances to have time to adjust to the external action. High frequencies give rise, for instance, to memory effects which fall outside the framework of thermodynamics of systems close to equilibrium, but which can be studied in statistical mechanics.

Finally, if an *isolated* system is initially not in global equilibrium, it will return to it through a local equilibrium regime, provided the departure from global equilibrium is not too large. This return to global equilibrium is a macroscopic *relaxation* process. Because the microscopic mechanisms underlying response and relaxation are the same, these two regimes are related to one another, even for processes far from equilibrium (Exerc.14b).

In § 1.4.5 we gave an example, namely, that of *magnetic resonance* – electron paramagnetic resonance or nuclear magnetic resonance. Equation (1.44) governs the dynamics of the total magnetic moment M in a field B if we can neglect the spin-spin interactions and the interactions of the spins with the other degrees of freedom. These interactions are responsible (Exerc.2a) for the *relaxation* of the magnetic moment towards the direction of the fixed field B and towards the equilibrium value (1.37). An experimental study of the relaxation thus gives us information about the interactions which operate within the substance and is a valuable investigative method. Nuclear magnetic resonance of the hydrogen nuclei and their relaxation, for instance, enable us to measure the hydrogen density in a biological tissue; chemistry similarly uses this effect to determine the structure of molecules. The principle of the measurements is simple. We observe the magnetization as function of time in the presence of a fixed field B and a varying field at right angles to B. If this variation is periodic with a frequency which sweeps the Larmor frequency, we observe resonance and its shape is related to the relaxation (Exerc.14b). We can also observe the relaxation directly, if we change the field suddenly.

In the case where the spin-spin interactions are more efficient than their interactions with the other degrees of freedom the relaxation takes place in two stages. The first stage leads towards a state of equilibrium of the spins by themselves, which is characterized by a *spin temperature*. This can differ for several hours from the temperature of the rest of the substance and can even be negative (Exerc.1a). The other interactions – called the *"spin-lattice"* interactions, because they include the interactions of the spins with the phonons, that is, with the lattice vibrations – eventually lead to a complete relaxation with an equalization of all temperatures. Spin echo experiments (§ 15.4.5) rely on the existence of different relaxation times for spin-spin and spin-lattice processes.

14.1.3 Relations Between Fluxes, Currents, and Densities

We want to determine the change with time of the macroscopic variables $A_i(a)$. Their temporal evolution will be controlled by the dynamics of *transfers* from one subsystem to another of each of the conservative quantities A_i. We thus define the *flux*,

$$\Phi_i(a \longrightarrow b) = -\Phi_i(b \longrightarrow a), \tag{14.1}$$

associated with the quantity i as the amount which is *transferred from a to b per unit time*. When a, b, ... denote discrete subsystems, the exchanges take generally place between neighbours because of the short range of the interactions, so that the fluxes (14.1) only exist if a and b are close to one another.

The conservation laws for each of the quantities i, such as the energy, the particle numbers, or the momentum components, enable us to express the change with time of the $A_i(a)$ quantities for each subsystem a as *function of the fluxes* towards the subsystems b with which a is coupled: using the definition (14.1) of the fluxes we have the *macroscopic balance*

$$\frac{dA_i(a)}{dt} + \sum_b \Phi_i(a \longrightarrow b) = 0. \tag{14.2}$$

The conservation equation (14.2) assumes that the system is isolated from the exterior. If it interacts with *external sources* – for instance, if one of its subsystems a exchanges heat with a thermostat – we should either include those sources amongst the subsystems b in (14.2), or isolate their contribution in the form of an extra term,

$$\frac{dA_i(a)}{dt} + \sum_b \Phi_i(a \longrightarrow b) = \Phi_i(\text{sources} \longrightarrow a). \tag{14.3}$$

In the example of a thermostat, and if A_i is the energy, the right-hand side of (14.3) is the heat flux given up by the thermostat to the subsystem a.

A similar situation occurs in neutron transport problems for the particle number A_i: if the medium in which the neutrons move and are diffused absorbs them, it gives rise to a violation of the conservation, but if its effect is known, we can consider it as a source and introduce its, negative, contribution into the right-hand side of (14.3); the same applies when we wish to take the finite lifetime of the neutrons into account. More generally, a *nearly conservative* variable which evolves slowly according to known laws obeys a balance equation of the type (14.3). The source term, which may be positive or negative, in that case does not describe a coupling with the exterior, but accounts for creation or decay processes associated with the violation of the conservation law.

In the case of chemical kinetics and similar reactions – electron-hole annihilation in a semiconductor, for instance – we take for the A_i the numbers N_j of molecules of the various species (§ 6.6.3). (Chemists usually take the number of moles.) It is convenient to parametrize these numbers by the *degrees of progress* $M^{(k)}$ of each of the possible processes (6.74), namely

$$\sum_j \nu_j^{(k)} X_j \overset{\rightarrow}{\underset{\leftarrow}{=}} 0$$

for each k. The dynamics are then characterized by the *reaction speeds*, which are the time derivatives of the degrees of progress $M^{(k)}$; these speeds can be identified with the *chemical fluxes* Φ_k. We adopt the sign convention that Φ_k is positive when the reaction (6.74) is proceeding from left to right, that is, (6.75) or (6.76) from right to left. The changes in the numbers N_j can then be calculated exactly as we did in (6.77) for the infinitesimal displacements $dM^{(k)}$ of each reaction in the vicinity of chemical equilibrium. At all times we thus find

$$\frac{dN_j}{dt} + \sum_k \nu_j^{(k)} \Phi_k = 0. \tag{14.4}$$

The dynamical equations (14.4) take into account all the conservation laws characterized by the values of the stoicheiometric coefficients $\nu_j^{(k)}$ which appear in the various possible reactions (6.74).

In the point of view, which is suitable for reactions in the gas phase or in solution, where we treat each set of molecules X_j as a different subsystem j, the $A_i(j)$ variables are the number of atoms of each species in each of the subsystems, for instance, $N(H/H_2O)$, $N(H/H_2)$, $N(O/H_2O)$, and $N(O/O_2)$ for the reaction $2H_2 + O_2 \rightarrow 2H_2O$, which is the irreversible version of the equilibrium (6.75). The flux Φ_H of H from H_2 to H_2O satisfies the law (14.2), which here expresses the conservation of the total number of H atoms; the same is true for Φ_O. Moreover, the identity $N(H/H_2O) = 2N(O/H_2O)$ implies that $\Phi_H = 2\Phi_O$. These relations are clearly equivalent to (14.4).

When the macroscopic state variables vary continuously, the division of space into volume elements ω, which are sufficiently large for the matter to be almost at equilibrium within each of them and sufficiently small so that the matter inside is almost homogeneous, enables us to regard ω as infinitesimal on the macroscopic scale even though its dimensions remain large compared to all typical microscopic lengths. The subsystems are then volume elements; we characterize them by the three spatial coordinates r which vary continuously and we get rid of the elementary volume by redefining the variables, no longer as the extensive quantities $A_i(a)$ themselves, which are proportional to ω, but as their values per unit volume, denoting them by $\varrho_i(r) = A_i(a)/\omega$. Depending on what we are dealing with, $\varrho_i(r)$ will thus be an energy, particle, charge, (one of three) momentum (component) *density per unit volume*. Of course, the hypothesis of local equilibrium implies that ϱ_i changes slowly in space and time.

Similarly, the flux (14.1) between two neighbouring volume elements is *proportional to the area* which separates them when we go over to the continuum limit and when the transfer takes place locally. More precisely, the flux across a fixed surface σ with an oriented normal n, which is a *scalar*, can be derived from a *vector* field $J_i(r,t)$ as follows:

$$\Phi_i(\sigma, t) = \int_\sigma d\sigma \; (n \cdot J_i(r,t)) \, . \tag{14.5}$$

The vector J_i, which points in the direction of the flow of the quantity i at the point r, is the *current density* associated with that quantity. It has three components for the energy, the number of particles of a given kind, or the charge, and 3×3 components for the momentum, as i itself has in that case three possible values. Although properly speaking the flux is (14.5) we shall often employ the usual terminology of "flux" to denote the current density J_i.

The conservation law (14.3) applied to a fixed volume Ω bounded by a surface σ with external normal n gives according to (14.5)

$$\frac{d}{dt} \int_\Omega d^3r \; \varrho_i(r,t) + \int_\sigma d\sigma \; (n \cdot J_i(r,t)) = 0 \, .$$

From Green's formula it then follows that

$$\int_\Omega d^3r \left[\frac{\partial}{\partial t} \varrho_i(r,t) + \operatorname{div} J_i(r,t) \right] = 0$$

for any volume Ω, so that we find at any point

$$\boxed{\frac{\partial \varrho_i}{\partial t} + \operatorname{div} J_i = 0} \, . \tag{14.6}$$

This form of a conservation equation for an inhomogeneous system, in terms of a density ϱ and a current density $\boldsymbol{J}$, is often called the *continuity equation* associated with the conserved quantity i.

As in the case of (14.3) the existence of exchanges with the outside, added to the exchanges between neighbouring elements which are described by the currents $\boldsymbol{J}_i$, is taken into account by means of source terms to be added to the right-hand side of (14.6). If, for instance, i denotes a *momentum* component P_α, the conservation law describes the change with time of the momentum contained in an infinitesimal volume; this change is due, on the one hand, to the fact that particles enter and leave the volume concerned, and, on the other hand, to the *forces* exerted on the particles inside that volume. Indeed, the change in the momentum of one single particle during a time dt is equal to the impulse $\boldsymbol{f}\,dt$ that is receives, where $\boldsymbol{f}$ is the total force exerted on that particle. Thus, if there are *forces applied* from the outside, such as electrical or gravity fields, we must *add to the right-hand side* of (14.6) the contribution per unit volume from those forces. Their work also contributes to the right-hand side of the local *energy* balance (14.6). On the other hand, we shall see that the viscosity forces, which are due to the short-range interactions between particles, are included in $\boldsymbol{J}_i$. More generally, the introduction of source terms is necessary for systems which are not isolated or for which the conservation laws are violated in a controlled manner, as we have indicated above in the example of neutron transport.

The local conservation laws (14.2) or (14.6) of non-equilibrium thermodynamics, which are the *dynamic* form of the First Law, are more detailed than the global conservation laws of thermostatics. Those, in fact, concern a complete, isolated system for which it follows from (14.1) and (14.2) that

$$\frac{d}{dt}\sum_a A_i(a) = 0. \tag{14.7}$$

Moreover, in thermostatics we are only interested in comparing values at the initial and the final times and we use in that case only the integral of (14.7) rather than (14.7) itself.

The local character of the conservation laws (14.2) or (14.6) gives rise to some difficulties when the system contains parts which interact with one another through macroscopic-range forces. In particular, the attribution of the energy associated with the interaction field to some specific parts of the substance becomes ambiguous, if charged or magnetized substances are present. We have seen in the thermostatics framework (§ 6.6.5) how one can, depending on the procedure, arrive at different, but, of course, equivalent, formulations.

We shall return in §§ 14.3.1 and 14.3.2 to a justification of the local conservation laws in the framework of statistical mechanics.

14.2 Response Coefficients

We have introduced the macroscopic, local, *extensive A_i variables* (or the corresponding *densities ϱ_i*) and the *fluxes Φ_i* (or the *current densities J_i*) which characterize the rates of exchanging the A_i quantities, and we have written down the conservation laws which connect them. The fluxes are produced by the deviations of the system from equilibrium; we shall see that these deviations are measured by the *affinities*. We shall write down the empirical equations which relate the fluxes to the affinities and we shall study their general properties. This will eventually lead to a systematic method for applying thermodynamics to some problem or other.

14.2.1 Local Equations of State

We know (§ 6.2.2) that the global equilibrium of an isolated system is determined by writing down that the intensive quantities γ_i of the same nature are equal for all subsystems which can exchange the quantity A_i. In the *local equilibrium regime* each subsystem a is at any time almost at equilibrium. It is thus natural to start by introducing its intensive variables $\gamma_i(a)$ which are calculated as if the subsystem did not interact with its neighbours. The situation is then the same as in Chap.6: the exchanges are inhibited and all $A_i(a)$ variables are quenched.

This enables us to define the *instantaneous thermodynamic entropy* of the system as the sum of the thermostatic entropies of each of its parts:

$$S = \sum_a S_a(\{A_i(a)\}). \tag{14.8}$$

In this definition we have neglected the coupling and the correlations between the subsystems; this is legitimate if the coupling is weak and the fluxes small, which are the conditions necessary to ensure quasi-equilibrium. The $A_i(a)$ variables take the values which are observed at the time considered, so that S will depend on the time. As in equilibrium, the entropy (14.8) is *additive*. The *intensive variables which are conjugate* to the conserved quantities A_i are, for each subsystem, given by

$$\gamma_i(a) = \frac{\partial S_a}{\partial A_i(a)} = \frac{\partial S}{\partial A_i(a)}; \tag{14.9}$$

these relations are the *equations of state*. They are the same as (6.6) at equilibrium; the only difference with thermostatics consists in the fact that the γ_i of neighbouring subsystems can here be different, whether exchanges are permitted or not.

For instance, if A_i is the energy, $\gamma_E(a) = 1/T(a) = k\beta(a)$ is the *local temperature* which can vary from one part to another. For the number of particles of a given kind, $\gamma_N = -\mu/T = -k\alpha$ is similarly related to the *local chemical potential* $\mu(a)$. Let us look for the interpretation of the intensive variables $\gamma_{P\alpha}$ which are associated with the momentum components P_α (§ 4.3.3 and Exerc.4e). To do this, we consider a system of mass M at equilibrium and at rest in a given Galilean reference frame and let $S_0(U_0)$ be its entropy as function of the internal energy U_0; its total momentum is equal to zero. Let us assume that this Galilean frame moves uniformly with a velocity $\boldsymbol{u}$ with respect to a second frame which is taken to be fixed. With respect to this new frame the system is still in equilibrium, but with a momentum $\boldsymbol{P} = M\boldsymbol{u}$, an energy $U_0 + P^2/2M$, and an unchanged entropy. This statement is intuitive, but we shall derive the result from statistical mechanics in § 14.4.4. As function of the constants of the motion, that is, the total energy U and the total momentum $\boldsymbol{P}$, we have thus for the entropy

$$S(U, \boldsymbol{P}) \;=\; S_0(U_0) \;=\; S_0\left(U - \frac{P^2}{2M}\right),$$

so that the intensive variable $\gamma_{P\alpha}$ which is the conjugate of P_α is given by

$$\gamma_{P\alpha} \;=\; \frac{\partial S}{\partial P_\alpha} \;=\; \frac{\partial S}{\partial U}\left(-\frac{P_\alpha}{M}\right) \;=\; -\frac{1}{T}\,u_\alpha. \tag{14.10}$$

It is thus directly related to the velocity $\boldsymbol{u}$ of the system with respect to the fixed reference frame. In non-equilibrium thermodynamics we shall be dealing with the *local velocity*, which for the subsystem a is equal to $\boldsymbol{u}(a) = \boldsymbol{P}(a)/M(a)$ by virtue of the Galilean invariance.

When the subsystems form a *continuum* with slowly varying densities $\varrho_i(\boldsymbol{r}, t)$, the summation over a in (14.8) is replaced by an integral over space

$$S \;=\; \int d^3r \; s\left(\{\varrho_i(\boldsymbol{r}, t)\}\right), \tag{14.11}$$

where s denotes the *entropy density*, that is, the entropy per unit volume of thermostatics. Because of the extensivity of substances at equilibrium the latter – the entropy per unit volume – is a function of the densities ϱ_i of the conservative quantities. The volume elements are fixed, but the particles can enter or leave them so that each of them is an *open system*. The *local intensive variables* are thus, at all times, defined as the functional derivative of S,

$$\gamma_i(\boldsymbol{r}) \;=\; \frac{\partial s(\boldsymbol{r})}{\partial \varrho_i(\boldsymbol{r})} \;=\; \frac{\delta S}{\delta \varrho_i(\boldsymbol{r})}, \tag{14.12}$$

which is equivalent to the identity

$$dS \;=\; \int d^3r \; \gamma_i(\boldsymbol{r}) \, d\varrho_i(\boldsymbol{r}) \tag{14.13}$$

for arbitrary changes $d\varrho_i(r)$ of the densities at each point. In this way we can describe substances the temperature or the chemical potential of which are functions of the coordinates; this we had not done until now.

For chemical type of reactions in the gas phase where we introduce subsystems which are superimposed upon one another in space (§ 14.1.2) the thermodynamic entropy is again the sum of those of each of the species of molecules (§ 8.2.2) so that (14.8) remains valid.

We shall not follow here the tradition of thermodynamicists which consists in exchanging the rôles of the energy and the entropy. We submitted to that practice for the study of systems in equilibrium (Chaps.5 and 6) even though the energy plays the same rôle as the other constants of the motion and though the entropy follows from statistical physics as the most natural thermodynamic potential. However, now there is a risk of misunderstandings and it is better to use the natural variables γ and afterwards to translate the results in terms of the traditional variables T, μ, or u.

The equations of state (14.10) or (14.12) give us a set of relations which at each time connect the A_i (or ϱ_i) variables with the γ_i variables for the same subsystem (or at the same point in space). These relations are the same as in equilibrium and once we know one equilibrium thermodynamic potential – it does not matter which one – we can write them down. The instantaneous quasi-equilibrium state can thus be characterized by giving the variables of one kind or of another kind, for which it now remains to obtain the evolution equations.

14.2.2 Responses of the Fluxes to the Affinities

When the system reaches global equilibrium the γ_i variables of the same nature associated with two subsystems which are in communication with one another are equal (§ 6.2.2). The deviation from equilibrium is thus characterized by the *affinities* which are in general defined as the differences,

$$\Gamma_i(a,b) \equiv \gamma_i(b) - \gamma_i(a), \qquad (14.14a)$$

between the intensive variables of neighbouring subsystems. For the chemical reactions (6.74) the equality of the γ_i at equilibrium is replaced by the conditions (6.78) which are linear relations between the chemical potentials $\mu_j = -T\gamma_j$ of different species X_j. It is thus natural to introduce for each possible chemical reaction (6.74) a *chemical affinity*

$$\Gamma_k \equiv \sum_j \nu_j^{(k)} \left(-\frac{\mu_j}{T}\right), \qquad (14.14b)$$

which is the counterpart of the definition (14.4) of the chemical fluxes Φ_k. These affinities measure in how far the equilibrium conditions (6.78), that is, $\Gamma_k = 0$, are violated. Finally, in the case of a continuum, a difference between neighbouring volume elements becomes a gradient and the affinity about i is defined as

$$\nabla \gamma_i(\boldsymbol{r}, t). \tag{14.14c}$$

Again, it describes the local deviation from equilibrium where $\gamma_i(\boldsymbol{r}, t)$ would be uniform in space. All affinities (14.14) are directly connected with differences or gradients of the temperature, the chemical potential, the velocity, or the pressure: depending on which exchanges we are considering, γ_i is $1/T$, or $-\mu/T$, or $-\boldsymbol{u}/T$, or $\mathcal{P}/T$.

Let us consider an isolated system; we thus include in it the sources, for instance, the thermostats, with which the system studied is possibly coupled. In global equilibrium two subsystems a and b which are in contact do not exchange anything and all fluxes $\Phi_i(a \to b)$ vanish; moreover, their affinities $\Gamma_i(a, b)$ vanish for all quantities i which may be exchanged. The existence of non-vanishing affinities is a perturbation of the relative equilibrium situation between a and b, and the system *"responds"* to it by the creation of fluxes which *tend to reestablish equilibrium*. These fluxes which are created between a and b are functions of the variables $\gamma_i(a)$ and $\gamma_i(b)$; they vanish at the same time as the affinities $\gamma_i(b) - \gamma_i(a)$. In order to simplify the discussion we shall assume that the latter, and hence also the fluxes, are sufficiently small that a linearized approximation is justified. We thus define the *linear responses* $L_{ij}(a, b)$ through the relations

$$\Phi_i(a \longrightarrow b) = \sum_j L_{ij}(a, b)\, \Gamma_j(a, b), \tag{14.15}$$

which express the fluxes as functions of the affinities for each pair of subsystems. These responses are empirical coefficients from the macroscopic point of view but, in principle, can be calculated in statistical mechanics. They depend on the two subsystems a and b and their coupling and are functions of the intensive variables γ_i, which in the linear approximation considered should be regarded as being practically the same for a and b. As both the fluxes (14.1) and the affinities (14.14a) are antisymmetric in the exchange of a and b we have $L_{ij}(a, b) = L_{ji}(b, a)$. Of course, the more difficult the exchanges, the smaller the coefficients L are, and some of them vanish if the corresponding transfer is forbidden. For instance, if a and b cannot exchange particles, the flux $\Phi_N(a \to b)$ is always zero, so that the coefficients $L_{Nj}(a, b)$ vanish. The existence of coefficients L_{ij} with $i \neq j$ reflects the possibility that one kind of affinity can produce a flux of another kind, for instance, a temperature difference producing an electric current. In the applications to *chemical type* of kinetics the indices i and j in (14.15) include the indices k of the possible reactions, the chemical affinities (14.14b) appear amongst the Γ_j, and the chemical fluxes Φ_k amongst the Φ_i; if there are several homogeneous phases, they are referred to by the subsystem index a.

Two complications appear for *continuous* inhomogeneous systems. On the one hand, the fluxes J_i, like the affinities (14.14c), are vectors. The response coefficients L which connect the components of the one to the components of the other are thus *tensors* with two indices α and β, one of which is associated with J_i and the other with $\nabla \gamma_i$. On the other hand, if we want to study a substance which moves macroscopically, as in the dynamics of deformable solids or in hydrodynamics, we must take into account *non-vanishing equilibrium currents*. In fact, in an equilibrium situation where the three constants of the motion P_α have non-zero average values, the material shows a uniform motion with velocity u. There exist therefore between the given subsystems, which are here a set of fixed volume elements, fluxes of particles, energy, and momentum, and the corresponding currents J_i^0 can be calculated starting from the properties of the matter at rest. For instance, the particle current J_N^0 equals $\varrho_N u$; the other equilibrium currents are given in a classical fluid by (14.116). In a non-equilibrium state we must thus single out for each current $J_i(r,t)$ a contribution $J_i^0(r,t)$. The latter is calculated for the case *where there are no gradients*, as if the intensive variables γ_i were uniform, everywhere taking their value $\gamma_i(r,t)$ at the position and the time under consideration, for instance, $J_N^0(r,t) = \varrho_N(r,t) u(r,t)$. The difference $J_i - J_i^0$, which vanishes with the affinities, is related linearly to them through the response coefficients L. Altogether, the linear response equations have here the general form

$$
J_i^\alpha(r,t) = J_i^{0\alpha}(r,t) + \sum_{j,\beta} L_{ij}^{\alpha\beta} \frac{\partial}{\partial r_\beta} \gamma_j(r,t) \quad ,
\tag{14.16}
$$

which includes terms of zeroth and first order in the affinities. The response coefficients $L_{ij}^{\alpha\beta}$ depend on the values of the intensive variables γ_i at the point r and the time t.

We now have the complete set of equations which can be used to find a *practical solution of any problem of macroscopic transport or relaxation*: the equations of state (14.9) or (14.12) couple the conserved quantities with their conjugated intensive variables; the conservation equations (14.2) or (14.6) give us the change in time of the conserved quantities as function of the fluxes; finally, the linear response equations (14.15) or (14.16) express the fluxes as functions of the affinities, that is, as functions of the local intensive variables. This set of equations refer to an isolated system; if there are couplings to the outside, the exchanges with the sources of heat, work, or particles can suitably be taken into account as in (14.3), and one should write down response relations like (14.15) for the fluxes coming from these sources. We can also use the right-hand side of (14.3) to deal with a small violation of a conservation law.

In what follows we shall illustrate by examples the abstract general method which we have just sketched and which we shall summarize in § 14.2.6.

The principal problem which remains to be solved, apart from the mathematical treatment of the equations which we have found, is the *determination of the response coefficients L*. Experiments give us empirical results; moreover, the principles of thermodynamics which we shall now enounce help us considerably to simplify that determination with the aid of theoretical arguments.

Whereas the conservation equations are valid in any regime, even far from equilibrium, the equations of state, which define the intensive variables, are only of interest in quasi-equilibrium regimes. The response equations presuppose, moreover, that the affinities are sufficiently small to justify a linear approximation. Experiments show, nevertheless, that one hardly exceeds the linear domain in practice: Ohm's law, the Navier-Stokes equations, Fourier's law, or the equations of chemical kinetics remain valid for electric fields, velocity gradients, temperature gradients, or chemical affinities which are quite large.

As we have already mentioned, there does not exist a standard terminology to denote the entities occurring in non-equilibrium thermodynamics. Depending on the context, the *affinities* are also called *thermodynamic forces* or *tensions*. We must watch out and not confuse them with the forces X_α which appear in thermostatics in the definition (5.11) of work; the "thermodynamic forces" Γ_i or $\nabla\gamma_i$ only appear when there is no equilibrium. Chemists often use the name affinities for the $-T\Gamma_k$, which are associated with the use of the energy rather than the entropy as fundamental thermodynamic potential, rather than the quantities (14.14b). Similarly, one sometimes calls the linear response equations (14.15) or (14.16) *complementary equations, or laws*, to contrast them with the equations of state (14.9) or (14.12). In their continuous version (14.16) they are also called macroscopic *diffusion* or *transport equations*, as they characterize the way each of the conserved variables A *diffuses from one element to the neighbouring one due to spatial variations* of the intensive variables γ. Finally, it is important to distinguish between the linear responses *of fluxes to affinities* which are introduced here and the linear responses to static (Exerc.4a) or time-dependent (Exerc.14b) *external perturbations*, which also play an essential rôle in many practical applications of statistical mechanics, for instance, the reaction of an electric circuit to a periodic potential applied to its terminals. This absence of systematic conventions and terminology reflects the unclear origins of thermostatics and the fact that too great an accent was put on mechanical and energy concepts; undoubtedly it is also a consequence of the long period elapsed between the start of the practical and intuitive use of non-equilibrium thermodynamics, at the beginning of the nineteenth century, and the working out of its general principles, which was first done by De Donder in 1927 for chemical kinetics.

14.2.3 Some Common Transport Coefficients

All dynamic effects in a quasi-equilibrium regime, from thermal to mechanical properties of continuous media, from chemistry to electromagnetism in matter, can be described in the above framework. However, experimental results are usually expressed in terms of variables such as the temperature or the flow velocity, rather than in terms of the variables, better suited for a theoretical analysis, which occur in our general equations. It is therefore expedient to relate the transport coefficients used in practice to the response coefficients L occurring in the general theory.

A simple example is *heat diffusion* across a solid which is governed by *Fourier's law* (1811)

$$J_E = -\lambda \nabla T. \tag{14.17}$$

This empirical law defines the *heat conductivity* λ as the coefficient of proportionality between the thermal gradient and the heat flux, which is given by (14.5) in terms of the current density J_E; the solid is at rest and the transported energy here consists only of heat. A comparison with (14.16), where $\gamma_E = 1/T$, leads to the following identification:

$$L_{EE}^{\alpha\beta} = \delta_{\alpha\beta}\lambda T^2. \tag{14.18}$$

The heat flow equation is obtained by eliminating the fluxes from (14.17) and the energy conservation equation (14.6), and then using the definition of the specific heat per unit volume, C, which gives us

$$\frac{\partial \varrho_E}{\partial t} = \text{div}(\lambda \nabla T) = \text{div}\left(\frac{\lambda}{C} \nabla \varrho_E\right) = C\,\frac{\partial T}{\partial t}. \tag{14.19}$$

This equation covers different situations, for instance, transport problems where the matter is coupled to two thermostats at different temperatures, or relaxation problems where we are trying to find out how long it will take an isolated sample with a non-uniform temperature, which is thus removed from equilibrium, to return to it.

At the start of the nineteenth century heat flow was an open problem which appeared so important that the French Académie des Sciences made it the subject of a competition. Joseph Fourier (Auxerre 1768–Paris 1830) was the Prize winner, after several years of trying. First of all, he found the equations for the problem, but he still needed to solve, assuming that λ/C is constant, a partial differential equation, the *heat conduction equation*

$$\frac{\partial \varrho_E}{\partial t} - \frac{\lambda}{C}\,\nabla^2 \varrho_E = 0,$$

with appropriate boundary and initial conditions. An unavailing attempt led to the invention of Fourier series. The final solution was obtained thanks to the discovery of the Fourier integral transformation.

Diffusion is a formally analogous phenomenon, in which the energy is replaced by particles of a given kind. It consists in the macroscopic drift of those particles under the action of a spatial variation of their density. Different materials can be involved: a solute in a liquid solvent, impurities in a solid – with applications to "doping", where the motion of these impurities is only significant at rather high temperatures – gas mixtures, neutrons in substances which make up a nuclear reactor, and so on. In the case of a gas with only one kind of particles, the density also tends to become uniform, but the mechanism of that effect, which is called self-diffusion, is less simple, as it involves a global motion of matter (§ 14.4.6). In the case of diffusion proper, macroscopic velocities are negligible, as is the momentum density ϱ_P; the temperature, moreover, is uniform so that the only relevant response coefficient is L_{NN}. In general, diffusion obeys *Fick's law* (1855; Adolf Fick was a German physiologist), an empirical law stating that the current of the particles considered is proportional to the gradient of their density:

$$J_N = -D \nabla \varrho_N. \tag{14.20}$$

The proportionality coefficient, D, the *diffusion coefficient*, can be expressed as a function of L_{NN} by comparing (14.20) and (14.16) and using the equations of state. To do this it is sufficient to know $\partial \varrho_N / \partial \gamma_N = -T \partial \varrho_N / \partial \mu$ for constant $\gamma_E = 1/T$; the methods of § 6.3.5 give us $\partial \varrho_N / \partial \mu = \kappa \varrho_N^2$, where κ is the compressibility (6.42) of the gas of diffusing particles, and hence we have

$$L_{NN}^{\alpha\beta} = \delta_{\alpha\beta} D \kappa \varrho_N^2 T. \tag{14.21}$$

The *diffusion equation* following from the particle conservation law (14.6) and (14.20) has the same form as the heat equation (14.19).

Similarly, *osmotic diffusion*, and the flow of a fluid across a *porous wall*, which belong to the category of transport phenomena between two *discrete* subsystems, are governed by *Darcy's law* (1856; Henri Darcy was an engineer in charge of the city water system in Dijon). The flux passing through per unit area is proportional to the pressure difference, a variable which is related to the affinity Γ_N, that is, to the difference between the values of $-\mu/T$. One usually writes the coefficient in the form $K\varrho_N/l\eta$, where l is the thickness of the wall and η the viscosity of the fluid; this defines the *permeability coefficient* K of the porous medium.

The oldest historical example of transport phenomena is undoubtedly Newtonian *viscosity* (14.134) which connects the constraints on a fluid flowing in the x-direction – with a velocity u_x which varies with z – to the velocity gradient $\partial u_x / \partial z$ (§ 7.4.6). As we shall see in § 14.4.6, the viscosity coefficient η is related to $L_{PP}^{\alpha\beta}$ which is the tensor describing *momentum transport*.

In *chemical kinetics* and similar reaction phenomena it is natural to express, as in the case of diffusion, the variables $\gamma_N = -\mu/T$ associated with the various kinds of particles in terms of the corresponding densities ϱ_N. The affinities Γ_k, defined by (14.14b), then become functions of these densities. Taking into account the conditions of the reaction, for instance, constant pressure and constant temperature, we are thus led to a law for chemical kinetics in terms of directly measurable quantities, where (14.15) provides the fluxes or reaction speeds Φ_k as functions of the partial densities of the different constituents, and where (14.4) relates in turn the time-derivatives of the densities to the fluxes.

All those examples describe *direct* effects where an affinity Γ_i produces the flux Φ_i of the same nature. However, if, for chemical type kinetics, several reactions are coupled, the flux Φ_k for each one may depend through (14.15) on the affinities Γ_l for the others. More generally, in what follows we shall meet with other *indirect* effects in which an affinity produces not only a flux of the quantity to which it is conjugated, but also a flux of another quantity. For instance, a temperature gradient can produce in a fluid not only heat flow, but also matter transport as the particles can move in bulk when the substance is not in equilibrium. This phenomenon, *thermal diffusion*, is associated with the off-diagonal matrix element L_{NE} of the linear response matrix. When thermal diffusion is allowed, the expression for the heat conductivity becomes complicated (see Eq.(14.89)) as λ is defined in conditions when heat flows without matter being transported.

In § 14.4 we shall return to the above transport phenomena which we shall study more systematically together with other examples, such as electric conduction, thermoelectric effects, or hydrodynamics. In all cases, whether we are dealing with macroscopic transport or with relaxation phenomena, we must determine the relations between the directly observed quantities of interest and the conservative variables, or their conjugate variables, and make clear what quantities remain fixed during the process under consideration.

14.2.4 Curie's Principle and Onsager's Relations

The elements of the linear response matrix L are functions of the intensive variables which locally characterize the state of the substance. From a macroscopic point of view we are dealing with phenomenological coefficients which must be determined from a comparison with experiments. Nevertheless, before any measurement, thermodynamics provides a certain amount of information which partially lifts the arbitrariness of these coefficients. We shall, first of all, use *symmetry* and *invariance* properties which are important as they enable us to reduce the number of independent matrix elements. We shall state here the general principles, postponing to § 14.4 their application to specific examples. Pierre Curie (Paris 1859–1906) was the first to stress the importance of the *symmetry principle* after he discovered piezoelectricity – the indirect effects which relate to each other electric fields and mechanical constraints in materials such as quartz – and after he studied magnetism.

Let us assume that the substance has symmetry properties under some spatial transformation, for instance, rotational invariance or parity invariance, that is, symmetry with respect to a point. We should consider it to be an active transformation (§ 2.1.5), operating on the substance itself, but it is also equivalent to a passive change of coordinates under which *all equations must remain unchanged*. As far as the *equations of state* are concerned, they are obtained by taking derivatives of the entropy; the symmetries of the substance are thus reflected in writing down the invariance of the entropy as function of the various densities ϱ_i. For instance, for an isotropic fluid these densities are ϱ_E, ϱ_N, and the three components of ϱ_P, ϱ_{Px}, ϱ_{Py}, and ϱ_{Pz} ; the first two are scalars, that is, quantities which are invariant under rotation, and the last three form a vector: under a rotation they transform by combining linearly. The entropy density s is a scalar and the only scalar which can be built from the vector ϱ_P is $\varrho_P^2 = \sum_\alpha \varrho_{P\alpha}\varrho_{P\alpha}$ so that s is a function of the three variables ϱ_E, ϱ_N, and ϱ_P^2. We shall similarly use the Galilean invariance (§ 14.4.4) and the invariance of the entropy under a change in the energy zero (§ 14.4.2) to set constraints on the form of the function $s(\{\varrho_i\})$ for a fluid.

The relations between invariances and *conservation laws* are fundamental, but subtle. We shall, in fact, see in § 14.3.1 that the conservation laws themselves follow from the symmetry properties of the microscopic equations of motion.

The consistency of the *response coefficients* with the symmetry properties provides valuable information about them, as we shall see in the examples of § 14.4. We shall here illustrate this point on one example, namely, that of an isotropic substance, for the response coefficients $L_{ij}^{\alpha\beta}$ associated with scalar variables i and j, for instance, the energy or the number of particles. Under a rotation (14.16) must remain invariant; the quantities $\boldsymbol{J}_i$ and $\nabla\gamma_j$ which occur in it transform as vectors. As a result, $L_{ij}^{\alpha\beta}$ must transform as a *tensor* with two indices, that is, as a vector for each of the indices α and β. If, moreover, it is only a function of scalars (such as the temperature or the chemical potential) and independent of any vector (such as the local velocity) $L_{ij}^{\alpha\beta}$ itself must be invariant under rotation. As we shall see in what follows this implies that $L_{ij}^{\alpha\beta}$ must be of the form $L_{ij}\delta_{\alpha\beta}$ so that for the variables i, j considered the response equation can be written in the form

$$\boldsymbol{J}_i = \sum_j L_{ij}\, \nabla\gamma_j. \tag{14.22}$$

To prove this it is sufficient to write down the invariance of $L^{\alpha\beta}$ under an infinitesimal rotation around an arbitrary direction δ, which gives us

$$0 = \sum_\gamma \left(\varepsilon_{\alpha\delta\gamma} L_{ij}^{\gamma\beta} + \varepsilon_{\beta\delta\gamma} L_{ij}^{\alpha\gamma} \right).$$

When $\delta = \alpha$, we find that $L^{\alpha\beta} = 0$ for $\alpha \neq \beta$; when $\delta \neq \alpha$, we find that $L^{\beta\beta} = L^{\alpha\alpha}$ for $\alpha \neq \beta$. Hence, the isotropic tensor $L^{\alpha\beta}$ must be proportional to the unit matrix $\delta_{\alpha\beta}$.

A similar calculation enables us to check that the only isotropic tensor with three indices is proportional to the completely antisymmetric unit tensor

$$\varepsilon_{\alpha\beta\gamma}, \quad \text{with } \varepsilon_{123} = 1 = -\varepsilon_{213} = \dots, \tag{14.23a}$$

and that the only isotropic tensors with 4 indices are proportional to

$$\delta_{\alpha\beta}\delta_{\gamma\delta}, \quad \delta_{\alpha\gamma}\delta_{\beta\delta}, \quad \delta_{\alpha\delta}\delta_{\beta\gamma}. \tag{14.23b}$$

Curie's symmetry principle can be used in the same way for other invariances. For instance, a crystalline substance, even though it is not isotropic, is invariant under certain symmetries and rotations of a discrete group which characterizes its structure (§ 11.1.2). A study of the transformations of the $L_{ij}^{\alpha\beta}$ under the action of this group is an efficient way to simplify the description of transport properties by reducing the number of independent coefficients, as we did in § 6.6.4 for thermostatics. In particular, in a cubic crystal and in the case where the quantities i and j are scalars, $L_{ij}^{\alpha\beta}$ may a priori depend on the three basic vectors of the crystal lattice but it must remain invariant under any permutation or sign change of them. This again leads to the relation (14.22): notwithstanding the anisotropy the transport of scalar quantities in a cubic crystal takes place as in an isotropic substance. On the other hand, in single crystals which have less simple crystal structures transport properties such as the electric conductivity can be anisotropic and thus involve several response coefficients. As a further example, piezoelectricity cannot exist in substances which have a centre of symmetry; for practical applications such as the generation of regular oscillations by resonance in watches, one uses a single crystal of quartz, which involves a helical microscopic structure with a given parity.

Another symmetry principle for the response coefficients L is known as the *Onsager reciprocal relations*. Below we shall give its general form. The simplest situation concerns the special case of scalar quantities i and j which remain invariant under time reversal, such as the energy, the number of particles, or chemical variables. In such a case we have

$$L_{ij} = L_{ji}. \tag{14.24}$$

These relations which can be proved starting from statistical mechanics are an important general property of the thermodynamics of non-equilibrium processes. They have been referred to as a Fourth Law of thermodynamics.

Onsager's reciprocal relations concern the off-diagonal elements of the L_{ij} matrix which describe *indirect effects in the approach to equilibrium*. In this way a temperature gradient produces energy flow, but it can also produce a flow of particles, even though the variable $-\mu/T$ which is conjugated to the number of particles is uniform. Inversely, even when the temperature is uniform, a gradient of the chemical potential, or, what amounts to the same, of the pressure or of the density, can produce a heat flow at the same time as a particle flow. Denoting by $\boldsymbol{J}_E$ and $\boldsymbol{J}_N$ the energy and particle current densities, those four effects are for an isotropic substance described by the transport equations

$$\boldsymbol{J}_E = L_{EE} \, \nabla\left(\frac{1}{T}\right) + L_{EN} \, \nabla\left(-\frac{\mu}{T}\right), \tag{14.25a}$$

$$J_N = L_{NE} \, \nabla \left(\frac{1}{T} \right) + L_{NN} \, \nabla \left(-\frac{\mu}{T} \right). \tag{14.25b}$$

In this case the Onsager relations state that

$$L_{EN} = L_{NE}. \tag{14.26}$$

The two reciprocal effects, namely, particle transport produced by a temperature gradient or heat transport produced by a mechanical gradient, are thus related to each other. Similarly, for coupled chemical reactions with fluxes Φ_k, Φ_l, ... and affinities Γ_k, Γ_l, ..., we have $L_{kl} = L_{lk}$.

Another classical example of an application of the Onsager relations concerns the thermoelectric effects, either in a homogeneous substance or in a circuit with junctions. For instance, the heat which is produced at a junction when a current passes through it (Peltier effect) is related (§ 14.4.3) to the electromotive force generated in a thermocouple by a temperature difference (Seebeck effect). These effects, discovered in 1834 by Jean Charles Peltier (Ham 1785–Paris 1845) and in 1821 by Thomas Seebeck (Tallinn 1770–Berlin 1831), were studied by William Thomson (later Lord Kelvin) in 1854 from a thermodynamic point of view. One had to wait to 1931, however, before Lars Onsager (Oslo 1911–Miami 1976) gave a correct proof of the second relation written down by Thomson – the first one gave the energy conservation. The most general proof of the Onsager relations was given in 1945 by Hendrik B.G. Casimir (The Hague 1909).

The Onsager relations are useful for reducing the number of independent transport coefficients for various phenomena, from chemistry to mechanics of solids or of anisotropic fluids; they cover many indirect effects which mix heat, magnetism, and electricity. They follow from the symmetry of the microscopic equations of motion when we reverse the sign of the time in them. In general, they remain invariant; however, we can for the sake of greater generality consider cases where they are changed. For instance, when there is an external magnetic field present, time reversal leaves the equations of motion unchanged only if one at the same time reverses the magnetic field. We shall denote by $\widehat{H}^{\mathrm{T}}$ the Hamiltonian which is transformed in that way (see the end of § 2.1.5). To write down the Onsager relations in their general form, we must also find the behaviour of the quantities ϱ_i and γ_i under time reversal. The energy, particle, or charge densities are even, that is, they remain unchanged under a change in the sign of t; for them we thus have $\varrho_i^{\mathrm{T}} = \varrho_i$. On the other hand, the momentum density changes sign with the time, so that $\varrho_{P\alpha}^{\mathrm{T}} = -\varrho_{P\alpha}$; the same is true for the electric current, angular momentum, or magnetization densities. We write, in general, $\varrho_i^{\mathrm{T}} = \varepsilon_i \varrho_i$ with $\varepsilon_i = \pm 1$. When an index i refers to a chemical affinity, ε_i equals $+1$. The value of the entropy is independent of the direction of the time so that we have $\gamma_i^{\mathrm{T}} = \varepsilon_i \gamma_i$: the temperature and the chemical potentials remain unchanged and the velocity changes sign. The response coefficients $L_{ij}^{\alpha\beta}$ are functions of the set of intensive variables $\{\gamma_k\}$ at the point considered; they also depend on the temporal evolution governed by the Hamiltonian $\widehat{H}$. To indicate those dependences we write them as $L_{ij}^{\alpha\beta}(\{\gamma_k\}, \widehat{H})$. The *general reciprocal Onsager relations* can then be written in the form

$$L_{ij}^{\alpha\beta}(\{\gamma_k\}, \widehat{H}) = \varepsilon_i \varepsilon_j L_{ji}^{\beta\alpha}(\{\varepsilon_k \gamma_k\}, \widehat{H}^{\mathrm{T}}). \tag{14.27}$$

These relations are valid both in the case of a continuum at a given point and for the responses $L_{ij}(a, b)$ defined by (14.15) for a pair of discrete subsystems a and b; in that case we have also clearly $L_{ij}(a, b) = L_{ij}(b, a)$.

In this general form the Onsager relations connect the transport coefficients of two systems which differ from one another in the direction of the local velocities, through changes in the γ_k, and in the direction of a possible applied magnetic field, through $\widehat{H}$. If there is no magnetism and if the substance is at rest, the same system occurs on both sides of (14.27) which then simply expresses the symmetry or antisymmetry of the linear response matrix elements. The interchange of the indices i and j, as of α and β, in (14.27) reflects the fact that, if one reverses the direction of the time in the equations, the *causes and effects interchange*.

14.2.5 Dissipation

The *rate of change of the entropy* of a subsystem a is given by

$$\frac{dS_a}{dt} = \sum_i \frac{\partial S_a}{\partial A_i(a)} \frac{dA_i(a)}{dt} = \sum_i \gamma_i(a) \frac{dA_i(a)}{dt}$$

$$= -\sum_{i,b} \gamma_i(a) \, \Phi_i(a \longrightarrow b), \tag{14.28}$$

where we have used, successively, the definition (14.9) of the intensive variables γ and the conservation laws (14.2).

Let us write the right-hand side of (14.28) as a sum of two contributions, one antisymmetric and the other symmetric under an interchange of the subsystems a and b. The first contribution,

$$\Phi_S(a \longrightarrow b) \equiv \frac{1}{2} \sum_i \left[\gamma_i(a) + \gamma_i(b) \right] \Phi_i(a \longrightarrow b), \tag{14.29}$$

which is antisymmetric thanks to (14.1), can be *interpreted as an entropy flux* from a to b, which is the opposite of the flux from b to a as should be the case. This definition of entropy flux is also justified by the fact that when $\gamma_i(a) = \gamma_i(b)$ expression (14.29) reduces to the change in the entropy of b in a *reversible* transformation where a gives off to b the quantities Φ_i. In particular, if a and b only exchange heat energy, and if that exchange is reversible, the temperatures of a and b are very close to one another and (14.29) expresses that the heat flow Φ_E equals $T\Phi_S$.

We should, however, not push the analogy of Φ_S with the fluxes Φ_i of conservative quantities too far. As the entropy has a nature which is different from that of the extensive variables, its flux plays a rôle which is different from that of the fluxes associated with those variables, which entered the conservation equations (14.2). In fact, on the right-hand side of (14.28) there occurs a second contribution which is symmetric under an interchange of

a and b: if we use the definition (14.14) of the affinities and the definition (14.29) of the entropy flux, we can rewrite (14.28) in the form

$$\frac{dS_a}{dt} + \sum_b \Phi_S(a \longrightarrow b) = \frac{1}{2} \sum_{i,b} \Gamma_i(a,b)\, \Phi_i(a \longrightarrow b). \qquad (14.30)$$

The left-hand side reminds us of the conservation laws. However, there occurs here a right-hand side which is a sum of terms. Each of these terms is symmetric under an interchange of a and b and can be interpreted as a *rate of entropy production at the interface between a and b*, associated with the *irreversible* transfer of the quantity i. The symmetry of $\frac{1}{2}\Gamma_i\Phi_i(a \to b)$ under an interchange of a and b implies that the same term appears also in the entropy balance for the subsystem b so that the entropy created at the a, b interface is divided equally between these two subsystems. On the other hand, because it is antisymmetric, the term Φ_S on the left-hand side of (14.30) does not contribute to the global entropy balance of a and b since the contributions $\Phi_S(a \to b)$ and $\Phi_S(b \to a)$ are the opposite of each other: the entropy fluxes Φ_S cancel one another pairwise in a global balance. We shall use the term *dissipation* in the general meaning of entropy production – and not in the often used restricted meaning of energy degradation, for instance, transformation of heat into work.

In its dynamic form, the *Second Law* can be expressed through the *Clausius-Duhem inequality*

$$\sum_i \Gamma_i(a,b)\, \Phi_i(a \longrightarrow b) \geq 0, \qquad (14.31)$$

which must be satisfied whatever the affinities. The fluxes are functions of the latter and the Second Law indicates thus that they cannot be arbitrary functions: the right-hand side of (14.30) must under all circumstances reflect the *existence of positive dissipation*. The Clausius-Duhem inequality is more detailed than the Second Law in its thermostatics version: in fact, it must be satisfied at all times and for any pair of subsystems a, b, whereas in thermostatics one is only concerned with comparing the initial and the final entropies of an isolated system which both initially and finally is at equilibrium. Nonetheless, the Second Law of thermostatics is more general inasmuch as it makes no assumptions about the intermediary states, whereas (14.31) is only concerned with systems which evolve in a quasi-equilibrium regime: at all times an infinitesimal irreversible transformation takes place, while the subsystems a and b remain always close to equilibrium.

This analysis is valid for any evolution in a quasi-equilibrium regime. If the regime, moreover, is *linear*, we can use (14.15) to express the dissipation (14.30) in terms of the local intensive variables γ in the form

$$\frac{dS_a}{dt} + \sum_b \Phi_S(a \longrightarrow b) = \frac{1}{2} \sum_{i,j,b} \Gamma_i(a,b)\, L_{ij}\, \Gamma_j(a,b). \qquad (14.32)$$

The dissipation is then a quadratic form of the affinities $\Gamma_i(a,b) \equiv \gamma_i(b) - \gamma_i(a)$. According to the principle (14.31) it must be positive whatever the affinities. The symmetric part, $L_{ij} + L_{ji}$, of the *linear response* is thus a positive matrix in i, j. Thermodynamics thus imposes *constraints on the transport coefficients* by predicting that they must obey a certain number of inequalities. We shall, in particular, see that the fact that the diagonal elements of the L matrix are positive implies that quantities such as thermal or electrical conductivities, diffusion coefficients, or viscosities, also must be positive. Of course, if L_{ij} has an antisymmetric part, the fact that the dissipation is positive does not give any information about $L_{ij} - L_{ji}$. By virtue of the Onsager relations (14.27) this may occur if i and j do not behave in the same way under time reversal, or if there is a magnetic field present.

For an *isolated* system the summation over the subsystems a in (14.30) or (14.32) leads to the cancellation of the entropy currents, and for the expression of the *total dissipation* we are left with

$$\frac{dS}{dt} = \sum_{i,a>b} \Gamma_i(a,b)\, \Phi_i(a \longrightarrow b) = \sum_{i,j,a>b} \Gamma_i(a,b)\, L_{ij}\, \Gamma_j(a,b). \quad (14.33)$$

One should, however, note that if the system considered is coupled to *sources*, we must in the global balance (14.33) include possible exchanges of entropy with these sources. Moreover, if we bear in mind that the dissipation (14.31) is associated not with a subsystem, but with a pair of subsystems, the interfaces between the system proper and its sources are places where entropy is produced, if the affinities do not vanish there.

For a given value of the transfers Φ_i, (14.31) and (14.33) show that the increase in the total entropy of an isolated system, which measures the irreversibility of the transformation, is small like the affinities $\Gamma_i(a,b)$. In order for a transformation to be *almost reversible* it is thus necessary that the *intensive variables* of the parts of the system between which exchanges take place be *almost equal*. However, the form (14.15) of the response equations itself entails that the fluxes are then small and that the transformation is *very slow*. Significant jumps, or gradients, in the temperature, the chemical potential, or the pressure are necessary to *reduce the duration* of the processes. However, the price which we must pay for that greater speed is an *increase in the dissipation*, hence a loss of efficiency. The art of the engineer often consists in looking for compromises, aiming to face up to this *contradiction between reasonably high efficiency and reasonably short delay* of the process.

We now extend the preceding ideas to *continuous systems*. If we use (14.12) and (14.6), we see that in the case of continuous variations in space the local entropy density $s(\boldsymbol{r}, t)$ per unit volume evolves as follows:

$$\frac{\partial s}{\partial t} = \sum_i \gamma_i \frac{\partial \varrho_i}{\partial t} = -\sum_i \gamma_i \operatorname{div} \boldsymbol{J}_i. \quad (14.34)$$

Using the identity $-\gamma \operatorname{div} \boldsymbol{J} = -\operatorname{div}\gamma\boldsymbol{J} + (\boldsymbol{J} \cdot \nabla)\gamma$ and the same procedure as above, we can separate the right-hand side of (14.34) into two parts, which leads to

$$\frac{\partial s}{\partial t} + \operatorname{div}\left(\sum_i \gamma_i \boldsymbol{J}_i \right) = \sum_i (\nabla\gamma_i \cdot \boldsymbol{J}_i). \tag{14.35}$$

In the case where the fluxes $\boldsymbol{J}_i$ vanish together with the affinities $\nabla\gamma_i$, (14.35) is the required entropy balance, its right-hand side defining the dissipation, as in (14.30). However, when the response equations (14.16) contain terms of zeroth order, we expect that the contribution from the equilibrium currents $\boldsymbol{J}_i^0$ to the right-hand side of (14.35) should be eliminated through integration over the volume so that the total entropy would remain constant if $\boldsymbol{J}_i = \boldsymbol{J}_i^0$. For this it is necessary that $\sum (\nabla\gamma_i \cdot \boldsymbol{J}_i^0)$ be the divergence of a vector field, which we shall then include in the definition of the entropy flux. In fact, we shall show in (14.149) that for a fluid

$$\sum_i (\nabla\gamma_i \cdot \boldsymbol{J}_i^0) \equiv -\operatorname{div} \frac{\mathcal{P}}{T} \boldsymbol{u},$$

where the index i takes 5 values, referring to the energy E, the particle number N, and the momentum $\boldsymbol{P}$, and where $\mathcal{P}$, T, and $\boldsymbol{u}$ are the pressure, the temperature and the local velocity at the point $\boldsymbol{r}$, expressed in terms of the γ_i as if the substance were at equilibrium. It is thus natural to define the *entropy current density* through

$$\boldsymbol{J}_S \equiv \sum_i \gamma_i \boldsymbol{J}_i + \frac{\mathcal{P}}{T} \boldsymbol{u}. \tag{14.36}$$

This expression is the continuous analogue of (14.29). The extra term it contains has again the same form $\gamma_\Omega \boldsymbol{J}_\Omega$ and refers to the volume. In fact, one can identify $\mathcal{P}/T$ with the intensive variable $\gamma_\Omega = \partial S(E, N, \boldsymbol{P}, \Omega)/\partial\Omega$, which is the conjugate of the volume, while $\boldsymbol{u}$ plays the rôle of the current associated with the volume Ω of a fluid element, since its flux across the surface of that element is the rate of expansion $d\Omega/dt$ of the fluid when it is followed in its macroscopic motion.

Substituting the definition (14.36) into (14.35) we get the *local balance equation for the entropy*

$$\frac{\partial s}{\partial t} + \operatorname{div} \boldsymbol{J}_S = \sum_i \left(\nabla\gamma_i \cdot (\boldsymbol{J}_i - \boldsymbol{J}_i^0) \right). \tag{14.37}$$

In the linear regime (14.16) and (14.37) give

$$\frac{\partial s}{\partial t} + \operatorname{div} \boldsymbol{J}_S = \sum_{\substack{i,j \\ \alpha,\beta}} \frac{\partial\gamma_i}{\partial r_\alpha} L_{ij}^{\alpha\beta} \frac{\partial\gamma_j}{\partial r_\beta}. \tag{14.38}$$

The Clausius-Duhem inequality here implies that (14.38) is positive, whatever the $\nabla \gamma_i$. It thus imposes constraints on $L_{ij}^{\alpha\beta}$, considered to be a matrix in i, j and also in the tensorial indices α, β relative to the three spatial directions: the symmetric part of that matrix must be positive.

The right-hand side of (14.37) or of (14.38) can be interpreted as the *rate of dissipation at the point* $\boldsymbol{r}$: the entropy production occurs *in the bulk* of the substance. To illustrate this point let us consider *heat transport* in a permanent regime, from a thermostat a at a temperature T_a to a second thermostat b at a temperature $T_b < T_a$, across a bar of length l and cross-section σ which consists of matter with a conductivity λ. We assume that there are no jumps in the temperature at the contacts between the thermostats and the bar so that there is no entropy production at those contacts. If an amount of heat Q is transported from the one source to the other during a time t, thermostatics gives us the *global* entropy balance over that period: the entropy of the thermostat b has increased by Q/T_b, that of a has decreased by Q/T_a, while that of the bar has not changed – as its state remains unchanged in a permanent regime. However, the *detailed* balance (14.37) of thermodynamics shows that entropy is created, not in the thermostats, which evolve reversibly, but *in the conductor* according to

$$\frac{ds}{dt} + \operatorname{div} \boldsymbol{J}_S = \left(\nabla \left(\frac{1}{T} \right) \cdot \boldsymbol{J}_E \right). \tag{14.39}$$

In a permanent regime we have $ds/dt = 0$ and ∇T is given by (14.17). The heat flux $\boldsymbol{J}_E$ equals $Q/t\sigma$ in the propagation direction, which we take to be the x-direction; it is uniform: energy conservation. On the other hand, the entropy flux defined by (14.36), $J_S = J_E/T$, increases from J_E/T_a to J_E/T_b from the left end of the bar at $x = 0$ to its right end at $x = l$. According to (14.39), in the section of the bar between x and $x + dx$ an amount of entropy is *created*, equal to

$$\frac{d}{dx} \frac{1}{T(x)} \frac{Q}{t} \, dt$$

during the time dt; this section *receives* at its left-hand side an amount of entropy

$$J_S(x) \, \sigma \, dt = \frac{1}{T(x)} \frac{Q}{t} \, dt,$$

and *gives off* to its right-hand side a larger amount,

$$J_S(x + dx) \, \sigma \, dt = \frac{1}{T(x + dx)} \frac{Q}{t} \, dt.$$

The entropy flux leaving the section is larger than that entering it, because this flux carries permanently the created excess. The integral over dt and over dx of the dissipation yields again the total increase in the entropy $Q/T_b - Q/T_a$ given by thermostatics. However, our arguments show what is the dissipation mechanism and where it is localized: an entropy flux J_E/T_a leaves a, accompanying the heat flux J_E; it is directed towards b, while *gradually increasing* like J_E/T as the temperature decreases. This increase is produced by the irreversibility of the processes taking place in each volume element of the conducting substance. The final balance is supported by the sources; however, b has received more entropy than a has given up only because there was dissipation all along the bar.

In (14.30) as in (14.37) the dissipation terms have the form of the product of an *affinity* with the corresponding *flux*: these quantities thus appear as *conjugates for the dynamics* of the system, in the same way as the intensive γ and the extensive A variables in the case of equilibrium. There exists, in fact, a formal analogy between (14.33) and the expression for the change in the entropy between equilibrium states,

$$dS = \sum_{i,a} \gamma_i(a) \, dA_i(a),$$

which is equivalent to (14.9). Similarly, in the case of a continuum (14.37) resembles (14.13). We note, however, that dS is a total differential, which implies that the partial derivatives $\partial \gamma_i/\partial A_j$ and $\partial \gamma_j/\partial A_i$ are equal; this is not true in general for (14.33). The analogy becomes less superficial in the particular cases where there is a "dissipation potential", a function of the fluxes (or of the affinities) from which we can find the affinities (or the fluxes) through differentiation, in the same way as the knowledge of a "thermodynamic potential" produces through differentiation the pairs of conjugated A_i, γ_i variables.

14.2.6 Summary: Macroscopic Approach to Dynamic Phenomena

The principles stated in §§ 14.1 and 14.2 provide us with a general framework in which a large number of problems from irreversible thermodynamics and, more generally, of macroscopic dynamics, enter. One meets with such problems in all branches of science and technology: transport of heat, matter, or electricity, hydrodynamics, chemical or nuclear kinetics, astrophysics, energetics, physiology, and so on. The questions which we put, the models which we construct to explain actual facts, the chosen level of our description, either macroscopic or microscopic at a more or less finite scale, are clearly specific for each problem and present a great variety. However, we find in many cases a number of common stages through which the solution of the problem passes, and a general scheme of equations (Fig.14.1).

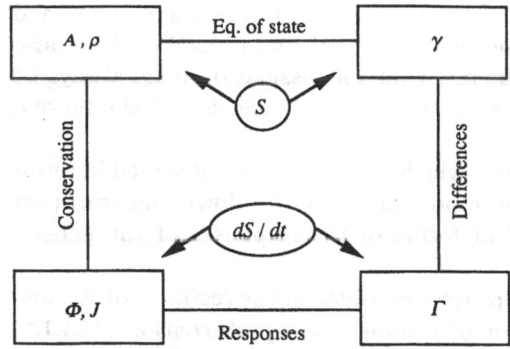

Fig. 14.1. *The general scheme of equations for a thermodynamic problem near equilibrium.* The local extensive variables $A_i(a)$ and the densities $\varrho_i(a)$ are related to the local intensive variables γ_i through the entropy of thermostatics. They are related to the fluxes $\Phi_i(a \to b)$ or to the current densities $J_i(r)$ through the conservation laws. The affinities $\Gamma_i(a, b)$ are differences between the $\gamma_i(a)$ variables or gradients of the γ_i. In a linear regime, the fluxes are linear functions of the affinities, and the dissipation dS/dt is a bilinear form of fluxes and affinities. The coefficients obey constraints imposed by the symmetries and by the Clausius-Duhem inequality

- Analyze, while building a model, the system under study in terms of *subsystems* which are nearly at equilibrium and between which exchanges can take place (§ 14.1.1). These subsystems may be discrete, they may be volume elements of a continuum, or they may be superimposed upon one another in space (§ 14.1.2).
- State which are the *conserved quantities* and the variables that at any time characterize the state of each of the subsystems (§§ 14.1.1 and 14.1.2).
- Determine the *relations between conjugated variables*, such as E and $1/T$, or N and $-\mu/T$. These relations are in a microscopic approach obtained through differentiation of a partition function. When they are given by experiments, check that they are compatible with the existence of a thermodynamic potential, such as the entropy as function of the extensive variables, from which they can be produced through differentiation (§ 14.2.1). This step ensures that the Second Law of thermostatics is satisfied.
- Write down the *conservation equations* which connect the rates of change in the conservative quantities of each subsystem with the fluxes between the subsystems (§ 14.1.3). Include source terms, if the system is not isolated or if the conservation laws are not rigorous. This step corresponds to the First Law and its extensions.
- Express the fluxes in terms of the affinities – the deviations between intensive variables – through *response relations*, for which the linear approximation usually is sufficient (§ 14.2.2). The coefficients, which are intrinsic to the substance, may be empirical and given by experiments, or they may be calculated using a microscopic theory.

- Reduce the number of independent coefficients by using *symmetry* and *invariance* properties, and the *Onsager relations* (§ 14.2.4). Check also the constraints provided by the fact that the dissipation must always be *positive* (§ 14.2.5); this property is the dynamical version of the Second Law.
- Identify the *quantities of interest* which can be actually observed in terms of the natural variables of thermodynamics. Write down the transport coefficients usually employed in terms of the responses of the general theory (§ 14.2.3).
- Specify the *conditions* for the process considered: the regime – of stationary transport, of relaxation, or of response to a perturbation (§ 14.1.2) – whether the system is open or not, isolated or not, the boundary and initial conditions, the fixed quantities.
- *Solve* either algebraically or numerically the coupled equations, namely, the equations of state, the conservation equations, and the response equations, under the conditions of interest. This stage, which is the object of macroscopic sciences, is often the most difficult one.
- *Compare* the results with facts, possibly improve them by including new variables and starting again, and so on.

Most of these points will be met with again in the applications studied in § 14.4, and the reader will recognize them when he analyzes the approach to many other problems. They will, however, appear in an arbitrary order imposed by the inner logic of the problem considered. The present general scheme is useful as a guide to a macroscopic phenomenology. It gives us guidelines to prevent the macroscopic theory from being incompatible with information coming from microscopic physics, for instance, when the symmetry principles force one or other of the coefficients to vanish, or when the Second Law imposes some inequality. It also enables us to connect with one another phenomena which are apparently independent, as we shall see later on for diffusion and electrical conduction, or for the thermoelectric effects.

Let us, however, remind ourselves that the method described here is only suitable for processes which are sufficiently close to equilibrium. In that case a purely macroscopic description is often sufficient even though interesting extra information could be provided by a simultaneous more microscopic study. If we are far from equilibrium, there is no general method; one may proceed either by empirical methods which are more or less reliable, or by more fundamental microscopic approaches which, however, because of their complexity have a restricted range of applicability.

14.3 Microscopic Approach

The point of view of statistical physics is needed for many purposes: to derive from microscopic physics the various properties posed above as general principles, to understand the significance on the microscopic scale of the new quantities which we introduced, such as the fluxes and the currents, and to express them in terms of the particles which make up the matter, and also to calculate the transport coefficients using microscopic dynamics. It is also indispensable for demarcating the domain of validity of thermodynamics, the neighbourhood of equilibrium, and, most of all, for going beyond that domain. For instance, we must have recourse to statistical mechanics whenever we want to make a theory about processes where some macroscopic quantities change rapidly, either in space – shock waves – or in time – electrodynamics at high frequencies (Exerc.14b). We shall restrict ourselves to giving below a few hints about the justification of the laws of thermodynamics and to present in Chap.15 Boltzmann's and Lorentz's microscopic approach to the dynamics of almost perfect gases.

14.3.1 Invariance and Conservation

We want to use statistical physics to justify the local conservation laws which we gave as principles in § 14.1.3. To do this, we shall need to identify, in terms of the elementary constituents, those quantities which occur in the local laws. For instance, to prove the conservation (14.6) of the energy, we must express the density ϱ_E and the current density J_E at a point r as functions of the microscopic variables, namely, the positions r_j and the momenta p_j of the various particles. The main difficulty comes from the fact that ϱ_E and J_E should be local quantities defined at a point r, whereas the potential energy, even though it has a short range, is non-local and depends on the coordinates r_j and r_k of particles in pairs. An extra complication arises in quantum mechanics: in a definition such as (14.49) which we shall give for the particle flux at the point r, the positions r_j and the momenta $p_j = mv_j$ of the various particles should be replaced by operators which do not commute. To simplify matters we shall restrict ourselves to proving the local conservation laws in *classical* statistical mechanics.

We have already indicated that the existence of an *invariance* in the dynamics implies that of a *conservation law* (§ 2.1.5). We have thus associated with each invariance or symmetry a *global* conservation law. For instance, invariance of the equations of motion under translation in space leads to conservation of the total momentum; invariance under translation in time leads to conservation of the energy, and invariance under rotation to that of the angular momentum. In the Hamiltonian formalism this is just the result of the fact that $\widehat{H}$ commutes with the infinitesimal generator of the invariance considered. The *local* conservation laws (14.6) also follow from the same invariances, combined with the short range of the dynamic effects. For instance, if the number of particles is conserved, we have globally $dN/dt = 0$; moreover, particles situated in a given volume element at time t remain in each other's neighbourhood at time $t + dt$. Due to this weak non-locality in the evolution, $d\varrho_N/dt$ does not vanish, but can on the macroscopic scale be expressed in the form of transfers between neighbouring volume elements. Our aim in what follows is to justify the form $-\operatorname{div} \boldsymbol{J}$ for such transfers and to construct a microscopic expression for the flux $\boldsymbol{J}$.

In the classical Lagrangian formalism (§ 2.3.3) the connection between invariance and conservation is the subject of *Noether's theorem* (Amalie Emmy Noether, Erlangen 1882–Bryn Mawr 1935), which provides us with a systematic method for constructing conserved quantities and their associated fluxes and which can be proved as follows. For a *global* conservation law we start from the fact that the action is the time-integral (2.60) of the Lagrangian $L\{q, \dot{q}, t\}$ and that it is stationary along a trajectory, and we assume that the equations of motion remain unchanged under some change in the coordinates q_k which depends on a continuous parameter. We therefore carry out an *infinitesimal* change in coordinates,

$$\delta q_k = \varepsilon g_k\{q, \dot{q}, t\}. \tag{14.40}$$

The velocities are changed from $\dot{q}_k$ to $\dot{q}_k + \delta\dot{q}_k$ where

$$\delta\dot{q}_k = \varepsilon \sum_l \left(\frac{\partial g_k}{\partial q_l} \dot{q}_l + \frac{\partial g_k}{\partial \dot{q}_l} \ddot{q}_l \right) + \varepsilon \frac{\partial g_k}{\partial t}. \tag{14.41}$$

The equations of motion of $q + \delta q$ are the same as those of q, if the Lagrangian $L\{q, \dot{q}, t\}$ remains invariant under the transformation (14.40) or, more generally, if its change under that transformation,

$$\delta L \equiv \varepsilon \sum_k \left(\frac{\partial L}{\partial q_k} \delta q_k + \frac{\partial L}{\partial \dot{q}_k} \delta\dot{q}_k \right) = \varepsilon \frac{d}{dt} \Lambda\{q, \dot{q}, t\}, \tag{14.42}$$

has the form of the total derivative with respect to the time of some function Λ; in fact, $d\Lambda/dt$ does only contribute boundary terms to the action, $\Lambda(t_1) - \Lambda(t_2)$, which do not affect the equations of motion. Using the Lagrangian equations (2.61) we find from (14.42) that

$$\frac{1}{\varepsilon} \delta L = \frac{d}{dt} \left[\sum_k \frac{\partial L}{\partial \dot{q}_k} \delta q_k \right] = \frac{d\Lambda}{dt}, \tag{14.43}$$

and, if we use (14.40), we find the conservation law

$$\sum_k \frac{\partial L}{\partial \dot{q}_k} g_k - \Lambda = \text{constant}. \tag{14.44}$$

As exercises one could check the conservation of *energy* for a time-independent Lagrangian (in which case $\delta q_k = q_k(t + \varepsilon) - q_k(t) = \varepsilon\dot{q}_k$ and $\Lambda = L$) and, in the case of a Lagrangian for interacting particles of the form,

$$L \equiv \frac{1}{2} m \sum_j \dot{r}_j^2 - \frac{1}{2} \sum_{j \neq k} W(|r_j - r_k|), \tag{14.45}$$

the conservation of *total momentum* (when $\delta r_j = a$ and $\delta L = 0$) and the conservation of *angular momentum* (when $\delta r_j = [\omega \times r_j]$ and $\delta L = 0$). In a *Galilean transformation* the positions r_j in the fixed frame become $r_j - ut$ in the frame moving with velocity u and the change in the Lagrangian,

$$\delta L = - \sum_j m(\dot{r}_j \cdot u) + \tfrac{1}{2} N m u^2, \tag{14.46}$$

is again a time derivative; the conservation law (14.44) associated with this Galilean invariance expresses that the velocity of the centre of mass remains constant.

The extension of Noether's theorem to establishing *local* conservation laws is easy in *field theory* where the action is the integral of a Lagrangian density $l(r, t)$ over time and space. For instance, in the case of a vibrating string, the Lagrangian (11.136) is given by a Lagrangian density

$$l(x, t) = \frac{\tau}{2} \left[\frac{1}{c^2} \left(\frac{\partial \varphi}{\partial t} \right)^2 - \left(\frac{\partial \varphi}{\partial x} \right)^2 \right]. \tag{14.47}$$

The dynamical q variables are now the values of the field φ at each point and the set of space-time coordinates play here the same rôle as the time did earlier. Accordingly, the equations of motion become partial differential equations, and the conservation equations which generalize (14.43) do involve not only a time derivative, but also partial derivatives with respect to the spatial coordinates; this gives them the required form (14.6). In the example (14.47), the invariance under translation in time corresponds to $\delta\varphi(x, t) = \varepsilon \partial\varphi/\partial t$, which leads to the variation $\delta l = \varepsilon dl/dt$. If we then use the equations of motion (11.134) to get an expression for this variation, as in (14.42) and (14.43), we find

$$\frac{d}{dt} \left(\frac{\partial l}{\partial \dot{\varphi}} \delta\varphi \right) + \frac{\partial}{\partial x} \left(\frac{\partial l}{\partial \nabla\varphi} \delta\varphi \right) = \varepsilon \frac{dl}{dt}, \qquad \delta\varphi = \varepsilon \dot{\varphi},$$

which gives us the energy conservation in the form (14.6); the energy density and flux can, respectively, be identified with

$$\varrho_E = \dot{\varphi} \frac{\partial l}{\partial \dot{\varphi}} - l = \frac{\tau}{2} \left[\frac{1}{c^2} \left(\frac{\partial \varphi}{\partial t} \right)^2 + \left(\frac{\partial \varphi}{\partial x} \right)^2 \right],$$

$$J_E = \frac{\partial l}{\partial \nabla\varphi} \dot{\varphi} = - \tau \frac{\partial \varphi}{\partial x} \frac{\partial \varphi}{\partial t}.$$

If the field φ is given through a probability law, these expressions should be replaced by their expectation values.

The same method, when applied to an *electromagnetic field* in vacuo, gives, for instance, the local energy conservation equation (14.6), with the expectation value of (13.15) as the energy density ϱ_E^{em} of the field, and with the expectation value of the Poynting vector (13.16) as the radiative flux $J_E^{em} = \langle N \rangle$. If, however, the electromagnetic field is coupled to particles which have a charge q, the quantities which are conserved no longer pertain solely to the field, since its energy, momentum, or angular momentum can be exchanged with those of the particles. In particular, one checks easily that the right-hand sides of the Maxwell equations (13.2) give, as in (14.3), rise to a source term in the energy balance equation for the radiation:

$$\frac{\partial \varrho_E^{em}}{\partial t} + \operatorname{div} \langle \boldsymbol{N} \rangle = - \langle (\boldsymbol{E} \cdot \boldsymbol{J}_{el}) \rangle.$$

This term, which involves the electric current density $\boldsymbol{J}_{el} = q \boldsymbol{J}_N$ of the particles, can be interpreted as work done by the latter; it occurs, of course, also in the energy balance equations for the particles themselves, with the opposite sign.

14.3.2 Microscopic Expression for the Fluxes in a Fluid

The above method cannot be applied as systematically to the problems in which we are interested. As an illustration we shall deal with the case of a *classical fluid* the dynamics of which is produced by the Lagrangian (14.45). The global conservation laws deal with the number of particles, the momentum, and the energy; we wish to associate with them local conservation laws (14.6), for which we shall try to identify the densities ϱ_N, ϱ_P, and ϱ_E, and the currents $\boldsymbol{J}_N$, $\boldsymbol{J}_P$, and $\boldsymbol{J}_E$ in terms of the $\boldsymbol{r}_j$ and $\boldsymbol{p}_j$ variables of the N particles. If we could write down the Lagrangian (14.45) as the integral of a Lagrangian density with the same properties as the Lagrangian of a field, this construction would be automatic, but this is impossible because of the finite range of the interparticle potentials. We must thus have recourse to a direct analysis.

Note, to begin with, that the dynamics of a macroscopic variable $\langle A \rangle = \operatorname{Tr} DA$, which is the average, at time t, of a function A of the coordinates $\boldsymbol{r}_j$ and the momenta $\boldsymbol{p}_j$, can be produced in two ways, corresponding in quantum mechanics to the Schrödinger and the Heisenberg pictures (§ 2.1.5). On the one hand, one can consider A to be a random variable which is fixed in time: the density in phase then changes according to the Liouville equation (2.68). On the other hand, one can keep the state fixed – and we shall do this here – and A, like the $\boldsymbol{r}_j$ and the $\boldsymbol{p}_j$, evolves according to the Hamiltonian equation $dA/dt = \{A, H\}$, or the Lagrangian equation (2.61). Note that $\boldsymbol{r}$ denotes the fixed point in space for which we are writing down the conservation law, whereas the particle coordinates $\boldsymbol{r}_j$ are functions of the time.

The local conservation (14.6) of the *number of particles* can then be checked easily by writing

$$\varrho_N(\boldsymbol{r}, t) = \left\langle \sum_j \delta^3 (\boldsymbol{r}_j - \boldsymbol{r}) \right\rangle = \int d^3 p \, f(\boldsymbol{r}, \boldsymbol{p}, t), \qquad (14.48)$$

$$\boldsymbol{J}_N(\boldsymbol{r}, t) = \left\langle \sum_j \boldsymbol{v}_j \delta^3 (\boldsymbol{r}_j - \boldsymbol{r}) \right\rangle = \int d^3 p \, \boldsymbol{v} \, f(\boldsymbol{r}, \boldsymbol{p}, t), \qquad (14.49)$$

where we have used the notation $\boldsymbol{v}_j = \dot{\boldsymbol{r}}_j = \boldsymbol{p}_j/m$ and the definition (2.81) of the reduced single-particle density $f(\boldsymbol{r}, \boldsymbol{p}, t)$. In fact, the time-derivative of ϱ_N is

$$\left\langle \sum_j \left(\dot{r}_j \cdot \nabla_j \delta^3 (r_j - r) \right) \right\rangle = -\operatorname{div}_r \left\langle \sum_j v_j \delta^3 (r_j - r) \right\rangle .$$

To justify the local *momentum* conservation we start by defining, through

$$\varrho_P(r,t) = \left\langle \sum_j p_j \, \delta^3 (r_j - r) \right\rangle = \int d^3p \, p \, f(r,p,t), \qquad (14.50)$$

the components $\varrho_{P\beta}$ ($\beta = 1, 2, 3$) of the momentum density at the point r, which is a vector. We want to express the time-derivative of $\varrho_{P\beta}$ as the divergence with respect to r of some vector, in order to prove the continuity equation (14.6) for the momentum and to construct the current $J_{P\beta}^\alpha$. We find

$$\frac{\partial}{\partial t} \varrho_{P\beta} = \left\langle \sum_{j\alpha} p_{j\beta} v_{j\alpha} \frac{\partial}{\partial r_{j\alpha}} \delta^3 (r_j - r) + \sum_j f_{j\beta} \delta^3 (r_j - r) \right\rangle , \qquad (14.51)$$

where f_j is the force,

$$f_{j\beta} = - \sum_{k \neq j} \frac{\partial}{\partial r_{j\beta}} W \left(|r_j - r_k| \right) = \sum_{k \neq j} r'_\beta \frac{1}{r'} \frac{dW}{dr'} \bigg|_{r' \equiv r_k - r_j} , \qquad (14.52)$$

exerted on the j-th particle by the other particles. The first term in (14.51) can immediately be written as a divergence with respect to r. Using (14.52), the symmetry between particles j and k, and the notation $r' \equiv r_k - r_j$, we transform the second term to read

$$\left\langle \frac{1}{2} \sum_{j,k} r'_\beta \frac{1}{r'} \frac{dW}{dr'} \left[\delta^3 (r_j - r) - \delta^3 (r_k - r) \right] \right\rangle .$$

There now remains only to express the difference between the two δ-functions as a divergence with respect to r; to do this, we use the identity

$$\delta^3 \left(r_j - r \right) - \delta^3 \left(r_j - r' - r \right) = - \int_0^1 d\lambda \, \frac{d}{d\lambda} \, \delta^3 \left(r - r_j - \lambda r' \right)$$

$$= \sum_\alpha \int_0^1 d\lambda \, r'_\alpha \frac{\partial}{\partial r_\alpha} \delta^3 \left(r - r_j - \lambda r' \right). \qquad (14.53)$$

From (14.51), (14.52), and (14.53) we then get the conservation equation

$$\frac{\partial}{\partial t} \varrho_{P\beta} + \operatorname{div} J_{P\beta} = 0, \qquad (14.54)$$

with the components of the momentum current density

$$
J_{P\beta}^{\alpha} = \left\langle \sum_j p_{j\beta} v_{j\alpha} \delta^3 (r_j - r) - \frac{1}{2} \sum_{j,k} \int_0^1 d\lambda \, \frac{r_\beta' r_\alpha'}{r'} \frac{dW}{dr'} \delta^3 (r - r_j - \lambda r') \right\rangle
$$

$$
= \int d^3p \, \frac{p_\alpha p_\beta}{m} f(r, p, t) - \frac{1}{2} \int d^3p \, d^3p' \, d^3r'
$$

$$
\times \int_0^1 d\lambda \, \frac{r_\alpha' r_\beta'}{r'} \frac{dW}{dr'} f_2 (r - \lambda r', p, r + r' - \lambda r', p').
\tag{14.55}
$$

In (14.55) we have introduced the two-particle reduced density f_2, defined by (2.82), which is equivalent to

$$
f_2 (r, p, r', p') \equiv \sum_{j \neq k} \left\langle \delta^3 (r_j - r) \delta^3 (p_j - p) \delta^3 (r_k - r') \delta^3 (p_k - p') \right\rangle.
\tag{14.56}
$$

In § 14.4.5 we shall see how J_P, which governs the change with time of the momentum density, can be interpreted in terms of the *stress tensor* in the liquid.

We can also use Noether's method to obtain the same results (14.50), (14.54), and (14.55) for the local momentum balance. Under an infinitesimal, uniform translation $\delta r_j = a$ the change δL in the Lagrangian vanishes and that invariance implies the conservation of the total momentum. Let us carry out a transformation $\delta r_j = a(r_j, t)$ where now the vector a is a field $a(r, t)$ which *depends on the position and time*. The variation of the action is an integral over r and t, with an integrand which contains terms proportional to a and its partial derivatives. We know, however, that δL vanishes when a is uniform and independent of the time so that only the terms in $\partial a / \partial t$ and $\partial a / \partial r_{j\alpha}$ can remain. The variation of the action δS must therefore necessarily have the form

$$
\delta S = \int d^3r \, dt \left[\sum_\beta \varrho_{P\beta}(r, t) \frac{\partial a_\beta(r, t)}{\partial t} + \sum_{\alpha\beta} J_{P\beta}^{\alpha} \frac{\partial a_\beta}{\partial r_\alpha} \right],
\tag{14.57}
$$

which defines the coefficients ϱ_P and J_P. The actual calculation of these, starting from (14.45), gives the same expressions (14.50) and (14.55) as above. (The evaluation of J_P involves $a(r_j) - a(r_k)$ which, as in (14.53), can be transformed into an integral of the gradient $\partial a_\beta(r, t)/\partial r_\alpha$.) It is then sufficient to integrate (14.57) by parts and after that write down that δS vanishes for any $a(r, t)$ to get the momentum conservation equation (14.54).

In expression (14.55) the force exerted by particle k on particle j has been transferred to the point r, lying between r_j and r_k, where we want to strike the momentum balance. Similarly, in order to write down the local *energy* conservation we define the energy density by distributing the potential energy over the segments connecting the particles in pairs, the point r being a barycentre of r_j and r_k. We thus write

$$\varrho_E(r,t) = \left\langle \sum_j \frac{p_j^2}{2m} \delta^3(r_j - r) + \frac{1}{2} \sum_{j,k} \int_0^1 d\lambda \ W(r') \chi'(\lambda) \delta^3 \left(r - r_j - \lambda r' \right) \right\rangle$$

$$= \int d^3p \ \frac{p^2}{2m} \ f(r,p,t) + \frac{1}{2} \int d^3p \, d^3p' \, d^3r'$$

$$\times \int_0^1 d\lambda \ W(r') \chi'(\lambda) \ f_2(r + r' - \lambda r', p, r - \lambda r', p') , \qquad (14.58)$$

where we have used the notation of (14.55). The weight $\chi' \equiv d\chi/d\lambda$ is produced by some function $\chi(\lambda)$ which increases from $\chi(0) = -\frac{1}{2}$ to $\chi(1) = \frac{1}{2}$ and which satisfies the relation $\chi(1 - \lambda) = -\chi(\lambda)$; for instance, a uniform distribution corresponds to $\chi' = 1$, $\chi = \lambda - \frac{1}{2}$, and a distribution at the two ends to $2\chi' = \delta(\lambda) + \delta(1 - \lambda)$, $\chi = 0$. Through a calculation analogous to that of J_P one can prove the equation of continuity (14.6) for the energy and construct the energy flux,

$$J_E = \left\langle \sum_j \frac{p_j^2}{2m} \ v_j \, \delta^3(r_j - r) \right.$$

$$+ \frac{1}{2} \sum_{j,k} \int_0^1 d\lambda \ W(r') \chi'(\lambda) \left(v_j - \lambda v_j + \lambda v_k \right) \delta^3 \left(r - r_j - \lambda r' \right)$$

$$\left. - \frac{1}{2} \sum_{j,k} \int_0^1 d\lambda \ \frac{r'}{r'} \frac{dW}{dr'} \left(r' \cdot \left[(\frac{1}{2} - \chi) v_j + (\frac{1}{2} + \chi) v_k \right] \right) \delta^3 \left(r - r_j - \lambda r' \right) \right\rangle$$

$$= \int d^3p \ \frac{p^2}{2m} \ v \ f(r,p,t)$$

$$+ \frac{1}{2} \int d^3p \, d^3p' \, d^3r' \int_0^1 d\lambda \ f_2(r - \lambda r', p, r + r' - \lambda r', p')$$

$$\times \left\{ W(r') \chi' \left[(1 - \lambda)v + \lambda v' \right] - \frac{r'}{r'} \frac{dW}{dr'} \left(r' \cdot \left[(\frac{1}{2} - \chi) v + (\frac{1}{2} + \chi) v' \right] \right) \right\}.$$

$$(14.59)$$

The fact that there is still some arbitrariness in the definitions of ϱ_E and J_E is connected with the fact that we had to assign to the point r properties connected with particles at the points r_j and r_k. In the case of *long-range* forces, such as Coulomb forces, the natural choice is to assign half of the potential energy to the point r_j and half to the point r_k, by taking $2\chi' = \delta(\lambda) + \delta(1 - \lambda)$, $\chi = 0$ in the energy density (14.58); however, the flux (14.59) contains even in that case contributions coming from the segment which joins r_j and r_k. Nonetheless, in the case of *short-range* forces, r' is small in (14.55), (14.58), and (14.59) so that only particles close to r contribute to the densities and the fluxes at that point. In this case f_2 is, close to equilibrium, hardly at all affected if we simultaneously translate its two arguments over a distance of order r' smaller than the range of the forces, and it therefore hardly at all depends on λ. As a result it does not matter how we choose χ and we find altogether

$$\varrho_E \simeq \int d^3p \, \frac{p^2}{2m} \, f + \frac{1}{2} \int d^3p \, d^3p' \, d^3r' \, W(r') \, f_2 \left(r + \tfrac{1}{2}r', p, r - \tfrac{1}{2}r', p' \right),$$

$$J_E \simeq \int d^3p \, \frac{p^2}{2m} \, v f + \frac{1}{2} \int d^3p \, d^3p' \, d^3r' \tag{14.60}$$

$$\times f_2 \left(r + \tfrac{1}{2}r', p + \tfrac{1}{2}p',, r - \tfrac{1}{2}r', p - \tfrac{1}{2}p' \right) \left[v \, W(r') - \frac{r'}{r'} \frac{dW}{dr'} \, (r' \cdot v) \right].$$

The conservation of *angular momentum* is connected with the invariance under rotation. The Lagrangian (14.45) is invariant under an infinitesimal global rotation $\delta r_j = [\omega \times r_j]$. As in the case of the translation in (14.57), the change in the action under a transformation $\delta r_j = [\omega(r_j, t) \times r_j]$, where the infinitesimal rotation vector ω is replaced by a field $\omega(r, t)$ which varies in space and time, has the form

$$\delta S = \int d^3r \, dt \left[\sum_\gamma \varrho_{L\gamma}(r, t) \, \frac{\partial \omega_\gamma(r, t)}{\partial t} + \sum_{\alpha\gamma} J^\alpha_{L\gamma} \, \frac{\partial \omega_\gamma}{\partial r_\alpha} \right], \tag{14.61}$$

since δL vanishes when ω is uniform and constant. After we have identified the coefficients ϱ_L and J_L in (14.61) we shall obtain the local angular momentum conservation law by requiring that δS vanishes for any choice of ω:

$$\frac{\partial}{\partial t} \varrho_{L\gamma} + \operatorname{div} J_{L\gamma} = 0. \tag{14.62}$$

However, the variation δr_j in the trajectories under a "local rotation" $\omega(r, t)$ is nothing but a special case, with $(u \cdot r) = 0$, of the variation $\delta r_j = u(r_j, t)$ under a "local translation", which produces the change (14.57) in the action; to see this we only need to identify $u(r, t)$ with $[\omega(r, t) \times r]$. As a result, we expect, in contrast to what occurs in the case of the global laws, that the *local angular momentum conservation law* (14.62) is a *consequence* of the local momentum conservation, just as the form of the densities and the fluxes which occur in it. As a matter of fact, the calculation of (14.61) follows that of (14.57) closely; the only new feature is the use of the invariance of the Lagrangian under a global rotation for writing down that the term in ω vanishes. The result can be written as

$$\delta S = \int d^3r \, dt \left[\sum_{\beta\gamma\delta} \varrho_{P\beta} \, \varepsilon_{\beta\gamma\delta} \, \frac{\partial \omega_\gamma}{\partial t} r_\delta + \sum_{\alpha\beta\gamma\delta} J^\alpha_{P\beta} \, \varepsilon_{\beta\gamma\delta} \, \frac{\partial \omega_\gamma}{\partial r_\alpha} r_\delta \right], \tag{14.63}$$

where $\varepsilon_{\alpha\beta\gamma}$ is the completely antisymmetric third rank unit tensor (14.23a). Comparing (14.61) with (14.63) we find the required expressions,

$$\varrho_{L\gamma} = \sum_{\delta\beta} \varepsilon_{\gamma\delta\beta} \, r_\delta \, \varrho_{P\beta}, \qquad J^\alpha_{L\gamma} = \sum_{\delta\beta} \varepsilon_{\gamma\delta\beta} \, r_\delta \, J^\alpha_{P\beta}, \tag{14.64}$$

without having had to use the specific form of ϱ_P or of J_P. If we use (14.64) and (14.54), we then find from the local angular momentum conservation (14.62) that

$$\sum_{\alpha\beta} \varepsilon_{\gamma\alpha\beta} J_{P\beta}^{\alpha} = 0,$$

which means that the J_P tensor is symmetric:

$$J_{P\beta}^{\alpha} = J_{P\alpha}^{\beta}. \tag{14.65}$$

Even without having evaluated J_P we could thus have expected to find a symmetric tensor if the Lagrangian is invariant under rotation. One can check this directly from (14.55).

On the other hand, a Lagrangian invariant under translation, but not under rotation, such as, for instance,

$$L = \frac{1}{2} \sum_{j\alpha} m_\alpha v_{j\alpha}^2,$$

which describes non-interacting particles with a mass tensor with different eigenvalues m_α, again gives rise to a momentum conservation law (14.54); however, in this case the current,

$$J_{P\beta}^{\alpha} = \left\langle \sum_j p_{j\beta} v_{j\alpha} \delta^3(r_j - r) \right\rangle = \int d^3p \, \frac{p_\alpha p_\beta}{m_\alpha} \, f(r, p, t),$$

is not symmetric in α and β. The fact that there is no invariance under rotation leads in this case to an extra term in (14.63) proportional to ω which remains even when ω is uniform. An example of such a situation is provided by conduction electrons in a semiconductor with an energy spectrum (11.58).

Note also that the property (14.65) is based upon the fact that the molecules which make up the fluid have no structure and are assumed to be frozen into their ground state, so that they can be modelled as interacting point particles. In the case of molecules which can rotate we must add to the Lagrangian (14.45) terms depending on the internal degrees of freedom. Their contribution to the angular momentum is small as compared to that from the motions of the molecules. However, if we want be rigorous, we note that the orientations of the molecules are coupled to their relative positions by the interactions, and invariance of the Lagrangian under rotation requires not only that we shift the molecules over $\delta r_j = [\omega \times r_j]$, but also that we let them rotate over ω around their centre of mass. This gives extra contributions to the angular momentum density and current (14.64) and involves a weak violation of the symmetry (14.65). Let us finally remember that we have only studied a *classical* liquid. For helium at low temperatures, which is a quantum fluid, the quantization of the angular momentum gives rise to particular kinds of effects which can be observed in macroscopic hydrodynamics (§ 12.3.3).

The proof of the conservation laws, starting from microscopic dynamics, is general and can be applied even to situations which are *far from local equilibrium*. In a regime where the local equilibrium is practically reached at all points and all times we can make a major simplification by taking into account that the densities ϱ_i *suffice to characterize* the state of the system. For instance, in a classical liquid in local equilibrium the distribution of the momenta p_j is practically Maxwellian and depends only on ϱ_E, ϱ_N, and ϱ_P – or on T, μ, and u; the one- and two-particle densities f and f_2 occurring in the various expressions (14.48), (14.49), (14.50), (14.55), (14.58), or (14.59) for the densities and the currents then take their equilibrium forms, which follow from the grand canonical distribution described by T, μ, and $\boldsymbol{u}$. Far from equilibrium the dynamics does not depend only on those variables and we must follow the evolution of f and f_2 for the evaluation of the densities and the currents.

We have found the microscopic expression for the currents $\boldsymbol{J}_i$ for a substance where the various quantities $i = E, N, \boldsymbol{P}$ change continuously from one point to another. For systems where exchanges occur between homogeneous macroscopic parts, the fluxes Φ_i occurring in (14.2) are expressed as the surface integrals (14.5). As a result, at the microscopic scale the fluxes Φ_i across a surface σ can be written as functions of the dynamical variables of particles situated near the surface σ, in a layer with a thickness of the order of magnitude of the range of the forces between the particles. For instance, the integration of (14.55) over a closed surface σ leads to the flux of $\boldsymbol{J}_P$ entering σ; using (14.54) we find that this flux is the time-derivative of the momentum inside σ. It thus contains, apart from a contribution associated with the momenta of the particles leaving or entering σ, the *macroscopic force* exerted by the outside on the inside. As one might have expected, (14.55) expresses this force as a surface integral of *pressure forces* applied at the points $\boldsymbol{r}$ of σ, and it shows that only particle pairs with one particle inside and the other outside σ contribute to the pressure force.

14.3.3 Relevant Entropy

Let us now look for the microscopic meaning of the conjugate macroscopic quantities $A_i(a)$ and $\gamma_i(a)$, and also that of the entropy S of non-equilibrium thermodynamics which we introduced in § 14.2.1 by extrapolating the laws of thermostatics. We shall work in the quantum framework and divide the system into discrete subsystems. These involve independent degrees of freedom and the Hilbert space of the global system is the direct product of the spaces relating to the subsystems. We shall proceed as in the theory of canonical equilibria. To do this we work at a fixed time and assume that the only quantities which are known at that time are the values of the relevant variables $A_i(a)$. To fix the ideas we can imagine that those variables represent the energies of various bodies in thermal contact. What density operator $\widehat{D}_0$ must we assign under those conditions to the system in order to make reasonable predictions about the other quantities, which are *irrelevant* at the macroscopic scale? The answer to this question comes from the *maximum statistical entropy principle* (§ 4.1.3): the least biased density operator $\widehat{D}_0$ is the one for which the uncertainty in the sense of information theory is the largest, when we take into account that the $A_i(a)$ are given.

We identify these data with the statistical expectation values of some observables $\widehat{A}_i(a)$, which are operators acting in the Hilbert space associated with the subsystem a, so that $\widehat{D}_0$ is constrained to satisfy the conditions

$$\text{Tr } \widehat{D}_0 \widehat{A}_i(a) = A_i(a). \tag{14.66}$$

If, for instance, $A_i(a)$ is an average particle number, $\widehat{A}_N(a)$ is the operator with as eigenvalues the number of particles in the volume of the subsystem a. In the case of an energy, $\widehat{A}_E(a)$ represents that part of the Hamiltonian which is associated with the region a. The complete Hamiltonian contains, apart from $\sum_a \widehat{A}_E(a)$, *coupling terms* $\widehat{V}_{ab}$ between the subsystems, with negligible expectation values. The macroscopic additivity of the energy is satisfied in this way, but the coupling terms, albeit small, will be essential for producing the fluxes and for determining the dynamics of the system.

As in §4.2.1 the maximum of $S(\widehat{D})$ under the constraints (14.66) is reached for a *generalized canonical* distribution

$$\widehat{D}_0 = \frac{1}{Z} \exp\left[- \sum_{i,a} \frac{1}{k} \gamma_i(a)\widehat{A}_i(a) \right]. \tag{14.67}$$

Anticipating the result we are looking for, we have introduced Boltzmann's constant k in the definition of the Lagrangian multipliers $\gamma_i(a)/k$. Those multipliers are determined by the conditions (14.66) which give us

$$-k \frac{\partial}{\partial \gamma_i(a)} \ln Z = A_i(a), \tag{14.68}$$

and the maximum of the statistical entropy is, according to (4.7), equal to

$$S(\widehat{D}_0) = k \ln Z - \sum_{i,a} k \, \gamma_i(a) \frac{\partial}{\partial \gamma_i(a)} \ln Z. \tag{14.69}$$

Considered as a *function of the* $A_i(a)$ *variables*, (14.69) defines the *relevant entropy* relating to those variables (Exerc.3c): it characterizes the *missing information when we know only the macroscopic variables* $A_i(a)$. Mathematically speaking, changing through (14.69) from $k \ln Z$, as function of the γ, to $S(\widehat{D}_0)$, as function of the A, is a Legendre transformation (§6.3.1). The symmetry of this transformation implies that (14.68) is equivalent to

$$\frac{\partial}{\partial A_i(a)} S(\widehat{D}_0) = \gamma_i(a). \tag{14.70}$$

When the $A_i(a)$ variables are the constants of motion of the subsystem a, Z can be factorized into contributions associated each with a subsystem, and calculated exactly as in equilibrium. As a result, the statistical entropy (14.69), the sum of the thermostatic entropies of the various parts, *can be identified with the thermodynamic entropy* of §14.2.1. A comparison of (14.70) with (14.9) then enables us to identify the *multipliers* $\gamma_i(a)$ with the macroscopic *local intensive variables*.

The microscopic justification which we have just given for the local equations of state is a general one and can be applied to any set of observables $\widehat{A}_i(a)$. *At any time* it associates with the $A_i(a)$ variables a set of conjugated $\gamma_i(a)$ variables as well as a relevant entropy. *The latter depends on the choice of variables*; it becomes smaller when we enlarge this choice, as in that case the system is better determined. At the same time we have solved a *statistical inference* problem: what expectation value should we assign to an irrelevant quantity $\widehat{B}$, if we have the $A_i(a)$ as data? The answer is: $\mathrm{Tr}\,\widehat{D}_0\widehat{B}$.

Even though it provides us with the equations of state of §14.2.1 this construction remains a formal one. The $\gamma_i(a,t)$ variables are defined at all times starting from the $A_i(a,t)$ through a definite procedure based upon information theory. However, their relevance to dynamics is not yet clear: we must still close the thermodynamic equations by completing the square of Fig.14.1 and show by using a microscopic theory how the $\gamma_i(a,t)$ determine the fluxes and, hence, the dynamics of the $A_i(a,t)$.

14.3.4 Macroscopic, Microscopic, and Mesoscopic Descriptions

On a *macroscopic* scale the evolution is characterized by the changes in the $A_i(a,t)$ quantities. On a *microscopic* scale it is governed, in the case when the system is isolated, by the Liouville-von Neumann equation (2.49) for the density operator $\widehat{D}(t)$, completed by assigning to the system a density operator $\widehat{D}(0)$ at the initial time $t=0$. The solution of this problem is given by (2.27) and (2.50), that is,

$$\widehat{D}(t) \;=\; \mathrm{e}^{-\mathrm{i}\widehat{H}t/\hbar}\,\widehat{D}(0)\,\mathrm{e}^{\mathrm{i}\widehat{H}t/\hbar}. \tag{14.71}$$

In general, the initial data are the values of the relevant macroscopic $A_i(a,0)$ variables. As a result we must choose for $\widehat{D}(0) = \widehat{D}_0(0)$ the canonical form (14.67) and use (14.70) to determine the multipliers $\gamma_i(a,0)$ in it in terms of the $A_i(a,0)$.

With increasing t there is no reason that (14.71) should retain the special generalized canonical form (14.67). Nevertheless we shall now see that a density operator $\widehat{D}_0(t)$ of that form appears naturally. The macroscopic quantities $A_i(a,t)$ can be derived from $\widehat{D}(t)$ through

$$A_i(a,t) \;=\; \mathrm{Tr}\,\widehat{D}(t)\,\widehat{A}_i(a). \tag{14.72}$$

Let us assume that at time t *we throw out all information except that contained in* $A_i(a,t)$; using the procedure of §14.3.3 we should then assign to the system a density operator $\widehat{D}_0(t)$ of the form (14.67), with parameters $\gamma_i(a,t)$ characterized by the data

$$A_i(a,t) = \text{Tr } \widehat{D}_0(t) \, \widehat{A}_i(a). \tag{14.73}$$

This leads us to associate with $\widehat{D}(t)$, which evolves according to (14.71), a density operator $\widehat{D}_0(t)$ *which accompanies it* all the time while *retaining a canonical form*. The state $\widehat{D}_0(t)$ defines the *mesoscopic description* of the system, which is intermediate between the macroscopic and the microscopic descriptions, and which is also called the *contracted or reduced description*. It has the interest of involving directly the thermodynamic variables $\gamma_i(a,t)$.

As far as the macroscopic quantities $\widehat{A}_i(a)$ are concerned, the three descriptions are equivalent, since (14.72) and (14.73) are satisfied. The macroscopic description gives no information about other quantities $\widehat{B}$ which are irrelevant on the macroscopic scale, whereas $\langle B \rangle$ can be evaluated as well from $\widehat{D}(t)$ as from $\widehat{D}_0(t)$. However, the result is not the same. In fact, according to the definition of $\widehat{D}_0(t)$ itself, the mesoscopic description gives for $\langle B \rangle$ predictions which are the least biased ones that we can make at time t by induction from the values of the $A_i(a,t)$ *at the same time*. The mesoscopic density operator $\widehat{D}_0(t)$ thus takes into account only the *instantaneous macroscopic information*. In contrast, the microscopic description through $\widehat{D}(t)$ *takes the history of the system into account*, as the mean value of $\langle B \rangle$ which it gives depends not only on the $A_i(a,t)$, but also on the $A_i(a,0)$ through (14.71).

A statistical entropy balance shows that $\widehat{D}(t)$ contains more information than $\widehat{D}_0(t)$. In fact, we have seen in § 3.2.3 that $S[\widehat{D}(t)]$ *remains constant* in time when $\widehat{D}(t)$ evolves according to (14.71): the information existing initially about the $\widehat{A}_i(a)$ is transferred to other variables, but its total remains constant. In contrast, the mesoscopic description retains the full information about the $\widehat{A}_i(a)$, but it loses all information about the other quantities. This loss can be expressed by the inequality

$$S[\widehat{D}_0(t)] > S[\widehat{D}(t)] = S[\widehat{D}_0(0)], \tag{14.74}$$

which is a consequence of the definition itself of $\widehat{D}_0$ through the maximum entropy principle. As $S(\widehat{D}_0)$ can be identified with the macroscopic entropy, the *dissipation* is equivalent to a *transfer of information* to irrelevant degrees of freedom, where this information becomes *inaccessible*.

We now have to face the problem in the following terms. We want to obtain the equations of motion for the $A_i(a,t)$ or, what amounts to the same, for the mesoscopic density operator $\widehat{D}_0(t)$ which through (14.67) and (14.70) is characterized by giving the $A_i(a,t)$. To do this we have available the microscopic Liouville-von Neumann equation which governs the evolution of

$$\widehat{D}(t) \equiv \widehat{D}_0(t) + \widehat{D}_1(t). \tag{14.75}$$

However, $\widehat{D}(t)$, which contains all the information about the evolution of the microscopic quantities, is too detailed to be manageable. We therefore want to *eliminate* its irrelevant part $\widehat{D}_1(t)$ in order to write down the evolution of

$\widehat{D}_0(t)$ only. This procedure, which is called the reduction or *contraction of the description*, is the object of the *projection method*,[1] which is a powerful tool of non-equilibrium statistical physics; we shall be satisfied to just sketch the basic ideas of this method.

14.3.5 The Projection Method

Our final aim is to justify the macroscopic equations of motion. In particular, the theory should show how the reduction of the description through the elimination of $\widehat{D}_1$ has as a result a *qualitative* change in the dynamics. The origin of some of the differences between the microscopic and macroscopic scales are easy to understand. The *probabilistic* nature of microphysics is not obvious at our scale if the statistical fluctuations of the $\widehat{A}_i(a)$ remain small at all times. This can be seen rather generally for time-dependent macroscopic systems, as in thermostatics; a notable exception is *turbulence* where small initial fluctuations can with time develop to such an extent that they prevent deterministic predictions on a macroscopic scale. The *non-linearity* of thermodynamics, which is in contrast to the linearity of the Liouville-von Neumann equation, comes from that of the relations between the A_i and the γ_i. Other differences, however, are less simple to explain. The *irreversible* nature of the macroscopic equations is in contrast to the *invariance* of the underlying Liouville-von Neumann equations under time reversal. Moreover, the macroscopic motion in a quasi-equilibrium regime is *Markovian*, that is, $A_i(a, t + dt)$ depends only on the $A_j(a,t)$; in fact, it is given by differential equations which are of first order in t, and the system retains no memory of the past, except through the conservation laws. In contrast, the form of (14.71) shows that the elimination of $\widehat{D}_1$ gives us a mesoscopic density operator $\widehat{D}_0(t + dt)$ which depends *a priori* not only on $\widehat{D}_0(t)$, but also on the *previous history* of the system which is memorized in $\widehat{D}_1(t)$. Statistical physics must thus show that in the limit where the macroscopic processes are slow and where the systems are large, this *memory effect disappears*; in other words, the information lost between times 0 and t to the irrelevant variables does not affect the later macroscopic evolution.

In order to get an idea how statistical mechanics approaches this problem we must work in the Liouville representation which was sketched in §§ 2.1.7 and 2.2.7. It consists in considering the matrix elements A_{mn} (or D_{nm}) of the observables $\widehat{A}$ (or of the states $\widehat{D}$) as the coordinates of a *vector* characterized by the *pair of indices* m, n. This enables us to introduce "operators in Liouville space" or

---

[1] S. Nakajima, Progr. Theor. Phys. **20** (1958) 948; R. Zwanzig, J. Chem. Phys. **33** (1960) 1338; Ann. Rev. Phys. Chem. **16** (1965) 67; H. Mori, Progr. Theor. Phys. **33** (1965) 423; H. Grabert, *Projection Operator Techniques in Nonequilibrium Statistical Mechanics*, Springer Tracts in Modern Physics, Vol. 95 (1982); R. Balian, Y.Alhassid, and H.Reinhardt, Phys. Repts. **131** (1986) 1.

"superoperators" which linearly transform the observables into one another (or the states into one another) by operating on the pair of indices. For instance, the operation which, for a given $\widehat{H}$, associates $[\widehat{H}, \widehat{D}]/i\hbar$ with $\widehat{D}$ can be thought of as the action $\mathcal{L}\widehat{D}$ of the *Liouvillian* $\mathcal{L}$ on $\widehat{D}$, so that we can write the Liouville-von Neumann equation as

$$\frac{d\widehat{D}}{dt} = \mathcal{L}\widehat{D} \equiv \frac{1}{i\hbar} [\widehat{H}, \widehat{D}]; \tag{14.76}$$

the superoperator $\mathcal{L}$ has two pairs of indices and operates as follows:

$$[\mathcal{L}\widehat{D}]_{mn} = \sum_{m'n'} \mathcal{L}_{mn,m'n'} D_{n'm'},$$

$$\mathcal{L}_{mn,m'n'} = \frac{1}{i\hbar} \left(H_{mn'}\delta_{m'n} - \delta_{mn'}H_{m'n} \right).$$

In this language an expectation value $\operatorname{Tr} \widehat{D}\widehat{A} = \sum_{mn} A_{mn}D_{nm}$ occurs as the scalar product of two vectors with coordinates A_{mn} and D_{nm}. The relation which associates a density operator $\widehat{D}_0$ with each $\widehat{D}$ according to the rules given in § 14.3.4 can in the Liouville representation be represented as a Nakajima-Zwanzig *projection*

$$\widehat{D}_0 = \mathcal{P}\widehat{D}, \qquad \widehat{D}_1 = \mathcal{Q}\widehat{D}, \tag{14.77}$$

where the projector $\mathcal{P}$ and also its complementary projector $\mathcal{Q}$ are superoperators. We shall not need to write down their explicit form, which clearly depends on the choice of the observables $\widehat{A}_i(a)$ as well as on $\widehat{D}_0$, and thus on the time.

The evolution of $\widehat{D}_0$ and of $\widehat{D}_1$ is, according to (14.76) and (14.77), governed by coupled equations. The equation for $\widehat{D}_0$ is equivalent to the Ehrenfest equations for the relevant variables,

$$\frac{dA_i(a,t)}{dt} = \frac{1}{i\hbar} \operatorname{Tr} [\widehat{H}, \widehat{D}_0] \widehat{A}_i(a) + \frac{1}{i\hbar} [\widehat{H}, \widehat{D}_1] \widehat{A}_i(a), \tag{14.78}$$

and we can write the equation for $\widehat{D}_1$ as

$$\frac{d\widehat{D}_1}{dt} - \mathcal{Q}\mathcal{L}\mathcal{Q}\widehat{D}_1 = \mathcal{Q}\left(\mathcal{L}\mathcal{P}\widehat{D}_0 - \frac{d\widehat{D}_0}{dt} \right) \equiv \widehat{\Delta}. \tag{14.79}$$

Let us for the sake of simplicity assume in the following that all the operators $\widehat{A}_i(a)$ commute with one another. The first term of (14.78) then vanishes, since in that case $\widehat{D}_0$ commutes with the $\widehat{A}_i(a)$. In the second term all parts $\widehat{H}_b$ of $\widehat{H}$ commute with $\widehat{A}_i(a)$ and hence do not contribute; we can thus replace $\widehat{H}$ by $\sum_{a>b} \widehat{V}_{ab}$, where $\widehat{V}_{ab}$ is the interaction between the subsystems a and b. As a result we can identify (14.78) with the conservation equation (14.2) and we now get a microscopic expression,

$$\Phi_i(a \rightarrow b) = \frac{1}{i\hbar} \operatorname{Tr} \widehat{D}_1 \left[\widehat{V}_{ab}, \widehat{A}_i(a) \right], \tag{14.80}$$

for the *fluxes*. Using (14.79) to calculate $\widehat{D}_1$ then will give us the fluxes. Since the right-hand side $\widehat{\Delta}$ of Eq.(14.79) itself depends on the fluxes through $d\widehat{D}_0/dt$ we should, in principle, proceed iteratively. We shall not give these calculations here, but we shall in Chap.15 study a similar example, the Chapman-Enskog method. Here we shall restrict ourselves to a formal solution, exhibiting the qualitative features of the results. Noting that (14.79) is a linear equation for $\widehat{D}_1$ we introduce its resolvent kernel $\mathcal{W}(t, t')$; this is the superoperator which is defined for $t > t'$ as the solution of Eq.(14.79) without the right-hand side and with the initial condition $\mathcal{W}(t, t) = \mathcal{Q}(t)$. The right-hand side $\widehat{\Delta}(t)$, defined by (14.79), is proportional to the deviation of $d\widehat{D}_0/dt$ as compared to the Liouville-von Neumann equation; it can be expressed as a function of the intensive or extensive macroscopic variables $\gamma_i(a, t)$ or $A_i(a, t)$. We thus obtain the solution of (14.79) with the initial condition $\widehat{D}_1(0) = 0$ as

$$\widehat{D}_1(t) = \int_0^t dt' \, \mathcal{W}(t, t') \, \widehat{\Delta}(t'). \tag{14.81}$$

We have altogether formally achieved the required elimination of $\widehat{D}_1$, and (14.80) and (14.81) express the *fluxes at time t as functions of the $\gamma_i(a, t')$ variables at earlier times t'*. As we have not made any approximations we have not yet obtained the equations of thermodynamics where the fluxes depend on the affinities, and hence on the γ, *at the same time*.

The superoperator $\mathcal{W}$, called the *memory kernel*, describes, according to (14.79), an evolution produced by the effective Liouvillian $\mathcal{QLQ}$ which operates in the space spanned by the projector $\mathcal{Q}$. According to (14.77) it is, like $\widehat{D}_1$, associated with the irrelevant quantities, whereas $\mathcal{P}$ and $\widehat{D}_0$ are associated with the relevant quantities, the $A_i(a)$. As a result, $\mathcal{W}$ characterizes the *evolution of the irrelevant variables*. According to Eqs.(14.80) and (14.81) the relevant variables are at all times $t' < t$ coupled to the others through $\widehat{\Delta}(t')$; the latter, irrelevant variables, evolve from t' to t and are then recoupled back to the relevant quantities through the fluxes (14.80). The macroscopic equations of motion (14.80) and (14.81) thus describe a *retarded transfer of information* amongst the relevant variables, carried out through their coupling with the other variables, which evolve according to the memory kernel $\mathcal{W}$.

If we have made an adequate choice of relevant variables $\widehat{A}_i(a)$ we expect that the remaining, microscopic, non-conservative variables change very rapidly with time. We shall show that this enables $\mathcal{W}$ to be significant only during a short time interval $t - t'$. This kernel, which is a solution of Eq.(14.79) without its right-hand side, is a sum of exponentials $e^{i\omega(t-t')}$ associated with the different eigenvalues of $\mathcal{QLQ}$ – we neglect here the slower t-dependence of $\mathcal{Q}$. The frequencies ω which contribute to (14.81) are very numerous and they can be replaced by a continuum for a macroscopic system. Moreover, their characteristic values Ω are very large as $\mathcal{W}$ describes microscopic motions. These features are illustrated by the example

$$\int d\omega \, e^{i\omega(t-t')} \frac{1}{\omega^2 + \Omega^2} = \frac{\pi}{\Omega} \, e^{-\Omega|t-t'|},$$

which enables us to understand that the characteristic memory time $1/\Omega$ of $\mathcal{W}$ is of the order of the characteristic microscopic periods (see also the note in Exerc.14b).

Let us therefore assume that our system has a *short memory*, that us, that the contribution to the integral (14.81) is concentrated in a time interval of t'-values close to t, where $\widehat{\Delta}(t')$ *is close to* $\widehat{\Delta}(t)$. With this approximation we find that

$$\widehat{D}_1(t) = \mathcal{K}(t)\,\widehat{\Delta}(t), \qquad \mathcal{K} \equiv \int_0^t dt'\, \mathcal{W}(t,t'), \tag{14.82}$$

where the integral which defines the superoperator $\mathcal{K}$ is only over the neighbourhood of t. Equations (14.80) and (14.82) then determine the fluxes solely from the *macroscopic variables* $\gamma_i(a,t)$ *at time* t.

In this way statistical mechanics has justified the scheme of thermodynamics close to equilibrium provided the short-memory approximation is valid for the kernel $\mathcal{W}$. This property is connected with the possibility to separate the variables into two families, the relevant variables which evolve slowly and the others with short characteristic times. In fact, the existence of a quasi-equilibrium regime involving *two time scales*, a microscopic one characterizing the *approach to local equilibrium* for the irrelevant variables, and a macroscopic one governing the *changes in the macroscopic quantities* $A_i(a)$, is the result of the *conservation laws*. Those, in fact, force the $A_i(a)$ variables to evolve only under the influence of the macroscopic fluxes from one subsystem to another. Such transfers are hindered by the weakness of the couplings, a *bottleneck* which makes the characteristic times for the evolution of the A_i macroscopic.

In the projection method it is interesting to identify the respective rôles of the two parts, $\widehat{D}_0$ and $\widehat{D}_1$, of the density operator. Local equilibrium is described by $\widehat{D}_0$, which is the dominant contribution to $\widehat{D}$. However, the deviation $\widehat{D}_1$ from the local equilibrium is, according to (14.80), solely responsible for the macroscopic dynamics and the approach to global equilibrium of an isolated system. This deviation is itself determined by (14.82) in the case of a short-memory kernel, and $\widehat{D}_1(t)$ then depends only on $\widehat{D}_0(t)$, independently of the initial conditions, which it soon forgets. The very processes involved in the approach to *global* equilibrium thus prevent the establishing at all times of an exact *local* equilibrium, for which we would have $\widehat{D}_1(t) = 0$. In return, this small *local disequilibrium* $\widehat{D}_1$ *is necessary in order to instigate exchanges* between the subsystems. The small deviation $\widehat{D}_1$ between the microscopic description $\widehat{D}$ and the mesoscopic description $\widehat{D}_0$ is therefore essential since it *determines the fluxes while being completely determined by* $\widehat{D}_0$, even though the local equilibrium state is characterized by $\widehat{D}_0$ alone. We shall again encounter these rather subtle features in the Chapman-Enskog method in Chap.15.

Altogether, we have achieved our goals: (i) to show how from microphysics one can derive the equations which relate the various thermodynamic quantities to one another according to the scheme of Fig.14.1 and § 14.2.6, (ii) to understand what approximations are involved in the justification of thermodynamics, and (iii) to find, at least formally, explicit microscopic expressions for the thermodynamic equations. The conservation laws follow exactly, either from §§ 14.3.1 and 14.3.2 or from (14.78). The equations of state also can be derived exactly using the approach of § 14.3.3, but the latter should rather be regarded as a microscopic definition of the $\gamma_i(a)$ variables and of the thermodynamic entropy $S(\widehat{D}_0)$. The only approximation made has been the replacement of the exact, retarded equation (14.81) by the instantaneous equation (14.82); substituting this result into (14.80) provides a formal expression for the fluxes as functions of the $\gamma_i(a)$ variables, whereas the exact expressions (14.80) and (14.81) depend in principle on their past.

The closed set of equations thus obtained have the form expected from macroscopic phenomenology. They are valid in any *quasi-equilibrium regime*, which microscopically means a *short memory* for the kernel W describing the motion of the microscopic, irrelevant variables other than the $A_i(a)$. If we are sufficiently close to equilibrium we can make a second approximation by expanding (14.80) and (14.82) in powers of the affinities. Thus, if the *fluxes are sufficiently small* linearization gives us a microscopic basis for thermodynamics in the *linear regime*. The expressions which we obtain in this way for the response coefficients L should enable us to prove their properties which we postulated in §§ 14.2.4 and 14.2.5, especially the Onsager relations and the Clausius-Duhem inequalities. We shall not carry out this general task, but we shall work out the above ideas in Chap.15 for the special case of the kinetic theory of gases. Note that, starting from the microscopic dynamics (14.76) which is *invariant under time reversal* provided $\widehat{H}^{T} = \widehat{H}$, we have been able to find *irreversible* macroscopic equations. The key to this qualitative change is provided by the short-memory property of W which, through (14.82), has made it possible to close the macroscopic equations of motion; the fact that $\widehat{D}_1(t)$ then depends only on $\widehat{D}_0(t)$ is reflected by an irretrievable loss of information to the irrelevant variables.

Let us, in conclusion, stress the generality and the flexibility of the contraction of the description and the projection method which implements it. The choice of the relevant quantities is, in fact, *a priori* left to our own discretion. With each choice we associate a relevant entropy and get different dynamic equations which are more or less detailed, depending on the number of relevant variables which we have kept (§ 15.4.4). The most adequate choice is that which separates the slow from the fast variables; the short-memory approximation, which is necessary to simplify the equations of motion, is then legitimate.

14.4 Applications

In order to illustrate the general methods of thermodynamics (§ 14.2.6), complemented by hints from statistical physics, we shall analyze a few simple examples: diffusion, conduction, hydrodynamics. The study of most substances can be approached in a similar way with a few complications which are principally due to the larger number of macroscopic variables.

14.4.1 Particle and Heat Diffusion

Various problems can be modelled as follows. An inert, fixed medium contains particles which can move in it by colliding with the constituents of the medium. The system studied, which disregards the scattering substratum, consists solely of these mobile particles, the density of which may change from one point to another. We assume that the collisions are elastic so that the energy of the moving particles is conserved, but not their momentum. This assumption is justified provided the particles are much lighter than the centres by which they are scattered.

Such a situation occurs in the case of *neutron diffusion* in the various substances which make up a nuclear reactor: the density of these neutrons is so small – only 10^7cm^{-3} in the core of the reactor – that they can be treated as a classical gas without interactions; however, they collide with the nuclei of the material they pass through, which are much heavier (§ 15.2.5, Prob.17). The diffusion of *impurities in a solid*, which in general is slow but which can become appreciable at high temperatures (Exerc.15a, Prob.19), is a similar effect. In the case of a liquid solvent the model represents, provided this solvent is modelled as being fixed, diffusion of the *molecules of a solute*, which tends to make their density homogeneous. The *conduction of electricity and of heat* falls into the same framework (§§ 11.3 and 15.2, Prob.18). In that case the charge and energy carriers, namely, electrons and holes in semiconductors, electrons in metals, interact with the *crystal lattice imperfections*, which arise either from the displacements of nuclei from their equilibrium positions (phonons), from crystal defects, or from impurities. The existence of long-range Coulomb interactions, however, complicates matters and we shall discuss this in § 14.4.2. If the inertia of the substrate is small, which often is the case for neutral or ionic solutions, it easily collects energy and plays the rôle of a heat bath, so that the following considerations would be valid in such cases only for uniform T.

In the general case the macroscopic state of our system is characterized by giving locally two macroscopic conservative variables, the *energy* and the *number of particles*. (For a semiconductor we must introduce both the number of conduction electrons and the number of holes.) They are associated with two local intensive variables, $\gamma_E = 1/T$ and $\gamma_N = -\mu/T$, and two fluxes, J_E and J_N. *A priori*, the response coefficients form four 3×3 matrices, but invariance under rotation (14.22) reduces them to four scalars. We wrote down in (14.25) the transport equations connecting the fluxes and the affinities and we have seen that Onsager's relations give us the equality, (14.26), of L_{EN} and L_{NE}. This reduces the number of independent coefficients to three. Finally, the Clausius-Duhem principle (14.32) can be expressed through the *inequalities*

$$L_{NN} \geq 0, \qquad L_{EE} L_{NN} \geq \left(L_{EN}\right)^2. \tag{14.83}$$

The *diffusion coefficient* D is defined by (14.20) for processes where the temperature is uniform; this connects it, according to (14.25b), to L_{NN} through the relation

$$D = \frac{1}{T} L_{NN} \left. \frac{\partial \mu}{\partial \varrho_N} \right|_T, \tag{14.84}$$

or through (14.21). If the diffusing particles behave as a classical perfect gas, their density ϱ_N depends on μ only through the factor $e^{\mu/kT}$ of (7.32). As a result, the gradients of the density and of the chemical potential are, for constant T, related to each other through

$$\nabla \mu = \frac{kT}{\varrho_N} \nabla \varrho_N, \tag{14.85}$$

and (14.84) reduces to

$$D = \frac{k}{\varrho_N} L_{NN}. \tag{14.86}$$

More generally, it follows from (4.35) and (4.38) that $\partial \varrho_N / \partial \mu$ is always positive so that (14.83) and (14.84) imply that D is positive.

The *thermal conductivity* is defined by Fourier's empirical law,

$$\boldsymbol{J}_E = -\lambda \nabla T, \qquad \boldsymbol{J}_N = 0, \tag{14.87}$$

where it is important to note that the *mean particle flux vanishes*. On the microscopic scale the number of particles passing through a surface element in one direction or the opposite one are equal, but the particles moving in one direction carry more energy than those moving in the opposite one. The equation of continuity for the particle number then implies that the densities ϱ_N remain constant with time at each point during the process (14.87). It follows from (14.25b) that the chemical potential varies in space in order to prevent a flux $\boldsymbol{J}_N$ to occur:

$$0 = L_{NE} \nabla \left(\frac{1}{T} \right) + L_{NN} \nabla \left(-\frac{\mu}{T} \right), \tag{14.88}$$

and this variation determines that of ϱ_N. Eliminating $\nabla \mu$ between (14.25a) and (14.88) enables us to express $\boldsymbol{J}_E$ in terms of $\nabla(1/T)$ and, hence, to express the conductivity (14.87) in terms of the transport coefficients L,

$$\lambda = \frac{1}{T^2 L_{NN}} \left(L_{EE} L_{NN} - L_{EN}^2 \right). \tag{14.89}$$

It follows from (14.83) that λ is positive.

The fact that heat transport may here be accompanied by transport of particles is reflected in the difference between the expressions (14.18) and (14.89) for the conductivity. The first expression is adapted for an insulating solid where heat on the microscopic scale corresponds to lattice vibrations – or to motion of phonons, but as their number is not conserved there is no associated ϱ_N density. On the other hand, in a metal the, much more efficient, heat conduction is produced by the motion of the electrons and it satisfies (14.89).

We have indicated in § 14.2.4 that the coefficients L_{EN} and L_{NE} govern indirect thermo-mechanical effects which are symmetric with respect to one another: the flux of heat produced by a gradient of the chemical potential, or pressure, at uniform temperature, and the flux of particles produced by a temperature gradient. These effects, which are easier to observe in a charged substance (§ 14.4.3), are related to one another through the Onsager relation $L_{EN} = L_{NE}$.

14.4.2 Electrodynamics

When the moving particles are charged, the above ideas can be adapted to electric transport. In order to simplify the discussion let us assume that there is a single kind of carriers with a charge q, which is equal to $-e = -1.6 \times 10^{-19}$ C for the electrons in a metal or a strongly doped n-type semiconductor, or to $+e$ for the holes in a strongly doped p-type semiconductor. In the case of an ionic solution or of an intrinsic semiconductor we should add the contributions of the different carriers. The electric charge and current densities, $\varrho_{el} = q\varrho_N$, $J_{el} = qJ_N$, are related to each other through the conservation law of carriers. We note that we are dealing with macroscopic quantities and that their microscopic definition makes it necessary to use averages, as we saw in § 11.3.3 in the case of the electrostatic charge density. The long range of the forces on the microscopic scale implies that on our scale there still remains a *macroscopic electric potential* $\Phi(\boldsymbol{r})$ produced by the averaged charges; its expression as function of ϱ_{el} and of the external sources was given in § 11.3.3. In turn we must add the self-consistent macroscopic potential $q\Phi(\boldsymbol{r})$ to the single-particle Hamiltonian of the carriers. We shall take into account the force $q\boldsymbol{E}$ which is produced in this way. However, we shall disregard magnetic effects, neglecting the $\boldsymbol{B}$ field. We shall also ignore the dynamical effects associated with the polarization of the medium; in order to take these into account we would need to introduce a polarization density in addition to the two densities ϱ_E and ϱ_N to which we restrict ourselves.

The *electric conductivity* σ is, for a substance which is at a uniform temperature and macroscopically neutral, defined by *Ohm's empirical law*

$$J_{el} = \sigma \boldsymbol{E} = -\sigma \nabla \Phi. \tag{14.90}$$

On the microscopic scale one also defines the *electrical mobility* (11.72), which is the mean drift velocity acquired by the carriers in unit field. This quantity, μ_{el}, is directly related to the conductivity through the equation

$$\sigma = q\varrho_N \mu_{el}. \tag{14.91}$$

It is customary to use the same symbol μ for the mobility and for the chemical potential; we distinguish them through the index for μ_{el}. As we did in the case of the diffusion coefficient, we want to connect the conductivity to the coefficient L_{NN} of (14.25). To do this, we must analyze the effect of the presence of the potential $\Phi(r)$ on the equations of thermodynamics, and especially the spatial variations produced in the chemical potential μ. We shall proceed in the same spirit as in § 14.3.2 where we studied the local symmetry properties. We first exhibit the *invariance* associated with the *addition of a constant* $q\Phi$ to the single-particle Hamiltonian; afterwards we shall *let Φ vary* in space. In thermostatics the entropy density $s(\varrho_E, \varrho_N, \Phi)$ has an evident invariance property: it is connected to $s(\varrho_E, \varrho_N)$ when there is no potential through

$$s(\varrho_E, \varrho_N, \Phi) = s(\varrho'_E, \varrho_N), \qquad \varrho_E \equiv \varrho'_E + q\Phi\varrho_N. \tag{14.92}$$

As a result we see that the local equation of state (14.12) which defines μ depends as follows on the electric potential Φ:

$$\mu \equiv \mu(\varrho_E, \varrho_N, \Phi) = -T \frac{\partial}{\partial \varrho_N} s(\varrho_E, \varrho_N, \Phi)$$

$$= \mu(\varrho'_E, \varrho_N) + q\Phi. \tag{14.93}$$

This result was microscopically obvious in the grand canonical ensemble where nothing is changed when we simultaneously shift the single-particle energies and the chemical potential by $q\Phi$. It remains valid *at each point* for the *local* chemical potential, provided the electric potential $\Phi(r)$ changes slowly on the microscopic scale. The first term $\mu' \equiv \mu(\varrho'_E, \varrho_N)$ in (14.93) depends only on the local temperature and on the local density, but *not on the potential Φ*. For instance, if the carriers behave as a classical gas – semiconductors – the reduced chemical potential μ' varies with ϱ_N according to (14.85) whatever the changes in $\Phi(r)$.

This result has important consequences. The general theory indicates that *whatever its physical origin, a change in μ produces a flux*, equal at uniform temperature to

$$\boldsymbol{J}_N = -\frac{1}{T} L_{NN} \nabla\mu.$$

We shall see later that L_{NN} is independent of the potential $\Phi(r)$. Using (14.93) we obtain the flux produced, at uniform temperature, by *simultaneous changes in the density and the potential*, in the form

$$\boldsymbol{J}_N = -\frac{1}{T} L_{NN} \left.\frac{\partial\mu'}{\partial\varrho_N}\right|_T \nabla\varrho_N + \frac{q}{T} L_{NN} \boldsymbol{E}. \tag{14.94}$$

We have thus split the total particle flux into two contributions. The first one represents the *diffusion current*, the physical interpretation of which is the tendency of the system to make its density ϱ_N uniform. The second one represents the *current produced by the forces applied* to the volume element d^3r which would, in the absence of a scattering medium, tend to accelerate the particles into the direction $q\boldsymbol{E}$; the friction by the medium leads to a final drift velocity $\mu_{el}\boldsymbol{E}$. These two effects are governed by the *same microscopic mechanism*, which is reflected in that there occurs only a single response coefficient in (14.94). A comparison of the first term of (14.94) with Fick's law (14.20) has already provided us with the relation (14.84) between the diffusion coefficient D and L_{NN}. On the other hand, the electrical conductivity (14.90) is defined in a neutral substance, which implies a uniform carrier density ϱ_N, so that it is given by the second term in (14.94). Hence, we find for the conductivity σ

$$\sigma = \frac{q^2}{T} L_{NN}. \tag{14.95}$$

Diffusion and *conduction* are thus *related phenomena* since they are governed, according to (14.84) and (14.95), by the same response coefficient, L_{NN}: different causes produce the same effects, if the deviation from equilibrium which they bring about is *measured by the same* $\nabla\mu$. In the first case, when there are no applied forces, Φ vanishes and $\nabla\mu$ is given as function of the density variations by (14.85) for a perfect gas; in the second case the density is uniform and we get $\nabla\mu = q\nabla\Phi$ from (14.93). If the carriers constitute a classical gas this connection is reflected by *Einstein's relation* (1905) between the diffusion coefficient and the mobility,

$$D = \frac{kT}{\varrho_N q^2} \sigma = \frac{kT}{q} \mu_{el}, \tag{14.96}$$

where we have used (14.86), (14.91), and (14.95).

In the context of electromagnetism μ/q is often called the *electrochemical potential*. However, one should take care not to confuse it with the electrical potential Φ with which it is connected through (14.93). In particular, in a substance in electrostatic equilibrium (§ 11.3.3) the current vanishes, μ is uniform, but this is not true for Φ; the density ϱ_N follows the changes in Φ according to (14.93) which, if we take (14.85) into account, implies for a classical gas that

$$\varrho_N(\boldsymbol{r}) \propto \exp \frac{\mu - q\Phi(\boldsymbol{r})}{kT},$$

in agreement with (7.32). This equilibrium state can, if we use (14.94), be interpreted as the result of the cancellation of the diffusion current by the current which would be induced by the electrostatic forces, if they were operating in a homogeneous medium. Einstein used such a balance argument to derive the relation (14.96). More generally, if $\mu(\boldsymbol{r})$ changes from one point to another, we can interpret $q\mu(\boldsymbol{r}) - q\mu(\boldsymbol{r}')$ as the *electromotive force* which tends to create a current from r to r'. If there is a macroscopic charge density present, or a temperature gradient, this electromotive force is different from the potential difference $q\Phi(r) - q\Phi(r')$ so that Ohm's law (14.90) is no longer valid; we must go back to the general response equations (14.25) together with (14.93). Such situations often occur in semiconductors, but not in a metal where the screening confines the charges to the surface so that $\nabla \varrho_N = 0$ inside, and where thermal gradients are small.

Let us complete the study of the effects of an electrical potential on the various equations of thermodynamics. If Φ is an *external* potential, we should associate with the change (14.92) of the energy density a change in the energy flux, $\boldsymbol{J}_E = \boldsymbol{J}'_E + q\Phi \boldsymbol{J}_N$. The *local energy conservation law* then remains valid, even when the applied potential $\Phi(r)$ varies in space. This is easily checked microscopically by adding $q\Phi(r)$ to $p^2/2m$ in equations (14.58) to (14.60). (The momentum conservation would make it necessary to introduce into the right-hand side of (14.6) a source term $-q\varrho_N \nabla \Phi$, but it is anyway violated by the collisions with the scattering centres.) If the applied potential depends on the time we must add to the right-hand side of the energy conservation equation a source term $q\varrho_N \partial \Phi / \partial t$; even the total energy is then no longer conserved, since the Hamiltonian is time-dependent. However, the potential $\Phi(r,t)$ contains, in general, a *self-consistent* part produced by the charges situated at other points r' and depending on the time, because these charges change. If we include the whole of $q\Phi$ in the energy density and flux at r we must again introduce a source term in the evolution equation for ϱ_E; the effect of this term is cancelled in the calculation of the change in the total energy by the fact that the latter is then no longer equal to $\int d^3r\, \varrho_E$, but to $\int d^3r \left(\varrho_E - \frac{1}{2} q\varrho_N \Phi \right)$, because otherwise we would have counted the potential energy twice. Another way of proceeding consists in dividing the interaction potential energy between r and r' evenly between those two points when we define ϱ_E, as we did on the microscopic scale in (14.58) with $\chi = 0$; we must then change the definition of the flux $\boldsymbol{J}_E$ according to the macroscopic version of (14.59), and the local energy conservation law remains satisfied without sources; but the price we pay is the introduction of non-local terms in $\boldsymbol{J}_E$.

We must still determine the *response coefficients* L_{ij} when there is a potential $\Phi(r)$ present, assuming that we know the coefficients L'_{ij} when there is no such potential. Note, to begin with, that the L_{ij} depend on Φ, but not on its gradient so that we can calculate them, assuming that Φ is uniform. It then suffices to start from Eqs.(14.25) which relate the fluxes $\boldsymbol{J}'_i$ to the affinities $\nabla \gamma'_i$ in the absence of Φ, and to make the change of variables $T = T'$, $\mu = \mu' + q\Phi$, $\boldsymbol{J}_E = \boldsymbol{J}'_E + q\Phi \boldsymbol{J}_N$, $\boldsymbol{J}_N = \boldsymbol{J}'_N$, to get the relations between the $\boldsymbol{J}_i$ and the $\nabla \gamma_i$. The resulting coefficients are the required response coefficients L_{ij}. We find in this way how they *depend explicitly on* $\Phi(r)$:

$$L_{NN} = L'_{NN}, \tag{14.97a}$$

$$L_{NE} = L_{EN} = L'_{NE} + q\Phi L'_{NN}, \tag{14.97b}$$

$$L_{EE} = L'_{EE} + 2q\Phi L'_{NE} + q^2 \Phi^2 L'_{NN}. \tag{14.97c}$$

We check that L_{NN} is independent of Φ, as we assumed when we wrote down (14.95). Moreover, it follows from (14.97) that the presence of an *electrical potential does not affect the thermal conductivity* (14.89).

14.4.3 Thermoelectric Effects

The indirect response coefficients $L_{NE} = L_{EN}$ are involved in the thermo-electric effects which connect the two kinds of conservative quantities, charge and heat. The *Seebeck effect* (1822) is the production of an *electromotive force* by a *temperature gradient*. It introduces the *thermoelectric power* or Seebeck coefficient ϵ of a substance, a quantity which is defined as minus the electromotive force produced in an open circuit by unit temperature gradient:

$$\nabla \left(\frac{\mu}{q} \right) = -\epsilon \, \nabla T, \qquad \boldsymbol{J}_N = 0. \tag{14.98}$$

In this definition the electric charges and currents vanish so that the electro-motive force can be identified with the spatial variation of μ/q. We are in the same circumstances (14.88) as in the case of heat transport, but now we are interested in $\nabla \mu$; using (14.93) and (14.97) we thus obtain

$$\epsilon = \frac{1}{qT} \left(\frac{L_{NE}}{L_{NN}} - \mu \right) = \frac{1}{qT} \left(\frac{L'_{NE}}{L'_{NN}} - \mu' \right). \tag{14.99}$$

In order to use this property for applications and to observe it in a per-manent regime one must construct a circuit containing two kinds of different conductors, in order to prevent the electromotive forces produced by the sections of the circuit which are subject to opposite temperature gradients from cancelling one another. This gives us the principle of *thermocouples* (Fig.14.2). At the two junctions, which are at temperatures T_1 and T_2, both the chemical potential and the temperature between the conductors A and B are continuous. On the other hand, we have seen in § 11.3.5 that the elec-tric potential Φ is discontinuous; its macroscopic jump, which is created by a double layer of charges, has just the effect of making the chemical potentials at both sides of the junction equal to one another. However, the temperature and the chemical potential vary, according to (14.98), along the conductors. This produces at the points 3 and 4 of the simple circuit shown in Fig.14.2 an electromotive force,

$$\begin{aligned} \Phi_3 - \Phi_4 &= \frac{1}{q} \left[(\mu_3 - \mu_1) + (\mu_1 - \mu_2) + (\mu_2 - \mu_4) \right] \\ &= \epsilon_B (T_1 - T) + \epsilon_A (T_2 - T_1) + \epsilon_B (T - T_2) \\ &= (\epsilon_B - \epsilon_A)(T_1 - T_2), \end{aligned} \tag{14.100}$$

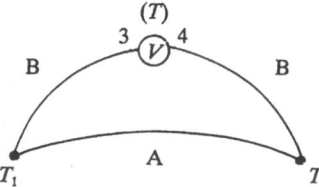

Fig. 14.2. Thermocouple

which is proportional to the difference between the temperatures of the two junctions. A voltmeter, with a large resistivity in order that the current is practically zero, makes it possible for us to determine this electromotive force, which is equal to the potential difference between the ends, and thus to *measure temperature differences* after calibration. The effect can be amplified by placing several similar circuits in series; it enables us to obtain a high sensitivity and great convenience.

The best known thermal effect of electricity is clearly the *Joule effect* (1841) which is connected with the passage of a current in uniform conduction regime where the temperature, electric field, and carrier density are uniform and the substance is neutral on the macroscopic scale. By eliminating $\nabla\mu$ between Eqs.(14.25) and using (14.93) we see that in this regime the particle flux is accompanied by an energy flux,

$$\boldsymbol{J}_E = -\frac{q}{T}\, L_{EN}\, \nabla\Phi = \frac{L_{EN}}{qL_{NN}}\, \boldsymbol{J}_{\mathrm{el}}. \tag{14.101}$$

Even though this energy flux is an indirect effect connected with the coefficient L_{EN}, the divergence of (14.101) is given by the term in $q\Phi$ in (14.97b), the only one which varies from one point to another. Energy conservation then gives an expression which does not involve L_{EN}:

$$\frac{\partial \varrho_E}{\partial t} = -\operatorname{div}\boldsymbol{J}_E = -\left(\nabla\Phi\cdot\boldsymbol{J}_{\mathrm{el}}\right) = \frac{J_{\mathrm{el}}^2}{\sigma}. \tag{14.102}$$

In the model considered the charge carriers do not exchange energy with the medium; the conduction is not a stationary regime, and the Joule power (14.102) produced at every point contributes to an increase in the energy of the carriers with time and thus to a rise in their temperature. However, in practice (§ 15.2.4) the inelasticity of the collisions enables the carriers to exchange energy with the lattice, and hence the heat (14.102) is carried off when the conductor is maintained at a constant temperature. Note that the heat produced depends only on σ, that is, on the *direct* response coefficient L_{NN}.

If (14.97) is taken into account, the entropy flux (14.36) reduces to

$$\boldsymbol{J}_S = \frac{1}{T}\boldsymbol{J}_E - \frac{\mu}{T}\boldsymbol{J}_N = \frac{q}{T^2}\left(\mu L_{NN} - L_{EN}\right)\nabla\Phi$$
$$= \frac{q}{T^2}\left(\mu' L'_{NN} - L'_{EN}\right)\nabla\Phi, \tag{14.103}$$

and its divergence vanishes. The local entropy balance (14.38) gives us the *dissipation rate*

$$\frac{\partial s}{\partial t} = \frac{q^2}{T^2}\, L_{NN}\left(\nabla\Phi\right)^2 = \frac{1}{\sigma T}\, J_{\mathrm{el}}^2, \tag{14.104}$$

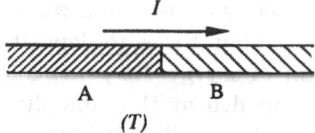

Fig. 14.3. The Peltier effect

where we have used (14.95). In fact, this dissipation of entropy is the basic phenomenon in the Joule effect; it is a consequence of the *irreversibility* of the charge transport, and the heating (14.102) follows from quasi-equilibrium thermodynamics.

The inverse indirect phenomenon of the Seebeck effect is the *Peltier effect* (1834), which involves the coefficient L_{EN}. It is also easily observed only when one uses two different conductors: when an electric current passes through a junction between A and B, one observes locally a heating or a cooling, depending on the direction of the current. Let us assume that the junction and its surroundings are maintained at a temperature T. The current passing through the conductor A is accompanied by an energy flux (14.101). However, whereas the electric current is conserved, the energy flux (14.101) in B differs from that in A. The local energy balance then implies that the thermostat which maintains the junction at the temperature T provides it with a power

$$W = \left[\left(\frac{L_{EN}}{qL_{NN}} \right)_{\mathrm{B}} - \left(\frac{L_{EN}}{qL_{NN}} \right)_{\mathrm{A}} \right] I, \tag{14.105}$$

which is proportional to the strength I of the current passing through the junction and which changes sign when the current changes sign. Bearing in mind the continuity of μ across the junction and expression (14.99) for the Seebeck coefficient, we find for the *Peltier coefficient*, which is the power absorbed for unit current from A to B,

$$\begin{aligned} \Pi_{\mathrm{AB}} &= \frac{W}{I} = \left(\frac{L_{EN}}{qL_{NN}} - \frac{\mu}{q} \right)_{\mathrm{B}} - \left(\frac{L_{EN}}{qL_{NN}} - \frac{\mu}{q} \right)_{\mathrm{A}} \\ &= T \left(\epsilon_{\mathrm{B}} - \epsilon_{\mathrm{A}} \right). \end{aligned} \tag{14.106}$$

This relation between two reciprocal thermoelectric effects is, in fact, a consequence of the Onsager relation $L_{EN} = L_{NE}$, as the response which occurs in the thermoelectric power is L_{NE}, whereas that which occurs in the energy flux is L_{EN}. Note that it was useless in these two cases to analyze in detail what happens at the junctions: it has been sufficient to consider the bulk transport in the two substances together with the continuity property of the chemical potential, even though complicated effects occur at the junction – notably, the electrical potential has a discontinuity which is connected with a localized charge distribution. Note also that, in contrast to the Joule effect which accompanies it in the conductors A and B, the Peltier effect is *reversible*: the entropy exchanges of the junction with A and B and with the thermostat are balanced, without dissipation; they merely change sign when the current changes sign. The Peltier effect is used in laboratories to deposit or take away at a given point a well defined amount of energy. Its practical use – for instance, to make small refrigerators – is impeded by the difficulty of finding substances with a large Peltier coefficient and a small resistivity as well as a small heat conductivity – to avoid a loss of efficiency through heat leaks. Moreover, one must restrict oneself to small intensities, as the Peltier effect is proportional to I, whereas the Joule effect is proportional to I^2.

As an exercise, one could analyze another thermoelectric effect, the *Thomson effect*, which occurs when an electric current passes through a substance in the presence of a temperature gradient. As in the Peltier effect, but here for a single substance, we have locally the creation or absorption of an amount of heat, proportional to $I\nabla T$, in addition to the Joule heat.

Let us finally stress that all the foregoing only concerns regimes where there is local equilibrium. In the case of effects which involve fast variations in space or time one tends to use empirical laws which define *non-local* or *retarded*, or, what amounts to the same, *wavelength- or frequency-dependent*, transport coefficients. For instance, the static permittivity ε, introduced in § 11.3.3, or the conductivity σ of (14.90) are replaced by functions of the frequency – complex permittivity or impedance – for rapidly oscillating phenomena. These go beyond the framework of thermodynamics as they imply either situations far from equilibrium or memory effects. Their study needs an appeal to statistical mechanics (Exerc.14b and Chap.15).

14.4.4 Local Equilibrium of a Fluid

From now on we shall be interested in a fluid consisting of one kind of molecules, to be treated as point particles, which evolves near equilibrium. We may be dealing with a *gas* – we shall come back to that case in Chap.15 – or with a *liquid*, if the interactions between the molecules are strong. Our aim is to derive the equations of hydrodynamics and heat transport, starting solely from the general principles of thermodynamics, and to identify the usual transport coefficients in terms of the theoretical coefficients L_{ij}. In particular, we shall classify the equations of fluid dynamics either as conservation or as response laws, and we shall make a systematic use of invariances and symmetries to reduce the number of independent L_{ij} coefficients to three. We shall always regard the transformations as passive (§ 2.1.5), that is, operating on the coordinate frame rather than on the system itself.

We start by writing down the constants of the motion. As in the case of the systems studied in § 14.4.1 they contain the *number of molecules* and the *energy* but we must also include here the three components of the *momentum* $\boldsymbol{P}$. The presence of these new conserved quantities makes it possible that quasi-equilibrium regimes are established where the fluid is macroscopically

in motion. This is not the case when particles interact as in § 14.4.1 with
a scattering medium, since collisions with scattering centres quickly bring
the momenta p_j of the particles to an isotropic distribution if there is no
external force present; then the system of molecules within a macroscopic
volume element cannot retain over a macroscopic time interval a global drift
velocity, unless there is an external force.

A local equilibrium state is characterized by giving at each point the five
densities of the conservative quantities, the particle density ϱ_N, the energy
density ϱ_E, and the momentum densities $\varrho_{P\alpha}$, for each of the three components P_α ($\alpha = x, y, z$). Thermostatics teaches that all equilibrium properties
can be derived by giving the entropy density s as function of these five variables. Especially, the partial derivatives of this function with respect to ϱ_E,
ϱ_N, and $\varrho_{P\alpha}$ define the local intensive variables $\gamma_i = 1/T$, $-\mu/T$, $-u_\alpha/T$.

We already started to use the *Galilean invariance* in (14.10) to identify
u with the local mean displacement velocity of the fluid. More precisely,
Galilean invariance implies that the thermodynamic potentials *depend only
on two independent variables* and not on the five variables associated with
E, N, and P. For instance, the entropy density has the form

$$s(\varrho_E, \varrho_N, \varrho_P) = s\left(\varrho_E - \frac{\varrho_P^2}{2m\varrho_N}, \varrho_N, 0\right), \tag{14.107}$$

which is valid for any fluid, whether it is a gas or a liquuid, classical or quantal.
In practice, we characterize the local equilibrium state by the *temperature T*,
the *mass density*

$$\varrho \equiv m\varrho_N, \tag{14.108}$$

and the *velocity* $u = \varrho_P/m\varrho_N$. Knowing the free energy per unit volume
$F(T, \varrho_N)$ then gives us the equations of state in the frame moving with a
velocity u, using the invariance (14.107) of s, just as (14.92) enabled us to
take advantage of the invariance under a change in the energy origin. Here
we find $T = T'$, $\mu = \mu' - \varrho_P^2/2m\varrho_N^2$, where μ' is the chemical potential in the
frame moving with the fluid at velocity u. Altogether we can, if we introduce
the *internal energy density* ϱ_U, write the *equations of state* in the form

$$\varrho_U = F - T\frac{\partial F}{\partial T} = F + Ts, \qquad \mu' = \frac{\partial F}{\partial \varrho_N}, \tag{14.109a}$$

$$\varrho_E = \varrho_U + \tfrac{1}{2}\varrho u^2, \qquad \mu = \mu' - \tfrac{1}{2}mu^2, \qquad \varrho_P = \varrho u, \tag{14.109b}$$

$$\mathcal{P} = T\left(s - \sum_i \gamma_i \varrho_i\right) = Ts - \varrho_U + \mu'\varrho_N. \tag{14.109c}$$

We have used the Gibbs-Duhem relation (6.9) or (5.79) to get expression
(14.109c) for the *equilibrium pressure $\mathcal{P}$*.

To justify (14.107), starting from microscopic physics, we use a technique which is of interest also for applications to other invariance problems. In the present case, when we make a Galilean change in the frame of reference, the coordinates and the momenta of the particles transform as

$$r'_j = r_j - ut, \qquad p'_j = p_j - mu. \qquad (14.110)$$

The relative velocity u of the two frames is arbitrary; as the transformations (14.110) form a group, it is sufficient to use the Galilean invariance for an *infinitesimal* u and then to proceed by integration. The total momentum and Hamiltonian operators then transform as

$$\widehat{P}' = \widehat{P} - mu\widehat{N}, \qquad \widehat{H}' = \widehat{H} - (u \cdot \widehat{P}). \qquad (14.111)$$

In the case of a finite transformation we must add to the energy a term $\frac{1}{2}mu^2\widehat{N}$. Let us assume that the system is in generalized canonical equilibrium, where we have given the expectation values of $\widehat{H}$, $\widehat{N}$, and $\widehat{P}$, with Lagrangian multipliers β, $-\alpha$, and $-\lambda$ (§ 4.3.3). In another Galilean frame its density operator is the same, but the equilibrium state which it describes is characterized by new multipliers β', $-\alpha'$, $-\lambda'$; the partition function may *a priori* also be changed. Using (14.111) we identify the two expressions

$$\widehat{D} = \frac{1}{Z} e^{-\beta\widehat{H}+\alpha\widehat{N}+(\lambda\cdot\widehat{P})} = \frac{1}{Z'} e^{-\beta'\widehat{H}'+\alpha'\widehat{N}'+(\lambda'\cdot\widehat{P}')},$$

which, to first order in u, gives

$$\beta = \beta', \qquad \alpha = \alpha' - m(u \cdot \lambda'), \qquad \lambda = \lambda' + \beta u, \qquad Z = Z'. \qquad (14.112)$$

Since $Z(\beta, \alpha, \lambda)$ is invariant under the infinitesimal transformation (14.112), it satisfies the equations

$$-m\lambda \frac{\partial Z}{\partial \alpha} + \beta \frac{\partial Z}{\partial \lambda} = 0. \qquad (14.113)$$

The solution of these partial differential equations shows that Z depends on α and λ only through the combination $\alpha + m\lambda^2/2\beta$ so that we can express $Z(\beta, \alpha, \lambda)$ in terms of the grand canonical partition function, obtained for the case when $\lambda = 0$, as follows:

$$Z(\beta, \alpha, \lambda) = Z_G\left(\beta, \alpha + \frac{m\lambda^2}{2\beta}\right). \qquad (14.114)$$

We can derive (14.107) from (14.114) using equation (4.14) which gives the entropy in the canonical formalism. We could also have noted that s is the Legendre transform of $k \ln Z$ per unit volume with respect to β, α, λ, so that (14.113) is equivalent to

$$m \frac{\partial s}{\partial \varrho_P} \varrho_N + \frac{\partial s}{\partial \varrho_E} \varrho_P = 0, \qquad (14.115)$$

which has (14.107) as a solution.

Here the response equations (14.16) involve non-zero *equilibrium currents* J_i^0, calculated for the case when there are no gradients of T, μ, or $\boldsymbol{u}$. Let us begin by calculating them for a fluid at rest ($\boldsymbol{u} = 0$). The rotational invariance implies that $\boldsymbol{J}_E^0$ and $\boldsymbol{J}_N^0$, which are vectors, vanish (§ 14.2.4); on the other hand, $J_{P\beta}^{0\alpha}$, which is a tensor with two indices, invariant under rotations, should be proportional to the unit tensor $\delta_{\alpha\beta}$. In order to determine its coefficient we note that the flux of $\boldsymbol{J}_{P\alpha}$ through a surface with a normal in the α direction represents the momentum which passes perpendicularly through it per unit time. For a fluid at rest in equilibrium this momentum transfer must be balanced by the force exerted by the walls; this enables us to identify $J_{P\beta}^{0\alpha} = \mathcal{P}\delta_{\alpha\beta}$ with the *hydrostatic pressure* $\mathcal{P}$ given by (14.109c). We confirm this result below, when we start from the microscopic expressions for the currents in a classical liquid and evaluate them at equilibrium. We shall thus prove the equivalence between the three definitions of the pressure: (i) the force exerted by the fluid on a wall, (ii) the momentum current density $\boldsymbol{J}_P^0$ within the fluid in equilibrium, and (iii) the intensive variable conjugate $\mathcal{P}$ to the volume in thermostatics.

Coming back to a fluid in motion we again use a Galilean transformation. According to equations (14.121) which are proved below we thus obtain the *equilibrium currents* in an arbitrary frame of reference:

$$\boldsymbol{J}_N^0 = \boldsymbol{u}\varrho_N, \quad \boldsymbol{J}_E^0 = \boldsymbol{u}\big(\varrho_E + \mathcal{P}\big), \quad J_{P\beta}^{0\alpha} = u_\alpha u_\beta \varrho + \mathcal{P}\delta_{\alpha\beta}. \tag{14.116}$$

In the case of a liquid at equilibrium and at rest the one- and two-particle densities f and f_2 which occur in the microscopic expressions of § 14.3.2 for the densities and the fluxes have, respectively, the form $\varrho_N g(p)$ and $\varrho_2(r')g(p)g(p')$, where $g(p)$ is the Maxwell distribution (7.17) and r' is the distance between the two points in f_2. As a result the only non-vanishing currents are given by the diagonal elements of (14.55) which are equal and reduce to

$$J_{P\alpha}^{0\alpha} = \varrho_N kT - \frac{1}{6}\int d^3r'\, r'\, \frac{dW}{dr'}\,\varrho_2(r'). \tag{14.117}$$

On the other hand, the grand potential (9.4) of the liquid depends on the volume Ω only through the integration domain. A change of coordinates, $r_j = s_j L$, where $L = \Omega^{1/3}$ gives us

$$A = -kT \ln\left[\sum_N \frac{f^N \Omega^N}{N!}\int d^3s_1 \cdots d^3s_N\, e^{-\beta\sum_{i>j} W_{ij}}\right],$$

where the integrations are over unit volume; the volume Ω appears both explicitly and through $W_{ij} \equiv W\big(|s_i - s_j|L\big)$. There are two terms, corresponding to these two functional dependences, in the derivative of A with respect to the volume which gives us the pressure:

$$\mathcal{P}\Omega = -\Omega\frac{\partial A}{\partial\Omega} = -kT\frac{\partial A}{\partial\mu} - \frac{L}{3}\left\langle\sum_{i>j}\frac{d}{dL}W\big(|s_i - s_j|L\big)\right\rangle. \tag{14.118}$$

In order to write down the first term in (14.118) we have used the fact that μ appears solely through the combination $f\Omega \propto e^{\mu/kT}\Omega$; for the second term we noted that $3\Omega\partial/\partial\Omega = L\partial/\partial L$ and recognized the fact that the result was a grand canonical average. Taking also into account that $\partial A/\partial\mu = -N$, returning to the r_j variables, and using the definition (14.56) of f_2 and its equilibrium form $\varrho_2 g(p)g(p')$, we obtain from (14.118) the *equilibrium pressure in the liquid* as

$$\mathcal{P} = \frac{N}{\Omega}kT - \frac{1}{6}\int d^3r'\, r'\, \frac{dW}{dr'}\, \varrho_2(r'), \tag{14.119}$$

which is just the same as (14.117).

This proof was based upon the definition of the pressure as the intensive variable which is the conjugate of the volume with respect to the energy in thermostatics. Another proof, based upon the virial theorem (9.17), relates J_P to the pressure $\mathcal{P}$ defined as the force per unit area from the wall. In fact, in (9.17) the total force F_i exerted on the i-th particle consists, on the one hand, of the contributions $-\nabla_i W_{ij}$ from the other particles, j, and, on the other hand, of the contribution from the external forces, which is localized near the walls, and the average of which defines $\mathcal{P}$. The summation in (9.17) over all particles i then provides us with

$$3NkT = -\sum_i \langle(F_i \cdot r_i)\rangle = \frac{1}{2}\Omega \int d^3r'\, r'\, \frac{dW}{dr'}\, \varrho_2(r') + \int d^2\sigma\, \mathcal{P}(n \cdot r),$$

where we have used the symmetry between pairs of interacting particles to write down the first term; the second term, which is the flux of $\mathcal{P}r$ leaving the volume Ω, equals $\int d^3r \operatorname{div}(\mathcal{P}r) = 3\Omega\mathcal{P}$ and we find again that (14.117) equals $\mathcal{P}$.

14.4.5 Conservation Laws in a Fluid

The rates of change of the densities ϱ_E, ϱ_N, and ϱ_P are connected with the corresponding fluxes through the conservation equations (14.6). We shall give for these a more familiar form which has a simpler interpretation, by writing them in terms of intrinsic quantities, independent of the motion, for instance, the internal energy density ϱ_U given by (14.109b) instead of ϱ_E.

A first remark is based upon the microscopic expressions (14.49) and (14.50) of the *mass current* mJ_N and the *momentum density* ϱ_P. As the velocity and the momentum – in the absence of a magnetic field – are proportional to one another, these quantities are equal:

$$mJ_N = \varrho_P. \tag{14.120}$$

Let us then express, as we have done in (14.109) for the variables which characterize local equilibrium, the current densities in a fixed frame of reference, at the point r_0 and at time t_0, as functions of the current densities at the same point, but measured in a Galilean frame moving with a velocity $u_0 = u(r_0, t_0)$, in which the fluid is locally and macroscopically at rest. As ϱ_P vanishes in that local frame, the same is true for J_N. We shall denote by J_U and by $\mathcal{P}_\beta^\alpha = J_{P\beta}^\alpha$ the *energy current* and the *momentum current in the frame moving with the fluid* at the point r_0. In the case of a fluid which is globally at equilibrium, J_U vanishes and $\mathcal{P}_\beta^\alpha$ is equal to $\mathcal{P}\delta_{\alpha\beta}$ as we saw in § 14.4.4. We shall show that the currents *in the fixed frame* are given by

$$J_N = u\varrho_N, \qquad J_E^\alpha = u_\alpha \varrho_E + J_U^\alpha + \sum_\beta \mathcal{P}_\beta^\alpha u_\beta,$$

$$J_{P\beta}^\alpha = u_\alpha u_\beta \varrho + \mathcal{P}_\beta^\alpha. \tag{14.121}$$

To do this we write down how the densities and currents transform under an arbitrary Galilean transformation, corresponding to a velocity u which once again we assume to be *infinitesimal*. It follows from (14.111) that

$$\varrho_E' = \varrho_E - (u \cdot \varrho_P), \qquad \varrho_N' = \varrho_N, \qquad \varrho_P' = \varrho_P - mu\varrho_N, \tag{14.122}$$

and that $s' = s$, $\mathcal{P}' = \mathcal{P}$. As far as the fluxes are concerned, the global motion u has two effects. On the one hand, the transported quantities, the energy and the momentum, are changed according to (14.122). On the other hand, the J_i define fluxes across surfaces which are fixed in the original frame, whereas the J_i' relate to surfaces bound to the frame which moves with the velocity u. Altogether, we get to first order in u

$$J_E' = J_E - \sum_\beta u_\beta J_{P\beta} - u\varrho_E, \qquad J_N' = J_N - u\varrho_N,$$

$$J_{P\beta}' = J_{P\beta} - mu_\beta J_N - u\,\varrho_{P\beta}. \tag{14.123}$$

We can check that the transformation (14.122), (14.123) keeps the conservation equations intact, as it should, if we take into account the fact that this transformation is accompanied by a change of coordinates, $r' = r - ut$; the partial derivatives $\partial \varrho_i'/\partial t$ must be evaluated at constant r', and not at constant r.

We need to write down the transformations (14.122) and (14.123) for a finite velocity u, which we shall set equal to u_0 at the point r_0 and at time t_0. To do this we must, as in (14.112), consider these equations as partial differential equations for the components of u, for instance,

$$\frac{\partial \varrho_E}{\partial u_\beta} = -\varrho_{P\beta}, \qquad \frac{\partial J_E^\alpha}{\partial u_\beta} = -J_{P\beta}^\alpha - \varrho_E\,\delta_{\alpha\beta},$$

and integrate them from 0 to u_0. Integration of (14.122) leads again to (14.109b). Integration of (14.123), with $J_E = J_U$, $J_N = 0$, $J_{P\beta}^\alpha = \mathcal{P}_\beta^\alpha$ in the frame with velocity u_0, gives (14.121), and also (14.120).

The forms (14.109b) and (14.122) of the densities and the currents are not only valid close to equilibrium, but follow from the general expressions (14.48)–(14.50), (14.55), (14.58), and (14.59) which were at the microscopic level established for any form of the one- and two-particle densities f and f_2. To see this, it is sufficient to prove that those expressions transform as (14.122) and (14.123) under an infinitesimal Galilean transformation. In fact, the only change produced by a change u in the frame velocity is the replacing of p by $p - mu$, which makes this check an easy one.

The variables in terms of which one usually writes down the equations of hydrodynamics are, on the one hand, the local *temperature* T, the *mass density* ϱ, and the *velocity* $\boldsymbol{u}$, and, on the other hand, the *internal energy flux* $\boldsymbol{J}_U$ and the *stress tensor* $-\mathcal{P}_\beta^\alpha$. We have used these variables in (14.109) and (14.121) to find expressions for the various quantities occurring in our theoretical scheme, and there remains for us to rewrite the general conservation and response equations in terms of them. Let us begin with the conservation laws, which are exact under all circumstances.

The conservation of the number of particles, $i = N$, gives us the *mass conservation* equation, which is often referred to by the name "equation of continuity",

$$\frac{\partial \varrho}{\partial t} + \operatorname{div} \varrho \boldsymbol{u} \; = \; 0. \tag{14.124}$$

This relation involves only the local equilibrium variables ϱ and $\boldsymbol{u}$, since the particle flux only depends on those, in contrast to the other fluxes.

If we use (14.109), (14.121), and (14.124) we find from the momentum conservation law, $\partial \varrho_{P\beta} + \operatorname{div} \boldsymbol{J}_{P\beta} = 0$, the equation of motion

$$\frac{\partial u_\beta}{\partial t} + (\boldsymbol{u} \cdot \nabla) u_\beta + \frac{1}{\varrho} \sum_\alpha \frac{\partial}{\partial r_\alpha}\, \mathcal{P}_\beta^\alpha \; = \; 0. \tag{14.125}$$

We recognize here, after multiplication by $d^3\boldsymbol{r}$, the *fundamental equation of dynamics* applied to the infinitesimal, but macroscopic fluid element $d^3\boldsymbol{r}$. In the language of hydrodynamics this is considered as a piece of continuous matter which, when it passes the point $\boldsymbol{r}$ at time t, moves with the velocity $\boldsymbol{u}(\boldsymbol{r}, t)$. As for all other equations that we shall write down, we used for Eqs.(14.124) and (14.125) the Eulerian description (§ 14.1.2), where $\boldsymbol{r}$ denotes a *fixed* point for fields such as $\boldsymbol{u}(\boldsymbol{r}, t)$, $\varrho(\boldsymbol{r}, t)$, $\mathcal{P}_\beta^\alpha$, and so on. Let us follow a piece of fluid, of volume $d^3\boldsymbol{r}$ and mass $\varrho\, d^3\boldsymbol{r}$, in its macroscopic motion. At the time t, its velocity is $\boldsymbol{u}$, and the first two terms in (14.125) can be interpreted as its *acceleration*. We must then identify

$$-\sum_\alpha \frac{\partial}{\partial r_\alpha}\, \mathcal{P}_\beta^\alpha\, d^3\boldsymbol{r}$$

as the β-component of the *force* exerted at the macroscopic level on this volume element by the remainder of the fluid. As a result $-\mathcal{P}_\beta^z\, dx\, dy$ is the force exerted in the β-direction by the layers of fluid situated above z on the surface element $dx\, dy$ of the fluid situated below. We can thus, apart from the sign, identify the components $\mathcal{P}_\beta^\alpha$ of the *momentum flux in the local frame* with those of the *stress tensor* of the mechanics of continuous media.

It is interesting to note that the *equation of motion of macroscopic hydrodynamics* has here been obtained by solely using the microscopic particle number and momentum *conservation equations* in connection with *transla-*

tional and Galilean invariance. If the fluid is subjected to *external forces* such as gravity, their value per unit mass occurs on the right-hand side of (14.125). In macroscopic dynamics, this just expresses the relation between the acceleration and the force for a volume elemnt. In the microscopic approach of § 14.3.2, the external potential contributes to equation (14.51) for $\partial \varrho_{P\beta}/\partial t$. Its gradient equals minus the applied force; since $\partial/\partial r_j$ in quantum mechanics is represented by $i\widehat{p}_j/\hbar$, the total external force is represented by the commutator

$$\widehat{F} = \frac{1}{i\hbar} \, [\widehat{P}, \widehat{H}],$$

so that $\widehat{F}$ can be interpreted as measuring the *lack of translational invariance* of the Hamiltonian, if we bear in mind that $\widehat{P}$ is the generator of translations (§ 2.1.5). This produces a *violation of the momentum conservation law* (14.54), and the external force per unit mass here enters as a *source term* taking care of that violation, as in (14.3).

The *energy conservation* equation $\partial \varrho_E/\partial t + \operatorname{div} J_E = 0$ can, in turn, be rewritten as

$$\frac{\partial \varrho_U}{\partial t} + \operatorname{div} \, (u \varrho_U + J_U) + \sum_{\alpha\beta} \mathcal{P}_\beta^\alpha \, \frac{\partial u_\beta}{\partial r_\alpha} = 0, \qquad (14.126)$$

where we have used (14.109b), (14.121), (14.124), and (14.125). If the last term were not there, this equation would express the conservation of *internal* energy, the internal energy flux in the fixed frame being the sum of one part, $u \varrho_U$, associated with the overall motion, as in (14.124), and of another part, J_U, which can be interpreted as a *heat flux*. The last term reflects the existence of energy transfer between the internal energy and the motion, which is interpreted as *work done by the stresses* on the volume element. The use of Galilean invariance has, by providing a natural decomposition of the densities and the fluxes, enabled us to identify the various kinds of energy exchange inside the fluid. The kinetic energy $\frac{1}{2}\varrho u^2$ occurring in (14.109b) has been split off, and the balance (14.126) involves in each point heat and work done by the internal forces in the fluid. Here again, the effect of an external potential would be expressed by a source term added to the energy balance (14.126).

We have so far ignored *conservation of angular momentum.* In principle, we should have introduced the three components of the angular momentum density, ϱ_L, and their associated quantities, the intensive variables and the fluxes, in addition to the densities ϱ_E, ϱ_N, and ϱ_P, and should have written down the corresponding dynamic equations. However, we saw in § 14.3.2 from a microscopic study that for the simple fluid discussed here those quantities can be derived trivially from the ones we have considered, as ϱ_L and J_L are given by (14.64); the intensive variable which is the conjugate of ϱ_L is $-\omega/T$, where ω is the *local angular velocity*, or *vorticity*, given by

$$\boldsymbol{\omega} = \tfrac{1}{2}\operatorname{curl}\boldsymbol{u}, \tag{14.127}$$

and it also is not an independent variable. Moreover, we saw that a consequence of the angular momentum conservation was the symmetry (14.65) of the $\boldsymbol{J}_P$ tensor, which implies the *symmetry of the stress tensor*

$$\mathcal{P}_\beta^\alpha = \mathcal{P}_\alpha^\beta. \tag{14.128}$$

This property, which must be added to the conservation laws (14.125) and (14.126), follows from the microscopic expression (14.55) for $\mathcal{P}_\beta^\alpha$. Provided it is satisfied, the dynamics of the fluid rotations follows from the equations which we have written down for the energy, the number of particles, and the momentum.

14.4.6 Equations of Hydrodynamics

We still have to close our set of equations by finding an expression for the non-equilibrium fluxes $\boldsymbol{J}_i$. We achieve this task approximately in the framework of quasi-equilibrium, to wit, we assume that the time-dependences are sufficiently slow and the gradients sufficiently weak that (i) memory effects are negligible, and (ii) the linear response equations (14.16), characterized by empirical coefficients L, are valid. This linear regime in local equilibrium defines the so-called *Newtonian fluids*, which are characterized by their independent response coefficients. As each of the fluxes $\boldsymbol{J}_i - \boldsymbol{J}_i^0$ and each of the gradients $\nabla \gamma_i$ in (14.16) has three components, the response matrix forms *a priori* a 15×15 matrix connecting the former to the latter, where each matrix element is a function of the five local state variables. We shall show that there is, in fact, far less arbitrariness, by using all available information and, especially, invariances.

Let us first of all note that the *particle flux* $\boldsymbol{J}_N$ reduces to the equilibrium contribution $\boldsymbol{J}_N^0 = \boldsymbol{u}(\boldsymbol{r})\varrho_N(\boldsymbol{r})$ and therefore does not contain any additional part of first order in the affinities. The coefficients L_{Ni} which are associated with it are therefore *all equal to zero*. This is also consistent with the equation of continuity (14.124) for ϱ, which solely involves the local equilibrium quantities ϱ and $\boldsymbol{u}$, without any non-trivial flux.

The *Onsager relations* then imply that matrix elements of the kind $L_{iN}^{\alpha\beta}$, which are either equal to or the opposite of the corresponding elements $L_{Ni}^{\beta\alpha}$, must also vanish. As a result, a *spatial variation of* $-\mu/T$, without variations in the other intensive variables $1/T$ and $-\boldsymbol{u}/T$, *will not produce any flux*.

The remaining 12×12 matrix connects the components of $\boldsymbol{J}_E - \boldsymbol{J}_E^0$ and $\boldsymbol{J}_P - \boldsymbol{J}_P^0$ with those of the gradients of $1/T$ and $-\boldsymbol{u}/T$. In order to reduce that matrix we shall again use the *Galilean invariance*. Let us consider a fixed point $\boldsymbol{r}_0$, a given time t_0, and the Galilean frame $\boldsymbol{u}_0 = \boldsymbol{u}_0(\boldsymbol{r}_0, t_0)$ in which the fluid is locally at rest. Let us assume that we know the response matrix in that frame, that is, the matrix connecting the fluxes to the affinities, all evaluated in the vicinity of $\boldsymbol{r}_0$ and expressed in that Galilean frame. It will be easy afterwards to obtain the response equations in the fixed frame, which is *independent of the point r_0 considered*: to do this we shall use the transformations (14.121) for the fluxes and the transformations (14.109) for the affinities to go back to the fixed frame. The latter transformations can be simplified as follows. The temperature is an invariant. The transformation from μ' to μ, given by (14.109b) is useless, as $\nabla \gamma_N$ does not occur in the response equations. Finally, denoting by $r' = r - (t - t_0)u_0$ and $u'(r',t) = u(r,t) - u_0$ the position and the velocity field in the Galilean frame which follows the motion of the fluid at point r_0, the affinities $\nabla \gamma_P$ in that frame can be written, at the point r_0, as

$$\frac{\partial}{\partial r'_\gamma} \left(-\frac{u'_\delta}{T} \right) = -\frac{\partial}{\partial r_\gamma} \left(\frac{u_\delta - u_{0\delta}}{T} \right) = -\frac{1}{T} \frac{\partial u_\delta}{\partial r_\gamma}, \tag{14.129}$$

where we used the fact that $u(r_0, t_0) = u_0$; going from one frame to the other has thus resulted in an *elimination of the temperature gradient*.

We must thus connect the components of the fluxes J_U and $\mathcal{P}^\alpha_\beta - \delta_{\alpha\beta}\mathcal{P}$ in the local frame with those of the affinities $\nabla(1/T)$ and $(-1/T)\nabla u$ in that frame, and this defines the responses in the local frame: L_{EE} (a 3×3 matrix), L_{EP} (a 3×9 matrix), L_{PE} (a 9×3 matrix), and L_{PP} (a 9×9 matrix). To simplify these matrices let us use *Curie's principle* (§ 14.2.4). The properties of the fluid must be *invariant under rotation*. Let us classify the various quantities according to their tensorial nature, that is, according to how their components transform into one another under rotation. Some of them, which are invariant, ϱ, T, ..., are scalars; others, which are vectors, u, J_U, $\nabla(1/T)$, ..., have one index on which the rotations operate; still others, $\mathcal{P}^\alpha_\beta$, $\partial u_\alpha / \partial r_\beta$, ..., are tensors with two indices. The responses L_{EE} which connect two vectors are second rank tensors, the responses L_{EP} and L_{PE} are third rank tensors, and the responses L_{PP} form a fourth rank tensor. These various responses are, in principle, functions of the five state variables ϱ, T, and u at the point considered. Our choice of frame made u equal to zero so that each of the components of the various tensors L, written down in the local frame, is, in fact, *only a function of the two variables ϱ and T*. As the latter are scalars there is no available vector to construct the tensors L; hence they must be *invariant under rotation*: otherwise a simultaneous rotation of the affinities and the fluxes would change the relation connecting them. However, we have seen in § 14.2.4 that the only second rank invariant tensor is $\delta_{\alpha\beta}$, apart from a scalar factor. The response $L^{\alpha\beta}_{EE}$ of J^α_E to $\partial(1/T)/\partial r_\beta$ thus reduces to $L_U \delta_{\alpha\beta}$, where L_U denotes a scalar function of ϱ and T. Similarly (see Eqs.(14.23)), the only available invariant third rank tensor to construct L_{EP} and L_{PE} is the completely antisymmetric tensor $\varepsilon_{\alpha\beta\gamma}$. Finally, for the fourth rank tensor L_{PP} we have available only the invariant linear combinations $\delta_{\alpha\beta}\delta_{\gamma\delta}$, $\delta_{\alpha\gamma}\delta_{\beta\delta}$, and $\delta_{\alpha\delta}\delta_{\beta\gamma}$. The isotropy of the medium thus considerably reduces the possible forms of the responses L_{ij} in the frame in which the fluid is locally at rest.

Invariance under reflection in space then gets rid of the responses $L^{\alpha\beta}_{EP\gamma}$ in that frame. In fact, if we use the earlier result which requires that those responses must be proportional to $\varepsilon_{\alpha\beta\gamma}$, we find using (14.129) that the energy flux must be proportional to curl u. However, when we change the direction of the axes, the components of the energy flux change sign, whereas those of the derivatives of u remain unchanged. Therefore, the proportionality constant can only be zero, and the responses L_{EP} must vanish in the frame fixed to the fluid. The *Onsager relations* then show that the responses L_{PE} are also equal to zero.

Finally, we have seen from the angular momentum conservation that the tensor $\mathcal{P}_\beta^\alpha - \mathcal{P}\delta_{\alpha\beta}$ was *symmetric*. Thus, among the three invariant tensors (14.23b) which can connect it with (14.129) only $\delta_{\alpha\beta}\delta_{\gamma\delta}$, with a scalar coefficient denoted as l, and $\delta_{\alpha\gamma}\delta_{\beta\delta} + \delta_{\alpha\delta}\delta_{\beta\gamma}$ can remain. Instead of the latter it is convenient to introduce the combination with zero trace,

$$\tfrac{1}{2}\left(\delta_{\alpha\gamma}\delta_{\beta\delta} + \delta_{\alpha\delta}\delta_{\beta\gamma}\right) - \tfrac{1}{3}\delta_{\alpha\beta}\delta_{\gamma\delta},$$

with a scalar coefficient denoted by L.

Altogether the response equations thus involve only *three independent coefficients* L_U, l, and L, which are functions of the two variables ϱ and T – instead of 15×15 functions of five variables, which we thought of *a priori*. The affinities have been reduced to $\nabla(1/T)$ and to (14.129), and the fluxes by use of (14.121) to $\boldsymbol{J}_U$ and $\mathcal{P}_\beta^\alpha - \mathcal{P}\delta_{\alpha\beta}$. We are left with only diagonal coefficients and the linear response equations can be written as

$$\boldsymbol{J}_U = L_U \nabla\left(\frac{1}{T}\right), \tag{14.130}$$

$$\mathcal{P}_\beta^\alpha - \mathcal{P}\delta_{\alpha\beta} = -\frac{l}{T}\delta_{\alpha\beta}\,\mathrm{div}\,\boldsymbol{u} - \frac{L}{T}\Delta_{\alpha\beta}, \tag{14.131}$$

where the tensor Δ, with zero trace, is defined by

$$\Delta_{\alpha\beta} \equiv \frac{1}{2}\left(\frac{\partial u_\beta}{\partial r_\alpha} + \frac{\partial u_\alpha}{\partial r_\beta}\right) - \frac{1}{3}\delta_{\alpha\beta}\,\mathrm{div}\,\boldsymbol{u}. \tag{14.132}$$

These equations, combined with the conservation equations and with equations (14.109) which connect the internal energy density ϱ_U and the pressure $\mathcal{P}$ at thermal equilibrium with the state variables ϱ and T, enable us to discuss all mechanical and thermal problems about the fluid considered under circumstances where the gradients are not too large; the fluid is then entirely *characterized by four functions* of two variables, namely, a thermodynamic potential, such as the free energy, and the responses L_U, l, and L. Notwithstanding that there are here five conservative variables instead of two, the number of independent transport coefficients remains the same as in § 14.4.1, but their natures are quite different.

Thanks to the simplifications brought about, notably by the Galilean invariance and the Onsager relations, the experimental variables occur here directly in the conservation and response equations. The connection between the general theory and the usual description of transport phenomena will therefore be particularly simple. Especially, going from the fixed frame to a local frame attached to the fluid has changed the affinities according to (14.129) so that the gradient of the thermostatic, natural, but unusual variable $-u/T$ has been replaced by $-(1/T)\nabla u$. On the other hand, *thermal and mechanical effects have been decoupled* in (14.130) and (14.131). Moreover, there are *no diffusion phenomena*: since we have simply $L_{Ni} = L_{iN} = 0$ for the particle flux $J_N = \varrho P/m$, a gradient of the chemical potential μ does not produce any flux. Finally, note that the *chemical potential* has been eliminated: the final equations only involve directly ϱ, T, and u. We have stressed several times in the present book the importance of the chemical potential, or of the multiplier α, a quantity which plays a similar rôle for particle exchanges as the temperature for energy exchanges, and which has to be used in diverse problems: the physics and technology of semiconductors and metals, chemical equilibria and reactions, adsorption, equilibria between phases, and so on. No doubt the fact, that the oldest sciences of matter – mechanics and thermodynamics of monatomic fluids and of solids – happened not to need the chemical potential, was the reason why μ is not systematically introduced in elementary thermodynamics books and why the terminology is so vague: "chemical potential" is doubly inappropriate, "Fermi level" is specialized to just fermions, and the natural variable $-\mu/T = -k\alpha$ does not even have a name.

Let us connect the coefficients L_U, l, and L with the *usual transport coefficients* of Newtonian fluids. The *heat conductivity* λ is defined empirically by (14.87) for a fluid *at rest* ($u = 0$). The equation of motion (14.125) then implies that the *pressure is uniform*. As the temperature varies from one point to another, this implies that the density must also vary, in such a way that the pressure gradient remains zero. The energy flux J_E, given by (14.121), reduces to the heat flux J_U and when we write

$$J_U = -\lambda \nabla T = L_U \nabla \frac{1}{T}, \tag{14.133}$$

we find that $\lambda = L_U/T^2$.

Note the simplicity of this result as compared to expression (14.89) which is, for instance, valid for heat conduction in metals. In fact, (14.89) accounts for dissipative *diffusion* effects, which here do not appear, as the fluid contains only one kind of molecules. A metal, on the other hand, contains mobile electrons interacting with a substratum, the lattice imperfections, which is practically at rest. In the two cases heat conduction on the microscopic scale appears as a superposition of two opposed fluxes which transport the same number of particles, but not the same energy: the flux from the hot region is made up from particles with a higher energy. However, whereas in a simple liquid thermalization is due solely to the collisions of the molecules with one another, in the case of the electron gas one must include collisions with the substratum, which take place at different energies for the two fluxes. The phenomenon thus involves effects of *relative diffusion* of the light and the heavy particles, which the extra term in (14.89) takes into account.

Newton defined at the end of the seventeenth century the *shear viscosity* η for a laminar flow with velocities $\boldsymbol{u}(\boldsymbol{r})$ which are parallel to the x-direction and which are functions of the height z. *Newton's equation*,

$$-\mathcal{P}_x^z = \eta \, \frac{du_x}{dz}, \tag{14.134}$$

connects empirically the velocity gradient to the force per unit horizontal area which is exerted in the x-direction by the layers of the fluid situated above z on the lower layers. We identified this force with $-\mathcal{P}_x^z$ when we interpreted the momentum conservation equation (14.125) as the macroscopic equation of motion for a fluid volume element. A comparison of (14.134) with (14.131) and (14.132) gives us

$$\eta = \frac{L}{2T}. \tag{14.135}$$

Such a flow also gives rise to a stress $\mathcal{P}_z^x = \mathcal{P}_x^z$ in the vertical direction, acting on the surface elements perpendicular to the flow.

Finally, the *volume viscosity* $\eta_{\rm v}$ is associated with the extra pressure $\delta\mathcal{P}$ created by a compressional motion of the fluid, and an empirical law indicates that this is proportional to the compression rate,

$$\delta\mathcal{P} = \eta_{\rm v} \left(-\frac{1}{\Omega} \frac{d\Omega}{dt} \right) = -\eta_{\rm v} \, {\rm div} \, \boldsymbol{u}. \tag{14.136}$$

We have used the form

$$\frac{\partial u_\alpha}{\partial r_\beta} = -\frac{1}{3\Omega} \frac{d\Omega}{dt} \, \delta_{\alpha\beta}$$

for the velocity field in a uniform compression. Comparing (14.136) with the general form (14.131) of the response laws, where $\Delta = 0$, we find

$$\eta_{\rm v} = \frac{l}{T}. \tag{14.137}$$

When the motions are sufficiently smooth in space and sufficiently slow in time so that we can not only use linearized response equations, but also neglect dissipative effects, we are dealing with a so-called *perfect fluid*, and the dynamics are completely given by the conservation laws. The latter reduce in that case to the equation of *continuity* (14.124), the *Euler equation* (1757),

$$\frac{\partial \boldsymbol{u}}{\partial t} + (\boldsymbol{u} \cdot \nabla)\boldsymbol{u} + \frac{1}{\varrho}\nabla\mathcal{P} = 0, \tag{14.138}$$

and the *energy* conservation equation

$$\frac{\partial \varrho_U}{\partial t} + \text{div}\left(\boldsymbol{u}\varrho_U\right) = -\mathcal{P}\,\text{div}\,\boldsymbol{u}, \tag{14.139}$$

which expresses the *reversible* transformation of the work done by the hydrostatic pressure into heat.

In the linear dissipative regime, the combination of the conservation equation (14.125) and the response equation (14.131) gives us the *Navier-Stokes equation* of hydrodynamics which, if we neglect a possible dependence of η and η_v on the temperature and the density, can be written as

$$\frac{\partial \boldsymbol{u}}{\partial t} + (\boldsymbol{u}\cdot\nabla)\boldsymbol{u} + \frac{1}{\varrho}\nabla\mathcal{P} = \frac{\eta}{\varrho}\nabla^2\boldsymbol{u} + \frac{1}{\varrho}\left(\frac{\eta}{3}+\eta_v\right)\nabla\,\text{div}\,\boldsymbol{u}. \tag{14.140}$$

These equations were for an incompressible liquid arrived at towards 1825 by Henri Navier, ingénieur des Ponts et Chaussées (Dijon 1785–Paris 1836), who already described the fluid using an atomistic model, and in 1845 by Sir George Gabriel Stokes, mathematician and physicist (Bornat Skreen, Ireland 1819–Cambridge 1903) whose treatment of the fluid as a continuum at that time seemed to be more realistic. This change, over twenty years, of the view taken of the microscopic nature of matter illustrates the eclipse of atomism during the nineteenth century; we have already several times indicated that the dominant trend between 1830 and 1890 brought scientists and philosophers to the idea that matter was continuous.

Recall that, in contrast to what happened in § 14.4.1, there is here no specific transport coefficient of the nature of L_{NN} associated with making the density of the fluid, originally inhomogeneous, uniform. In fact, the *self diffusion* effect which describes this approach to a uniform density is here a relatively complex process. Let us assume that initially the temperature is uniform, the velocity zero, and the density varying in space. This is reflected by the existence of a gradient in the chemical potential. However, we have seen that the *responses associated with the affinity* $\nabla(-\mu/T)$, and especially the direct response L_{NN} of the particle flux to this affinity, are *zero*. Initially no particle flux is therefore created and, more generally, no dissipation is produced immediately. Nonetheless, the system will start to evolve through the indirect effect of the *conservation equations*, since Eq.(14.125) at the initial time gives us

$$\frac{\partial \boldsymbol{u}}{\partial t} + \frac{1}{\varrho}\nabla\mathcal{P} = 0,$$

where the pressure follows the variations of the density. Motion starts, first in the form of the propagation of a wave through the coupling with the continuity equation (14.124). It is the *final damping of this wave* as a result of the viscosities which will ultimately lead *indirectly* to the dispersion of the non-uniformity of the density.

The situation is different in a fluid containing several kinds of particles. An analysis similar to the one we have just given shows that for a binary mixture there appear new independent response coefficients which describe the *relative diffusion* of the two components and the *indirect thermomechanical effects* associated with the coupling between thermal phenomena and diffusion. In particular, we find again Fick's law for the relative diffusion, in agreement with § 14.4.1 where, in fact, gas diffusion ocurred with respect to a fixed diffusive medium made of heavier particles.

We still must write down the *entropy balance*, which makes it necessary to introduce an entropy flux J_S. Equation (14.35) would suggest that one should choose for this $\sum \gamma_i J_i$, if there were no equilibrium fluxes J_i^0. To guess the actual form of J_S, already anticipated in (14.36), we approach this situation by first sitting in the Galilean frame which, at the point r considered, accompanies the fluid motion, and where we put

$$J'_S = \sum_i \gamma'_i J'_i = \frac{1}{T} J_U. \tag{14.141}$$

The other terms, $i \neq E$, vanish, as $J'_N = 0$ and $\gamma'_P \propto u' = 0$. In order to get back to the fixed frame, as we did in (14.121) for the fluxes J_i, we note that the entropy density s is invariant under a Galilean transformation; the entropy flux is thus changed only because the surface it passes through is moving, and hence we find that

$$J_S = J'_S + su = \frac{1}{T} J_U + su. \tag{14.142}$$

The *entropy flux* J_S defined by (14.142) is the sum of a flux, su, associated with the motion of the fluid as a whole, and the term J_U/T. This confirms the interpretation of J_U as *heat flux* which followed from (14.126). Quite remarkably, whereas the entropy (14.107) is a function of all variables E, N, and P, its flux contains only one non-trivial contribution, associated with the internal energy.

Using the equations of state (14.109), the conservation equations, and the definition (14.142) we find, in agreement with the general equation (14.37) for the *dissipation*, the local entropy balance

$$\frac{\partial s}{\partial t} + \operatorname{div} J_S = \left(\nabla \left(\frac{1}{T} \right) \cdot J_U \right)$$
$$+ \sum_{\alpha\beta} \left(-\frac{1}{T} \frac{\partial u_\beta}{\partial r_\alpha} \right) \left(\mathcal{P}_\beta^\alpha - \mathcal{P} \delta_{\alpha\beta} \right). \tag{14.143}$$

In the linear regime Eqs.(14.130) and (14.131) give us

$$\frac{\partial s}{\partial t} + \operatorname{div} J_S = L_U \left(\nabla \frac{1}{T} \right)^2 + l \left(\frac{1}{T} \operatorname{div} u \right)^2$$
$$+ L \sum_{\alpha\beta} \left(\frac{1}{T} \Delta_{\alpha\beta} \right)^2. \tag{14.144}$$

The Clausius-Duhem inequality which expresses that, whatever the gradients on the right-hand side of (14.144), the dissipation must be positive then means that the *three responses* L_U, l, and L *are positive* or zero; necessarily the same must then hold also for the conductivity λ and the viscosities η and η_v.

Equation (14.144) clearly separates the rôles of the three response coefficients in dissipation, and it associates the irreversibility with two causes: the transport of heat due to temperature gradients, and the existence of motions which make div $\boldsymbol{u}$ or $\Delta_{\alpha\beta}$ non-vanishing. One easily checks that a translation or a rotation of the system as a whole, where $\partial u_\beta / \partial r_\alpha = \varepsilon_{\alpha\beta\gamma} \omega_\gamma$ implies that div $\boldsymbol{u} = 0$, $\Delta_{\alpha\beta} = 0$, do not give rise to dissipation. The mechanical dissipation terms can also be found from the work done by the stresses on a volume element of the fluid; this work is obtained from the last term of the internal energy balance (14.126) and from expression (14.131) for the stresses, and is, per unit volume and unit time, equal to

$$-\mathcal{P} \operatorname{div} \boldsymbol{u} + \frac{l}{T} \left(\operatorname{div} \boldsymbol{u} \right)^2 + \frac{L}{T} \sum_{\alpha\beta} \left(\Delta_{\alpha\beta} \right)^2 . \tag{14.145}$$

The first term, $-\mathcal{P} d\Omega / \Omega dt$, is nothing but the *mechanical power which is produced reversibly* by the hydrostatic pressure forces, and it contributes to $-\operatorname{div} \boldsymbol{J}_S$. The others are *positive* work done by the viscous stresses, which is *irreversibly transformed into heat* δQ in the volume element; we check by comparison with (14.144) that $\delta Q = T\, dS$.

We shall give below a direct construction of the entropy current (14.142), which we derived from the Ansatz (14.141). At the same time, we shall establish the more general form (14.36) or (14.151) of $\boldsymbol{J}_S$, which can be extended, for instance, to the dynamics of fluid mixtures or of solids. According to the general method of § 14.2.5 the entropy flux must have the form

$$\boldsymbol{J}_S = \sum_i \gamma_i \boldsymbol{J}_i + \boldsymbol{X}, \tag{14.146}$$

where the vector $\boldsymbol{X}$ must satisfy the equation

$$\operatorname{div} \boldsymbol{X} = - \sum_i \left(\nabla \gamma_i \cdot \boldsymbol{J}_i^0 \right) . \tag{14.147}$$

To check these points we must show that the right-hand side of (14.147) is, indeed, the divergence of a vector field, construct the latter, and then compare (14.146) with expression (11.142) which had been introduced heuristically. The existence of $\boldsymbol{X}$ is not obvious, as the $\boldsymbol{J}_i^0$, defined by (14.116), change from one point to another in a local equilibrium situation. Separating in each $\boldsymbol{J}_i^0$ the contribution $\varrho_i \boldsymbol{u}$, due to the displacement, from the contribution coming from the equilibrium pressure, we find

$$- \sum_i \left(\nabla \gamma_i \cdot \boldsymbol{J}_i^0 \right) = - \sum_i \varrho_i \left(\boldsymbol{u} \cdot \nabla \right) \gamma_i - \mathcal{P} \left(\boldsymbol{u} \cdot \nabla \right) \frac{1}{T} - \mathcal{P} \operatorname{div} \left(-\frac{\boldsymbol{u}}{T} \right)$$

$$= - \left(\boldsymbol{u} \cdot \nabla \right) \left[\sum_i \gamma_i \varrho_i \right] + \sum_i \gamma_i \left(\boldsymbol{u} \cdot \nabla \right) \varrho_i + \frac{\mathcal{P}}{T} \operatorname{div} \boldsymbol{u} .$$

The second term is equal to $(\boldsymbol{u} \cdot \nabla)s$, if we use the definition (14.12) of the γ_i variables. In order to rewrite the first term, we use the Gibbs-Duhem relation,

$$\sum_i \gamma_i \varrho_i \;=\; s - \frac{P}{T} \;=\; \sum_i \gamma_i' \varrho_i', \tag{14.148}$$

which finally gives

$$-\sum_i \left(\nabla \gamma_i \cdot \boldsymbol{J}_i^0 \right) \;=\; (\boldsymbol{u} \cdot \nabla)\frac{P}{T} + \frac{P}{T} \operatorname{div} \boldsymbol{u} \;=\; \operatorname{div}\left(\frac{P}{T}\boldsymbol{u} \right). \tag{14.149}$$

We have thus proved (14.147) and found $\boldsymbol{X}$. To identify (14.146) with (14.142) we examine the effect of a Galilean transformation on $\sum \gamma_i \boldsymbol{J}_i$. Whereas, according to (14.148), $\sum \gamma_i \varrho_i$ remains invariant, we easily check, using (14.109) and (14.121), that the combination

$$\sum_i \gamma_i \left(\boldsymbol{J}_i - \varrho_i \boldsymbol{u} \right) \;=\; \sum_i \gamma_i' \boldsymbol{J}_i' \;=\; \frac{1}{T} \boldsymbol{J}_U \tag{14.150}$$

is the one which is invariant. Using (14.148) to rewrite the left-hand side of (14.150) we finally find

$$\boldsymbol{J}_S \;=\; \sum_i \gamma_i \boldsymbol{J}_i + \frac{P}{T}\boldsymbol{u} \;=\; \frac{1}{T}\boldsymbol{J}_U + s\boldsymbol{u}, \tag{14.151}$$

in agreement with both (14.36) and (14.142).

The equations which we have constructed are the just same ones which are introduced and worked out in hydrodynamics. Our aim, however, was not so much to reach this result, as to demonstrate the power of the general approach of non-equilibrium thermodynamics. We have in this way seen why in a simple fluid processes which *a priori* could be imagined, such as the existence of indirect responses connecting thermal and mechanical effects, or the presence of a diffusion coefficient in a Fick-type law, are precluded. In fact, one can generalize easily the above methods to other, less simple substances, mixtures of fluids, crystalline solids, liquid crystals, and so on, or to other, chemical, magnetic, electromechanical, ..., phenomena, and this leads to an enormous richness of results. In all such cases, the methods of thermodynamics are essential to build a sound phenomenology, as they restrict the number of independent parameters which one is allowed to consider in a given situation and which then can be determined empirically or by a microscopic theory.

Summary

The macroscopic study of a system evolving close to equilibrium states can be based upon the systematic approach of irreversible thermodynamics, the principles of which are summarized in § 14.2.6: identifying the processes involved – relaxation towards equilibrium, transport, forced regime, chemical kind of system, and so on; analyzing the system in terms of weakly coupled subsystems, each of which is practically at equilibrium; finding the macroscopic state variables and the conserved quantities; writing down the conservation equations, the local equations of state, and the equations describing the linear response of the fluxes to the affinities; using symmetry and invariance laws and Onsager relations and checking that the dissipation is positive; finally, solving the coupled evolution equations which we have obtained.

These elements, postulated by non-equilibrium thermodynamics, can be deduced from the microscopic equations of motion and the methods of statistical mechanics, provided the evolution is sufficiently slow. The proof of the conservation laws gives us the microscopic interpretation of the fluxes. The study of the dynamics involves, together with the microscopic description through a density operator $\widehat{D}$ which evolves according to the Liouville-von Neumann equation, a mesoscopic description through a simplified density operator $\widehat{D}_0$ which follows $\widehat{D}$ in its motion and which contains only information about the macroscopic, relevant variables. The relevant statistical entropy $S(\widehat{D}_0)$, which is the missing information associated with these variables, can be identified with the entropy of thermodynamics; its increase reflects a leak of information towards the other, irrelevant variables. The reduction, or contraction, of the description from $\widehat{D}$ to $\widehat{D}_0$ is the subject of the projection method; it associates the macroscopic quasi-equilibrium regime with a short-memory approximation for the dynamics of the irrelevant microscopic variables. The linear regime corresponds, moreover, to a weak coupling between subsystems.

The use of symmetries and invariances enables us to analyze the structure of the macroscopic dynamic equations, to reduce the number of independent response coefficients, and to connect various effects with one another. We illustrated the general method by studying in that spirit diffusion, heat or electric conduction, thermoelectric effects, and hydrodynamics. We proved, especially, the macroscopic laws governing the thermal and mechanical behaviour of Newtonian fluids, starting solely from the general principles of non-equilibrium thermodynamics.

Exercises

14a Design of an Isotope Separation Plant

The balance method, though simple, is quite efficient. This was illustrated on
the microscopic scale by studying effusion (Exercs.7g and 8a), an elementary
mechanism which enables us to separate the uranium hexafluoride molecules
with ^{238}U or ^{235}U according to their mass. Effusion is often called "gas diffusion" which should not be confused with the diffusion of § 14.4.1. Studying
a macroscopic balance will enable us, moreover, to understand the design of
plants using this process. One passage of the gas through a porous barrier
can only increase its molecular concentration C in ^{235}U by at most 4.3×10^{-3}
(Exerc.7g) so that one needs a large number of stages. The Eurodif plant in
Tricastin (Rhône valley) which treats one third of the world nuclear fuel and
feeds a hundred 1000 MW power stations has 1400 stages. We denote them
by p, with $-M < p \leq P$ and $M = 600$, $P = 800$. Each stage is a tower
of height between 16 and 23 m, containing from top to bottom a diffuser,
a heat exchanger, and a compressor. The gas, with a concentration C_p in
^{235}U, arrives at the bottom of the diffuser (Fig.14.4) and circulates there at
a pressure close to one atmosphere in vertical porous ceramic tubes, numbering several thousand. The pressure at the outside of the tubes is about
five times lower and the gas passing through their wall leaves enriched in the
lighter isotope ^{235}U with an average concentration C'_p. This gas is sucked
up and then compressed by the compressor at the bottom of the p-th stage
to be sent to the $p + 1$-st stage. The work done by the compression heats
the gas by a few tens of degrees (Exerc.8a); this heat is carried away by

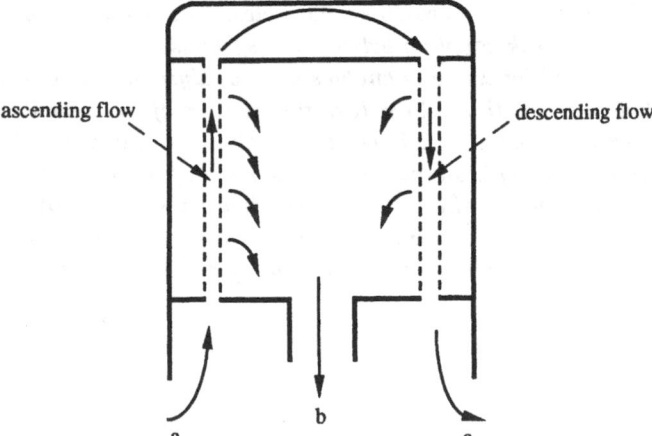

ascending flow   descending flow

Fig. 14.4. Sketch of a diffuser: (a) incoming mixture; (b) departing enriched gas;
(c) departing depleted gas. We show only two tubes: the mixture ascends in one of
them and descends in the other; in reality the cylindrical diffuser contains several
thousands of them

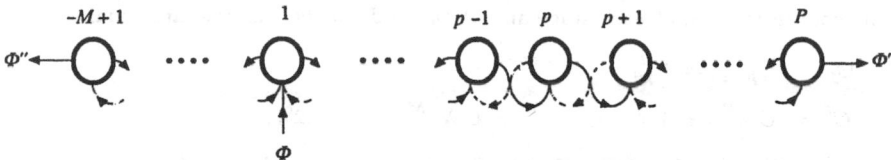

Fig. 14.5. The enrichment (solid lines) and depletion (dashed lines) cascades. The input (a) and outputs (b) and (c) in Fig.14.4 are schematically indicated here by the points below, to the right of, and to the left of each diffuser, respectively

the water flowing in the exchangers, and then ejected into the atmosphere by two cooling towers which dominate the site. Meanwhile, the flux in the tubes decreases gradually as the gas diffuses through their walls and as the richness in ^{235}U of the remaining gas decreases. The flow first is upwards and then, in fewer tubes, downwards, and the gas which leaves is depleted with a concentration C_p''. Calculate the molecular fluxes Φ_p' and Φ_p'' of the departing enriched and depleted gases in terms of the incoming flux Φ_p, assuming that $C_p'/C_p = C_p/C_p'' = \lambda = 1.002$.

The enriched flux Φ_p' leaving the p-th stage is sent to enter the $p + 1$-st stage and this gives rise to an enrichment cascade (Fig.14.5). The depleted gas coming from the p-th stage is recycled in the preceding $p - 1$-st stage, its concentration C_p'' being equal to C_{p-1}; to do that it is sent to the compressor of the $p - 2$-nd stage where it mixes with the flux Φ_{p-2}' to form the Φ_{p-1} flux. This gives rise to a depletion cascade (Fig.14.5). The whole chain is fed at the $p = 1$ stage by a flux Φ which has the natural concentration $C = 0.7\%$. A flux Φ' of enriched uranium with a concentration C', which will be used in power stations, leaves the $P = 800$ stage. The waste, a flux Φ'' of depleted uranium with concentration C'', leaves the stage $-M + 1 = -599$. Evaluate C' and C'', and also Φ' and Φ'' as functions of Φ. Calculate the fluxes Φ_p passing through the various stages. Especially, compare the maximum flux with the incoming flux Φ. What is the total flux passing through the compressors, per unit flux leaving the last enriching stage? Estimate the work consumed per kg of enriched uranium produced in this way.

Solution:

The conservation of total flux and of the ^{235}U flux in each diffuser gives us

$$\Phi_p = \Phi_p' + \Phi_p'', \qquad \Phi_p C_p = \Phi_p' C_p' + \Phi_p'' C_p'',$$

and hence

$$\frac{\Phi_p'}{C_p - C_p''} = \frac{\Phi_p''}{C_p' - C_p} = \frac{\Phi_p}{C_p' - C_p''}.$$

From this we get

$$\Phi_p' = \frac{\Phi_p}{\lambda + 1} = \frac{\Phi_p''}{\lambda}.$$

The concentrations, at each stage and at the ends of the cascade, are equal to

$$C_p = C'_{p-1} = C''_{p+1} = C\lambda^{p-1},$$

$$C' = C\lambda^P \simeq 3.5\%, \qquad C'' = C\lambda^{-M-1} \simeq 0.2\%;$$

in reality C' is equal to 3.2%. For the fluxes we get from the equation

$$\frac{\Phi'}{C - C''} = \frac{\Phi''}{C' - C} = \frac{\Phi}{C' - C''}$$

the relations

$$\Phi \simeq 6\Phi', \qquad \Phi'' \simeq 5\Phi'.$$

The balance at the entry to each stage can be written as

$$\Phi_p = \Phi'_{p-1} + \Phi''_{p+1},$$

except for $p = 1$, where we must add Φ, for $p = P$, where $\Phi''_{p+1} = 0$, and for $p = -M + 1$, where $\Phi'_{-M} = 0$. Hence we get the equation

$$(\lambda + 1)\Phi_p - \Phi_{p-1} - \lambda\Phi_{p+1} = 0, \qquad p \neq 1,$$

the general solution of which is the sum of two exponentials x^p of p, with either $x = 1$ or $x = \lambda^{-1}$. Using the boundary conditions at $p = 1$ and at the ends of the cascade, we get for $p \geq 1$

$$\Phi_p = \frac{\left(\lambda^{P+1-p} - 1\right)\left(\lambda^{M+1} - 1\right)(\lambda + 1)}{\left(\lambda^{P+M+1} - 1\right)(\lambda - 1)} \Phi \simeq 150\left(\lambda^{P+1-p} - 1\right)\Phi,$$

and for $p \leq 1$

$$\Phi_p = \frac{\left(1 - \lambda^{-M-p}\right)\left(\lambda^{P+M+1} - \lambda^{M+1}\right)(\lambda + 1)}{\left(\lambda^{P+M+1} - 1\right)(\lambda - 1)} \Phi \simeq 850\left(1 - \lambda^{-M-p}\right)\Phi.$$

The flux is a maximum for the input stage $p = 1$ and equals there $\Phi_1 \simeq 600\Phi$. The injected flux Φ represents only a tiny part of the fluxes which pass through each of the 1400 stages, except for the end stages, since the cycling and recycling make it necessary that the gas passes to and fro very many times along the cascade, especially near the centre. The total flux handled by the compressors is equal to

$$\Phi_{\text{tot}} = \sum_p \Phi_p = \frac{(\lambda + 1)\Phi'}{\lambda - 1}\left[(M + 1)\frac{\lambda^P - 1}{1 - \lambda^{-M-1}} - P\right] \simeq 2.6 \times 10^6\Phi'.$$

At each compression, the work done equals $kTN \ln \mathcal{P}/\mathcal{P}'$ where $T = 80\,°C$ and where $\mathcal{P}/\mathcal{P}' = 5$ for the compression of the enriched gas; that of the depleted gas is small so that one only needs to take into account half of Φ_{tot}. This gives us 50×10^9 J, or 7 MWh per kg. Our estimate agrees with the actual figures, since the effective consumption is 9 MWh per kg.

Note. According to the theory just given the stages should all differ from one another. In practice they are of three sizes: (a) 23 m, 150 tons, a 3.3 MW compressor;

(b) 19 m, 80 tons, 1.6 MW; (c) 16 m, 30 tons, 0.6 MW. One plays with small viola-
tions of the relations $C_p'/C_p = C_p/C_p''$ and $C_{p+1}'' = C_{p-1}'$, which we assumed above
to hold, to replace the ideal Φ_p curve which we evaluated by an, approximate, step
curve taking on only three values. The stages are combined in groups of 20; the 40
groups on the enriched side contain 16 groups (a), 14 groups (b), and 10 groups
(c), whereas the 30 groups at the depleted side contain 20 groups (a), 6 groups (b),
and 4 groups (c); these numbers have been chosen so as best to approximate the
Φ_p curve which we determined above.

14b Response to an External Perturbation

Many macroscopic – electromagnetic, electronic, mechanical, acoustic, thermal, ... – or microscopic effects – magnetic resonance, lasers, Josephson
effects in superconductors, elementary excitations in various substances,
neutron or light scattering, collisions in atomic, nuclear, or particle
physics, ... – can be described as the reaction of a system to an external
perturbation. The result is then a function of the time t, depending on the
strength of the perturbation at earlier times t'; in the linear regime this
dependence is characterized by a response function $\chi(t - t')$. In statistical
mechanics such a situation is described as follows. A system with Hamilton-
ian $\widehat{H}_0$, initially at time $t = -\infty$ in canonical equilibrium $\widehat{D}_0 \propto \exp(-\beta\widehat{H}_0)$,
is subjected to a weak time-dependent perturbation so that the Hamiltonian
has the form $\widehat{H}_0 + \sum_\alpha \widehat{B}_\alpha \widetilde{\lambda}_\alpha(t)$, where the $\widehat{B}_\alpha$ play the rôle of the "force"
observables $\widehat{X}$ of § 5.2.3 and the $\widetilde{\lambda}_\alpha$ those of the "position" variables ξ. The
final outcome is then the mean value $\langle A \rangle$ of some observable $\widehat{A}$, or more
precisely, $\Delta A = \langle A \rangle - \langle A \rangle_0$, $\langle A \rangle_0 = \text{Tr}\,\widehat{D}_0\widehat{A}$, which we want to determine as
function of the time for a given perturbation strength $\widetilde{\lambda}(t)$. As we are dealing
with a linear regime, it will be sufficient to take only one term $\widehat{B}\widetilde{\lambda}(t)$.

1. We expect to find, for small $\widetilde{\lambda}$, a relation between the input $\widetilde{\lambda}$ and the
output A of the form

$$\Delta A(t) = \int_{-\infty}^{+\infty} dt' \, \widetilde{\chi}_{AB}(t - t')\widetilde{\lambda}(t'), \qquad (14.152)$$

where $\widetilde{\chi}_{AB}$, which defines the *retarded response* of $\widehat{A}$ to the perturbation $\widehat{B}$,
should depend solely on the time difference $t - t'$ and should vanish when
$t' > t$. Determine $\widetilde{\chi}$ by solving the Liouville-von Neumann equation to first
order. Write $\widetilde{\chi}$ in the base in which $\widehat{H}_0$ and $\widehat{D}_0$ are diagonal with E_m and p_m
as their eigenvalues.

One should take care not to confuse the linear responses of thermodynamics (§ 14.2.2), which relate the fluxes to the affinities, with the linear
responses introduced here, which describe the effects of a time-dependent
perturbation applied to the system. Whereas thermodynamics deals only
with *quasi-equilibrium* states, the present theory can be applied even if $\widetilde{\lambda}$
and hence A *vary rapidly*, on microscopic time scales. As shown by (14.152),

we integrate the influence of the perturbation over the past history. The response function $\widetilde{\chi}$ thus takes *memory effects* into account, in contrast to the response coefficients L. However, the present theory is restricted by the fact that the state $\widehat{D}(t)$ differs only slightly from the initial *global* equilibrium state $\widehat{D}_0$, even though its variations are fast. In thermodynamics, the small difference was between $\widehat{D}(t)$ and its associated *local* equilibrium state $\widehat{D}_0(t)$ (§ 14.3.5).

2. Especially because of the practical interest in *periodic* perturbations – for instance, an alternating electric potential – it is convenient to perform a Fourier transformation,

$$\lambda(\omega) = \int dt\, e^{i\omega t}\, \widetilde{\lambda}(t), \qquad \widetilde{\lambda}(t) = \frac{1}{2\pi} \int d\omega\, e^{-i\omega t}\, \lambda(\omega),$$

$$\chi_{AB} = \int dt\, e^{i\omega t}\, \widetilde{\chi}_{AB}(t) \equiv \chi'_{AB} - i\chi''_{AB}(\omega).$$

What are the expressions for $\Delta A(t)$ and for its Fourier transform? Determine χ_{AB} and also

$$\chi''_{AB} \equiv \frac{i}{2} \left[\chi_{AB}(\omega) - \chi_{BA}(-\omega)\right].$$

Study the symmetry properties of $\widetilde{\chi}, \chi, \chi', \chi''$. Show that *causality* is reflected in the fact that $\chi(\omega)$ is the limit as $z \to \omega + i0$ of a function $\chi(z)$ defined in the upper z-half-plane, *analytic*, without singularities, and bounded at infinity. Show how one can get this function from $\chi''(\omega)$. We shall see below that χ'' is the dissipative part of the response.

3. Assume that $\chi(\omega)$ is a smooth function, or that $\widetilde{\chi}$ has a finite range in t. What is the expression for ΔA in the limit where the perturbation changes slowly? What are the expressions for the responses $\widetilde{\chi}$ or χ in that limit? Compare with thermodynamics.

4. In the case of a sinusoidal input $\widetilde{\lambda}(t) = \lambda \cos \omega_0 t$ the output is $\Delta A = a \cos(\omega_0 t - \varphi)$. Determine its amplitude a and its phase shift φ. Is it possible that a system operates as a perfect filter, that is, that the output $a(\omega_0)$ vanishes for all frequencies $\omega_0/2\pi$ over a certain range? Is it possible for $a(\omega_0)$ and $\varphi(\omega_0)$ to be constant over a certain range? Or for $\varphi(\omega_0)$ to be constant for all frequencies? What is the result $\Delta A(t)$ for a perturbation strength $\widetilde{\lambda}(t) = \lambda \cos \omega_0 t\, e^{\gamma t}$ and how can we interpret $\chi(z)$ for complex z in the upper half-plane?

5. Depending on the problem, the responses χ can be determined either theoretically or empirically – for instance, the *impedance* in an electrical circuit. In both circumstances, it often happens that they show *resonances*, which are sharp peaks and which are associated with poles $z = \Omega - i\Gamma$ of $\chi(z)$, when that function is analytically continued into the lower half-plane. Assume that χ is given by the one-resonance approximation,

$$\chi_{AB}(z) = \frac{Qe^{i\alpha}}{z - \Omega + i\Gamma} - \frac{Qe^{-i\alpha}}{z + \Omega + i\Gamma};$$

we need the two terms because of the symmetry $\chi(\omega) = \chi^*(-\omega)$, the parameters which occur are real, and we assume that $\Gamma \ll \Omega$. How do the amplitude a and the phase shift φ of $\Delta A(t)$ behave when ω_0 sweeps the vicinity of Ω in the case of a sinusoidal perturbation $\tilde{\lambda}(t) = \lambda \cos \omega_0 t$? This behaviour enables one to observe a resonance in a *forced oscillation* regime and to determine its position Ω and its width Γ. One can also observe it in a *relaxation* regime. In that case one subjects the system to an impulse $\tilde{\lambda}(t) = \lambda \delta(t)$; how does $\Delta A(t)$ behave for $t > 0$? One can also remove $\langle A \rangle$ slowly from its equilibrium position, then let it go at $t = 0$, using a perturbation $\tilde{\lambda}(t) = \lambda e^{\gamma t} \theta(-t)$ with $\gamma \to 0$; how does $\Delta A(t)$ behave in that case?

6. Express the change dU/dt in the energy $U \equiv \langle \widehat{H}_0 + \widehat{B}\tilde{\lambda}(t) \rangle$ with time, to second order in λ. Calculate the total energy received by a system when it has been subjected to a perturbation $\tilde{\lambda}(t)$ which is localized in time so that $\widehat{H}_0$ returns to its initial value. Show, by expressing this energy in terms of $\chi''_{BB}(\omega)$, that it is always positive. Calculate, within the same regularity hypotheses as under 3, the change in energy with time in the case of a sinusoidal perturbation $\tilde{\lambda}(t) = \lambda \cos \omega_0 t$ and interpret χ''.

7. Express this energy dissipation χ''_{BB}, or more generally, $\chi''_{AB}(\omega)$, in terms of the Fourier transform of the time-dependent correlations $\psi_{AB}(\omega)$ and $\psi_{BA}(\omega)$, which are defined by

$$\tilde{\psi}_{AB}(t) = \mathrm{Tr}\, \widehat{D}_0\, e^{i\widehat{H}_0 t/\hbar}\, \widehat{A}\, e^{-i\widehat{H}_0 t/\hbar}\, \widehat{B} - \langle A \rangle_0 \langle B \rangle_0.$$

This is *Kubo's relation* or the *fluctuation-dissipation theorem*. What happens to it in the classical limit? What happens, as $\hbar \to 0$ and $t = 0$? Use it to find the *noise* $\langle I(t)I(0)\rangle$ due to fluctuations in the current $I(t)$ in an open electrical resistor R; this is the *Nyquist theorem*.

Solution:

1. We can write the Liouville-von Neumann equation as

$$i\hbar \frac{d}{dt} \left(e^{i\widehat{H}_0 t/\hbar} \widehat{D}\, e^{-i\widehat{H}_0 t/\hbar} \right) = \tilde{\lambda}(t)\, e^{i\widehat{H}_0 t/\hbar} \left[\widehat{B}, \widehat{D} \right] e^{-i\widehat{H}_0 t/\hbar},$$

a form suggested by the solution for $\widehat{D}(t)$ when there are no perturbations. The right-hand side is small and to first order we can replace in it $\widehat{D}$ by $\widehat{D}_0$ whence we find after integration that

$$\widehat{D} \approx \widehat{D}_0 + \frac{1}{i\hbar} \int_{-\infty}^{t} dt'\, \tilde{\lambda}(t') \left[e^{-i\widehat{H}_0(t-t')/\hbar} \widehat{B}\, e^{i\widehat{H}_0(t-t')/\hbar}, \widehat{D}_0 \right].$$

We took into account the fact that $\hat{H}_0$ and $\hat{D}_0$ commute. If we now use the cyclic invariance of the trace, we get

$$\langle A \rangle = \text{Tr}\, \hat{D}(t)\hat{A} = \text{Tr}\, \hat{D}_0\hat{A} + \int_{-\infty}^{+\infty} dt'\, \widetilde{\chi}_{AB}(t - t')\widetilde{\lambda}(t'),$$

where

$$\widetilde{\chi}_{AB}(t) = \frac{\theta(t)}{i\hbar}\, \text{Tr}\, \hat{D}_0 \left[e^{i\hat{H}_0 t/\hbar}\, \hat{A}\, e^{-i\hat{H}_0 t/\hbar}, \hat{B} \right]$$

$$= \frac{\theta(t)}{i\hbar} \sum_{mn} \left(p_m - p_n \right) A_{mn} B_{nm}\, e^{i(E_m - E_n)t/\hbar}. \tag{14.153a}$$

2. Fourier transforming $\widetilde{\chi}$ and $\widetilde{\lambda}$ gives us

$$\Delta A(t) = \frac{1}{2\pi} \int d\omega\, e^{-i\omega t}\chi_{AB}(\omega)\lambda(\omega),$$

and the Fourier transform of $\Delta A(t)$ is $\chi_{AB}(\omega)\lambda(\omega)$. We have

$$\chi_{AB}(\omega) = \sum_{mn} \frac{(p_m - p_n)A_{mn}B_{nm}}{E_m - E_n + \hbar\omega + i0}, \tag{14.153b}$$

$$\chi''_{AB}(\omega) = \pi \sum_{mn} (p_m - p_n)A_{mn}B_{nm}\, \delta(E_m - E_n + \hbar\omega). \tag{14.153c}$$

Responses to Hermitean perturbations must be real, which is reflected in the general equations

$$\widetilde{\chi}_{AB}(t) = \widetilde{\chi}_{A^\dagger B^\dagger}(t)^*, \qquad \chi_{AB}(-\omega) = \chi_{A^\dagger B^\dagger}(\omega)^*,$$

$$\chi'_{AB}(-\omega) = \chi'_{BA}(\omega) = \chi'_{A^\dagger B^\dagger}(\omega)^*,$$

$$\chi''_{AB}(-\omega) = -\chi''_{BA}(\omega) = -\chi''_{A^\dagger B^\dagger}(\omega)^*.$$

When the perturbation $\sum \hat{B}_\alpha \widetilde{\lambda}_\alpha(t)$ includes several terms, each $\hat{B}_\alpha$ need not be Hermitean; for instance, if $\hat{B}_1 = \hat{B}_2^\dagger$, we must have $\widetilde{\lambda}_1 = \widetilde{\lambda}_2^*$. If $\hat{H}_0$ is invariant under time reversal, we have

$$\widetilde{\chi}_{A^T B^T}(t) = \widetilde{\chi}_{B^\dagger A^\dagger}(t), \qquad \chi_{A^T B^T}(\omega) = \chi_{B^\dagger A^\dagger}(\omega),$$

so that the retarded responses satisfy the same Onsager relations, with an extra t- or ω-dependence, as the responses L of thermodynamics. If $\hat{A}$ or $\hat{B}$ commutes with $\hat{H}_0$, χ_{AB} vanishes.

Causality means that $\Delta A(t)$ at time t only responds to values of $\widetilde{\lambda}(t')$ at earlier times t' so that $\widetilde{\chi}(t)$ vanishes for $t < 0$, as is shown by the factor $\theta(t)$ in (14.153a). The function

$$\chi_{AB}(z) \equiv \int_0^\infty dt\, e^{izt}\chi_{AB}(t)$$

is then analytic for $\text{Im}z \geq 0$, and it is the analytical continuation of $\chi_{AB}(\omega)$. Inversely,

$$\frac{1}{2\pi} \int_{-\infty}^{+\infty} d\omega\, \chi_{AB}(\omega)\, e^{-i\omega t}$$

vanishes for $t < 0$, as the contour can then be closed by a large semi-circle in the upper ω-half-plane. The definitions of χ and χ'' give

$$\chi_{AB}(z) = \frac{1}{\pi} \int_{-\infty}^{+\infty} \frac{d\omega}{z - \omega}\, \chi''_{AB}(\omega), \qquad \text{Im}z > 0, \qquad (14.154)$$

so that knowledge of χ'' is sufficient to determine χ'. This integral relation, called a *dispersion relation*, or a *Kramers-Kronig relation*, is often useful, for instance, to determine the high-frequency behaviour of χ.

3. If $\tilde{\lambda}(t)$ changes slowly, $\lambda(\omega)$ is concentrated at small values of ω and we can replace $\chi(\omega)$ by $\chi(0)$, which leads to

$$\Delta A(t) = \frac{1}{2\pi} \chi_{AB}(0)\lambda(t).$$

The average value $\langle A \rangle$ follows the perturbation at the time t itself, without any memory effect. We have found again the behaviour of quasi-equilibrium thermodynamics. The instantaneous response is equal to

$$\chi_{AB}(0) = \sum_{mn} \frac{(p_m - p_n)A_{mn}B_{nm}}{E_m - E_n + i0} = \int_{-\infty}^{t} dt'\, \tilde{\chi}_{AB}(t - t'),$$

as in (14.81), (14.82). Because of the term $i0$ in the denominator the terms with $p_m = p_n$ do not contribute. These results can also be derived directly from (14.152), when $\tilde{\lambda}(t')$ varies slowly over the range of $\chi(t - t')$.

4. The Fourier transform of $\Delta A(t)$ is

$$\pi\lambda \left[\delta(\omega - \omega_0)\chi_{AB}(\omega_0) + \delta(\omega + \omega_0)\chi_{AB}(-\omega_0)\right],$$

so that

$$\Delta A(t) = \text{Re}\left[\chi_{AB}(\omega_0)\, \lambda\, e^{-i\omega_0 t}\right], \qquad (14.155)$$

and hence $a = \lambda|\chi(\omega_0)|$, $\varphi = \arg\chi(\omega_0)$.

The distribution $\chi(\omega)$ is the boundary value at $z = \omega + i0$ of an analytic function $\chi(z)$; on the other hand, when one gives an analytic function on a continuum of z-values, it is determined everywhere. Hence the vanishing of $a(\omega_0)$ over some range of frequencies implies that $\chi = 0$: *causality forbids the existence of perfect filters*. Similarly, a and φ can be constant over some range of frequencies only if $\chi(z)$ is everywhere a constant, so that this occurs only when no incoming signal is deformed. Finally, if $\chi(\omega)e^{-i\varphi}$ has an imaginary part which vanishes for all ω, it again is a constant.

In the case when $\tilde{\lambda}(t) = \lambda \cos \omega_0 t\, e^{\gamma t}$, there is no Fourier transform $\lambda(\omega)$ and we go back to (14.152) for the evaluation of $\Delta A(t)$:

$$\Delta A(t) = \int_{-\infty}^{t} dt' \int \frac{d\omega}{2\pi} \chi_{AB}(\omega) \, e^{-i\omega(t-t')} \lambda \cos \omega_0 t' \, e^{\gamma t'}$$

$$= \operatorname{Re}\left[\chi_{AB}(\omega_0 + i\gamma)\lambda \, e^{-i\omega_0 t + \gamma t}\right].$$

The integral over ω was evaluated by closing the contour at infinity in the upper half-plane. The response $\chi(z)$, where z lies in the upper half-plane, can thus directly be interpreted, as it gives the amplitude and the phase of the output, for an incoming signal proportional to $\cos \omega_0 t \, e^{\gamma t}$.

5. The contribution from the second term in χ is negligible for $|\omega_0 - \Omega| \ll 2\Omega$ so that we have

$$a \simeq \frac{\lambda Q}{\sqrt{(\omega_0 - \Omega)^2 + \Gamma^2}}, \qquad \varphi \simeq \alpha - \arg(\omega_0 - \Omega + i\Gamma). \qquad (14.156a)$$

It is well known that the amplitude of the outgoing signal as function of ω_0 has a very sharp *maximum* at $\omega_0 = \Omega$ with a width of the order of Γ; the phase shift *increases fast* over that width from $\alpha - \pi$ to α. Measuring either the amplitude or the phase shift enables us to determine Ω and Γ.

After an impulse we get, using the fact that $\lambda(\omega) = \lambda$,

$$\Delta A(t) = \lambda \chi_{AB}(t) = \frac{\lambda}{\pi} \operatorname{Re} \int d\omega \, \frac{Q e^{i\alpha - i\omega t}}{\omega - \Omega + i\Gamma}$$

$$= -2\lambda Q \sin(\Omega t - \alpha) \, e^{-\Gamma t} \theta(t). \qquad (14.156b)$$

The variable A relaxes towards its initial value *over a time of order $1/\Gamma$, while oscillating with a frequency $\Omega/2\pi$*. For the relaxation after a progressive deviation we have $\lambda(\omega) = \lambda/(\gamma + i\omega)$ and

$$\Delta A(t) = \frac{1}{2\pi} \int \frac{\lambda \, d\omega}{\gamma + i\omega} \, e^{-i\omega t} \chi_{AB}(\omega)$$

$$= \lambda e^{\gamma t} \chi_{AB}(i\gamma)\theta(-t) - 2\lambda Q \, e^{-\Gamma t} \operatorname{Re} \frac{e^{i\Omega t - i\alpha}}{\Omega + i(\Gamma + \gamma)} \, \theta(t). \qquad (14.156c)$$

The first term describes the slow process which, as under 3 moves $\langle A \rangle$ away from its equilibrium value to lead, at $t = 0$, to $\Delta A \simeq -2\lambda Q \cos \alpha/\Omega$, in the case when $\gamma \ll \Gamma \ll \Omega$; the second term is a relaxation of the form

$$\Delta A \simeq -\frac{2\lambda Q}{\Omega} \, \cos(\Omega t - \alpha) \, \theta(t),$$

which is comparable with the relaxation after an impulse, apart from a phase shift $\frac{1}{2}\pi$ and a factor Ω. This factor, which reduces the higher harmonics for a system such as a vibrating string having a series of resonances $\Omega_n = n\Omega$, is partly responsible for the difference in tone between a piano where the string relaxes after having been hit and a harp or a harpsichord where the string relaxes after having been plucked.

6. In the case of a time-dependent Hamiltonian the Liouville-von Neumann equation gives us

$$\frac{dU}{dt} = \frac{d}{dt} \, \mathrm{Tr} \, \widehat{D}(t)\widehat{H}(t) = \frac{1}{i\hbar} \, \mathrm{Tr} \, \big[\widehat{H},\widehat{D}\big] \, \widehat{H} + \mathrm{Tr} \, \widehat{D}(t) \, \frac{\partial \widehat{H}}{\partial t} = \left\langle \frac{\partial \widehat{H}}{\partial t} \right\rangle.$$

It therefore suffices to use the response of B to itself to first order to obtain the evolution of the energy to second order:

$$\frac{dU}{dt} \approx \frac{d\widetilde{\lambda}}{dt} \left[\langle B \rangle_0 + \int dt' \, \widetilde{\chi}_{BB}(t-t')\widetilde{\lambda}(t') \right]$$

$$= \frac{d\widetilde{\lambda}}{dt} \langle B \rangle_0 + \frac{1}{2\pi} \frac{d\widetilde{\lambda}}{dt} \int d\omega \, e^{-i\omega t} \chi_{BB}(\omega)\lambda(\omega).$$

Extending this result to a Hamiltonian $\widehat{H}_0 + \sum \widehat{B}_\alpha \widetilde{\lambda}_\alpha(t)$ would involve a double summation over α in the second term, as in (14.157a) below.

If the Hamiltonian returns to its initial value, ΔA or ΔU also returns to zero to first order in λ as $t \to +\infty$. We can use a Fourier transformation to get to second order for the energy

$$U(+\infty) - U(-\infty) = \langle B \rangle_0 \int_{-\infty}^{+\infty} dt \, \frac{d\widetilde{\lambda}}{dt} + \frac{1}{(2\pi)^2} \int_{-\infty}^{+\infty} dt \, d\omega' \, d\omega \, (-i\omega')$$

$$\times \, e^{-i(\omega+\omega')t} \lambda(\omega') \chi_{BB}(\omega)\lambda(\omega)$$

$$= \frac{i}{2\pi} \int d\omega \, \omega \lambda(-\omega) \chi_{BB}(\omega)\lambda(\omega).$$

Using the definition of χ'' and the symmetry property $\chi_{BB}(-\omega) = \chi^*_{BB}(\omega)$ we find

$$U(+\infty) - U(-\infty) = \frac{1}{\pi} \int_0^\infty d\omega \, \omega \, \chi''_{BB}(\omega) \, |\lambda(\omega)|^2,$$

or more generally

$$U(+\infty) - U(-\infty) = \frac{1}{\pi} \sum_{\alpha,\beta} \int_0^\infty d\omega \, \omega \, \chi''_{B_\alpha^\dagger B_\beta}(\omega) \, \lambda_\alpha^*(\omega)\lambda_\beta(\omega). \tag{14.157a}$$

Expression (14.153c) shows that $\chi''_{B^\dagger B}(\omega)$ has the same sign as ω so that any perturbation *in a closed cycle increases the energy* of the system. More precisely, we have

$$U(+\infty) - U(-\infty) = \frac{1}{2\hbar^2} \sum_{mn} (p_m - p_n)(E_n - E_m)|B_{mn}|^2$$

$$\times \left| \lambda \left([E_n - E_m]/\hbar \right) \right|^2, \tag{14.157b}$$

where $p_m \propto e^{-\beta E_m}$ decreases with increasing E_m, so that each term is positive.

In the case when $\lambda(\omega) = \pi\lambda \big[\delta(\omega - \omega_0) + \delta(\omega + \omega_0) \big]$, one obtains, separating the terms in χ' and χ'',

$$\frac{dU}{dt} = -\langle B \rangle_0 \lambda\omega_0 \sin \omega_0 t$$

$$+ \lambda^2 \omega_0 \sin \omega_0 t \, \big[\chi''_{BB}(\omega_0) \sin \omega_0 t - \chi'_{BB}(\omega_0) \cos \omega_0 t \big]$$

$$= -\langle B \rangle_0 \lambda \omega_0 \sin \omega_0 t - \tfrac{1}{2} \lambda^2 \omega_0 \, |\chi_{BB}(\omega_0)| \, \sin(2\omega_0 t - \varphi)$$
$$+ \tfrac{1}{2} \lambda^2 \omega_0 \chi''_{BB}(\omega_0). \tag{14.157c}$$

The last term describes a linear increase in the energy so that $\chi''(\omega)$ is proportional to the *energy dissipation* in a system subject to a perturbation of frequency $\omega/2\pi$. The preceding terms describe oscillations in U with the imposed or twice the imposed frequency; their time-average vanishes. Both the energy dissipation and the amplitude of the energy oscillations are sharply peaked at a resonance.

Expression (14.157c) for the dissipation can be interpreted in terms of *Fermi's Golden Rule* which gives the probability Q_m that a system in a state m makes per unit time a transition to one of the states n of energy $E_m + \hbar \omega_0$ as the effect of a perturbation $\hat{B} \lambda \cos \omega_0 t$:

$$Q_m(\omega_0) = \frac{\pi \lambda^2}{2\hbar} \sum_n |B_{mn}|^2 \, \delta\big(E_m - E_n + \hbar \omega_0\big). \tag{14.158}$$

For transitions involving energies $\pm \hbar \omega_0$ we then get by striking a balance

$$\frac{dU}{dt} = \hbar \omega_0 \sum_m p_m \, [Q_m(\omega_0) - Q_m(-\omega_0)] = \tfrac{1}{2} \lambda^2 \omega_0 \chi''_{BB}(\omega_0).$$

7. We can calculate $\widetilde{\psi}$ in the same way as $\widetilde{\chi}$ in the first section, and we find

$$\widetilde{\psi}_{AB}(t) = \sum_{mn} p_m A_{mn} B_{nm} \, e^{i(E_m - E_n)t/\hbar} - \langle A \rangle_0 \langle B \rangle_0.$$

Its Fourier transform equals

$$\psi_{AB}(\omega) = 2\pi \hbar \sum_{mn} p_m A_{mn} B_{nm} \, \delta\big(E_m - E_n + \hbar \omega\big) - 2\pi \langle A \rangle_0 \langle B \rangle_0 \delta(\omega).$$

On the other hand, as $\hbar \omega = E_n - E_m$, the factor p_n of $\chi''_{AB}(\omega)$ equals $p_m e^{-\beta \hbar \omega}$ so that

$$\chi''_{AB}(\omega) = \frac{1}{2\hbar} \left(1 - e^{-\beta \hbar \omega}\right) \psi_{AB}(\omega) = \frac{1}{2\hbar} \left(e^{\beta \hbar \omega} - 1\right) \psi_{BA}(-\omega).$$

Thanks to the presence of the term with $\langle A \rangle_0 \langle B \rangle_0$ in $\widetilde{\psi}$, this function tends to zero at infinity; in the Fourier transform the contributions with $\delta(\omega)$ cancel each other. This enables us to write the *fluctuation-dissipation relation* or *Kubo relation* in the form

$$\psi_{AB}(\omega) = \frac{2\hbar}{1 - e^{-\beta \hbar \omega}} \, \chi''_{AB}(\omega), \tag{14.159}$$

even in the vicinity of $\omega = 0$, where $\chi'' \to 0$.

In the classical limit we find

$$\psi_{AB}(\omega) = \psi_{BA}(-\omega) = \frac{2kT}{\omega} \, \chi''_{AB}(\omega).$$

At $t = 0$ the Kramers-Kronig relation then gives

$$\widetilde{\psi}_{AB}(0) = \frac{1}{2\pi} \int d\omega \, \frac{2kT}{\omega} \, \chi''_{AB}(\omega) = -kT\chi_{AB}(0). \tag{14.160}$$

In the case of a classical system the correlation or the fluctuation $\widetilde{\psi}(0)$ in equilibrium is proportional to the response $\chi(0)$ to a static perturbation (Exercs.4a and 5d).

When we apply at the ends of a resistor R a periodic potential difference $V \cos \omega_0 t$, the dissipated Joule power equals $V^2/2R$. This potential plays the rôle of $\lambda(t)$ and in the Hamiltonian it is coupled to the charge Q transferred from one end of the sample to the other. We can thus identify $\chi''_{QQ}(\omega)$ with $1/R\omega$. This result can also be obtained by noting that the relation $\widetilde{V}(t) = R\widetilde{I}(t)$ means that $\chi_{IQ}(\omega) = 1/R$; as $\widetilde{I}(t) = -d\widetilde{Q}(t)/dt$ it follows that $I(\omega) = i\omega Q(\omega)$ and hence that $\chi_{QQ}(\omega) = -i/R\omega$ which implies that $\chi'_{QQ}(\omega) = 0$, $\chi''_{QQ}(\omega) = 1/R\omega$. As a result we have $\chi''_{II}(\omega) = \omega/R$ and hence $\psi_{II}(\omega) = 2kT/R$. Altogether we find from the Kubo relation the *Nyquist theorem*

$$\langle I(t)I(0) \rangle = \frac{1}{2\pi} \int d\omega \, e^{-i\omega t} \, \frac{2kT}{R} = \frac{2kT}{R} \, \delta(t).$$

The noise is "white noise" with a very short memory; in other words, its Fourier transform is a constant. Its amplitude is proportional to the temperature and to the conductivity (Exerc.5c).

Notes:

In studying under 3 slow processes and under 4, 5, and 6 periodic processes, we have implicitly assumed that $\widetilde{\chi}(t)$ tends to zero as $t \to \infty$, or that $\chi''(\omega)$ is a regular function of ω. However, for a *finite system* the spectrum of $\widehat{H}_0$ is discrete, so that (14.153a) defines $\widetilde{\chi}(t)$ as an enumerable sum of periodic terms; one can show that such a function, which is called almost periodic, returns for sufficiently large t to values close to its initial value and, hence, cannot tend to zero. A finite system thus has an *infinitely long memory*. Strictly speaking, one cannot apply here the concepts of thermodynamics, and $\Delta A(t)$ retains a recollection of past history even in the case of a very slowly varying perturbation $\widetilde{\lambda}(t)$. The difficulty crops up in a different form for the Fourier transform where (14.153c) gives $\chi''(\omega)$ as a sum of Dirac distributions, and not as a regular function. The response (14.155) or (14.157c) to a periodic perturbation is therefore pathological in the case of a finite system: it diverges if $\hbar\omega_0$ equals a difference $E_n - E_m$ between two levels of $\widehat{H}_0$. In other words, a finite system leads to *infinitely narrow and sharp resonances*. According to (14.156) it relaxes *without damping* ($\Gamma = 0$).

Nevertheless, for the *macroscopic* systems studied in statistical physics, the spectrum of $\widehat{H}_0$ is extremely *dense*: we have seen that typical distances ΔE between neighbouring levels decrease exponentially with the particle number. As a result, the recurrence times after which $\widetilde{\chi}(t)$ returns to the neighbourhood of its initial value are huge, of the order of $1/\hbar\Delta E$, much larger than the duration of our experiments or even than the age of the Universe. On a reasonable time scale, $\Delta A(t)$ therefore follows $\widetilde{\lambda}(t)$ without any memory effect, as long as $\widetilde{\lambda}(t)$ evolves slowly. On the other hand, even the smallest uncertainty in the frequency ω_0 suffices to smooth (14.155) or (14.157c), which means that in the expressions (14.153) for the response we can replace the sums by integrals. Once we have made this change, the distribution $\chi''(\omega)$ becomes a continuous function, and $\widetilde{\chi}(t)$ becomes a function which tends to zero as $t \to \infty$; this makes the results obtained above valid.

In atomic, nuclear, or particle physics the number of elementary constituents is often small, but the objects studied are not confined to a finite volume so that we are dealing with *continuous* spectra; this thus makes $\chi''(\omega)$ a regular function and enables it, for instance, to show resonances with a finite width Γ, as under 5. In this case, it is the possibility for radiating or for emitting fragments *into the surrounding space* which leads to irreversibility and dissipation. The finite lifetime of excited atomic states can then be identified with the relaxation time Γ^{-1} in (14.156c). On the other hand, recent experiments where an atom was confined to a small cavity have shown that the levels again become discrete: preventing the photon radiation to escape suppressed dissipation, and one observed, for example, *undamped* oscillations between atomic levels.

14c Desorption

We want to study the dynamics of the degassing of an adsorbing wall in the very schematic model of Exerc.4b. We assume that the wall, on the one hand, and the gas, on the other hand, are at equilibrium with, respectively, the variables α' and α, and β' and β. Show that the four response coefficients associated with particle and energy exchanges are related to one another. If the gas is at a temperature T and a pressure $\mathcal{P}$, write down the evolution equation for the number of adsorbed molecules and discuss the effect of $\mathcal{P}$ and T.

Hints. The particle and energy fluxes Φ_N and Φ_E from the wall to the gas are connected through $\Phi_E \equiv -u\Phi_N$, which leads to $L_{EN} = -uL_{NN}$, $L_{EE} = -uL_{NE} = u^2 L_{NN}$ (Onsager). It then follows from (14.15) and the equations of state of the gas and of the wall that

$$\frac{dn}{dt} = -kL_{NN} \ln\left(\frac{n}{N-n}\, \frac{\mathcal{P}_0(T)}{\mathcal{P}}\right),$$

where $\mathcal{P}_0(T) = kT\,e^{-u/kT}\lambda_T^{-3}$. The approach of n to the Langmuir equilibrium value is slow because of the logarithmic behaviour of the flux.

15. Kinetic Equations

"... des centaines de mille d'Égyptiens dont les costumes
blancs ou bigarrés de couleurs vives papillotaient au soleil,
dans ce fourmillement perpétuel qui caractérise la multitude,
même lorsqu'elle semble immobile. ... Aussitôt qu'Aharon
eut fait le geste, surgirent des millions de grenouilles; mues
comme par des ressorts, elles bondissaient entre les jambes,
à droite, à gauche, en avant, en arrière. A perte de vue, on
les voyait clapoter, sauteler, passer les unes sur les autres."

Théophile Gautier, Le Roman de la Momie

"A notre avis, le caractère métaphysique le plus profond de
la théorie cinétique des gaz, c'est qu'elle réalise une transcendance de la qualité, en ce sens qu'une qualité n'appartenant
pas aux composants appartient cependant au composé. C'est
contre cette transcendance que protestent sans fin les esprits
logiques."

Gaston Bachelard, Le Nouvel Esprit Scientifique, 1934

Whereas there exists a unified microscopic approach for equilibrium problems, the methods used in non-equilibrium statistical physics are varied and
often adjusted to the questions that are treated. We have given the main
features of two examples, the projection method (§ 14.3.5) and the response
theory (Exerc.14b). We have also used rough balance methods in §§ 7.4, 11.3,
and 13.3, where we met with kinetic theory in an elementary and qualitative
form. Here we return to this kinetic theory in a more detailed manner[1]. It
applies to systems such as classical gases, charge carrier gases in semiconductors and metals, or neutrons in a nuclear reactor, all of which can be
described as sets of particles interacting through forces with a range, short
as compared to their distances apart. This makes it possible to distinguish
at the microscopic level two time scales, the very short duration of each collision and the time between successive collisions of a particle. The latter is
itself much shorter than the characteristic evolution times of non-equilibrium
thermodynamics. We shall construct and study the kinetic equations which
satisfactorily describe the dynamics for times, long compared to the duration
of the collisions. One gets these equations by striking the *balance* of the dis-

---

[1] An extensive bibliography can be found in James A.McLennan, *Introduction to
Nonequilibrium Statistical Mechanics*, Prentice Hall, Englewood Cliffs, NJ, 1989.

placements of particles in the *single-particle phase space*, under the effects of their collisions and of their free motion between collisions.

We start by studying the *Lorentz gas* where non-interacting particles collide with scattering centres which are fixed at random positions (§ 15.1). This simple model is suitable for a microscopic study of phenomena such as diffusion or conduction (§ 15.2.1); it can easily be generalized for a treatment of the motion of charge carriers in a semiconductor (§ 15.2.2) or in a metal (§ 15.2.3), of energy exchanges through collisions between electrons and imperfections in a crystal (§ 15.2.4), or of the physics of neutron flow in matter (§ 15.2.5). We then study the *Boltzmann equation* (§ 15.3) which describes in a realistic manner the properties of non-equilibrium classical gases if intermolecular collisions are taken into account. When there are several kinds of molecules, the Boltzmann equation also provides us with a microscopic approach to the mutual diffusion phenomenon (§ 15.3.4). We take up the theory of an important limiting case, *Brownian motion*, that is, the random motion of a heavy particle plunged into a medium consisting of light particles (§ 15.3.5).

When analyzing the solutions of the kinetic equations near equilibrium by the so-called *Chapman-Enskog* method (§ 15.1.5) we shall for classical gases obtain a microscopic justification for the general principles of thermodynamics which we stated in Chap. 14. At the same time, we shall express the transport coefficients, such as the thermal conductivity or the viscosity for a Boltzmann gas, in terms of the *cross-section* for scattering of molecules by one another (§ 15.3.3). However, the kinetic equations cover a domain much wider than that of transport in a local equilibrium regime (§ 15.1.4). In fact, they also enable us to study situations *far from equilibrium*: hyperfrequencies and micro-miniaturization in electronics, shock waves and boundary layers in aerodynamics, neutron physics, and so on. One thus needs to use them in the many cases where the macroscopic phenomenology of thermodynamics does not work for applications to pure science or to technology.

We end with a discussion of the *irreversibility paradox* in the framework of kinetic theory (§ 15.4). The key to this problem lies in the existence of several, more or less detailed, descriptions of the same process, ranging from the microscopic to the macroscopic scale, and passing through one or more mesoscopic descriptions (§§ 15.1.2 and 15.4.1). The entropy appears then, not as a unique function of the state of the system, but as a *relative* concept; its growth reflects an increase in disorder or a loss of information, not in an absolute sense,but *at the level of the chosen description* (§§ 15.4.2 to 15.4.4). The thermodynamic entropy is a special case, associated with the macroscopic description. This relative nature of the entropy may appear surprising; for that reason we discuss it using the example of *spin echo* experiments which exhibit the subtle nature of macroscopic irreversibility (§ 15.4.5).

15.1 The Lorentz Model

15.1.1 Gases with Scattering Centres

The first examples of systems, the properties of which we are going to study starting from the microscopic structure, are sets of *classical non-interacting* particles with a dynamics controlled by *collisions with scattering centres*, considered to be exterior to the system, which therefore is *not isolated*.

Such a model satisfactorily represents a *semiconductor* : we have shown in § 11.3.2 that the negative and positive charge carriers, that is, electrons in the conduction band and holes in the valence band, behave like two classical gases. We saw that the definition itself of these charge carriers takes into account the Coulomb potentials between the elementary constituents of the crystal, that is, the electrons and the nuclei which are in fixed positions on a regular lattice. We can thus regard the carriers as practically non-interacting; there remains only the slowly varying mean electric field produced by possible macroscopic charge densities. However, this description rests on band theory and assumes that the potential seen by the electrons is periodic. The presence of crystal *defects*, of *impurities*, or of *phonons*, that is, displacements of the nuclei, is a perturbation of the periodic reference potential. Our model represents this perturbation by an effective potential, which is localized at each of the imperfections and which can scatter the charge carriers. The collisions of the carriers by the imperfections are the microscopic mechanism underlying the electrical resistivity. In a perfect crystal, the carriers would be freely accelerated by the applied field, just like electrons in a vacuum.

The *neutrons* circulating in one or other of the parts of a nuclear reactor, for instance, in the uranium bars, also constitute a classical gas with a very low density, typically $10^7 \mathrm{cm}^{-3}$. Their interactions with one another are negligible, but this is not true for their interactions with the material they pass through, which can be thought of as a system of scattering centres.

In the Lorentz model one assumes the scattering centres to be *infinitely heavy* compared to the non-interacting particles in the gas under consideration. Moreover, they are *randomly distributed in space*. These assumptions are well justified for the two examples studied above. The large mass of the scattering centres justifies the assumption that they are *fixed*: their energies are of the order of kT so that their velocities are low; after a collision, their recoil velocity remains small. Finally, we neglect the effects of the internal structure of the scatterers so that we may consider the scattering of each particle to be *elastic*.

The fact that the scattering centres are placed randomly rather than on a regular lattice guarantees that successive scattering processes do not produce any coherence effects. In fact, the interaction of the carriers with a periodic lattice potential gives rise to bands in a semiconductor, but does not produce any friction. The latter phenomenon, necessary to account for the electric resistivity, appears on the macroscopic scale when the scattering centres are distributed irregularly.

Hendrik A. Lorentz (Arnhem 1853–Haarlem 1928) who gave his name to the space-time transformation associated with a change of frame in special relativity (1903) also made major contributions to statistical mechanics: electron theory of metals, interaction of matter with radiation, kinetic theory. The model defined here, which he introduced to describe the electrons in a metal, is actually only applicable to the carriers in semiconductors, substances which were poorly known at the time. In metals one must make changes to account for the Pauli principle (§ 15.2.3).

Apart from its intrinsic interest, a study of the Lorentz model will prepare us for an investigation of the Boltzmann equation for ordinary classical gases, which is more complicated but based on the same ideas.

15.1.2 Successive Contractions of the Description

We must in the whole of this chapter distinguish carefully *several levels of description*. The most detailed *microscopic description* rests upon the use of the phase density D, which represents the state of the system in the N-particle phase space and which evolves according to the Liouville equation (§ 2.3). At the other extreme, the empirical *macroscopic description*, or *hydrodynamic description*, which is suitable for the near-equilibrium regime involves solely two variables at each point: the energy and particle local densities ϱ_E and ϱ_N. Following the general ideas of § 14.3.4 we shall below introduce the associated mesoscopic description; to do this we assign at any time a phase density D_0 to the system, equivalent to D as regards $\varrho_E(\boldsymbol{r}, t)$ and $\varrho_N(\boldsymbol{r}, t)$, but containing no more information.

However, as the gas is not dense, it is natural to work at an intermediate level, that of *Boltzmann's description* which plays the rôle of a macroscopic level as regards D, but of a microscopic level as regards ϱ_E and ϱ_N (Fig.15.1). At the Boltzmann level, we discard any correlations which might exist between particles; this is legitimate as one may expect these correlations to remain weak at low densities. A state is then characterized by the *single-particle reduced density* $f(\boldsymbol{r}, \boldsymbol{p})$ which at any time t is defined by (§ 2.3.5)

$$f(\boldsymbol{r}, \boldsymbol{p}, t) = \left\langle \sum_j \delta^3[\boldsymbol{r} - \boldsymbol{r}_j(t)]\, \delta^3[\boldsymbol{p} - \boldsymbol{p}_j(t)] \right\rangle; \tag{15.1}$$

the quantity $f\, d^3\boldsymbol{r}\, d^3\boldsymbol{p}$ represents the number of particles situated on average in the volume element $d^3\boldsymbol{r}\, d^3\boldsymbol{p}$ of the *single-particle phase space* and can be derived from the phase density D through (2.81). In what follows (§ 15.4.1) it will be useful to associate with f a mesoscopic phase density D_B which differs from D by suppressing all correlations which D might contain. The macroscopic densities ϱ_E and ϱ_N can be expressed in terms of f as follows:

$$\varrho_N(\boldsymbol{r}, t) = \int d^3\boldsymbol{p}\, f(\boldsymbol{r}, \boldsymbol{p}, t), \qquad \varrho_E(\boldsymbol{r}, t) = \int d^3\boldsymbol{p}\, f(\boldsymbol{r}, \boldsymbol{p}, t)\, \varepsilon. \tag{15.2}$$

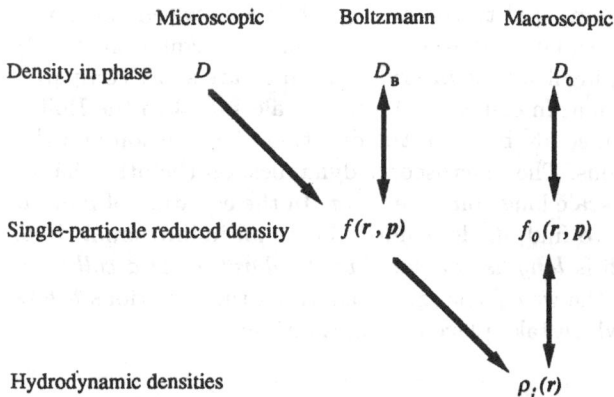

Fig. 15.1. Levels of description of a gas. The arrows indicate how the various densities follow from one another. Two successive *projections* (§ 14.3.5) lead from D to D_B then to D_0

The reason for restricting ourselves for ϱ_E to the kinetic energy $\varepsilon = p^2/2m$ and neglecting the interactions of the particles with one another and with the scattering centres is our low density assumption: even though these interactions are essential in controlling the dynamic processes, they hardly at all contribute to the energy, since the duration of the collisions is very short and between collisions the energy is purely kinetic energy. At any given time the fraction of particles which are interacting with a scattering centre is small, and the contribution to ϱ_E from their potential energy is negligible.

Altogether (Fig.15.1), starting from the microscopic description in terms of D, *two successive contractions*, leading through (15.1) to f and then through (15.2) to the ϱ_i, eliminate the non-relevant variables. At each stage the maximum statistical entropy principle enables us to reconstruct mesoscopic phase densities D_B and D_0 which, respectively, contain no more information than f and than the ϱ_i. By construction the associated statistical entropies $S(D_B)$ and $S(D_0)$ satisfy at all times the inequalities

$$S(D) \leq S(D_B) \leq S(D_0). \tag{15.3}$$

The description which is best suited for a gas on the microscopic scale is Boltzmann's, which has already eliminated the correlations between particles. However, before we use it, we must write down the evolution equations for $f(r, p, t)$ or, what amounts to the same, for D_B. To do this we could start from the Liouville equation for D and thence, using a method similar to the one of § 14.3.5, derive the dynamics of $f(r, p, t)$. It is more expedient to establish the equation of motion for f directly by noting that there are *different time scales* associated with the three levels of description. In fact, since the scattering centres have a *short range* δ, a particle never interacts with more than one centre at the same time and it moves freely between them. Each encounter with a scattering centre behaves as a collision with a duration which is *short* compared to the average time τ between collisions. The microscopic Liouville equation (2.68), or the equivalent equations (2.89) for the reduced densities, involve time intervals dt which are short compared to the duration of a collision. In contrast, the time scale suited to the Boltzmann description is intermediate between the duration of a collision and the interval τ between collisions. The macroscopic dynamics, on the other hand, is associated with a time scale long compared to τ. In the equation of motion for df/dt which we want to find, dt does not indicate an infinitesimal time, but a time interval which is *long as compared to the duration of a collision*. We must therefore study the way f changes, comparing the situations *before* and *after* the collisions which take place during the time dt.

15.1.3 Evolution of the Single-Particle Reduced Density

For simplification, let us assume that each scatterer has spherical symmetry. The classical trajectory of an incident particle is then characterized by its *impact parameter* b, that is, the distance from the centre of the scattering potential at which this trajectory would pass by, if it were not scattered (Fig.15.2). Conservation of the energy ε, which is equal to $p^2/2m$ outside the range of the potential, implies that the initial momentum $\boldsymbol{p}$ and the final momentum $\boldsymbol{p}'$ have the same absolute magnitude. The global effect of the scattering is thus a deflection of the trajectory over an angle θ, defined by $\cos\theta = (\boldsymbol{p}\cdot\boldsymbol{p}')/pp'$. The magnitude $\theta(b)$ of this angle is a function of the impact parameter, and it depends on the form of the potential.

However, the range of this potential, a few Å for a defect in a solid and a few fm for a nucleus scattering a neutron, is very short, not only as compared to the macroscopic distances involved in local equilibria, but also as compared to the mean free path, which is 10^2 to 10^4 Å, and even as compared to the distances between the particles or the scatterers. Over such distances the reduced density f is practically a constant so that we are not so much interested in the deflection angle as function of the impact parameter, as in the *distribution* of the angles of deflection for the case of a *uniform flux* of incident particles. More precisely, the scattering is characterized by the differential *cross-section* $d\sigma(p,\theta)$; this quantity is defined as the *number of*

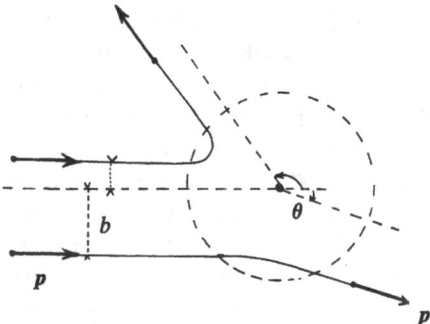

Fig. 15.2. Collision with a scatterer

particles scattered per unit time into a solid angle $d^2\omega$ around the direction of $\boldsymbol{p}'$, in the case of an incident beam, parallel to $\boldsymbol{p}$, with a *uniform unit flux density*. Equivalently, $d\sigma(p,\theta)$ represents in a plane at right angles to $\boldsymbol{p}$ the area $b\, db\, d\varphi$ covered by those incident trajectories which after scattering will point in a direction within the solid angle $d^2\omega = \sin\theta\, d\theta\, d\varphi$. This interpretation justifies the expression "cross-section". It also gives us for a given $\theta(b)$ an expression for $d\sigma/d\omega$, in the form

$$\frac{d\sigma}{d\omega} = \sum_k \left| \frac{b_k\, db_k}{\sin\theta\, d\theta} \right|, \tag{15.4}$$

where the b_k are the values of the impact parameter for which $\theta(b_k) = \theta$; we have $b > 0$, $0 < \theta < \pi$, $0 < \varphi < 2\pi$, the incident azimuthal angle is either equal to that of $\boldsymbol{p}'$, φ, or to $\varphi \pm \pi$, and (15.4) contains both kinds of contributions.

The classical approximation which we have just used is, in fact, valid only (§ 7.1.3) over distances which are much larger than the de Broglie wavelength h/p. Usually, the distances between the scattering centres easily satisfy $d \gg \hbar/p$, but this is not the case for the size δ of each centre. For instance, in a semiconductor at room temperatures when the effective mass of the carriers is $m/10$, typical values of $\hbar/p$ are 50 Å, whereas δ is of the order of a few Å. Similarly, neutrons are scattered by nuclei with a size of the order of fm (10^{-15} m), whereas, depending on their energy, their thermal lengths λ_T lie between 10^{-14} and 10^{-10} m. Hence, while it is legitimate to use the classical formalism on the Boltzmann scale where the characteristic distances are d or the mean free path l, the idea of a trajectory has no meaning for the description of a collision on the scale δ; the above analysis in terms of an impact parameter is not correct. All the same, the concept of a *cross-section remains valid for quantum scattering* even though $d\sigma/d\omega$ is no longer given by (15.4). In what follows we shall therefore assume that the cross-section $d\sigma/d\omega$ which characterizes the microscopic scattering has been provided either by quantum theory or from experiments. We must use this *quantum mechanical* quantity to write down at the Boltzmann level the *classical* equation of motion for $f(\boldsymbol{r},\boldsymbol{p},t)$.

If, for instance, one takes *hard spheres* of radius δ as a model for the scattering centres, for $p\delta \gg \hbar$ the classical expression (15.4) gives an isotropic cross-section,

$$\frac{d\sigma}{d\omega} = \frac{1}{4}\delta^2. \tag{15.5a}$$

The total cross-section, obtained by integrating (15.5a) over all possible scattering directions gives us the total number of particles scattered per unit time for unit incident flux. In the present case it is equal to $\sigma_t = \pi\delta^2$, that is, as one might have expected, the area of the circle, in which form the scattering sphere is seen by the incident particles. For smaller values of $p\delta$ the calculation of $d\sigma/d\omega$ needs the solution of the scattering wave equation. In the limit as $p\delta \ll \hbar$ we find

$$\frac{d\sigma}{d\omega} \approx \delta^2 + \frac{p^2\delta^4}{\hbar^2}\left(2\cos\theta - \frac{1}{3}\right), \tag{15.5b}$$

still isotropic, but with a four times larger coefficient.

We look for the change $df(r, p)$ in the single-particle reduced density over a time dt which is long as compared to the duration of a collision, but short as compared to τ. To do this we strike a *detailed balance* between the number of particles entering and leaving the volume element $d^3r\, d^3p$ of single-particle phase space. If there are no collisions, the value of p remains constant for each particle, and the flow in phase space is simply produced by the uniform motion of each particle with a velocity $v \equiv p/m$ in r space. The corresponding contribution to $\partial f / \partial t$ is then obtained in the same way as the conservation (14.6) of the number of particles or, as in (2.88), from the Liouville theorem. This gives us an equation of the form

$$\frac{\partial f}{\partial t} + \left(v \cdot \nabla_r\right)f = \mathcal{I}(f), \tag{15.6}$$

where the left-hand side describes the dynamics of non-interacting particles, while the right-hand side describes the effect of collisions. To simplify the notation we shall in what follows write $v \equiv p/m$ and $\varepsilon \equiv p^2/2m$ even though the only independent variables are the components of p. This will also allow our equations to remain valid for charged, or relativistic, particles, in spite of changes in the relations between v, p, and ε (§ 15.2.1 and Exerc.15d).

In order to write down the *collision term* $\mathcal{I}$ we note that $\mathcal{I}(f)\, d^3r\, d^3p\, dt$ is the average increase during a time dt in the number of particles situated in the volume element $d^3r\, d^3p$ of the single-particle phase space, due only to collisions. Since f is a function which changes slowly on the scale of the scatterers, we can choose the dimensions of the element d^3r large as compared to the range of the potentials and as compared to the de Broglie wavelength. A classical or quantum collision has the effect of changing the momentum of a particle, while leaving the particle inside d^3r. We must thus consider two kinds of processes, both produced by those scatterers which are situated inside d^3r. On the one hand, one of the particles with a momentum p inside d^3p is scattered into p', which lies outside d^3p, and this reduces $f\, d^3r\, d^3p$ by unity. On the other hand, inversely, a particle situated inside d^3r with a momentum p', acquires as the result of a collision a momentum p inside d^3p, which increases $f\, d^3r\, d^3p$ by unity.

Let us evaluate the first effect in terms of the cross-section and the density ϱ_{sc} of the scatterers. The flux density of the $f\, d^3r\, d^3p$ particles considered is $v\, f\, d^3p$. In unit time, therefore, each centre scatters a number $v\, f\, d^3p\, d\sigma(p, \theta)$ of those particles into the solid angle $d^2\omega$, in agreement with the definition of $d\sigma$ itself; the number of centres involved equals $\varrho_{sc}\, d^3r$. Writing $d^3p' \equiv d^2\omega'\, p'^2\, dp'$ and $d^3p \equiv d^2\omega\, p^2\, dp$ for the volume elements in momentum space, we thus find the total number of particles lost from $d^3r\, d^3p$ during the time dt

$$v\, f(r, p, t)\, d^3p\, \varrho_{sc}\, d^3r\, dt \int \frac{d\sigma(p, \theta)}{d\omega}\, d^2\omega',$$

which gives a contribution

$$-v\, f(\boldsymbol{r},\boldsymbol{p},t)\, \varrho_{\text{sc}} \int \frac{d\sigma}{d\omega}\, d^2\omega'$$

to $\mathcal{I}(f)$.

To calculate the second effect we note that scattering conserves the absolute magnitude of the momentum so that the extra particles liable to appear in the region $d^3r\, d^3p$ are those within the volume element d^3r which have momenta $\boldsymbol{p}'$ with absolute magnitude between p and $p+dp$ and arbitrary directions. The number of such incident particles is $f(\boldsymbol{r},\boldsymbol{p}')\, d^3r\, d^2\omega'\, p^2\, dp$, and ultimately one must integrate over the direction $d^2\omega'$ of $\boldsymbol{p}'$. From those we must count the ones which will be scattered into the solid angle $d^2\omega$ around $\boldsymbol{p}$; their current density is $v'\, f(\boldsymbol{r},\boldsymbol{p}')\, d^2\omega'\, p^2\, dp$ so that the number of such particles scattered by each centre into $d^2\omega$ equals

$$v\, f(\boldsymbol{r},\boldsymbol{p}',t)\, d^2\omega'\, p^2\, dp\, \frac{d\sigma(p',\theta)}{d\omega}\, d^2\omega,$$

where we have again used the definition of the cross-section. Note that the angle θ for the deflection from $\boldsymbol{p}'$ to $\boldsymbol{p}$ is the same as for the deflection from $\boldsymbol{p}$ to $\boldsymbol{p}'$ and that the cross-section occurring here is *the same* as the one corresponding to the direct process. This property, often called the *detailed balancing principle* or *microreversibility*, will enable us to regroup the two terms and will have important consequences. Integrating over the direction of $\boldsymbol{p}'$ and summing over the $\varrho_{\text{sc}}\, d^3r$ active scattering centres, we find the total number of particles gained in $d^3r\, d^3p$ during the time dt:

$$v\, d^3p\, \varrho_{\text{sc}}\, d^3r\, dt \int \frac{d\sigma(p,\theta)}{d\omega}\, f(\boldsymbol{r},\boldsymbol{p}',t)\, d^2\omega'.$$

Combining all this, we find finally for the collision term of (15.6) in the Lorentz model

$$\mathcal{I}(f) \;=\; \int d^3p'\, W_\theta(p)\, \delta(\varepsilon - \varepsilon')\, \big[f(\boldsymbol{r},\boldsymbol{p}',t) - f(\boldsymbol{r},\boldsymbol{p},t)\big], \qquad (15.7)$$

where we have used the relation

$$\delta(p - p') \;=\; \frac{p}{m}\, \delta(\varepsilon - \varepsilon')$$

to introduce explicitly the conservation of kinetic energy,

$$\varepsilon \;\equiv\; \frac{p^2}{2m} = \varepsilon' \equiv \frac{p'^2}{2m},$$

and where we have for $p = p'$ defined the function

$$W_\theta(p) \;\equiv\; \frac{1}{m^2}\, \varrho_{\text{sc}}\, \frac{d\sigma(p,\theta)}{d\omega}, \qquad (15.8)$$

which depends on the absolute magnitude of p and the angle θ between p and p'.

The particular form (15.7) of $\mathcal{I}(f)$ in the Lorentz model shows a general property of the collision term: it is *local and instantaneous*. It involves the reduced density at the same point and the same time as the ones ocurring on the left-hand side of the transport equation. These two characteristics result from the approximations which we have made when working at the Boltzmann scale, both regards the time interval dt, assumed to be long as compared to the duration of a collision, and as regards the dimensions of the volume element d^3r, assumed to be large as compared to the range δ. However, each collision can considerably change the momentum, with the energy fixed. This is reflected by the integration over p' occurring in $\mathcal{I}(f)$, which is often called the *collision integral*.

The expression

$$d^3p'\, W_\theta(p)\, \delta(\varepsilon - \varepsilon')$$

can be interpreted as the *transition probability per unit time* for a particle to go from an initial momentum p into the volume element d^3p' as the result of collisions. The probability for the inverse process is the same, due to the reversibility of the microscopic collision equations. Using this interpretation we could have easily striken the detailed balance leading to the form (15.7) of the collision term. Our approach, moreover, provided us with the explicit expression (15.8) for the transition probability.

15.1.4 Ballistic and Local Equilibrium Regimes

The dynamics of the Lorentz gas is governed by the kinetic equation (15.6), (15.7) which is an integro-differential equation. This equation is valid at low densities where the correlations between the particles remain negligible and where therefore the reduced Boltzmann description is adequate. The evolution of the various macroscopic quantities will follow once we know $f(r, p, t)$; for instance, the densities ϱ_N and ϱ_E are given by (15.2) at any time.

On the microscopic scale, the *conservation laws* are valid for *each collision* in which *neither the number of particles, nor their energy* is changed. This is reflected by the identities

$$\int d^3p\, \mathcal{I} \equiv 0, \qquad \int d^3p\, \mathcal{I}\, \varepsilon \equiv 0, \tag{15.9}$$

which must be satisfied by the collision term. In fact, the two integrals (15.9) represent the rate of increase, due only to collisions, of the particle and energy densities at the point r. One can easily check that they vanish for the collision term (15.7) of the Lorentz model: the first identity follows from the antisymmetry of the integrand under an exchange of p and p'; the second in addition uses the relation

$$\varepsilon \, \delta(\varepsilon - \varepsilon') \;=\; \tfrac{1}{2} \, (\varepsilon + \varepsilon') \, \delta(\varepsilon - \varepsilon').$$

Integrating the kinetic equation (15.6), possibly multiplied by ε, over p and using (15.9), we find the macroscopic conservation laws

$$\frac{\partial \varrho_N}{\partial t} + \operatorname{div} \boldsymbol{J}_N \;=\; 0, \qquad \frac{\partial \varrho_E}{\partial t} + \operatorname{div} \boldsymbol{J}_E \;=\; 0, \tag{15.10}$$

provided we define the densities by (15.2) and the fluxes by

$$\boldsymbol{J}_N \;=\; \int d^3 \boldsymbol{p} \, \boldsymbol{v} \, f, \qquad \boldsymbol{J}_E \;=\; \int d^3 \boldsymbol{p} \, \boldsymbol{v} \, f \, \varepsilon. \tag{15.11}$$

These expressions are special cases of (14.48), (14.49), (14.58), and (14.59) which were established for the dynamics produced by the Liouville equation for the density in phase D. Passing to the kinetic equation (15.6) for the reduced density f is accompanied by the total disppearance of the contributions from the potentials in the definitions of ϱ_E and $\boldsymbol{J}_E$. It is therefore consistent to use the Boltzmann description for the collisional dynamics of the gas and at the same time to treat the latter as being perfect when we calculate ϱ_E and $\boldsymbol{J}_E$.

The evolution of f is governed by the interplay of the left- and right-hand sides of the kinetic equation (15.6). On the one hand, the *drift term* $(\boldsymbol{v} \cdot \nabla) f$ vanishes for a *spatially uniform* distribution f; on the other hand, the *collision term* vanishes,

$$\mathcal{I}(f_0) \;=\; 0, \tag{15.12}$$

for a *local equilibrium* distribution which has the *Maxwellian form* (7.31) at each point:

$$f_0(\boldsymbol{r}, \boldsymbol{p}, t) \;=\; \frac{1}{h^3} \, e^{\alpha(\boldsymbol{r},t) - \beta(\boldsymbol{r},t)\varepsilon}. \tag{15.13}$$

The parameters $1/k\beta(\boldsymbol{r},t)$ and $\alpha(\boldsymbol{r},t)/\beta(\boldsymbol{r},t)$ can be interpreted as a local temperature and a local chemical potential; for spin-s particles, we must include in (15.13) a factor $2s + 1$. More generally, (15.12) is satisfied for a reduced density f which is *isotropic in p at each point* of space. Of course, a *global* equilibrium distribution for which f is both uniform and isotropic in p is stationary. However, if initially f has a *local* equilibrium form f_0, it evolves due to the drift term $(\boldsymbol{v} \cdot \nabla) f$ and it *loses this special form* with time; as a result, the collision term starts again to play a rôle. The existence of strong heterogeneities reinforces the effect of the drift term; a large deviation from local equilibrium, or local isotropy, makes the collision term the dominant one.

To estimate the relative orders of magnitude of the two effects we note that the characteristic length associated with $(\boldsymbol{v} \cdot \nabla)f$ is the distance over which f changes significantly, and this can be a macroscopic length. The correponding time is the one needed for a particle with a typical speed v to feel a spatial variation of the properties of the gas. This time can be macroscopic; over a time shorter than it the evolution of f is governed by the collision term (15.7). The order of magnitude of the latter is characterized by the *total transition probability* per unit time,

$$W(p) \equiv \int d^3\boldsymbol{p}' \, W_\theta(p) \, \delta(\varepsilon - \varepsilon') = v \, \varrho_{sc} \, \sigma_t \equiv \frac{1}{\tau(p)}, \qquad (15.14)$$

which we have evaluated in terms of the cross-section, starting from the definition (15.8) of $W_\theta(p)$. The total cross-section σ_t, which is the integral of $d\sigma/d\omega$, is of the order of magnitude of the area $\pi\delta^2$ of a scattering centre, and may depend on p. The length

$$l \equiv \frac{1}{\varrho_{sc}\sigma_t} \qquad (15.15)$$

can be interpreted as the *mean free path* between two successive collisions of the same particle, which we have previously introduced in § 7.4.5. The time $\tau(p)$, defined by (15.14), is thus the *average time between successive collisions*. For the charge carrier gases in semiconductors the speed v is of the order of 4×10^5 m s^{-1}; if the scattering centres have an effective radius δ of 7 Å, and if their number is one in a cube with 70 Å edgelength, we find a mean free path $l \simeq 0.2$ μm and a time $\tau \simeq 0.5 \times 10^{-12}$ s. The *large value of W* in macroscopic units implies that the collision term $\mathcal{I}(f)$ modifies f rapidly, over a time of the order of τ, as long as the two terms of (15.7) do not cancel each other.

This discussion shows up two extreme regimes for the dynamics of f. In the *ballistic regime* (§ 7.4.4) the left-hand side of (15.6) dominates and the collisions play only a minor rôle, notwithstanding the large value of W. This occurs as long as the sample is *small as compared to the mean free path*; for instance, the base of a transistor (§ 11.3.5) is rather thin and collisions are not sufficiently efficient to produce even local equilibrium. This occurs also when the sample is subject to external perturbations which *vary rapidly*, over times comparable to τ, which again prohibits relaxation to local equilibrium (Exerc.15b). Conversely, in the *local equilibrium regime* or *hydrodynamic regime* which we shall study in § 15.1.5 the two terms of $\mathcal{I}(f)$ nearly cancel each other, as f remains close to a local equilibrium distribution at all times. The drift term $(\boldsymbol{v} \cdot \nabla)f$ controls the evolution, but the collision term continues to play a rôle. In fact, if

$$f = f_0 + f_1 \qquad (15.16)$$

deviates little from a local equilibrium distribution (15.13), $\mathcal{I}(f)$ reduces to $\mathcal{I}(f_1)$ by virtue of (15.12) and the smallness of f_1 is compensated for by the large value of W.

In the local equilibrium regime the densities (15.2) associated with f_0 can be evaluated exactly like the expressions (7.32) and (7.44) which we had found for the gas in global equilibrium, since those expresions followed simply from the Maxwellian form of f_0. In terms of $\alpha(\boldsymbol{r}, t)$ and $\beta(\boldsymbol{r}, t)$ we have thus

$$ \varrho_N = \frac{1}{h^3} e^\alpha \left(\frac{2\pi m}{\beta} \right)^{3/2}, \qquad \varrho_E = \frac{1}{2\beta} \varrho_N. \tag{15.17} $$

Inverting these relations we can equally well characterize the local equilibrium state by the densities $\varrho_N(\boldsymbol{r}, t)$ and $\varrho_E(\boldsymbol{r}, t)$. Thus we recover microscopically the local equations of state of § 14.2.1 which connect the ϱ_i and the γ_i. The entropy density, given by (7.29) and (7.41) equals

$$ s = \varrho_N k \left(\tfrac{5}{2} - \alpha \right). \tag{15.18} $$

If initially f has an arbitrary value, far from equilibrium and without any very large spatial variations, the dynamics is dominated by the collision term which is much larger than $(\boldsymbol{v} \cdot \nabla) f$, at least for sufficiently short times. To simplify the study of the corresponding *transient regime* let us assume that each scattering process is isotropic so that $W_\theta(p)$ is independent of the angle θ, as in the examples (15.5). Using the definition (15.14) of W, the collision term then reduces to

$$ \mathcal{I}(f) = W (\langle f \rangle - f), \tag{15.19} $$

where $\langle f \rangle$ denotes the average of $f(\boldsymbol{r}, \boldsymbol{p}, t)$ over the directions of $\boldsymbol{p}$:

$$ \langle f(\boldsymbol{r}, p, t) \rangle = \int \frac{d^2\omega}{4\pi} f(\boldsymbol{r}, \boldsymbol{p}, t). \tag{15.20} $$

Over times which are sufficiently short so that we can neglect $(\boldsymbol{v} \cdot \nabla) f$ as compared to $\mathcal{I}(f)$, the kinetic equation

$$ \frac{\partial f}{\partial t} \simeq -W (f - \langle f \rangle) $$

has as its solution

$$ f(\boldsymbol{r}, \boldsymbol{p}, t) \simeq \langle f(\boldsymbol{r}, p, 0) \rangle + [f(\boldsymbol{r}, \boldsymbol{p}, 0) - \langle f(\boldsymbol{r}, \boldsymbol{p}, 0) \rangle] \, e^{-Wt}. \tag{15.21} $$

In a time of the order of the *time between collisions* τ the distribution *becomes isotropic* at each point. After a few collisions the particles lose any memory of the direction of their momentum in any volume element of dimensions larger than the mean free path. This reduces the collision term until it becomes comparable with the drift term. The evolution then proceeds in a local equilibrium regime, where f is almost isotropic in $\boldsymbol{p}$ at each point, on a macroscopic time scale which is long as compared to τ.

The Lorentz model, in fact, is pathological: its microscopic conservation laws include not only those of the number of particles and the energy, (15.9), but also the conservation of any function of the absolute magnitude of p, which remains unchanged in the collisions. The local conservative variables therefore do not reduce to (15.17), as we tacitly assumed. In actual fact, the distributions which are locally isotropic in p play the rôle of the local equilibrium distributions; we should have denoted those by f_0 in (15.12), (15.16), and in § 15.1.5. The distributions which are independent of r and isotropic in p play the rôle of global equilibrium distributions for the Lorentz dynamics (15.6), (15.7). The evolution toward a true local Maxwellian equilibrium requires energy exchanges with the scatterers (§ 15.2.4).

The analysis of the microscopic evolution has shown up a *double rôle* played by the collisions. The *high frequency* of the latter, which is reflected in the large value of the transition probability per unit time W occurring in the collision term, results in establishing on a microscopic time scale τ a local equilibrium where $\mathcal{I}(f)$ *becomes small*. The evolution then proceeds on a macroscopic time scale, with the state at all times remaining close to a local equilibrium. In Chap.14 we noted that this involves, on the one hand, an efficient mechanism for the evolution toward equilibrium inside any subsystem and, on the other hand, a coupling between subsystems which is sufficiently weak to allow a slow evolution of the system as a whole. The collision term by itself here plays these two rôles: its effectiveness is very large in the transient regime when it brings every macroscopic volume element to local equilibrium, but it becomes small as regards the coupling between these volume elements as soon as f gets close to an f_0 form. The *collision term alone* controls the *rapid relaxation* toward a local equilibrium; *the interplay of the collision term and the drift term* on the left-hand side controls the *slow relaxation* toward global equilibrium, or the transport in a permanent regime.

15.1.5 The Chapman-Enskog Method

Starting from the kinetic equation (15.6), (15.7) of the Lorentz model in a local equilibrium dynamic regime, we look for a *justification* of the various evolution equations postulated by *macroscopic thermodynamics* (§ 14.2.6). Also we want to *calculate the response coefficients* which we introduced empirically in § 14.4.1. We have already established the *conservation laws* (15.10) and the *equations of state* (15.17). However, the latter were written down, not starting from f, but starting from a reduced density f_0 which has the special form (15.13). Near local equilibrium we want to split f at all times, according to (15.16), into a part f_0 corresponding to local equilibrium and a small correction f_1. However, such a splitting is *a priori* not unique. To determine f_0 unambiguously from f, we must at every point write down two constraints to determine $\alpha(r,t)$ and $\beta(r,t)$. However, the densities ϱ_N and ϱ_E follow from f through (15.2). It is therefore natural to require that f and f_0 be *equivalent for the calculation of these local densities*, that is, to force f_1 to satisfy the *constraints*

$$\int d^3p\, f_1(\boldsymbol{r},\boldsymbol{p},t) \;=\; 0, \qquad \int d^3p\, f_1(\boldsymbol{r},\boldsymbol{p},t)\,\varepsilon \;=\; 0. \tag{15.22}$$

Equations (15.2), to which f_1 does not contribute, and (15.17) then *define* α and β at each point and at all times: thus, the equations of state are automatically satisfied just by the way they are constructed. As to the fluxes, the situation is the opposite; the isotropy of f_0 implies that f_0 does not contribute to (15.11) so that

$$\boldsymbol{J}_N \;=\; \int d^3p\, \boldsymbol{v}\, f_1, \qquad \boldsymbol{J}_E \;=\; \int d^3p\, \boldsymbol{v}\, f_1\, \varepsilon. \tag{15.23}$$

Our aim is to determine the dynamics of f_0, that is, of the local equilibrium Maxwell distribution which *follows f in its motion* under the constraints (15.22).

The method of Chapman and Enskog (1912–17) which is based upon the splitting (15.16), (15.13), and (15.22) of the reduced density and which we shall work out below is nothing but an adaptation of the *projection method* (§ 14.3.5) to kinetic theory. Here, the reduced density f which is governed by the kinetic equation (15.6) replaces the density operator $\widehat{D}$ which is governed by the Liouville-von Neumann equation, and the decomposition (15.16) replaces (14.75). The densities $\varrho_N(\boldsymbol{r},t)$ and $\varrho_E(\boldsymbol{r},t)$ are the macroscopic variables $A_i(a,t)$, defined by (14.72). The constraints (15.22) which express that f and f_0 give us the same densities ϱ_N and ϱ_E play the rôle of the constraints (14.73) on $\widehat{D}_0$; the maximum statistical entropy condition, which led in § 14.3.3 to the determination of $\widehat{D}_0$ in the local equilibrium form (14.67), here imposes upon f_0 the Maxwellian form (15.13). Finally, $\alpha(\boldsymbol{r},t)$ and $\beta(\boldsymbol{r},t)$ are the multipliers $\gamma_i(a,t)$ of the general theory. We shall see that the Chapman-Enskog method proceeds by determining the dynamics of f_0 and f_1 through iteration. This procedure will, in fact, amount to implementing a short-memory approximation which is similar to that of (14.82) in the general projection method. That is the reason why it will end up with instantaneous equations which connect the fluxes with the affinities, as postulated by thermodynamics.

To simplify the discussion of the dynamics in the local equilibrium regime we restrict ourselves to an isotropic cross-section $d\sigma/d\omega$ as would be the case, according to (15.5), for scattering by hard spheres in the limits $p\delta \gg \hbar$ or $p\delta \ll \hbar$. Equation (15.19) shows that the kinetic equation (15.6) can then be written in the form

$$\frac{\partial f}{\partial t} + (\boldsymbol{v}\cdot\nabla)f \;=\; W\left(\langle f\rangle - f\right). \tag{15.24}$$

The coefficient $W(p)$, defined by (15.14), is large in macroscopic units. We shall therefore solve (15.24) in two stages, first neglecting the deviation $f_1 = f - f_0$ from the local equilibrium state (15.13) and treating f_0 as a quantity of zeroth order in $1/W$, and then determining f_1, considered to be of first order in $1/W$. *In each stage* we must take two precautions. First of all, we must satisfy to each order the *conservation equations* (15.10). Secondly, we also wish to retain the *equations of state* (15.17) to each order, which is guaranteed by (15.22).

Therefore, we replace f by $f_0 + f_1$ in (15.24) and expand. To lowest order in $1/W$ this equation reduces to

$$0 = W\left(\langle f_0 \rangle - f_0\right), \tag{15.25}$$

that is, to (15.12), which is automatically satisfied. The fluxes (15.23) are of order $1/W$ so that to zeroth order the transport coefficients vanish and the conservation equations simply become

$$\frac{\partial \varrho_N}{\partial t} = 0, \qquad \frac{\partial \varrho_E}{\partial t} = 0. \tag{15.26}$$

Hence, if we take the equations of state (15.17) into account, it follows that $\partial \alpha / \partial t = 0$, $\partial \beta / \partial t = 0$. The definition (15.13) then shows that f_0 *does not change* with time to lowest order. One should note that the *dynamics* of f_0 did not follow from the kinetic equation (15.24) itself, but from the *conservation laws* which are consequences from it, a fact which we shall also find in the next order. The absence of transport is not as surprising as it might seem at first sight, if we remember the double rôle played by the collision term (see the end of § 15.1.4): in the limit as $W \to \infty$ each volume element immediately reaches local equilibrium, while at the same time the coupling between neighbouring volume elements vanishes, so that the spatial variations in $\alpha(\boldsymbol{r}, t)$ and $\beta(\boldsymbol{r}, t)$ do not entail any flux generation.

The fluxes appear in the *next order*. The smallness of their magnitude, which is proportional to $1/W$, is in accordance with experiments: as we have often noted in Chap.14, the smallness of the coupling between different volume elements entails the slowness of the transport of conserved quantities. Let us therefore write (15.24), and also the constraints (15.22) and the conservation laws, to next order. For (15.24) we retain the terms of zeroth order in $1/W$, which involve the zeroth and first orders of f_0 and f_1. The left-hand side must be evaluated using f_0, but we have seen that consistency with the conservation laws implies that to the order considered we put $\partial f_0 / \partial t = 0$. The right-hand side must be evaluated, if we take (15.25) into account, using f_1; the constraints (15.22) on f_1 are satisfied if we put $\langle f_1 \rangle = 0$, so that to the order considered equation (15.24) gives

$$\left(\boldsymbol{v} \cdot \nabla_r\right) f_0 = -W f_1.$$

This equation, which expresses the competition between the drift and the collision terms in the local equilibrium regime, can be directly solved as

$$f = f_0 + f_1 \simeq f_0 - \frac{1}{W}\left(\boldsymbol{v} \cdot \nabla_r\right) f_0. \tag{15.27}$$

From (15.27) we can check that our assumption $\langle f_1 \rangle = 0$ was consistent.

Using the value of f_1 from (15.27) we get, to order $1/W$, an approximate value for the fluxes (15.23):

$$\boldsymbol{J}_N = -\int d^3p \, \frac{\boldsymbol{v}}{W} \, (\boldsymbol{v} \cdot \nabla f_0), \qquad \boldsymbol{J}_E = -\int d^3p \, \frac{\boldsymbol{v}}{W} \, (\boldsymbol{v} \cdot \nabla f_0) \, \varepsilon;$$

The isotropy of f_0 and W enables us to integrate over the direction of $\boldsymbol{p}$, and using the relation

$$\int d^2\omega \, v_\alpha v_\beta = \frac{4\pi}{3} \, \delta_{\alpha\beta} v^2,$$

where the factor of $\delta_{\alpha\beta}$ is found by summing over $\alpha = \beta$, we get

$$\left.\begin{array}{l} \boldsymbol{J}_N = -\dfrac{4\pi}{3} \displaystyle\int_0^\infty p^2 \, dp \, \dfrac{v^2}{W} \, \nabla f_0, \\[4mm] \boldsymbol{J}_E = -\dfrac{4\pi}{3} \displaystyle\int_0^\infty p^2 \, dp \, \dfrac{v^2}{W} \, \varepsilon \, \nabla f_0. \end{array}\right\} \tag{15.28}$$

Using the explicit form (15.13) of f_0 and the relations between the multipliers α and β, on the one hand, and the intensive local variables μ and T, on the other hand, we find

$$\left.\begin{array}{l} \boldsymbol{J}_N = \dfrac{4\pi}{3k} \displaystyle\int_0^\infty p^2 \, dp \, f_0 \, \dfrac{v^2}{W} \left[\nabla \left(-\dfrac{\mu}{T} \right) + \varepsilon \nabla \left(\dfrac{1}{T} \right) \right], \\[4mm] \boldsymbol{J}_E = \dfrac{4\pi}{3k} \displaystyle\int_0^\infty p^2 \, dp \, f_0 \, \dfrac{v^2 \varepsilon}{W} \left[\nabla \left(-\dfrac{\mu}{T} \right) + \varepsilon \nabla \left(\dfrac{1}{T} \right) \right], \end{array}\right\} \tag{15.29}$$

If one substitutes these results into the conservation laws (15.10) and uses the equations of state (15.17), one obtains the equations of motion for ϱ_N and ϱ_E, or for μ and T. Using (15.13) one could then, to order $1/W$, find $\partial f_0/\partial t$, and substitute it into the kinetic equation to continue the iteration to the next orders in $1/W$; however, the first order suffices, if we want to find the linear response coefficients. Note that this method does not provide us with a simple series expansion, as the coupling between f_0 and f_1 through the conservation laws changes f_0 at each iteration.

Equations (15.29) *prove the phenomenological relations between the affinities and the fluxes* from § 14.4.1. In particular, they imply *Ohm's, Fick's, and Fourier's laws* and this provides us with a microscopic justification of the near-equilibrium thermodynamics for the Lorentz gas. At the same time they give us an expression for the *macroscopic response coefficients* in terms of the microscopic scattering cross-section. In the particular case of isotropic scattering which we are considering, we find, in fact, from (15.14) and (15.29) that

$$L_{NN} = \frac{4\pi}{3k} \int_0^\infty p^2 \, dp \, f_0 \, \frac{v}{\varrho_{sc}\sigma_t}, \qquad L_{EE} = \frac{4\pi}{3k} \int_0^\infty p^2 \, dp \, f_0 \, \frac{v\varepsilon^2}{\varrho_{sc}\sigma_t},$$

$$L_{NE} = L_{EN} = \frac{4\pi}{3k} \int_0^\infty p^2 \, dp \, f_0 \, \frac{v\varepsilon}{\varrho_{sc}\sigma_t}, \tag{15.30}$$

where σ_t is the total scattering cross-section. The microscopic results (15.30) also allow us to prove here the principles of thermodynamics. We easily check the *Onsager relation* $L_{NE} = L_{EN}$ for this problem. Finally, L_{NN} and L_{EE} are clearly positive; the same is true for $L_{NN}L_{EE} - L_{NE}^2$, as a consequence of the Schwartz inequality. This proves the positivity of the response matrix, and hence of the *dissipation*, that is, the increase in entropy with time (§ 14.2.5).

We can easily finish the calculation of the response coefficients when the cross-section σ_t, and hence the mean free path $l = 1/\varrho_{sc}\sigma_t$, are independent of p. To simplify the notation we introduce the *mean time τ between collisions*, weighted by $f_0 v^2$:

$$\tau \equiv \frac{\int d^3\boldsymbol{p} \, f_0 \, v^2 \, \tau(p)}{\int d^3\boldsymbol{p} \, f_0 \, v^2} = \frac{2l}{3} \left(\frac{2m}{\pi kT} \right)^{1/2} = \frac{8l}{3\pi\bar{v}}, \tag{15.31}$$

where $\tau(p) = l/v = 1/W(p)$ was defined by (15.14), and where $\bar{v}$ denotes the mean particle speed at thermal equilibrium. Using (15.2) and formulae from the end of the book we then get from (15.30)

$$L_{NN} = \frac{\tau}{m} \varrho_N T, \qquad L_{EE} = \frac{6\tau}{m} \varrho_N k^2 T^3,$$

$$L_{NE} = L_{EN} = \frac{2\tau}{m} \varrho_N kT^2. \tag{15.32}$$

The Chapman-Enskog method shows that the condition for the existence of the local equilibrium regime is that the time between collisions τ be small as compared to the characteristic times for the macroscopic evolution. This requires either each collision to be efficient, that is, the cross-section to be large, or a large density of scatterers. If these conditions are fulfilled, the technique used here enables us to *contract the description* from the dynamics of the reduced density f to that of the macroscopic densities ϱ_N and ϱ_E of § 14.4.1, and thence to justify the empirical approach of thermodynamics.

Like the projection method, the Chapman-Enskog method has a remarkable characteristic. One is interested only in the evolution of macroscopic quantities, or, equivalently, of the mesoscopic reduced density f_0. However, even though f_1 has been eliminated, it is *just this deviation* f_1 from the local equilibrium which *governs the dynamics* of the local equilibrium quantities.

A surprising fact, called *Hilbert's paradox*, remains to be explained. The general solution, $f(\boldsymbol{r}, \boldsymbol{p}, t)$, of the microscopic transport equation depends on an *initial condition* $f(\boldsymbol{r}, \boldsymbol{p}, 0)$, which is a *function of 6 variables*. However, the iterative solution which we were able to construct, using the Chapman-Enskog method, is completely determined, once we give the initial condition for f_0. One sees this from the result (15.27), where f_1, to order $1/W$, is given explicitly in terms of f_0 at the same time, and where f_0 evolves according to the conservation equations governed by the fluxes (15.29). To higher order, these same characteristics are retained so that the Chapman-Enskog solution $f(r, p, t)$ is obtained as a power series in $1/W$, the convergence of which can be proved, and which depends solely on $f_0(r, p, 0)$. In fact, we noted when discussing (15.21) that a correct mathematical analysis of the kinetic equation (15.6), (15.7) requires that we take for the densities f_0 not the local equilibrium distributions (15.13), but, more generally, *isotropic* functions of p. As a result the Chapman-Enskog solutions depend only on *a function of 4 variables*, r and p. It therefore appears as if we have lost some of the solutions of the original equation! However, *experience tells us that the local equilibrium regime is general*. In order to understand the significance of the particular solutions which we have obtained and which have forms close to f_0 we must find the relation between the particular Chapman-Enskog solutions and the general solution and answer the question why in actual cases only those solutions are observed.

The key to this problem is the following. Let us assume that initially $f(r, p, 0)$ is very different from a Chapman-Enskog form, due to a large anisotropy in p or very fast variations in r. The evolution then takes place in two stages. First, on a microscopic time scale of the order of the *time between collisions*, f rapidly approaches a form which is at each point isotropic in p, in accordance with (15.21). The details of $f(r, p, t)$ are important on that time scale. However, subsequently we can forget about them, as the new initial condition, from which we start at the end of the first, transient stage, is a distribution close to the f_0 type. The evolution on *macroscopic time scales* will then to a good approximation follow a local equilibrium regime for *all* solutions of the microscopic equation. The particular Chapman-Enskog solutions are thus the *asymptotic form* of the general solution for large times. Moreover, the behaviour of f as $\exp(-Wt)$ in the transient regime (15.21) could not have been obtained by the Chapman-Enskog method, which through its very construction produces only solutions which can be expanded in powers of $1/W$. One can rigorously prove that any solution of equation (15.6) is the sum of a Chapman-Enskog type solution and a correction, possibly large initially, but decreasing exponentially as $\exp[-t/\tau(p)]$.

15.2 Applications and Extensions

15.2.1 Diffusion and Conduction

The Lorentz model is sufficiently realistic to describe semiconductors in the exhaustion regime (§ 11.3.4) where there is practically only one kind of carriers, with charge $q = \pm e$, and the number of which is conserved. For those substances (14.86), together with (15.32), gives us the *diffusion coefficient*

$$D = \frac{\tau}{m} kT, \tag{15.33}$$

in terms of the time between collisions, or in terms of the mean free path, through (15.31), or, alternatively, in terms of the cross-section and the density of scattering centres through (15.14). According to Fick's law, D governs the average motion of the carriers toward the less dense regions.

One also finds from (15.32) the thermal conductivity (14.89),

$$\lambda = 2\varrho_N k^2 T \frac{\tau}{m}, \tag{15.34}$$

the electrical conductivity (14.95),

$$\sigma = \varrho_N q^2 \frac{\tau}{m}, \tag{15.35}$$

and the thermoelectric coefficients (§ 14.4.3).

These expressions, obtained towards the start of the twentieth century by Drude and Lorentz, and others, in order to explain the properties of *metals*, were the first successes of statistical mechanics applied outside its original field, the physics of gases. In particular, it follows from (15.34) and (15.35) that the *ratio of the thermal and electrical conductivities*,

$$\frac{\lambda}{\sigma} = \frac{2k^2 T}{q^2} = 1.5 \times 10^{-8} T \qquad \text{in SI units,} \tag{15.36}$$

is independent of the substance, and varies proportionally to the temperature. This property, the *Wiedemann-Franz law*, had been known experimentally since 1853 and was thus proved by statistical mechanics. However, the numerical coefficient occurring in (15.36) is smaller than the one experimentally found for metals. In fact, we shall see in § 15.2.3 that the theory of Drude and Lorentz, although conceived originally for metals, could not be applied to them, but only to semiconductors.

Our evaluation of the conductivity leaves somewhat to be desired, since it is implicitly based upon the macroscopic arguments of § 14.4.2. The physical origin of the conduction is an *external electric field* applied to the substance, and we have not included that into our considerations. We must therefore repeat the earlier calculations and include into the microscopic kinetic equation an extra term coming from the electric potential in which the carriers are moving. That will allow us to justify expression (15.35) for the conductivity on a completely microscopic basis. Denoting the external potential by $V(r) = q\Phi(r)$ and using the same collision term as in the simplified Lorentz model (15.24) we can write the kinetic equation as

$$\frac{\partial f}{\partial t} + \left(v \cdot \nabla_r \right) f - \left(\nabla_r V \cdot \nabla_p \right) f = W \left(\langle f \rangle - f \right). \tag{15.37}$$

The extra term on the left-hand side was obtained as in (2.88). It describes the drift of the particle momenta due to the applied force $-\nabla V$. Its presence, while leaving the conservation laws (15.10) unchanged, changes the expressions (15.2) and (15.11) for the energy flux and the energy density, as well as expressions (15.22) and (15.23) which follow from them. One can easily check, by integrating (15.37) over p after multiplying by $p^2/2m + V(r)$, that the only change we must make in all the equations is the replacement of $\varepsilon = p^2/2m$ by

$$\varepsilon \equiv \frac{p^2}{2m} + V(r). \tag{15.38}$$

The same substitution must, of course, be made in the reduced local equilibrium density f_0, (15.13), since β is associated with the *total, kinetic and potential, energy* of the particles. The effect of this is to change $\mu = \alpha/\beta$ in expressions (15.17) and (15.18) for the particle and entropy densities to

$$\mu'(r,t) \equiv \mu(r,t) - V(r), \tag{15.39}$$

in agreement with the macroscopic definition (14.93) of μ'.

The solution of (15.37) by the Chapman-Enskog method now gives us, instead of (15.27),

$$f_1 = -\frac{1}{W}\left(v \cdot \nabla_r\right)f_0 + \frac{1}{W}\left(\nabla V \cdot \nabla_p\right)f_0 = -\frac{1}{W}\left(v \cdot \left\{\nabla_r f_0 + \beta f_0 \nabla_r V\right\}\right)$$

$$= \frac{1}{W}f_0\left(v \cdot \left\{-\nabla\alpha + \varepsilon\nabla\beta\right\}\right). \tag{15.40}$$

In deriving (15.40) we have taken into account, when differentiating f_0 with respect to r, that not only α and β, but also ε has a spatial dependence (Eqs.(15.13) and (15.38)). Equation (15.40), as well as Eqs.(15.29) and (15.30) which follow from it, are formally the same as when there is no V, provided V is introduced in the definition (15.38) of ε. Altogether, the form of the response equations remains the same, but the response coeffcients (15.32) are changed to

$$L_{NN} = \frac{\tau}{m}\varrho_N T,$$

$$L_{NE} = L_{EN} = \frac{\tau}{m}\varrho_N T\left(2kT + V\right),$$

$$L_{EE} = \frac{\tau}{m}\varrho_N T\left(6k^2T^2 + 4kT V + V^2\right). \tag{15.41}$$

These expressions are in agreement with the general form (14.97) which we found by macroscopic arguments. The fact that neither L_{NN}, nor $L_{NN}L_{EE} - L_{EN}^2$ depend on V justifies expressions (15.35) and (15.34) for the electrical and thermal conductivities. We must, however, rely on the following remark. The conditions for the applicability of Ohm's law, namely, *uniform temperature* and *electrical neutrality* in a homogeneous substance, imply that μ', which is a function of ϱ_N and T, must be uniform, and hence, if we recall (15.39), that

$$\nabla\left(-\frac{\mu}{T}\right) = -\frac{1}{T}\nabla V = \frac{q}{T}E; \tag{15.42}$$

the response equation $J_N = L_{NN}\nabla(-\mu/T)$ then gives us *Ohm's law* with (15.35) for the conductivity.

The elimination of μ by using the equation of state (15.17) with $\alpha = \beta\mu'$ and (15.39) gives, for the case of a uniform temperature, the following particle flux

$$ \boldsymbol{J}_N \;=\; L_{NN} \left(\frac{q}{T}\, \boldsymbol{E} - k\, \frac{\nabla \varrho_N}{\varrho_N} \right) \;=\; \frac{\sigma}{q} \left(\boldsymbol{E} - \frac{kT}{q\varrho_N}\, \nabla \varrho_N \right). \qquad (15.43) $$

The two terms in (15.43) represent, respectively, the contributions associated with the *electric force* and with the *inhomogeneity*. We saw in § 14.4.2 that the proportionality of their coefficients is the *Einstein relation* (14.96).

The *electric mobility*, defined by (14.91), equals here

$$ \mu_{\mathrm{el}} \;=\; q\,\frac{\tau}{m} \;=\; ql \left(\frac{8}{9\pi mkT} \right)^{1/2}. \qquad (15.44) $$

It is the average velocity acquired by a particle of, positive or negative, charge q being subjected to unit field. It is directly observed in *electrophoresis*: if some ions of charge q and mass m initially form a drop in a colloidal medium, the latter plays the rôle of the scattering centres of the Lorentz model and the ions can diffuse and migrate under the influence of an electric field, according to (15.43). One can show (Exerc.15c) that the drop spreads rather little as compared to its average displacement with a velocity given by (15.44). This phenomenon is the basis of a method of analysis which enables one to separate ions according to their mobility by the simple application of an electric field. Its applications are essential in *biology*. For instance, one can separate and dose in this way albumin and globulins from blood serum, which have different mobilities.

Electrophoresis is also commonly used to *read the genetic information* contained in the deoxyribonucleic acid (DNA). DNA has the famous double helix structure. One of the two strands (1) is a long polymer chain along which four bases, denoted by T, C, A, and G, are positioned in succession, in a well defined order which determines the genetic information. Our aim is to determine the sequence. The second strand of DNA contains the same information, but in a complementary form. In fact, chemical bonds can occur, on the one hand, between the T and A bases and, on the other hand, between the C and G bases. The strands (1) and (2) are thus complementary sequences, where, on the one hand, T and A and, on the other hand, C and G, face one another so that the strands may be bonded all along their length.

We start from a solution containing the DNA which we want to analyse. Molecular biology methods allow us to synthesize chains which are complementary to the strand (2), that is, which are analogous to the strand (1), by using (2) as a matrix along which the complementary bases will order themselves. To do this, one incorporates in the solution the elementary constituents, T, C, A, and G, a polymerization enzyme, and *radioactively marked* special molecules for the beginning of the chain. One of the latter fixes itself to start with at the beginning of the strand (2) of the DNA. Starting from this initial element, the bases which are drawn from the solution place themselves successively face to face with their complements in the strand (2) of the DNA. Through chain polymerization one has thus reconstructed a copy of the strand (1) with a radioactive marker at its beginning. However, we also have an elementary constituent T′, similar to T, which can replace the latter during the chain polymerization process, while *blocking it*. If we put a little T′ in our solution, the copy breaks off randomly at one or other of the T sites of the strand (1) of the DNA which we are trying to analyze. We thus produce a batch of molecules, marked radioactively at their beginning, with different *lengths* which reflect accurately the *positions of the* T *bases* along the strand (1). Let us call this the T batch. Incorporating in the solution inhibitors C′, A′, or G′ instead of T′, we similarly produce C, A, and G batches.

We then analyze the four batches through electrophoresis: each of them is placed at one end of a strip of gel, a porous material in which the molecules can diffuse, and between the two ends of this strip we apply a potential difference during a certain time interval. The charge and mass of the molecules are proportional to their length, but their mobility is in this case not given by (15.44); the diffusion mechanism is different because of the compact nature of the gel and the great length of the molecules. However, one can show that the mobility decreases, logarithmically, with increasing size of the molecules. The partial copies of the strand (1) migrate therefore the further, the shorter they are. One determines the distances travelled by the various radioactive fragments by photographically detecting the radioactivity of the strips at each point. For each of the four strips, for instance T, we thus obtain a series of spots, each of which visualizes the size of one type of molecules in the T batch, that is, the position of one of the T bases in the strand (1) which we analyzed. By arranging the spots in the inverse order to the distance travelled and classifying them according to the T, C, A, or G batch to which they belong, we can thus *read directly the sequence of the bases* of the original DNA. This procedure, which is convenient and efficient, is universally used in biology laboratories. By a single experiment it enables one to determine the sequence of several hundred bases.

In semiconductors the mobilities at room temperature are of the order of $0.1 \ \mathrm{m^2 s^{-1} V^{-1}}$. They increase as the temperature decreases, following (15.44) and also because the density of scatterers decreases, which increases the mean free path. In metals, they are smaller by two orders of magnitude. The corresponding drift velocities of the charge carriers are very small as compared to their thermal random velocities, which are of the order of $10^6 \ \mathrm{m \ s^{-1}}$ in a metal, and about half that in a semiconductor.

15.2.2 Semiconductors

The Lorentz model has the merit of explaining the conduction mechanism in a way which is less qualitative than the arguments of § 11.3. However, coming from the pre-quantum electron theory, it has more or less serious defects. We shall discuss it in order to find ways of improving it for applications to some substance or another.

One of the major approximations made consists in treating the charge carriers as a *classical gas*, which is justified as long as their density is low. The theory therefore needs to be profoundly altered for *metals*, where the Pauli principle plays an essential rôle (§ 15.2.3). Nevertheless, we have seen at the start of § 15.2.1 that the model can be successfully applied to semiconductors, at least under circumstances when the number of majority charge carriers is conserved and that of the minority carriers is negligible. Another example of systems which behave like classical gases is provided by dilute solutions (Exerc.9f) and it is natural to try to apply the Lorentz model to dilute solutions of *electrolytes*. However, in that case we must take into account at least two kinds of ions, positive and negative ones, and above all, the properties of the solvent. In fact, the model assumes that the charge carriers move in a medium consisting of scattering centres at large distances from one another and practically fixed. Neither the one nor the other of these two assumptions is satisfied since the solvent is a liquid, the molecules of which are relatively close to one another and have masses of the same order of magnitude of, if not smaller than, those of the ions. A microscopic theory of electrolytes, even a rather rough one, therefore calls for complicated methods. The simplest case is that of insulators or *semiconductors*, for which rather simple changes are sufficient to make the Lorentz model realistic.

We must first, of course, attribute to the carriers the *effective masses* which govern their dynamics (§ 11.3.2). In the charge balance we must include the donor and acceptor centres and the carriers which are trapped in them, and we must add to the external electrical potential the self-consistent *potential produced by the charge distribution*. We must also take into account the *polarization* effects of the medium (§ 11.3.3) which reduce the electrostatic forces, to first approximation dividing them by the dielectric constant. The electron *spin* introduces a factor 2 into (15.13).

The major change concerns the presence of two kinds of carriers. Let us first give a *macroscopic* analysis. We saw in §§ 11.3.2 and 11.3.4 that a semiconductor behaves as a mixture of *two very dilute classical gases*, one consisting of the, negative, electrons in the conduction band, the other of the, positive, holes in the valence band. At equilibrium the density of each of them depends solely on the temperature and the doping, in the case of a neutral substance; if the semiconductor is charged, it also depends on the local density of the total charge or, what amounts to the same, on the difference $\mu' = \mu - V$ between the local chemical potential and the macroscopic potential $V = -e\Phi$. In a *local equilibrium* regime we must, instead of a single carrier density ϱ_N, introduce the two local densities n and p of the conduction electrons and the holes, as well as the densities n_d and p_a of electrons and holes bound to donor and acceptor centres, respectively. We wrote in (11.101) the equilibrium expressions for these various quantities; this gives here, in particular, rise to the introduction at each point of *two chemical potentials* μ_n and μ_p which are associated, respectively, with the filling of the conduction band with negative carriers and the valence band with positive carriers. Their difference $(\mu_n - \mu_p)/T$ which vanishes at equilibrium defines a *chemical kind of affinity* (14.14b); when it is positive, there is an *excess* of holes and electrons with respect to the equilibrium density and the system reacts by a *recombination* process where the pairs of carriers with opposite signs annihilate each other with a release of energy. When this difference is negative, we have, on the other hand, *pair creation*, that is, the excitation of an electron from the valence band into the conduction band. The flux associated with the affinity $(\mu_n - \mu_p)/T$ is the recombination rate per unit time.

Apart from this chemical kind of thermodynamic effect, the *spatial variations of μ_n and μ_p* give rise to the two affinities $\nabla(-\mu_n/T)$ and $\nabla(-\mu_p/T)$, with which two negative and positive carrier currents are associated. However, the continuity equations (15.10) for the two kinds of carriers contain source terms, equal to the rate of pair creation. If we add to all that the thermal effects and the possibilities for trapping in and releasing from the donor and acceptor sites, we expect from a macroscopic analysis a large number of response coefficients.

At the microscopic level we explain the above features as follows. The two carrier gases are described by *two reduced densities* $f_n(r, p, t)$ *and* $f_p(r, p, t)$ which each, in local equilibrium, have a Maxwellian form (15.13) provided the Pauli principle does not come into play, which is most often the case. As in equilibrium, one must include in the energy ε the local potential energy and change the signs of the energies and of μ for the hole density f_p. The equations of motion for the quasi-particles (§ 11.3.2) show that f_n and f_p to first approximation evolve according to a kinetic equation like (15.6) with a collision term (15.7) describing the interaction of the carriers with the crystal imperfections. As a result, each of the populations thermalizes *independently of the other*, over a short time, of the order of the time τ between collisions, and the number of each kind of carrier is conserved on that time scale. This enables us to understand why over longer times we observe local equilibrium regimes where the two electron and hole gases evolve almost independently, while being superimposed upon each other in space.

However, there also exist microscopic processes, rarer than the simple scattering of carriers by lattice imperfections, which can be interpreted as the *annihilation or the creation of a pair* of carriers with opposite signs. In fact, when an electron in the valence band acquires through a collision an energy just above the width of the forbidden band it can make a transition into the conduction band; this manifests itself as pair creation, to use the language of quasi-particles. The inverse process also exists and is described as pair annihilation. These two processes give in the kinetic equations rise to extra collision terms which *couple* f_n *and* f_p and which *do not conserve the number* of carriers. However, these terms, describing the probability for the transition of an electron from one band to another, involve matrix elements between wavefunctions of the valence and the conduction bands. We know that the corresponding Wannier functions χ_v and χ_c are orthogonal to each other, which considerably reduces these matrix elements and hence the corresponding collision term – we encountered a similar situation in § 11.3.4 where for a similar reason the valence band was not involved in the calculation of the binding energy of an impurity level around a donor centre. Moreover, the necessity to provide or to absorb an energy of the order of the forbidden band width requires coupling with other degrees of freedom, for instance, with the phonons. This again reduces the probability for chemical type reactions, that is, pair creation or annihilation, as compared to the collision probability $W = 1/\tau$ for a carrier. The typical characteristic recombination times are of the order of 10^{-4} s, which should be compared with the order of magnitude of 10^{-12} s for τ; this justifies the use of the two gas model for intermediate times.

Relaxation toward equilibrium also involves other microscopic processes, for instance, when an electron impinges upon an unoccupied donor centre and gets *trapped*. Those are processes which give rise to the evolution of the densities n_d and p_a of electrons and holes bound to donor and acceptor centres, respectively. Finally, we shall see in § 15.2.4 that we must also include *inelastic* processes which are responsible for the thermalization.

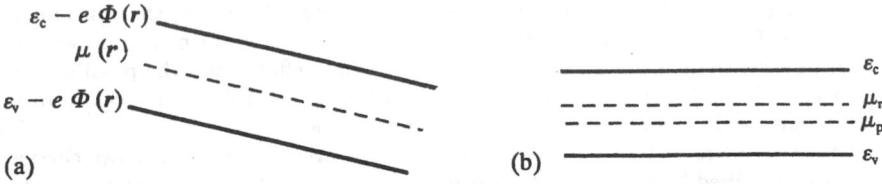

Fig. 15.3a,b. Semiconductor close to equilibrium. (a) Ohmic conduction; (b) pair excitation

In semiconductor technology (§ 11.3.5) one usually depicts the local energies of the band edges across the various parts of a device. At equilibrium, the chemical potential μ is uniform and represented by a horizontal line whereas the band edges follow the spatial variation of the potential $-e\Phi$. In this way, Fig.11.15 gave a schematic picture of a p-n junction at equilibrium. In a local equilibrium regime the chemical potentials μ_n and μ_p, associated with the two kinds of carrier, may be split and they may vary in space. Figure 15.3a, for example, shows a regime where a globally neutral semiconductor *conducts an electric current*. The chemical potential varies parallel to $-e\Phi$ in order to ensure neutrality. However, as it is not constant, the charges move in the direction which would tend to reestablish a uniform μ, namely, to the left for the p carriers and to the right for the n carriers; this produces in a stationary regime a current proportional to $\nabla\mu$, that is, proportional to $\nabla\Phi$, as in § 15.2.1, but now with both n and p carriers.

Another situation of interest (Fig.15.3b) corresponds to an excess of both kinds of carriers at the same time; this occurs, for instance, when the semiconductor is *illuminated*: the energy of the incident photons is absorbed and creates pairs. In the stationary regime which is set up, the carriers of either sign are practically in an equilibrium, which is controlled by the ordinary collision term, between electrons or between holes, and the distribution of each set of carriers is Maxwellian. However, the *recombination* coefficient is so small that the chemical potentials μ_n and μ_p get stuck at different values, the more different, the stronger the intensity of the illuminating flux in the range of frequencies associated with the width δ of the forbidden band. The excess of carriers, produced in this way, gives rise to a pronounced *photoconductivity* effect, since the conductivity (15.35) is proportional to the carrier density. In § 11.3.4 we listed some technical applications of this effect.

We also indicated in § 11.3.5 the use of p-n junctions as *photocells*. Figure 15.4 shows this schematically. The two p and n layers are joined together along the plane of the junction which is subjected to solar radiation. The layer on the illuminated side is sufficiently thin to let the radiation pass through. When there is no light, the potential is represented by the dot-dash curve and the chemical potential is constant, as in Fig.11.15. Illumination creates pairs and produces a positive difference $\mu_n - \mu_p$ between the chemical potentials in the junction region. For the case of an open illuminated circuit and a permanent regime, we have shown the potential by a full drawn curve and the chemical potentials by dashed curves. Far from the junction, recombination takes place locally and this reduces $\mu_n - \mu_p$ practically to zero; the carrier densities are close to their equilibrium values, and $\mu + e\Phi$ takes on both sides the same value as when there is no illumination. A more complicated situation occurs in the junction region. The difference $\mu_n - \mu_p$ remains rather large, since recombination does not occur immediately and locally: indeed, its character-

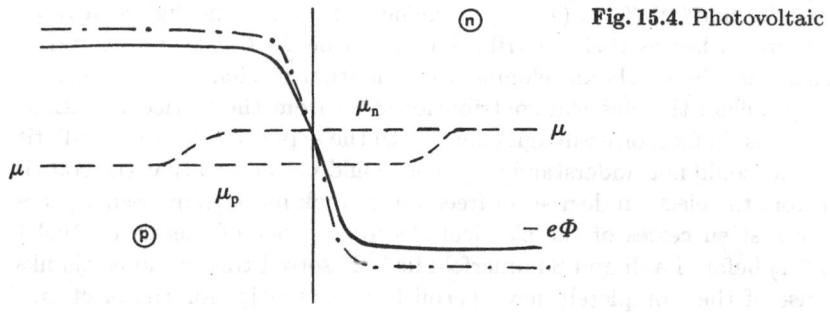

Fig. 15.4. Photovoltaic effect

istic time of 10^{-4} s is rather long and the inhomogeneity causes the carriers to migrate during that time. As compared to the equilibrium situation there is on the p side an excess of holes which is more marked than the excess of electrons, since the latter are on that side the minority carriers. The opposite is true on the n side. As a result the double layer of charges in the vicinity of the junction is reduced by the illumination. This entails a reduction of the spontaneous potential difference; this reduction is equal to the difference between the chemical potentials and thus to the *electromotive force* from p to n which one would observe, if one closed the circuit. A complete theory would have to write down (i) the self-consistent electro-static equations, (ii) the continuity equations (15.10) for the two kinds of carrier, including their source terms which describe the pair creation by the light and the recombination, and (iii) the relations between the two carrier fluxes and the gradients of the chemical potentials, in which the conductivities (15.35) are proportional to the local carrier density.

If we consider very small samples, for instance, a transistor base (Fig.11.18), the thickness of which of the order of a micron is not large as compared to the mean free path of the carriers, or if we deal with applications to fast electronics, the quasi-equilibrium assumption is not justified; we may find ourselves in a regime intermediate between the local equilibrium and the *ballistic* regimes (§ 15.1.4). We must then go back to the microscopic kinetic equations and look for their solutions, without making the Chapman-Enskog approximation, which is inoperative here since it assumes that the collisions are sufficiently efficient to ensure local equilibrium.

The study of semiconductors thus appeals to a rather complicated theoretical arsenal. It is essential to note that the various points listed above are commonly used not only in pure science, but above all in *applied* science. As in the case of other up-to-date technologies, the mastering of the scientific facts and the use of advanced methods are indispensable for a solution of industrial problems which range here from the design of electronic devices to the capture of solar energy by photocells.

15.2.3 Metals

We saw in §§ 10.4.3 and 11.3.1 that *quantum mechanics* is indispensable for a correct description of the behaviour of the electrons in a metal. In particular, their contribution (10.75) or (11.52) to the heat capacity is negligible at room temperatures, whereas their contribution would be $\frac{3}{2}Nk$ if they behaved as a classical gas. Before the development of quantum mechanics, this posed a difficult problem: the classical contribution $3Nk$ from the lattice vibrations (§ 11.4.3) was, in fact, often in agreement with the experimental Dulong-Petit law and one could not understand why one should not add to that the contribution from the electron degrees of freedom. It took more than twenty years after the first successes of the classical electron theory of metals (§§ 10.4.1 and 15.2.1) before Pauli and Sommerfeld in 1927 solved this paradox, thanks to the use of the, completely new, Fermi-Dirac statistics for the electrons. The introduction of quantum mechanics is equally indispensable for the understanding of transport phenomena in metals. We shall therefore repeat the above approach, adapting it for the *fermion gas* of the electrons in the metal.

First of all we note that several of the features of the Lorentz model are fairly well satisfied here. Solid state theory shows (§§ 11.3.1 and 11.3.3) that to a good approximation one can consider the conduction electrons as noninteracting particles, provided one introduces a macroscopic electric potential to account for possible macroscopic charge densities, which are due to the lack of balance between the electron and nuclear densities. The scattering centres, that is, nuclei displaced from their equilibrium positions and defects, have a large inertia as compared to the electrons, which therefore are a *quantum Lorentz gas*.

The use of a reduced density $f(r, p, t)$ does not pose too many problems in quantum mechanics provided, as will be the case for transport problems, this function varies slowly (§ 10.3.4b). It is true that, because $\hat{r}$ and $\hat{p}$ do not commute, one cannot give the position of a particle in phase space with a precision better than a volume of size h^3. Let us, however, divide the single-electron phase space into volume elements much larger than h^3. This condition enables us to define the single-electron quantum micro-states in the volume $d^3r\, d^3p$; their number is large and, if we take the spin into account, equal to $2\, d^3r\, d^3p/h^3$ (§ 2.3.4). We shall then, as in classical statistical mechanics, define $f\, d^3r\, d^3p$ as the *average number of electrons within the region* $d^3r\, d^3p$ of phase space, that is, occupying the set of micro-states localized within that region.

In thermal equilibrium the theory of § 10.3.3 can be applied to the volume element d^3r, considered to be macroscopic. The average occupation number for each micro-state is the Fermi factor (10.31) so that at equilibrium f is given by

$$f\, d^3r\, d^3p = \frac{2}{h^3} \frac{1}{e^{\beta\varepsilon - \alpha} + 1} d^3r\, d^3p.$$

Close to equilibrium we can treat the points r and p situated at the centre of the various volume elements $d^3r\, d^3p \gg h^3$ as a continuum, provided the quantities involved vary little from one volume element to the next. We thus introduce the *local equilibrium reduced density*

$$f_0(\boldsymbol{r}, \boldsymbol{p}, t) \;=\; \frac{2}{h^3} \, \frac{1}{e^{\beta(\boldsymbol{r},t)\varepsilon - \alpha(\boldsymbol{r},t)} + 1}, \tag{15.45}$$

where α and β are slowly varying functions. The energy ε, which in § 10.4.3 reduced to the kinetic energy $p^2/2m$, may also include a potential $V(\boldsymbol{r})$ as in (15.38). However, it is necessary that this potential varies slowly, in order that we can neglect the discrete nature of the energy levels (§ 10.3.4b). To be more precise, the variations of α, β, and V must be sufficiently slow that f_0 remains practically unchanged within a cell $d^3r \, d^3p \gg h^3$, which means that

$$\lambda_T \, |\nabla(\beta\varepsilon - \alpha)| \;\ll\; 1. \tag{15.46}$$

Note the difference in normalization of the Fermi factor f_q, which refers to a single micro-state, and of f_0 which is associated with the volume $d^3r \, d^3p$.

Farther from equilibrium, provided the electron gas in the metal shows no other correlations than those following from the Pauli principle, we can still characterize the states by $f(\boldsymbol{r}, \boldsymbol{p}, t)$, which has no longer the form (15.45). The Fermi-Dirac statistics impose the condition that $f \leq 2h^{-3}$. The densities and the fluxes can still be expressed by (15.2) and (15.11) in the non-interacting gas model. The next stage consists in obtaining the equation of motion for f. We show in what follows that the microscopic dynamics of the gas of electrons in a metal remains to a good approximation described by the *same kinetic equation* (15.6), (15.7) as in the case of the classical Lorentz gas, notwithstanding the essential rôle played by quantum mechanics.

A complete proof would need a more precise definition of f than we have just given. To do this one should use the Wigner representation (§ 2.3.4) and strike the balance of collisions in the slow variation limit of § 10.3.4b, while all the time satisfying Pauli's principle. Here we shall be satisfied with heuristic arguments.

As regards the left-hand side of the kinetic equation (15.6) or (15.37), which describes the motion of a population of non-interacting electrons, possibly situated in a slowly variable potential $V(\boldsymbol{r})$, we recall that it is the direct result of the classical equations of motion of a particle. However, according to Ehrenfest's theorem (2.29) and the relation (2.78), the centre of a wavepacket evolves according to the classical equations provided its spread $\Delta r \Delta p$ is large compared to h. A sufficiently slow variation of $f(\boldsymbol{r}, \boldsymbol{p}, t)$ in phase space guarantees that f describes a population of electrons, each of which has such a spread. As a result, the two drift terms in (15.37) retain their classical form.

The *collision term* (15.7), however, must in principle be changed in order that we can account for the Pauli exclusion principle. The electrons are fermions and one cannot put more than one in each of the single-particle micro-states. We did not take this into account in the evaluation of the second term

$$-\int d^3p' \, W_\theta(p) \, \delta(\varepsilon - \varepsilon') \, f(r, p, t) \tag{15.47}$$

of (15.7), which describes the scattering of electrons from the volume element $d^3r\, d^3p$ around r, p to the volume element $d^3r\, d^3p'$ around r, p'. Our balance remains correct when the final micro-states are not occupied; however, transitions are *forbidden* toward a micro-state which *is already occupied.* Denoting by $\mathcal{P}_q$ the probability that a single-particle micro-state q is occupied, we must thus weight the term, that describes the collisions from $d^3r\, d^3p$ around r, p to a state q in the volume element $d^3r\, d^3p'$ around r, p', by the probability $1 - \mathcal{P}_q$ that the latter state is empty. The number of states q within that volume element is $2\, d^3r\, d^3p'/h^3$, and the average number of electrons contained in it is $f(r, p', t)\, d^3r\, d^3p'$ so that, on average, $\mathcal{P}_q$ equals $\frac{1}{2} h^3 f(r, p', t)$. The contribution (15.47) to the collision term must thus be replaced by

$$- \int d^3p'\, W_\theta(p)\, \delta(\varepsilon - \varepsilon')\, f(r, p, t) \left[1 - \tfrac{1}{2} h^3\, f(r, p', t) \right]. \tag{15.48}$$

The correction is far from negligible. In fact, if f is close to a low-temperature Fermi distribution, (15.47) would involve all the momenta p and p' *inside* the Fermi sphere, whereas (15.48) contains only collisions where p and p' are close to the *surface* of that Fermi sphere. The effect of the collisions is thus greatly reduced by the Fermi-Dirac statistics.

At the same time, the first term in (15.7) must also be changed to

$$\int d^3p'\, W_\theta(p)\, \delta(\varepsilon - \varepsilon')\, f(r, p', t) \left[1 - \tfrac{1}{2} h^3\, f(r, p, t) \right]. \tag{15.49}$$

In fact, no collision can induce an electron within the volume element $d^3r\, d^3p'$ around r, p' to make a transition to one of the micro-states within $d^3r\, d^3p$ around r, p, if the latter is already occupied. Altogether, combining (15.48) and (15.49), we note that the two corrections which we have introduced, although very large, *cancel exactly,* if we take the symmetry of W_θ under an exchange of p and p' (*microreversibility*) into account. The collision integral retains its form (15.7), notwithstanding the exclusion effects. This cancellation is an accident, due to the special properties of the Lorentz model. In general, the collision terms are not the same for a classical gas as for a gas of fermions (see Eqs.(15.74) and (15.157)).

The difference between the dynamics of the electrons in a metal in a local equilibrium regime and that of a classical Lorentz gas comes from the fact that one must now look for solutions of (15.6) which are *close to a local equilibrium f_0 of fermions*, (15.45), rather than close to a classical f_0, (15.13). To do this one needs change very little in the formalism of §§ 15.1.6 and 15.2.1. It suffices, in fact, to use everywhere the new definition of f_0 instead of the old one, especially in the densities and the fluxes; the equations of state (15.17) of the perfect gas must also be replaced by those of the Fermi gas, (10.42). In the Chapman-Enskog method nothing is changed until (15.28), but the new form of f_0 changes the expression for the gradients of f_0. Those can be expressed in terms of the derivative

$$g \equiv \frac{\partial f_0}{\partial \alpha} = \frac{1}{2h^3 \cosh^2 \frac{1}{2}(\beta\varepsilon - \alpha)}, \tag{15.50}$$

whereas for the classical gas we had $\partial f_0/\partial\alpha = f_0$. We must thus in (15.29) and (15.30) replace f_0 by g. Similarly, when there is a potential $V(\boldsymbol{r})$ present, (15.40) becomes

$$f \;=\; f_0 + \frac{1}{W}\,g\left(\boldsymbol{v}\cdot\{-\nabla\alpha + \varepsilon\nabla\beta\}\right),\qquad\qquad(15.51)$$

which again justifies the V dependence (14.97) of the responses.

A remarkable effect appears. From the fact that g is localized in a region with a width of the order of kT around the Fermi surface $\varepsilon = \alpha/\beta$, *only electrons with energies close to the Fermi surface, within a margin kT, will contribute to the transport.* The electrons with lower energies are *frozen in* and can be neglected. Especially, the heat and electricity conductions are, like the equilibrium properties of the metals (Chaps.10 and 11) governed by the single-electron states close to the Fermi surface. When one applies an electric field, one slightly and anisotropically changes, according to (15.51), the populations of the single-particle states of momenta $\boldsymbol{p}$ with absolute magnitudes close to p_{F}; this change, which, however, affects only a very small fraction of the electrons, produces an electric current. The large *conductivity* of a metal is not due to the acceleration, between collisions, of *all* the electrons in the electric field, as for the carriers in a semiconductor, but to a *marginal effect* on the electrons situated, in momentum space, close to the surface of the Fermi sphere: those electrons, close to the Fermi level, can be excited and can create a current while absorbing a very small amount of energy. On the other hand, the electrons with momenta well inside the Fermi sphere do not feel any effect, due to the Pauli principle, as the neighbouring states into which they might have been scattered are already occupied. To appreciate how weak a perturbation of the motion of the electrons the current is, recall that the Fermi velocity is of the order of 10^6 m s^{-1}, whereas the mean drift velocity is only 10^{-4} m s^{-1} for a current density of 1 A mm^{-2}. In the same units of velocity, the thickness of the Fermi surface at room temperature is of the order of 10^4 m s^{-1}.

One can easily complete the calculation of the linear responses L and the corresponding transport coefficients, by taking the derivative of the Sommerfeld expansion (10.64) which is valid in a metal at room temperature; this gives

$$g \;\approx\; \frac{2}{h^3\beta}\,\delta(\varepsilon - \mu) + \frac{\pi^2}{3h^3\beta^3}\,\delta''(\varepsilon - \mu).\qquad\qquad(15.52)$$

In particular, the *electrical conductivity* (14.95), where $q = -e$ is the electron charge,

$$\sigma \;=\; \frac{e^2}{T}\,L_{NN} \;=\; \frac{4\pi e^2}{3kT}\int_0^\infty p^2\,dp\,g\,\frac{v^2}{W},$$

can be calculated using the first term in (15.52), which yields

$$\sigma = \frac{2}{h^3} \frac{4\pi}{3} p_F^3 \frac{e^2}{mW(p_F)} = \varrho_N e^2 \frac{\tau}{m}. \tag{15.53}$$

We have used the value (10.69) of the electron density and defined the time between collisions as

$$\tau = \frac{1}{W(p_F)}, \tag{15.54}$$

in agreement with the fact that the only active collisions are those of the electrons at the Fermi surface. The evaluation of the *thermal conductivity* λ and of the thermoelectric power ϵ, defined in § 14.4.3, is slightly less simple. In what follows we shall perform it in a more general framework; the expressions which we shall then obtain reduce for the Fermi gas considered here to

$$\lambda = \frac{\pi^2}{3} \varrho_N k^2 T \frac{\tau}{m}, \tag{15.55}$$

$$\epsilon = -\frac{\pi^2}{3} \frac{k^2 T}{e} \frac{2m}{p_F^2} \left[1 + \frac{W}{2} \frac{d}{dp} \left(\frac{p}{W} \right) \right]; \tag{15.56}$$

the derivative of $p/W(p) = m/\varrho_{sc}\sigma_t$, to be taken at $p = p_F$, vanishes in the case of hard spheres in the two limits (15.5). We then find for the ratio of the conductivities

$$\frac{\lambda}{\sigma} = \frac{\pi^2}{3} \frac{k^2}{e^2} T = 2.5 \times 10^{-8} T \quad \text{in SI units,} \tag{15.57}$$

in excellent agreement with the experimental *Wiedemann-Franz law*, in contrast to (15.36) where the coefficient was too small by a factor $\pi^2/6$ for metals.

The fact that the correct expressions (15.53), (15.55), and (15.57) hardly differ from the corresponding classical expressions (15.35), (15.34), and (15.36) explains why, by a happy coincidence, classical electron theory appeared satisfactory for explaining the electrical and thermal conductivities of metals. However, the *thermoelectric powers* in the two theories are very *different*; one sees easily that expression (15.56) which holds for metals gives much lower numerical values, which are in reasonable agreement with experiments, than the classical expression which is valid for semiconductors. In fact, according to (14.99), (15.32), and (15.41), the thermoelectric power of an n-type semiconductor, neglecting the holes, equals

$$\epsilon = \frac{1}{qT} \left(\frac{L_{NE}}{L_{NN}} - \mu \right) = -\frac{k}{e} \left(2 + \frac{\varepsilon_c - \mu}{kT} \right). \tag{15.58}$$

We have here for a direct application of the Lorentz model taken the energy origin to coincide with the bottom ε_c of the conduction band. As $(\varepsilon_c - \mu)/kT$

is of the order of unity, the ratio of (15.58) to (15.56) is of the order of ε_F/kT, where ε_F is the Fermi energy of the metal, that is, of the order of a factor 100. In effect, semiconductors have much more pronounced thermoelectric properties than metals. Note finally that comparing σ or λ with experiments is far less testing than such a comparison for λ/σ or ϵ; in fact, these latter quantities depend hardly or not at all on the mean time τ between collisions, a microscopic quantity which cannot be estimated very accurately.

For a more realistic theory of metals, when we take the *band structure* (§ 11.3.1) into account, we must interpret p as the quasi-momentum and ε as the associated energy. On the left-hand side of (15.6), as in the definition of the fluxes, the velocity v denotes $\nabla_p\varepsilon$, if we use the equations of motion of § 11.3.2. Neglecting for the sake of simplicity the anisotropy of ε and introducing the density of states $\mathcal{D}(\varepsilon)$, we get, instead of (15.30), for the response coefficients

$$L_{NN} = \frac{1}{3k} \int d^3p\, g\, \frac{v^2}{W} = \frac{h^3}{6k} \int d\varepsilon\, \frac{\mathcal{D}(\varepsilon)}{\Omega}\, g\, \frac{v^2}{W},$$

$$L_{NE} = L_{EN} = \frac{h^3}{6k} \int d\varepsilon\, \frac{\mathcal{D}(\varepsilon)}{\Omega}\, g\, \frac{v^2\varepsilon}{W}, \qquad (15.59)$$

$$L_{EE} = \frac{h^3}{6k} \int d\varepsilon\, \frac{\mathcal{D}(\varepsilon)}{\Omega}\, g\, \frac{v^2\varepsilon^2}{W}.$$

Hence we get for the *electrical conductivity* (14.95), in the low-temperature approximation (15.52),

$$\sigma = \frac{q^2}{T} L_{NN} \sim \frac{e^2}{3} \frac{\mathcal{D}(\varepsilon_F)}{\Omega} v_F^2\tau, \qquad (15.60)$$

where we have again defined τ by (15.54). In the box model, using (15.60) and (10.43), we find again (15.53). However, (15.60) shows better the fact that the electrical conduction depends only on the immediate neighbourhood of the Fermi surface. Expression (15.53) involved, by a false analogy with the conductivity (15.35) for semiconductors, the *total* electron density ϱ_N. In fact, we see that this was solely because, *by accident*, in the box model, where $\varepsilon = p^2/2m$, we have

$$\frac{1}{3} \frac{\mathcal{D}(\varepsilon_F)}{\Omega} v_F^2 = \frac{\varrho_N}{m}.$$

In the calculation of the *thermal conductivity* (14.89) the dominant terms from $L_{NN}L_{EE} - L_{EN}^2$ cancel. One must therefore use the second term in (15.52). To do this it is convenient to rearrange the terms coming from (15.59) in the form of a double integral:

$$\lambda = \frac{1}{T^2 L_{NN}} \frac{h^6}{36k^2} \int d\varepsilon\, d\varepsilon'\, \frac{\mathcal{D}(\varepsilon)}{\Omega} \frac{\mathcal{D}(\varepsilon')}{\Omega}\, g\, g'\, \frac{v^2}{W} \frac{v'^2}{W'} \frac{(\varepsilon - \varepsilon')^2}{2},$$

and to use the identity

$$\delta(\varepsilon' - \mu)\, \delta''(\varepsilon - \mu)\, (\varepsilon - \varepsilon')^2 = 2\delta(\varepsilon' - \mu)\, \delta(\varepsilon - \mu),$$

which yields

$$\lambda = \frac{\pi^2}{9} k^2 T \frac{\mathcal{D}(\varepsilon_{\mathrm{F}})}{\Omega} v_{\mathrm{F}}^2 \tau. \tag{15.61}$$

Remarkably, the ratio λ/σ continues to satisfy the *Wiedemann-Franz law* (15.57), notwithstanding the band structure. Experimental and theoretical studies of the Wiedemann-Franz law are thus a useful investigatory means for showing more subtle effects than the ones we have considered, such as the lattice thermal conduction, or anisotropy or retardation effects in the collisions.

Finally, in the calculation of the *thermoelectric power* (14.99) it is again useful to regroup the two terms in

$$\epsilon = \frac{1}{qT} \left(\frac{L_{NE}}{L_{NN}} - \mu \right) = -\frac{1}{eTL_{NN}} \frac{h^3}{6k} \int d\varepsilon \frac{\mathcal{D}(\varepsilon)}{\Omega} g \frac{v^2}{W} (\varepsilon - \mu).$$

Only the second term of (15.52) contributes, and using the relation

$$\delta''(\varepsilon - \mu)(\varepsilon - \mu) = -2\delta'(\varepsilon - \mu),$$

we find

$$\epsilon = -\frac{\pi^2}{3} \frac{k^2 T}{e} \frac{d}{d\varepsilon} \left[\ln \frac{\mathcal{D}(\varepsilon)}{\Omega} \frac{v^2}{W} \right]_{\varepsilon = \varepsilon_{\mathrm{F}}}. \tag{15.62}$$

For $\varepsilon = p^2/2m$, expression (15.56) follows.

15.2.4 Thermalization of Electrons in a Conductor

The electron theory of semiconductors and metals, based upon the Lorentz model, does not describe the thermalization processes where the carrier gas exchanges energy with the other degrees of freedom. We saw in § 14.4.3 that the explanation of the *Joule effect* requires the existence of such exchanges. Moreover, the Lorentz model does not enable us to understand why the reduced density f can at local equilibrium tend to a *Maxwell* distribution for semiconductors and to a *Fermi* distribution for metals, whereas the *same kinetic equation* governs the two cases. In fact, in this model the collisions with *fixed* and *structureless* centres conserve p and thus do not change the initial kinetic energy distribution over the particles, as is shown by (15.21). Nevertheless, in reality, even if the interactions between the carriers are negligible, their interactions with the scattering centres, regarded as an external reservoir, may produce not only exchanges of momentum, as we saw, but also energy exchanges. Here and in § 15.2.5 we shall examine these mechanisms and the physical consequences of such exchanges, which involve either a change in the *internal state* of the scatterer, or its *recoil*.

Once again, the electrons are scattered by the imperfections of the crystal lattice. The inertia of the scatterers is huge as compared to that of the carriers – the mass of a nucleon is 1836 times the electron mass – so that their recoil energy is negligible. In fact, when two particles with masses $m_a \ll m_b$ undergo an *elastic* collision, the total momentum and energy are the same before and after the collision,

$$\boldsymbol{p}_a + \boldsymbol{p}_b = \boldsymbol{p}_a' + \boldsymbol{p}_a', \qquad \varepsilon_a + \varepsilon_b = \varepsilon_a' + \varepsilon_b';$$

this produces an energy transfer

$$\varepsilon_a - \varepsilon_a' = \frac{1}{2m_b} \left(\{2\boldsymbol{p}_b - \boldsymbol{p}_a' + \boldsymbol{p}_a\} \cdot \{\boldsymbol{p}_a' - \boldsymbol{p}_a\} \right). \tag{15.63}$$

Close to equilibrium, ε_a and ε_b are of the same order of magnitude as kT so that the relative change in ε_a in a collision is of the order $\sqrt{m_a/m_b}$. Elastic collisions are thus inefficient for thermalization. However, we have left out of the construction of the Lorentz model the fact that scattering centres have an *internal structure* which may give rise to *inelastic* collisions with energy exchanges, notwithstanding the absence of recoil.

As regards *impurities* and crystal *defects*, these inelastic processes are the ones which are mainly responsible for energy exchanges, since the effective masses of those defects are large and they hardly propagate, while a defect can have energy levels similar to molecular levels and the carriers can induce transitions from one such level to another. Another kind of imperfection is the potential created by the displacement of an ion in the lattice from its equilibrium position during the thermal vibrations of the lattice. Both classically and quantally, energy can in this way be exchanged between the carriers and the crystal lattice, the vibrational state of which is changed. We have indicated in § 11.4.2 that in quasi-particle terms such an interaction is described by the scattering of a carrier with the *creation or absorption of a phonon* which removes or supplies the momentum and energy necessary for the conservation. For instance, the process of the absorption of a phonon of momentum $\boldsymbol{p}'' = \boldsymbol{p}' - \boldsymbol{p}$ by a carrier with momentum $\boldsymbol{p}$ increases the energy of the latter by

$$\varepsilon' - \varepsilon = u|\boldsymbol{p}' - \boldsymbol{p}| \simeq 2up \sin \tfrac{1}{2}\theta, \tag{15.64}$$

where u is the, supposedly isotropic, sound velocity and θ the deflection angle when $\boldsymbol{p}$ changes to $\boldsymbol{p}'$. The relative change in energy resulting from (15.64) is of the order of the ratio of the sound speed, typically $5000\,\mathrm{m\,s^{-1}}$, to the carrier velocity, which in semiconductors is of the order of $4 \times 10^5\,\mathrm{m\,s^{-1}}$ and in metals of the order of $10^6\,\mathrm{m\,s^{-1}}$. In contrast to what happens to the momentum itself, the distribution of which becomes isotropic after a few collisions, *little energy is transferred* in each collision so that the time scale associated with thermalization will be large as compared to τ.

Let us go back to the evaluation of the collision term $\mathcal{I}(f)$ which must now replace (15.7). A complete detailed balance would depend on the kind of imperfection by which the charge carrier is scattered; the contribution from the absorption of a phonon of momentum $\boldsymbol{p}''$, for instance, is proportional to the density of phonons of momentum $\boldsymbol{p}''$ at the point considered, and it contains the constraint (15.64). Similarly, we shall in § 15.3.4 study the effect of the recoil of structureless scatterers with a large, but not infinite, mass. We shall content ourselves here with a more rudimentary and global approach, which follows from the interpretation of $d^3\boldsymbol{p}' \, W_\theta \, \delta(\varepsilon - \varepsilon')$ as a transition probability (end of § 15.1.3). This quantity is here replaced by $d^3\boldsymbol{p}' \, Y(\boldsymbol{p}, \boldsymbol{p}')$, that is, the probability per unit time for a transition of a charge carrier with initial momentum p to a final momentum p' within a margin d^3p'. The length of p' is no longer constrained to be the same as that of p, but, in general, $Y(p, p')$ will be localized in a narrow region $\varepsilon \simeq \varepsilon'$. Treating the carriers as a classical gas, for the semiconductor case, and repeating the arguments of § 15.1.3, we see that (15.7) is replaced by

$$\mathcal{I}(f) = \int d^3p' \left[Y(p', p) f(r, p', t) - Y(p, p') f(r, p, t) \right].$$ (15.65)

Of course, this collision term no longer conserves energy. Besides, $Y(p, p')$ is no longer symmetric under an exchange of the initial state p and the final state p' of the carrier. More precisely, recall that the processes which do not conserve energy operate only on a time scale which is long as compared to the collision time τ. We expect that the medium also has a rather short proper thermalization time. It is thus natural to assume that the *scattering medium is at thermal equilibrium*, at least locally, at a temperature T. Let α denote the micro-states of one of the scatterers (impurity, defect, or phonon) considered and E_α the corresponding energy. The density $\varrho_{sc}(\alpha)$ of scatterers which are in the state α is proportional to $\exp[-E_\alpha/kT]$. Consider the probability $\mathcal{Y}(p, \alpha; p', \alpha')$ for the transition per unit time from a state p, α *for the carrier and scatterer* to the state p', α'. Microscopic physics tells us that *this processs and the inverse process have the same probability* if we take the reversibility of the microscopic equations of motion into account. This property, the symmetry of $\mathcal{Y}$, which generalizes the symmetry of $W_\theta(p)$, is a new form of the *detailed balance principle* or of the *microreversibility*. We find thus, assuming that $\mathcal{Y}$ refers to unit volume,

$$Y(p, p') = \sum_{\alpha, \alpha'} \varrho_{sc}(\alpha) \, \mathcal{Y}(p, \alpha; p', \alpha'),$$

since the global transition probability for the carrier itself must be weighted by the density of scattering centres which are in each of the possible initial states α, and summed over all accessible final states α'. The thermal equilibrium of the scattering medium and conservation of energy imply that

$$\frac{\varrho_{sc}(\alpha)}{\varrho_{sc}(\alpha')} = e^{-(E_\alpha - E_{\alpha'})/kT} = e^{(\varepsilon - \varepsilon')/kT}.$$

Using, finally, the symmetry of $\mathcal{Y}$ we find, for those p, p' values for which $Y \neq 0$,

$$\frac{Y(p, p')}{Y(p', p)} = e^{(\varepsilon - \varepsilon')/kT}.$$ (15.66)

We know that the condition (15.12) for the vanishing of the collision integral characterizes the local equilibrium states. In the original Lorentz model the absence of energy exchanges between carriers and scatterers gave for solutions of $\mathcal{I}(f_0) = 0$ all isotropic functions of p, and not only the Maxwell distributions, so that there was no thermalization. Now, the condition that (15.65) be equal to zero means, in general, that for all pairs of momenta p, p' which can be related through collisions, we have

$$f_0(r, p, t) \, e^{\varepsilon/kT} = f_0(r, p', t) \, e^{\varepsilon'/kT} = \text{constant},$$

where we have used (15.66). As a result f_0 depends on p solely through the Maxwell factor. The system thus evolves to a local equilibrium

$$f_0(r, p, t) = \frac{2}{h^3} e^{\alpha(r,t) - \varepsilon/kT}, \tag{15.67}$$

where *the temperature is determined by the medium*, since $Y(p, p')$ gradually couples all the momenta.

Let us estimate the characteristic time for thermalization by studying how f, initially arbitrary, relaxes locally to the Maxwellian (15.67). As we did when finding (15.21), we can then neglect the drift term. An expansion in spherical harmonics shows now again that, *first of all*, f becomes *isotropic* over a time of order τ, defined by

$$\frac{1}{\tau} = \int d^3p' \, Y(p, p'), \tag{15.68}$$

provided Y depends not too strongly on the angle between p and p'. After that the isotropic part of f at the point r, which we denote by $\varphi(\varepsilon, t) e^{-\varepsilon/kT}$, *continues to evolve*, by virtue of (15.65) and (15.66), according to

$$\frac{\partial}{\partial t} \varphi(\varepsilon, t) \simeq \frac{1}{\tau} \int d\varepsilon'' \, y(\varepsilon, \varepsilon'') \left[\varphi(\varepsilon + \varepsilon'', t) - \varphi(\varepsilon, t)\right], \tag{15.69}$$

where we have written

$$y(\varepsilon, \varepsilon'') \equiv \tau \int d^3p' \, \delta\left(\varepsilon + \varepsilon'' - \frac{p'^2}{2m}\right) Y(p, p'). \tag{15.70}$$

The function y satisfies the relations

$$\int d\varepsilon'' \, y(\varepsilon, \varepsilon'') = 1, \tag{15.71a}$$

$$y(\varepsilon, \varepsilon'') = y(\varepsilon + \varepsilon'', -\varepsilon'') e^{-\varepsilon''/kT} \sqrt{1 + \frac{\varepsilon''}{\varepsilon}};$$

it varies slowly with ε, but is strongly peaked around a transferred energy $\varepsilon'' \simeq 0$, so that its width, defined by

$$\Delta(\varepsilon) \equiv \left[\frac{1}{2} \int d\varepsilon'' \, y(\varepsilon, \varepsilon'') \varepsilon''^2\right]^{1/2}, \tag{15.71b}$$

is small as compared to kT and as compared to typical particle energies ε. It is thus legitimate to expand $\varphi(\varepsilon + \varepsilon'', t)$ in (15.69) in powers of ε'', which, if we use (15.71), yields

$$\tau \frac{\partial \varphi}{\partial t} \simeq \frac{\partial}{\partial \varepsilon} \Delta^2 \frac{\partial \varphi}{\partial \varepsilon} + \Delta^2 \left(\frac{1}{2\varepsilon} - \frac{1}{kT}\right) \frac{\partial \varphi}{\partial \varepsilon}. \tag{15.72}$$

We look for the time t after which the initially arbitrary function φ becomes almost constant in a region of ε of order kT. Even without solving (15.72), a homogeneity argument shows that this *characteristic time for thermalization* is of order

$$ t \sim \tau \left(\frac{kT}{\Delta}\right)^2. \tag{15.73}$$

As we had anticipated it is considerably longer than τ, by a factor $(kT/\Delta)^2$ which is of order 10^4 for the case of electron-phonon interactions. Over this time range one is still far from equilibrium and macroscopic thermodynamics cannot be applied.

Note that the result (15.73) could have been guessed by interpreting (15.69) as an equation describing the following stochastic process: in each time interval τ a collision takes place, which makes the energy increase or decrease by an amount of the order of Δ. After $N = t/\tau$ independent steps, the variance of the energy has increased by $N\Delta^2$; the calculation is formally the same as for (1.10) or in Exerc.15a. This variance becomes equal to $(kT)^2$ when t is of the order (15.73).

In the case of a *metal* the collision term (15.65) must be modified as in § 15.2.3 to take account of the exclusion principle in the final state (see (15.48) and (15.49)). We thus get

$$ \mathcal{I}(f) = \int d^3 p' \left\{ Y(p',p) \, f(r,p',t) \left[1 - \tfrac{1}{2}h^3 f(r,p,t)\right] \right.$$
$$ \left. - Y(p,p') \, f(r,p,t) \left[1 - \tfrac{1}{2}h^3 f(r,p',t)\right] \right\}. \tag{15.74}$$

The quantum corrections do not cancel each other here, as the kernel Y does not have the required symmetry. The condition $\mathcal{I}(f_0) = 0$ then gives for the electrons in a metal

$$ \frac{f_0(r,p,t) \, e^{\varepsilon/kT}}{1 - \tfrac{1}{2}h^3 f_0(r,p,t)} = \frac{f_0(r,p',t) \, e^{\varepsilon'/kT}}{1 - \tfrac{1}{2}h^3 f_0(r,p',t)} = \text{constant}, $$

or

$$ f_0(r,p,t) = \frac{2}{h^3} \, \frac{1}{e^{\varepsilon/kT - \alpha(r,t)} + 1}. \tag{15.75}$$

We thus find here the Fermi factor *dynamically*, as a *consequence of the particular form of the collision term* (15.74), or, more precisely, as a consequence of the fact that an electron is forbidden to be scattered into a state which is already occupied.

Once thermalization has taken place, according to (15.67) for semiconductors and according to (15.75) for metals, the approximation of replacing $Y(p,p')$ by $W_\theta(p)\delta(\varepsilon - \varepsilon')$, which makes the collision terms (15.65) and (15.74) the same, is in general justified. Depending on the case, it leads to the local equilibrium regimes of §§ 15.1.5 or 15.2.3. However, for thermal phenomena, and especially for the *Joule effect* (§ 14.4.3), the increase in the kinetic energy of the carriers between two collisions due to the action of an applied field is, in a permanent regime, compensated for by a transfer to the other degrees of freedom, especially phonons, through collisions with $\varepsilon \neq \varepsilon'$. These exchanges are altogether important since an electron, moving in such a way that its electric potential changes by 1 V, gains a 1 eV energy which is comparable with the Fermi energy and large compared to kT. However, even though the inelasticity of each collision is very small, the high frequency of the collisions makes such energy exchanges possible. In fact, these exchanges are, in general, followed by a release of energy to the ambient medium through conduction or radiation.

15.2.5 Neutron Physics

The design and operation of a nuclear reactor make it necessary to have a theoretical mastery and advanced practical experience of neutron distributions in energy and space. A fissile nucleus, usually ^{235}U, can split into two parts, for instance, ^{140}Xe and ^{94}Sr, carrying away a large amount of kinetic energy, of the order of 200 MeV. This fission reaction is initiated by a neutron colliding with the nucleus; it produces 2 or 3 neutrons which in turn may produce a new fission, but which may also escape or be absorbed, that is, be lost. Moreover, the efficiency of nuclear reactions depends on the energy of the incident neutron. In order that a chain reaction proceeds regularly in a stationary regime it is essential that on average neither more nor less than *one* of the neutrons liberated by each fission gives rise to a new fission. This needs a detailed control of the reduced density $f(\boldsymbol{r}, \boldsymbol{p}, t)$ of the neutrons.

The Lorentz model is not sufficient for this since the *recoil effects of the scattering centres* are important, even more than in § 15.2.4. Indeed, the neutrons can lose their energy only through such recoil. The neutrons produced by the fission of ^{235}U nuclei start out with kinetic energies of the order of MeV. Like the other fission products they thermalize with the other elements of the core of the reactor, thus contributing to the heating of the circuit, usually a water circuit, which feeds the turbines; this contribution is small, of the order of 2.5% of the heat conveyed by the other fission fragments. However, the thermalization of the neutrons is important for another reason. On economic grounds the fuel usually consists of rods of slightly enriched uranium, containing only 3.2% of the ^{235}U isotope. In such a medium a neutron can either produce another fission of a ^{235}U nucleus, or it can be absorbed by a ^{238}U nucleus. The fission cross-section, which characterizes the efficiency of a collision of a neutron with a ^{235}U nucleus for producing fission, varies greatly with the neutron energy: it is about 2 b for 1 MeV neutrons, but reaches 600 b for *"thermal" neutrons* with energies of the order of $\frac{1}{40}$ eV (1 b, barn, the standard unit for nuclear cross-sections, is equal to 10^{-28} m^2). On the other hand, the neutron absorption cross-section of the ^{238}U isotope varies little: it varies from 1 to 3 b for the same change in energies. Under those conditions, because of the low density of ^{235}U, it is necessary to *slow down the neutrons* since absorption dominates for the fast neutrons produced by fission, whereas thermal neutrons more easily initiate the chain reaction.

The slowing down of the neutrons is mainly due to the recoil of the particles by which they are elastically scattered, and it is then described by the collision term (15.130). It starts in the uranium rods through elastic collisions with ^{238}U nuclei which are the majority. However, due to the large mass ratio this process is not very efficient. In fact, the momentum p_a of 1 MeV neutrons is very much larger than the momentum p_b of the thermalized uranium nuclei; under those conditions Eq.(15.63) gives us for the energy loss of a neutron in a collision

$$0 < \frac{\varepsilon_a - \varepsilon_a'}{\varepsilon_a} < 4\,\frac{m_a}{m_b} \simeq 1.7\%, \qquad (15.76)$$

which has a small upper bound. This is the reason why it is necessary to separate the fuel elements by a so-called *moderator* in which the neutrons become thermalized through collisions without being absorbed. Having passed through the moderator they reach another rod with kinetic energies of the order of $\frac{1}{40}$ eV, which enables them easily to produce new fissions. Usually the moderator is water which circulates between the uranium rods where the nuclear reactions take place, and which serves at the same time as the coolant. Even though the collisions with the scattering centres are elastic, the difference in mass between these centres, the H and O nuclei, and the neutrons is sufficiently small for their recoil to be large so that an important amount of kinetic energy is lost by a fast neutron during a collision. When the neutron kinetic energy is sufficiently small, part of it can also be given off to the water molecules in the form of vibrational or rotational energy. The neutrons acquire in this way "thermal" kinetic energies during times of the order of µs. Whatever the thermalization mechanism we must note that the perturbation introduced by the neutrons to the moderator is negligible, since the density of the gas of neutrons, 10^7 cm^{-3}, is very small as compared to that of water, 3×10^{22} cm^{-3}. The moderator can thus be regarded as being in thermal equilibrium so that the analysis of § 15.2.4 remains valid and the slowing down can be described by the collision term (15.65), (15.66). Various approximation methods have been devised for the solution of such a kinetic equation for the reactor geometry (Prob.17).

After thermalization the neutron velocities have a Maxwellian distribution at the temperature of the water, and the macroscopic thermodynamic approach, based upon the diffusion equation (14.20), becomes adequate to describe the flux of thermal neutrons; note that neutron physicists call the quantity $D/\langle v\rangle$ the diffusion coefficient and denote it by D. The characteristic time which elapses before the neutrons react again with uranium is much longer, by one or two orders of magnitude, than the thermalization time.

Another important effect which must be taken into account is the *absorption* of neutrons in collisions with ^{238}U nuclei in the fuel rods, and with Cd and B, contained in the control rods. By plunging the latter more or less deeply into the reactor core, one can fine control the neutron fluxes and hence the reaction rate. Strong absorption can also occur in the security rods which fall into the core in case of an incident. On the other hand, neutrons *disappear* and are *created* by fission reactions. All these effects are taken into account by modifying the kinetic equation in the fuel elements and in the control rods, through the addition of source or sink terms which do not conserve the number of neutrons.

The absorption of neutrons by ^{238}U is unlucky in so far as it reduces the number of neutrons available for maintaining the chain reaction. However, it is indirectly useful, since it leads to the production of ^{239}Pu. Apart from for military purposes, the latter can serve as fuel in *fast neutron reactors* where it undergoes chain reactions. The moderator is here useless since the absorption by ^{238}U is no longer a problem. Indeed, reactors of this kind, the most powerful prototype of which, Superphénix in Creys-Malville produces 1200 MW electric for 3000 MW thermal power, function as *superregenerators* or *breeders*: the core is surrounded by a blanket of depleted uranium, the residue of the isotope separation (Exerc.8a), so that the neutrons which escape again produce Pu. The same amount of natural uranium thus provides 50 times more energy than in an ordinary reactor, and there are fewer waste products.

Mastering the operation of a reactor thus requires a detailed study of the transport and the slowing down of neutrons on the microscopic scale. As in the case of microelectronics or solar collectors, nuclear energy technology appeals fundamentally to advanced methods from statistical physics.

15.3 The Boltzmann Equation

Transport phenomena in *classical gases* are governed by collisions between molecules which Boltzmann's kinetic equation takes into account. The ideas and methods are the same as in the Lorentz model and we shall therefore not repeat the explanations of § 15.1. The main changes to be made are technical complications due to the non-linearity of the equation and to the existence of five conserved macroscopic variables, rather than two, namely, the energy, the number of particles, and the three components of the momentum. This gives rise to macroscopic motions. For the local equilibrium regime we shall here again justify the results of macroscopic thermodynamics and hydrodynamics (§ 14.4.6) and we shall express the response coefficients of the gas in terms of the microscopic cross-sections[2].

15.3.1 The Boltzmann Collision Term

We shall, as in the case of the Lorentz gas, neglect in our description the correlations between the gas molecules and look for the dynamics of the *single-particle reduced density* (15.1), working at the Boltzmann scale (§ 15.1.2). This is possible since here again the duration of the intermolecular collisions is short as compared to the average time τ between collisions. In § 7.4.5 we estimated the latter to be 10^{-9} s for a gas under normal conditions, with a mean free path between 0.1 and 1 µm. On Boltzmann's time and distance scales each collision is treated as an instantaneous and point collision and we shall get the equation of motion of $f(r,p,t)$ as in § 15.1.3 by a detailed balance method. The left-hand side, including its drift term, is the same as in (15.6). The right-hand side $\mathcal{I}(f)$ describes how many particles appear during a time interval dt in the volume element $d^3r\, d^3p$ around the point r,p in the single-particle phase space and how many disappear, due to *elastic binary collisions* within the volume d^3r. Those collisions change quasi-instantaneously and quasi-locally the momenta of two particles by letting particles enter into or leave from d^3p. The result, to be derived below, is the following

---

[2] As reference book we mention C.Cercignani, *Theory and Application of the Boltzmann Equation*, Scottish Academic Press, Edinburgh, 1975.

$$\mathcal{I} = \frac{1}{2} \int d^3 \boldsymbol{p}_2 \, d^3 \boldsymbol{p}_3 \, d^3 \boldsymbol{p}_4 \; W(\boldsymbol{p}_3, \boldsymbol{p}_4; \boldsymbol{p}, \boldsymbol{p}_2) \, \delta(\varepsilon + \varepsilon_2 - \varepsilon_3 - \varepsilon_4)$$

$$\times \, \delta^3(\boldsymbol{p} + \boldsymbol{p}_2 - \boldsymbol{p}_3 - \boldsymbol{p}_4) \{ f(\boldsymbol{r}, \boldsymbol{p}_3, t) f(\boldsymbol{r}, \boldsymbol{p}_4, t) - f(\boldsymbol{r}, \boldsymbol{p}, t) f(\boldsymbol{r}, \boldsymbol{p}_2, t) \}.$$
$$(15.77)$$

The various factors in this collision term have obvious meanings. The function $W(\boldsymbol{p}_1, \boldsymbol{p}_2; \boldsymbol{p}_3, \boldsymbol{p}_4)$ represents, apart from a normalization factor, the probability per unit time that a pair of particles with momenta $\boldsymbol{p}_1$ and $\boldsymbol{p}_2$ acquires momenta $\boldsymbol{p}_3$ and $\boldsymbol{p}_4$ through a collision. Below we shall express it in terms of the *cross-section* for elastic scattering of a pair of particles, a quantity which we shall define exactly and which can be measured or calculated in microscopic physics. Since the particles are indistinguishable, W is symmetric under the exchanges $\boldsymbol{p}_1 \leftrightarrow \boldsymbol{p}_2$ and $\boldsymbol{p}_3 \leftrightarrow \boldsymbol{p}_4$; the factor $\frac{1}{2}$ is associated with the indistinguishability of particles 3 and 4 over which we integrate. Moreover, W is invariant under rotation, under parity and under the Galilean transformation $\boldsymbol{p} \mapsto \boldsymbol{p} - m\boldsymbol{u}$, as is the interaction between the particles. Finally, W is symmetric under the exchange $\boldsymbol{p}_1, \boldsymbol{p}_2 \leftrightarrow \boldsymbol{p}_3, \boldsymbol{p}_4$ (*"microreversibility"*) as the pair collision process is not altered by inverting the motion. The first term in $\mathcal{I}$, associated with the scattering of two particles $\boldsymbol{p}_3, \boldsymbol{p}_4$ into $\boldsymbol{p}, \boldsymbol{p}_2$ is weighted with the average pair density $f(\boldsymbol{r}, \boldsymbol{p}_3) \, f(\boldsymbol{r}, \boldsymbol{p}_4)$ in the initial state; it describes an increase in $f(\boldsymbol{r}, \boldsymbol{p})$ when one of the particles in the final state has acquired a momentum $\boldsymbol{p}$. Inversely, the second term corresponds to the scattering of a particle $\boldsymbol{p}$ by another, with arbitrary momentum $\boldsymbol{p}_2$, into an arbitrary final state; again it is weighted by the average pair density $f(\boldsymbol{r}, \boldsymbol{p}) \, f(\boldsymbol{r}, \boldsymbol{p}_2)$ in the initial state considered and it describes a decrease in $f(\boldsymbol{r}, \boldsymbol{p})$. Conservation of energy and momentum in each elastic collision is explicitly expressed by the δ-functions which occur in (15.77); W is defined only for the momenta which are compatible with these conservation laws.

The kinetic equation

$$\frac{\partial f}{\partial t} + (\boldsymbol{v} \cdot \nabla) f = \mathcal{I}(f),$$
$$(15.78)$$

with the right-hand side given by (15.77), is the famous *Boltzmann equation* (1872). It made it possible to formalize the kinetic gas theory, started, among others, by Maxwell in 1860, and it played an essential rôle in creating statistical mechanics: it provided a more efficient calculation method for the transport properties of gases than those of § 7.4.6 and it led Boltzmann a few years later (1877) to identify the entropy as a microscopic statistical quantity (§§ 3.4.2 and 15.4.1).

The analysis of the *scattering of two particles* rests in classical mechanics upon the concept of the impact parameter and, more generally, both in classical and in quantum mechanics, on that of the scattering *cross-section*. We want to extend

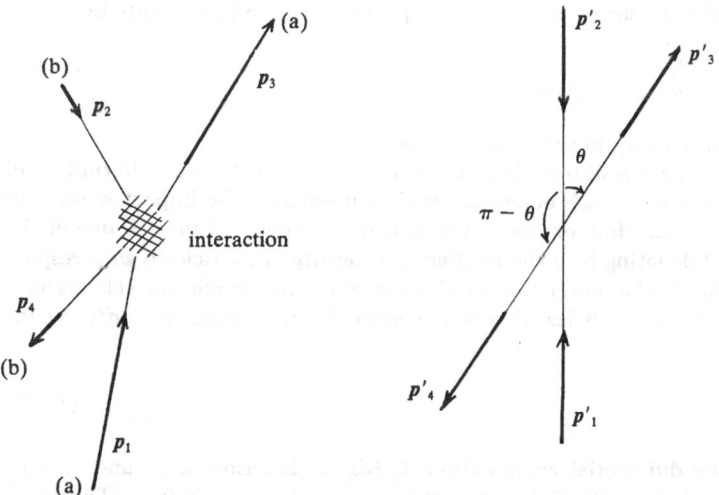

Fig. 15.5. Elastic scattering of two particles in the fixed and the centre-of-mass frame of reference

the definition of cross-section, which we gave in § 15.1.3 for the special case where particle b was very heavy and could be replaced by a fixed scattering centre, to the scattering of two particles a and b with, at the moment, arbitrary masses m_a and m_b; we thus analyze the kinematics of the scattering in a Galilean frame of reference with a velocity v_{cm} which is that of the centre of mass of the pair (Fig.15.5). Denoting by p' the momenta in that new frame, which for each particle are related to the momenta in the fixed frame by $p = mv_{cm} + p'$, we have

$$p_1 + p_2 = p_3 + p_4 = (m_a + m_b)v_{cm},$$

$$p'_1 = -p'_2, \quad p'_3 = -p'_4, \quad |p'_1| = |p'_3|, \qquad (15.79)$$

where the last equality expresses conservation of energy. The indices 1 and 2 refer to the particles a and b before the collision, 3 and 4 to them after the collision (Fig.15.5). The scattering of one particle by another does not involve more parameters than the scattering by a fixed centre. These parameters are the *energy* in the centre of mass frame,

$$\varepsilon_0 = \varepsilon_1 + \varepsilon_2 - \tfrac{1}{2}(m_a + m_b)v_{cm}^2$$

$$= \varepsilon'_1 + \varepsilon'_2 = \frac{m_a m_b v_{12}^2}{2(m_a + m_b)} = \frac{m_a m_b v_{34}^2}{2(m_a + m_b)}, \qquad (15.80)$$

and the *deflection angle* θ between p'_1 and p'_3, which in the fixed frame is given by

$$\cos \theta = \frac{(v_{12} \cdot v_{34})}{v_{12}^2}. \qquad (15.81)$$

We have denoted the relative velocity of the particles in the initial state by

$$v_{12} = \frac{p_1}{m_a} - \frac{p_2}{m_b} = \frac{m_a + m_b}{m_a m_b} p'_1, \tag{15.82}$$

and their relative velocity in the final state by v_{34} ($v_{12} = v_{34}$).

The *differential cross-section* $d\sigma$ is, as in § 15.1.3, defined for a uniform flux of particles directed toward each other; in classical mechanics the impact parameter is completely random. More precisely, working in the centre of mass frame of the two particles and denoting by j the *relative flux density* of particles b with respect to particles a, that is the product of the density of b and the relative velocity v_{12}, the *average number of particles* of type b *scattered into a solid angle* $d^2\omega$ *in the time dt* is

$$j \frac{d\sigma}{d\omega} d^2\omega \, dt, \tag{15.83}$$

which defines the differential cross-section $d\sigma/d\omega$ as function of ε_0 and θ. This definition reduces to that of § 15.1.3 when the mass m_a becomes infinite. The cross-section is an area: the average number of particles deflected per unit time into $d^2\omega$ is equal to the average number of incident particles crossing per unit time an area $d\sigma$, in the centre of mass frame. In the case of indistinguishable particles, we must, apart from the simplification $m_a = m_b$, take into account that the deflections θ and $\pi - \theta$ must be regarded as the same process; besides, in quantum mechanics, it is impossible to distinguish the two final states p_3, p_4 and p_4, p_3. Therefore, if in classical mechanics we calculate a cross-section by taking the average over impact parameters, we must include both contributions; a correct quantum calculation does this automatically. On the other hand, even in the case of distinguishable particles, the definition (15.83) implies that the cross-section is symmetric in the exchange of a and b.

If we take *classical hard spheres* as a model for the particles, and denote the sum of the radii of the two colliding spheres by δ, the calculation of the cross-section will be the same as in § 15.1.3, since we can essentially forget the motion of the particle a. In particular, from the fact that the cross-section refers to a *relative* unit incident flux, it follows that the number of collisions per unit time for an impact parameter between b and $b + db$ remains unchanged and equal to $2\pi b \, db$. The differential cross-section is thus again equal to $d\sigma = \frac{1}{4}|d^2\omega|\delta^2$; this number must be multiplied by a factor 2 for the case of indistinguishable particles.

We now turn to the calculation of the number of collisions undergone during the time dt by the $f \, d^3r \, d^3p$ particles situated within a single-particle phase space volume element; that will give us the second, negative, contribution to $\mathcal{I} \, d^3r \, d^3p \, dt$. To be more general, we assume that these collisions take place with particles b which may be different from the particles a considered. Let us consider the collisions of one of the particles a with particles b of momentum p_2, within d^3p_2; the relative velocity is $v_{12} = (p/m_a) - (p_2/m_b)$, and the density of particles of type b is $f_b(r, p_2) \, d^3p_2$ so that their relative flux density is $v_{12} f_b(p_2) \, d^3p_2$. The number of collisions undergone by each of the particles a under consideration of momentum p is thus

$$dt \int d^2\omega \int d^3p_2 \, v_{12} \, f_b(r, p_2) \, \frac{d\sigma}{d\omega}, \tag{15.84}$$

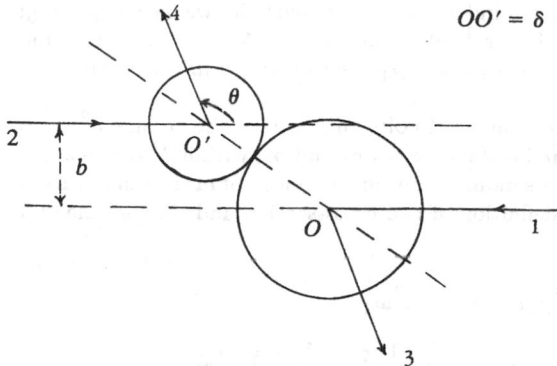

$$OO' = \delta$$

Fig. 15.6. Collision between classical hard spheres

where we have used the definition (15.83) of the cross-section; this gives us the following contribution to $\partial f_a/\partial t$:

$$- f_a(\boldsymbol{r}, \boldsymbol{p}, t) \int d^2\omega \, d^3 p_2 \, v_{12} \, f_b(\boldsymbol{r}, \boldsymbol{p}_2, t) \, \frac{d\sigma}{d\omega}.$$

Using expression (15.81) for the angle θ we can replace the integration over $d^2\omega$ by an integration over v_{34}, by writing

$$\int d^2\omega \; = \; \int d^3 v_{34} \, \frac{1}{v_{12}^2} \, \delta(v_{34} - v_{12}).$$

It follows from (15.80) that

$$\frac{1}{v_{12}} \, \delta(v_{34} - v_{12}) \; = \; \delta\left(\frac{1}{2} v_{34}^2 - \frac{1}{2} v_{12}^2\right) \; = \; \frac{m_a m_b}{m_a + m_b} \, \delta(\varepsilon_3 + \varepsilon_4 - \varepsilon - \varepsilon_2),$$

and from (15.82) and (15.79) that

$$\int d^3 v_{34} \; = \; \int \left(\frac{m_a + m_b}{m_a m_b}\right)^3 d^3 p_3'$$

$$= \; \int \left(\frac{m_a + m_b}{m_a m_b}\right)^3 d^3 p_3 \, d^3 p_4 \, \delta^3(\boldsymbol{p}_3 + \boldsymbol{p}_4 - \boldsymbol{p} - \boldsymbol{p}_2),$$

so that the contribution (15.84) of the losses of particles a through collisions with particles b can finally be expressed as

$$- \int d^3 p_2 \, d^3 p_3 \, d^3 p_4 \, \delta^3(\boldsymbol{p}_3 + \boldsymbol{p}_4 - \boldsymbol{p} - \boldsymbol{p}_2) \, \delta(\varepsilon_3 + \varepsilon_4 - \varepsilon - \varepsilon_2)$$

$$\times \left(\frac{1}{m_a} + \frac{1}{m_b}\right)^2 \frac{d\sigma}{d\omega} \, f_a(\boldsymbol{r}, \boldsymbol{p}, t) \, f_b(\boldsymbol{r}, \boldsymbol{p}_2, t).$$

Writing

$$W(\boldsymbol{p}_1, \boldsymbol{p}_2; \boldsymbol{p}_3, \boldsymbol{p}_4) \; \equiv \; \left(\frac{1}{m_a} + \frac{1}{m_b}\right)^2 \frac{d\sigma}{d\omega}, \tag{15.85}$$

we get the second term in (15.77). For indistinguishable particles we have $m_a = m_b$; moreover, we must introduce the factor $\frac{1}{2}$ which occurs in (15.77) in order to avoid counting twice the same process when p_3 and p_4 are exchanged in the integration, using the definition of $d\sigma/d\omega$.

We can similarly evaluate the number of collisions, within the volume d^3r and the time dt, between particles a and b of momenta p_3 and p_4, within d^3p_3 and d^3p_4, which produce an a particle with its momentum, in the centre of mass frame, within the solid angle $d^2\omega_1'$. Using the definition of the cross-section and the formulae for the change of frames, we find

$$v_{34}\, f_a(r,p_3)\, d^3p_3\, f_b(r,p_4)\, d^3p_4\, d^3r\, dt\, \frac{d\sigma}{d\omega}\, d^2\omega_1'$$

$$= f_a(p_3)\, d^3p_3\, f_b(p_4)\, d^3p_4\, d^3r\, dt\, \frac{d\sigma}{d\omega}\, \delta\left(\frac{1}{2}v_{12}^2 - \frac{1}{2}v_{34}^2\right)\, d^3v_{12}$$

$$= f_a(p_3)\, d^3p_3\, f_b(p_4)\, d^3p_4\, d^3r\, dt\, \frac{d\sigma}{d\omega}\, \delta(\varepsilon_1 + \varepsilon_2 - \varepsilon_3 - \varepsilon_4)$$

$$\times \delta^3(p_1 + p_2 - p_3 - p_4)\left(\frac{1}{m_a} + \frac{1}{m_b}\right)^2\, d^3p_2\, d^3p_1.$$

Among these collisions those for which p_1 is equal to p, within d^3p, are the ones which contribute to the increase in $f_a\, d^3r\, d^3p$. For indistinguishable particles their contribution gives just the first term in (15.77).

We have thus justified the Boltzmann collision term (15.77). The function W, defined by (15.85) in terms of the differential cross-section for elastic scattering, is a function of the two variables ε_0 and θ, defined through (15.80) and (15.81). One can easily check that it has the above mentioned symmetry and invariance properties. For classical indistinguishable hard spheres with diameter δ,

$$W = \frac{2\delta^2}{m^2} \tag{15.86}$$

is a constant.

Boltzmann's equation has some simple properties which are a generalization of those of the Lorentz model. The *conservation laws* for particle number, energy, and momentum for each collision entail the *identities*

$$\int d^3p\, \mathcal{I} \equiv 0, \qquad \int d^3p\, \mathcal{I}\, \varepsilon \equiv 0, \qquad \int d^3p\, \mathcal{I}\, p \equiv 0, \tag{15.87}$$

which are satisfied by the Boltzmann collision term for any distribution f. One can easily prove these identities, using the explicit form (15.77) of $\mathcal{I}$ and the symmetry between the four momenta, which is fully reinstated by the integration over p. The particle number conservation, $\int d^3p\, \mathcal{I} = 0$ thus follows from the antisymmetry of the integrand $\mathcal{I}$ under the exchange $p, p_2 \leftrightarrow p_3, p_4$. Using that antisymmetry and the symmetry of $\mathcal{I}$ under the exchanges $p \leftrightarrow p_2$ and $p_3 \leftrightarrow p_4$, we have

$$\int d^3p\, \mathcal{I}\, \varepsilon = \frac{1}{4}\int d^3p\, \mathcal{I}\, (\varepsilon + \varepsilon_2 - \varepsilon_3 - \varepsilon_4),$$

and this last expression vanishes, since it contains

$$(\varepsilon + \varepsilon_2 - \varepsilon_3 - \varepsilon_4)\,\delta(\varepsilon + \varepsilon_2 - \varepsilon_3 - \varepsilon_4).$$

Similarly, the last identities (15.87) for the conservation of the momentum components follow from the symmetry of the integrand and the fact that they contain the factor $\delta^3(\boldsymbol{p} + \boldsymbol{p}_2 - \boldsymbol{p}_3 - \boldsymbol{p}_4)$.

The microscopic conservation equations (15.87) are *local*, which is a consequence of the fact that collisions change the momenta of the particles, but leave their positions practically unchanged. The *macroscopic conservation equations* (14.6) follow directly from the Boltzmann equation (15.78) and the microscopic conservation equations (15.87), provided we define the *densities* by

$$\varrho_N(\boldsymbol{r}, t) \;=\; \int d^3p\, f(\boldsymbol{r}, \boldsymbol{p}, t), \qquad \varrho_E(\boldsymbol{r}, t) \;=\; \int d^3p\, f(\boldsymbol{r}, \boldsymbol{p}, t)\,\varepsilon,$$

$$\varrho_P(\boldsymbol{r}, t) \;=\; \int d^3p\, f(\boldsymbol{r}, \boldsymbol{p}, t)\,\boldsymbol{p}, \tag{15.88}$$

and the *fluxes* by

$$\boldsymbol{J}_N(\boldsymbol{r}, t) \;=\; \int d^3p\, \boldsymbol{v}\, f(\boldsymbol{r}, \boldsymbol{p}, t), \qquad \boldsymbol{J}_E(\boldsymbol{r}, t) \;=\; \int d^3p\, \boldsymbol{v}\, f(\boldsymbol{r}, \boldsymbol{p}, t)\,\varepsilon,$$

$$J_{P_\beta}^\alpha(\boldsymbol{r}, t) \;=\; \int d^3p\, v_\alpha\, f(\boldsymbol{r}, \boldsymbol{p}, t)\, p_\beta. \tag{15.89}$$

In fact, it is sufficient to integrate (15.78) over $\boldsymbol{p}$, after having multiplied it, successively, by 1, ε, and $\boldsymbol{p}$: the right-hand side vanishes, the first term on the left-hand side produces $\partial \varrho_i / \partial t$, and the second term, which we can also write in the form div $(\boldsymbol{v} f)$, produces the divergence of $\boldsymbol{J}_i$.

Like for the Lorentz model, expressions (15.88) and (15.89) can be obtained by retaining only the kinetic part of the Hamiltonian in the general forms of § 14.3.2. Using the Boltzmann equation to take account of the dynamic effects of the interactions should thus be on a par with *completely* neglecting these interactions in calculating the densities and the fluxes.

15.3.2 Densities and Fluxes in a Gas

The Boltzmann collision term has another simple and important property. It *vanishes*,

$$\mathcal{I}(f_0) \;=\; 0, \tag{15.90}$$

for any local equilibrium distribution of the form

$$f_0 \;=\; \frac{1}{h^3}\, e^{\alpha(\boldsymbol{r}, t) - \beta(\boldsymbol{r}, t)\varepsilon + (\lambda(\boldsymbol{r}, t) \cdot \boldsymbol{p})}. \tag{15.91}$$

The existence of the five local conservative variables (15.88) implies here the presence of five associated Lagrangian multipliers, α, β, and λ. One can check (15.90) immediately: substituting (15.91) into (15.77) the combination

$$f_0(\boldsymbol{r},\boldsymbol{p})\,f_0(\boldsymbol{r},\boldsymbol{p_2}) - f_0(\boldsymbol{r},\boldsymbol{p_3})\,f_0(\boldsymbol{r},\boldsymbol{p_4})$$

appears; it vanishes provided $\varepsilon + \varepsilon_2 = \varepsilon_3 + \varepsilon_4$, $\boldsymbol{p} + \boldsymbol{p_2} = \boldsymbol{p_3} + \boldsymbol{p_4}$ so that the integrand of $\mathcal{I}(f_0)$ itself vanishes. (In (15.91), as in (15.13), we dropped the spin factor $2s + 1$.) We shall see in § 15.4.2 that, conversely, *all* solutions of (15.90) have in general the form (15.91).

When f is arbitrary, far from the form (15.91), one can estimate the importance of the collision term by the order of magnitude of $f/\mathcal{I}(f)$, which has the dimensions of a time. Expression (15.84) and the definition of the cross-section σ show that this time τ is of the order of $1/\sigma \varrho_N v$, that is, the average time between successive collisions of one of the particles. On a time scale small as compared to τ or over distances small as compared to the mean free path $l \simeq 1/\sigma \varrho_N$, the collision term does hardly come into play and we are in the *ballistic regime* (§ 15.1.4). This occurs at very *low densities*, for instance, for transport properties in the upper atmosphere; the gas may also evolve far from local equilibrium in regions with *large gradients* such as shock waves within flows or boundary layers at their boundary.

Inversely, transport phenomena with typical times large as compared to τ and typical distances large as compared to l are studied in the *local equilibrium* regime by putting $f = f_0 + f_1$, where f_1 is small as compared to f_0. As in (15.22), in order unambiguously to associate to each f a local equilibrium distribution f_0 which accompanies it in its motion, we constrain f_1 to satisfy

$$\int d^3p\, f_1(\boldsymbol{r},\boldsymbol{p},t) = 0, \quad \int d^3p\, f_1\,\varepsilon = 0, \quad \int d^3p\, f_1\,\boldsymbol{p} = 0, \quad (15.92)$$

so that the densities (15.88) can equivalently be expressed in terms of f or of f_0. This gives us the equations of state which connect at each point the densities ϱ_i with the multipliers α, β, and λ of (15.91). Putting

$$\beta \equiv \frac{1}{kT}, \qquad \lambda \equiv \beta u, \qquad \alpha = \alpha' - \tfrac{1}{2}mu^2\beta, \qquad (15.93)$$

we can rewrite (15.91) in the form

$$f_0 = \frac{1}{h^3}\, e^{\alpha' - \beta(\boldsymbol{p}-m\boldsymbol{u})^2/2m}, \qquad (15.94)$$

which enables us to identify $\boldsymbol{u}(\boldsymbol{r},t)$ with the macroscopic *local flow velocity* and $\mu'(\boldsymbol{r},t) \equiv \alpha'/\beta$ with the chemical potential in the Galilean frame in which the fluid is locally at rest. Hence we find the explicit forms of the *local equations of state* for the gas:

$$\left. \begin{aligned} \varrho_N &= \frac{1}{h^3}\, e^{\alpha'} \left(\frac{2\pi m}{\beta}\right)^{3/2}, & \varrho_E &= \varrho_U + \tfrac{1}{2}mu^2 \varrho_N, \\ \varrho_U &= \tfrac{3}{2}\varrho_N kT, & \varrho_P &= \varrho_N m u, \\ \mathcal{P} &= \varrho_N kT, & s &= \varrho_N k\big(\tfrac{5}{2} - \alpha'\big); \end{aligned} \right\} \tag{15.95}$$

they are a special case of the general form (14.109) which we found macroscopically for any fluid.

The fluxes (15.89) contain two contributions, associated one with f_0 and the other with f_1. The first we identify with the *equilibrium fluxes* $\boldsymbol{J}_i^0$ for which we found the general form (14.116) from macroscopic invariance arguments. Their explicit form is here obtained by using (15.89) and (15.94) which give

$$\left. \begin{aligned} \boldsymbol{J}_N^0 &= \boldsymbol{u}\varrho_N = \frac{1}{m}\,\varrho_P, & \boldsymbol{J}_E^0 &= \boldsymbol{u}\varrho_N \left(\tfrac{5}{2}kT + \tfrac{1}{2}mu^2\right), \\ J_{P\beta}^{0\alpha} &= m\varrho_N u_\alpha u_\beta + \varrho_N kT \delta_{\alpha\beta}. \end{aligned} \right\} \tag{15.96}$$

These results are in agreement with expressions (14.116), if we bear in mind the form (15.95) of the energy density ϱ_E and of the equilibrium pressure $\mathcal{P}$. Similarly, the contributions from f_1 to the fluxes,

$$\left. \begin{aligned} \boldsymbol{J}_N^1 &= \int d^3p\,\boldsymbol{v}\,f_1 = 0, \\ J_{P\beta}^{1\alpha} &= \int d^3p\,v_\alpha f_1 p_\beta \equiv \mathcal{P}_\beta^\alpha - \mathcal{P}\,\delta_{\alpha\beta}, \\ J_E^{1\alpha} &= \int d^3p\,v_\alpha f_1 \varepsilon \equiv J_U^\alpha + \sum_\beta \left(\mathcal{P}_\beta^\alpha - \mathcal{P}\,\delta_{\alpha\beta}\right) u_\beta, \end{aligned} \right\} \tag{15.97}$$

have the form (14.121) which is valid for any fluid. They define the *heat flux*,

$$\boldsymbol{J}_U = \int d^3p\,\boldsymbol{v}'f_1\varepsilon', \tag{15.98}$$

and the *stress tensor*,

$$\mathcal{P}_\beta^\alpha = \mathcal{P}\,\delta_{\alpha\beta} + \int d^3p\,v_\alpha' f_1 p_\beta', \tag{15.99}$$

where we have taken the velocities $\boldsymbol{v}' \equiv \boldsymbol{v} - \boldsymbol{u}$, the momenta $\boldsymbol{p}' \equiv m\boldsymbol{v}'$, and the kinetic energies $\varepsilon' \equiv p'^2/2m$ of the particles in the local rest frame. As in all fluids we find $\varrho_P = m\boldsymbol{J}_N$, and $\mathcal{P}_\beta^\alpha$ is symmetric. Moreover, we get from (15.88), (15.89), (15.92), and (15.97) a new property,

$$\varrho_E = \tfrac{1}{2} \sum_\alpha J_{P\alpha}^\alpha, \qquad \sum_\alpha \left(\mathcal{P}_\alpha^\alpha - \mathcal{P}\right) = 0, \tag{15.100}$$

which is *specific for dilute gases*.

Since we have again obtained the general expressions (14.121) for the fluxes we have at the same time justified the equations which follow from them, namely: the *equation of continuity* (14.124), the *equation of motion* (14.125), and the *energy conservation* equation (14.126). We have found here these properties as consequences of the microscopic conservation laws (15.87) and we did not make any approximations beyond the ones needed to derive the Boltzmann equation.

We must still derive the response equations. To do that, we note that, as in the Lorentz model (§ 15.1.4), if the density is sufficiently high, the collisions will *in a microscopic time τ lead to local equilibrium*. After that their effect is less and gives rise to a *slow relaxation* to global equilibrium in a linear response regime. As in § 15.1.5, we can calculate that relaxation by pushing the Chapman-Enskog method to order f_1. Recall that the latter consists in treating W as a quantity of order -1 whereas f_0 and f_1 are of order 0 and 1, respectively; moreover, in each stage of the iteration which follows from this idea, one must satisfy, on the one hand, the constraints (15.92), that is, the equations of state (15.95) and, on the other hand, the conservation laws.

However, in contrast to what happens for the Lorentz gas, here we get already a non-trivial macroscopic dynamics *to lowest order*. In fact, the Boltzmann equation (15.78) reduces in that order to (15.90), the solution of which is a local equilibrium distribution. In conformity with the general Chapman-Enskog strategy we must also write down the *conservation equations* to lowest order, where only the equilibrium fluxes (15.96) remain, since the contributions (15.97) coming from f_1 are of higher order. As in (15.26), it is not the kinetic equation itself which produces the equations of motion for the densities (15.88) or, what amounts to the same, for the intensive variables given by (15.95), or for f_0 given by (15.91), but the conservation laws, which are its consequences. Using the form of (15.95) and (15.96) and putting $\varrho = m\varrho_N$, we get the *equation of continuity*

$$\frac{\partial \varrho}{\partial t} + \operatorname{div} \varrho \boldsymbol{u} = 0, \tag{15.101}$$

the *Euler equation*

$$\frac{\partial \boldsymbol{u}}{\partial t} + (\boldsymbol{u} \cdot \nabla)\boldsymbol{u} + \frac{1}{\varrho} \nabla \mathcal{P} = 0, \tag{15.102}$$

and the *energy conservation*

$$\frac{\partial \varrho U}{\partial t} + \operatorname{div} \varrho_U \boldsymbol{u} + \mathcal{P} \operatorname{div} \boldsymbol{u} = 0. \tag{15.103}$$

The Chapman-Enskog approximation thus gives to lowest order the equations of a non-dissipative *perfect fluid* which we encountered already in § 14.4.6. In particular, Eq.(15.103) describes the *reversible* transformation of the internal energy of the gas into work done by the quasi-static pressure forces.

15.3.3 Transport Coefficients of a Gas

If we want to work out the Chapman-Enskog method of § 15.1.5 to order f_1 we must follow a programme, the principles of which are relatively simple, but which gives rise to rather cumbersome calculations. First of all we must write down the left-hand side of the Boltzmann equation to order f_0, taking care to satisfy *the lowest order conservation laws*; to do that we must start by using the equations of state to express the derivative $\partial f_0/\partial t$, which will involve $\partial \alpha/\partial t$, $\partial \beta/\partial t$, and $\partial \lambda/\partial t$, in terms of the derivatives $\partial \varrho_i/\partial t$; then we must eliminate these time derivatives by using Eqs.(15.101), (15.102), and (15.103) of non-dissipative hydrodynamics to replace them by the equilibrium fluxes J_i^0; the left-hand side of the Boltzmann equation will then be expressed solely in terms of local equilibrium variables and their gradients. On the other hand, we must write out the right-hand side $\mathcal{I}(f)$ by expanding it *to first order in* f_1. After that we must solve the equation obtained and *express f_1 as function of f_0 and the affinities*, taking all the time account of the *constraints* (15.92) which ensure that the relations between the local extensive and intensive variables remain unaltered. Finally, we must evaluate the *dissipative parts of the fluxes* using their expressions (15.97) in terms of f_1 and hence derive the linear responses.

Let us start by using the conservation equations of non-dissipative hydrodynamics to transform the first term in the Boltzmann equation $\partial f_0/\partial t$. It will be convenient to replace the local state variables ϱ_i by ϱ, β, and $\boldsymbol{u}$ so that we shall, instead of the energy conservation equation (15.103), use the equation for β which follows from it:

$$\frac{\partial \beta}{\partial t} + (\boldsymbol{u} \cdot \nabla)\beta - \frac{2\beta}{3} \operatorname{div} \boldsymbol{u} = 0. \tag{15.104}$$

Starting from the form (15.94) for f_0 and using (15.95) to express the intensive variables α', β, and $\boldsymbol{u}$, which depend on $\boldsymbol{r}$ and t, as functions of ϱ, β, and $\boldsymbol{u}$, we get

$$\frac{\partial f_0}{\partial t} = f_0 \left[\frac{\partial \alpha'}{\partial t} - \varepsilon' \frac{\partial \beta}{\partial t} + \beta \left(\boldsymbol{p}' \cdot \frac{\partial \boldsymbol{u}}{\partial t} \right) \right]$$

$$= f_0 \left[\frac{1}{\varrho} \frac{\partial \varrho}{\partial t} + \left(\frac{3}{2\beta} - \varepsilon' \right) \frac{\partial \beta}{\partial t} + \beta \left(\boldsymbol{p}' \cdot \frac{\partial \boldsymbol{u}}{\partial t} \right) \right].$$

Using the relation $\mathcal{P} = \varrho/m\beta$ and the conservation equations (15.101), (15.102), and (15.104) we then get

$$\frac{1}{f_0} \frac{\partial f_0}{\partial t} = -\frac{1}{\varrho} (\boldsymbol{u} \cdot \nabla)\varrho - \operatorname{div} \boldsymbol{u} - \left(\frac{3}{2\beta} - \varepsilon' \right) (\boldsymbol{u} \cdot \nabla)\beta$$

$$+ \left(\frac{3}{2\beta} - \varepsilon' \right) \frac{2\beta}{3} \operatorname{div} \boldsymbol{u} - \beta \big(\boldsymbol{p}' \cdot (\boldsymbol{u} \cdot \nabla)\boldsymbol{u} \big) - \frac{\beta}{\varrho} (\boldsymbol{p}' \cdot \nabla)\mathcal{P}.$$

Combining this with the second term $(\boldsymbol{v} \cdot \nabla)f_0$, which we transform in the same way as $\partial f_0/\partial t$ in terms of the gradients of ϱ, β, and $\boldsymbol{u}$, we obtain after some calculations for the *left-hand side* of the Boltzmann equation

$$\frac{\partial f_0}{\partial t} + (\boldsymbol{v} \cdot \nabla) f_0 = f_0 \left[\left(\frac{5}{2\beta} - \varepsilon' \right) (\boldsymbol{v}' \cdot \nabla) \beta + \beta m \sum_{\alpha, \beta} v'_\alpha v'_\beta \Delta_{\alpha\beta} \right]. \quad (15.105)$$

Note that $\nabla \varrho$ has been eliminated and that the gradients of the local velocity $\boldsymbol{u}$ only appear through the symmetric combination with zero trace

$$\Delta_{\alpha\beta} \equiv \frac{1}{2} \left(\frac{\partial u_\beta}{\partial r_\alpha} + \frac{\partial u_\alpha}{\partial r_\beta} \right) - \frac{1}{3} \delta_{\alpha\beta} \operatorname{div} \boldsymbol{u}, \quad (15.106)$$

which was already introduced in (14.132).

The expansion of the *collision term* (15.77) to order 1 in f_1 gives an expression of the form

$$\mathcal{I}(f_0 + f_1) \sim - f_0(\boldsymbol{r}, \boldsymbol{p}, t) \int d^3 p_1 \, \mathcal{K}(\boldsymbol{p}, \boldsymbol{p}_1) \, f_1(\boldsymbol{r}, \boldsymbol{p}_1, t), \quad (15.107)$$

where the *collision kernel* $\mathcal{K}(\boldsymbol{p}, \boldsymbol{p}_1)$ is defined by

$$\mathcal{K}(\boldsymbol{p}, \boldsymbol{p}_1) \equiv - \int d^3 p_2 \, d^3 p_4 \, W(\boldsymbol{p}_1, \boldsymbol{p}_4; \boldsymbol{p}, \boldsymbol{p}_2) \, \delta(\varepsilon + \varepsilon_2 - \varepsilon_1 - \varepsilon_4)$$

$$\times \, \delta^3(\boldsymbol{p} + \boldsymbol{p}_2 - \boldsymbol{p}_1 - \boldsymbol{p}_4) \, e^{\beta(\varepsilon' - \varepsilon'_4)} + \frac{1}{2} \int d^3 p_3 \, d^3 p_4 \, W(\boldsymbol{p}_3, \boldsymbol{p}_4; \boldsymbol{p}, \boldsymbol{p}_1)$$

$$\times \, \delta(\varepsilon + \varepsilon_1 - \varepsilon_3 - \varepsilon_4) \, \delta^3(\boldsymbol{p} + \boldsymbol{p}_1 - \boldsymbol{p}_3 - \boldsymbol{p}_4) + \frac{1}{2} \delta^3(\boldsymbol{p} - \boldsymbol{p}_1) \int d^3 p_2 \, d^3 p_3 \, d^3 p_4$$

$$\times \, W(\boldsymbol{p}_3, \boldsymbol{p}_4; \boldsymbol{p}, \boldsymbol{p}_2) \, \delta(\varepsilon + \varepsilon_2 - \varepsilon_3 - \varepsilon_4) \, \delta^3(\boldsymbol{p} + \boldsymbol{p}_2 - \boldsymbol{p}_3 - \boldsymbol{p}_4) \, e^{\beta(\varepsilon' - \varepsilon'_2)}. \quad (15.108)$$

Using the Galilean invariance of W and the δ-functions which express energy and momentum conservation we can rewrite Eqs.(15.107) and (15.108) in terms of the $\boldsymbol{p}' = \boldsymbol{p} - m\boldsymbol{u}$ momenta rather than the $\boldsymbol{p}$. The velocity $\boldsymbol{u}$ which occurs in the energies $\varepsilon' = (\boldsymbol{p} - m\boldsymbol{u})^2/2m$ in (15.108) will then occur only implicitly through the $\boldsymbol{p}'$ momenta.

Altogether, if we equate (15.105) with (15.107), the Boltzmann equation to first order of the Chapman-Enskog method gives us

$$\mathcal{K} f_1 \equiv \int d^3 p_1 \, \mathcal{K}(\boldsymbol{p}, \boldsymbol{p}_1) \, f_1(\boldsymbol{p}_1) = h, \quad (15.109)$$

where $h(\boldsymbol{p})$ is defined by

$$h \equiv \left(\varepsilon - \frac{5}{2\beta} \right) (\boldsymbol{v} \cdot \nabla) \beta - \beta \sum_{\gamma\delta} v_\gamma p_\delta \Delta_{\gamma\delta}, \quad (15.110)$$

and where we wrote $\boldsymbol{p} = m\boldsymbol{v}$ and ε instead of $\boldsymbol{p}' = m\boldsymbol{v}'$ and ε'. We shall henceforth use this simplified notation. Neither will it be necessary to write explicitly the variables $\boldsymbol{r}$ and t, which now only appear through the temperature, both explicitly and through $\mathcal{K}$. The quantities ϱ and $\boldsymbol{u}$ have disappeared.

The next stage consists in *solving* the linear integral equation (15.109) for f_1. We shall do this formally in order to point out the properties of the solution f_1. The integral kernel $\mathcal{K}$ has the following properties which one can check from (15.108). It is *real* and *symmetric*; to see this we use the microreversibility $W(\boldsymbol{p}_1, \boldsymbol{p}_4; \boldsymbol{p}, \boldsymbol{p}_2) = W(\boldsymbol{p}, \boldsymbol{p}_2; \boldsymbol{p}_1, \boldsymbol{p}_4)$ and the relation $\varepsilon' - \varepsilon_4' = \varepsilon_1' - \varepsilon_2'$ for the first term in (15.108). It *depends only on the temperature*, and on the interactions between the particles through the cross-section. It is *invariant under rotations* of the two momenta on which it depends. Moreover, the microscopic conservation properties (15.87) of the Boltzmann collision term imply that

$$\mathcal{K} f_0 = 0, \qquad \mathcal{K} f_0 \varepsilon = 0, \qquad \mathcal{K} f_0 \boldsymbol{p} = 0, \tag{15.111}$$

so that the kernel $\mathcal{K}$ has a five-fold degenerate *zero eigenvalue* with f_0, $f_0\varepsilon$, and $f_0\boldsymbol{p}$ as eigenfunctions. This property implies that the inhomogeneous equation (15.109) has a solution only provided h is orthogonal to these eigenfunctions, or, if we define a scalar product by $\{f, g\} \equiv \int d^3\boldsymbol{p}\, fg$, only provided

$$\{f_0, h\} = 0, \qquad \{f_0\varepsilon, h\} = 0, \qquad \{f_0\boldsymbol{p}, h\} = 0. \tag{15.112}$$

From the explicit form (15.110) of h one sees that these conditions are satisfied; this was, in fact, just the reason why we determined the approximate left-hand side $-f_0 h$ of the Boltzmann equation in such a way that $\partial f_0/\partial t$ satisfied the conservation equations. Finally, $\mathcal{K}$ is a *non-negative kernel*

$$\{f, \mathcal{K}f\} \geq 0, \qquad \forall\ f, \tag{15.113}$$

and all its eigenvalues, bar those associated with (15.111) are positive.

To prove that we use the symmetry properties of W and find from (15.108) that

$$\{f, \mathcal{K}f\} = \frac{1}{8} \int d^3\boldsymbol{p}_1\, d^3\boldsymbol{p}_2\, d^3\boldsymbol{p}_3\, d^3\boldsymbol{p}_4\, W(\boldsymbol{p}_3, \boldsymbol{p}_4; \boldsymbol{p}_1, \boldsymbol{p}_2)$$
$$\times\, \delta(\varepsilon_1 + \varepsilon_2 - \varepsilon_3 - \varepsilon_4)\, \delta^3(\boldsymbol{p}_1 + \boldsymbol{p}_2 - \boldsymbol{p}_3 - \boldsymbol{p}_4)\, e^{-\beta(\varepsilon_1 + \varepsilon_2)} .$$
$$\times\, \big[\varphi(\boldsymbol{p}_1) + \varphi(\boldsymbol{p}_2) - \varphi(\boldsymbol{p}_3) - \varphi(\boldsymbol{p}_4)\big]^2, \tag{15.113'}$$

where $\varphi(\boldsymbol{p}) \equiv f(\boldsymbol{p})\, e^{\beta\varepsilon}$. This expression is clearly positive or zero. Provided the scattering is not pathological, the function W is positive for all values compatible with energy and momentum conservation and the vanishing of (15.113') implies that

$$\varphi(\boldsymbol{p}_1) + \varphi(\boldsymbol{p}_2) - \varphi(\boldsymbol{p}_3) - \varphi(\boldsymbol{p}_4) \equiv 0,$$

whatever the values of $\boldsymbol{p}_1, \boldsymbol{p}_2, \boldsymbol{p}_3, \boldsymbol{p}_4$ which satisfy the conditions $\varepsilon_1 + \varepsilon_2 - \varepsilon_3 - \varepsilon_4 = 0$ and $\boldsymbol{p}_1 + \boldsymbol{p}_2 - \boldsymbol{p}_3 - \boldsymbol{p}_4 = 0$; this problem is solved by the Lagrangian multiplier method and leads to

$$\varphi(\boldsymbol{p}) = \xi + \eta\varepsilon + (\boldsymbol{\zeta} \cdot \boldsymbol{p}),$$

that is,

$$f(\boldsymbol{p}) \propto \big(\xi + \eta\varepsilon + (\boldsymbol{\zeta} \cdot \boldsymbol{p})\big) f_0.$$

The only eigenfunctions of $\mathcal{K}$ which are associated with the eigenvalue zero are thus those given by (15.111). All other eigenvalues are positive.

Let Π be the projection operator onto the space associated with the non-vanishing eigenvalues of $\mathcal{K}$. In that subspace the kernel $\mathcal{K}$ *can be inverted* and this defines a kernel $\mathcal{L}$ characterized by

$$\mathcal{L}\mathcal{K} = \mathcal{K}\mathcal{L} = \Pi, \qquad \mathcal{L}\Pi = \Pi\mathcal{L} = \mathcal{L}. \qquad (15.114)$$

One checks easily that $\mathcal{L}$ has the same properties as those we just listed for $\mathcal{K}$. Constructing $\mathcal{L}$ enables us to solve (15.109) and its general solution has the form

$$f_1 = \mathcal{L}h - \left(\frac{\beta}{2\pi m}\right)^{3/2} e^{-\beta\varepsilon}\left(\xi + \eta\varepsilon + (\boldsymbol{\zeta}\cdot\boldsymbol{p})\right). \qquad (15.115)$$

This solution depends on 5 arbitrary parameters, associated with the eigenspace of $\mathcal{K}$ and $\mathcal{L}$ corresponding to the eigenvalue 0. The constraints (15.92), imposed upon f_1, then enable us to determine these constants by writing

$$\{1, f_1\} = 0, \qquad \{\varepsilon, f_1\} = 0, \qquad \{\boldsymbol{p}, f_1\} = 0,$$

which gives

$$\xi = \left\{\frac{5}{2} - \beta\varepsilon, \mathcal{L}h\right\}, \qquad \eta = \left\{\frac{2}{3}\beta^2\varepsilon - \beta, \mathcal{L}h\right\}, \qquad \boldsymbol{\zeta} = \left\{\frac{\beta\boldsymbol{p}}{m}, \mathcal{L}h\right\}. \qquad (15.116)$$

To obtain the responses of the gas to the perturbations $\nabla\beta$ and $\nabla\boldsymbol{u}$ which occur in h, we must still substitute (15.115) for f_1 in expressions (15.98) and (15.99) for the fluxes. The result of this substitution is an integral over the momenta $\boldsymbol{p}$ and $\boldsymbol{p}_1$ of the kernel $\mathcal{L}$. As all expressions are written in the local Galilean frame, the integrand must be even under a sign change in $\boldsymbol{p}$ and $\boldsymbol{p}_1$. The kernel $\mathcal{L}$ itself is even, and the two parts of h defined by (15.110) have opposite parities; these two parts thus become decoupled. Only the second one contributes to the momentum flux and only the first one to the heat flux.

We thus get for the momentum flux (15.99)

$$\mathcal{P}_\beta^\alpha - \mathcal{P}\delta_{\alpha\beta} = -\beta \sum_{\gamma\delta} \left\{v_\alpha p_\beta, \mathcal{L}v_\gamma p_\delta\right\} \Delta_{\gamma\delta}$$

$$+ \left(\frac{\beta}{2\pi m}\right)^{3/2} \left\{v_\alpha p_\beta, e^{-\beta\varepsilon}(\xi + \eta\varepsilon)\right\};$$

if we use the fact that $(\varepsilon, f_1) = 0$, we see that the last term equals

$$\frac{2}{3}\delta_{\alpha\beta}\left(\frac{\beta}{2\pi m}\right)^{3/2}\int d^3\boldsymbol{p}\,\varepsilon\,e^{-\beta\varepsilon}(\xi + \eta\varepsilon) = \frac{2}{3}\delta_{\alpha\beta}\{\varepsilon, \mathcal{L}h\},$$

and combining this with the first term we find

$$\mathcal{P}_\beta^\alpha - \mathcal{P}\delta_{\alpha\beta} = -\frac{\beta}{m^2} \sum_{\gamma\delta} \left\{p_\alpha p_\beta - \frac{1}{3}p^2\delta_{\alpha\beta}, \mathcal{L}p_\gamma p_\delta\right\} \Delta_{\gamma\delta}. \qquad (15.117)$$

Taking into account that $\sum_\gamma \Delta_{\gamma\gamma} = 0$, we can make (15.117) more symmetric, subtracting $\frac{1}{3}p^2\delta_{\gamma\delta}$ from $p_\gamma p_\delta$. The integral kernel $\mathcal{L}(\boldsymbol{p}, \boldsymbol{p}_1)$, which is invariant under rotation, only depends on the lengths of $\boldsymbol{p}$ and $\boldsymbol{p}_1$ and on the angle between them; after integration over $\boldsymbol{p}$ and $\boldsymbol{p}_1$, the coefficient of $\Delta_{\gamma\delta}$ will thus be a tensor with four indices, $\alpha, \beta, \gamma, \delta$, which is invariant under rotation, symmetric under the exchanges $\alpha \leftrightarrow \beta$, $\gamma \leftrightarrow \delta$, $\alpha\beta \leftrightarrow \gamma\delta$, and with zero trace both over $\alpha = \beta$ and over $\gamma = \delta$. The only tensor which has those properties is

$$\tfrac{1}{2} \left(\delta_{\alpha\gamma}\delta_{\beta\delta} + \delta_{\alpha\delta}\delta_{\beta\gamma} \right) - \tfrac{1}{3} \delta_{\alpha\beta}\delta_{\gamma\delta},$$

and the *response relation* (15.117) thus has the form

$$\mathcal{P}_\beta^\alpha - \mathcal{P}\,\delta_{\alpha\beta} \;=\; L\left(-\frac{1}{T}\,\Delta_{\alpha\beta} \right), \tag{15.118}$$

a special case of the general law (14.131) which was expected on the basis of macroscopic arguments. We also have obtained a *microscopic expression for the response coefficient*

$$L \;=\; \frac{2}{km^2}\,\{p_\alpha p_\beta, \mathcal{L}p_\alpha p_\beta\}, \qquad \alpha \neq \beta. \tag{15.119}$$

Similarly we obtain the heat flux (15.98), following from the first term of h. Using (15.116) we find

$$
\begin{aligned}
J_U &= \{\varepsilon v, \mathcal{L}h\} - \left(\frac{\beta}{2\pi m} \right)^{3/2} \{\varepsilon v, e^{-\beta\varepsilon}(\boldsymbol{\zeta}\cdot\boldsymbol{p})\} \\
&= \{\varepsilon v, \mathcal{L}h\} - \frac{5}{2\beta^2}\,\boldsymbol{\zeta} = \left\{ \left(\varepsilon - \frac{5}{2\beta} \right) v, \mathcal{L}h \right\} \\
&= \left\{ \left(\varepsilon - \frac{5}{2\beta} \right) v, \mathcal{L}\left(\varepsilon - \frac{5}{2\beta} \right) v \right\} \cdot \nabla\beta.
\end{aligned}
\tag{15.120}
$$

Here again, the invariance of $\mathcal{L}$ under rotation enables us to justify the *response relation*

$$J_U \;=\; L_U\,\nabla\left(\frac{1}{T} \right), \tag{15.121}$$

in agreement with (14.133), and to express the *response coefficient in terms of microscopic physics* through

$$L_U \;=\; \frac{1}{k}\left\{ \left(\varepsilon - \frac{5}{2}kT \right) v_\alpha, \mathcal{L}\left(\varepsilon - \frac{5}{2}kT \right) v_\alpha \right\}. \tag{15.122}$$

Altogether, we have, starting from the Boltzmann equation and solving it in a regime close to local equilibrium, proved for gases *all general properties* of the responses which were established in § 14.4.6 on the basis of the postulates of macroscopic thermodynamics, such as symmetries, invariances or Onsager relations. In particular, the *Onsager relations*, which imply the vanishing of several coefficients, derive here from microreversibility, which leads to the symmetry of the kernels $\mathcal{K}$ and $\mathcal{L}$ under interchange of the momenta and which itself is the result of the invariance of the microscopic motion under time reversal. The *absence of diffusion* is connected with the vanishing of J_N^1, which follows from (15.92). The fact that the response coefficients are *independent of the velocities* u follows, as on the macroscopic scale, from the Galilean invariance which here led to the disappearance of u in $\mathcal{L}$. Furthermore, the fact that the *kernel $\mathcal{L}$ is positive* implies that the responses (15.119) and (15.122) are *positive* so that the evolution properly gives rise to dissipation.

The explicit expressions (15.119) and (15.122) moreover supply us with a number of *properties specific to low-density gases* to which the Boltzmann equation is applicable. We found in connection with the vanishing (15.100) of the trace of $\mathcal{P}_\beta^\alpha - \mathcal{P}\delta_{\alpha\beta}$, which is automatically true here, whatever the affinities, that the response l associated with a compression or dilatation rate was zero. Hence, the *volume viscosity η_v of a gas vanishes*. This property, specific for gases of weakly interacting particles, does not hold for denser fluids for which the interactions between molecules contribute to the energy and momentum densities and fluxes.

On the other hand, the two other responses, L and L_U, which are given by (15.119) and (15.122) in the form of integrals over the two momenta of the kernel $\mathcal{L}$, depend, like it, *only on the temperature* and the scattering cross-section for a pair of molecules. According to the general macroscopic theory of § 14.4.6, we also expected to find a density dependence. The fact that the *viscosity $\eta = L/2T$ and the heat conductivity $\lambda = L_U/T^2$ do not depend on the density*, which is proved exactly here, had already been indicated by the rough microscopic calculation of § 7.4.6. These results are with good accuracy confirmed experimentally. Because they are not intuitive, their original observation by Maxwell was a major success for statistical mechanics which was then in its infancy.

In order to obtain more precise results and, in particular, to calculate how the two transport coefficients η and λ depend on the temperature, we must work with special models describing the intermolecular collisions. Let us especially consider the model of *classical hard spheres* with a diameter δ (Eq.(15.86)) for which $W = 2\delta^2/m^2$ is independent of the p. From dimensional arguments it follows that the kernel $\mathcal{K}$, defined by (15.108), is for $u = 0$ a distribution of the form

$$\frac{\delta^2}{\sqrt{m\beta}}\,\varphi\left(p\sqrt{\frac{\beta}{m}}, p_1\sqrt{\frac{\beta}{m}}\right),$$

where φ is a dimensionless function. The inverse kernel $\mathcal{L}$ is therefore equally a distribution of the form

$$\left(\frac{\beta}{m}\right)^3 \frac{\sqrt{m\beta}}{\delta^2}\,\psi\left(p\sqrt{\frac{\beta}{m}}, p_1\sqrt{\frac{\beta}{m}}\right);$$

the coefficient $(\beta/m)^3$ comes from the fact that the kernels $\mathcal{K}$ and $\mathcal{L}$ are each other's inverse in p-space and not in $p\sqrt{\beta/m}$-space. Evaluating (15.119) and (15.122) then leads to

$$L = a\,\frac{\sqrt{m\beta}}{k\delta^2\beta^2}, \qquad L_U = b\,\frac{\sqrt{m\beta}}{k\delta^2 m\beta^3},$$

where a and b are numbers which a detailed calculation based on the explicit form of φ gives; one finds that $a = 0.358$ and $b = 0.678$. We thus find for the viscosity and the thermal conductivity

$$\eta = \frac{1}{2}\,a\,\frac{\sqrt{mkT}}{\delta^2}, \qquad \lambda = b\,\frac{k}{m}\,\frac{\sqrt{mkT}}{\delta^2}, \tag{15.123}$$

in accordance with our estimates of § 7.4.6. Both *increase with temperature as* $\sqrt{T}$, and their *ratio*

$$\frac{\eta}{\lambda} = \frac{a}{2b}\,\frac{m}{k}$$

is independent of the molecular radius. In order to compare this number with experiments, note that

$$c_v = \frac{3k}{2m} \tag{15.124}$$

is the specific heat at constant volume, per unit mass, so that *the theory gives the universal relation*

$$\lambda = \frac{4b}{3a}\,\eta\,c_v = 2.53\,\eta\,c_v. \tag{15.125}$$

For the inert gases, to which one may reasonably expect that the model of structureless classical hard spheres, used here, would be applicable, the experimental values lie between 2.45 and 2.58. The agreement is not satisfactory for other gases: 2.02 for H_2 and 1.96 for air. Moreover, as early as between 1861 to 1873 experiments by Oscar Emil Meyer have shown that the viscosity increases faster with temperature than the $\sqrt{T}$ law of (15.123). Such a behaviour can be understood by noting that the potential for the interaction between the molecules is less stiff than that between hard spheres; as a result, the molecules can get closer to one another as their energies increase, so that the average cross-section decreases with temperature. This fact is empirically reflected in a decrease in the denominator δ^2 in (15.123), which explains the increase of $\eta/\sqrt{T}$ with T.

The mathematical complications of solving the Boltzmann equation have led practitioners to replace the collision term by simple approximations, more manageable than the Boltzmann collision integral. For instance, one often uses the *relaxation time approximation* which is based upon the approximate collision term

$$\frac{\partial f}{\partial t} + \left(v \cdot \nabla_r\right)f = \mathcal{I} = -\frac{f - f_0}{\tau}, \tag{15.126}$$

where f_0 is the local equilibrium distribution around which one tries to construct a solution and where τ has the dimension of a time. This expression occurred directly in (15.24) for the Lorentz gas when $W = 1/\tau$ was independent of the angles. The time τ can be interpreted as a *microscopic relaxation time* towards local equilibrium, since in the case when $\nabla f = 0$ the solution of (15.126) tends exponentially to f_0 as $e^{-t/\tau}$. It is of the order of the average time between successive collisions of one of the molecules in the gas. The approximation (15.126) satisfies the condition $\mathcal{I}(f_0) = 0$ for the *particular* local equilibrium distribution considered, but not for any arbitrary f_0, in contrast to the Boltzmann expression. On the other hand, it violates the microscopic conservation equations (15.87) for arbitrary f; however, these conservation equations are identical with the constraints (15.92), if one equates $f - f_0$ with f_1. By imposing the conservation equations as conditions on f_0 and the constraints (15.92) on f_1, we can thus use (15.126) effectively to simplify transport problems. For instance, as an exercise one could calculate *directly* the thermal conductivity and the viscosity, starting from (15.126), in the spirit of Exerc.15b. The main interest of (15.126) lies, however, in studies of ballistic regimes.

Let us illustrate the Chapman-Enskog method by applying it to the collision term (15.126). The kernel $\mathcal{K}$ defined by (15.108) can here be written as

$$\mathcal{K}(\boldsymbol{p}, \boldsymbol{p}_1) = \frac{1}{\tau f_0(\boldsymbol{p})} \delta^3(\boldsymbol{p} - \boldsymbol{p}_1).$$

It has the same general properties as the ones we listed above, except the existence of zero eigenvalues (15.111) associated with the conservation laws, which we now have to impose, as they are not consequences of the simplified Boltzmann equation. The kernel $\mathcal{K}$ depends only on the temperature, as a result of the fact that f_0 and $1/\tau$ are both proportional to the density. The kernel $\mathcal{L}$ is equal to

$$\mathcal{L}(\boldsymbol{p}, \boldsymbol{p}_1) = \tau f_0(\boldsymbol{p}) \delta^3(\boldsymbol{p} - \boldsymbol{p}_1).$$

Using (15.119) we get for the viscosity

$$\eta = \frac{L}{2T} = \frac{\tau}{kTm^2} \int d^3 p \ p_x^2 p_y^2 f_0(\boldsymbol{p}) = \varrho_N \tau kT. \tag{15.127}$$

Similarly, we obtain the thermal conductivity from (15.122):

$$\lambda = \frac{L_U}{T^2} = \frac{2\tau}{3kT^2 m} \int d^3 p \ f_0(\boldsymbol{p}) \varepsilon \left(\varepsilon - \frac{5}{2\beta} \right)^2 = \frac{5}{2} \varrho_N \tau \frac{k^2 T}{m}. \tag{15.128}$$

Since τ is proportional to $1/\varrho_N \sqrt{T}$, we find again the behaviour predicted by (15.123). Using (15.124), we get, instead of (15.125),

$$\lambda = \tfrac{5}{3} \eta c_v = 1.7 \eta c_v. \tag{15.129}$$

The results obtained are qualitatively correct, notwithstanding the gross character of the approximation.

15.3.4 Gas Mixtures

The use of the Boltzmann equation for a simple monatomic gas in a local equilibrium regime has given us important theoretical results, but rather too few numerical consequences to compare with experiments. In such a gas, in fact, only two transport coefficients, η and λ, come into play. The theoretical predictions given earlier are based on a rather badly known number, the molecular diameter δ. If one considers the latter as an adjustable parameter, the associated microscopic phenomenology will be hardly more predictive than the macroscopic phenomenology based upon adjusting two transport coefficients to experimental data. However, the Boltzmann equation becomes an indispensable tool in other circumstances, especially, if one is interested in situations far from equilibrium, involving large gradients or fast variations, but also for less simple gases consisting of molecules which can absorb and emit energy, or angular momentum, or gas mixtures containing several kinds of molecules. In those cases, even close to equilibrium, there are many and varied macroscopic phenomena, and it becomes profitable to interconnect these through an economic microscopic theory.

As an example of such phenomena we mention the *Dufour effect* (1872). Assume that we maintain a concentration gradient in a mixture, for instance, by connecting two reservoirs with each other which contain gas mixtures or liquid solutions with different concentrations. One notices, apart from the diffusion which tends to make the system homogeneous, a heat flow. This effect is a new instance of an indirect transport phenomenon, an energy flux being produced by a chemical potential gradient accompanying the concentration gradient. There exists also an inverse effect, *thermal diffusion*, which in liquids is called the *Soret effect* (1879). In that case, a temperature gradient produces in a mixture a flux of matter which differs for each of the components, hence affecting the local concentration. The two effects are related through an Onsager relation which has been checked experimentally. The Soret effect is interesting as it produces, starting from a homogeneous mixture, concentration variations thanks to a thermal gradient which partially separates the molecules. It is the basis of an isotope separation method of technological and medical interest.

The solution of the Boltzmann equation for gas mixtures goes beyond the framework of the present book. We shall restrict ourselves to an examination of the form of the right-hand side for a few special cases. Consider a binary mixture of two kinds of particles, a and b. We must write down a balance equation for each of the reduced single-particle densities f_a and f_b. For f_a the right-hand side contains, apart from the collision term (15.77) for particles a colliding with one another, a mixed term which we derived in § 15.3.1:

$$\mathcal{I}_{ab} = \int d^3p_2 \, d^3p_3 \, d^3p_4 \, W_{ab}(\boldsymbol{p}_3, \boldsymbol{p}_4; \boldsymbol{p}, \boldsymbol{p}_2) \, \delta(\varepsilon + \varepsilon_2 - \varepsilon_3 - \varepsilon_4)$$

$$\times \, \delta^3(\boldsymbol{p} + \boldsymbol{p}_2 - \boldsymbol{p}_3 - \boldsymbol{p}_4) \left[f_a(\boldsymbol{p}_3) f_b(\boldsymbol{p}_4) - f_a(\boldsymbol{p}) f_b(\boldsymbol{p}_2) \right] . \tag{15.130}$$

The momenta and energies $\boldsymbol{p}, \boldsymbol{p}_3$ and $\varepsilon, \varepsilon_3$ refer to an a particle of mass m_a, whereas $\boldsymbol{p}_2, \boldsymbol{p}_4$ and $\varepsilon_2, \varepsilon_4$ refer to b; we have not indicated explicitly the r and t dependence of the distributions f_a and f_b. The coefficient $W_{ab} = W_{ba}$ is connected with the cross-section for scattering of a particles by b particles, or *vice versa*, through (15.85). The collision terms $\mathcal{I}_{ab}$ and $\mathcal{I}_{ba}$, which couple the two Boltzmann equations for f_a and f_b, are responsible for the effects connected with the relative diffusion of the two gases in the mixture considered here.

In what follows we shall assume that the a gas in which we are interested is much more rarefied than the b gas. For the latter, the collision term $\mathcal{I}_{ba}$ is negligible as compared to $\mathcal{I}_{bb}$, and we assume that the term $\mathcal{I}_{bb}$ has been sufficiently efficient to produce for f_b a thermal equilibrium Maxwell distribution,

$$f_b(\boldsymbol{p}) = \varrho_b \left(\frac{\beta}{2\pi m_b}\right)^{3/2} e^{-\beta p^2/2m_b}, \tag{15.131}$$

where ϱ_b is the number of b particles per unit volume. In contrast, for the a gas we can restrict ourselves to considering only collisions with b particles so that the right-hand side of the Boltzmann equation reduces to (15.130) with f_b known. We want to analyze that equation for the a particles. Its collision term involves only the integral

$$Y(\boldsymbol{p}, \boldsymbol{p}') \equiv \int d^3 p_2 \, d^3 p_4 \, W_{ab}(\boldsymbol{p}', \boldsymbol{p}_4; \boldsymbol{p}, \boldsymbol{p}_2) \, \delta(\varepsilon + \varepsilon_2 - \varepsilon' - \varepsilon_4)$$
$$\times \delta^3(\boldsymbol{p} + \boldsymbol{p}_2 - \boldsymbol{p}' - \boldsymbol{p}_4) \, f_b(\boldsymbol{p}_2). \tag{15.132}$$

Energy conservation results in

$$e^{-\beta\varepsilon} f_b(\boldsymbol{p}_2) = e^{-\beta\varepsilon'} f_b(\boldsymbol{p}_4)$$

for the equilibrium distribution (15.131). From this relation and the microreversibility of W_{ab} it follows that Y satisfies (15.66). The collision term $\mathcal{I}_{ab}$, defined by (15.130), can then be identified with the one we introduced through (15.65) to describe the *thermalization* of a gas of neutrons or electrons. We there assumed that the particles in such a gas were scattered by exterior centres which could take up momentum or energy. The model of § 15.2.4 is recovered here as a consequence of the Boltzmann equation for two gases, of which one, b, which is *much denser* than a, is rapidly thermalized and after that plays for a the rôle of the scattering medium.

The *Lorentz model* also follows from the Boltzmann equation, if one assumes moreover that the b particles have not only a much larger density, but also a *much larger mass* than the a particles. In fact, the presence of the Maxwell factor f_b in (15.132) forces p_2 to be large as $\sqrt{m_b kT}$. As a result

$$\varepsilon_4 - \varepsilon_2 = \frac{1}{2m_b} \left((\boldsymbol{p} - \boldsymbol{p}') \cdot (\boldsymbol{p} - \boldsymbol{p}' + 2\boldsymbol{p}_2)\right)$$

is negligible as compared to $\varepsilon - \varepsilon' = (p^2 - p'^2)/2m_a$ in the integration domain, in the limit where $m_b \gg m_a$. The factor W_{ab} in (15.132), defined by (15.85), only depends on the variables (15.80) and (15.81), which tend to $\varepsilon = \varepsilon'$ and to the angle between $\boldsymbol{p}$ and $\boldsymbol{p}'$, respectively. The integration over the momenta in (15.132) then yields

$$Y(\boldsymbol{p}, \boldsymbol{p}') = \delta(\varepsilon - \varepsilon') \int d^3 p_2 \, f_b(\boldsymbol{p}_2) \frac{1}{m_a^2} \frac{d\sigma(p, \theta)}{d\omega}$$
$$= \delta(\varepsilon - \varepsilon') \, W_\theta(p), \tag{15.133}$$

where we have replaced W_{ab} in (15.132) by its expression (15.85) and used expression (15.8) for W_θ. The collision term (15.130) with (15.132) and (15.133) thus reduces finally to the Lorentz term (15.7) when the b scatterers are heavy.

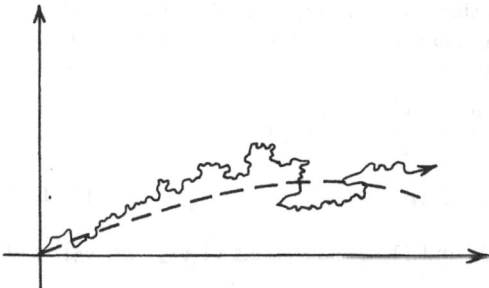

Fig. 15.7. Brownian motion

15.3.5 Brownian Motion

Brownian motion corresponds to the opposite limit $m_b \ll m_a$, where one is interested in *relatively heavy* a particles being scattered by b particles of a gas in thermal equilibrium and at rest. Brownian motion was observed for the first time in 1827 by Robert Brown, a naturalist, who in his microscope noted the erratic motion of grains of pollen, and later of dust particles, in suspension in his preparations (Fig.15.7). This effect was for a long time the subject of controversies, until Einstein in 1905 and Langevin in 1908 gave its theory. The b molecules of the fluid in which the Brownian particles move produce through their collisions the observed motions. Boltzmann's constant is involved in the effect; this enabled Jean Perrin in 1908 to determine directly and accurately the Boltzmann constant and hence the Avogadro number.

When the a particles, the motion of which we are studying, are heavy, each collision which they undergo through interacting with the b particles of the scattering medium has only a small effect so that the magnitude of their momentum is relatively little changed. This is reflected in the fact that in the collision term (15.130) p and p_3 lie close to one another. Below we shall prove that in the limit as $m_b/m_a \to 0$, their difference becomes infinitesimal so that the integration over p_3 disappears and the collision term gets a differential form. Brownian motion is thus governed by the *Fokker-Planck equation* in phase space:

$$\frac{\partial f}{\partial t} + (v \cdot \nabla_r)f + (\varphi \cdot \nabla_p)f = \gamma \operatorname{div}_p v f + \gamma kT \nabla_p^2 f, \qquad (15.134)$$

where we have dropped the subscript of the distribution function f_a of the Brownian particles. On the left-hand side we have included a term reflecting the action of an external force $\varphi = -\nabla V$, which might, for instance, be exerted by gravity, or by an electric field, if the particles considered are charged. Equation (15.134) was actually found by Kramers and by Chandrasekhar; the Fokker-Planck equation proper deals with the momentum distribution only, and it results from (15.134) by integrating over r, which results just in the suppression of the second term.

To derive (15.134) we start by rewriting the collision term (15.130) as an integral over the variables v_{cm}, v_{12}, and v_{34}, defined by (15.79) and (15.82). After we integrate over v_{cm} and use (15.85), we find

$$
\mathcal{I}_{ab} = \int d^3 v_{12}\, d^3 v_{34}\, m_b^3\, \frac{d\sigma(\varepsilon_0, \theta)}{d\omega}\, \delta\left(\frac{1}{2}v_{12}^2 - \frac{1}{2}v_{34}^2\right)
$$

$$
\times \left[f(p+q) f_b(p_2 - q) - f(p) f_b(p_2) \right],
\tag{15.135}
$$

where ε_0 and θ are defined by (15.80) and (15.81) in terms of v_{12} and v_{34}, and where

$$
q \equiv \frac{m_a m_b}{m_a + m_b} \left(v_{34} - v_{12} \right), \qquad p_2 \equiv \frac{m_b}{m_a} p - m_b v_{12}.
$$

As f_b is Maxwellian and $m_b \ll m_a$, the integral (15.135) is over momenta p_2 and q which are small as compared to p, whereas, on the other hand, the relative velocities v_{12} and v_{34} are dominated by the b particles. We expand $f(p+q)$ in the square bracket of (15.135) in powers of q and expand the f_b around the Maxwellian distribution associated with the velocity $v^2 = v_{12}^2 = v_{34}^2$. We need retain only the terms which are even in v_{12} and v_{34}, since the remainder of the integrand is invariant under a change in sign of the velocities. Apart from a factor $m_b^2 f_b(m_b v)$, the square bracket then gives to lowest order

$$
\frac{\beta^2}{2m_a^2} \left[\left(p \cdot v_{34}\right)^2 - \left(p \cdot v_{12}\right)^2 \right] f(p) + \frac{\beta}{m_a} \left[v^2 - \left(v_{12} \cdot v_{34}\right) \right] f(p)
$$

$$
+ \frac{\beta}{m_a} \left(p \cdot v_{34}\right) \left(\left(v_{34} - v_{12}\right) \cdot \nabla \right) f
$$

$$
+ \frac{1}{2} \sum_{\alpha\beta} \left(v_{34} - v_{12}\right)_\alpha \left(v_{34} - v_{12}\right)_\beta \frac{\partial^2 f}{\partial p_\alpha \partial p_\beta}.
$$

We need retain only the part which is symmetric under an exchange of v_{12} and v_{34} as in the remainder of the integrand; this yields

$$
\frac{1}{2} \sum_{\alpha\beta} \left(v_{34} - v_{12}\right)_\alpha \left(v_{34} - v_{12}\right)_\beta \left[\frac{\beta}{m_a} \left(\delta_{\alpha\beta} f + p_\alpha \frac{\partial f}{\partial p_\beta} \right) + \frac{\partial^2 f}{\partial p_\alpha \partial p_\beta} \right].
$$

Taking into account the invariance of the rest of the integral under a simultaneous rotation of v_{12} and v_{34} we see that only the terms $\alpha = \beta$ contribute, and their values are the same. We thus get

$$
\mathcal{I}_{ab} \sim \frac{1}{6} \int d^3 v_{12}\, d^3 v_{34}\, m_b^5\, \frac{d\sigma}{d\omega}\, \delta\left(\frac{1}{2}v_{12}^2 - \frac{1}{2}v_{34}^2\right) f_b(m_a v_{12})
$$

$$
\times \left(v_{34} - v_{12}\right)^2 \left[\beta \operatorname{div} \left(\frac{p}{m_a} f \right) + \nabla^2 f \right],
$$

which can be identified with the right-hand side of (15.134), provided we put

$$\gamma \equiv \frac{\beta}{6m_b^2} \int d^3p\, d^3p'\, \frac{d\sigma}{d\omega}\, \delta(\varepsilon - \varepsilon')\, f_b(p)\, (p - p')^2$$

$$= \frac{\sqrt{\pi}}{3} \left(\frac{2\beta}{m_b} \right)^{5/2} \varrho_b \int p^5\, dp\, d(\cos\theta)\, \frac{d\sigma}{d\omega}\, e^{-\beta\varepsilon}\, \sin^2 \frac{1}{2}\theta. \qquad (15.136)$$

We have thus justified the form (15.134) of the collision term in the limit $m_a \gg m_b$ and obtained an expression for the *coefficient* γ *in terms of the cross-section* $d\sigma/d\omega$ for the scattering of b particles by a particles. In (15.136) the a particle is fixed, p and p' denote the momenta of b before and after scattering, while $d\sigma/d\omega$ depends on the energy $\varepsilon = p^2/2m_b$ and the angle θ between p and p'. In the case where the b particles are very small and where the a particles can be treated as hard spheres of radius δ, the cross-section (15.5) equals $\frac{1}{4}\delta^2$ and the coefficient γ given by (15.136) equals

$$\gamma = \tfrac{8}{3}\, \varrho_b\, \delta^2\, \sqrt{2\pi m_b kT}.$$

The *local equilibrium* regime for the Fokker-Planck equation (15.134) corresponds to a reduced density f which makes $v\, f + kT \nabla_p f$ vanish, that is, to a Maxwellian distribution at each point. The density may vary from point to point, but the *temperature T is determined* by the scattering medium: the latter plays the rôle of a *thermostat* for the heavier particles in suspension. One can study this regime using methods similar to those used for the Boltzmann equation; one also obtains interesting results in situations *far from equilibrium*. Given an arbitrary distribution f, one can, for instance, find from (15.134) the evolution of the expectation values of observables such as

$$\langle r \rangle \left(\equiv \int d^3r\, d^3p\, f\, r \right), \qquad \langle r^2 \rangle, \qquad \langle p \rangle, \qquad \dots\,.$$

In particular, one finds

$$\frac{d\langle r \rangle}{dt} = \frac{\langle p \rangle}{m},$$

$$\frac{d\langle p \rangle}{dt} = \langle \varphi \rangle - \gamma \frac{d\langle r \rangle}{dt}, \qquad (15.137)$$

where only the first part of the collision term contributes to $d\langle p \rangle/dt$. These equations are valid even for a single a particle, in which case f, normalized to 1, describes the probability distribution for the position and the momentum of this particle. They show that the average statistical motion of a particle differs from the free motion through the presence of an *effective friction force* which is proportional to the velocity. The coefficient γ, calculated in (15.136), depends on the properties of the scattering medium, namely, the mass of its particles, their density, and the temperature, and also on the size of the scattered particles. It characterizes this mean friction force; the latter is due to the asymmetry of the effect of the collisions which are more efficient at the leading face of the moving particle than at the back. One can easily measure γ, which is equal to the ratio of the force $\boldsymbol{\varphi}$ to the velocity $d\langle r \rangle/dt$ in a regime where a uniform velocity is acquired as the result of an external force.

The collisions with the medium have also another effect on the trajectory of a particle. They not only decelerate it, but their randomness tends to give it an erratic motion, characterized by the dispersion $\langle r^2 \rangle - \langle r \rangle^2$. Assume, for instance, that a particle is originally at rest at the origin; its probability distribution has the initial value

$$f(\boldsymbol{r}, \boldsymbol{p}, 0) = \delta^3(\boldsymbol{r})\, \delta^3(\boldsymbol{p}).$$

If there were neither external potentials nor collisions, its state would remain unchanged; the collisions spread this distribution out progressively. There is no approach to equilibrium when there is no potential since the particle has a tendency to become uniformly distributed in space. We show in what follows that after a transition period of order m/γ, during which the momentum becomes thermalized, the *statistical fluctuation of r increases as $\sqrt{t}$*:

$$\langle r^2 \rangle \sim \frac{6kT}{\gamma} t. \tag{15.138}$$

The collision term (15.134) thus plays two complementary rôles: its first part, which has a form similar to the contribution from an external force, describes the *deceleration* of the particle; its second part, which depends on the temperature of the medium is, in contrast, associated with a *random force* which tends to *agitate* the particle and which causes its diffusion. The thermalization of the momenta results in a competition between these two effects. The indefinite drift (15.138) of the position is the result of the absence of a restoring force which would counterbalance the tendency for diffusion in r-space.

In order to prove (15.138) we integrate (15.134) after having multiplied it successively with r^2, $(\boldsymbol{p} \cdot \boldsymbol{r})$, and $\boldsymbol{p}^2$; this leads in the case where there is no force $\boldsymbol{\varphi}$ and with f normalized to 1 to

$$\frac{d}{dt} \langle r^2 \rangle = \frac{2}{m} \langle (\boldsymbol{p} \cdot \boldsymbol{r}) \rangle,$$

$$\frac{d}{dt} \langle (\boldsymbol{p} \cdot \boldsymbol{r}) \rangle = \frac{1}{m} \langle \boldsymbol{p}^2 \rangle - \frac{\gamma}{m} \langle (\boldsymbol{p} \cdot \boldsymbol{r}) \rangle, \tag{15.139}$$

$$\frac{d}{dt} \langle \boldsymbol{p}^2 \rangle = -\frac{2\gamma}{m} \langle \boldsymbol{p}^2 \rangle + 6\gamma kT.$$

The solution of these equations for a particle which lies at $r = p = 0$ at $t = 0$ is

$$\langle \boldsymbol{p}^2 \rangle = 3mkT \left(1 - \mathrm{e}^{-2\gamma t/m} \right),$$

$$\langle (\boldsymbol{p} \cdot \boldsymbol{r}) \rangle = \frac{3mkT}{\gamma} \left(1 - \mathrm{e}^{-\gamma t/m} \right)^2, \tag{15.140}$$

$$\langle r^2 \rangle = \frac{6kT}{\gamma} t - \frac{3mkT}{\gamma^2} \left(3 - \mathrm{e}^{-\gamma t/m} \right) \left(1 - \mathrm{e}^{-\gamma t/m} \right).$$

This solution shows a *relaxation time of order* m/γ. The limit of $\langle p^2 \rangle$ is in accordance with the *equipartition* theorem, and the behaviour of $\langle r^2 \rangle$ for $t \gg m/\gamma$ is given by (15.138).

As an exercise one could construct the solution of the Kramers-Chandrasekhar equation in the absence of a field with the initial condition $f(r, p, 0) = \delta^3(r)\delta^3(p - p_0)$. The result is

$$ f = C \exp \left\{ - \frac{[\gamma r - (p + p_0) \tanh \tau]^2}{8mkT(\tau - \tanh \tau)} - \frac{(p - p_0 e^{-2\tau})^2}{2mkT(1 - e^{-4\tau})} \right\}, $$

$$ C^{-1} = (4\pi mkT)^3 \gamma^{-3} (\tau - \tanh \tau)^{3/2} (1 - e^{-4\tau})^{3/2}, \qquad \tau = \frac{\gamma t}{2m}. \quad (15.141) $$

Starting from (15.141) one can easily check the various properties which we described earlier. This expression again shows the time m/γ for the relaxation towards the local equilibrium regime.

15.4 Microscopic Reversibility
vs Macroscopic Irreversibility

15.4.1 The Boltzmann Description

We relied in § 15.1.2 on the ideas of the projection method (§ 14.3.4) to define the Boltzmann description: at all times we retain only the information about the single-particle quantities, summarized by our knowledge of the reduced density $f(r, p)$ at the time considered. In the N-particle phase space we are thus led to associate with the density in phase D a simpler density, its *projection* D_B, in the sense of § 14.3.4. The reduced macro-state D_B is the *mesoscopic* density in phase defined in § 14.3.4, for the contraction of the description to the *single-particle quantities* $f(r, p)$.

To construct it we must look for the *maximum of the statistical entropy*

$$ S(D_B) = -k \sum_N \int d\tau_N \, D_B \ln D_B \tag{15.142} $$

under the constraints that D_B gives us the same values (15.1) for the reduced density,

$$ \sum_N \int d\tau_N \, D_B \sum_j \delta^3(r - r_j) \, \delta^3(p - p_j) = f(r, p), \tag{15.143} $$

as D. As in § 2.3.6, D_B denotes a set of functions of the $6N$ variables r_j, p_j, the number N being arbitrary; $f(r, p)$ is for each value of r, p the expectation value of the observable $\sum_j \delta^3(r - r_j) \delta^3(p - p_j)$. One determines D_B using the general method of Chap.4, introducing a continuum of Lagrangian multipliers for each constraint (15.143), together with one multiplier associated with the normalization of D_B. This gives for the maximum-entropy distribution D_B the usual exponential form, where each observable involved in (15.143) occurs with its appropriate multiplier $\lambda(r, p)$:

$$
D_B = \frac{1}{Z} \exp \int d^3r\, d^3p\, \lambda(r, p) \sum_j \delta^3(r - r_j) \delta^3(p - p_j)
$$

$$
= \frac{1}{Z} \exp \sum_{j=1}^{N} \lambda(r_j, p_j). \tag{15.144}
$$

This expression determines each of the N-components of D_B as a product of factors associated with each particle j. Its form implies that the reduced 2-, 3-, ... particle densities (2.82) can also be *factorized* so that the projection of D onto D_B has the effect of *eliminating the correlations* which might occur in D.

To complete the determination of D_B we must express its normalization in the form

$$
Z = \sum_N \frac{1}{N!} \int \prod_{j=1}^{N} \left[\frac{d^3r_j\, d^3p_j}{h^3}\, e^{\lambda(r_j, p_j)} \right],
$$

which gives

$$
\ln Z = \frac{1}{h^3} \int d^3r\, d^3p\, e^{\lambda(r, p)}, \tag{15.145}
$$

and we must eliminate the Lagrangian multipliers from (15.143) and (15.144); we finally obtain

$$
f(r, p) = \frac{1}{Z} \sum_N \frac{1}{N!} \int \prod_{j=2}^{N} \left[\frac{d^3r_j\, d^3p_j}{h^3}\, e^{\lambda(r_j, p_j)} \right]
$$

$$
\times N \int \frac{d^3r_1\, d^3p_1}{h^3}\, e^{\lambda(r_1, p_1)}\, \delta^3(r - r_1)\, \delta^3(p - p_1),
$$

that is,

$$
f(r, p) = \frac{1}{h^3}\, e^{\lambda(r, p)}. \tag{15.146}
$$

In agreement with the scheme of Fig.15.1, §15.1.2, giving f or giving D_B are equivalent. It follows from (15.145) and (15.146) that $\ln Z$ is equal to the average number of particles.

The above calculations are analogous to those we performed in §7.2.4 for the perfect gas in grand canonical equilibrium. Indeed, f there took the particular form

$$\frac{1}{h^3}\, e^{\alpha - \beta p^2 / 2m};$$

the density in phase (15.144) reduced to the grand canonical density of the perfect gas, and Z to its grand partition function. Expressions (15.144), (15.145), and (15.146) are their generalizations, defining a non-equilibrium macro-state where the particles remain uncorrelated.

The *Boltzmann entropy* S_B, which is the maximum of (15.142), follows directly from (15.144), whence

$$S_B = k \ln Z - k \left\langle \sum_j \lambda(r_j, p_j) \right\rangle.$$

Using (15.145) and (15.146) we get

$$S_B = k \int d^3r\, d^3p\, f(r, p)\, [1 - \ln h^3 f(r, p)], \tag{15.147}$$

where $h^3 f \ll 1$ in the classical limit with which we are here concerned. The Boltzmann entropy is a special case of the *relevant entropies* introduced in Exerc.3c and § 14.3.3; the quanties $A_i(a)$ are here replaced by $f(r, p)$, and the observables $\widehat{A}_i(a)$ by $\sum_j \delta^3(r - r_j)\delta^3(p - p_j)$. This entropy characterizes the uncertainty associated with *knowing only* the expectation values of the *single-particle quantities.*

It is natural to introduce also a *local Boltzmann entropy density* defined by

$$s_B = k \int d^3p\, f\, [1 - \ln(h^3 f)], \tag{15.148}$$

and the associated *flux* defined by

$$J_{SB} = k \int d^3p\, v\, f\, [1 - \ln(h^3 f)]. \tag{15.149}$$

In a local equilibrium regime where f has the form f_0 of (15.91) or (15.94), the entropy density (15.148) reduces automatically to the Sackur-Tetrode value (15.95) for the perfect gas at equilibrium. Taking into account the definitions (15.89) of the conservative fluxes, we find for the flux density (15.149)

$$\begin{aligned}
J_{SB} &= k u \varrho_N - k \alpha J_N + k \beta J_E + k \sum \lambda_\alpha J_{P\alpha} \\
&= \frac{P}{T} u + \sum_i \gamma_i J_i. \tag{15.150}
\end{aligned}$$

It can thus be identified with expression (14.36) for the entropy flux density of macroscopic thermodynamics. It is, though, important to note that *far from equilibrium* the definitions (15.148) and (15.149) which follow from statistics continue to have a meaning whereas neither the macroscopic entropy, nor the local intensive variables, such as the temperature, exist.

Historically, Boltzmann followed exactly the opposite road to the one we travelled. Knowing only the entropy of a perfect gas at equilibrium he associated by inference a number H with an arbitrary distribution f; he defined it by $- \int f \ln f$, that is, apart from multiplying and additive constants, by (15.147). He meant the symbol H to stand for the capital of η. He proved, as we shall do in what follows, that this number increases with time and he recognized that, at equilibrium, it was the same as the thermodynamic entropy (1872). This enabled him to extend the entropy concept to non-equilibrium distributions and after that, in 1877, to associate the equilibrium entropy through $S = k \ln W$ with the microcanonical density in phase. It took more than half a century before the interpretation of entropy as lack of information was first guessed, then put into mathematical terms, along the approach of our Chap.3. Thus, Boltzmann's H-theorem which stated the increase of the variable H has played an important rôle in the history of the entropy (§ 3.4): it helped to bring out the idea that a quantity H which is *statistical* by nature was, in fact, an extension of the macroscopic entropy concept introduced by thermodynamicists solely at equilibrium and close to equilibrium.

15.4.2 The H-Theorem

Let us strike the *local Boltzmann entropy balance* for the evolution of f generated by the Boltzmann equation

$$\frac{\partial f}{\partial t} + (\boldsymbol{v} \cdot \nabla_r)f - (\nabla V \cdot \nabla_p)f = \mathcal{I};$$

we have for greater generality introduced an external potential V. After multiplying by $-k \ln(h^3 f)$, integrating over $\boldsymbol{p}$, using the definition (15.148) of s_B, and noting that for any variation of f

$$\delta\{f[1 - \ln(h^3 f)]\} = -\ln(h^3 f)\,\delta f, \tag{15.151}$$

we obtain

$$\frac{\partial s_B}{\partial t} + k \int d^3 p \, (\boldsymbol{v} \cdot \nabla_r)\left[f(1 - \ln(h^3 f))\right]$$
$$- k\left(\nabla V \cdot \int d^3 p \, \nabla_p[f(1 - \ln(h^3 f))]\right) = -k \int d^3 p \, \mathcal{I} \ln(h^3 f).$$

The left-hand side can be simplified, if we use (15.149) and integrate by parts; moreover, replacing the collision term by its form (15.77) we get

$$\frac{\partial s_B}{\partial t} + \operatorname{div} \boldsymbol{J}_{SB} = -\frac{1}{2}k \int d^3 p \, d^3 p_2 \, d^3 p_3 \, d^3 p_4 \, \ln[h^3 f(\boldsymbol{p})]$$
$$\times W(\boldsymbol{p}_3, \boldsymbol{p}_4; \boldsymbol{p}, \boldsymbol{p}_2)\delta(\varepsilon + \varepsilon_2 - \varepsilon_3 - \varepsilon_4)\delta^3(\boldsymbol{p} + \boldsymbol{p}_2 - \boldsymbol{p}_3 - \boldsymbol{p}_4)$$
$$\times [f(\boldsymbol{p}_3)f(\boldsymbol{p}_4) - f(\boldsymbol{p})f(\boldsymbol{p}_2)],$$

where again we have not written out explicitly the r- and t-dependence of the f's. Using the symmetry of W under the exchanges $p \leftrightarrow p_2$, $(p, p_2) \leftrightarrow (p_3, p_4)$, $p_3 \leftrightarrow p_4$ we can rewrite the right-hand side as an average of four equal contributions in the form

$$
\frac{\partial s_B}{\partial t} + \operatorname{div} J_{SB} = \frac{1}{8} k \int d^3 p_1 \, d^3 p_2 \, d^3 p_3 \, d^3 p_4 \, W(p_3, p_4; p_1, p_2)
$$
$$
\times \delta(\varepsilon_1 + \varepsilon_2 - \varepsilon_3 - \varepsilon_4) \, \delta^3(p_1 + p_2 - p_3 - p_4)
$$
$$
\times \left[f(p_1)f(p_2) - f(p_3)f(p_4) \right] \ln \frac{f(p_1)f(p_2)}{f(p_3)f(p_4)}, \tag{15.152}
$$

Apart from factors which are clearly non-negative, the integrand of (15.152) contains an expression of the form $(x - 1) \ln x$ which is also non-negative for all $x \equiv f(p_1)f(p_2)/f(p_3)f(p_4)$, and hence the Boltzmann entropy satisfies the *growth property*

$$
\frac{\partial s_B}{\partial t} + \operatorname{div} J_{SB} \geq 0. \tag{15.153}
$$

For a gas enclosed in a box described by the potential V, the integral over the whole of space of $\operatorname{div} J_{SB}$ vanishes, and hence

$$
\frac{dS_B}{dt} \geq 0. \tag{15.154}
$$

This inequality is *Boltzmann's H-theorem: the entropy of a gas without correlations, the distribution of which evolves according to the Boltzmann equation, increases with time.*

It is not impossible that $dS_B/dt = 0$. However, for that to happen it is necessary that the integral of (15.152) over r vanishes; this implies, if the cross-section $d\sigma/d\omega$ is everywhere positive, that at each point r we have

$$
\ln \frac{f(p_1)f(p_2)}{f(p_3)f(p_4)} = 0
$$

for all values of p_1, p_2, p_3, p_4 which satisfy the energy and momentum conservation equations. We shall prove in what follows that this property is satisfied only if f at each point has the general Maxwellian form f_0 of (15.91). Hence, the *time-derivative of the Boltzmann entropy vanishes only if f is a local equilibrium distribution* at each point at the time considered. The same proof also implies that, if the *collision term is ineffective* at a given point r for any p, the reduced density f has at that point a *local equilibrium form*. We have already met with this property in § 15.3.2.

The proof is similar to the one of (15.113). Associating, at the point r and the time t, Lagrangian multipliers β, λ with each of the conservation properties, we must write down the condition that

$$\ln \frac{f(\boldsymbol{p}_1)f(\boldsymbol{p}_2)}{f(\boldsymbol{p}_3)f(\boldsymbol{p}_4)} + \beta(\varepsilon_1 + \varepsilon_2 - \varepsilon_3 - \varepsilon_4) - \left(\boldsymbol{\lambda} \cdot [\boldsymbol{p}_1 + \boldsymbol{p}_2 - \boldsymbol{p}_3 - \boldsymbol{p}_4]\right)$$

be stationary for *arbitrary* variations of the $\boldsymbol{p}$. Hence we find for all $\boldsymbol{p}$ that

$$\nabla_{\boldsymbol{p}}\left[\ln f(\boldsymbol{p}) + \beta \frac{p^2}{2m} - (\boldsymbol{\lambda} \cdot \boldsymbol{p})\right] = 0, \tag{15.155}$$

which means that the square bracket is a constant with respect to $\boldsymbol{p}$. Denoting this constant by $\alpha - \ln h^3$, we find that f has the form (15.91).

It is easy to extend the H-theorem to other collision terms. For the Lorentz model (15.7) we find

$$\frac{\partial s_B}{\partial t} + \mathrm{div}\, \boldsymbol{J}_{SB} = \frac{1}{2}k \int d^3p\, d^3p'\, W_\theta(\boldsymbol{p})\, \delta(\varepsilon - \varepsilon')$$

$$\times \left[f(\boldsymbol{p}) - f(\boldsymbol{p}')\right] \ln \frac{f(\boldsymbol{p})}{f(\boldsymbol{p}')}, \tag{15.156}$$

and the right-hand side vanishes only if f is an isotropic function of $\boldsymbol{p}$. A more interesting example is that of a *gas of fermions* with interactions between the particles which are weak and represented by a collision term. This is a rough model for *liquid helium 3* or for *nuclear matter*; for the electron gas in solids the Lorentz model is generally to be preferred (§ 15.2). The *Landau-Uhlenbeck collision term* – also named the Uehling-Uhlenbeck collision term – which describes such situations must, as in § 15.2.3, take the exclusion principle into account. In fact, two particles with momenta $\boldsymbol{p}_1$ and $\boldsymbol{p}_2$ can be scattered into states with momenta $\boldsymbol{p}_3$ and $\boldsymbol{p}_4$ only if beforehand both these states are unoccupied. As in the terms (15.48) and (15.49) which described the scattering of a single fermion, we must weight here the transition probabilities by a factor $[1 - \frac{1}{2}h^3 f]$ for each final state; this factor describes the probability that it is empty. The Landau-Uhlenbeck collision term follows thus from the Boltzmann collision term (15.77) by replacing $f(\boldsymbol{p}_3)f(\boldsymbol{p}_4) - f(\boldsymbol{p})f(\boldsymbol{p}_2)$ by

$$f(\boldsymbol{p}_3)f(\boldsymbol{p}_4)\left[1 - \tfrac{1}{2}h^3 f(\boldsymbol{p})\right]\left[1 - \tfrac{1}{2}h^3 f(\boldsymbol{p}_2)\right]$$

$$- f(\boldsymbol{p})f(\boldsymbol{p}_2)\left[1 - \tfrac{1}{2}h^3 f(\boldsymbol{p}_3)\right]\left[1 - \tfrac{1}{2}h^3 f(\boldsymbol{p}_4)\right]. \tag{15.157}$$

Moreover, the entropy of a gas of fermions, which have no correlations other than those coming from the statistics, can be written as

$$S = -k \int d^3r\, d^3p \left[f \ln\left(\tfrac{1}{2}h^3 f\right) + \left(\tfrac{2}{h^3} - f\right) \ln\left(1 - \tfrac{1}{2}h^3 f\right)\right]. \tag{15.158}$$

This expression follows from (10.37) which we wrote down for the grand canonical equilibrium of non-interacting fermions. In fact, for spatial variations which are sufficiently slow (§ 10.3.4) we can treat d^3r as a large volume and $f(\boldsymbol{r}, \boldsymbol{p})\, d^3r\, d^3p$, the number of particles in that volume with momenta within d^3p, is equal to $2f_p\, d^3r\, d^3p/h^3$, where f_p is the Fermi factor (10.62), and where the factor 2 comes from the spin. Note that expressions (15.157) and (15.158) are valid only in situations where we neglect spin effects; a more accurate theory must replace f_p by a 2×2 density matrix in spin space, and $f(\boldsymbol{r}, \boldsymbol{p})$ would be proportional to its trace. We can check that the Boltzmann entropy (15.147) is the classical limit of (15.158) for $h^3 f \ll 1$ and that we also get the Boltzmann collision term from (15.157) in the same approximation. We define the Landau-Boltzmann entropy density and flux by modifying (15.148) and (15.149), along the same lines as in (15.158); the factor (15.151) coming from the variation of s_B then becomes

$$\delta \left[-f \ln \left(\frac{1}{2} h^3 f \right) + \left(\frac{2}{h^3} - f \right) \ln \left(1 - \frac{1}{2} h^3 f \right) \right] = \ln \frac{f}{2/h^3 - f} \, \delta f. \quad (15.159)$$

The extension of the H-theorem to fermions can now be proved by multiplying the Landau-Uhlenbeck kinetic equation by (15.159) and integrating over p. We find an equation similar to (15.152) where the last two factors are replaced, respectively, by (15.157) and by

$$\ln \frac{f(p_1) f(p_2) \left[1 - \frac{1}{2} h^3 f(p_3) \right] \left[1 - \frac{1}{2} h^3 f(p_4) \right]}{\left[1 - \frac{1}{2} h^3 f(p_1) \right] \left[1 - \frac{1}{2} h^3 f(p_2) \right] f(p_3) f(p_4)}.$$

The fact that $\partial s / \partial t + \operatorname{div} J_S$ is *positive* for the Landau-Boltzmann entropy (15.158) is then clear if we replace everywhere $\ln h^3 f$ by (15.159). The *vanishing of the dissipation* requires the same replacement in (15.155), which becomes

$$\nabla_p \left[\ln \frac{f}{2/h^3 - f} + \beta \frac{p^2}{2m} - (\boldsymbol{\lambda} \cdot \boldsymbol{p}) \right] = 0;$$

hence local equilibrium is characterized for fermions by a distribution of the form

$$f = \frac{2}{h^3} \frac{1}{e^{\beta \varepsilon - \alpha - (\boldsymbol{\lambda} \cdot \boldsymbol{p})} + 1}, \quad (15.160)$$

where β, α, and $\boldsymbol{\lambda}$ are functions of $\boldsymbol{r}$. Just as we found the Maxwell factor by starting from the Boltzmann equation and requiring that the collision term vanishes, the Landau-Uhlenbeck form (15.157) describing collisions between fermions *leads to the Fermi factor*. The latter is thus found as a consequence of the *dynamics* and the prohibition of a fermion to scatter into an already occupied state. We have already encountered this property when studying the thermalization of electrons in a metal, but the temperature in (15.75) was imposed by the scattering medium, mainly phonons in thermal equilibrium.

Another kind of extension of the H-theorem concerns the kinetic equations where the particles interact with external scatterers which play the rôle of a *thermostat* at a temperature T. In this case, it is not the Boltzmann entropy S_B which increases, but, as in similar situations in macroscopic thermodynamics, the *free energy* $F = U - T S_B$, calculated using rules adapted to the Boltzmann description, which decreases. This holds, for instance, when the collision term has the form (15.65), or the form (15.74) for electrons. Verifying this is similar to the calculation leading to (15.152) and for (15.65) it is based upon

$$\frac{\partial \varrho_F}{\partial t} + \operatorname{div} J_F = -\frac{1}{2} kT \int d^3 p \, d^3 p' \, e^{-\beta \varepsilon} \, Y(p, p')$$

$$\times \ln \frac{e^{\beta \varepsilon} f(p)}{e^{\beta \varepsilon'} f(p')} \left[e^{\beta \varepsilon} f(p) - e^{\beta \varepsilon'} f(p') \right] \leq 0, \quad (15.161)$$

where $\varepsilon = p^2/2m + V(r)$, $\varrho_F = \varrho_E - Ts$, and where $T = 1/k\beta$ is the temperature of the scattering medium; (15.161) vanishes only if $f \propto e^{-\beta\varepsilon}$. Similarly, the *Fokker-Planck* equation (15.134) leads to

$$\frac{\partial \varrho_F}{\partial t} + \operatorname{div} \boldsymbol{J}_F = -\gamma \int d^3p \, f \left[kT \, \nabla_p \ln f + \nabla_p \varepsilon\right]^2 \leq 0, \qquad (15.162)$$

and the right-hand side of this equation vanishes again only in the case of a local equilibrium distribution *at the temperature imposed by the bath*.

15.4.3 Irreversibility of the Boltzmann Equation

The historical importance of the H-theorem for the clarification of the entropy concept has been enormous (§ 3.4.2). For the first time a microscopic quantity emerged which increased with time. Nonetheless, we shall see in § 15.4.4 that S_B is the same as the thermodynamic entropy S_{th} only in a local equilibrium regime. In a ballistic regime, we *must* resort to statistical mechanics even for macroscopic processes: the Boltzmann entropy S_B, and not S_{th}, is the useful quantity.

In particular, the H-theorem has important implications for the *asymptotic behaviour* of an *arbitrary* solution f of the Boltzmann equation as $t \to \infty$. Consider a gas sample, confined in space by a potential V, for instance, a box potential. The average number of particles N and the total energy U remain constant with time when f evolves. In this case the total momentum is not a constant of the motion because of the interactions with the walls and this changes the momentum conservation equation into

$$\frac{\partial \varrho_P}{\partial t} + \operatorname{div} \boldsymbol{J}_P = -\varrho_N \nabla V, \qquad (15.163)$$

with a source term which describes the applied force localized at the walls. Let S_{eq} be the *grand canonical equilibrium* entropy, associated with the given values of N and U. By definition, S_{eq} is larger than the entropy of any macrostate with the same energy and the same number of particles. It thus gives an *upper bound* for the Boltzmann entropy (15.147) at all times,

$$S_B(t) < S_{eq}. \qquad (15.164)$$

The function $S_B(t)$ which, according to (15.154), is non-decreasing and bounded thus tends as $t \to \infty$ to a finite value $S(\infty)$ which is smaller than or equal to S_{eq}. However, the vanishing of dS_B/dt implies that the Boltzmann gas is in local equilibrium at all points. In general, if this condition is satisfied at a given moment, it will not remain so eventually, and as a result the entropy will again increase; we refer to the discussion of the self-diffusion effect in § 14.4.6 after Eq.(14.140). The evolution thus goes on until S_B is practically equal to $S(\infty)$ and the asymptotic behaviour of f for large t is thus a local equilibrium solution of the Boltzmann equation, without its right-hand side. We construct in what follows all solutions such that f retains a form f_0 with time. It will follow that, if the gas is confined to a volume Ω, the only possible solutions are the *global* equilibrium distributions. This finally proves that $S(\infty) = S_{\mathrm{eq}}$ and hence that *any solution of the Boltzmann equation tends to the global equilibrium distribution* associated with the constants of motion, the energy and the number of particles, which are determined by the initial conditions.

If the gas is not confined, its entropy may increase without bounds at the same time as its density decreases, and we never reach equilibrium; besides, the equilibrium entropy itself increases without bounds as $\Omega \to \infty$.

The distributions of the form f_0 which satisfy the Boltzmann equation are characterized by 5 functions, α', β, λ, of r and t which are solutions of

$$\left(\frac{\partial}{\partial t} + \left(\frac{p}{m} \cdot \nabla_r\right) + \left(\varphi \cdot \nabla_p\right)\right)\left(\alpha' - \beta\,\frac{p^2}{2m} + (\lambda \cdot p)\right) = 0. \tag{15.165}$$

Arranging this in powers of p we get a set of partial differential equations for α', β, λ, φ, the general solution of which in the case when there is no external force φ is

$$\left.\begin{aligned}
\alpha' &= \alpha_0 - m\left(\lambda_1 \cdot r\right) - \tfrac{1}{2}\beta_2 r^2, \\
\beta &= \beta_0 + 2\beta_1 t + \beta_2 t^2, \\
\lambda &= \lambda_0 + \lambda_1 t + \beta_1 r + \beta_2 t r + [\omega \times r].
\end{aligned}\right\} \tag{15.166}$$

The various parameters are independent of r and t; λ_0, β_1, and ω describe, respectively, a *uniform translation*, a *uniform dilatation* with cooling, and a *uniform rotation*, while λ_1 and β_2 describe a *uniformly accelerated translation* and a *uniformly accelerated dilatation*; β_0 and α_0 are associated with the initial temperature and the initial density at the origin. The evolution is reversible, without dissipation. For an arbitrary time-independent potential, we always have global equilibrium solutions where $\alpha = \alpha'(r) + \beta V(r)$, β, and λ are constants. However, in the case of some special forces we also find solutions f_0 differing from global equilibrium and depending on the time, for instance, for a harmonic central force, $\varphi = -\kappa r$, where f_0 can represent a drop of gas which periodically extends and contracts. However, if the gas is enclosed *in a box* the solution should be given by (15.166) with as boundary conditions the vanishing of the normal component of the velocity $u = \lambda/\beta$ at the wall. This condition implies that $\lambda_0 = \lambda_1 = 0$, $\beta_1 = \beta_2 = 0$. For a vessel of arbitrary shape it also implies that $\omega = 0$ so that the only possible solutions of the Boltzmann equation with the form f_0 are the global equilibrium distributions $\alpha' = \alpha = \alpha_0$, $\beta = \beta_0$, $\lambda = 0$ with a uniform temperature and a uniform density, the gas being at rest.

A notable exception concerns the case where the gas is enclosed in a vessel which has *axial symmetry*, such as the cylinder of Exerc.7b, or a spherical shape. The solution (15.166) with $\omega \neq 0$ then describes a gas in uniform rotation with a density which increases with the distance from the axis of rotation (Exerc.7b). In fact, the *angular momentum* around the axis of symmetry is in this case a constant of the motion together with N and U. If its initial value is non-vanishing, the gas tends, as $t \to \infty$, to a *global rotating equilibrium* state.

The preceding analysis can also be applied to a *Landau-Boltzmann* gas. Here the entropy (15.158) tends to a constant value and f to a Fermi factor f_0. Notwithstanding this change, Eq.(15.165) remains valid as well as the conclusions we have drawn from it.

The Boltzmann equation thus gives a new justification, based upon *dynamics*, for the Boltzmann-Gibbs equilibrium distribution in the case of a gas, whereas the justification given in Chap.4 was mainly based upon *statistical* considerations. The global equilibrium macro-state appears not only as the *most disordered and thus the most likely macro-state* but also as the *final state towards which any evolution will lead*, starting from an arbitrary initial state. There is thus a *convergence of the two approaches*, the purely statistical and the dynamical approaches, to the theory of thermal equilibrium.

The H-theorem clearly shows the *irreversibility* of the dynamics described by the Boltzmann equation: if one there changes t to $-t$ and p to $-p$, the left-hand side of that equation changes sign, whereas the collision term $\mathcal{I}(f)$ remains unchanged. However, the Liouville equation which governs the *exact* evolution of the gas is *invariant* under time reversal; the statistical entropy $S(D)$ remains *constant* in time, according to (3.29). In more detail, the reduced single-particle density f, two-particle density f_2, ..., evolve according to a hierarchy of exact equations (§ 2.3.5). The first equation

$$\frac{\partial f}{\partial t} + (\boldsymbol{v} \cdot \nabla)f = \int d^3 r' \, d^3 p' \, \left(\nabla_r V(|\boldsymbol{r} - \boldsymbol{r}'|) \cdot \nabla_p f_2(\boldsymbol{r}, \boldsymbol{p}, \boldsymbol{r}', \boldsymbol{p}') \right),$$

(15.167)

has the same left-hand side as the Boltzmann equation, but its right-hand side changes sign with t and thus cannot be equal to the collision term!

The *irreversibility paradox* (§§ 3.4.3 and 4.1.5) is based upon the apparent contradiction between these two kinds of behaviour. Admittedly the increase in the entropy which is reflected in the H-theorem looks like being in agreement with macroscopic experience. However, for an isolated gas, governed by a well defined Hamiltonian, the Liouville equation shows us that if $D(t)$ is a solution, so is $D^{\mathrm{T}}(-t)$, which is obtained by changing the signs of all momenta and the direction of the time. If $D(t)$ gives us a reduced density $f(\boldsymbol{r}, \boldsymbol{p}, t)$ with which we associate an increasing entropy S_{B}, the solution $D^{\mathrm{T}}(-t)$ will, symmetrically, lead to a decreasing entropy S_{B} so that $f(\boldsymbol{r}, -\boldsymbol{p}, -t)$ cannot be a solution of the Boltzmann equation, although it satisfies the exact equation (15.167). Notwithstanding the existence of solutions $D^{\mathrm{T}}(-t)$ of the exact Liouville equation, one never observes such a behaviour – bar exceptionally (§ 15.4.5); why is this so? The paradox is even more striking if one notes that the derivation of the Boltzmann equation in §§ 15.1.3 and 15.3.1 does not seem to be based upon any approximation, apart from the fact that we are dealing with a rarefied gas, so that its consequences should not contradict those of the exact Liouville evolution equation.

In fact, there was an approximation, albeit a fairly well hidden one. Recall the reasoning of §§ 15.1.3 and 15.3.1. The description of a *single collision* is rigorous. It involves a transition probability $W_\theta(p)$ or $W(p_1, p_2; p_3, p_4)$ which is invariant under time reversal, that is, under substitution of $-p'$ for p or $-p_3, -p_4$ for p_1, p_2. Moreover, we noted that this invariance, which constitutes the principle of detailed balancing or the "*microreversibility*" of the Boltzmann equation, was essential in order that the transport coefficients derived from this equation satify the Onsager relations. The approximation which gives rise to the irreversibility was made in the next stage where we constructed the collision term $\mathcal{I}(f)$ using W. In fact, we implicitly assumed that *before the collision* the particles *are uncorrelated*. Classically, this amounts to saying that all impact parameters have the same probability; quantum mechanically, it means that the incident flux is uniform. In the Boltzmann equation (15.77) the factor $f(r, p_3)f(r, p_4)$, for instance, represents the probability density that the two incident particles have momenta p_3 and p_4. Its factorization reflects the absence of correlations; moreover, this form assumes implicitly that f varies little over distances of the order of the range δ of the forces. This *random collision hypothesis* is traditionally called by its German name "*Stosszahlansatz*". Nevertheless, even if we assume that before the collision the two particles were actually statistically independent, their distribution *during* and *after* the collision *contains correlations* which are just connected with the kinetics of the collision itself. In the Lorentz model we are dealing with correlations between the position R of the centre of the scatterer and the coordinates r', p' of the emerging particle in phase space: the vector p' must be along $r' - R$ for $|r' - R| \gg \delta$. In the Boltzmann equation, the positions and momenta of the emerging particles are also correlated, as by reversing their momenta they must collide. The Stosszahlansatz has thus introduced an asymmetry in time: particles which were uncorrelated before a collision become correlated afterwards.

However, the reasoning which leads to the Boltzmann equation is *valid notwithstanding this creation of correlations*. Two particles which have already interacted with each other have practically no chance of colliding again; they will collide with *other* particles so that the correlations which have been created are innocuous and the random collision hypothesis will remain justified for each of the subsequent collisions. Little by little there is thus produced a set of complex correlations occurring in the two-, three-, ..., N-, ... particle reduced densities, but the Boltzmann equation is not interested in them. These correlations involve at the same time the momenta and positions of a large number of particles and this in an extremely subtle manner, as they retain a memory of all the collisions which the particles have undergone earlier. They involve very fine details, since a simple change, of the order of the range δ of the forces, in the position of one particle would radically change them, for instance, by preventing a collision to occur. Nevertheless, they have no effect on the kinetic equations. As regards the Lorentz model, it derives from the hypothesis we made about a *random distribution of the scattering centres*; taking an average over their positions amounts to taking an average over a uniform incident particle flux. As to the Boltzmann equation, one can show that it becomes *exact for finite times* and finite geometries in the limit as $\varrho_N \to \infty$, provided that the particle mass tends to zero with a fixed mass density, and that the cross-section σ tends to zero in such a way that $\varrho_N \sigma$ remains finite. This condition implies that

$$l \gg \varrho_N^{-1/3} \gg \delta, \tag{15.168}$$

which expresses the fact that the relative volume $\varrho_N \delta^3$ occupied by the molecules is negligible, or equivalently, that the mean free path l is large as compared to the intermolecular distances. For the values realized in Nature, we should expect times long as compared to the age of the Universe before we see the appearance of deviations from the Boltzmann equation.

Let us make a few extra remarks about the validity of the latter. It was derived on time scales long as compared to the duration η of a single collision so that $\mathcal{I}(f)$ must be identified with the *average* of the right-hand side of (15.167) over times long as compared to η, under initial conditions without correlations, rather than with that right-hand side itself. Whereas (15.167) involves a *true* time-derivative, $\partial f / \partial t$ in the Boltzmann equation stands for a change in f over a time ∂t much longer than η. On the other hand, $\mathcal{I}(f)$ is the result of the elimination of the two-particle and higher reduced densities. The situation is the same as in § 14.3.5: this elimination is here realized by projecting the density in phase D onto the Boltzmann density D_B, and the memory time is of the order of η, the duration of a single collision. There was therefore no reason for being surprised that over times long as compared to η we found an equation of motion for f which is qualitatively different from the exact equation, to wit, *of first order in t, non-linear*, and *irreversible* – since all those characteristics are present in the general expressions (14.78) and (14.81).

The difference in behaviour of the two entropies $S(D)$ and S_B simply reflects the fact that D_B differs from D only through the suppression of the correlations between molecules. During the exact Hamiltonian evolution which is described by D, the total missing information $S(D)$ remains constant. The quantity $S_B - S(D)$, which represents the information about the correlations contained in D, increases; this means that an ever increasing part of the total information is *transferred to the correlations* which are continuously created in D. In the Boltzmann description D_B which is incomplete, this information is *lost*; the increase in S_B with time reflects the fact that the *reduced density f becomes more and more disordered* as the order escapes towards the correlation degrees of freedom, in such a way that the total disorder $S(D)$ does not increase.

We still must understand *why one never observes the inverse processes*, even though they are allowed by the Liouville equation. Let $C(\{\boldsymbol{r}, \boldsymbol{p}\}, t)$ denote the set of correlations between all the particles; the simplest of them can be described in terms of the one- and two-particle reduced densities by $f_2(\boldsymbol{r}, \boldsymbol{p}; \boldsymbol{r}', \boldsymbol{p}') - f(\boldsymbol{r}, \boldsymbol{p}) f(\boldsymbol{r}', \boldsymbol{p}')$. Giving f and the C is equivalent to giving the density in phase D, whereas D_B corresponds to f and $C = 0$. Let us assume that initially at $t = -t_0$ we have $C(\{\boldsymbol{r}, \boldsymbol{p}\}, -t_0) \simeq 0$, and $f(\boldsymbol{r}, \boldsymbol{p}, -t_0)$ is arbitrary; we can associate with the corresponding solution $D(t)$ of the Liouville equation a solution $f(\boldsymbol{r}, \boldsymbol{p}, t)$ of the Boltzmann equation and a set of correlations $C(\{\boldsymbol{r}, \boldsymbol{p}\}, t)$ which become more and more important. The entropy $S(D)$ remains constant and close to the entropy $S_\mathrm{B}(-t_0)$ of the distribution $f(\boldsymbol{r}, \boldsymbol{p}, -t_0)$ whereas the Boltzmann entropy $S_\mathrm{B}(t)$ resulting from $f(\boldsymbol{r}, \boldsymbol{p}, t)$ is larger than either for $t > -t_0$. Let us now consider the solution $D'(t) = D^\mathrm{T}(-t)$ obtained by time reversal. Its initial conditions are $f'(\boldsymbol{r}, \boldsymbol{p}, -t_0) = f(\boldsymbol{r}, -\boldsymbol{p}, t_0)$ and $C'(\{\boldsymbol{r}, \boldsymbol{p}\}, -t_0) = C(\{\boldsymbol{r}, -\boldsymbol{p}\}, t_0) \neq 0$; its reduced density $f'(t)$ *does not obey the Boltzmann equation* but an equation, obtained by changing the sign of the collision term. The entropy $S(D')$ is at all times equal to $S(D)$, but the associated Boltzmann entropy $S_\mathrm{B}(f')$ shows a pathological behaviour, since it *decreases*. To observe such a solution $f'(t)$ we should *realize the initial conditions $f'(-t_0)$ and $C'(-t_0)$*. However, the correlations $C'(-t_0)$ are extremely complicated and subtle; they are very sensitive to the small initial correlations $C(-t_0)$ which do not affect $f(t)$; besides, their average over distances, large as compared to δ, is almost zero. Producing a gas showing these correlations is out of the question. Only a thought experiment, unrealizable in practice, which would consist in reversing suddenly all momenta would allow us to do this (see § 15.4.5). This *practical impossibility*, but not the properties of the system itself, prohibits the Boltzmann entropy from decreasing.

Statistical considerations enable us better to estimate the extent of this impossibility. Recall that the quantity $\exp[S(D)/k]$ gives the order of magnitude of the *number of micro-states* involved in the macro-state D, in the classical limit where the volume $d\tau_N$ of phase space measures a number of quantum micro-states. If we want to see the Boltzmann entropy decrease, we should prepare a macro-state $D'(-t_0)$ containing the correlations $C'(-t_0)$; the entropy $S(D')$ of this macro-state is *smaller* than the Boltzmann entropy S_B associated with the reduced density $f'(-t_0)$. The fact that this difference in entropy is *macroscopic* implies that the number of micro-states described by $D'(-t_0)$ is considerably smaller than the total number of micro-states characterized by giving only $f'(-t_0)$, by a factor estimated to be

$$\exp\big\{ [S_\mathrm{B} - S(D')]/k \big\}. \tag{15.169}$$

The number (15.169) is huge since macroscopic entropies are of the order of J K^{-1} while $k = 1.38 \times 10^{-23}$ J K^{-1}. Moreover, since S_B is the maximum of the entropy when $f'(-t_0)$ is given and since it is reached when $C'(-t_0) = 0$, the very large value of (15.169) shows that *practically all micro-states associated with $f'(-t_0)$ do not show correlations*. The preparation of an initial macro-state has thus every chance of *leading to an uncorrelated or little correlated state*, which then will evolve with an increasing Boltzmann entropy;

the very large relative rarity (15.169) of micro-states contributing to the correlated macro-state $D'(-t_0)$ implies that the latter cannot be constructed in practice. A violation of the irreversibility is thus not forbidden as a matter of principle, but because it is *highly improbable*.

15.4.4 Boltzmann and Thermodynamic Entropies

In § 15.1.2 we have shown that the macroscopic, thermodynamic or hydrodynamic, description of a gas could be derived from the Boltzmann description through the projection method (Fig.15.1). We thus associate with any density in phase D or D_B at all times a thermodynamic density in phase D_0 which has the following properties: (i) all three give the same values for the local densities ϱ_N, ϱ_E, and ϱ_P; (ii) the entropy $S(D_0)$ is the maximum one. The second condition implies that D_0 has the factorized form (15.144), (15.146) where, moreover, f is a particular local equilibrium distribution f_0, given by (15.91) and quadratic in p. The first condition can be expressed by the constraints (15.92) which must be satisfied by $f_1 \equiv f - f_0$. The thermodynamic entropy $S_{\text{th}} = S(D_0)$ associated with f_0 is, according to (15.95), equal to

$$
S_{\text{th}} = k \int d^3r\, d^3p\, f_0(r,p)\left[1 - \ln\left(h^3 f_0(r,p)\right)\right]
$$

$$
= k \int d^3r\, \varrho_N(r)\left\{\tfrac{5}{2} - \ln\left[\varrho_N(r)\lambda_T(r)^3\right]\right\}, \tag{15.170}
$$

where $\lambda_T^2 \equiv 2\pi\hbar^2 \beta(r)/mk$. As expected, it is just the sum of the thermostatic entropies of each volume element. Several entropies have thus been introduced for a given state of the gas, and we shall now compare them.

In the *local equilibrium* regime where f_1 is small as compared to f_0, the difference between S_B and S_{th} is found by expanding S_B in powers of f_1, which gives

$$
S_B \simeq S_{\text{th}} - \int d^3r\, d^3p\, f_1(r,p)\, \ln\left[h^3 f_0(r,p)\right]
$$

$$
- \frac{1}{2} \int d^3r\, d^3p\, \frac{[f_1(r,p)]^2}{f_0(r,p)}.
$$

The first-order term in f_1 vanishes by virtue of (15.92), and this allows us to check that

$$
S_{\text{th}} - S_B \sim \frac{1}{2} \int d^3r\, d^3p\, \frac{f_1^2}{f_0} \tag{15.171}
$$

is positive as it should be. Its value would follow from (15.115), but it is negligible to first order in f_1, while to this order the change with time of S_B or S_{th} is given by (14.144). We have thus

$$\frac{dS_{\rm B}}{dt} \sim \frac{dS_{\rm th}}{dt} \sim \int d^3 r \left[\frac{\lambda}{T^2} (\nabla T)^2 + \frac{2\eta}{T} \sum_{\alpha\beta} \Delta_{\alpha\beta}^2 \right]. \qquad (15.172)$$

As a result, in this regime and to this order, the *H-theorem is the same as the increase in the thermodynamic entropy* due to the irreversible thermal fluxes and viscous flows. We have thereby found a microscopic justification for the Clausius-Duhem inequality for gases.

However, outside the local equilibrium regime the difference $S_{\rm th} - S_{\rm B}$ can be macroscopic. Nothing even guarantees that $S_{\rm th}$ grows, whereas the *H*-theorem ensures that $S_{\rm B}$ increases. Let us, for instance, imagine two drops of gas, each initially in local equilibrium, which are projected towards one another with large average velocities $\pm u$, much larger than the thermal velocities; let us assume that their densities are sufficiently low so that intermolecular collisions are rare. This ensures that $S_{\rm B}$ grows only slowly during the whole process, including when the two gas masses pass through one another. Before the encounter, $S_{\rm th} \simeq S_{\rm B}$ shows the same behaviour. However, during the overlap the local momentum distribution becomes a superposition of two Maxwellians centred on mu and $-mu$; it has thus rapidly moved far from a local equilibrium distribution. As a result, $S_{\rm th}$ has suddenly increased, but after the encounter each drop comes back practically to its earlier state so that $S_{\rm th}$ *falls back* to practically its initial value. The fact that we retained in the description only the variables ϱ_N, ϱ_E, and ϱ_P has thus given rise to a *memory* effect: the value of these quantities at time $t + dt$ depends on their past history before the encounter, and not only on their values at time t. The two drops did not have enough time during their overlap to thermalize each other. They separate while *creating order in the thermodynamic variables* at the expense of information hidden in the non-Maxwellian $f(r, p)$. Of course, if the system is enclosed in a vessel, it will end up reaching first a local and then a global equilibrium regime (Fig.15.8). However, the corresponding characteristic time is the longer, the lower the density.

The hierarchy of entropies $S(D)$, $S_{\rm B}$, $S_{\rm th}$, $S_{\rm eq}$ associated with descriptions becoming less and less detailed is shown in Fig.15.8. They correspond, respectively, to D, to $f(r, p)$, to the local densities ϱ_i, and to the global conservative variables. As long as the gas is away from the local equilibrium regime, the usual macroscopic entropy $S_{\rm th}$ is of little interest because of memory effects which persist if we follow the evolution of the ϱ_i densities only; the Boltzmann equation is best suited to describe the evolution. On the contrary, later on, the thermodynamic description becomes adequate while the Boltzmann description is uselessly complicated.

This discussion shows up a general conceptual problem: there does not exist just *one* entropy, but *several* ones, *depending on the choice of relevant variables*, the dynamics of which one follows with time. For a gas, depending on the regime and the time scale, we have thus been forced to consider two levels of description, the *thermodynamic* level in terms of the densities ϱ_N,

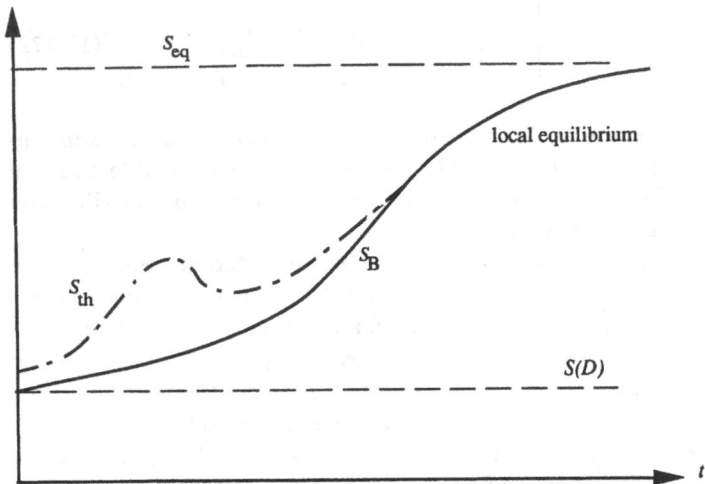

Fig. 15.8. Evolution of the various entropies in a gas. The most detailed entropy $S(D)$ remains constant. If the molecules are uncorrelated at the start, the Boltzmann entropy, S_B, parts company from $S(D)$ and increases by virtue of the H-theorem. The thermodynamic entropy S_{th}, which is larger than S_B, rejoins S_B in the local equilibrium regime. Both tend to the global equilibrium entropy S_{eq} if the gas is enclosed in a vessel

ϱ_E, and ϱ_P, and the more detailed *Boltzmann* description in terms of $f(r, p)$, and we associated with each of them an entropy. We have seen also that the question of irreversibility should be put in different terms in the two cases.

The *choice of relevant variables* can pose delicate problems. If we had tried to use the normal thermodynamic description in the above example, the impossibility to make reasonable predictions in agreement with experiments would have led us to introduce *new variables* which were *hidden* in the – too macroscopic – thermodynamic description. The relevant dynamic variables, for instance $f(r, p)$ for a gas in a ballistic regime, can thus involve quantities which are not accessible to macroscopic experiments; they constitute in general the smallest set of variables necessary to represent the dynamics by equations *without memory* (§ 14.3.5). The same kind of question arose already in thermostatics as in the example of the specific heats of hydrogen (§ 8.4.5). In fact, the appearance of several specific heats could not be explained except by enlarging the set of variables needed to describe the – metastable – equilibrium macro-state; these variables must include not only the number of molecules, the energy, and the volume, but also the proportion of para- and ortho-hydrogen, which is a hidden variable. The discussion of the Gibbs paradox (§ 8.2.1) also appeals to several entropies, which depend on the coarseness of the description and which may thus present a *relative* and anthropocentric character. When there is some natural choice of thermodynamic variables, for instance, the local conservative variables in slow dynamics, the entropy associated with these variables will play a privileged rôle. However, it is important to know how to extend the description by including new variables and introducing a different entropy when this is necessary; examples include gases in a ballistic regime, systems in which an order parameter appears, plastic substances, shape memory materials, or amorphous substances.

As an example, slightly a caricature, but instructive, of a system where hidden variables play an important rôle, let us consider a perfectly elastic spring with a very great length L. Let $\xi(x,t)$ be the elongation which is a function of the abscissa, $0 < x < L$, and the time. We assume the spring to be enclosed in a "black box" with only its $x = 0$ end being accessible. Let us try to describe the macroscopic state by a single variable, the displacement $\xi(0,t)$ of the free end, on which we can exert a force $F(t)$. Let k denote the elastic constant and u the speed of the propagation of perturbations along the spring; we then have

$$F(t) = -k \frac{\partial \xi(0,t)}{\partial x}. \qquad (15.173)$$

Starting at $t = 0$ from a situation where the spring is at rest, $\xi(x,t)$ is a function solely of $x - ut$ on a time scale $t < 2L/u$. As a result, we find from (15.173) that

$$F(t) = \frac{k}{u} \frac{\partial \xi(0,t)}{\partial t}. \qquad (15.174)$$

The force exerted is proportional to the speed, so that the system behaves as a perfect *damper*! Over short times. nothing thus distinguishes the end of the spring from the rod of a piston acting on a viscous medium. The apparent dissipation is due to the elimination of the degrees of freedom $\xi(x,t)$ for $x > 0$ which play here the rôle of microscopic variables. However, we must, of course, reintroduce these hidden variables $\xi(x,t)$ to describe the dynamics of $\xi(0,t)$ when $t > 2L/u$; after that time the initial perturbation returns to the origin after having propagated and having been reflected at the $x = L$ end.

15.4.5 Spin Echoes

Our discussion of the significance of irreversibility for a macroscopic evolution applies to the relaxation towards equilibrium of most physical systems: usually the *flight of order towards inaccessible degrees of freedom is final*; this order which hides in the complex degrees of freedom does not give rise to any detectable physical effect and everything happens, exactly as if the disorder increases. There are, however, experiments, for instance, on spin systems, the so-called "spin echo" experiments, where order which is initially obvious is transferred to unobservable degrees of freedom; this order which seems to be lost can, however, reappear later on in easily accessible degrees of freedom, thanks to a manipulation which is equivalent to reversing the time in the equations of motion. We shall in what follows describe the simplest of these experiments. Their existence is conceptually important: it shows clearly that *irreversibility is not an absolute concept*, but is partly subjective by nature, depending on the complexity of the systems and on the details and ingenuity of our observations.

Let us consider a paramagnetic salt of the kind studied in Chap.1, in which the N elementary magnetic moments interact very little with one another or with the other degrees of freedom of the crystal. The microscopic dynamics of the non-interacting spins in a magnetic field B is governed by Eq.(1.44) which can be written as

$$\frac{d\boldsymbol{\mu}_i}{dt} = \frac{2\mu_{\mathrm{B}}}{\hbar} \left[B \times \boldsymbol{\mu}_i \right], \tag{15.175}$$

where $\mu_{\mathrm{B}} = e\hbar/2m$ is the Bohr magneton and where $\boldsymbol{\mu}_i$ denotes the expectation value of the magnetic moment $-\mu_{\mathrm{B}}\widehat{\boldsymbol{\sigma}}_i$ of the spin $\widehat{\boldsymbol{\sigma}}_i$; the factor 2 in (15.175) is associated with the magnetic moment of the spin of an electron and should possibly be changed, if the magnetic moment $\boldsymbol{\mu}_i$ has an orbital origin (§ 1.4.6). Equation (15.175) describes the *Larmor precession* motion: each vector $\boldsymbol{\mu}_i$ rotates around B with an angular velocity $\omega = 2\mu_{\mathrm{B}}B/\hbar$, proportional to the magnitude B of the field; its length remains fixed and its end describes a circle in a plane at right angles to B. Moreover, the interactions of the spins with one another and with the lattice are responsible for their relaxation (§ 14.1.2). They are sufficiently weak for the relaxation time, say of 0.1 s, to be very long as compared to the duration of the experiment; during the latter we can neglect these interactions as compared to the dynamics described by (15.175). They are, nonetheless, involved in the preparation of the initial state, which is an equilibrium state in a magnetic field B_0 along the x-axis; this field forces the $\boldsymbol{\mu}_i$ to take at $t = 0$ their equilibrium average value (1.37), that is,

$$\mu_{\mathrm{B}} \tanh \frac{\mu_{\mathrm{B}}B_0}{kT}, \tag{15.176}$$

and to be directed along the x-axis (Fig.15.9a). We assumed in the figure that B_0 was sufficiently large that the spins were nearly perfectly ordered.

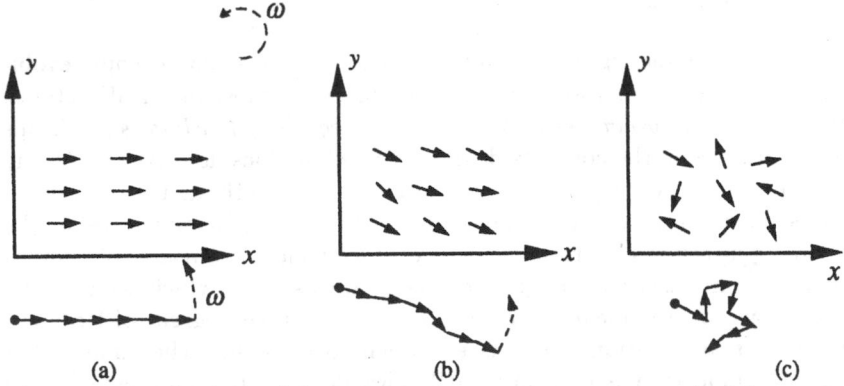

(a)   (b)   (c)

Fig. 15.9a–c. Behaviour of the individual magnetic moments and of the total magnetic moment in a slightly inhomogeneous field B along the z-axis

At time $t = 0$ we replace the field $\boldsymbol{B}_0$ by a permanent field $\boldsymbol{B}$ *in the z-direction*. The vectors $\boldsymbol{\mu}_i$ start to rotate parallel to one another in the xy-plane with the Larmor angular velocity ω. We choose the field B sufficiently strong so that this velocity is large, say 10^7 rotations per second. (One can even easily obtain 10^{10} s^{-1}.) The total magnetization $\boldsymbol{M}$ of the sample rotates with the same velocity and can easily be detected by induction in a coil. However, the field $\boldsymbol{B}$ cannot be strictly homogeneous and some spins feel a field which is slightly stronger, and others a field which is slightly weaker. Since the angular velocity ω_i is proportional to the field B_i, the former spins will rotate a bit faster and the latter a bit more slowly than the average. With each revolution the spread will become more and more pronounced. The total magnetization gets progressively shorter, while rotating (Fig.15.9b). If the heterogeneity of the field B is of the order of 0.1%, one thousand revolutions are sufficient to lead to the spins pointing in all directions in the xy-plane. After that time, that is, after 0.1 ms, their orientations remain *completely disordered* and the substance no longer retains any magnetization (Fig.15.9c). The irregularities in the orientation of $\boldsymbol{B}$ can be neglected, as they do not have the same cumulative effect as the irregularities in its length.

So far nothing unexpected has occurred. The process illustrates a new irreversibility mechanism. On the microscopic scale, the motion is deterministic and reversible: reversing the magnetic field $\boldsymbol{B}_i$ seen by each atom would make its spin rotate the other way, according to (15.175), as if one reversed the time. However, on the macroscopic scale the disorder has increased, as shown by the demagnetization of the material. In fact, it is the slightly random nature of $\boldsymbol{B}$ in space which has been transferred to the spins through the lack of synchronism of their rotations. The macroscopic equation of motion for the case of a Gaussian distribution of B,

$$\frac{d\boldsymbol{M}}{dt} = \frac{2\mu_B}{\hbar} \left[\overline{\boldsymbol{B}} \times \boldsymbol{M} \right] - \left(\frac{2\mu_B}{\hbar} \right)^2 \Delta B^2 \, t \, \boldsymbol{M}, \tag{15.177}$$

can easily be found, solving (15.175) and averaging over B_i; its last term describes the damping of M. We have a measure of the macroscopic irreversibility by looking at the time-dependence of the *thermodynamic spin entropy*, which follows from (1.17) for a homogeneous sample with total magnetization $M \equiv \mu_B N \varrho$,

$$S_{\text{th}} = kN \left(\frac{1+\varrho}{2} \ln \frac{2}{1+\varrho} + \frac{1-\varrho}{2} \ln \frac{2}{1-\varrho} \right); \tag{15.178}$$

it *increases* as M decreases. On the other hand, the entropy

$$S = k \sum_i \left(\frac{1+\varrho_i}{2} \ln \frac{2}{1+\varrho_i} + \frac{1-\varrho_i}{2} \ln \frac{2}{1-\varrho_i} \right), \tag{15.179}$$

which is associated with the detailed microscopic description (15.175), remains constant. In fact, the expectation value $\mu_i = \mu_B \varrho_i$ of each magnetic moment conserves the value (15.176) during the precession.

At a time t_0 of the order of milliseconds the spins have thus been disoriented for a long time, the total magnetic moment is zero, and the signal which it induces has disappeared. The system seems fully disordered, "dead". Note, however, that the mechanisms for the relaxation through interactions have not yet had time to come into play and to thermalize the moments parallel to $\boldsymbol{B}$. At that time, one now

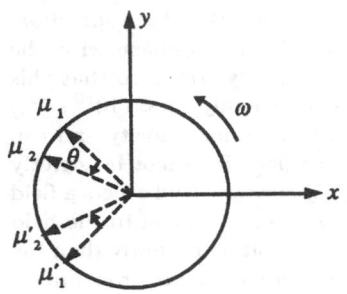

Fig. 15.10. Rotation of the spins over π around the x-axis: each moment $\boldsymbol{\mu}_i$ suddenly goes to its symmetric image $\boldsymbol{\mu}_i'$

applies, without touching the field $\boldsymbol{B}$, for a very short time τ a field $\boldsymbol{B}'$ in the x-direction. One chooses the values of τ and of B' such that $\omega'\tau = \pi$, where ω' is the Larmor frequency associated with B'. As a result of this pulse all the moments $\boldsymbol{\mu}_i$ suddenly make a *semi-revolution around the x-axis* (Fig.15.10). In this operation, if a moment $\boldsymbol{\mu}_1$ was *behind* another moment $\boldsymbol{\mu}_2$ by an angle $\theta = (\omega_2 - \omega_1)t_0$ because $\omega_1 < \omega_2$, its transformed moment $\boldsymbol{\mu}_1'$ gets *ahead* of $\boldsymbol{\mu}_2'$ by the same angle θ. One should bear in mind that the field $\boldsymbol{B}$ is still present. Starting from the time t_0 the spins thus continue to rotate, always in the same direction, each one with its own local Larmor frequency ω_i. However, because they changed position at time t_0, *at time $2t_0$ they find themselves again all* oriented in the x-direction. During the time interval $t_0, 2t_0$ the *total magnetic moment increases* and the *macroscopic entropy* (15.178) *decreases*; the evolution is the symmetric counterpart of the one observed between 0 and t_0. The reappearance of the initial signal at time $2t_0$ is the so-called "*spin echo*". Nothing prevents the production of successive echos, as long as the other relaxation mechanisms have not yet been effective.

This experiment has a surprising characteristic. *The order which seemed to have been lost* during the first phase, *reemerges* during the second one. Irreversibility and increase of the thermodynamic entropy turn out not to be all-conquering and between the times t_0 and $2t_0$ the system behaves *as if the time were reversed*. All this illustrates the *relative* nature of the irreversibility and entropy concepts. For a fresh observer the state of the spins at time t_0 is disordered and this disorder is characterized by the thermodynamic entropy (15.178). However, the experimenter who has prepared this state *knows* that it contains correlations between the local field $\boldsymbol{B}_i$ and the orientation of the moment $\boldsymbol{\mu}_i$. These correlations are complex and invisible in the thermodynamic description. However, in contrast to what happens for a gas, this information is *not irretrievably lost*; there exists in this case an operation, the rotation over π around the x-axis, which enables us to bring back the order hidden in the correlations and to transform it into a macroscopic order shown by the total magnetic moment. One sees thus that the *disorder is not an intrinsic property* of the objects. Its definition involves the *observer* in agreement with the points of view of information theory.

Summary

In order to study the dynamics of gases of weakly interacting particles we work within the Boltzmann description which disregards the correlations between molecules. The kinetic equation (15.6) then deals with the reduced density f in the single-particle phase space. Its right-hand side, the collision term, can take on various forms depending on whether the particles are scattered by one another, (15.77), or by other scatterers which are fixed, (15.7), or which can absorb energy, (15.65), or whether they are scattered by other, light particles in equilibrium, (15.134), as in the Brownian motion, or whether Fermi statistics plays a rôle, (15.74) and (15.157). One can find this term as a function of the cross-sections by striking a balance of the effect of collisions on the particle distribution f in phase space. It satisfies the microreversibility property and the microscopic conservation equations (15.9) or (15.87), and its vanishing characterizes the local equilibrium distributions (15.13) or (15.91).

The kinetic equation describes a ballistic regime if the collisions are not very efficient and a local equilibrium regime in the opposite limit. In the second case it can be solved by the Chapman-Enskog method: we use the maximum entropy method to associate with f a local equilibrium distribution f_0 which follows it in its motion and which is equivalent to it for calculating the conservative macroscopic quantities (15.2) or (15.88). This enables us to justify all macroscopic laws of thermodynamics and hydrodynamics and to calculate the transport coefficients.

The Lorentz model, which describes collisions with fixed scatterers, is well suited for a study of the gases of carriers in semiconductors or metals, and also for neutron physics. In regimes close to equilibrium it can account for diffusion, thermal or electrical conduction, and thermoelectric effects; it enables us to connect these properties with each other and with the mean free path. It can also be extended to thermalization and be applied to processes far from equilibrium. For metals transport involves only the electrons which are close to the Fermi surface.

For ordinary gases the Boltzmann equation gives us in the local equilibrium regime the transport coefficients, the viscosity and the thermal conductivity, as functions of the scattering cross-section of the molecules. These coefficients are independent of the density and increase as $\sqrt{T}$; the volume viscosity vanishes.

Boltzmann's H-theorem expresses that the entropy associated with the reduced single-particle density increases; it proves the relaxation of the system towards equilibrium. The distinction between three levels of description, the microscopic, Boltzmann, and thermodynamic levels, enables us to discuss the problem of the irreversibility of the macroscopic evolution for a gas with a microscopic evolution which is reversible. The introduction of several entropies, each associated with a level of description, shows especially the relative nature of the entropy concept and even of the irreversibility concept.

Exercises

15a Diffusion in Solids; Doping

If the temperature is sufficiently high, impurity atoms in a crystal can jump from one lattice site to another and thus migrate. This diffusion process is applied in industry for doping semiconductors (§ 11.3.4). One can study it experimentally by letting marked radioactive atoms diffuse and observing their concentration at each point as a function of time; this gives some information about the structure of the material. A convenient method for marking the atoms is to irradiate at the initial time one point of the solid with a fast particle beam.

To understand this effect we study the following one-dimensional model which simulates a copper wire heated to a few hundreds of degrees. Initially we mark the atom Cu*, placed at $x = 0$. Each atom can perform jumps $\pm l$ along the x-axis, where l is the cell size of Cu; the jumps are independent and the time between two successive jumps of the same atom is τ. This time τ, which at room temperatures is practically infinite, is a rapidly decreasing function of the temperature (Prob.19). Calculate the probability that the marked Cu* atom lies after a time t between x and $x + dx$ $(dx \gg l, t \gg \tau)$. Starting from a given distribution of marked Cu* atoms, localized near $x = 0$, find the concentration as function of x and t. Determine the diffusion coefficient.

Solution:

After a time $t = n\tau$, the Cu* atom has jumped n_r times to the right and n_l times to the left, where n_r and $n_l = n - n_r$ are random. The corresponding probability is (binomial law)

$$p(n_r, n_l) = \frac{1}{2^n} \frac{n!}{n_r! n_l!};$$

one obtains it simply by counting the number of different ways of performing these jumps. It has a sharp maximum for $n_r \sim n_l$ and with $\Delta n \equiv n_r - n_l \ll n$ reduces in the Stirling approximation to the Gaussian law

$$p(n_r, n_l) \sim \sqrt{\frac{2}{\pi n}} e^{-\Delta n^2 / 2n}.$$

The distance travelled by x equals $l\Delta n$ and the number of possible values of Δn associated with the range $x, x + dx$ equals $dx/2l$. The required probability $p(x) dx$ is thus equal to

$$p(x) dx = \sqrt{\frac{\tau}{2\pi l^2 t}} e^{-x^2 \tau / 2l^2 t} dx.$$

If the initial concentration is $n_0 \delta(x)$, where n_0 is the number of Cu* atoms created near the origin, the concentration is $n_0 p(x)$. Diffusion covers a region of magnitude $l\sqrt{t/\tau}$ which increases as $\sqrt{t}$. This random walk model gives the same result as the macroscopic diffusion theory which is based upon the solution of the equation

$$\frac{\partial \varrho}{\partial t} = - \operatorname{div} \boldsymbol{J} = D \nabla^2 \varrho.$$

If we identify ϱ with $n_0 p(x)$ we find

$$D = \frac{l^2}{2\tau}.$$

Note. The manufacture of electronic devices requires a detailed control of the impurity concentration at each point. In practice one starts from Si which is as pure as possible. Using masks one introduces on the surface of the required regions a controlled amount of n or p impurities, which one lets migrate towards specially selected positions in the interior of the material through heating for a carefully chosen period and at a carefully chosen temperature. This makes it possible to have in an extremely small volume a large number of different components which have predetermined functions, such as diodes, transistors, and so on. The progress of microelectronics thus relies on the mastering of doping techniques at smaller and smaller scales.

15b Impedance in the Lorentz Model

1. Use the Lorentz model with W assumed to be constant to write down the microscopic equation which describes a classical gas of charged particles in a uniform field in a permanent and uniform regime. Use this equation to derive directly the mobility, that is, the ratio of the mean drift velocity $\boldsymbol{u}$ to the field, and prove Ohm's law.

2. Show that the result is the same as the one obtained by an elementary calculation where one assumes that each particle is accelerated between two collisions and that each collision makes the momentum distribution isotropic.

3. Use the Lorentz model to find the equation which connects the electric current $\boldsymbol{J}(t)$ with a given time-dependent uniform applied field $\boldsymbol{E}(t)$. Integrate that equation with $\boldsymbol{E}(-\infty) = \boldsymbol{J}(-\infty) = 0$. What is the value of the impedance $Z(\omega)$, that is, the ratio of the Fourier transforms of $\boldsymbol{E}$ and $\boldsymbol{J}$ (see Exerc.14b)? Compare this with the predictions of thermodynamics for a local equilibrium regime.

4. Assume that we have an arbitrary applied field $\boldsymbol{E}(\boldsymbol{r}, t)$ which is sufficiently small that it can be treated as a perturbation. Neglect the effect of the associated magnetic field. Expanding f in the vicinity of the unperturbed global equilibrium distribution f_0, and Fourier transforming $\boldsymbol{E}$ and $f - f_0$, which replaces t and $\boldsymbol{r}$ by ω and $\boldsymbol{k}$, write down and solve the equation for $f - f_0$. Use this result to find the current as a function of the field, distinguishing between the components $\boldsymbol{E}_\perp$ and $\boldsymbol{E}_\parallel$ at right angles and parallel to $\boldsymbol{k}$. Discuss its behaviour as $\omega \to 0$, $k \to 0$. For what values of ω and k do we get a local and instantaneous response in agreement with Ohm's law?

Hints:

1. Start from (15.37) with $\partial f/\partial t = 0$, $\nabla_r f = 0$. Multiplying by p and integrating, by parts, over p, we get

$$-qE\varrho_N = -WmJ_N = -Wmu\varrho_N.$$

Hence we find the mobility (15.44) if $W = 1/\tau$ and the conductivity (15.35).

2. The calculation follows Eqs.(11.70)–(11.73).

3. Multiplying the equation

$$\frac{\partial f}{\partial t} + q\big(E(t)\cdot\nabla_p\big)f = W\big(\langle f\rangle - f\big),$$

by qp/m and integrating over p we find

$$\frac{\partial J}{\partial t} - \frac{q^2}{m}E\varrho_N + WJ = 0.$$

The density is constant as a consequence of the flux conservation equation. Integration gives

$$J(t) = \sigma E(t) - \sigma \int_{-\infty}^{t} dt'\, e^{-W(t-t')}\frac{dE}{dt'},$$

where σ is the static conductivity $\varrho_N q^2/mW$. The second term describes a retardation effect which is not included in near-equilibrium thermodynamics. The complex impedance is equal to

$$Z(\omega) = \frac{1}{\sigma}\left(1 - \frac{i\omega}{W}\right),$$

and the ballistic effects become sizeable when the frequency is not small as compared to τ^{-1}.

4. Writing

$$f(r,p,t) \equiv f_0(p) + \frac{1}{(2\pi)^4}\int d\omega\, d^3k\, e^{-i\omega t + i(k\cdot r)} f_1(k,p,\omega),$$

$$E(r,t) \equiv \frac{1}{(2\pi)^4}\int d\omega\, d^3k\, e^{-i\omega t + i(k\cdot r)} E_1(k,\omega),$$

we have, to first order,

$$-i\omega f_1 + i(v\cdot k)f_1 - \beta q\big(v\cdot E_1\big)f_0 = W\big(\langle f_1\rangle - f_1\big),$$

which is solved by

$$f_1 = \frac{\beta q\big(v\cdot E_1\big)f_0 + W\langle f_1\rangle}{W - i\omega + i(v\cdot k)}.$$

Averaging over the orientation of $p \equiv mv$ leads to

$$\langle f_1 \rangle = -i\beta q \, \frac{(k \cdot E_1)}{k^2} \, \frac{W - i\omega}{W\xi - i\omega} \, \xi f_0,$$

$$\xi \equiv \left\langle \frac{i(v \cdot k)}{W - i\omega + i(v \cdot k)} \right\rangle = 1 - \frac{1}{\eta} \arctan \eta, \qquad \eta \equiv \frac{vk}{W - i\omega}.$$

The current is given by

$$J(\omega, k) = q \int d^3 p \, v \, f_1$$

$$= \frac{\beta q^2}{W - i\omega} \int d^3 p \, f_0(p) \, v^2 \left[\frac{1}{2}\left(1 - \xi - \frac{\xi}{\eta^2}\right) E_\perp + \frac{\omega}{\omega + iW\xi} \frac{\xi}{\eta^2} E_\parallel \right]$$

$$\simeq \frac{q^2 \varrho_N (W - i\omega)}{m(W - i\omega)^2 + k^2/\beta} E_\perp + \frac{4q^2 \varrho_N}{m\sqrt{\pi}(W - i\omega)} \int_0^\infty dx \, e^{-x} x^{3/2}$$

$$\times \left[3 + \frac{2k^2 \left(iW + \frac{9}{5}\omega\right)x}{\beta m \omega (W - i\omega)^2} \right]^{-1} E_\parallel,$$

where the approximation leading to the last expression is valid for $k^2 \ll \beta m (W^2 + \omega^2)$. The longitudinal part is singular, but bounded, as both $\omega \to 0$, $k \to 0$. Its coefficient is the same as that of the transverse part, if we first take $k \to 0$, and then $\omega \to 0$; however, it tends to zero, if we first take $\omega \to 0$, and then $k \to 0$, in which case it behaves as

$$-iq^2 \varrho_N \beta \frac{\omega}{k^2},$$

and thus is independent of the collisions. The local equilibrium regime holds provided the response coefficients are constants as far as ω and k are concerned, and equal to $q^2 \varrho_N / mW$, which means that

$$\omega \ll W, \qquad k^2 \ll \beta m W^2, \qquad k^2 \ll \beta m W \omega.$$

The temporal variations of the field must be slow as compared to τ; its spatial variations must not only be slow as compared to the mean free path, but they also must be the smaller, the lower the frequency. The first two conditions were expected, but the third one is also necessary to ensure the validity of Ohm's law for the component of E parallel to k.

15c Dynamics of a Lorentz Gas Particle

Initially a particle with momentum p_0 is placed at the origin. This particle is subjected to a uniform force field φ and moves in a scattering medium described by the simplified Lorentz model (15.37).

1. Study the evolution of the expectation values of its coordinates $\langle r \rangle$ and $\langle p \rangle$ in phase space. Interpret the result for the evolution of $\langle r \rangle$.

2. Study, for the case where there is no external force, the diffusion, that is, the evolution of the dispersion of r. Do this by solving the equations which couple $\langle r^2 \rangle$, $\langle p^2 \rangle$, and $\langle (p \cdot r) \rangle$. Compare the result with Brownian motion.

3. Compare the result with the macroscopic Fick law.

Hints:

1. As in the case of the solution of the Fokker-Planck equation, we multiply (15.37) successively by r and p and integrate over both r and p, and then we solve the differential equations which we have obtained; the result is

$$\langle p \rangle = p_0 \, e^{-Wt} + \frac{\varphi}{W} \left(1 - e^{-Wt} \right),$$

$$\langle r \rangle = \left(\frac{p_0}{m} - \frac{\varphi}{Wm} \right) \frac{1 - e^{-Wt}}{W} + \frac{\varphi}{Wm} \, t.$$

The collisions with the scattering medium are equivalent to a friction force $-Wm\langle v \rangle$, which is proportional to the average velocity. After a rather short time $1/W$ there is relaxation towards a uniform motion with a velocity φ/Wm, and the initial momentum has been forgotten after that time.

2. Similarly, if there is no force φ we find

$$\langle p^2 \rangle = p_0^2, \qquad \langle (p \cdot r) \rangle = \frac{p_0^2}{Wm} \left(1 - e^{-Wt} \right),$$

$$\langle r^2 \rangle = \frac{2}{m} \int_0^t dt \, \langle (p \cdot r) \rangle = \frac{2p_0^2}{Wm^2} \left[t - \frac{1 - e^{-Wt}}{W} \right].$$

After a time long as compared to $1/W$ the position of the particle becomes less and less well defined; it is statistically spread out as $\sqrt{t}$. The fluctuations Δp and Δr increase with time; one can check that they satisfy the Schwartz inequality

$$\Delta p^2 \, \Delta r^2 - \left[\langle (p \cdot r) \rangle - (\langle p \rangle \cdot \langle r \rangle) \right]^2$$

$$= \frac{2p_0^4}{W^2 m^2} \left(1 - e^{-Wt} \right)^2 \left[\frac{Wt}{\tanh \frac{1}{2} Wt} - 2 \right] > 0.$$

The motion of $\langle r \rangle$ and of $\langle p \rangle$ is the same as in Brownian motion (15.137), if we identify W with γ/m. This analogy is, however, superficial, since the scattering medium is fixed for the Lorentz model, while it consists of very light particles for the Brownian motion. There is now no thermalization of $\langle p^2 \rangle$, but $\langle r^2 \rangle$ has the same form as (15.140) for times which are long as compared to the relaxation time $1/W$, if p_0^2 has the thermal equilibrium value $3mkT$.

3. The linearity of the equation of motion for f makes it possible for it to describe equally well a set of particles as the statistical distribution of a single particle in phase space, normalized initially to $\delta^3(r) \, \delta^3(p - p_0)$. For $t \gg 1/W$ we find again the local equilibrium regime of § 15.1.5, but f_0 is replaced by

$$f_0 = \frac{1}{2\pi p_0} \, \delta\left(p^2 - p_0^2\right) \varrho_N(r), \qquad \int d^3 r \, \varrho_N(r) = 1.$$

The analogue of (15.29) is

$$J_N = \frac{1}{3} \varrho_N \frac{p_0^2}{m^2 W} \nabla(-\ln \varrho_N).$$

Fick's law follows:

$$\frac{\partial \varrho_N}{\partial t} = -\operatorname{div} J_N = D \nabla^2 \varrho_N,$$

with the diffusion coefficient

$$D = \frac{p_0^2}{3W m^2}.$$

The solution of this macroscopic diffusion equation with the required initial condition is

$$\varrho_N = (4\pi Dt)^{-3/2} e^{-r^2/4Dt},$$

and it leads to $\langle r^2 \rangle \sim 6Dt$, as under 2. We also regain this result by multiplying the diffusion equation for ϱ_N by r^2 and integrating over r.

15d Hall Effect

A semiconductor, described as a gas of carriers with charge q in the Lorentz model with W assumed to be constant, is subjected to uniform static electric and magnetic fields. Assuming that a homogeneous stationary regime has been established find the relation between the current J and the fields. Write down the matrix of the responses of J_x and J_y to E_x and E_y for the case where the magnetic field B is along the z-axis and the electric field lies in the xy-plane. Check the Onsager relations.

Each of the four sides $\pm x$, $\pm y$ of a rectangular film in the xy-plane is connected with a terminal, and the film is subjected to a field B along the z-axis. What happens if one applies a potential difference between the $\pm x$ terminals, while keeping the $\pm y$ terminals open? What, if one then shorts the $\pm y$ terminals? How can one use the Hall effect to find the sign of the majority carriers?

Solution:

The distribution f is governed by (see note below)

$$\frac{\partial f}{\partial t} + (v \cdot \nabla_r)f + q\left(\{E + [v \times B]\} \cdot \nabla_p\right)f = W\left(\langle f \rangle - f\right),$$

where the first two terms vanish in a homogeneous stationary regime. Multiplying by $-qv$ and integrating over $p = mv$ we find

$$WJ = \frac{q^2 \varrho_N}{m} E + \frac{q}{m} [J \times B].$$

Denoting the conductivity in the absence of the field by $\sigma \equiv q^2 \varrho_N \tau / m$ and $\tau \equiv 1/W$, we then find

$$\sigma \begin{pmatrix} E_x \\ E_y \end{pmatrix} = \begin{pmatrix} 1 & -\dfrac{q\tau}{m} B \\ \dfrac{q\tau}{m} B & 1 \end{pmatrix} \begin{pmatrix} J_x \\ J_y \end{pmatrix},$$

$$\begin{pmatrix} J_x \\ J_y \end{pmatrix} = \sigma \left(1 + \dfrac{q^2 \tau^2 B^2}{m^2} \right)^{-1} \begin{pmatrix} 1 & \dfrac{q\tau}{m} B \\ -\dfrac{q\tau}{m} B & 1 \end{pmatrix} \begin{pmatrix} E_x \\ E_y \end{pmatrix}.$$

According to (14.27) the response matrix must remain unchanged if one transposes it and if one changes the Hamiltonian by time reversal, that is, if one changes B to $-B$; one can check that this is the case for the above matrix.

If the current J_y is zero, a *field*

$$E_y = \frac{q\tau}{m} B E_x = \frac{B}{q\varrho_N} J_x$$

is created in the *direction at right angles to the current*: at the $\pm y$ terminals we see appear a potential difference, *the sign of which depends on that of the carriers*. The field E_y is due to an accumulation of opposite charges on the $\pm y$ sides of the film. The relation $J_x = \sigma E_x$ remains unchanged. If one shorts the $\pm y$ terminals, the field E_y vanishes; a current

$$J_y = -\frac{q\tau}{m} B J_x = -\frac{\sigma m q \tau B}{m^2 + q^2 \tau^2 B^2} E_x$$

circulates in the film in the y-direction, which is collected at the $\pm y$ terminals. The direction of this current depends on the sign of the carriers.

The contributions from the two kinds of carriers must be subtracted from one another and the sign of the Hall effect is thus governed by the majority carriers.

Notes. In the above $p = mv$ denotes not the *momentum*, but the *kinetic momentum*. In terms of the momentum p', which in contrast to p depends on the choice of gauge, the Hamiltonian can be written as

$$H(r, p') = \frac{1}{2m} \left(p' - qA \right)^2 + q\Phi,$$

where we may take $A = \frac{1}{2}[B \times r]$ and $\Phi = -(r \cdot E)$ to describe uniform fields. The velocity is given by $p = mv = p' - qA$; it, and not p'/m, occurs in the definition of the current J. The reduced density $f'(r, p') \equiv f(r, p)$ satisfies the equation

$$\frac{\partial f'}{\partial t} + \left(\nabla_{p'} H \cdot \nabla_r \right) f' - \left(\nabla_r H \cdot \nabla_{p'} \right) f' = \mathcal{I}(f).$$

On the right-hand side the collisions conserve the kinetic energy and the angular integration is over p, rather than over p'. On the left-hand side $\nabla_r f'$ contains not only $\nabla_r f$, which vanishes in a homogeneous state, but also contributions with $\nabla_p f$; those must be combined with the third term to lead to the gauge-invariant equation for f which we started from without justification.

It is not consistent to assume, as we did, that $\partial f/\partial t = 0$ since the Lorentz gas is heated following $\partial \varrho_E/\partial t = (\boldsymbol{J} \cdot \boldsymbol{E})$. This, however, does not affect our calculations.

If W depends on p, for instance linearly as in (15.14) for scattering by hard spheres, σ will involve the average of W^{-1} over $f_0 p^2$ according to (15.31). The Hall coefficient will involve the average of W^{-2} over $f_0 p^2$ which will introduce a numerical factor $3\pi/8$ in front of τ in the above calculations.

15e Absorption, Stimulated and Spontaneous Emission

Einstein established Planck's radiation law by a balance method similar to the one which leads to the Maxwell or the Fermi distributions starting from the Boltzmann or the Landau-Uhlenbeck equation (§ 15.4.3). The idea consists in deriving the equilibrium of the photons in an enclosure from that of atoms contained in that vessel, which can absorb or emit photons. To simplify the argument we assume that there exists only one mode, of frequency ν, containing an average number $f(\nu)$ of photons. We also assume that each of the N atoms has two levels E_1 and E_2 with $E_2 - E_1 = h\nu$. A more realistic discussion would introduce all the modes, with a quasicontinuous ν spectrum, and levels E lying densely for the matter which can absorb or emit radiation.

The probability per unit time that an atom in the state E_1 is excited by absorbing a photon in the ν mode has the form $P_{12} = Af(\nu)$ where A defines the Einstein coefficient for *induced absorption*. Similarly one introduces the *spontaneous emission* coefficient B: the probability per unit time that an excited atom in the state E_2 falls back into the state E_1 while emitting a ν photon is $P_{21} = B$; this process *a priori* seems independent of the state of the radiation in the enclosure.

1. Using this hypothesis, write down the equations governing the evolution of the numbers N_1 and N_2 of atoms with energies E_1 and E_2. What is the relation between N_1, N_2, and $f(\nu)$ as $t \to \infty$? What is the resulting temperature dependence of $f(\nu)$ in thermal equilibrium?

2. The discrepancy between this result and experiments on thermal radiation led Einstein to suggest the existence of a supplementary process, *stimulated emission*: the presence of a ν photon in the vicinity of the atom induces it to emit a new ν photon. As in the case of induced absorption, the probability per unit time for this process has thus the form $P'_{21} = B'f(\nu)$. Write down the equations governing the dynamics of the three processes. Determine the new relation between N_1, N_2, and $f(\nu)$ as $t \to \infty$. What must be the values of the Einstein coefficients A, B, and B' in order that in thermal equilibrium $f(\nu)$ be compatible with Planck's law? Show that the relations obtained in this way result from the principle of detailed balancing.

3. One lets impinge on N atoms a beam of photons with the same direction, the same polarization, and the same frequency. A model for this situation consists in assuming that the average number $f(\nu)$ in a single mode is large. Show by solving the equations of motion for $f(\nu)$ for given N_1 and N_2 that the exit flux can be larger than the incident flux, provided one suit-

ably chooses the populations N_1 and N_2. This property, *superradiance*, gives the principle of *lasers*.

Answers:

1. We have

$$\frac{dN_1}{dt} = -\frac{dN_2}{dt} = -Af(\nu)N_1 + BN_2.$$

If there is no other source for emission or absorption of ν photons, we also have $df(\nu) = dN_1$. As $t \to \infty$ we find that all solutions become stationary with

$$\frac{N_2}{N_1} = \frac{A}{B} f(\nu).$$

If the gas is in thermal equilibrium, it follows from $N_2/N_1 = \exp[-(E_2 - E_1)/kT]$ that the number of photons in the mode ν in the stationary regime must be

$$f(\nu) = \frac{B}{A} e^{-h\nu/kT};$$

the A and B coefficients, which are related to quantum processes, can be expressed by (14.158) in terms of matrix elements of the transition operator between the initial and final states and are thus independent of the temperature. We find for f a Boltzmann distribution, an incorrect result for photons.

2. We now have

$$\frac{dN_1}{dt} = -Af(\nu)N_1 + \left[B + B'f(\nu)\right]N_2.$$

In the stationary regime reached as $t \to \infty$, the vanishing of the right-hand side gives

$$\frac{N_2}{N_1} = \frac{Af(\nu)}{B + B'f(\nu)}.$$

If the gas is in equilibrium, we have thus

$$f(\nu) = \frac{B}{Ae^{h\nu/kT} - B'},$$

which reduces to the Bose factor, provided

$$A = B = B'.$$

The probability for the transition from a state with n photons and an atom of energy E_1 to a state with $n-1$ photons and the atom with the energy E_2 is An. The probability for the inverse process is $B + B'(n-1)$. Putting these two numbers equal to one another for all n gives $A = B = B'$. The *Bose distribution* for photons thus arises here from the *microreversibility* of emission and absorption processes, which itself *implies the existence of stimulated emission*.

3. Neglecting the time changes of N_1 and N_2, and the coupling between the photons and the other degrees of freedom, we find that the evolution equation,

$$\frac{df}{dt} = A(N_2 - N_1)f + AN_2,$$

has the solution

$$f = \left(f_0 - \frac{N_2}{N_1 - N_2} \right) e^{-A(N_1 - N_2)t} + \frac{N_2}{N_1 - N_2}.$$

If $N_1 > N_2$, f tends to the stationary value $N_2/(N_1 - N_2)$. However, if there is a *population inversion* in a *non-equilibrium* atomic state with $N_2 > N_1$, f will increase exponentially, whatever its initial value. In practice, this increase is limited because photons are absorbed by other degrees of freedom and because the atoms are de-excited, especially by spontaneous emission to other modes of the same frequency which are neglected here. One thus ends up with a stationary state where the number of photons f in the ν mode is saturated at a very large value. This is a good model for the behaviour of a laser where an inversion in population of some atomic levels can give rise to a luminous coherent beam with a very high intensity.

15f Heat Transfer in Stars

1. The energy produced by thermonuclear fusion reactions in the core of the main-sequence stars such as the Sun is transferred by radiation, convection, and conduction from the core to the surface, where it is emitted in the form of radiation. In such stars the dominant process of heat transport is radiation: the photons propagate and are scattered, absorbed, and re-emitted by the electrons and nuclei, mainly hydrogen, which are the constituents of the star. Assume that this process is governed by a Lorentz equation for the photons, in which the scattering cross-section does not depend on the frequency; this is true for the nearly elastic Thomson scattering of photons by electrons, where $\sigma_{\text{Th}} = \frac{8}{3}\pi(e^2/4\pi\varepsilon_0 mc^2)^2$. Prove directly that the energy density ϱ_E and energy current density $\boldsymbol{J}_E$ of the photons within the star are related by $\boldsymbol{J}_E = -D\nabla\varrho_E$, and find an expression for D. Assuming that the star has a sharp surface with radius R, across which the photons pass suddenly from a local thermal equilibrium (LTE) regime to a ballistic regime, use the Lorentz equation to show that the star radiates as a black body with temperature $T_S = T(R)$.

2. Denote the mass density by $\varrho(r)$, and describe the matter as a fully ionized hydrogen plasma, behaving as a locally neutral mixture of classical perfect gases of protons and electrons. Denote by $q(r)$ the heat power production per unit volume, which is significant only in the core of the star where the temperature is sufficiently high to induce fusion. Write down the equations which determine the temperature $T(r)$, the density $\varrho(r)$, the pressure $\mathcal{P}(r)$ and the luminosity L of the star in a stationary, LTE, regime. (The luminosity is the total radiated power.) Discuss which parameters govern the state of the star.

3. Take as a model of the Sun a sphere with a uniform density, and with an active core of radius $R_C = 0.1\,R_\odot$ in which q is uniform. Determine $T(r)$

in terms of $M_\odot, L_\odot$ and $R_\odot$. Using the data from the end of this volume, determine numerically the temperature at the surface and at the centre of the Sun. How long does it take for a photon produced at the centre of the Sun and diffusing through it to escape from its surface? Write down an additional relation expressing self-gravitational equilibrium in this model, and derive from it the luminosity of a star as function of its mass.

4. Heat is produced in the core of the star by a sequence of nuclear reactions, typically

$$p + p \longrightarrow {}^2H + e^+ + \nu,$$
$$e^+ + e^- \longrightarrow \gamma,$$
$$p + {}^2H \longrightarrow {}^3He + \gamma,$$
$${}^3He + {}^3He \longrightarrow {}^4He + 2p.$$

Fusion of 4 protons into helium thus produces an energy of 27 MeV. We focus on the first reaction, with a cross-section $\sigma_f(\varepsilon)$ depending on the energy ε of the two protons in the centre of mass frame. Denoting by $f(r, p)$ the (Maxwellian) reduced number density of the protons, write down the number of fusions n_f of proton pairs per unit volume and unit time. Integrate over the centre of mass motion and over the angular variables. The cross-section $\sigma_f(\varepsilon)$ behaves as $A \, \varepsilon^{-1} \exp\left[-(\varepsilon_B/\varepsilon)^{1/2}\right]$, where A is a slowly varying function of ε and where $\varepsilon_B = \pi^2 \alpha^2 m_p c^2 \simeq 0.5\,\mathrm{MeV}$; the small exponential factor, easily obtained by the WKB method, describes the quantum tunnelling through the Coulomb barrier which is needed to bring the protons close together, at a distance of the order of fermis where the nuclear forces begin to operate. How does the heat production rate depend on ϱ and on T? Show that the energy of the protons which contribute to fusion has a sharp maximum (*Gamow peak*). Show that the stationary state of the star is stable.

5. Consider now a model of a white dwarf, with mass $0.6\,M_\odot$ and radius R, which has an interior part of fully ionized carbon with radius R_0, and an envelope with negligible relative mass and energy. Check that the electron gas in the interior is degenerate and non-relativistic, while the C nuclei constitute a classical gas (see Prob.9). Assume also for the sake of simplicity that the interior is homogeneous, with a uniform density. Determine R_0. White dwarfs are stars which have exhausted their nuclear fuel, and thus evolve slowly without heat creation. The main heat transfer mechanism in their interior is through conduction by electrons, which scatter on the C nuclei. Describe the Coulomb scattering classically and derive the *Rutherford formula* for the differential cross-section,

$$\frac{d\sigma}{d\omega} = \frac{\sigma_R}{\sin^4(\theta/2)}, \qquad \sigma_R \equiv \left(\frac{Ze^2}{8\pi\varepsilon_0 m v^2}\right)^2.$$

To simplify the resulting Boltzmann equation, neglect the recoil of the C nuclei as well as the angular dependence of the cross-section and take σ_R as an effective total cross-section. Hence estimate the time it takes for a heat pulse to spread out in the whole star and show that the internal temperature T_0 is uniform.

6. Take as a model of the envelope $R_0 < r < R$ of the white dwarf a classical perfect gas of non-ionized atomic hydrogen. In it, the heat transfer is dominated by radiative processes (photoionization and bremsstrahlung) which produce an opacity κ of the form

$$\kappa = \frac{W}{c\varrho} \equiv \frac{\sigma}{m_{\mathrm{p}}} = a\varrho T^{-7/2}, \qquad a = 2.5 \times 10^{19} \text{ SI},$$

where ϱ and T are the mass density and the temperature at the point considered. (Astrophysicists define the *opacity* of a medium as $\kappa \equiv 1/l\varrho$ where l is the mean free path for photons and ϱ the mass density.) Write down the temperature and pressure gradients dT/dr and dP/dT as functions of T, P and the luminosity L, in a LTE, nearly stationary regime. Solve the equation for dP/dT, assuming that both the density and the temperature are negligible at the surface $r = R$, and relate $\varrho(r)$ to L and $T(r)$. Find the dependence of T on r. In order to match the interior and the envelope in spite of the qualitative differences between the models which describe them, assume that P, ϱ and T are continuous across the sphere $r = R_0$ and drop in this question the assumption of a uniform density in the interior $r \leq R_0$. From the equations of state on both sides, find a relation between $\varrho(R_0)$ and $T(R_0)$. Express the luminosity in terms of T_0, and make numerical estimates of T_0, of the effective surface temperature T_{S} associated with the emitted radiation and of the relative thickness of the envelope for a white dwarf with mass $M = 0.6\ M_\odot$ and luminosity $L = 10^{-3}\ L_\odot$.

7. Write down the internal energy of the star, keeping only the ground state contribution U_0 of the electrons and the perfect gas contribution U_1 of the C nuclei. White dwarfs are formed by contraction of the less massive main-sequence stars when thermonuclear fusion reactions are achieved; at that initial stage they have a large luminosity, larger than that of the Sun, because the radiation and the contraction have produced heating (Exerc.6e). Using the above results, relate the age of a white dwarf to its luminosity.

Answers and Hints:

1. In the Lorentz equation (15.24) and in Eqs. (15.2), (15.23) we have here $\varepsilon = cp = h\nu$, $v = cp/p$ and $W = c\sigma_{\mathrm{Th}}\, n$ where $n = \varrho/m_{\mathrm{p}}$ is the number of electrons (or protons) per unit volume. Here the opacity is just $\sigma_{\mathrm{Th}}/m_{\mathrm{p}}$. In the LTE regime prevailing within the star, we have $f_1 = -(v \cdot \nabla f_0)/W$, which yields, by integration over p after multiplication by $v\varepsilon$, the required diffusion equation $J_E = -(c^2/3W)\nabla\varrho_E$. If we integrate the Lorentz equation over r in a thin shell lying astride the surface, we get from the dominant term $\int d^3r \ \mathrm{div}\ (vf) = 0$; as a consequence, the non-equilibrium distribution f of photons just outside equals the equilibrium distribution $f_0(T_{\mathrm{S}})$ for outgoing momenta, and vanishes for ingoing

momenta. Integration over p then yields for the radiated flux $J_E = \frac{1}{4}c\, \varrho_E(T_S) = \sigma T_S^4$. Note that $J_E(r) \neq \sigma T^4(r)$ within the star.

2. The equation of state is at each point $\mathcal{P} = 2(\varrho/m_p)kT$. The hydrostatic equilibrium equation (Exerc.6e) reads

$$\frac{d\mathcal{P}}{dr} = -\frac{G\varrho(r)}{r^2} 4\pi \int_0^r dr'\, r'^2 \varrho(r').$$

The above diffusion equation, where $\sigma T^4 = \frac{1}{4}c\varrho_E$, takes the form

$$-\frac{4m_p\sigma}{3r^2} \frac{d}{dr}\left(\frac{r^2}{\sigma_{\mathrm{Th}}\varrho} \frac{dT^4}{dr}\right) = q(r).$$

The luminosity is

$$L = 4\pi r^2 J_E(r) = 4\pi \int_0^r dr'\, r'^2 q(r') = 4\pi R^2 \sigma T^4(R),$$

for any r outside the core. The solution of these coupled equations depends *only on the total mass* of the star. However, the presence of nuclei other than hydrogen influences the processes of photon transfer and of nuclear reactions; it thus modifies D and q and can give rise to slight differences between stars.

3. For $r \leq R_{\mathrm{C}}$, the solution of the heat transfer equation is

$$\sigma T^4 = -\frac{\sigma_{\mathrm{Th}}\varrho q}{8m_p} r^2 + C, \quad \varrho = \frac{3M_\odot}{4\pi R_\odot^3}, \quad r \leq R_{\mathrm{C}}.$$

Matching it with the solution for $R_c \leq r \leq R_\odot$ yields

$$\sigma T^4 = \frac{\sigma_{\mathrm{Th}}\varrho q R_{\mathrm{C}}^3}{4m_p}\left(\frac{1}{r} - \frac{3}{2R_{\mathrm{C}}}\right) + C, \quad r \geq R_{\mathrm{C}},$$

and the constant C is determined by the boundary condition

$$\sigma T^4(R_\odot) = \frac{L_\odot}{4\pi R_\odot^2}, \quad L_\odot = \frac{4}{3}\pi R_{\mathrm{C}}^3 q.$$

Numerically, this yields $T_S = 5750$ K and

$$\sigma T^4(0) = \sigma T_S^4 + \left(\frac{3R_\odot}{2R_{\mathrm{C}}} - 1\right)\frac{9\sigma_{\mathrm{Th}} M_\odot L_\odot}{64\pi^2 R_\odot^4 m_p}$$

$$\simeq \left(\frac{3R_\odot}{2R_{\mathrm{C}}} - 1\right)\frac{9\sigma_{\mathrm{Th}} M_\odot}{16\pi R_\odot^2 m_p}\sigma\, T_S^4,$$

that is, $T(0) \simeq 5 \times 10^6$ K. The actual surface temperature is slightly higher, as the luminosity is reduced by absorption in the solar atmosphere. Our calculation underestimates the internal temperature by a factor 4, which is not too bad in view of the coarseness of our approximation. In actual fact the Thomson scattering is not the sole process for photon diffusion and absorption, so that the opacity is significantly larger than σ_{Th}/m_p; moreover, the density and hence W decrease from the centre of the star to its surface, which hinders the diffusion of photons in the central part; both effects enhance the central temperature.

The probability distribution of a photon which leaves the centre of the Sun at $t = 0$ spreads out as $\exp(-r^2/4Dt)$ (§ 15.3.5 and Exerc. 15c). The expectation value of the time it takes to leave the Sun is

$$\int_0^\infty dt\, t\, \frac{d}{dt} 4\pi \int_R^\infty dr\, r^2 \frac{e^{-r^2/4Dt}}{(4\pi Dt)^{3/2}}$$

$$= \frac{1}{(4\pi D^3)^{1/2}} \int_0^R dr\, r^2 \int_0^\infty \frac{dt}{t^{3/2}}\, e^{-r^2/4Dt} = \frac{R_\odot^2}{2D}.$$

Even if we take only the Thomson scattering into account, this time is about 5000 years; the length travelled by the photon in its Brownian motion is 3×10^{28} times the Sun's radius!

The equation of state and the hydrostatic equation are replaced in this model by $2U = 3\,\mathcal{P}\Omega = -E_G$, that is

$$\int_0^{R_\odot} dr\, r^2 T(r) = \frac{GM_\odot m_\mathrm{p} R_\odot^2}{30k}.$$

The average temperature within the star is proportional to its central temperature evaluated above. Hence we find the *mass-luminosity relation*,

$$L \simeq \frac{G^4 m_\mathrm{p}^5}{400\hbar^3 c^2 \sigma_{\mathrm{Th}}}\, M^3,$$

where the numerical coefficient 400 is a rough estimate, corresponding to $\langle T \rangle \simeq 0.4\,T(0)$.

4. As in Eq. (15.84) we find

$$n_\mathrm{f} = \frac{1}{2} \int d^3 p_1\, d^3 p_2\, v_{12}\, \sigma_\mathrm{f}(\varepsilon) f(r, p_1)\, f(r, p_2)$$

$$= \frac{2\varrho^2}{\sqrt{\pi (kT)^3 m_\mathrm{p}^5}} \int_0^\infty d\varepsilon\, \varepsilon\, e^{-\varepsilon/kT} \sigma_\mathrm{f}(\varepsilon).$$

This quantity, and hence q, is proportional to ϱ^2 and rapidly increases with T. The integral can be evaluated by the steepest descent method (Exerc. 5b); the integrand is concentrated around the Gamow energy

$$\varepsilon_G = 2^{-2/3}(kT)^{2/3}\varepsilon_\mathrm{B}^{1/3},$$

with a relative fluctuation $(4kT/27\varepsilon_\mathrm{B})^{1/6}$, and we find

$$n_\mathrm{f} \simeq \frac{8A\varrho^2}{\sqrt{6(kT)^3 m_\mathrm{p}^5 \varepsilon_\mathrm{B}^{1/2}}} \exp\left[-3\left(\frac{\varepsilon_\mathrm{B}}{4kT}\right)^{1/3}\right].$$

The stationary state of the star is characterized by an *exact balance* between the luminosity L and the heat production, $\int d^3 r q(r)$, which increases very rapidly with

T. If more heat is produced than radiated, the total energy $U + E_G$ increases. However, the restoration of gravitational equilibrium, when $2U = -E_G$, implies an increase of E_G and a decrease of U, so that the star expands and cools down, thus reducing q. Conversely, if $\int d^3 r q(r) < L$, the evolution enhances q until a stationary regime is reached.

5. From Exerc. 6e, we find

$$E_G = -\frac{3}{5}\frac{GM^2}{R_0} = -\frac{1}{2}U = -\frac{3}{10}\left(\frac{6M}{12m_p}\right)\varepsilon_F,$$

and hence $R_0 = 8600$ km. In classical Coulomb scattering, the impact parameter b and the scattering angle θ, φ are related by

$$b\tan\frac{\theta}{2} = \frac{Ze^2}{4\pi\varepsilon_0 m v^2}, \quad \varphi = \pi,$$

and the Rutherford formula results from (15.4). The thermal conductivity is obtained from (15.14), (15.55) as

$$\lambda = \frac{2\pi^2 k^2 T}{p_F \sigma_R}.$$

A heat pulse diffuses with a coefficient $D = \lambda/C$, where C is the electronic specific heat per unit volume,

$$C = \frac{\pi^2}{2}\frac{k^2 T}{\varepsilon_F}\frac{N_{el}}{\Omega}.$$

The characteristic time $R^2/2D$ is thus

$$\frac{3^{1/3}\sigma_R m}{16\pi^{4/3}\hbar}\left(\frac{M}{m_p}\right)^{2/3} \simeq 200\,000 \text{ years},$$

small compared to the age of the star and to its cooling time determined below.

6. The diffusion equation $J_E = -(c/3\kappa\varrho)\nabla\varrho_E$, together with $L = 4\pi r^2 J_E$, $c\varrho_E = 4\sigma T^4$ and $\mathcal{P} = \varrho kT/m_p$, yields

$$\frac{dT}{dr} = -\frac{3am_p^2 L\mathcal{P}^2}{64\pi\sigma k^2 r^2 T^{17/2}}.$$

The hydrostatic equilibrium is expressed by

$$\frac{d\mathcal{P}}{dr} = -\frac{GMm_p\mathcal{P}}{kr^2 T}.$$

The pressure, density and temperature at each point of the envelope are related by

$$\mathcal{P}^2 = \frac{256\pi\sigma GMk}{51am_p L}T^{17/2} = \frac{k^2}{m_p^2}\varrho^2 T^2,$$

and the solution of the differential equation for $T(r)$ is

$$T = \frac{4GMm_p}{17k}\left(\frac{1}{r} - \frac{1}{R}\right).$$

Matching the degeneration pressure of the electron gas for $r = R_0 - 0$ with the perfect gas pressure of hydrogen for $r = R_0 + 0$ yields

$$\frac{\hbar^2}{10m} \left[\frac{3\pi^2 \varrho(R_0)}{2m_{\rm p}} \right]^{2/3} = kT_0.$$

Elimination of $\varrho(R_0)$ provides

$$L - \frac{24\pi^5 \hbar^6 GM\sigma}{17k^4 am_{\rm p}(5m)^3} \, T_0^{7/2} = 4\pi R^2 \sigma T_{\rm S}^4,$$

from which one finds $T_0 \simeq 1.7 \times 10^7 {\rm K}$, $T_{\rm S} \simeq 8900$ K and $R - R_0 \simeq R/15$.

7. The internal energy is

$$U_0 + U_1 = \frac{3}{5} \left(\frac{M}{2m_{\rm p}} \right) \varepsilon_{\rm F} + \frac{3}{2} \left(\frac{M}{12m_{\rm p}} \right) kT_0.$$

Since $U_1 \ll U_0$, the star slowly cools down (see solution of the last part of Prob.9), according to

$$\frac{Mk}{8m_{\rm p}} \frac{dT_0}{dt} = -L(t).$$

Integration and elimination of $T_0(t)$ provides

$$L = \frac{k^3 M}{80\pi^2 \, m_{\rm p}} \left(\frac{425am^3}{3\hbar^6 \sigma G} \right)^{2/5} t^{-7/5}.$$

Assuming that most white dwarfs have masses of the order of 0.6 $M_\odot$, *their age is related to their luminosity* by

$$\log_{10} \frac{t}{\theta} = \frac{5}{7}x, \quad x \equiv \log_{10} \frac{L_\odot}{L}, \quad \theta = 5 \times 10^6 \text{ years.}$$

Note. This result suggests a method for determining the *age of our Galaxy*. One represents the statistics of luminosities of white dwarfs on a plot with abscissa x; dividing x into equal intervals Δx, one estimates the number ΔN of white dwarfs having a luminosity in each interval, and one draws a histogram of $y = \log_{10} \Delta N$ as function of x. For $0 < x < 4.5$, one observes a linear increase with a slope of about 0.7. This shape is explained by the above theory, which indicates that the number ΔN of white dwarfs in the interval x, $x + \Delta x$ is

$$\Delta N = \frac{dN}{dt} \left(\frac{5}{7} \ln 10 \right) t \, \Delta x,$$

where dN/dt is the rate of creation of such stars in our Galaxy at the time $-t$. If we reasonably assume that this rate has remained constant, we see that ΔN is proportional to $L^{-5/7}$ in agreement with observations. The shape of the age - luminosity curve thus explains why there are, for instance, 60 times more faint white dwarfs with $10^{-4.5} < L/L_\odot < 10^{-3.5}$ than brighter ones with $10^{-2} < L/L_\odot < 10^{-1}$. Observations show, however, a sudden drop of y for x beyond 4.5: no white dwarf with luminosity less than $2.5 \times 10^{-5} \, L_\odot$ has been seen, although much fainter objects can be detected. Hence we get a strong indication that our Galaxy was formed at a time corresponding to that luminosity. A rather reliable estimate of the age of our Galaxy is thus obtained as

$$(2.5 \times 10^{-5})^{5/7}\theta \simeq 10^{10} \text{ years.}$$

16. Problems

"L'examen n'était pas bien difficile; son cousin l'avait passé
sans peine: on exigeait, croyait-il, un peu de calcul, la con-
naissance du français et une bonne écriture."

Anatole France, La vie en fleur

"The answer is yes, but what is the question?"

Anonymous

"Oh! quelle est l'imagination malsaine, le cerveau dépravé où
germent ces problèmes révoltants dont on nous torture? Je
les exècre! Et les ouvriers qui se divisent en deux escouades
dont l'une dépense 1/3 de force de plus que l'autre, tandis
que l'autre, en revanche, travaille deux heures de plus! Et
le nombre d'aiguilles qu'une couturière use en 25 ans quand
elle se sert d'aiguilles à 0 fr. 50 le paquet pendant 11 ans,
et d'aiguilles à 0 fr. 75 pendant le reste du temps, mais que
celles de 0 fr. 75 sont ... etc ..., etc .... Et les locomotives
qui ... ! Odieuses suppositions, hypothèses invraisemblables,
qui m'ont rendue réfractaire à l'arithmétique pour toute ma
vie!"

Colette, Claudine à l'école

"On donne un trapèze convexe ABCD ... On le donne, le
trapèze convexe, mais, sitôt donné, on le reprend ...
On-do-nne-un-tra-pè-ze-con-ve-xe.
Ondo-nuntra-pèzecon-vexe."

Jacques Audiberti, Monorail

In this chapter we study in the form of problems a selection of different
subjects. These problems, numbered 16.1 to 16.19, are all independent of one
another; they each are applications of the main text and roughly speaking are
placed in the same order as the preceding chapters. Some concern solid state
physics, some classical or quantum fluids, and others astrophysics; several
deal with phase transitions and some of the later ones are related to applied
sciences. In most cases the statement of the problem is followed by more
or less short hints for solving them and by comments. The most difficult
questions are accompanied by detailed solutions.

16.1 Paramagnetism of Spin Pairs

Consider two fixed electrons with spins S_1 and S_2. These electrons are placed in a magnetic field B which is parallel to the z-axis, whence we get a contribution $\mathcal{H}_0$ to the Hamiltonian:

$$\mathcal{H}_0 = \frac{2\mu_B B}{\hbar}(S_{1z} + S_{2z}).$$

Moreover, they interact with one another according to a Hamiltonian

$$\mathcal{H}_1 = \frac{J}{\hbar^2}(S_1 \cdot S_2),$$

where J is a positive, constant energy. One can easily use quantum mechanics to determine the eigenstates of the complete Hamiltonian $\mathcal{H}_0 + \mathcal{H}_1$. Indeed, $\mathcal{H}_0$ and $\mathcal{H}_1$ commute and the eigenstates are those of $S_1^2 = S_2^2 = \frac{3}{4}\hbar^2$; of $(S_1 + S_2)^2$, which has the eigenvalues 0 for the singlet state $|s\rangle$ and $2\hbar^2$ for the triplet states $|t_-\rangle$, $|t_0\rangle$, and $|t_+\rangle$; and of $S_{1z} + S_{2z}$ which has the eigenvalues $-\hbar$, 0, $+\hbar$ for $|t_-\rangle$, $|t_0\rangle$, $|t_+\rangle$, respectively. One checks easily that the energies of the eigenstates $|s\rangle$, $|t_-\rangle$, $|t_0\rangle$, and $|t_+\rangle$ are given, respectively, by

$$E_s = -\frac{3J}{4}, \quad E_- = \frac{J}{4} - 2\mu_B B, \quad E_0 = \frac{J}{4}, \quad E_+ = \frac{J}{4} + 2\mu_B B.$$

Such a two-spin system will be called a spin pair.

1a. Write down the partition function Z_1 for a spin pair in equilibrium with a thermostat at a temperature T.

b. Use Z_1 to find the average energy U_1 and the entropy S_1 of this pair.

c. To what simpler expressions do Z_1, U_1, and S_1 reduce in the limit as $J = 0$? Give the physical reason for this.

2. We consider a piece of matter containing N pairs. Each pair occupies a given position in the sample. The pairs are far enough from one another that we can completely neglect the interactions between the spins of different pairs.

a. Write down the partition function Z of this system of N pairs.

b. Use Z to derive the average energy U and the entropy S of this system.

3. Show the behaviour of $U(T)$, $S(T)$, and $S(U)$, first for the case $B = 0$, and then for increasing values of B. Determine for each curve the limiting values, the extrema, and the derivatives in those points, using either simple physical considerations, or the equations obtained under 2. Bear in mind that $\beta = 1/kT$ can vary from $-\infty$ to $+\infty$ for spin systems.

4a. Evaluate the average magnetic moment M of the system as function of B and T.

b. Use that expression to find the magnetic susceptibility $\chi = (1/\Omega) \lim_{B \to 0} M/B$ where Ω is the volume.

c. Compare χ with the magnetic susceptibility χ_0 of $2N$ non-interacting spins (limiting case $J = 0$) and give a physical interpretation of the result, in particular for the extreme limits when $J/kT \gg 1$ and when $J/kT \ll -1$.

5a. Sketch the $M(B)$ curve for a temperature $T \gg J/k$. How does this curve change with T?

b. Let $J = 10^{-4}$ eV. For what value of B is the average magnetic moment half its maximum value for the given conditions?

c. Sketch the $S(B)$ curve for the same conditions, and for various values of T.

d. Show that it is possible to find values B_1 and B_2 of the magnetic field such that an adiabatic change from B_1 to B_2 lowers the temperature to a very low value. Specify the possible values for B_1 and B_2. What is the lowest temperature which, in principle, one might obtain in this way? What physical effects prevent in practice the attainment of this minimum?

Comments:

There are two reasons which prevent us from reaching a temperature $T = 0$ (last question). Firstly, the entropy related to the other degrees of freedom of the system must be added to the entropy evaluated here. Secondly, the degeneracy of the $|s\rangle$ and the $|t_-\rangle$ states when $B = J/2\mu_B$ will in practice be lifted through interactions which have been neglected in the present model, such as dipolar interactions or interactions between pairs or between spins and nuclei.

One often meets with spin-spin interactions of the $\mathcal{H}_1$-type in the case of fermions. They originate from orbital (Coulomb) interactions and the Pauli exclusion principle and they are called "exchange interactions" (§ 10.1.4). For electrons in atoms, molecules, or magnetic solids J is typically of the order of electron Volts, which makes fields of the order of $J/2\mu_B$ inaccessible in a laboratory since $(1 \, \text{eV})/2\mu_B \simeq 10^4$ T. Nevertheless there exist systems where the two electrons are sufficiently distant from one another that J is several orders of magnitude smaller, for instance, organic biradicals such as $(2,2',6,6'$-tetramethyl-piperidine-4-ol-1-oxy)-carbonate for which $J \simeq 10^{-7}$ eV, or two donors in a semi-conductor such as two phosphor impurities at a distance of 100 Å in silicon for which $J \simeq 10^{-4}$ eV. Such systems are well described by the model used here, but their use for obtaining low temperatures is purely hypothetical.

When $J < 0$, the ground state of the model considered here is threefold degenerate: the entropy does not tend to 0 in the low temperature limit, but to $Nk \ln 3$. The third law of thermodynamics is not satisfied. In this case also interactions neglected in the present model may lift this degeneracy and lead to a decrease in S at very low temperatures.

16.2 Elasticity of a Polymer Chain

We shall try to explain the elastic properties of materials which on a micro-scopic scale consist of polymer macromolecules, such as rubber and natural or synthetic textile fibres. To do this we shall study the elasticity of a single such molecule which we describe by a simplified model.

We give a schematic picture of a polymer molecule in the form of a long chain of N identical monomers (see Fig.16.1). These monomers form the links in the chain and have a fixed length a. We take the origin at one end of the chain and assume that the n-th link lies between the points r_{n-1} and r_n; its state is characterised by its direction, $\omega_n \equiv (r_n - r_{n-1})/a$. The relative orientation of two successive links can vary; to simplify the discussion we assume that the links can be treated like classical objects and that their orientation is completely random. All possible directions of ω_m have thus the same probability. To calculate a sum over the configurations we must integrate over the ω_n with a weight $d^2\omega_n$.

At the end $r_N = r$ of the chain we exert a force f in the direction of the z-axis. If we now include in the definition of the system studied a weight exerting the force f, as in Exerc.5a, we get the Hamiltonian $\mathcal{H} = -(f \cdot r)$, where we have neglected both the kinetic energy of the chain and the potential energy corresponding to the interaction between the links.

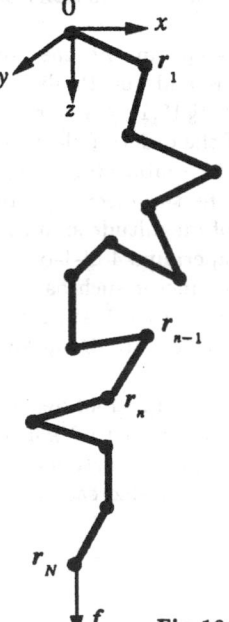

Fig. 16.1. Model of a polymer molecule

1. Calculate the canonical partition function. To simplify the notation put $x = \beta f a$.

2. If L is the average length between the ends of the chain, which is subject to a tension f and is in thermal equilibrium, find the equation of state for the chain, that is, find L as function of f and T. Sketch the isotherms $L(f)$ for fixed T.

3. What is the behaviour of the elasticity coefficient $\partial L/\partial f$ for small f ? And for large f ? How does the elasticity coefficient change with temperature in the linear region where L is proportional to f ?

4. What is the behaviour of the coefficient of linear expansion $\alpha = L^{-1}(\partial L/\partial T)$ for a given value of the tension f ? What important difference can you note between the polymer and ordinary solids?

5. Either by imagining what the form of the expression for the entropy will be, or by evaluating it explicitly, find out what happens if one releases adiabatically the tension exerted on the polymer. Assume that the chain does not exchange any heat with its surroundings during this shift in equilibrium.

6. Sketch how the specific heat at constant tension changes with temperature. What is the specific heat at constant length?

7. Evaluate the mean square length $\sqrt{\langle r^2 \rangle}$ of the polymer molecule when there is no tension.

16.3 Crystal Surfaces

In this problem we shall study some geometric characteristics of the faces of a crystal in equilibrium with its vapour. To do this we describe the crystal as a stack of N cubes with edge length a which are arranged on a simple lattice. Each cube represents an atom of mass m and zero spin; we neglect its internal structure. The centres of the cubes can occupy *sites* with coordinates $x = (q + \frac{1}{2})a$, $y = (q' + \frac{1}{2})a$, $z = (q'' + \frac{1}{2})a$, where $q, q', q'' \in \mathbb{Z}$. We take the energy of each isolated atom at rest to be zero. In particular, in the vapour, the energy reduces to the sum of the kinetic energies $p^2/2m$ of the atoms. In the crystalline phase we neglect the motion of the atoms, but take their interactions into account. The atoms repel one another at very short distances and this repulsion is simply represented in the model by the impenetrability of the cubes which represent the atoms. At larger distances, the atoms attract one another and this produces the cohesion of the crystal. In the present model we shall represent this attraction by a potential energy $-V$ which we associate with *each pair of neighbouring cubes* which have a face in common. It will be convenient in the evaluation of the energy of a configuration, characterized by having each site either occupied or not by a cube, to associate the energy $-V$ *with each face* which separates two occupied sites; remember that such a face is common to two cubes. We neglect all other interactions between two cubes, for the cases when they touch just along an edge, at a vertex, or not at all.

In the *numerical applications* we take the atomic mass to be 100 g mol^{-1}, the cell size $a = 2.5$ Å, and the potential energy $V = 0.2$ eV. The fundamental constants are given at the end of this book. We are only interested in orders of magnitude.

1. Bulk Properties of the Solid

a. Show that the cohesive energy per unit volume, ϱ, defined as the ratio of the ground state energy of a crystal to its volume Ω when $\Omega \to \infty$, equals $-3Va^{-3}$.

b. Use this result to find the chemical potential μ_0 of a crystal at zero temperature (and zero pressure).

c. At non-zero temperatures the crystal, still assumed to be infinite, contains a number of vacancies, that is, unoccupied sites. We shall assume that there are so few vacancies that they practically never touch one another. What is the energy needed to create a vacancy, assuming that the cube we remove is just annihilated? We consider a fixed volume $\Omega = \mathcal{N}a^3$ which is cut from an infinite crystal in our imagination – to avoid boundary effects – and which contains a large number of sites, $\mathcal{N}$; each of these can become a vacancy. Let ℓ be the number of vacancies and $N = \mathcal{N} - \ell$ be the number of the atoms which together occupy the $\mathcal{N}$ sites. Give the energy of a configuration as function of ℓ and N.

d. Assuming the vacancy density to be low, use the above result to calculate the grand partition function of the region Ω as function of β and $\alpha = \beta\mu$ where μ is the chemical potential of the atoms. Find an expression for the pressure $\mathcal{P}_s$ as function of T and μ.

e. Now write down the probability for a vacancy to exist at a given site in equilibrium at a temperature T and a pressure $\mathcal{P}_s$. How does it vary with T and $\mathcal{P}_s$?

f. Justify the hypothesis made sub d by giving a numerical upper bound for this probability at $T = 300$ K.

In what follows we shall systematically neglect the possibility for the creation of vacancies in the bulk of the crystal.

2. Saturated Vapour

We assume that the crystal is surrounded by its vapour and that it can exchange atoms and energy with the vapour so that an equilibrium is reached. The system is controlled by fixing the pressure $\mathcal{P}$ and the temperature T of the vapour which is assumed to be a monatomic perfect gas.

a. Give an expression for the chemical potential μ of the vapour as function of $\mathcal{P}$ and T. Assuming that the chemical potential of the crystal remains fixed at its value μ_0 when the temperature and the pressure increase, and that the saturated vapour is in equilibrium with the crystal, find an expression for the vapour pressure $\mathcal{P}_0(T)$. This is the saturated vapour pressure.

b. Sketch the solid-vapour phase diagram in the $T, \mathcal{P}$-plane. Evaluate the vapour presssure at room temperature, $\mathcal{P}_0(300 \text{ K})$.

c. Give a justification for the assumption made sub a that the chemical potential may be replaced by μ_0.

3. Smooth Crystal Face

We consider a very large crystal with one of its faces in the plane $z = 0$. The solid is in the $z < 0$ region and the vapour in the $z > 0$ region. At zero temperature the faces are perfectly smooth so that the energy is a minimum.

In what follows we reckon all energies from the energy of this configuration in which all the sites with $z < 0$ are occupied and all the sites with $z > 0$ are empty. At a finite temperature T exchange of atoms and of energy with the vapour may create irregularities at the crystal surface. In this section we restrict ourselves to the two simplest kinds of surface defects: a *surface vacancy* represented by a missing cube with its centre in the $z = -\frac{1}{2}a$-plane, and an *adsorbed atom* represented by an excess cube with its centre in the $z = \frac{1}{2}a$-plane.

a. Give the energies for these two kinds of defects, reckoned from the energy of a perfectly smooth face as zero.

b. The crystal is in grand-canonical equilibrium with its vapour at a pressure $\mathcal{P}_0(T)$. Let p_0, p_{-1}, and p_1 be, respectively, the probabilities that at a given point on the surface there is no defect, there is a surface vacancy, or there is an adsorbed atom. Find the ratios p_{-1}/p_0 and p_1/p_0 and show that p_{-1} and p_1 are equal, assuming that there are sufficiently few such defects that they are almost never neighbours. Check this assumption by calculating p_{-1} and p_1 for $T = 300$ K.

c. At a fixed temperature T we change the vapour pressure to a value $\mathcal{P}$ which is different from (either larger or smaller than) $\mathcal{P}_0(T)$. The crystal surface again comes to equilibrium with the vapour, which entails a change in its chemical potential. Find the new probabilities p_{-1} and p_1 as functions of the degree of supersaturation $\lambda \equiv \mathcal{P}/\mathcal{P}_0(T)$ and give a numerical value for p_1 for $\lambda = 10$.

d. What can one infer about the sublimation, that is, the evaporation, of the crystal and about the condensation of its vapour?

4. Growth of a Step on a Face

Adsorption of isolated atoms on a smooth surface thus cannot explain why a crystal will grow, if surrounded by its supersaturated vapour. The theory of crystal growth by Burton, Cabrera, and Frank[1] starts by assuming the presence of a *monatomic layer, bounded by a step* of height a on the face of

---

[1] Philos. Trans. Roy. Soc. **A243** (1950) 299.

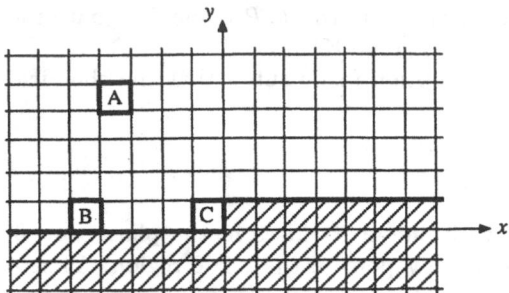

Fig. 16.2. A step with a kink

the crystal. We indicate the cubes which reach the $z = a$-plane by hatching while leaving the bare, smooth face in the $z = 0$-plane white (Fig.16.2)

a. Consider a step consisting of two straight sections ($x < 0$, $y = 0$ and $x > 0$, $y = a$), separated by a simple *discontinuity* or *kink* ($\Delta y = a$) at $x = 0$. Let A, B, C be sites with $z = \frac{1}{2}a$ which are, respectively, far from the step, at the step boundary, and at the corner of the kink (Fig.16.2). We assume that the surface is in contact with the supersaturated vapour which has a temperature $T = 300$ K and a pressure $\mathcal{P} = \lambda\mathcal{P}_0(T)$ where we take $\lambda = 10$. Neglect the formation of surface vacancies. Let p_A, p_B, p_C be, respectively, the probabilities of the configurations with an extra atom at A, at B, or at C (p_A is identical with the probability p_1 of question 3c, if $\lambda \gg 1$). Write down the ratios $p_A/(1 - p_A)$, $p_B/(1 - p_B)$, and $p_C/(1 - p_C)$, and use those expressions to find the numerical values of p_A, p_B, and p_C.

b. In order to find out whether there are many kinks or not along a step, we first estimate the energy of a single kink. To do this, we must compare two configurations with the same number of atoms, but differing in the number of kinks (Fig.16.3): the first configuration (1) contains two opposite straight steps; while in the second configuration (2) each of the steps has a discontinuity $\Delta y = +1$ at $x = 0$. Find the energy difference between these two configurations. Hence find the energy to be associated with a single kink.

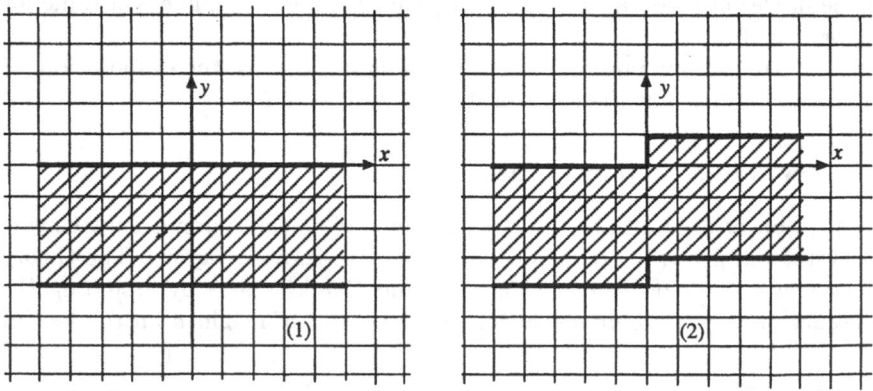

Fig. 16.3. Evaluating the energy of a kink

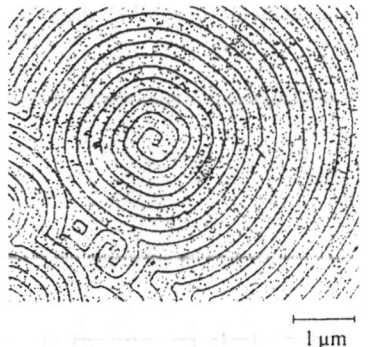

Fig. 16.4. A face of an NaCl crystal

1 μm

c. Consider a step parallel to the x-axis on a face in equilibrium with the saturated vapour $\mathcal{P}_0(T)$. To simplify the calculations we assume that this step contains only simple discontinuities ($\Delta y = +1$ or $\Delta y = -1$). We also assume that these kinks lie far from one another and that they may be treated by the methods of statistical physics as independent objects with an energy equal to the value found sub b. Give a numerical estimate for the probability p_D that there is a discontinuity, with either $\Delta y = +1$ or $\Delta y = -1$, at a given point of the step.

d. An atom coming from the vapour has little chance to hit just the edge of a step when it collides with a crystal face, and even less chance to hit the corner of a kink. On the other hand, it is natural to assume that the atoms are very mobile on a crystal surface and that they move without changing their energy. Use the values, obtained above, for p_A, p_B, p_C, and p_D and the fact that p_C, in particular, is very large, to imagine an efficient mechanism for crystal growth from the supersaturated vapour; assume that the degree of supersaturation is $\lambda = 10$.

The photograph shown in Fig.16.4 was taken by an electron microscope of a crystal face of NaCl. Can you interpret it?

5. Shape of a Step

We want to study the average shape of a step on the $z = 0$-face in the general case where this step is *not parallel* to the x-axis and where the degree of saturation $\lambda = \mathcal{P}/\mathcal{P}_0(T)$ *is not equal to* 1. In the x, y-plane the step goes from the point 0, 0 to the point Ka, Ma (Fig.16.5). To fix its ends, we assume that all sites in the column $x = -\frac{1}{2}a$, $y < 0$, $z = \frac{1}{2}a$ and all sites of the column $x = (K + \frac{1}{2})a$, $y < Ma$, $z = \frac{1}{2}a$ are occupied, like all sites far behind the step $0 < x < KA$, $y < -La$, $z = \frac{1}{2}a$ (these regions are indicated by heavy hatching) The, fixed, number L is chosen sufficiently large so that the step practically never approaches a y-coordinate less than $-La$. We characterize the step by a function $y(x)$ which is defined in discrete points. We thus neglect all configurations containing lateral protrusions (1), depressions behind the step (2), or higher islands in front of it (3). This

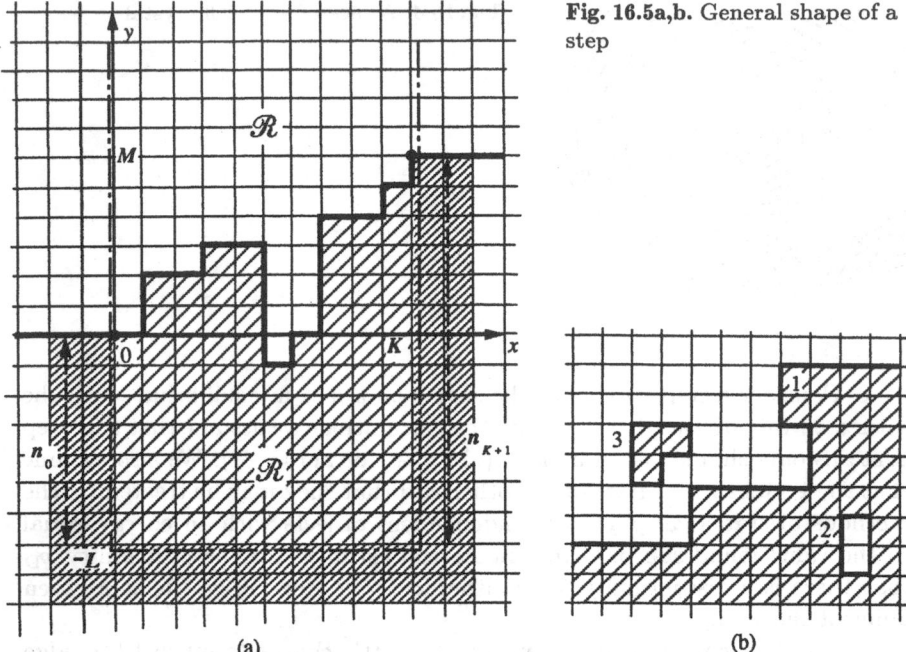

Fig. 16.5a,b. General shape of a step

(a)

(b)

is justified at low temperatures and for a step which does not have a large average slope $(M < K)$ in which case such accidental occurrences are unlikely. We thus characterize each configuration by a set of K integers, n_k, $1 \leq k \leq K$, such that each of them is the *number of cubes stacked in the column* $x = (k - \frac{1}{2})a$ between $y = -La$ and $y = (-L + n_k)a$. Equivalently, this configuration is characterized by the set $\{m_k\}$ of $K + 1$ integers (Fig.16.6)

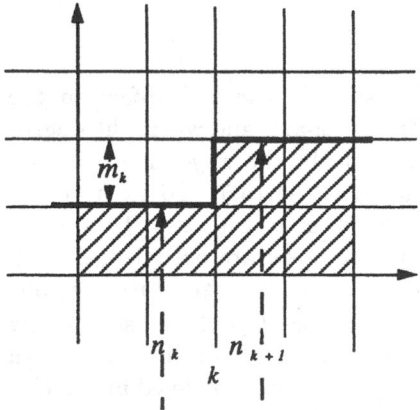

Fig. 16.6. Coordinates labelling a step edge

$$m_k = n_{k+1} - n_k, \qquad 0 \le k \le K, \qquad m_k = 0, \pm 1, \pm 2, \cdots ;$$

each of these integers represents the *algebraic value of the discontinuity* at the point of the step with abscissa $x = ka$ (at the ends we have $n_0 = L$ and $n_{K+1} = L + M$).

We shall now study the properties of the system $\mathcal{R}$ which is the region defined by $0 \le x \le Ka$, $y \ge -La$, $0 \le z \le a$, and bounded in Fig.16.5 by the dash-dot line. This system $\mathcal{R}$ does contain not only the atoms situated in the region, but also the faces in the $x = 0$, $x = Ka$, $y = -La$, and $z = 0$ planes; one assigns to these faces an energy $-V$ whenever they separate two atoms. The energy is reckoned from the energy of the region $\mathcal{R}$ without any atoms as zero.

a. Prove that for a given configuration $\{m_k\}$ the number N of atoms situated within the region $\mathcal{R}$ equals

$$N = KL + \sum_{k=0}^{K} (K - k) m_k.$$

b. Prove that for the configuration $\{m_k\}$ the energy associated with all the faces inside the region $\mathcal{R}$ equals

$$E = -3NV + \frac{V}{2} \sum_{k=0}^{K} |m_k| - V \left(L + \frac{1}{2} M \right),$$

where $\sum_{k=0}^{K} |m_k| a$ is the total length of the discontinuities.

This form of the energy shows the advantage of using the $K + 1$ variables $\{m_k\}$ rather than the K independent variables $\{n_k\}$. Nevertheless, the evaluation of the grand partition function for the region $\mathcal{R}$ is made more difficult because of the existence of a constraint,

$$\sum_{k=0}^{K} m_k = M,$$

which expresses that the end-point of the step is Ma for $x = Ka$. (In principle, there are other constraints following from the requirement that $n_k \ge 0$, but, provided L is sufficiently large, they do not come into play and we may let m_k vary from $-\infty$ to $+\infty$.) In order that we can sum freely over all the $K + 1$ numbers $m_k (0 \le k \le K)$, we impose the constraint on their total sum only *on average*, in the form

$$\sum_{k=0}^{K} \langle m_k \rangle = \langle M \rangle,$$

and we introduce a Lagrangian multiplier γ to take this constraint into account. The corresponding Boltzmann-Gibbs distribution thus gives us for each configuration $\{m_k\}$ a probability which is proportional to $e^{\alpha N - \beta E - \gamma M}$.

c. Prove that the partition function $Z(\alpha, \beta, \gamma)$ associated with the above-defined ensemble equals, apart from a multiplicative factor, the product of $K+1$ functions $Z_k(\alpha, \beta, \gamma)$, each of which is associated with a single abscissa $x = ka$ where a discontinuity may appear. Give an expression for Z_k. You may use the identity $(\xi > |\eta|)$

$$\sum_{m=-\infty}^{+\infty} e^{-\xi|m| - \eta m} = \frac{\sinh \xi}{\cosh \xi - \cosh \eta}.$$

d. Find $\langle m_k \rangle$, which is the average gradient dy/dx of the step in the x, y-plane at the point with abscissa $x = ka$, as function of α, β, γ.

e. Discuss the concavity of the step in its dependence on the degree of saturation λ.

6. Condensation Islands

Consider an island composed of a condensed monatomic layer on top of a face which is in equilibrium with a slightly supersaturated vapour ($\lambda \gtrsim 1$). This island is bounded by a step which is in the shape of a closed curve. We shall study the geometry of such islands (Fig.16.7).

a. Use the result of question 5d to determine γ and K as functions of λ and T such that the step has an average slope of 45° at the point $x = y = 0$ and an average zero slope at the point $x = K$, as shown in Fig.16.7.

b. Find $\langle M \rangle$, assuming that on the scale considered the edge of the step behaves as a continuum so that the summations over k reduce to integrals. This can be done starting from question 6a. It can also be done without

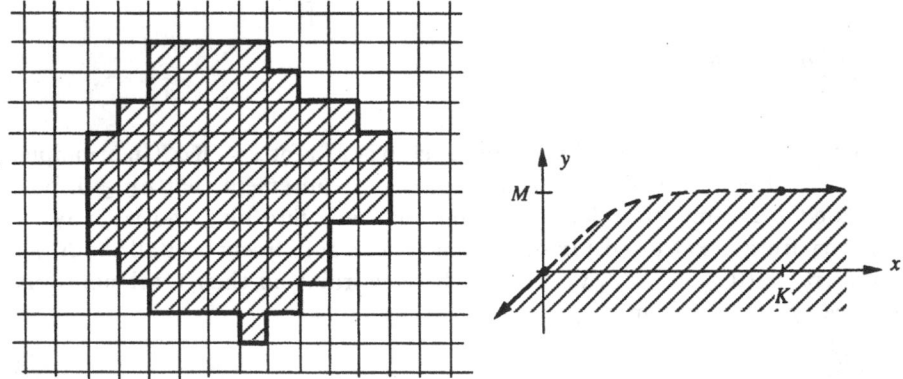

Fig. 16.7. An island on a crystal face. In equilibrium its boundary consists of eight segments, each of which is characterized by the parameters K and M of Fig. 16.5a

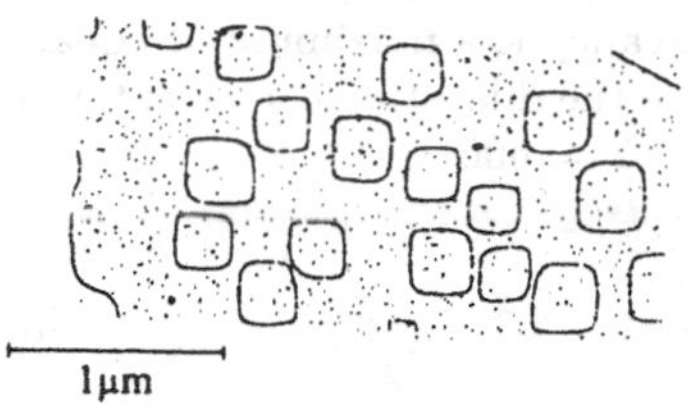

Fig. 16.8. Surface of an undersaturated crystal

calculations, if one first proves that the *thermodynamic potential* $\ln Z_k$ *is directly related to the shape* $y(x)$ *of the step.*

c. Find the shape and the average size of an island for $T = 300$ K and $\lambda = 1.02$. How do these geometric characteristics vary with the excess pressure λ? and with the temperature?

Discuss briefly the microphotograph of a NaCl crystal face shown in Fig. 16.8.

Solution

1a. We get the ground state for a compact packing. In a large volume Ω there are then $\mathcal{N} = \Omega/a^3$ cubes which are occupied and $3\mathcal{N}$ faces – six faces per cube, each of which is common to two cubes – so that the (negative) cohesive energy per unit volume is $\varrho = -3Va^{-3}$ (numerically, $\varrho = -6$ kJ cm^{-3}).

b. As $T = \mathcal{P} = 0$, the equation $dU = TdS - \mathcal{P}d\Omega + \mu dN$ reduces to $dU = \mu dN$. We find $U = -3VN$, and hence $\mu_0 = -3V$.

c. It costs an energy $6V$ to suppress a cube, as six faces common to two neighbouring atoms disappear from the energy sum. A configuration with ℓ vacancies thus has an energy $-3\mathcal{N}V + 6\ell V = -3NV + 3\ell V$.

d. From the previous answer we find that one must assign an energy $+3V$ to a vacant site and an energy $-3V$ to an occupied one. The $\mathcal{N}$ sites may be treated as being independent of one another, whence we find

$$Z_G(\alpha, \beta) = \left(e^{-3\beta V} + e^{3\beta V + \alpha}\right)^{\mathcal{N}}.$$

Hence we get for the pressure

$$\mathcal{P}_s = \frac{kT}{\Omega} \ln Z_G = \frac{kT}{a^3} \ln \left[e^{-3V/kT} + e^{(3V+\mu)/kT}\right]. \tag{1}$$

e. From the grand canonical Boltzmann-Gibbs distribution for a site we get for the probability that this site is unoccupied the expression $e^{-3\beta V}/\left[e^{-3\beta V} + e^{3\beta V + \alpha}\right]$. Using (1) to eliminate the chemical potential we find for this probability for a vacancy

$$\exp\left[-\frac{3V + \mathcal{P}_s a^3}{kT}\right].$$

This probability increases with increasing temperature, or increasing energy, and decreases with increasing pressure, or decreasing volume.

f. At zero pressure when the probability for a vacancy is a maximum, it equals $e^{-3V/kT} = 10^{-10}$ at 300 K. This quantity, and hence the dilatation coefficient and the specific heat, is negligible. It is still only 10^{-5} at $T = 600$ K.

2a. Chapter 7 gives

$$\mathcal{P} = -\frac{A}{\Omega} = kT\, e^{\mu/kT} \left(\frac{2\pi mkT}{h^2}\right)^{3/2}, \tag{2}$$

or

$$\mu = kT \ln\left[\mathcal{P}(kT)^{-5/2}\left(\frac{2\pi\hbar^2}{m}\right)^{3/2}\right].$$

At equilibrium, the chemical potential of the vapour equals $\mu_0 = -3V$ and hence

$$\mathcal{P}_0(T) = (kT)^{5/2}\, e^{-3V/kT}\left(\frac{m}{2\pi\hbar^2}\right)^{3/2}.$$

b. The expression for $\mathcal{P}_0(T)$ gives both the diagram shown in Fig.16.9 and the value $\mathcal{P}_0(300\text{ K}) = 300$ Pa. The curve rises very rapidly and already at 400 K one finds 2×10^5 Pa $= 2$ atm.

c. The low density of the vacancies entails, on the one hand, that the entropy per unit volume remains practically zero and, on the other hand, that the energy and the volume are practically proportional to the number of atoms. We thus have for the whole crystal

$$dU \simeq -3V\, dN = T\, dS + \mu\, dN - \mathcal{P}\, d\Omega \simeq \left(\mu - \mathcal{P}a^3\right)dN,$$

so that $\mu \simeq \mu_0 + \mathcal{P}a^3$. Microscopically, by inverting (1) one finds

$$\mu = -3V + \mathcal{P}a^3 + kT \ln\left(1 - e^{-(3V+\mathcal{P}a^3)/kT}\right).$$

If we insert this expression into (2) and put the chemical potentials and the pressures in the two phases equal to one another, we get the saturation curve $\mathcal{P}_0(T)$. However, the last term (which even at 600 K only equals 5×10^{-7} eV) as well as the term $\mathcal{P}a^3$ (which equals 10^{-5} eV at atmospheric pressure) are negligible compared to $\mu_0 = -0.6$ eV along this curve, so that it can be obtained simply by replacing μ by μ_0 in (2).

3a. As compared to a smooth surface, if we remove a cube from it, we get rid of 5 interaction faces, each with an energy $-V$, so that a surface vacancy has an energy $5V$. Adding a cube means adding one interaction face, so that an adsorbed atom has an energy $-V$.

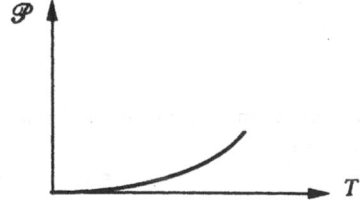

Fig. 16.9. Crystal-vapour phase diagram (the crystal phase is above and the vapour phase below the curve)

Note. The total energy of the crystal does not just consist of the volume terms calculated sub 1a and the above contribution from surface defects. In fact, consider as reference system a parallelepipedal crystal with edgelengths $L_x a$, $L_y a$, $L_z a$ and perfectly smooth faces. In the x-direction it contains L_x-1 lattice planes which each contribute $-VL_yL_z$ to the cohesive energy. As the crystal volume is $\Omega = L_xL_yL_z a^3$ and the area of the faces $S = 2(L_yL_z + L_zL_x + L_xL_y)$ the reference energy is not equal to $\varrho\Omega$, but to $\varrho\Omega + \sigma S$, where $\sigma = V/2a^2 \simeq 0.25$ J m^{-2}. The surface contribution σS increases with increasing surface area and should be a minimum for equilibrium at zero temperature. This is the reason why crystal surfaces must then be perfectly smooth.

b. According to the grand canonical Boltzmann-Gibbs equilibrium distribution the ratio of the probabilities p and p' of two configurations with energies ε and $\varepsilon + \Delta\varepsilon$ and atom numbers N and $N + \Delta N$, respectively, equals

$$\frac{p'}{p} = e^{-\beta\Delta\varepsilon+\alpha\Delta N}. \tag{3}$$

For the two kinds of defects discussed we have $\Delta\varepsilon + \mu\Delta N = 2V$ so that $p_1/p_0 = p_{-1}/p_0 = e^{-2V/kT}$. This is a small number, equal to 2×10^{-7} for $T = 300$ K, so that $p_0 = 1 - p_1 - p_{-1} \simeq 1$ and the required probabilities $p_1 = p_{-1}$ equal 2×10^{-7}. Moreover, this justifies the hypothesis which we have made implicitly that the sites may be treated independently of one another: there are practically never two defects which are side by side.

c. According to (2) the degree of saturation equals

$$\lambda \equiv \frac{\mathcal{P}}{\mathcal{P}_0(T)} = e^{(\mu-\mu_0)/kT} = e^{\alpha-\alpha_0} = e^{\alpha}e^{3V/kT}.$$

Substituting this new value of α into (3) and using the fact that $p_0 \simeq 1$, we get the probabilities $p_{-1} = e^{-2V/kT}/\lambda$ for the surface vacancies and $p_1 = \lambda e^{-2V/kT} = 2 \times 10^{-6}$ for adsorbed atoms, respectively.

d. If the vapour is supersaturated $(\lambda > 1, \mathcal{P} > \mathcal{P}_0(T))$, there are λ^2 more adsorbed atoms than surface vacancies. This excess of atoms indicates a tendency for the vapour to condense on the smooth crystal surface. On the other hand, if there is undersaturation $(\mathcal{P} < \mathcal{P}_0(T))$ there is a tendency for sublimation. Nevertheless, the overabundance or shortage of atoms *remains small* in the present theory, which thus does not yet enable us to understand why the equilibrium with under- or supersaturated vapour is unstable. In fact, there are configurations, to be considered later, which are more complicated than the surface point defects; they have large probabilities at equilibrium but are more difficult to produce. The equilibrium considered here, smooth surfaces with point defects, is thus *metastable*: only the crystal alone is truly stable when $\lambda > 1$ and only the vapour when $\lambda < 1$.

4a. The binding energies are, respectively, $-V$, $-2V$, and $-3V$ in A, B, and C, whereas the chemical potential is $-3V + kT\ln\lambda$. We thus find $p_A/(1 - p_A) = \lambda e^{-2\beta V}$; $p_B/(1 - p_B) = \lambda e^{-\beta V}$; $p_C/(1 - p_C) = \lambda$. Hence $p_A \simeq \lambda e^{-2V/kT} \simeq 2 \times 10^{-6}$, $p_B \simeq \lambda e^{-V/kT} \simeq 4 \times 10^{-3}$, and $p_C = \lambda/(1+\lambda) \simeq 0.9$.

b. As far as the number of faces in contact is concerned, configuration (2) differs from (1) only for the faces in the $x = 0$ plane: there is one less. The energy of (2) is thus higher than that of (1) by an amount V and we should attribute an energy $\frac{1}{2}V$ to a kink.

c. For each kind of discontinuity the probability is proportional to $e^{-\beta V/2}$ with a normalization factor $Z = 1 + 2e^{-\beta V/2} \simeq 1$. Hence, we find $p_D = 2e^{-V/2kT} = 4\,\%$.

d. For a vapour which is exactly saturated ($\lambda = 1$) we have $p_C = \frac{1}{2}$. A C site in the corner of a kink thus has an equal chance of being occupied as of being unoccupied; this means that there is no preference for the position of the kink, as one would have expected. However, if the vapour is supersaturated, a site in the corner of a kink has λ times more chance of being occupied than being unoccupied when there is equilibrium with the vapour. The large value $p_C = 0.9$ indicates that a step with a kink is unstable in the presence of supersaturated vapour: the kink has a tendency to move (leftward in Fig.16.2) so as to make the step progress. However, the atoms in the vapour must still reach C, which takes several stages and a certain amount of time, so that the step is actually metastable. In fact, the atoms from the vapour can only become part of the crystal after they have collided with the face; and there is practically no chance that the collisions will occur at a C, or even a B point, at the edge of the step. On the other hand, they easily end up in A sites. Thermal motion will see them depart again, but in the resulting metastable equilibrium there remains a proportion $p_A = 2 \times 10^{-6}$ per surface site. Two-dimensional diffusion of these atoms sitting on the smooth face then allows them to meet with a step, most often in a B point, more rarely in a C point (in 4 % of the cases, as $p_D = 4\,\%$). We thus gradually reach an average value $p_B = 4 \times 10^{-3}$. One-dimensional diffusion along the step finally lets the atoms arrive at C points where they have 9 chances out of 10 to become fixed. The efficiency of this process is based upon the fairly large number of kinks: 4 % of the sites along the step. The populations in A or B points would tend to get smaller, but new exchanges with the vapour and between one another maintain their populations at 2×10^{-6} and 4×10^{-3} while the step grows. An oblique step contains more kinks and thus grows more rapidly. It thus has a tendency to align itself parallel to an x- or y-axis. This is the microscopic dynamical process which explains why macroscopic crystals have simple geometric shapes.

Steps may be formed spontaneously on a face as the boundaries of condensation islands – which grow and can join onto others. However, there is a mechanism which *automatically* creates steps. In fact, crystals contain *screw dislocations*; these are linear defects along which the lattice is distorted. In the centre of the photograph of Fig.16.4 there is such a dislocation which is at right angles to the face; if we try to draw around it a square counterclockwise following the atoms in the lattice, we do not return at the starting point, as we would do in the case of perfect stacking, but we arrive in a site which is just above or just below it ($\Delta z = \pm 1$). The existence of this dislocation implies the existence on the surface of a step which starts from the end of the dislocation. In a supersaturated situation this step advances with a speed which is practically constant and at right angles to its local direction. This entails that the step will wind itself up and produce the spiral shape which is shown. Note that the spiral is macroscopic on the atomic scale: we count 6 windings over 1 µm, which means that there are about 700 lattice cells between two successive steps in the photograph. We see also two other screw dislocations with the same sign of Δz. We see, moreover, that the spiral flattens in the x- and y-directions which are oblique in the photograph; this comes about because the rate of growth of the step is larger when it is not along those directions, as we noted above. We pass between the centre and the edge of the photograph through about twenty steps, which means that we go down by about 50 Å.

5a. The number of atoms in $\mathcal{R}$ is

$$N = n_1 + n_2 + \cdots n_K$$

$$= (n_0 + m_0) + (n_0 + m_0 + m_1) + (n_0 + m_0 + m_1 + m_2)$$

$$+ \cdots + (n_0 + \cdots + m_{K-1})$$

$$= KL + \sum_{k=0}^{K}(K - k)m_k.$$

b. There are in the region $\mathcal{R}$, in the directions parallel to the $z=0$ and $y=0$ planes, as many faces with interactions $-V$ as atoms present; this leads to a contribution $-2NV$. In each vertical plane $x = k$ ($0 \le k \le K$) which separates n_k and n_{k+1} cubes, the number of faces contributing $-V$ is the smallest of the two numbers n_k and n_{k+1}, that is,

$$\tfrac{1}{2}(n_k + n_{k+1}) - \tfrac{1}{2}|n_{k+1} - n_k| = \tfrac{1}{2}(n_k + n_{k+1}) - \tfrac{1}{2}|m_k|.$$

Summing over k we get

$$\frac{1}{2}\sum_{k=0}^{K}(n_k + n_{k+1}) = \frac{1}{2}n_0 + N + \frac{1}{2}n_{K+1} = N + L + \frac{1}{2}M,$$

which gives us the total energy of the configuration.

c. Using the expressions for N, E, and M and summing over all configurations $\{m_k\}$ we have

$$Z(\alpha, \beta, \gamma) \equiv \mathrm{Tr}\, e^{\alpha N - \beta E - \gamma M}$$

$$= \sum_{\{m_k\}} \exp\left[(\alpha + 3\beta V)\left(KL + \sum_{k=0}^{K}(K - k)m_k \right) - \frac{1}{2}\beta V \sum_{k=0}^{K}|m_k| \right.$$

$$\left. + \beta V L + \left(\frac{1}{2}\beta V - \gamma\right)\sum_{k=0}^{K}m_k \right].$$

Apart from the factor

$$\exp\left[\alpha KL + \beta V L(3K + 1)\right],$$

this expression is the product of the partition functions

$$Z_k(\alpha, \beta, \gamma) \equiv \sum_{m=-\infty}^{+\infty} \exp\left\{ \left[(\alpha + 3\beta V)(K - k) + \left(\frac{1}{2}\beta V - \gamma\right)\right]m - \frac{1}{2}\beta V|m| \right\},$$

which are associated with each abscissa ($0 \le k \le K$) where the step may have a discontinuity.

Each function Z_k has the form

$$\sum_{m=-\infty}^{+\infty} e^{-\xi|m|-\eta m} = \sum_{m=0}^{+\infty}\left[e^{-(\xi+\eta)m} + e^{-(\xi-\eta)m}\right] - 1$$

$$= \frac{1}{1 - e^{-\xi-\eta}} + \frac{1}{1 - e^{-\xi+\eta}} - 1$$

$$= \frac{1 - 2e^{-2\xi}}{1 + e^{-2\xi} - e^{-\xi+\eta} - e^{-\xi-\eta}} = \frac{\sinh\xi}{\cosh\xi - \cosh\eta},$$

so that

$$\ln Z(\alpha,\beta,\gamma) = \alpha KL + \beta VL(3K+1) + (K+1)\ln\sinh\beta V/2$$

$$- \sum_{k=0}^{K} \ln\left[\cosh(\beta V/2) - \cosh\eta_k\right],$$

$$\eta_k \equiv \gamma - \beta V/2 - (\alpha + 3\beta V)(K-k) = \gamma - \beta V/2 - (K-k)\ln\lambda.$$

d. We get $\langle m_k \rangle$, according to the definition of Z_k, through taking a derivative:

$$\langle m_k \rangle = -\frac{\partial}{\partial\gamma}\ln Z_k = \frac{\partial}{\partial\gamma}\ln\left[\cosh(\beta V/2) - \cosh\eta_k\right]$$

$$= -\frac{\sinh\eta_k}{\cosh(\beta V/2) - \cosh\eta_k}.$$

e. The slope $\langle m_k \rangle$ is constant when $\lambda = 1$ since η_k is then independent of k. Therefore, if the vapour is exactly saturated, the step connects its ends, assumed to be fixed, on average through a straight line.

The function $\langle m_k \rangle$ decreases with increasing η_k since its derivative is

$$\frac{d\langle m_k \rangle}{d\eta_k} = -\frac{\cosh(\beta V/2)\cosh\eta_k - 1}{\left[\cosh(\beta V/2) - \cosh\eta_k\right]^2}.$$

As η_k increases with k when $\lambda > 1$ and decreases when $\lambda < 1$, the step is convex $(d^2y/dx^2 < 0)$ when the vapour is supersaturated and concave $(d^2y/dx^2 > 0)$ when it is undersaturated. The straight-line steps are unstable when $\lambda \neq 1$. This direction of the concavity indicates the tendency of a step, the ends of which are fixed, to advance when $\lambda > 1$ and to retreat when $\lambda < 1$. The curvature increases as $|\ln\lambda|$.
Notes. The formalism used above assumes that $|\eta_k| < \beta V/2$, which restricts the possible values of γ to

$$(K-k)\ln\lambda < \gamma < \beta V + (K-k)\ln\lambda.$$

These inequalities must be satisfied for all $0 \leq k \leq K$, whence

$$|2\gamma - \beta V - K\ln\lambda| < \beta V - K|\ln\lambda|.$$

The factor γ must be adjusted such that M takes its desired value, but that is impossible if

$$K > \frac{V}{kT|\ln \lambda|}.$$

This restricts step lengths which are possible when the vapour is super- or under-saturated. In fact, as $|\ln \lambda|$ increases, the curvature of the metastable steps which we have just studied increases, so that they cannot be very long.

One should also note that the existence of accidents which have here been neglected – lateral protrusions of the step, depressions behind it, islands in front of it – is essential to explain why the step is not stable, but only metastable when $\lambda \neq 1$.

Finally, we have $\langle m_k \rangle = 0$ for a step parallel to the x-axis in equilibrium with its saturated vapour ($\lambda = 1$), and hence $\eta_k = 0$ and $\gamma = \beta V/2$. The corresponding Boltzmann-Gibbs distribution gives for the probability of a discontinuity $\Delta y = ma$

$$p_m = \frac{\cosh(\beta V/2) - 1}{\sinh(\beta V/2)} \exp\left(-\frac{1}{2}\beta V|m|\right).$$

Using the fact that $e^{-\beta V/2} \simeq 2 \times 10^{-2} \ll 1$ we find that the hypotheses used in 4c are justified and recover the value $p_D = 4\,\%$.

6a. We must have

$$1 = \langle m_0 \rangle = \frac{\sinh(\beta V/2 + K \ln \lambda - \gamma)}{\cosh(\beta V/2) - \cosh(\beta V/2 + K \ln \lambda - \gamma)}$$

and

$$0 = \langle m_k \rangle = \frac{\sinh(\beta V/2 - \gamma)}{\cosh(\beta V/2) - \cosh(\beta V/2 - \gamma)},$$

which implies that $\gamma = \beta V/2$ and that

$$1 = \frac{\sinh(K \ln \lambda)}{\cosh(\beta V/2) - \cosh(K \ln \lambda)},$$

or

$$K = \frac{\ln \cosh(\beta V/2)}{\ln \lambda}.$$

b. Treating k as a continuous variable we get

$$M = \sum_K \langle m_k \rangle \simeq -\int_0^K dk \, \frac{\sinh \eta_k}{\cosh(\beta V/2) - \cosh \eta_k},$$

where one must substitute for η_k its value

$$\eta_k = -\ln \cosh(\beta V/2) + k \ln \lambda.$$

Using as variable $t = \cosh \eta_k$ we can integrate straight away and the result is

$$M = \frac{1}{\ln \lambda} \int_1^X \frac{dt}{\cosh(\beta V/2) - t} = \frac{1}{\ln \lambda} \ln \frac{\cosh(\beta V/2) - 1}{\cosh(\beta V/2) - X},$$

where $X = \cosh\left[\ln\cosh(\beta V/2)\right] = \frac{1}{2}\left[\cosh(\beta V/2) + \{\cosh(\beta V/2)\}^{-1}\right]$ so that, finally,

$$M = \frac{1}{\ln\lambda}\ln\frac{2\cosh(\beta V/2)}{\cosh(\beta V/2) + 1}.$$

Note. We could have obtained this result without calculations. In fact, we note that $\ln Z_k$ depends on the variables α, β, γ, and k only through β and the combination η_k. A first consequence of this is that

$$\frac{\partial}{\partial\alpha}\ln Z_k = -(K-k)\frac{\partial}{\partial\gamma}\ln Z_k = (K-k)\langle m_k\rangle,$$

which, after summation over k, gives us again the expression for N from 5a. We obtain a more remarkable consequence when we consider k as a continuous variable, x/a. We then find

$$\frac{\partial}{\partial x}\ln Z_k = \frac{\ln\lambda}{a}\frac{\partial}{\partial\gamma}Z_k = -\frac{\ln\lambda}{a}\langle m_k\rangle = -\frac{\ln\lambda}{a}\frac{dy}{dx};$$

after determining the integration constant at $x = 0$, we obtain

$$y = -\frac{a}{\ln\lambda}\ln\frac{Z_k}{Z_0}, \qquad x = ka.$$

We can thus, when $\lambda \neq 1$, apart from additive and multiplying constants, interpret the logarithm of the *partition function* $Z_k(\alpha, \beta, \gamma)$ as *the equation of the edge of the step* in the $y(x)$-plane. Explicitly, we have

$$\frac{y}{a} = \frac{1}{\ln\lambda}\ln\frac{\cosh(\beta V/2) - \cosh\eta_k}{\cosh(\beta V/2) - \cosh\eta_0}, \tag{4}$$

which for $k = K$ gives again M. Finally, integrating the equation for $y(x)$ from $k{=}0$ to $k{=}K$ shows that $\ln Z(\alpha, \beta, \gamma)$ is *itself* directly related to the area behind the step, that is, to the *number of atoms*, N. We can find the thermodynamic interpretation of the multiplier γ through identifying the Legendre transform

$$-kT\left[\ln Z + \gamma M\right] = U - TS - \mu N \equiv A$$

with the grand potential, which is a function of T, μ, and M, so that

$$-kT\gamma = \left.\frac{\partial A}{\partial M}\right|_{T,\mu}$$

is the *deformation energy* of the edge of the step in equilibrium with a vapour with given T and $\mathcal{P}$, when M changes by unity for fixed K.

c. If we neglect $e^{-\beta V/2} \simeq 2\times 10^{-2}$ as compared to 1, we can replace K and M by

$$K = \frac{\beta V}{2\ln\lambda} - \frac{\ln 2}{\ln\lambda}, \qquad M = \frac{\ln 2}{\ln\lambda} \simeq 35,$$

whence we have

$$K + M = \frac{\beta V}{2 \ln \lambda} \simeq 200.$$

The $y(x)$-curve determined above has a slope of 1 near $x = 0$ and soon turns down in such a way as to reach a rather long almost horizontal part; when $x = Ka/2$ the slope

$$\langle m_k \rangle = \frac{\sinh(K \ln \lambda / 2)}{\cosh(\beta V/2) - \cosh(K \ln \lambda / 2)} \simeq 0.1$$

is already small, and

$$M - \frac{y}{a} \simeq \frac{1}{\ln \lambda} \ln \frac{\cosh(\beta V/2) - 1}{\cosh(\beta V/2) - \cosh(K \ln \lambda / 2)} \simeq 3.5$$

is small compared to $M \simeq 35$.

To describe an island we must join up 8 such step sections, derived from one another using the symmetry of a square (Fig.16.10). The island thus has the form of a square with rounded angles, with an edgelength $2(K + M)a \simeq 1000\,\text{Å} = 0.1\,\mu\text{m}$; this value is the same we found at the end of 5 as upper bound for K for a stable step – a limit reached for an edge bounding half of the island.

As the supersaturation increases, the size of the island, which we can also write as $Va/(\mu - \mu_0)$ decreases as $\mu - \mu_0 = kT \ln \lambda$. As the instability becomes larger and larger, the metastable islands become smaller and smaller. However, the shape of the islands, characterized by the ratio $K/M = \beta V/(2 \ln 2)-1$, does not change with pressure. In fact, Eq.(4) for the edge $y(x)$ depends on the coordinates and on λ only through the combinations $x \ln \lambda / a$ and $y \ln \lambda / a$. On the other hand, increasing the temperature with λ constant makes K/M decrease and also makes $K+M$ decrease so that the islands become more rounded and they shrink when the temperature rises. At very low temperatures ($\beta V \gg 1$) they are nearly square.

In the photograph (Fig.16.8) we observe the various characteristics of the islands which we have studied here: square shapes with rounded angles and practically constant dimensions. The edges of the squares indicate the x- and y-directions of the cubic lattice; the curved parts represent many discontinuities at a scale of $a \simeq 2.5\,\text{Å}$. (In actual fact, the photograph corresponds to $\lambda < 1$ and the closed curves are the boundaries of hollow dips in the face and not raised islands; the above theory gives the same shape for λ as for $1/\lambda$, with convex steps being replaced by concave steps.)

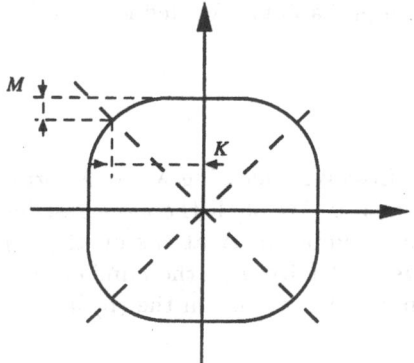

Fig. 16.10. Geometry of an island

16.4 Order in an Alloy

In this problem we consider order-disorder transitions in cubic alloys of the AB-type, such as CuZn.

To do this we consider a set of N points which constitute the sites of a cubic lattice; this lattice is designated by (a). We also consider another set of N points constituting the sites of another cubic lattice, (b). The two lattices are such that each site of the lattice (a) is the centre of a cube of the lattice (b) and *vice versa*. We see thus that each site of (a) is surrounded by 8 "nearest-neighbour" sites belonging to (b) and *vice versa*. We assume that the number N is sufficiently large that boundary effects due to the fact that the lattices are finite can be neglected. We also assume that a number N of A atoms and a number N of B atoms are situated on the lattice sites of the (a) and (b) lattices so that they form an AB alloy which has equal numbers of the two constituent atoms.

We could imagine that the alloy is perfectly ordered, that is, that the A atoms occupy the sites of one of the two lattices, say the (a) lattice, while the B atoms occupy the (b) lattice. We could also imagine a situation where the A and B atoms occupy the sites of the two lattices randomly: the alloy is then completely disordered. Between these two extreme cases there are intermediate, partially ordered, situations. We shall study the order of this alloy as a function of the temperature. We assume that the lattice is incompressible so that the pressure does not play a rôle.

We define a given configuration of the alloy, that is, a micro-state of the system considered, by placing on n well defined sites of the (a) lattice n A atoms and on $N-n$ well defined sites of the (b) lattice $N-n$ A atoms; the sites occupied by the B atoms are then also well defined. We now assume that the range of the interatomic forces is so short that the energy of each configuration depends solely on the binding energies of the nearest neighbours so that we only take those energies into account. All the A atoms, like all the B atoms, are indistinguishable from one another, and behave as structureless particles.

We consider the set $\{\lambda\}$ of the configurations corresponding to a given value of n ($0 \le n \le N$) and we denote by $\mathcal{D}_\lambda$ the statistical distribution such that all these configurations have the same probability. We define an order parameter λ through the relation

$$n = \frac{N}{2}(1 + \lambda),$$

where *a priori* λ lies between -1 and $+1$. However, changing λ to $-\lambda$ corresponds to interchanging the (a) and (b) lattices. We shall therefore restrict ourselves to values $\lambda \ge 0$. Let n_{Aa} be the number of A atoms on the (a) lattice, n_{Ab} the number of A atoms on the (b) lattice, n_{Ba} the number of B atoms on the (a) lattice, and n_{Bb} the number of B atoms on the (b) lattice; we see immediately that

$$n_{\mathrm{Aa}} = n_{\mathrm{Bb}} = n = \frac{N}{2}(1 + \lambda),$$

$$n_{\mathrm{Ab}} = n_{\mathrm{Ba}} = N - n = \frac{N}{2}(1 - \lambda).$$

We denote by N_{AA}, N_{BB}, and N_{AB}, respectively, the number of bonds between nearest neighbours of the AA, BB, and AB type for a given configuration.

1. Show that for a given value of λ between 0 and 1 the numbers N_{AA}, N_{BB}, and N_{AB} are not determined by just giving λ only. On the other hand, the averages $\overline{N}_{\mathrm{AA}}$, $\overline{N}_{\mathrm{BB}}$, and $\overline{N}_{\mathrm{AB}}$ of the numbers N_{AA}, N_{BB}, and N_{AB}, respectively, evaluated over the macro-state $\mathcal{D}_\lambda$, depend solely on λ. Evaluate these averages and show that

$$\overline{N}_{AA} = \overline{N}_{BB} = 2N(1 - \lambda^2),$$

$$\overline{N}_{AB} = 4N(1 + \lambda^2).$$

(One can calculate these values by finding the probability that two given nearest neighbour sites are both occupied by A atoms.)

In order to prepare for the last question show that for $\lambda = 0$ there exist micro-states which differ greatly from the average configuration, for instance, such micro-states that N_{AB}/N is very small.

2. We denote by W_{AA} the, negative, binding energy of two A atoms which are nearest neighbours; the energy $-W_{\mathrm{AA}}$ is the energy one needs to break the bond and separate these two atoms, assuming that this is the only bond involved. Similarly, we denote by W_{BB} and W_{AB} the energies of the B–B and A–B bonds. We assume that those energies are known and that they are independent of the temperature.

There are a large number of micro-states i of the system, with energy $E_i(\lambda)$, for a given value of λ. Write down the canonical partition function of the system in the form

$$Z = \sum_\lambda z(\lambda).$$

What is the form of $z(\lambda)$? Can this function be calculated exactly?

As it is impossible to use directly the Boltzmann-Gibbs distribution, we try to describe the system by the trial distribution $\mathcal{D}_\lambda$. We shall then determine the parameter λ such as to simulate thermal equilibrium in the best possible way. Calculate the average energy $U(\lambda)$ associated with the trial distribution $\mathcal{D}_\lambda$ as a function of W_{AA}, W_{BB}, W_{AB}, N, and λ. Show that $U(\lambda)$ has the form $U(\lambda) = U_0 - \lambda^2 U_1$, and write down expressions for U_0 and U_1.

3. Calculate the entropy $S(\lambda)$ associated with the trial distribution $\mathcal{D}_\lambda$. Sketch the behaviour of $S(\lambda)$.

4. Give a variational criterion for choosing amongst the distributions $\mathcal{D}_\lambda$ that one which gives the best possible approximation for the canonical partition function Z.

5. Use the above result to derive for each temperature T a value $\lambda_0(T)$ which approximately describes the equilibrium macro-state of the alloy as a function of the temperature. Distinguish the cases $U_1 > 0$ and $U_1 < 0$. What can one say about the order in the alloy as function of temperature? What happens at the temperature $T_c = U_1/Nk$?

6. Use the same approximation to sketch for the case $U_1 > 0$ the curves for the entropy and the specific heat as functions of the temperature. Indicate a simple physical measurement to obtain the experimental value of T_c.

7. Assume now that U_1 is negative and consider what happens at zero temperature. Show that there are stable configurations which are not included in the above analysis. Comment on this result.

Notes on Question 7. For $2W_{AB} > W_{AA} + W_{BB}$ one finds the minimum energy configurations when the A and the B atoms are separated in space. The trial density $\mathcal{D}_\lambda$, because it is translationally invariant, is not adequate to describe these configurations which occur in it with a negligible weight. It therefore cannot describe the transition which at low temperatures leads to an equilibrium macro-state such that in space the alloy is split into two phases in which either almost purely A or almost purely B atoms are each other's neighbours. Many alloys show this *segregation* effect; however, often the evolution towards equilibrium which requires a mass migration of atoms is so slow that the homogeneous mixture persists below the theoretical transition temperature.

16.5 Ferroelectricity

Barium titanate, $BaTiO_3$, is a ferroelectric crystal with electrical properties which are similar to the magnetic properties of ferromagnetic solids. We shall study these properties here.

First of all, let us look at the structure of this crystal. We shall hardly invoke this structure but it is the basis for the approximations we shall make and it enables us better to understand the physics. The crystal has cubic symmetry. We can consider it as a stacking of cubes with a Ba^{++} ion at each corner, an O^{--} ion at the centre of each face, and a Ti^{4+} ion at the centre of the cube. The titanium ions are small and can move freely to some extent, although the electrostatic forces will tend to keep them at the centre of the cube. The features of these attractive forces will be important in what follows.

In the temperature range considered, between 0 and 200 °C, we can treat the motion of the titanium ions *classically*.

We denote the *local* electrical field felt by a titanium atom by E and the charge of that ion by q. We consider a sample of volume Ω containing

N titanium ions. We denote by $\boldsymbol{P}$ the polarization, that is, the total dipole moment per unit volume, and by $\boldsymbol{E}_0$ an *external* electrical field which may be imposed upon the sample. The local field $\boldsymbol{E}$ is the sum of the field $\boldsymbol{E}_0$ and the electrostatic field created by all the neighbouring ions.

A rather simple electrostatic analysis which we shall use without proof enables us to relate the imposed field, the local field, and the polarization, where we retain solely the contribution due to the displacement of the titanium ions from their equilibrium positions; we find the relation

$$\boldsymbol{E} \; = \; \boldsymbol{E}_0 + \frac{\boldsymbol{P}}{3\varepsilon_0}.$$

We remind ourselves that when an ion of charge q is displaced by an amount $\boldsymbol{r}$ from its equilibrium position, its dipole moment is $q\boldsymbol{r}$.

1. If we neglected all forces acting on the titanium ions, they would form a perfect gas of the same volume Ω; what would be the canonical partition function Z_0 of this gas?

2. Assume that the forces acting on a titanium ion which pull it back to its equilibrium position are weak. Due to the cubic symmetry, they can then be derived from an isotropic harmonic potential energy,

$$\phi_0 \; = \; \tfrac{1}{2} a (x^2 + y^2 + z^2),$$

when the ion is at a position $\boldsymbol{r}$ with components x, y, z with respect to its equilibrium position. Calculate the partition function Z_1 of this sample, still assuming that $\boldsymbol{E}$ is zero and assuming that x, y, and z are small as compared to the interionic distances. For which values of $\langle r^2 \rangle$ would Z_1 go over into Z_0?

3. If the local field $\boldsymbol{E}$ is no longer zero, but is assumed to be the same for all ions and to be constant over the small displacements of the ions, calculate the new partition function Z_2 after having found an expression for the potential energy of each ion. Treat the Ti sites as being independent and express Z_2 as function of the temperature and of $\boldsymbol{E}$.

4. Express the polarization $\boldsymbol{P}$ as a derivative of Z_2, calculate it as function of $\boldsymbol{E}$, and then calculate it as function of the external field $\boldsymbol{E}_0$. Does it depend on the temperature? Show that in this model the crystal structure would be unstable at all temperatures if a is smaller than some value, which is to be determined.

5. Assume now that the crystal potential is anharmonic and of the form

$$\phi \; = \; \phi_0 + \phi_1,$$

where ϕ_1 is a homogeneous quartic polynomial in x, y, z. For symmetry reasons ϕ_1 is of the form

$$\phi_1 \; = \; \tfrac{1}{4} b r^4 + \tfrac{1}{4} c (x^4 + y^4 + z^4).$$

Calculate the partition function Z_3 in this new case, as function of T and of the local electrical field $\boldsymbol{E}$.

Note. In the calculations we encounter integrals involving exponentials of polynomials. To calculate them, we shall assume that the coefficients b and c are sufficiently small that we can expand $e^{-\beta\phi_1}$ as a power series under the integral sign and retain in $\ln Z_3$ only the terms which are independent of and linear in b and c. Take into account that

$$
\int\limits_{-\infty}^{+\infty} e^{-pz^2}\, z^{2n}\, dz \;=\;
\begin{cases}
\dfrac{(2n-1)(2n-3)\cdots 3\cdot 1}{(2p)^n}\sqrt{\dfrac{\pi}{p}}\,, & (n \geq 1) \\[3mm]
\sqrt{\dfrac{\pi}{p}}\,, & (n = 0),
\end{cases}
$$

or use the equipartition theorem (§ 8.4.2).

6. Calculate the polarization for this case as function of $\boldsymbol{E}$. Does it depend on the temperature?

7. We define the dielectric susceptibility χ of the medium through the relation

$$
\chi \;=\; \lim_{E_0 \to 0} \frac{1}{\varepsilon_0}\frac{\partial P}{\partial E_0}.
$$

Calculate χ and show that, if b and c are sufficiently small and satisfy a certain inequality, it is approximately of the form

$$
\chi \;=\; \frac{T_1}{T - T_0},
$$

provided $T_0 < T \ll T_1$; express T_1 and T_0 as functions of a, b, and c.

8. Knowing that $T_0 = 118\ °\text{C}$, $T_1 = 150\ 000\ \text{K}$, and that the size of an elementary cube of BaTiO_3 is 4 Å, evaluate a, b, and c using as units of energy and length the eV and the angstrom, and assuming that $c = 0.8b$. Remember that the electron charge is -1.602×10^{-19} C and that, if $kT = 1$ eV, $T = 11\ 605$ K.

9. To study the ferroelectric properties of the substance we take as thermodynamic variable the polarization $\boldsymbol{P}$. We then introduce the thermodynamic potential $F(T, \boldsymbol{P})$, defined by

$$
F(T, \boldsymbol{P}) \;=\; -kT \ln Z_3 + \Omega(\boldsymbol{P}\cdot\boldsymbol{E}_0) + \Omega\frac{P^2}{6\varepsilon_0},
$$

where $\boldsymbol{E}_0 = \boldsymbol{E} - \boldsymbol{P}/3\varepsilon_0$. Find the relation which connects $\boldsymbol{E}_0$ with the derivative $\nabla_{\boldsymbol{P}} F|_T$.

10. Calculate $F(T, \boldsymbol{P})$ to first order in b and c, and show that it is of the form

$$
F(T, \boldsymbol{P}) \;=\; F(T, 0) + \lambda(T - T_0)\boldsymbol{P}^2 + \mu(\boldsymbol{P}^2)^2 + \nu(P_x^4 + P_y^4 + P_z^4).
$$

It will not be necessary to give a detailed expression for $F(T, 0)$. Express λ, μ, and ν as functions of a, b, and c.

11. One says that there is *spontaneous polarization*, and the substance is then called a ferroelectric, when $\boldsymbol{P}$ can be non-vanishing although the external field $\boldsymbol{E}_0$ is zero. The expression found sub 7 for the susceptibility is no longer valid when there is spontaneous polarization. Calculate the spontaneous polarization and determine its possible directions, depending on b and c. Write down $\chi(T)$ in the various possible temperature ranges, for $b > c > 0$.

12. Give a numerical value for the displacement of the Ti ions due to ferroelectricity at $0\,°C$; compare it with the mean square displacement due to thermal motion; the atomic mass of Ti is $48.1\,\mathrm{g\,mol}^{-1}$. Using the numerical values found sub 8, discuss the approximations used.

13. What do you think gives rise to ferroelectricity? How will it show in practice?

Comments:

We have used several approximations, which we shall now justify, in order to be able to treat the Ti ions as *independent* objects, each of them moving in an external potential, and thus to be able to factorize Z_1, Z_2, and Z_3. To begin with, the fact that each ion remains close to a crystal site enables us to regard all ions as *distinguishable*; the *classical* approximation is justified for $kT \gg \hbar\sqrt{a/m}$.

We denote the ion situated near the origin by $\mathrm{Ti}^{(0)}$ and the other ions by $\mathrm{Ti}^{(n)}$, $1 \le n \le N$. The effective potential seen by $\mathrm{Ti}^{(0)}$ represents the average of the interactions to which it is subject from the Ba, O, and $\mathrm{Ti}^{(n)}$ ions, and we must take into account that this potential is a *self-consistent* one, as we did in §§ 11.2.1 and 11.3.3 for the electrons. In the above model ϕ describes the action on $\mathrm{Ti}^{(0)}$ due to the Ba, O, and $\mathrm{Ti}^{(n)}$ ions assumed to be *fixed* on the original lattice; in the language of § 11.4.1, it is the contribution to $W(\{\boldsymbol{R}_n\})$, to second and fourth order in the coordinates x, y, z of $\mathrm{Ti}^{(0)}$, evaluated by fixing the $\mathrm{Ti}^{(n)}$ ions at the centres of their respective cells. However, one should add to ϕ the contribution associated with the *displacement* of the $\mathrm{Ti}^{(n)}$ ions. Related to the original lattice, the displacement $\boldsymbol{r}^{(n)}$ of such an ion is equivalent to a dipole $q\boldsymbol{r}^{(n)}$. Implicitly we have approximately evaluated the potential produced near the origin by these $\mathrm{Ti}^{(n)}$ dipoles through replacing them by a uniform polarization $\boldsymbol{P} = Nq\langle \boldsymbol{r}^{(n)}\rangle/\Omega$ with an isotropic cut-off at short distances in order to take away the $\mathrm{Ti}^{(0)}$ ion. This potential equals $-q(\boldsymbol{P} \cdot \boldsymbol{r})/3\varepsilon_0$, and we have taken it into account by replacing the external field $\boldsymbol{E}_0$ by $\boldsymbol{E} = \boldsymbol{E}_0 + \boldsymbol{P}/3\varepsilon_0$. One can show that this continuous approximation is justified.

Nevertheless, even though the single-ion effective Hamiltonian obtained this way is correct, the interaction between $\mathrm{Ti}^{(0)}$ and $\mathrm{Ti}^{(1)}$, for example, is *counted twice* in the internal energy which one would find from Z_2 or Z_3, as it occurs once for $\mathrm{Ti}^{(0)}$ and once for $\mathrm{Ti}^{(1)}$. To correct this error, we must subtract the total dipole interaction energy $-\Omega P^2/6\varepsilon_0$. In this way we obtain the free energy $F^{(0)}(T, \boldsymbol{E}_0)$ defined in § 6.6.5 by Eq.(6.92) and associated with the crystal, including its interaction with the external field $\boldsymbol{E}_0$ occurring in the Hamiltonian:

$$F^{(0)}(T, \boldsymbol{E}_0) \;=\; -kT \ln Z_3 + \frac{\Omega}{6\varepsilon_0}\, P^2.$$

The same correction term was included in the definition of the thermodynamic potential $F(T, \boldsymbol{P})$ which we used for question 9; this potential, which is the Legendre transform of $F^{(0)}$ with respect to $\boldsymbol{E}_0$ is associated with the internal energy (6.95) of the dielectric, if the energy of the field is not included.

The double-counting error made in Z_2 and Z_3 has no consequences for the remainder of the problem. In fact, the expression for Z_3 for a single site,

$$Z_3^{(0)} = \int \frac{d^3r\, d^3p}{h^3}\, \exp\left[-\beta p^2/2m - \beta\phi + \beta q(\boldsymbol{E}\cdot\boldsymbol{r})\right],$$

shows that the polarization is correctly found by taking the derivative with respect to the total field $\boldsymbol{E}$. The situation is the same as for the electrons (end of §11.2.1): counting in the internal energy the interactions between pairs twice and neglecting the self-consistency of the effective potential leads to a wrong expression for the thermodynamic potentials, but these two errors cancel when one takes the derivatives of the latter.

We have seen (Exerc.11d) that when studying lattice vibrations it is incorrect to assume that the atoms are independent, as Einstein did; in that case the independent entities were the phonons rather than the atoms (§11.4.2). Here, it is legitimate to treat the Ti ions as being independent since they move in the fixed lattice of the Ba and O ions.

Interpreting the vector $\boldsymbol{P}$ as an *order parameter* with three components we see that the approximate expression,

$$F(T, \boldsymbol{P}) = -\frac{3}{2}NkT \ln \frac{mk^2T^2}{\hbar^2 a} + \frac{9NkT^2}{4T_1} + \frac{\Omega}{2\varepsilon_0}\frac{T-T_0}{T_1}\boldsymbol{P}^2$$

$$+ \frac{\Omega^4}{4N^3q^4}\left[b\left(\boldsymbol{P}^2\right)^2 - c(P_x^4 + P_y^4 + P_z^4)\right],$$

found in question 10, with

$$\frac{T_0}{T_1} = \frac{1}{3} - \frac{\varepsilon_0\Omega a}{Nq^2} \ll 1, \qquad kT_1 = \frac{3a^2}{5b-3c},$$

has exactly the form used by Landau in his theory of phase transitions (Exerc.6d). One could use it to extend to ferroelectricity the results obtained in Exerc.6d for Ising ferromagnetism.

16.6 Rotation of Molecules in a Gas

We shall study the thermodynamics of the rotational degrees of freedom of the diatomic molecules of a HCl gas. We assume that these rotational degrees of freedom are completely separated from the translational degrees of freedom of the molecules, which can be treated as in Chap.7. If we thus forget about the kinetic energy we shall assume in the whole of this problem that the Hamiltonian of a single molecule reduces to the rotational energy:

$$\widehat{h} = \frac{\widehat{L}^2}{2\mathcal{I}},$$

and that its micro-states are characterized by two quantum numbers l, m (integers with $|m| \leq l$). We also assume that we can treat the molecules as distinguishable from the point of view of their rotations. For an HCl molecule its moment of inertia is in energy units determined by

$$\varepsilon = \frac{\hbar^2}{2\mathcal{I}} = 1.3 \times 10^{-3} \text{ eV}.$$

Remember that room temperature of 300 K corresponds to an energy of $\frac{1}{40}$ eV. Remember also that the eigenvalues of $\widehat{L}^2$ are $\hbar^2 l(l+1)$.

1. Give for a single molecule in canonical equilibrium expressions for the rotational partition function Z_1, the average energy U_1, and the entropy S_1 as functions of the temperature. You do not have to sum the series which occur.

2. Now consider a mole of gas with $N = 6 \times 10^{23}$ independent molecules. Write down its canonical partition function Z and its rotational internal energy U and entropy S in terms of Z_1, U_1, and S_1.

3. Sketch the general behaviour of the function $S(U)$. Discuss the sign, slope and curvature of this curve.

4. Find the behaviour at very low temperatures ($T \ll 10$ K) of the specific heat per mole $C(T)$ associated with the rotation of the molecules. Give the numerical value of $C(T)$ at $T = 5$ K. How does the function $S(U)$ behave at the origin? Do we have to distinguish between the rotational specific heats at constant pressure and at constant volume? Why?

5. Evaluate $C(T)$ at room temperature (assume without mathematical proof that a series whose successive terms are very close behaves like an integral). Give its numerical value. Compare this with the result from classical statistical mechanics (equipartition theorem).

6. Sketch in detail the function $C(T)$ and indicate the numerical scales.

Notes. We have given in Chap. 8 a rigorous derivation of the hypotheses on which this problem is based. The translational degrees of freedom, which can be treated as if the molecules were classical point particles, contribute additively to the entropy and the energy, and thus to $\ln Z_C$. The fact that the molecules are indistinguishable must be taken into account when we deal with the translations but it turns out (§ 8.1.3) that the internal degrees of freedom of the molecules give a contribution which is simply proportional to N, as if they were distinguishable. Also, the electronic degrees of freedom and the molecular vibrations do not play a rôle except at very high temperatures and are frozen in at room temperatures (§§ 8.3.1, 8.4.1, and 8.4.4).

The rotational degrees of freedom are not coupled to external forces, in particular, not to the pressure since the molecular rotational energy is independent of the volume; the contribution from the rotations to the specific heat is thus the same for constant pressure as for constant volume; moreover, dU only contains the heat.

Detailed Solution of Question 6:

The $C(T)$-curve leaves the origin at low temperatures with a horizontal tangent and tends to the constant $C = Nk$ at high temperatures. On dimensional grounds we see that the scales are characterized by the quantities $\varepsilon/k = 15$ K along the abscissa and $Nk = 8.4$ J K^{-1} along the ordinate; in particular, the transition between the limiting cases of questions 4 and 5 occurs at temperatures of a few tens of degrees K.

The specific heat contains not only a rotational but also a translational contribution. The latter is at practically all temperatures given by the classical value $\frac{3}{2}Nk$. The specific heat of a diatomic gas thus goes from a value $\frac{3}{2}Nk$ at temperatures of a few K to a value of $\frac{5}{2}Nk$ at room temperatures.

If we wish to plot the $C(T)$-curve with some accuracy we must find correction terms at low and high temperatures. In the first case, keeping the terms $l = 0, 1,$ and 2 in Z, we find

$$\ln Z \approx N \ln\left(1 + 3\mathrm{e}^{-2\beta\varepsilon} + 5\mathrm{e}^{-6\beta\varepsilon}\right),$$

and hence, expanding in powers of $\mathrm{e}^{-\beta\varepsilon}$,

$$C \approx 12Nk(\beta\varepsilon)^2\mathrm{e}^{-2\beta\varepsilon}\left(1 - 6\mathrm{e}^{-2\beta\varepsilon} + 42\mathrm{e}^{-4\beta\varepsilon}\right).$$

Successive terms indicate where the expansion can be used ($kT < 0.8\varepsilon$) and indicate a rise and then a inflection of the $C(T)$-curve. To study the behaviour at high temperatures we need the calculation of the corrective terms in the approximation used under 5. Let us therefore turn again to the proof of that approximation. Putting

$$f(x) \equiv -\frac{1}{\beta\varepsilon}\,\mathrm{e}^{-\beta\varepsilon x(x+1)},$$

we find that the problem consists in evaluating

$$Z_1 = \sum_{l=0}^{\infty} f'(l)$$

in the limit as $\beta\varepsilon$ is small. We note that the successive derivatives of f with respect to x are for fixed l infinitesimal quantities in $\beta\varepsilon$ of increasing order, and that each of them can be calculated at the point $l+1$ by means of a Taylor expansion around the point l. The finite differences

$$\Delta_p(l) \equiv f^{(p)}(l+1) - f^{(p)}(l)$$

can thus be written as series in $f^{(q)}$ with $q > p$, and we use this to derive by a step by step inversion the Bernoulli series

$$f'(l) \approx \Delta_0(l) - \frac{1}{2}\Delta_1(l) + \frac{1}{6}\frac{\Delta_2(l)}{2!} - \frac{1}{30}\frac{\Delta_4(l)}{4!},$$

which expresses the function f' as a finite difference expansion of increasing orders in $\beta\varepsilon$. The summation of $f'(l)$ over l retains on the right-hand side only the extreme terms ($l = 0$ and $l = \infty$), which leads to the Euler-Maclaurin summation formula (see end of volume):

$$\sum_{l=0}^{\infty} f'(l) \approx -f(0) + \frac{1}{2}f'(0) - \frac{1}{6}\frac{f''(0)}{2!} + \frac{1}{30}\frac{f^{(4)}(0)}{4!}.$$

This gives us an asymptotic expansion in powers of $\beta\varepsilon$:

$$Z_1 \approx \frac{1}{\beta\varepsilon} + \frac{1}{2} - \frac{1}{6}\left(1 - \frac{\beta\varepsilon}{2}\right) + \frac{1}{30}\left(-\frac{\beta\varepsilon}{2} + \frac{(\beta\varepsilon)^2}{2} - \frac{(\beta\varepsilon)^3}{4!}\right)$$

$$\approx \frac{1}{\beta\varepsilon} + \frac{1}{3} + \frac{\beta\varepsilon}{15},$$

which leads to

$$\ln Z \approx -N \ln \beta\varepsilon + \frac{N}{3}\beta\varepsilon + \frac{N}{90}(\beta\varepsilon)^2,$$

whence we get

$$U \approx NkT - \frac{N\varepsilon}{3} - \frac{N\varepsilon^2}{45kT}, \qquad C \approx Nk + \frac{N\varepsilon^2}{45kT^2}.$$

The curve (Fig.8.3) tends to its asymptote from above and thus has a *maximum*. As the two limiting expressions match each other, the full curve can be determined through interpolating. It lies between the specific heat of an harmonic oscillator which *increases* from zero to the classical value Nk (Einstein model, Fig.11.20, or molecular vibrations, Fig.8.4) and that of a two-level system which shows a *peak* (paramagnet, Fig.1.5). This also agrees with the intermediate nature of the energy spectrum.

16.7 Isotherms and Phase Transition of a Lattice Gas

We are going to study the effect of the intermolecular interactions in a fluid in order to understand the shape of its isotherms and the existence of a transition between the liquid and vapour phases. Figure 16.11 gives us a typical experimental diagram in the $v, \mathcal{P}$-plane, where $v = 1/\varrho = \Omega/N$ is the reciprocal of the density and $\mathcal{P}$ the pressure. We want to explain this diagram qualitatively, starting from a simplified microscopic theory. To do this we evaluate the grand partition function $Z_G(\alpha, \beta, \Omega)$, using a model and approximations which we shall state as they become necessary. The molecules are treated as structureless point particles of mass m; this is justified for monatomic fluids such as helium or argon up to temperatures of the order of 10^5 K, and remains valid in a large temperature range for simple fluids, at least as far as the equation of state is concerned. We assume that the temperature is sufficiently high and the density sufficiently low that we may treat the molecules as classical particles; we shall not be interested here in

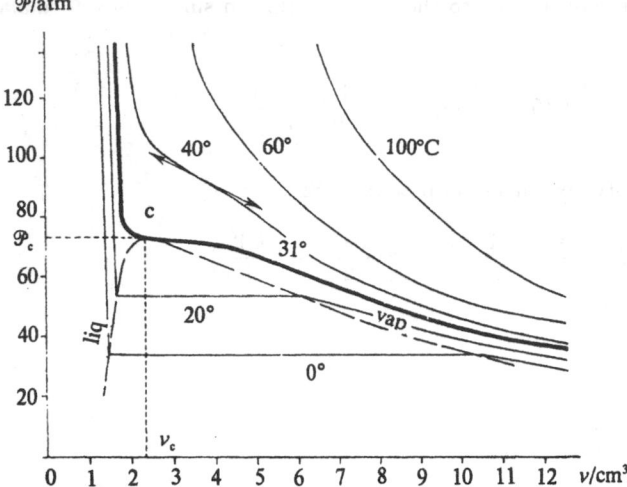

Fig. 16.11. Isotherms of 1 g of CO_2

high pressures or low temperatures such that crystallization – or for helium superfluidity – effects come into play. Let $r_1, \cdots, r_i, \cdots, r_N$ be the positions and $p_1, \cdots, p_i, \cdots, p_N$ the momenta of the particles. These particles interact with one another through an isotropic potential $V(|r_i - r_j|)$ which is known either from experiments on molecular collisions or from a theoretical approach based on the structure of the molecules. We show in Fig.16.12 the behaviour of this potential $V(r)$; it is practically infinite when the molecules are at the same position, strongly repulsive at short distances apart, and becomes attractive at larger distances apart: it changes sign for $r = r_0$ of the order of a few Å, and for a slightly larger r it has a minimum of the order of 10^{-2} eV while rapidly tending to zero at still larger distances apart.

1. Show that the grand partition function has the form

$$Z_G(\alpha, \beta, \Omega) = \sum_N \zeta^N Y_N, \tag{1}$$

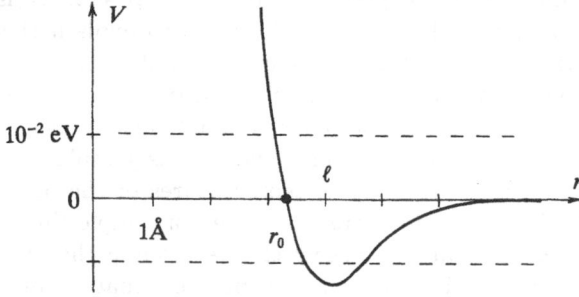

Fig. 16.12. The intermolecular potential

where Y_N is the multiple integral

$$Y_N \equiv \frac{1}{N!} \int_\Omega d^3 r_1 \cdots d^3 r_N \exp\left[-\beta \sum_{\substack{i,j=1 \\ i>j}}^{N} V(|r_i - r_j|) \right], \tag{2}$$

and where ζ is a function of α and β which must be written down.

As it is impossible to evaluate Y_N exactly we shall construct a model which retains the qualitative features of the problem and which is based upon the following trick. We replace the integral Y_N by a finite Riemann sum where each of the coordinates $x_1, y_1, z_1, \cdots, x_N, y_N, z_N$ takes on discrete integer values, multiplied by a length l. Instead of letting l go to zero, we fix it at a finite value which we choose slightly larger than r_0 and close to the minimum of the potential $V(r)$. This choice enables us to carry out the calculations and it will be justified by the agreement between the results and experiments. Space is thus quantized into a lattice with each cell having a volume $b = l^3$ and with $Q = \Omega/l^3$ sites which are denoted by the index $q = 1, 2, \cdots, Q$; the fluid is enclosed in a cubic vessel of volume Ω. We assume that the replacement for each particle of an integration over the continuum $r = (x, y, z)$ by a summation over the discrete points $R_q \equiv (x_q, y_q, z_q)$ of this lattice does not change the qualitative properties of (1), even though l is not small compared to the characteristic lengths of the potential. This approximation is commonly called the "*lattice gas model*".

2. Prove that in this approximation the expression (1), (2) can be written as a sum

$$Z_G(\alpha, \beta, \Omega) = \sum_{\{n\}} (b\zeta)^{n_1 + \cdots + n_Q} e^{-\beta W(\{n\})}, \tag{3}$$

over occupation numbers $n_1, \cdots, n_q, \cdots, n_Q$ associated with the sites, which each take on the values 0 and 1. The function W of these occupation numbers is defined as

$$W(\{n\}) \equiv W(n_1, \cdots, n_q, \cdots, n_Q) \equiv \frac{1}{2} \sum_{\substack{q,q'=1 \\ (q \neq q')}}^{Q} V(|R_q - R_{q'}|) n_q n_{q'}. \tag{4}$$

What is the physical property which corresponds to the restriction of the permitted values to 0 and 1 for the occupation numbers?

First of all, in questions 3 to 5 we try to estimate the rôle played by the short-range repulsions between the particles. To do this, we neglect the attractive forces by retaining solely that part of the potential which corresponds to distances apart less than r_0 (Fig.16.13). We denote by Z_1 the approximate grand partition function and use the index 1 to indicate quantities corresponding to this approximation.

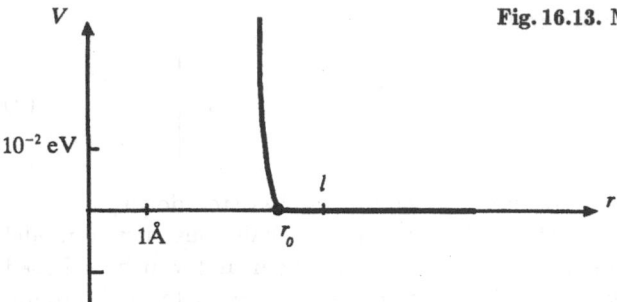

Fig. 16.13. Model repulsive potential

3. Evaluate for this case expression (3) for Z_1 and give the grand potential, A_1, and the average number of particles on a particular site, $\langle n_q \rangle_1$, as functions of α and β. To simplify the notation retain ζ as a parameter.

4. Show that there is a formal analogy between the model we have just introduced and a system of non-interacting fermions, the energy characteristics of which must be given. Use this analogy to find again Z_1 and $\langle n_q \rangle_1$.

5. Use the results of 3 to derive the set of isotherms $\mathcal{P}_1(v, T)$. Discuss briefly the limiting cases of low and high densities.

For the remainder of this problem we return to the realistic potential of Fig.16.12. From now on expression (3) for Z_G contains the attractive potentials between particles on neighbouring sites, q and q'. Notwithstanding the approximations we have made, it is still not possible to evaluate (3) exactly and we shall introduce a further approximation of a variational kind. We note that (3) has the same form as the partition function associated with a fermion density operator defined in the space of the occupation numbers, $n_1, \cdots, n_q, \cdots, n_Q$; we shall model this one by a simpler density,

$$D_0(\{n\}) \equiv D_0(n_1, \cdots, n_q, \cdots, n_Q) \equiv \frac{1}{Z_0}\, \lambda^{n_1 + \cdots n_q + \cdots n_Q}, \tag{5}$$

depending on an adjustable parameter λ and normalized by a factor Z_0. We introduce the function, which need not be evaluated explicitly for question 6,

$$\Psi(\alpha, \beta, \Omega; \lambda) \equiv -\sum_{\{n\}} D_0 \ln D_0 - \beta \sum_{\{n\}} D_0\, W(\{n\})$$

$$+ \ln(b\zeta) \sum_{\{n\}} D_0 \sum_q n_q. \tag{6}$$

6. Adapt the discussion of §4.2.2 to prove the inequality

$$\ln Z_G(\alpha, \beta, \Omega) > \Psi(\alpha, \beta, \Omega; \lambda). \tag{7}$$

This inequality will be the basis of the final approximation. The trial density D_0 which models best the properties of the system at equilibrium will be the one which leads to the smallest possible difference between $\Psi(\lambda)$ and $\ln Z_G$ for given values of the thermodynamic variables α, β, and Ω. Using

(7) we find that the best possible choice for λ is the one which makes $\Psi(\lambda)$ a maximum, and we therefore shall use the approximation

$$\ln Z_G(\alpha, \beta, \Omega) \simeq \max_{\lambda} \Psi(\alpha, \beta, \Omega; \lambda). \tag{8}$$

In order to get an explicit expression for Ψ we note that we can neglect boundary effects in the limit of large volumes, Ω. Thus, except for a small number of sites q near the surface of the sample which make a relatively negligible contribution, the expression

$$b \sum_{\substack{q'=1 \\ (q' \neq q)}}^{Q} V(|\boldsymbol{R}_q - \boldsymbol{R}_{q'}|) \equiv -2a, \tag{9}$$

which simulates in our model the integral $\int_{r > r_0} d^3 r\, V(r)$, for the attractive part of the potential, is a negative constant which does not depend on the site q.

7. Prove by a direct calculation that the function $\Psi(\alpha, \beta, \Omega; \lambda)$ is equal to

$$\Psi(\alpha, \beta, \Omega; \lambda) = \frac{\Omega}{b}\left\{ \ln(1 + \lambda) + \frac{\lambda}{1 + \lambda} \ln\left(\frac{b\zeta}{\lambda}\right) + \frac{\beta a}{b}\left(\frac{\lambda}{1 + \lambda}\right)^2 \right\}.$$

8. Write down the condition $\partial \Psi/\partial \lambda = 0$ which $\Psi(\lambda)$ must satisfy at its maximum. Use this condition to derive the equation of state $\mathcal{P}(v, T)$, first in parametric form in terms of λ and then explicitly. The distinction between minima and maxima of $\Psi(\lambda)$ will not be discussed until we come to question 10.

9. Sketch the isotherms $\mathcal{P}(v, T)$ obtained in the previous question. Show that there exists a critical temperature T_c for which the isotherm has an inflexion with horizonal tangent at the point $v_c, \mathcal{P}_c$. Evaluate for the case of the potential given in Fig.16.12 the order of magnitude of the critical temperature T_c, the critical molar volume $N_A v_c$, and the critical pressure $\mathcal{P}_c$. Use the numerical values indicated in Fig.16.12 and the data from the end of the book.

Not all the isotherms constructed here are in agreement with experiments. Moreover, they contain parts which do not correspond with the problem as we set it, that is, with finding the absolute maximum of $\Psi(\lambda)$. While the thermodynamic equilibrium state, characterized by T and μ, is described approximately by the absolute maximum of $\Psi(\lambda)$, we shall assume that a local maximum describes a metastable state.

10. Indicate the regions corresponding to stability or metastability in the $v, \mathcal{P}$-plane.

11. In order to retain only those parts of the isotherms which correspond to thermodynamical equilibrium states Maxwell proposed the construction indicated in Fig.16.14: we replace an oscillating part of the curve by a horizontal liquefaction plateau which, together with the original isotherm, bounds

Fig. 16.14. The Maxwell construction

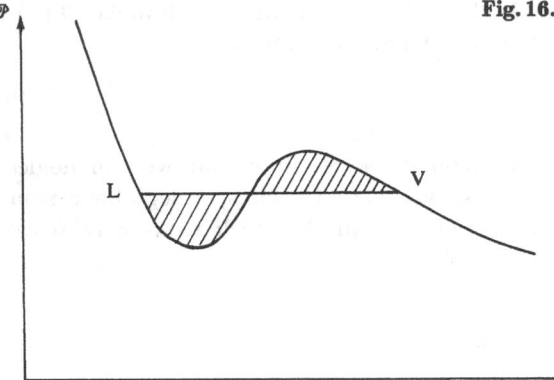

two equal areas. The ends L and V of the horizontal line correspond to the liquid and to its saturated vapour. Justify this construction, for instance, by tracing the variation of μ, $\mathcal{P}$, and v along an isotherm. Identify in the $v, \mathcal{P}$-diagram the region where a homogeneous stable phase exists.

12. Show that in the $T, \mathcal{P}$-plane the liquid phase and the saturated vapour phase coexist along a certain curve, $\mathcal{P}_\mathrm{s}(T)$; indicate how this curve can be determined in principle, without performing explicitly the necessary calculations. Show that the function $\mathcal{P}_\mathrm{s}(T)$ is an increasing function and that it terminates at the critical point determined in question 9. How are the metastable phases represented in the $T, \mathcal{P}$-plane?

13. Show that there exists a vaporization heat per mole, L, and that it satisfies the Clapeyron relation

$$L = T \frac{d\mathcal{P}_\mathrm{s}}{dT} N_\mathrm{A}(v_\mathrm{V} - v_\mathrm{L}),$$

where the derivative is calculated along the saturated vapour curve and where $N_\mathrm{A}v_\mathrm{V}$ and $N_\mathrm{A}v_\mathrm{L}$ denote the volumes of one mole of the fluid in the vapour and liquid phases.

Solution

1. The general expression of the grand partition function is in the classical limit

$$Z_\mathrm{G}(\alpha, \beta, \Omega) = \sum_{N=0}^{\infty} \mathrm{e}^{\alpha N} \int \frac{d^3 p_1 \cdots d^3 p_N d^3 r_1 \cdots d^3 r_N}{N! h^{3N}} \, \mathrm{e}^{-\beta H_N}.$$

After integrating over the p we get for each particle a factor

$$\zeta = \mathrm{e}^{\alpha} \int \frac{d^3 p}{h^3} \mathrm{e}^{-\beta p^2/2m} = \frac{\mathrm{e}^{\alpha}}{h^3} \left(\frac{2m\pi}{\beta}\right)^{3/2}, \tag{10}$$

which leads to expressions (1), (2) for Z_G.

2. In the lattice approximation the integration over each of the coordinates $r_1, \cdots, r_N$ gives a sum with a weight l for each dimension. We get for the integral Y_N

$$Y_N = \frac{1}{N!} l^{3N} \sum \exp\left[-\beta \sum_{\substack{i,j=1 \\ (i>j)}}^{N} V(|r_i - r_j|)\right],$$

where the sum is over the positions $r_1, \cdots, r_N$ of all the particles which each successively will occupy all the sites R_q of the lattice. Instead of analyzing this sum in terms of particles, let us analyze it in terms of sites. First of all, we note that configurations where a site is occupied by more than one particle do not contribute to Y_N as the potential $V(0)$ is infinite and thus $e^{-\beta V(0)}$ is zero. We can therefore restrict ourselves to configurations where N of the sites are occupied by a single particle and $Q-N$ are empty. Let us associate with each site q an occupation number n_q which can take on the values 0 and 1. Each of the configurations considered is characterized by giving a set of numbers n_q which are independent of one another, except for the constraint $\sum_q n_q = N$. In the summation over the positions of the particles, a given configuration will occur $N!$ times, where the $N!$ corresponding terms in Y_N are obtained by performing all the different permutations of the indices $1, \cdots, N$ of the particles. Finally, in calculating the potential energy associated with a configuration we have a contribution $V(|R_q - R_{q'}|)$ if both sites q and q' are occupied, and zero otherwise. The first situation corresponds to $n_q n_{q'} = 1$ and the others to $n_q n_{q'} = 0$, so that we have for the potential energy of a given configuration

$$W(\{n\}) = \sum_{\substack{q,q'=1 \\ (q>q')}}^{Q} V(|R_q - R_{q'}|)\, n_q n_{q'} = \frac{1}{2} \sum_{\substack{q,q'=1 \\ (q \neq q')}}^{Q} V(|R_q - R_{q'}|)\, n_q n_{q'}.$$

We get thus

$$Y_N = b^N \sum \exp\left[-W(\{n\})\right], \qquad (11)$$

where the sum is over those configurations for which $\sum_q n_q = N$. If we substitute (11) into (1), the constraint on the n_q is removed and we obtain (3).

The restriction to the values 0 and 1 of the numbers n_q comes from the fact that the potential V is infinite at the origin.

Note. If there were no potential, all values $0, 1, \cdots, +\infty$ would be possible for the n_q and each configuration would occur $N!/\prod(n_q!)$ times in the summation over the particle positions so that we would have

$$Z_G = \sum_{\{n\}} \prod_q \frac{(b\zeta)^{n_q}}{n_q!} = \prod_q e^{b\zeta} = e^{Qb\zeta} = e^{\Omega\zeta}.$$

We find again, even for finite grids, the well-known perfect-gas expression for Z_G, which we would not have obtained if we had restricted the values of the n_q to 0 and 1.

3. As two different sites will always be at a distance apart which exceeds r_0, W will be zero and the potential will not contribute to (3) – except implicitly through the restriction $n_q = 0$ or 1 which accounts for the repulsive part of the potential. We have thus

$$Z_1 = \sum_{\{n\}} \prod_{q=1}^Q (b\zeta)^{n_q} = \prod_{q=1}^Q \sum_{n=0,1} (b\zeta)^n = \prod_{q=1}^Q (1 + b\zeta) = (1 + b\zeta)^Q. \tag{12}$$

Hence we get the grand potential

$$A_1(\mu, T, \Omega) = -\frac{\Omega kT}{b} \ln(1 + b\zeta),$$

where ζ, which is given by (10), is a function of μ and T,

$$\zeta = \frac{e^{\mu/kT}}{h^3} (2m\pi kT)^{3/2}. \tag{13}$$

By differentiation we get the average number of particles per site

$$\langle n_q \rangle = -\frac{1}{Q}\frac{\partial A_1}{\partial \mu} = \frac{\Omega kT}{Qb}\frac{b\zeta}{1+b\zeta}\frac{1}{kT} = \frac{b\zeta}{1+b\zeta}. \tag{14}$$

4. The sites q play the same role as single-fermion states which can be either occupied ($n_q = 1$) or empty ($n_q = 0$). The probability for a configuration $\{n\}$ for non-interacting fermions is

$$\prod_{q=1}^Q \exp\left[-\frac{n_q(\varepsilon_q - \mu)}{kT}\right],$$

where ε_q is the energy of the single-fermion state q. Comparing (12) and (13) shows the formal analogy between the two probability distributions: our classical gas gives expressions which are the analogues of those for a gas of non-interacting fermions where all single-fermion states have the energy

$$\varepsilon = -\frac{3}{2}kT \ln\left(\frac{2\pi mkT}{h^2}b^{2/3}\right) \tag{15}$$

(one could also have assigned zero energy to those fermions, shifting at same time the chemical potential μ, since the only combination which occurs is $b\zeta = e^{-\beta(\varepsilon-\mu)}$).
Expression (12) for Z_1 is therefore the grand partition function

$$Z_1 = \left[1 + e^{-\beta(\varepsilon-\mu)}\right]^Q$$

for a gas of fermions and the mean occupation number (14) is the same as the Fermi factor

$$\frac{1}{e^{\beta(\varepsilon-\mu)}+1}.$$

Note that the repulsive potential at short distances apart for classical particles simulates the quantal exclusion principle according to which a site can be occupied by at most one fermion. One should not extend the analogy to thermal properties as the effective energy of the fermions is temperature dependent.

5. The density equals

$$\varrho = \frac{N_1}{\Omega} = \frac{\zeta}{1 + b\zeta} = \frac{1}{v},$$

and the pressure

$$\mathcal{P}_1 = -\frac{\partial A_1}{\partial \Omega} = \frac{kT}{b} \ln(1 + b\zeta).$$

For fixed T we get by using these two expressions to eliminate μ or, what amounts to the same, eliminate ζ

$$\mathcal{P}_1 = \frac{kT}{b} \ln \frac{v}{v - b}. \tag{16}$$

At low densities ($v \ll b$) we regain the perfect gas law

$$\mathcal{P}_1 = \frac{kT}{v},$$

but the repulsive forces lead to an increase in the pressure which, for large v/b, equals

$$\mathcal{P}_1 = \frac{kT}{v} \left(1 + \frac{b}{2v} + \cdots\right).$$

The pressure becomes infinite as $v \to b$: one cannot exceed a volume $1/b$ per particle. This behaviour (16) of the isotherms at low densities and high pressures agrees with experiment.

6. Expression (3) for Z_G is formally the same as the grand partition function of a fermion gas, as in question 4. Nonetheless, the energy of a configuration $\{n\}$ now includes not only the energy of the independent particles $\sum_q \varepsilon n_q$, but also the interaction energy W, expressed through (4) as a function of the occupation numbers.

The density operator $D_0(\{n\})$ represents in this language a system of non-interacting fermions at equilibrium where all configurations have zero energy and where the multiplier α equals $\ln \lambda$. The entropy of that state is $-k \operatorname{Tr} D_0 \ln D_0$. The second term in Ψ then represents the average $-\beta\langle W\rangle_0 \equiv -\beta \operatorname{Tr} D_0 W$ and the last term represents $\alpha'\langle N\rangle_0 = \alpha' \operatorname{Tr} D_0 N$, where $\alpha' \equiv \ln(b\zeta)$. Equation (4.10) of the main text, in which the trial density is D_0, while the conserved quantities are the energy W and the number of particles with associated multipliers β and α', reduces to (7).

7. The expression for Z_0 is formally the same as (12) which was evaluated for Z_1 in question 3, apart from replacing $b\zeta$ by λ. We have thus $Z_0 = (1 + \lambda)^Q$. At the same time the average $\langle n_q\rangle_0$ of n_q over D_0 is given by (14), whence

$$\langle n_q\rangle_0 = \frac{\lambda}{1 + \lambda}. \tag{14'}$$

Hence we get

$$\frac{S_0}{k} = -\operatorname{Tr} D_0 \ln D_0 = \left\langle \ln Z_0 - (\ln \lambda) \sum_q n_q \right\rangle_0$$

$$= Q \left[\ln(1+\lambda) - (\ln \lambda) \frac{\lambda}{1+\lambda} \right],$$

an expression which we could also have obtained by a straightforward calculation. To calculate the second term in Ψ we note that since D_0 is factorized, we have

$$\langle n_q n_{q'} \rangle_0 = \langle n_q \rangle_0 \langle n_{q'} \rangle_0 = \left(\frac{\lambda}{1+\lambda} \right)^2.$$

Hence we get, using the definition (9),

$$\operatorname{Tr} D_0 W = \sum_{\{n\}} D_0(\{n\}) \frac{1}{2} \sum_{\substack{q,q'=1 \\ (q \neq q')}}^{Q} V(|\mathbf{r}_q - \mathbf{r}_{q'}|) n_q n_{q'} = -\frac{Qa}{b} \left(\frac{\lambda}{1+\lambda} \right)^2.$$

Altogether we get for Ψ the expression

$$\Psi(\alpha,\beta,\lambda) = \frac{\Omega}{b} \left\{ \ln(1+\lambda) - \frac{\lambda}{1+\lambda} \ln \lambda + \frac{\beta a}{b} \left(\frac{\lambda}{1+\lambda} \right)^2 + \frac{\lambda}{1+\lambda} \ln(b\zeta) \right\}, \quad (17)$$

where ζ is the function of α and β defined in (10).

8. The extrema of Ψ satisfy the equation

$$0 = \frac{b}{\Omega} \frac{\partial \Psi}{\partial \lambda}, \quad \text{or}$$

$$\ln \frac{b\zeta}{\lambda} + 2 \frac{\beta a}{b} \frac{\lambda}{1+\lambda} = 0, \quad (18)$$

which determines λ as function of α and β.

The pressure is related to $A = kT \max \Psi$ through the equation

$$\mathcal{P} = -\frac{\partial A}{\partial \Omega} = -\frac{A}{\Omega} = \frac{kT}{\Omega} \Psi,$$

where Ψ is given by (17) and where λ, α, and β are related through (18). Using (17) and (18) to eliminate α (or ζ) we get

$$\mathcal{P} = \frac{kT}{b} \ln(1+\lambda) - \frac{a}{b^2} \left(\frac{\lambda}{1+\lambda} \right)^2. \quad (19)$$

We obtain the density either by differentiating A with respect to μ or directly by starting from (14'):

$$\varrho = \frac{1}{v} = -\frac{1}{\Omega} \frac{\partial A}{\partial \mu} = \frac{1}{\Omega} \left[\frac{\partial \Psi}{\partial \alpha} + \frac{\partial \Psi}{\partial \lambda} \frac{\partial \lambda}{\partial \alpha} \right] = \frac{1}{\Omega} \frac{\partial \Psi}{\partial \alpha}$$

$$= \frac{1}{\Omega} \zeta \frac{\partial \Psi}{\partial \zeta} = \frac{\langle n_q \rangle_0}{b} = \frac{1}{b} \frac{\lambda}{1+\lambda}. \quad (20)$$

Equations (19) and (20) give a parametric representation of the equation of state. We can eliminate λ by writing (20) in the form $\lambda = b/(v - b)$, which gives us

$$\mathcal{P} = \frac{kT}{b} \ln \frac{v}{v - b} - \frac{a}{v^2}. \tag{21}$$

9. The equation of state (21) is similar to the van der Waals equation and it gives us a similar family of isotherms in the $v, \mathcal{P}$-plane (Fig.16.15). At high temperatures, low densities, or high pressures, we get again (15). However, the isotherms which we have now found do not decrease everywhere, as the derivative

$$\frac{\partial \mathcal{P}}{\partial v} = -\frac{kT}{b} \left(\frac{1}{v - b} - \frac{1}{v} \right) + \frac{2a}{v^3} \tag{22}$$

can, for fixed v, become positive, provided the temperature is sufficiently low. The critical point is obtained from the condition that

$$\frac{\partial^2 \mathcal{P}}{\partial v^2} = \frac{kT}{b} \left(\frac{1}{(v - b)^2} - \frac{1}{v^2} \right) - \frac{6a}{v^4}$$

vanishes at the same time as (22), while (21) is satisfied. This yields

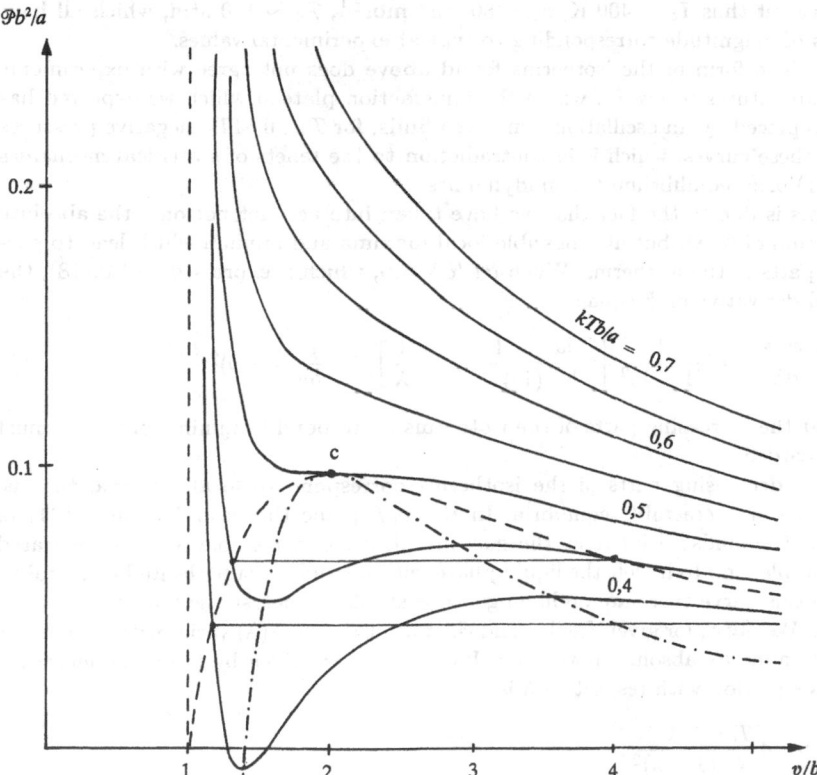

Fig. 16.15. The isotherms for the lattice gas

$$v_c = 2b, \qquad kT_c = \frac{a}{2b}, \qquad \mathcal{P}_c = \frac{a}{4b^2}(2\ln 2 - 1). \qquad (23)$$

For a fixed value of $v > b$ the pressure increases with increasing temperature so that the isotherms sweep the plane in a continuous way. When $T > T_c$, they decrease with increasing v; the critical isotherm for $T = T_c$ has an inflexion point with a horizontal tangent; when $T < T_c$ the isotherms have a maximum and a minimum given by equating (22) to zero. The curve on which the maxima and minima of $\mathcal{P}(v)$ lie is obtained by using (21) and (23) to eliminate T with $\partial \mathcal{P}/\partial v = 0$:

$$\mathcal{P}_m(v) = \frac{2a(v-b)}{bv^2}\ln\frac{v}{v-b} - \frac{a}{v^2}. \qquad (24)$$

This curve is represented by the dash-dot line in Fig.16.15; it has a maximum at the critical point.

The value of l taken from Fig.16.12 is 4 Å, so that $b = l^3 = 64 \times 10^{-30}\,\mathrm{m}^3$. Equation (9) gives

$$a = -\frac{b}{2}\left[6V(l) + 12V(l\sqrt{2}) + 8V(l\sqrt{3}) + \cdots\right],$$

where the numbers 6, 12, 8, $\cdots$ represent the number of sites on the lattice at distances l, $l\sqrt{2}$, $l\sqrt{3}$, $\cdots$ from the origin. This yields $a/b = 6.3\times10^{-2}$ eV. Using (23) we get thus $T_c \sim 400$ K, $v_c \sim 80$ cm^3 mol^{-1}, $\mathcal{P}_c \sim 150$ atm, which all have orders of magnitude corresponding to typical experimental values.

10. The form of the isotherms found above does not agree with experiments at temperatures below T_c, where the liquefaction plateau which we expected has been replaced by an oscillation. One even finds, for $T < 0.81T_c$, negative pressures along these curves, which is in contradiction to the tenets of statistical mechanics ($Z > 1$) or of equilibrium thermodynamics.

This is due to the fact that we have taken into account not only the absolute maximum of $\Psi(\lambda)$, but also possible local maxima and minima which lead to parasitic parts of the isotherms. When $\partial\Psi/\partial\lambda = 0$, which is expressed by Eq.(18), the second derivative of Ψ equals

$$\frac{b}{\Omega}\frac{\partial^2\Psi}{\partial\lambda^2} = \frac{1}{(1+\lambda)^2}\left[\frac{2\beta a}{b}\frac{1}{(1+\lambda)^2} - \frac{1}{\lambda}\right] = \frac{\beta}{bv}(v-b)^4\frac{\partial\mathcal{P}}{\partial v},$$

so that the increasing parts of the isotherms correspond to minima of Ψ and must be discarded.

The decreasing parts of the isotherms correspond to local maxima, that is, to stable or metastable equilibria. In the $v, \mathcal{P}$-plane the dash-dot curve (24) of Fig.16.15 bounds to its right the gaseous phase, either stable or supersaturated metastable, and to its left the liquid phase, either stable or superheated metastable. Inside this curve there are no homogeneous stable or metastable fluid states.

11. We must, for given fixed α and β, study the way $\Psi(\lambda)$ varies with λ, in order to determine its absolute maximum. Its extrema are given by (18); the derivative of this equation with respect to λ is

$$\frac{1}{\lambda} - 4\frac{T_c}{T}\frac{1}{(1+\lambda)^2},$$

which is positive everywhere when $T > T_c$ and changes sign twice when $T < T_c$.

Hence (18) increases when $T > T_c$ and has first a maximum and then a minimum when $T < T_c$: $\Psi(\lambda)$ has a single maximum for $T > T_c$, and it has either a single maximum, or two maxima separated by a minimum when $T < T_c$. The latter has already been placed on the rising part of the isotherm; we still have to eliminate the lower maximum.

To do this we fix T below T_c and let μ change from $-\infty$ to $+\infty$, while tracing the shape of $\Psi(\lambda)$ and the values of $\mathcal{P}$ and v associated with its maxima and minima. The equations we need are thus (10), (18), (20), and (21) which we write in differential form:

$$\frac{d\mu}{kT} = -d\lambda \left(\frac{1}{\lambda} - \frac{4T_c}{T} \frac{1}{(1+\lambda)^2} \right), \tag{25}$$

$$dv = -\frac{b\,d\lambda}{\lambda^2}, \tag{26}$$

$$d\mathcal{P} = -\frac{kT\lambda^2\,dv}{vb} \left(\frac{1}{\lambda} - \frac{4T_c}{T} \frac{1}{(1+\lambda)^2} \right). \tag{27}$$

Equation (27) is the same as (22); using (25), (26), and (27) along an isotherm we have

$$v\,d\mathcal{P} = d\mu. \tag{28}$$

We could also have obtained (28) directly by differentiating (17) and using the fact that $\partial\Psi/\partial\lambda = 0$, that $\beta\mathcal{P} = \partial\Psi/\partial\Omega = \Psi/\Omega$, that $N = \partial\Psi/\partial\alpha$, and that $U = \partial\Psi/\partial\beta$; hence we get $d\Psi = d(\beta\mathcal{P}\Omega) = \beta\mathcal{P}d\Omega + Nd\alpha - Ud\beta$ and $\Omega\,d\mathcal{P} = N\,d\mu$ for $\beta = 0$. Equation (28) is not unexpected as it can be derived without approximation in an equilibrium state from the Gibbs-Duhem relation $Nd\mu = \Omega d\mathcal{P} - SdT$. Nonetheless, the above proof shows that it is valid here even on the non-physical parts of the isotherms.

When v decreases from $+\infty$ to b, λ increases from 0 to $+\infty$, $\mathcal{P}$ increases, then decreases, and again increases along the isotherm, and μ follows the changes in $\mathcal{P}$, starting from $-\infty$ and ending up at $+\infty$. As the value of $\Psi(\lambda)$ in the stationary points equals $\mathcal{P}\Omega/kT$ we must choose from among the one to three $v, \mathcal{P}$ points associated with the same value of μ the one which corresponds to the maximum value of $\mathcal{P}$. We shall make this choice graphically by drawing in the same diagram (Fig.16.16) the isotherm considered and the associated $\mu(v)$ curve, which satisfies (28).

For values of μ less than the minimum μ_1 of $\mu(v)$ or larger than its maximum μ_2 there is only a single solution which of course we must keep. For intermediate values there are three solutions $v_1 > v_3 > v_2$, where v_1 and v_2 correspond to the maxima of $\Psi(\lambda)$ and v_3 to its minimum. As the values of μ are the same in the three points v_1, v_2, v_3, equation (28) then implies the relation

$$\int_{v_2}^{v_1} dv\, v\, \frac{d\mathcal{P}}{dv} = 0, \tag{29}$$

and the analogous equation for v_2 and v_3. This equation (29) expresses that the total algebraic area between the isotherm, from the point v_1 to the point v_2, the straight lines $\mathcal{P} = \mathcal{P}_1, \mathcal{P} = \mathcal{P}_2$, and the $v = 0$ axis must be zero (this area has been hatched from right to left in Fig.16.16 for the points v_2 and v_3).

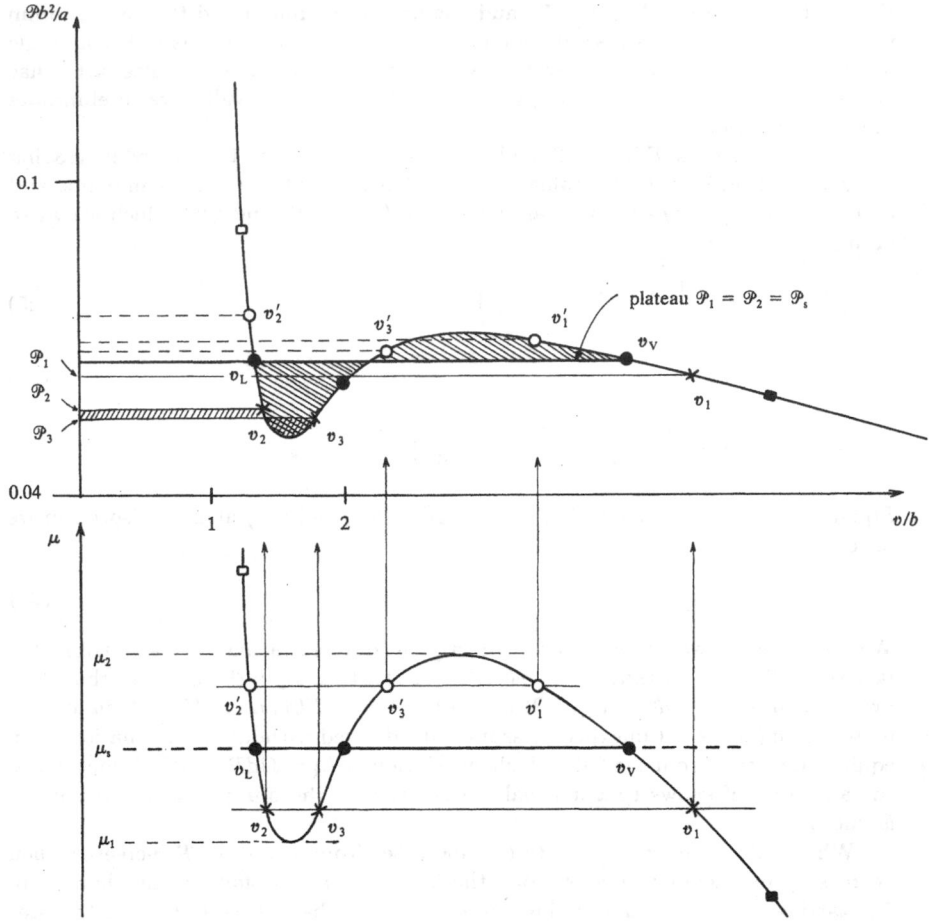

Fig. 16.16. Variational proof of the Maxwell construction. If several values of v correspond to a given chemical potential μ, we must choose the one which leads to the largest pressure $\mathcal{P}$

The two maxima of $\Psi(\lambda)$ have the same height when $\mathcal{P}_1 = \mathcal{P}_2 > \mathcal{P}_3$. The corresponding points (black circles v_V and v_L in Fig.16.16) are obtained by using (29). Hence the *Maxwell construction* follows, which expresses the equality of the areas, hatched from left to right in Fig.16.16. Let $\mu_s(T)$ be the value of the chemical potential corresponding to that situation and $\mathcal{P}_s$ the common value of the maximum $\mathcal{P}_1 = \mathcal{P}_2 = \mathcal{P}_s$. Algebraically, these values are obtained by writing down that in the points v_V and v_L the pressures (21) and the chemical potentials, given by (18) and (20), are equal, that is,

$$\mathcal{P}_s = \frac{kT}{b} \ln \frac{v_L}{v_L - b} - \frac{a}{v_L^2} = \frac{kT}{b} \ln \frac{v_V}{v_V - b} - \frac{a}{v_V^2}, \tag{30}$$

$$\mu_s + \frac{3}{2} kT \ln \frac{2\pi mkT}{h^2} = -kT \ln(v_L - b) - \frac{2a}{v_L} = -kT \ln(v_V - b) - \frac{2a}{v_V}. \tag{31}$$

Using (30) and (31) to eliminate v_L and v_V gives us $\mu_\mathrm{s}(T)$ as well as the value of $\mathcal{P}_\mathrm{s}(T)$ corresponding to the plateau produced by the Maxwell construction.

In order to conclude the discussion let us invoke continuity, letting μ increase from $-\infty$ to $+\infty$ for fixed T. We start with a single solution, $v_1, \mathcal{P}_1$, situated on the right-hand part of the isotherm and rising from right to left along this part. When μ passes μ_1, a supplementary pair, $v_2, \mathcal{P}_2$ and $v_3, \mathcal{P}_3$, appears at the minimum of the isotherm. Condition (29) shows that $\mathcal{P}_1 > \mathcal{P}_2 = \mathcal{P}_3$ so that the point $v_1, \mathcal{P}_1$ remains the correct one. From (28) it follows that $v_1\, d\mathcal{P}_1 = v_2\, d\mathcal{P}_2 = v_3\, d\mathcal{P}_3$ so that the point $v_2, \mathcal{P}_2$ is the one which rises the fastest and the point $v_1, \mathcal{P}_1$ the one which rises the most slowly (crosses in Fig.16.16). The cross-over occurs at the value $\mu = \mu_\mathrm{s}$ which we have already determined, where $\mathcal{P}_1 = \mathcal{P}_2 = \mathcal{P}_\mathrm{s}$. Thereupon the representative point jumps from 1 to 2. Beyond μ_s the point v_2' corresponds to the absolute maximum (open circles in Fig.16.16). After that, when μ passes μ_2 the points v_1' and v_3' meet and disappear at the maximum of the isotherm and the point v_2' continues alone to climb along the left-hand part of the isotherm.

Altogether, the useful isotherm consists of the two extreme parts separated by the horizontal segment between v_V and v_L. When μ passes $\mu_\mathrm{s}(T)$ the volume undergoes a discontinuity from v_V corresponding to the saturated vapour to v_L corresponding to the liquid, while the pressure remains constant. As the temperature changes, the points v_V and v_L describe in the $v, \mathcal{P}$-plane the saturation curve inside which there *cannot exist a homogeneous stable phase* (dashed line in Fig.16.15). Qualitatively, the model therefore describes the liquid-vapour transition satisfactorily.

12. Let us again trace the isotherms, letting μ increase from $-\infty$ to $+\infty$. As long as $T > T_\mathrm{c}$ the pressure increases from 0 to $+\infty$ and the volume decreases continuously from $+\infty$ to b; there is only a single phase. When $T < T_\mathrm{c}$ the pressure first increases from 0 to a value $\mathcal{P}_\mathrm{s}(T)$ corresponding to liquefaction and determined by using (30) and (31) to eliminate v_L and v_V; we are thus in the vapour phase where the vapour becomes saturated at the pressure $\mathcal{P}_\mathrm{s}$. After that, at the same pressure, the density increases suddenly without the representative point in the $T, \mathcal{P}$-plane changing; we switch to the liquid phase. Thereafter the pressure increases again. The $\mathcal{P}_\mathrm{s}(T)$-curve separates the representative regions of the two homogeneous, vapour and liquid, stable phases (which are continuously related by passing through temperatures $T > T_\mathrm{c}$). This curve stops at the critical point, $T_\mathrm{c}, \mathcal{P}_\mathrm{c}$. The function $\mathcal{P}_\mathrm{s}(T)$ increases, since the isotherms corresponding to decreasing values of T, and especially their horizontal segments, move from up to down in the $v, \mathcal{P}$-plane.

Note. To find the exact shape of the saturation curve (Fig.16.17) it is useful to rewrite equations (30), (31) which determine that curve, using (23), as

$$\ln \frac{v_\mathrm{V}}{v_\mathrm{L}} = \ln \frac{v_\mathrm{V} - b}{v_\mathrm{L} - b}\left(1 - \frac{b}{2}\frac{v_\mathrm{V} + v_\mathrm{L}}{v_\mathrm{V} v_\mathrm{L}}\right), \tag{32}$$

$$\frac{T_\mathrm{c}}{T} = \frac{v_\mathrm{V} v_\mathrm{L}}{4b(v_\mathrm{V} - v_\mathrm{L})}\ln \frac{v_\mathrm{V} - b}{v_\mathrm{L} - b}, \tag{33}$$

$$\frac{\mathcal{P}_\mathrm{c} - \mathcal{P}_\mathrm{s}}{\mathcal{P}_\mathrm{c}} = \frac{1}{2\ln 2 - 1}\left\{2\ln 2 - \frac{T}{T_\mathrm{c}}\ln \frac{v_\mathrm{V} v_\mathrm{L}}{(v_\mathrm{V} - b)(v_\mathrm{L} - b)}\right.$$

$$\left. + \frac{2b^2}{v_\mathrm{V}^2 v_\mathrm{L}^2}\left(v_\mathrm{V}^2 + v_\mathrm{L}^2\right) - 1\right\}. \tag{34}$$

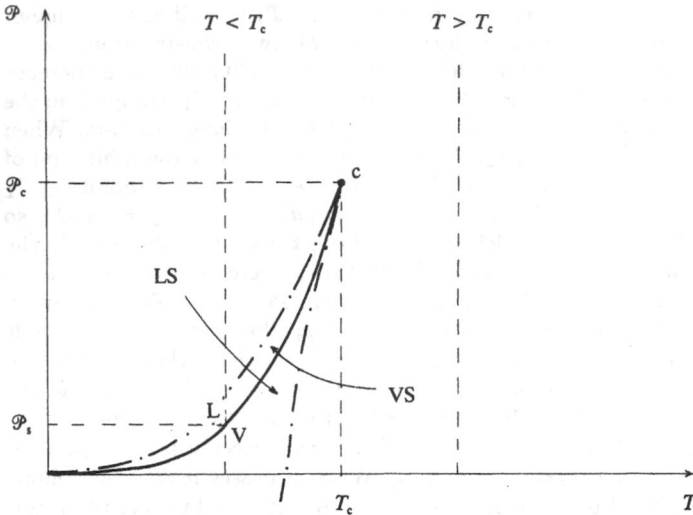

Fig. 16.17. The phase diagram in the $T, \mathcal{P}$-plane. The full drawn curve separates the stable phases: the liquid L and the vapour V; the latter are continued towards the metastable regions bounded by the dash-dot lines: superheated liquid LS and supersaturated vapour VS

Equations (33) and (34) together give us a parametric representation of the $\mathcal{P}_s(T)$-curve in terms of the variables v_V and v_L which are related through (32). If we want to determine how the curve behaves at its critical end, we can expand them:

$$v_V - v_c \sim v_c - v_L \sim b\varepsilon,$$

$$\frac{T_c - T}{T_c} \sim \frac{\varepsilon^2}{12}, \qquad \frac{\mathcal{P}_c - \mathcal{P}}{\mathcal{P}_c} \sim \frac{2\ln 2}{2\ln 2 - 1} \frac{T_c - T}{T}, \tag{35}$$

while at the other end $(T = 0)$ we have

$$v_V \sim \frac{b}{\varepsilon}, \qquad v_L \sim b(1 + \varepsilon),$$

$$\frac{T}{T_c} \sim \frac{2}{-\ln \varepsilon}, \qquad \frac{\mathcal{P}}{\mathcal{P}_c} \sim \frac{4}{2\ln 2 - 1} \frac{\varepsilon}{-\ln \varepsilon}. \tag{36}$$

In the $v, \mathcal{P}$-plane (Fig.16.15) the metastable states are bounded by the saturation curve (dashed curve) and the curve (24) of the minima of the isotherms (dash-dot curve). We have already drawn the first in the $T, \mathcal{P}$-plane (Fig.16.17). We get the second by using (21) and $\partial \mathcal{P}/\partial v = 0$ given by (22) to eliminate v. This leads to the parametric representation

$$kT = 2a\frac{v - b}{v^2}, \qquad \mathcal{P} = \frac{2a}{b}\frac{v - b}{v^2}\ln\frac{v}{v - b} - \frac{a}{v^2} \tag{37}$$

of the dash-dot curve which bounds the metastability regions in Fig.16.17. This curve has two branches starting from the critical point, one above (for $v \geq v_c$) and one below (for $v \leq v_c$) the saturation curve; these three curves bound two regions. In the VS region the stable phase is the liquid phase which we have already considered, but for the same values of T and $\mathcal{P}$ there can exist also a metastable supersaturated vapour phase. Similarly, in the LS region the same point represents a stable vapour phase and a metastable superheated liquid . The LS and VS metastability regions can be considered as the extensions on another sheet of the L and V stability regions.

Note. Using expansions we can give more details about the curve (37). In the vicinity of the critical point, writing $v = b(2 + \varepsilon)$ we get

$$\frac{T_c - T}{T_c} \approx \frac{\varepsilon^2}{4} - \frac{\varepsilon^3}{4}, \qquad \frac{\mathcal{P}_c - \mathcal{P}}{\mathcal{P}_c} \approx \frac{2\ln 2}{2\ln 2 - 1}\frac{\varepsilon^2}{4} - \frac{3\ln 2 + 1}{2\ln 2 - 1}\frac{\varepsilon^3}{6},$$

so that the curve bounding the metastability region shows a cusp at the critical point with a tangent which is the same as that of the saturation curve (Fig.16.17). As $v \to \infty$, we have

$$\frac{\mathcal{P}}{\mathcal{P}_c} \sim \frac{1}{2\ln 2 - 1}\left(\frac{T}{2T_c}\right)^2 \to 0.$$

As $v \to b$, we have

$$\frac{\mathcal{P}}{\mathcal{P}_c} \to -\frac{4}{2\ln 2 - 1}, \qquad T \to 0.$$

13. In the transition from liquid to vapour, at constant chemical potential, pressure, and temperature, there is an increase $\Delta v = v_V - v_L$ in volume, per molecule, and an increase ΔS in the entropy. The latter is given by the relation

$$S = -\frac{\partial A}{\partial T} = \frac{\partial}{\partial T}(kT\Psi),$$

where for Ψ we must take its absolute maximum. Using the stationarity of Ψ with respect to λ we need only consider the explicit dependence of (17) on T through β and $\zeta(T, \mu)$ given by (10). We get

$$S = k\Psi - \frac{\Omega a}{b^2}\left(\frac{\lambda}{1+\lambda}\right)^2\frac{1}{T} + \frac{kT\Omega}{b}\frac{\lambda}{1+\lambda}\left(-\frac{\mu}{kT^2} + \frac{3}{2T}\right).$$

Using (20) and the relation $\Psi = \mathcal{P}\Omega/kT$ we can write this expression in the form

$$S = \frac{\mathcal{P}\Omega}{T} - \frac{\Omega a}{Tv^2} + \frac{k\Omega}{v}\left(\frac{\mu}{kT} + \frac{3}{2}\right), \tag{38}$$

so that we get for the entropy per molecule

$$s = \frac{S}{N} = \frac{\mathcal{P}v}{T} - \frac{a}{Tv} - \frac{\mu}{T} + \frac{3}{2}k.$$

At the vaporization this entropy increases by

$$s_V - s_L = \frac{\mathcal{P}}{T}(v_V - v_L) - \frac{a}{T}\left(\frac{1}{v_V} - \frac{1}{v_L}\right),$$

so that the change in state is accompanied by an absorption of an amount of heat per mole equal to

$$L = TN_A(s_V - s_L) = \left(P + \frac{a}{Tv_L v_V}\right)N_A\left(v_V - v_L\right). \tag{39}$$

To prove the *Clapeyron relation* we note that the preceding calculation implies that

$$s = v\frac{\partial}{\partial T}\left(\frac{kT}{\Omega}\Psi\right) = v\frac{\partial P}{\partial T}\bigg|_\mu, \tag{40}$$

where $\mathcal{P}(T, \mu; \lambda)$, expressed through (17), satisfies the relations

$$\frac{\partial P}{\partial \mu} = \frac{1}{v}, \qquad \frac{\partial P}{\partial \lambda} = 0.$$

The derivative of $\mathcal{P}$ along the saturation curve is given by the formula

$$\begin{aligned}
\frac{d\mathcal{P}}{dT} &= \frac{\partial \mathcal{P}_V}{\partial T} + \frac{\partial \mathcal{P}_V}{\partial \mu}\frac{d\mu}{dT} + \frac{\partial \mathcal{P}_V}{\partial \lambda}\frac{d\lambda}{dT} \\
&= \frac{\partial \mathcal{P}_V}{\partial T} + \frac{1}{v_V}\frac{d\mu}{dT} = \frac{\partial \mathcal{P}_L}{\partial T} + \frac{1}{v_L}\frac{d\mu}{dT},
\end{aligned} \tag{41}$$

and we note that $\mathcal{P}$, T, and μ are continuous across the curve, but that this is not the case for the derivatives $\dfrac{\partial P}{\partial T}$ or $\dfrac{\partial P}{\partial \mu}$. Combining (40) and (41) we find

$$\begin{aligned}
s_V &= v_V\frac{\partial \mathcal{P}_V}{\partial T} = v_V\frac{d\mathcal{P}_s}{dT} - \frac{d\mu}{dT}, \\
s_L &= v_L\frac{d\mathcal{P}_s}{dT} - \frac{d\mu}{dT},
\end{aligned} \tag{42}$$

and hence

$$L = T N_A \left(v_V - v_L\right)\frac{d\mathcal{P}_s}{dT}. \tag{43}$$

16.8 Phase Diagram of Bromine

Figure 16.18 represents the experimentally measured phase diagram of bromine: the temperatures are in degrees centigrade and the pressures on a logarithmic scale in mm Hg.

We want to explain the general trend of the solid-gas (sublimation) and the solid-liquid (melting) coexistence curves using a rudimentary model. We assume that the three phases, solid, liquid, and gas, consist of Br_2 molecules which we shall treat as point particles. In the gas phase these point particles are non-interacting and have three translational degrees of freedom – perfect classical monatomic gas approximation. In the solid phase the Br_2 molecules

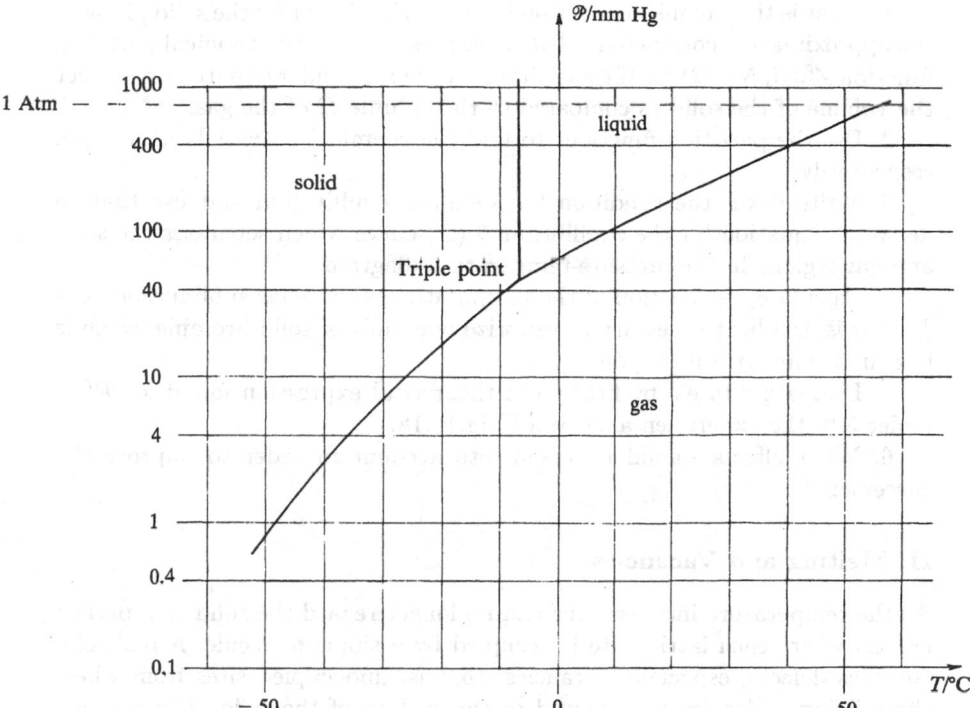

Fig. 16.18. The phase diagram of bromine. The triple point is given by $T_t = -7.3°$C and $\mathcal{P}_t = 43$ mm Hg

occupy the sites of a rigid lattice and have no degrees of freedom; we neglect the possibilities that they can vibrate around their equilibrium positions and that the size of the lattice mesh can change.

The necessary numerical data have been given in Fig.16.18 or can be found at the end of this volume. The atomic mass of bromine is 80 g mol^{-1}, its mass per unit volume in the solid state is 3.1 g cm^{-3}, while that of mercury under normal pressure and temperature is 13.6 g cm^{-3}.

The parts I and II of this problem are independent of each other.

I. Sublimation

In the solid-gas equilibrium region the temperature is low. We may thus assume that the solid phase is represented by a single micro-state such that each lattice site is occupied by a single Br$_2$ molecule which remains fixed to that site. We take the energy of that state as the energy origin and we denote by ε the energy necessary to remove one bromine molecule from the surface of the solid into the vapour. The indices "s" and "g" will refer to the solid and the gas.

1. What is the canonical partition function $Z_C^s(\beta, N^s)$ for the solid phase in the approximation considered? Give an expression for the canonical partition function $Z_C^g(\beta, N^g, \Omega)$ for the gas phase. In the remainder of part I we neglect the volume of the solid as compared to the volume Ω of the gas.

2. Use the partition functions to find the chemical potentials μ^s and μ^g, respectively.

3. Write down the condition for solid-gas equilibrium and use that to derive an equation for the equilibrium $\mathcal{P}(T)$-curve which separates the solid and gas regions in the pressure-temperature diagram.

4. Calculate, as function of the temperature, the molar sublimation heat L, that is, the heat necessary to vaporize one mole of solid bromine which is in equilibrium with its vapour.

5. Evaluate ε in eV by fitting the theoretical expression found for $\mathcal{P}(T)$ under 3 to the experimental curve of Fig.16.18.

6. What effects should be taken into account in order to improve the agreement?

II. Melting and Vacancies

As the temperature increases one can no longer regard the solid as a perfect crystal where each lattice site is occupied by a single molecule. A real solid contains defects, especially vacancies, that is, unoccupied sites from where the missing molecules have moved to the surface of the solid. The creation of a vacancy needs an energy ε' and induces a dilatation of the solid. We denote by v the volume occupied on average by each molecule in the perfect crystal. We assume that the vacancies do not interact with one another so that n vacancies have an energy $n\varepsilon'$ independent of their positions. The density α of the vacancies, that is, the fraction of unoccupied sites in the lattice in thermal equilibrium is a function of the temperature and the pressure. It increases with temperature and we shall assume that the cystalline structure becomes unstable, which means the melting of the solid, when this density α of the vacancies reaches a certain threshold $\alpha_0 = 10^{-3}$. To simplify the calculations we choose to describe the equilibrium solid in an isobaric-isothermal ensemble.

1. Give expressions for the isobaric-isothermal partition function Z_i and the free enthalpy G of the solid. Neglect the degeneracy connected with the number of arrangements of the displaced molecules on the surface of the crystal, so that the micro-states are simply characterized by giving the position of the vacancies.

2. Give an expression for the volume Ω of the solid in thermal equilibrium at a pressure $\mathcal{P}$.

3. Evaluate the density α of the vacancies at equilibrium as a function of the temperature and the pressure. Use the criterion for melting given above to find a theoretical expression for the melting curve $\mathcal{P}(T)$.

4. Determine the value of ε' in eV from a comparison with experimental data.

5. At a temperature of 20°C bromine is liquid under atmospheric pressure. In the model considered what pressure (in atmospheres) is necessary to solidify it at the same temperature?

6. We consider solid bromine at atmospheric pressure and a temperature close to the melting point. Calculate in the above model its coefficient of thermal expansion and its molar specific heat (at constant pressure).

7. There may be other effects apart from the thermal excitation of vacancies. In particular, the dilatation of a solid is not solely due to the creation of vacancies which we have considered above, but also to the change in the size of the crystalline lattice mesh. On the other hand, the specific heat contains not only the contribution from the vacancies, but also one from the vibrations of the molecules about their equilibrium positions. We also note that the formation and annihilation of vacancies needs migration of molecules, a process which may take a time of the order of minutes. Can you suggest an experimental technique to check the results obtained in 6, that is, to measure separately the contributions from the vacancies and those from other degrees of freedom to the dilatation and to the specific heat?

8. Which of the hypotheses used to formulate the above model for finding theoretically the solid-liquid equilibrium curve do you find most questionable? How would you set out to improve the model?

Solution

I. Sublimation

1. In the model considered the solid has only a single micro-state with zero energy so that

$$Z_C^s(\beta, N^s) = 1.$$

The canonical partition function of the gas is

$$Z_C^g(\beta, N^g, \Omega) = \frac{1}{N^g!} \left[\frac{\Omega}{h^3} e^{-\beta \varepsilon} \left(\frac{2m\pi}{\beta} \right)^{3/2} \right]^{N^g}.$$

2. The chemical potentials of the two phases are

$$\mu^s = -\frac{1}{\beta} \frac{\partial}{\partial N^s} \ln Z_C^s = 0,$$

$$\mu^g = -\frac{1}{\beta} \frac{\partial}{\partial N^g} \ln Z_C^g = \varepsilon - kT \ln \left[\frac{\Omega}{N^g h^3} (2m\pi kT)^{3/2} \right].$$

We have used here in the limit of large N

$$\frac{\partial}{\partial N} \ln N! \sim \ln N! - \ln (N-1)! = \ln N,$$

which amounts to using the Stirling formula for $N!$.

Note. In actual fact, if we do not neglect the volume of the solid which is assumed to be incompressible, μ^s does not vanish. This is clear, when we try to derive μ^s from the free enthalpy $G(T, N, \mathcal{P}) = F + \mathcal{P}\Omega$, which equals $\mathcal{P}Nv$, whence $\mu^s = \partial G / \partial N = \mathcal{P}v$. What is wrong with our calculation which uses a canonical ensemble? The free energy $F(T, N, \Omega)$ is not defined, unless when $\Omega = Nv$, in which case it is equal to zero; considered as a function of $T, N, \mathcal{P}$ it is nevertheless zero for all values of T, N, and $\mathcal{P}$. These variables are not those naturally associated with F, and we must thus change variables in the usual way by substituting $\mathcal{P}(T, N, \Omega)$ into $F(T, N, \mathcal{P})$ in order to find μ and $\mathcal{P}$ from $F(T, N, \mathcal{P})$. We thus obtain in the general case

$$\mu^s = \left.\frac{\partial F}{\partial N}\right|_{\mathcal{P}} + \left.\frac{\partial F}{\partial \mathcal{P}}\right|_N \left.\frac{\partial \mathcal{P}}{\partial N}\right|_\Omega, \qquad \mathcal{P} = -\left.\frac{\partial F}{\partial \mathcal{P}}\right|_N \left.\frac{\partial \mathcal{P}}{\partial \Omega}\right|_N,$$

whence we find again, using the extensivity, the Gibbs-Duhem relation

$$\mu^s = \left.\frac{\partial F}{\partial N}\right|_{\mathcal{P}} - \mathcal{P}\left.\frac{\partial \mathcal{P}}{\partial \mathcal{P}}N\right|_\Omega \bigg/ \left.\frac{\partial \mathcal{P}}{\partial \Omega}\right|_N = \frac{F}{N} + \mathcal{P}\frac{\Omega}{N}.$$

If the solid is incompressible, $\partial F/\partial \mathcal{P}$ vanishes, but $\partial \mathcal{P}/\partial N$ and $\partial \mathcal{P}/\partial \Omega$ are infinite, which explains why we find $\mu = \mathcal{P}v \neq 0$. In practice the density of the solid is so high that $\mathcal{P}v \ll \varepsilon$ and $v \ll \Omega/N^g$, and the error made by putting $\mu^s = 0$ is negligible.

3. Equilibrium between the two phases implies equality of temperatures and of chemical potentials. We can derive this property in the canonical ensemble by looking for the maximum of the total free energy $F^s + F^g$ for variations in N^s and N^g such that $dN^s + dN^g = 0$; it is useless to invoke equality of pressure, as the volume of the solid is neglected .

To characterize the solid-gas coexistence we thus get $\mu^g = 0$. If we use the fact that

$$\mathcal{P} = -\frac{\partial F^g}{\partial \Omega} = \frac{N^g kT}{\Omega},$$

the solid-gas coexistence curve in the $T, \mathcal{P}$-phase diagram is thus given by

$$\mathcal{P} = (kT)^{5/2} \left(\frac{m}{2\pi\hbar^2}\right)^{3/2} e^{-\varepsilon/kT}.$$

The validity of the perfect gas approximation for $\mu^g = 0$ requires that $\varepsilon \gg kT$.

4. The sublimation heat L, which is the amount of heat one must supply to transform one mole of solid into gas at fixed temperature and pressure, is

$$L = T\,\Delta S = T\,N_0 \left[\frac{\partial S^g}{\partial N^g} - \frac{\partial S^s}{\partial N^s}\right]\bigg|_{T,\mathcal{P}},$$

where N_0 is the Avogadro number. Using

$$S = -\left.\frac{\partial F}{\partial T}\right|_{N,\Omega},$$

we get $S^s = 0$ and

$$S^g = N^g k \left\{ \frac{5}{2} + \ln \left[\frac{\Omega}{N^g h^3} (2m\pi kT)^{3/2} \right] \right\}$$

$$= N^g \left[\frac{5}{2} k + \frac{\varepsilon - \mu^g}{T} \right].$$

As μ^g depends solely on T and $\mathcal{P}$ it does not make any difference whether we take the derivative of S^g for fixed T and $\mathcal{P}$ or for fixed T and μ^g and, hence, using the fact that $\mu^g = 0$, we find

$$L = N_0 \left[\frac{5}{2} kT + \varepsilon \right].$$

Another method consists in writing

$$L = \Delta U + \mathcal{P} \, \Delta \Omega = N_0 \left[\frac{\partial H^g}{\partial N^g} - \frac{\partial H^s}{\partial N^s} \right] \Bigg|_{T,\mathcal{P}},$$

where H is the enthalpy

$$H = U + \mathcal{P}\Omega = \left(-\frac{\partial}{\partial \beta} + \frac{\Omega}{\beta} \frac{\partial}{\partial \Omega} \right) \ln Z_{\mathrm{C}}(N, \Omega, \beta),$$

that is,

$$H^s = 0, \qquad H^g = N^g \left[\frac{5}{2} kT + \varepsilon \right].$$

We get again the above expression for L, which we could also have obtained from Clapeyron's formula, neglecting the volume occupied by the solid.

5. Numerically, the theoretical expression for $\mathcal{P}(T)$ gives, with $\mathcal{P}$ in mm Hg, ε in eV, and T in kelvins

$$\varepsilon = 8.614 \times 10^{-5} T \left[10.58 + \ln \frac{T^{5/2}}{\mathcal{P}} \right].$$

Recording the values of T and $\mathcal{P}$ on Fig.16.18 from a pressure of 1 mm Hg up to the triple point we get values for ε which increase regularly from 0.466 to 0.476 eV so that a value of

$$\varepsilon = 0.47 \pm 0.01 \text{ eV}$$

is compatible with the experimental curve.

6. Although the model is rather a rough one, it enables us to reproduce the experimental curve with a single adjustable parameter, ε, the value of which can be determined with a reasonable accuracy. Nonetheless, if we keep ε fixed to its value calculated for $\mathcal{P} = 1$ mm Hg, we get at the triple point a pressure which is too high by 50 %. The deviation is therefore significant. To improve the theory we must find corrections to $\mu^g - \mu^s$ of the order of the margin of 0.01 eV found above for ε. The effects which we can consider for this are twofold. One the one hand, in the solid phase the *vibrations* of the Br_2 molecules about their equilibrium positions and

the creation of *defects* – vacancies – can modify μ^s by an amount which increases with temperature. On the other hand, *internal motions*, rotations and vibrations, of each molecule contribute to F^g and to F^s. If the motions were the same their contributions to F^g and F^s, and thus to μ^g and μ^s would be equal and their effect could then not be seen on the $\mathcal{P}(T)$-curve. However, one should note that in the solid phase the presence of neighbouring molecules hinders the thermal excitation of molecular rotations. This leads to a contribution to the difference $\mu^g - \mu^s$ which varies with temperature and one should add it to ε.

II. Melting and Vacancies

1. For each micro-state with n vacancies the energy equals $E_n = n\varepsilon'$ and the crystal volume Ω_n equals $(N+n)v$. We get the isobaric-isothermal partition function by summing the expression $\exp[-\beta(E_n+\mathcal{P}\Omega_n)]$ over all possible micro-states and all possible values of the volume, which, if we take into account the number of possible configurations for each set of n vacancies, leads to

$$Z_i = \sum_n \frac{N!}{(N-n)!\,n!}\, e^{-\beta(\varepsilon'+\mathcal{P}v)n - \beta\mathcal{P}vN}$$

$$= \left[1 + e^{-\beta(\varepsilon'+\mathcal{P}v)}\right]^N e^{-\beta\mathcal{P}vN}.$$

The free enthalpy of the solid is thus (see § 5.6.6)

$$G(T,\mathcal{P},N) = -NkT \ln\left[1 + e^{-(\varepsilon'+\mathcal{P}v)/kT}\right] + \mathcal{P}vN.$$

The approximation is valid only for $n \ll N$, which requires $\varepsilon' + \mathcal{P}v \gg kT$.

2. The average volume of the solid in isobaric-isothermal equilibrium is given by the formula

$$\Omega = \left.\frac{\partial G}{\partial \mathcal{P}}\right|_{T,N} = Nv\left[\frac{1}{e^{(\varepsilon'+\mathcal{P}v)/kT} + 1} + 1\right].$$

3. As the volume Ω_n of a micro-state is connected with n by $\Omega_n = (N+n)v$, we get the value of α from the average volume Ω:

$$\alpha = \frac{\langle n\rangle}{N} = \frac{\Omega}{Nv} - 1 = \frac{1}{e^{(\varepsilon'+\mathcal{P}v)/kT} + 1}.$$

The melting curve is thus given by $\alpha(\mathcal{P},T) = \alpha_0$, or, using the fact that α_0 is small,

$$\mathcal{P} = \frac{kT}{v} \ln\frac{1}{\alpha_0} - \frac{\varepsilon'}{v}.$$

4. We have for the volume v

$$v = \frac{0.16}{3100}\frac{1}{N_0} = 8.5 \times 10^{-29}\ \mathrm{m}^3.$$

Using the values for $\mathcal{P}$ and T at the triple point we have

$$\varepsilon' = kT_t \ln \frac{1}{\alpha_0} - \mathcal{P}_t v = 0.16 \text{ eV}$$

(the contribution from the second term is negligibly small).

5. Bromine is solid at a temperature T, provided

$$\mathcal{P} > \frac{kT}{v} \ln \frac{1}{\alpha_0} - \frac{\varepsilon'}{v} = \frac{k(T - T_t)}{v} \ln \frac{1}{\alpha_0} + \mathcal{P}_t,$$

or, for $T - T_t = 27$ K,

$$\mathcal{P} > 300 \text{ atm}.$$

6. The expansion coefficient of the solid is

$$\frac{1}{\Omega} \frac{\partial \Omega}{\partial T}\bigg|_{\mathcal{P}} \simeq e^{-(\varepsilon' + \mathcal{P}v)/kT} \frac{\varepsilon' + \mathcal{P}v}{kT^2} \simeq \frac{\alpha}{T} \ln \frac{1}{\alpha},$$

where we have used the fact that α is small; this gives for $\alpha = \alpha_0$ and $T = 266$ K

$$\frac{1}{\Omega} \frac{\partial \Omega}{\partial T} = 2.6 \times 10^{-5} \text{ per degree}.$$

The entropy $S = -\dfrac{\partial G}{\partial T}$ equals in the same approximation

$$S \simeq N_0 k \alpha \frac{\varepsilon' + \mathcal{P}v}{kT}.$$

Hence we get the specific heat

$$C = \frac{dS}{dT}\bigg|_{\mathcal{P}} \simeq N_0 k \alpha \left(\frac{\varepsilon' + \mathcal{P}v}{kT} \right)^2 = N_0 k \alpha \left(\ln \frac{1}{\alpha} \right)^2,$$

which for $\alpha = \alpha_0$ gives

$$C = 0.4 \text{ J K}^{-1}.$$

7. As the formation and annihilation of vacancies is a rather slow process a rapid cooling down from a temperature T_1 to T_2 produces a quenching, that is, it freezes in the vacancies at the positions which they occupied initially; the density of the vacancies thus remains fixed at the value $\alpha(T_1)$ even though the temperature is T_2. On the other hand, thermal equilibrium is established rapidly with regards to the other degrees of freedom, lattice vibrations and size of the mesh, which are strongly coupled to one another and to the outside, but weakly coupled to the vacancies. This enables us to produce samples which for a time of the order of minutes behave as if the temperature were T_1 as far as the vacancies are concerned, but as if it were T_2 as regards the remainder of the sample. One can thus identify experimentally their respective contributions, provided one carries out the measurements sufficiently fast.

8. The hypothesis which is the least well justified is to assume that melting takes place when α reaches a fixed threshold α_0. It is true that the crystalline structure collapses when the vacancy density is too high. However, the correct criterion to

know whether or not the solid is in a stable thermal equilibrium form for given values of T and $\mathcal{P}$ is to compare (using the general variational principle of § 4.2.2) the free enthalpy $G^{s}(T, \mathcal{P}, N)$ of the solid with that of the liquid, $G^{l}(T, \mathcal{P}, N)$: the stable phase is the one for which the free enthalpy is the lower, that is, the entropy the higher for the given constraints on T and $\mathcal{P}$ and a fixed value of N. Moreover, the chemical potential equals

$$\mu = \frac{\partial G}{\partial N} = \frac{G}{N},$$

and the solid-liquid equilibrium curve is the one corresponding to $\mu^{s} = \mu^{l}$ in the $T, \mathcal{P}$-plane. Naturally, this programme presupposes that G^{l} is evaluated for a more or less approximate model of the liquid which takes into account the spatial disorder and the correlations between molecules due to their mutual repulsions.

Another hypothesis of the model which can be criticized is the assumption that the energy of n vacancies equals $n\varepsilon'$, independent of the positions of these vacancies. In actual fact, it needs less energy to create neighbouring vacancies. As soon as the number of vacancies becomes relatively large they tend to aggregate in groups, and to form bubbles in the solid which are the nuclei for the liquefaction. Taking this effect into account implies the calculation of corrections to G^{s} due to the interactions between vacancies.

Similarly, G^{s} contains a contribution due to the lattice vibrations.

Finally, we have assumed that the surface entropy can be neglected compared to the bulk entropy, that is, that the number of configurations due to the vacancies was equal to $N!/n!(N-n)!$. In actual fact, this number is larger, as the molecules which have migrated to the surface can take up different positions on the surface. There are $N' \sim aN^{2/3}$ such positions, which leads to a supplementary combinatory factor which has a maximum of the order of

$$\frac{N'!}{\left(\frac{1}{2}N'\right)!\left(\frac{1}{2}N'\right)!} \sim 2^{aN^{2/3}},$$

when half of the surface sites are occupied. The corresponding surface entropy is thus of the order of $kaN^{2/3}\ln 2$. The hypothesis consisting in neglecting this contribution as compared to extensive terms of the order of kN is thus well justified, except for small size solid grains, of the order of tens of lattice constants in each direction, for which $N^{1/3}$ would be of the order of unity.

16.9 White Dwarfs

White dwarfs are stars which have the remarkable property that their size is small, of the order of a few Earth radii, while their mass is comparable to that of the Sun. We shall study a model of them by first considering a gas of free electrons in the gravitational potential of the star, and then improving the model, making it more and more realistic.

1. Statistical Mechanics of an Electron Gas

In the first two, preliminary, parts we study a system of fermions consisting of N non-interacting spin-$\frac{1}{2}$ particles in a box of volume Ω. We first consider electrons with a number per unit volume equal to $N/\Omega = 10^{38}\,\text{m}^{-3}$.

a. Give the numerical value of the momentum p_F corresponding to the Fermi level, assuming the temperature of the system to be zero. Use this value to deduce that the electrons at the Fermi level are ultra-relativistic, that is, that $p_F \gg m_e c$, where m_e is the electron rest mass and c the speed of light. Calculate the position of the Fermi level $\varepsilon_F = k\Theta_F$ for $T = 0$. We take the rest mass energy of a particle as the zero of its kinetic energy so that this energy is $\varepsilon_p = \sqrt{p^2 c^2 + m^2 c^4} - mc^2$.

b. The system is at a non-zero temperature T ($kT = 10^4$ eV is the order of magnitude of the maximum temperature for a white dwarf). Show that it is legitimate to use the approximation $T = 0$ in calculating the pressure $\mathcal{P}$ and the average energy U of the electron gas. This pressure is often called the *degeneration pressure*.

c. Compare the orders of magnitude of the characteristic energies of the electron gas, the Fermi energy, the average energy, U/N, of the electrons, the temperature kT, with the ionization energy of a hydrogen atom (13.6 eV).

d. Write down the density of states at the Fermi surface and use its expression to find the specific heat of the electron gas at a temperature T. Compare this with the specific heat of a perfect gas at the same density.

2. Statistical Mechanics of a Proton Gas

Let now the fermions be protons of mass $m_p \simeq 2 \times 10^3 m_e$ with a number per unit volume again equal to $N/\Omega = 10^{38}\,\text{m}^{-3}$. Neglect again the charge.

a. Show that at the same temperature T, such that $kT = 10^4$ eV, the proton system behaves as a non-relativistic Maxwell-Boltzmann gas.

b. Compare the average energy U/N of the protons at a temperature T with that of the electrons at the same temperature. Compare the pressures exerted by the protons and the electrons on the walls of the box.

c. What approximations can we make when we consider a system of N electrons and N protons without interactions at the above temperature T? What happens when we replace the protons by nuclei, keeping the charge density the same?

3. Properties of White Dwarf Matter

White dwarfs are stars with a mass M which is a fraction of the solar mass, $M_\odot$, and which in contrast to the Sun have already "burned up" most of their hydrogen, except at the surface. There is thus little energy production through nuclear reactions and their radiation comes simply from cooling off. Their luminosity is thus very low, about 10^{-2} to 10^{-4} times that of the Sun.

Their interior consists of nuclei, from helium to iron, with varying atomic number A and charge Ze, where $\lambda \equiv A/Z \simeq 2$, and electrons, all nearly in thermal and gravitational equilibrium. It follows from the results of the previous questions that the interior is *completely ionized*: it is a *plasma*, that is, a superposition of gases of electrons and nuclei. The average kinetic energy of the electrons is sufficiently high that one can neglect that of the nuclei. The latter ensure the electrical neutrality and determine the mass of the star. The density of electrons is such that they can be treated as a "cold" Fermi gas and that we can make for them the $T = 0$ approximation. We examine two limiting situations: one where all electrons in the star are non-relativistic and one where the electrons at the Fermi level are ultra-relativistic ($p_F \gg m_e c$). We begin with neglecting gravitational and Coulomb interactions and treat the electrons as a gas with uniform density $n_{el} \equiv N/\Omega$.

a. Find for the non-relativistic case the kinetic internal energy density U/Ω and the pressure $\mathcal{P}$ of the stellar matter as functions of the electron density n_{el}.

b. Do the same assuming that all electrons are ultra-relativistic.

c. Give expressions for the density n_{el}, the internal energy density U/Ω, and the pressure $\mathcal{P}$ as functions of the dimensionless variable $\varphi \equiv p_F/m_e c$ which characterizes the Fermi momentum in the general (relativistic) case, where φ is neither small nor large as compared to unity. Check the preceding results. The integrals which appear can be evaluated by using the formula

$$8 \int x^2 \sqrt{x^2 + 1}\, dx = (2x^3 + x)\sqrt{x^2 + 1} - \ln(x + \sqrt{x^2 + 1}).$$

4. Typical Sizes and Limit Mass

We want to determine the equilibrium size of the star which we first assume to be homogeneous with a uniform density and to be electrically neutral. To do this we first of all consider a hypothetical situation where, with the mass M fixed, the radius $\tilde{R}$ takes on arbitrary values. We shall then vary $\tilde{R}$ to determine its equilibrium value R. The gravitational energy of the star is dominated by the mass of the nuclei, equal to $Am_p = \lambda Z m_p$; we bear in mind that this energy equals $-3GM^2/5\tilde{R}$ for a uniform density.

a. Write down the relation between the stellar mass M, its radius $\tilde{R}$, and the variable φ. What is the density ϱ_1 for which $\varphi = 1$?

b. Use the expression for the total energy of the star to find the equation from which we can determine φ and hence the equilibrium radius R. Show more generally that, if a star is treated as a homogeneous sphere, its mass, its radius, and its pressure, as given by the equation of state, are related to one another.

c. Discuss the equation for φ first by finding the behaviour of R as function of the mass M in the non-relativistic ($\varrho \ll \varrho_1$) and the ultra-relativistic

$(\varrho \gg \varrho_1)$ limits and after that by studying the variation of R and the stellar density with the mass M.

d. Show that there exists a maximum mass M_1, the so-called *Chandrasekhar limit*, above which there can be no white dwarf. Estimate $M_1/M_\odot$ for $\lambda = 2$ in the context of the approximations made; the solar mass $M_\odot$ is 2×10^{30} kg.

e. Estimate the radius and the mass density ϱ of a white dwarf with a mass which is half that of the Sun.

5. Mass and Charge Distributions

We shall now take into account the compressibility of the star, dropping the hypothesis of a uniform density made in 4. We should write down the equilibrium conditions, including the effects both of the (Newtonian) gravitational field produced almost completely by the nuclei and of the Coulomb field due to the charges of the electrons and the nuclei. The star is spherically symmetric so that quantities such as the mass density $\varrho(r)$, the gravitational field $W(r)$, connected to ϱ through the equation $\nabla^2 W = 4\pi G \varrho$, and the electron Fermi momentum $m_e c \varphi(r)$ are functions of the distance r from the centre of the star.

a. The energy of an electron contains, apart from the kinetic energy that we accounted for in part 3, a gravitational and a Coulomb contribution. As the electrical forces are much stronger than the gravitational forces $(e^2/4\pi\varepsilon_0 G(2m_p)^2 \simeq 3 \times 10^{35})$, there is a considerable tendency for the matter to remain neutral. This entails that, if we transport electrons from one point to another, they will drag along nuclei with on average a charge equal and opposite to that of the electrons. We shall therefore, in order to calculate the combined effects of electrostatics and gravitation, assume here that everything takes place as if the electrons had a mass $Am_p/Z \simeq 2m_p$ instead of m_e. Use that hypothesis, which will be justified under 5d, to give an expression for the chemical potential μ_{el} of the electrons in terms of $\varphi(r)$ and $W(r)$. Use this expression to find the differential equation which $\varphi(r)$ must satisfy and which therefore, in principle, determines $\varrho(r)$.

b. Show that the same result would have been obtained macroscopically by requiring that the pressure of the electron gas found in 3 should at each point in the star be balanced by the hydrostatic pressure.

c. Discuss, without solving, the equation for $\varphi(r)$. Show, in particular, that *the star has a sharp surface* beyond which the density is negligible and which determines its radius, and that all properties found under 4c and 4d for the stellar radius and mass remain unchanged, apart from some numerical factors which we shall not try to determine.

d. Treating the nuclei as a classical perfect gas in the gravitational field $W(r)$, show that they cannot be at equilibrium at the same time as the electrons under the hypotheses made above; to simplify the considerations assume that there exists only one type of nuclei of mass Am_p and charge

Ze. It is therefore necessary to reconsider the hypotheses of 5a, allowing the possibility of violating the electrical neutrality. The charge density $\sigma(r)$ which thus appears produces an electrical potential $\phi(r)$. Write down the chemical potentials μ_{el} and μ_{N} of the two gases (of electrons and of nuclei) in the potentials $W(r)$ and $\phi(r)$. Use them to find the differential equations which determine the mass and charge densities, $\varrho(r)$ and $\sigma(r)$. Taking care of orders of magnitude, show that in the interior of a star with a sufficiently large mass these equations give the same distribution $\varrho(r)$ as under 4c as well as a small charge density which is proportional to the density $\varrho(r)$. For which densities will significant corrections appear?

6. Electron Capture by the Nuclei

Due to the very large density of a white dwarf various nuclear reactions can take place. In particular, a proton from a nucleus can capture an electron, producing inside the nucleus a neutron and emitting a neutrino. The latter, which is very weakly coupled to the other particles, leaves the star rapidly and will be neglected in the thermodynamic balance. The nuclear reaction thus can be written schematically as

electron $+ (A, Z) \rightarrow (A, Z - 1)$.

We write $E(A, Z)$ for the binding energy of the ground state of the nucleus (A, Z) and $\varepsilon = E(A, Z - 1) - E(A, Z)$, which has a typical value of 2 MeV, for its change through electron capture.

a. Write down the equation satisfied by the chemical potentials of the electrons and the nuclei when they are in equilibrium under those reactions. Use this to derive the electron kinetic energy at the Fermi level in that equilibrium situation and the corresponding value of φ. What happens if φ is larger than this value? Or smaller?

b. Discuss qualitatively the consequences of the nuclear equilibrium realized through electron capture for the composition of white dwarfs, for the relation between their masses and radii, and for their limit mass.

7. Cooling of a White Dwarf

For the less massive white dwarfs there are no longer any nuclear reactions and the star cools down while radiating. A thermal gradient is established, as in any star, from the centre to the surface, in a nearly stationary non-equilibrium regime. Take the interior temperature T to be 10^8 K and the surface temperature T_s to be 10^4 K and consider a white dwarf with mass $M_\odot/2$ in the approximation of 4 with $Z = 8$, $A = 16$.

a. Estimate the specific heat of the stellar matter.

b. Give the order of magnitude of the time it will take for the stellar temperature to drop by 10 %.

Hints and Notes for the Latter Parts:

4. In the approximation where the matter of a star is treated as being uniform, the value R of its radius at equilibrium is found *variationally* by looking for the minimum of the total free energy $U - 3GM^2/5\tilde{R} - TS$ as function of the trial parameter $\tilde{R}$, for given M and T (see Exerc. 6e on self-gravitational equilibrium). The contribution $U - TS$, evaluated neglecting gravitation, is a function of the particle number, of T, and of the volume, such that $\partial(U - TS)/\partial\Omega = -\mathcal{P}$. We find thus at the minimum

$$0 = \frac{\partial}{\partial R}\left(U - \frac{3GM^2}{5R} - TS\right) = -\frac{3\Omega\mathcal{P}}{R} + \frac{3GM^2}{5R^2},$$

and hence the stellar radius can be found from the equation

$$\mathcal{P} = \frac{3GM^2}{20\pi R^4}.$$

For a white dwarf, $\tilde{R}$ is related to φ through the equation which gives an expression for the stellar mass

$$M = \lambda m_{\mathrm{p}} N = \lambda m_{\mathrm{p}} \frac{2\Omega}{h^3}\int_{p<p_F} d^3p = \frac{4\lambda m_{\mathrm{p}}}{9\pi}\left(\frac{\tilde{R}m_ec\varphi}{\hbar}\right)^3 \equiv \Omega\varrho_1\varphi^3.$$

The free energy of the nuclei practically reduces to the gravitational energy, while the entropy of the electrons is negligible as compared to U/T, where U is the kinetic energy of the electrons evaluated under 3. Thus, we just have to look for the minimum of the total energy $U - 3GM^2/5R$ as function of $\varphi \propto \tilde{R}^{-1}$. Thence we obtain the equation

$$\varphi^2\left(\frac{R}{R_1}\right)^2 = \left(\frac{M}{M_1}\right)^{2/3}$$

$$= \frac{3}{2\varphi^4}\ln\left(\varphi + \sqrt{1+\varphi^2}\right) + \left(\frac{1}{\varphi} - \frac{3}{2\varphi^3}\right)\sqrt{1+\varphi^2} \equiv f(\varphi),$$

where

$$R_1 \equiv \frac{3\sqrt{5\pi}}{4\lambda m_{\mathrm{p}}m_e}\left(\frac{\hbar^3}{cG}\right)^{1/2} \simeq 7500 \text{ km},$$

$$M_1 \equiv \frac{15\sqrt{5\pi}}{16(\lambda m_{\mathrm{p}})^2}\left(\frac{\hbar c}{G}\right)^{3/2} \simeq 3.5 \times 10^{30} \text{ kg} \simeq 1.7M_\odot,$$

$$\varrho_1 \equiv \frac{M_1}{\frac{4}{3}\pi R_1^3} = \frac{\lambda m_{\mathrm{p}}m_e^3c^3}{3\pi^2\hbar^3} \simeq 2 \times 10^6 \text{ g cm}^{-3}.$$

In the non-relativistic limit $\varphi \ll 1$ or $\varrho \ll \varrho_1$, we have $f(\varphi) \sim 4\varphi/5$, whence

$$\frac{R}{R_1} \sim \frac{4}{5}\left(\frac{M}{M_1}\right)^{-1/3} \sim \left(\frac{4}{5\varphi}\right)^{1/2} \gg 1;$$

in the ultra-relativistic limit $\varphi \gg 1$ or $\varrho \gg \varrho_1$ we have $f(\varphi) \sim 1 - \varphi^{-2}$, whence

$$\frac{R}{R_1} \sim \sqrt{1 - \left(\frac{M}{M_1}\right)^{2/3}} \sim \frac{1}{\varphi} \ll 1.$$

The function $f(\varphi)$ increases, whereas $\varphi^{-2}f(\varphi)$ decreases, and the density $\varrho = \varrho_1\varphi^3$ increases with φ. Therefore, the *density increases* and the *radius R decreases* with M.

When φ tends to infinity, M tends to M_1, the equilibrium radius R tends to zero and the density tends to infinity. A white dwarf with a mass larger than or equal to M_1 would be unstable and would implode. The existence of a limit mass was proved by Chandrasekhar[2].

The value $M = 0.5M_\odot = 0.3M_1$ is reached for $\varphi = 0.6$ and we have $R = 1.1R_1 = 8000$ km, a value comparable with the Earth radius and much smaller than the solar radius (700 000 km). The corresponding mass density is 4×10^5 g cm^{-3}, whereas we have only 1.4 g cm^{-3} for the Sun and 5.5 g cm^{-3} for the Earth. These values are comparable with the data found observationally for white dwarfs.

5. The solution of this part is based upon the same microscopic principles as we used for electrostatics in solids (§ 11.3.3). In the present case we must treat both the gravitational and electrostatic effects by a mean field method, replacing these *binary interactions* by a *single-particle self-consistent potential* (§ 11.2.1). As regards the single-particle properties, everything happens, as if the particles were independent, with a Hamiltonian which consists of the kinetic energy together with an effective potential which includes both a gravitational and a Coulomb part. Moreover, as this potential varies slowly in space, it is justified to treat it in each volume element as a *constant* (§ 10.3.4b). As a result, the Fermi factor is found at each point in space by adding to $\varepsilon_p - \mu$ the local potential, as we saw in § 11.3.3 for the case of the macroscopic electrostatic potential. We posed the question in such a way that this addition appears intuitively obvious. Nonetheless, as regards the global properties, the particles do not behave completely as independent entities. Especially, the gravitational energy of the star is not obtained simply by taking the sum of the potential energies of its constituent particles, since it is necessary to divide this sum by 2 (see end of § 11.2.1 and Prob.5). We have, of course, taken this into account when we wrote down the gravitational energy in question 4.

a. The mass density $\varrho(r)$ is related at each point to the electron Fermi momentum through $\varrho(r) = \varrho_1\varphi^3(r)$. From it we can derive the gravitational potential

$$W(r) = -G \int \frac{\varrho(r')\,d^3r'}{|r' - r|}.$$

Within a volume element in which $W(r)$ varies little, it simply produces a shift $\lambda m_\mathrm{p}W(r)$ in the single-electron energies ε_p. Thus, if we define $\mu_0(r)$ by

$$\mu_0(r) = \mu_\mathrm{el} + \lambda m_\mathrm{p}W(r),$$

and introduce μ_0 instead of μ_el in the Fermi factor, W drops out and the electron density is locally expressed in terms of $\mu_0(r)$ just as if there were no gravitation. We have therefore, as under 3,

---

[2] S.Chandrasekhar, Phil.Mag. **11** (1931) 592.

$$\mu_0(r) \;=\; m_e c^2 \left[\sqrt{1 + \varphi^2(r)} - 1 \right].$$

Equilibrium is expressed by stating that $T \simeq 0$ and μ_{el} are *uniform in space*. Then, by eliminating $W(r)$ and μ_{el} from the above *set of self-consistent equations*, we find a differential equation for $\varrho(r)$ or for $\varphi(r)$:

$$m_e c^2 \nabla^2 \sqrt{1 + \varphi^2} + 4\pi G \lambda m_p \varrho \;=\; 0.$$

In actual fact, the white dwarf is not at thermal equilibrium since it radiates (see last question). In the stationary regime which it reaches and which is governed by energy exchanges, mainly through radiation, the temperature decreases from the centre to the surface. However, the present results are valid since we set $T \simeq 0$.

b. The Gibbs-Duhem relation

$$\Omega \, d\mathcal{P} \;=\; N \, d\mu_0 + S \, dT,$$

which is valid for a volume element Ω if there is no potential, gives us

$$\nabla \mu_0 \;=\; \lambda m_p \frac{\nabla \mathcal{P}}{\varrho},$$

and hydrodynamic equilibrium implies that $\nabla \mathcal{P}/\varrho$ equals the force per unit mass $-\nabla W$. This leads again to the equation $\nabla \mu_{el} = 0$.

c. In terms of dimensionless quantities we can write the equation for $\varphi = (\varrho/\varrho_1)^{1/3}$ in the form

$$R_1^2 \left(\frac{d^2}{dr^2} + \frac{2}{r} \frac{d}{dr} \right) \sqrt{1 + \varphi^2} + \frac{15}{4} \varphi^3 = 0.$$

The characteristic size and density R_1 and ϱ_1 are the same as for question 4. This equation must be solved for positive values of $\varphi(r)$. Its solution depends on a dimensionless constant φ_c which is the value of $\varphi(r)$ at the centre of the star. All properties of the white dwarf and, in particular, the mass distribution are thus determined by a single parameter, say, the total mass M or the central density ϱ_c. We can integrate the equation for the function $\varphi(r)$ as:

$$\frac{d}{dr} \sqrt{1 + \varphi^2} = -\frac{15}{4R_1^2 r^2} \int_0^r r^2 \, dr \, \varphi^3(r) = -\frac{5 R_1 M(r)}{4 r^2 M_1},$$

where $M(r)$ is the mass within the sphere of radius r, so that near the origin $\varphi(r)$ behaves as

$$\varphi(r) \;=\; \varphi_c - \frac{5}{8} \varphi_c^2 \sqrt{1 + \varphi_c^2} \left(\frac{r}{R_1} \right)^2 + \cdots,$$

and decreases for any r while remaining positive. We could then imagine two scenarios. If φ becomes zero at some point R, the solution $\varphi(r)$ is acceptable for $r < R$, but its continuation beyond R has no physical significance; the density vanishes at $r = R$ and remains zero beyond that point. If φ remains positive up to infinity, the mass $M(r)$ tends to the total mass M as $r \to \infty$; the integro-differential equation for φ gives in this case after integration

$$\frac{\varphi^2}{2} \sim \frac{5R_1 M}{4r M_1}, \qquad r \to \infty;$$

however, the integral $\int_0^\infty r^2\,dr\,\varphi^3$, which must be proportional to the mass M, cannot then converge since $\varphi \propto r^{-1/2}$. The second scenario is thus excluded and the stellar mass is *completely* contained within a sphere of radius R. It equals

$$M = \frac{2M_1 R^2}{5R_1}\left(-\frac{d\varphi^2}{dr}\right)_{r=R},$$

so that $\varphi \propto (R-r)^{1/2}$ and the mass density tends to zero at the stellar surface as

$$\varrho(r) \sim \varrho_1 \left(\frac{5MR_1}{2M_1 R}\right)^{3/2}\left(1-\frac{r}{R}\right)^{3/2}.$$

In fact, in the *surface* region the *density is too low* for us to treat the electrons as a "cold" Fermi gas; the solution obtained is valid inside the region where $T \ll \Theta_F = (m_e c^2/2k)(\varrho/\varrho_1)^{2/3}$ while thermal effects would govern the atmosphere where $T > \Theta_F$.

For non-relativistic white dwarfs ($\varrho_c \ll \varrho_1$) the fact that the equation for φ is homogeneous implies that all its solutions are of the form $\varphi(r) = \varphi_c f_1(\varphi_c^{1/2} r/R_1)$ where $f_1(r/R_1)$ is a particular solution. Hence, the stellar radius and mass behave, respectively, like $R \propto R_1\varphi_c^{-1/2}$ and $M \propto M_1\varphi_c^{3/2}$. Similarly, in the core of an ultra-relativistic star, for values of r such that $\varrho(r) \gg \varrho_1$, the solutions are of the form $\varphi_c f_2(\varphi_c r/R_1)$; if φ_c is very large, such a solution extends over the larger part of the star so that $R \propto R_1\varphi_c^{-1}$ and M/M_1 takes on a value M_2/M_1, practically independent of φ_c; for slightly smaller values of φ_c an expansion shows that M is smaller than M_2 by an amount of order φ_c^{-2}. We thus get again the results of 4. A numerical solution of the equations[3] gives the coefficients $R \sim 1.33\,R_1\varphi_c^{-1/2}$, $M \sim 0.4M_1\varphi_c^{3/2}$, $\varrho_c \sim 6\bar\varrho$ in the non-relativistic case, and $R \sim 3.56\,R_1\varphi_c^{-1}$, $M_2 \simeq 0.83M_1 \simeq 1.4M_\odot$, $\varrho_c \sim 54\bar\varrho$ in the ultra-relativistic case. In particular, we find a limit mass M_2 for white dwarfs which is slightly smaller than in the approximation of a uniform density. Although the density changes considerably, the numerical results do not differ much from those which we obtained in that approximation.

d. In each point r the density $n_N(r)$ of the perfect gas of nuclei is related to the chemical potential μ_N of the nuclei through the equation

$$n_N(r) = \frac{2}{h^3}\left(2\pi A m_p kT\right)^{3/2}\exp\left[\frac{\mu_N - A m_p W(r)}{kT}\right].$$

The equilibrium of the nuclei requires that T and μ_N are uniform in space, which, together with the equation $\nabla^2 W = 4\pi G\varrho$, gives us an equation for $\varrho(r) = A m_p n_N(r)$. However, this equation does not have the same solution as the equation we got earlier for $\varrho(r)$ when we expressed the equilibrium for the electrons in neutral stellar matter.

---

[3] See, for instance, S. Weinberg, *Gravitation and Cosmology*, Wiley, New York, 1972, p.308.

If there are two potentials, the Newtonian and the Coulomb potential with $\nabla^2\phi = -\sigma/\varepsilon_0$, the chemical potentials are, for the (cold) electron gas

$$\mu_{\text{el}} = m_{\text{e}}c^2\left[\sqrt{1+\varphi^2(r)} - 1\right] + m_{\text{e}}W(r) - e\phi(r),$$

and for the gas of nuclei

$$\mu_{\text{N}} = kT\ln\left[\frac{n_{\text{N}}(r)}{2}\left(\frac{2\pi\hbar^2}{Am_{\text{p}}kT}\right)^{3/2}\right] + Am_{\text{p}}W(r) + Ze\phi(r).$$

For equilibrium we have

$$m_{\text{e}}c^2\nabla^2\sqrt{1+\varphi^2(r)} + 4\pi m_{\text{e}}G\varrho(r) + \frac{e}{\varepsilon_0}\sigma(r) = 0,$$

$$kT\nabla^2\ln n_{\text{N}}(r) + 4\pi Am_{\text{p}}G\varrho(r) - \frac{Ze}{\varepsilon_0}\sigma(r) = 0,$$

with

$$n_{\text{el}}(r) = \frac{1}{3\pi^2}\left(\frac{m_{\text{e}}c}{\hbar}\right)^3\varphi^3(r),$$

$$\varrho(r) = Am_{\text{p}}n_{\text{N}}(r) + m_{\text{e}}n_{\text{el}}(r),$$

$$\sigma(r) = Zen_{\text{N}}(r) - en_{\text{el}}(r).$$

Using the fact that $m_{\text{e}} \ll Am_{\text{p}}$ we get from the equations for $\varrho(r)$ and $\sigma(r)$

$$\frac{e}{\varepsilon_0}\sigma(r) = 4\pi\lambda m_{\text{p}}G\varrho(r) + \frac{kT}{Z}\nabla^2\ln\varrho(r),$$

$$m_{\text{e}}c^2\nabla^2\sqrt{1+\varphi^2(r)} + 4\pi\lambda m_{\text{p}}G\varrho(r) + \frac{kT}{Z}\nabla^2\ln\varrho(r) = 0.$$

In the second equation, the ratio of the last term to the first one is of order $3kT/m_{\text{e}}c^2Z\varphi$ in the ultrarelativistic limit, which for $T = 10^4$ eV gives a small number of order $0.06/Z\varphi$, and of order $3kT/m_{\text{e}}c^2Z\varphi^2$ in the non-relativistic limit, which remains small as long as the density is sufficiently high for our approximation $T \ll \Theta_{\text{F}}$ to be valid, that is, as long as $\varrho \gg 10^{-2}\varrho_1/Z^{3/2}$. Except near the stellar surface, where a correction due to the nuclei must be added to the thermal effects on the electrons, the second equation reduces to its first two terms; this justifies the results of 5c. Another result is that the last term in the first equation is small compared to the preceding one so that

$$\sigma(r) \simeq \frac{4\pi\varepsilon_0 Am_{\text{p}}G}{Ze}\varrho(r).$$

The stellar interior has a positive charge: the nuclei are attracted to the centre by gravitation which has little effect on the, lighter, electrons. The *charge* which is thus produced, and *not gravity*, retains the electrons. However, neutrality is nearly ensured locally since the Coulomb forces are much stronger than gravitation, as the smallness of the ratio

$$\frac{Zn_{\mathrm{N}}(r) - n_{\mathrm{el}}(r)}{n_{\mathrm{el}}(r)} \simeq \frac{4\pi\varepsilon_0 A^2 m_{\mathrm{p}}^2 G}{Ze^2} \simeq 3 \times 10^{-36} \, Z$$

shows. The charge in the interior of a white dwarf of one solar mass is only 300 C.

6. As in chemical equilibrium, we must have

$$\mu_{\mathrm{el}} + \mu(A, Z) \ = \ \mu(A, Z - 1).$$

It makes no difference whether or not we include in this equation the gravitational and Coulomb terms as they are the same on the two sides. The kinetic contributions to the chemical potentials of the nuclei differ only by an amount $kT \ln[n(A, Z - 1)/n(A, Z)]$ which is negligible compared to the dominant contributions $E(A, Z)$ from the energies of the nuclei at rest, so that in equilibrium $\mu_{\mathrm{el}} = \varepsilon \simeq 4m_e c^2$, whence we have $\varphi \simeq 5$. When $\varphi > 5$ the electrons are captured until their density is reduced to a value $\varrho_2 = \varrho_1 \, 5^3$. When $\varphi < 5$ nothing happens as the inverse reaction, which includes the production of anti-neutrinos, would require the presence of nuclei with a large neutron excess and these are just created only by the electron capture for dense white dwarfs.

Electron capture affects only the heaviest white dwarfs for which $\varphi_c > 5$. In this case the electron density at the centre decreases at nuclear equilibrium until $\varphi(r) \leq 5$ for all r. The sufficiently heavy white dwarfs thus contain neutron-rich nuclei, especially in the central region. The increase in the ratio $\lambda = A/Z$ implies a decrease in $M_1 \propto \lambda^{-2}$ and in $R_1 \propto \lambda^{-1}$, and an increase in $\varrho \propto \lambda$. For a fixed value of the mass, the radius is smaller and the density higher than when there is no capture; moreover, the decrease in the electron density reduces the kinetic pressure, which makes the star contract. The instability is thus increased. The limiting mass decreases for two reasons: on the one hand, M_1 and M_2 in the less coarse model of 5 decrease due to the increase in λ; on the other hand, M can no longer reach the value M_2, which corresponded to $\varphi_c \to \infty$, as φ remains everywhere bounded by 5. Chandrasekhar has shown by a realistic calculation that, in fact, the limiting mass for stability of white dwarfs is 1.2 $M_\odot$. The radius itself remains bounded from below, due to the bound on φ. The most massive white dwarfs thus have a radius of 4000 km.

7. The total specific heat of the star is

$$C \ = \ \frac{3}{2}kn_{\mathrm{N}}\Omega + \frac{\pi^2}{3}\mathcal{D}(\varepsilon_{\mathrm{F}})k^2 T \ = \ \frac{Mk}{\lambda m_{\mathrm{p}}}\left(\frac{3}{2Z} + \pi^2 \frac{kT}{m_e c^2}\frac{\sqrt{\varphi^2 + 1}}{\varphi^2}\right),$$

where $\mathcal{D}(\varepsilon_{\mathrm{F}})$ is the density of states at the Fermi surface. We have made the low-temperature approximation for the electrons, which is well justified as $T/\Theta_{\mathrm{F}} \simeq 0.1$. Nevertheless, the electron contribution dominates, by a factor 3, that of the nuclei, thanks, in particular, to the large value of Z. The stellar temperature decreases with time, due to the loss of energy through radiation from the surface, as

$$\frac{dT}{dt} \ = \ \frac{4\pi R^2 \sigma T_{\mathrm{s}}^4}{C} \ \simeq \ 1.5 \times 10^{-10} \ \mathrm{K\,s}^{-1},$$

and thus decreases by 10% in 2×10^9 years. Even though they have hardly any energy source, the white dwarfs can retain their luminosity over very long periods. In fact, the statistics of white dwarfs as function of their luminosity L shows a clear cut-off below $L = 2.5 \times 10^{-5} L_\odot$ which reflects the absence of very old, cooled white dwarfs. From that observation and from a theoretical study of cooling we can estimate the age of our Galaxy to be about 10^{10} years (Exerc. 15f).

In the above evaluation of dT/dt, we have implicitly disregarded the contraction of the star which results from the simultaneous shift in the self-gravitational equilibrium. A more correct calculation relies on Exerc. 6e, question 3. For non-relativistic matter, the self-gravitational energy E_G and the internal energy U are related by $2E_G + U = 0$. Hence, when an energy $-L\,dt = dE_G + dU$ is radiated, E_G decreases by $-2L\,dt$ whereas U *increases* by $L\,dt$. Let us denote by $U \equiv U_0 + U_1 + U_2$ its three contributions written above, which depend on R and T as $U_0 \propto R^{-2}$, $U_1 \propto T$, $U_2 \propto T^2 R^2$, respectively, while $E_G \propto R^{-1}$. From $dR/R = -dE_G/E_G = -dU/U$ and $dU = L\,dt$, we get

$$(U_1 + 2U_2)\frac{dT}{T} = L\,dt + (2U_0 - 2U_2)\frac{dR}{R} = -L\,dt\left(1 - \frac{2U_1 + 4U_2}{U}\right) \simeq -L\,dt,$$

or equivalently $C_v dT = -L\,dt$. It was therefore legitimate to completely forget about the contraction of the white dwarf and the changes that it induces in U and E_G, since $U_1 \ll U_0$ and $U_2 \ll U_0$. Note, however, that $dU_0 = 2L\,dt\,U_0/U$ is not small as compared to $dU_1 + dU_2$, but that it is nearly cancelled by the lowering $dE_G = -2L\,dt$ of the self-gravitational energy which accompanies radiation. This situation should be contrasted to the formation of stars, in which case the matter is dilute and behaves as a classical gas with $U \sim C_v T$; loss of energy by radiation then results in a *heating* $C_v dT = +L\,dt$ rather than in a cooling (Exerc. 6e). Compare this also with the stars of the main sequence such as the Sun, which remain in a stationary regime for periods of the order of 10^{10} years, as the emission is exactly compensated for by the production of thermonuclear heat in the core (Exerc. 15f). When the nuclear fuel gets exhausted, the thermal pressure can no longer balance the self-gravitational attraction, the star contracts and heats up again. If its mass lies below $1.4\,M_\odot$, it eventually ends up as a white dwarf; if its mass is larger, it implodes, its outer shells are expelled (supernova) while the residue may transform into a neutron star.

A theoretical study of the temperature gradients within the star would involve a balance equation (Chap. 15) describing *emission, transport, and absorption* of radiation inside the stellar matter. The solution of that equation determines the temperatures in the core of the star and at its surface, which we have assumed to be given (Exerc. 15f).

16.10 Crystallization of a Stellar Plasma

The central region of many white dwarfs, which are dense stars, is constituted mainly, one thinks, of carbon 12 nuclei and electrons, which together are electrically neutral and thus are a plasma. We take the order of magnitude of the temperature and of the mass density to be, respectively, $T_0 \simeq 10^8$ K and $\varrho_0 \simeq 10^{11}$ kg/m^3. We shall see that these data are close to the conditions where the plasma may appear as a crystal of carbon nuclei embedded in an electron Fermi sea, thus producing a system which is similar to a metal. The aim of the present problem is to study the conditions under which such a crystalline state exists and to find its phase diagram.

Note. All numerical values used can be found at the end of this book. The numerical results must be given in S.I. units, and you may use "astrophysical accuracy", that is, a single significant figure, everywhere. We shall use the following notation: m for the electron mass, M for the mass of a ^{12}C nucleus, $Z = 6$ for the charge of a ^{12}C nucleus, ϱ for the mass density, N for the number of ^{12}C nuclei in the given volume Ω, and N' for the number of electrons in this volume. The three parts of the problem are independent, the answers to the questions can always be given in a few lines, the algebraic calculations are easy, and the numerical applications are essential, within the accuracy asked for.

1a. Give a numerical value for the binding energy of a single electron by a carbon nucleus, using the hydrogen atom data.

b. Use these data to conclude that for the temperature of the star all carbon atoms are completely ionized.

2a. From 1 it follows that the electrons are a Fermi gas. Neglect the interactions of the electrons with one another and with the carbon nuclei. Give an expression for the Fermi temperature Θ_F as a function of the electron density N'/Ω.

b. Give a numerical estimate for Θ_F.

c. Use this value to conclude that the only part played by the electron gas is to ensure electrical neutrality at equilibrium.

3. The above results allow us to describe stellar matter as a system of ^{12}C nuclei embedded in a continuous uniform electron background which neutralizes the average charge. This matter is crystallized, if the carbon nuclei have equilibrium positions arranged on a regular lattice, around which they oscillate with an amplitude which increases with temperature. For the order of magnitude of the distance between neighbouring sites we take $d = (\Omega/N)^{1/3}$.

When a nucleus suffers a small displacement r from its equilibrium position, the Coulomb forces due to its neighbours and the continuous background of the electrons bring it back to this position and we shall assume that this restoring force has the form $F = -\alpha r/d^3$. For dimensional reasons α has the form

$$\alpha = \lambda \frac{Z^2 e^2}{4\pi\varepsilon_0},$$

where λ is a numerical constant which we shall put equal to 1.5.

There is an empirical criterion which we shall use to decide whether the crystalline phase is stable, the *Lindemann rule*: melting occurs when the mean square displacement $\Delta = \sqrt{\langle r^2 \rangle}$ exceeds a certain fraction γ of the intersite distance d. We assume here that this ratio will be $\gamma = 0.3$, which is much higher than the values close to $\frac{1}{8}$ found for ordinary crystals (§ 11.4.1).

a. Considering each nucleus to be an isotropic three-dimensional harmonic oscillator of frequency ω, which must be determined, write down its Hamiltonian H, its energy levels, and its partition function $Z_1(\beta, \omega)$, expressing the latter in terms of hyperbolic functions.

b. Use $Z_1(\beta, \omega)$ to calculate Δ; you may express $\langle r^2 \rangle$ as a derivative of $\ln Z_1$ by using the explicit expression for H.

c. Use the Lindemann rule to write down the melting curve equation which in the T, ϱ-plane bounds the region where the plasma is crystallized. It will be convenient for what follows to write the equation for this curve in a numerical form:

$$f(A\varrho^a T^b) = (B\varrho)^c,$$

where f is a hyperbolic function, a, b, and c are rational exponents, and A and B numerical constants which must be evaluated in the "astrophysical approximation".

d. Characterize in the T, ϱ-plane the region where classical statistics can be applied: $\beta\hbar\omega \ll 1$. Sketch the melting curve in this classical domain.

e. Indicate in the T, ϱ-plane the point corresponding to the limit $\beta\hbar\omega \to \infty$. Use continuity to outline the behaviour of the curve and indicate the nature of the phases in the various domains.

f. In which phase is the matter of the star considered here?

g. Assuming the plasma pressure $\mathcal{P}$ to be essentially that of the electron Fermi gas, outline qualitatively in the $T, \mathcal{P}$-plane the bounds of the region where the plasma is crystallized.

h. What striking difference exists between the phase diagram obtained here for the crystallization region and the one of ordinary matter at laboratory pressures and temperatures? Explain the origin of this difference.

Answers to the Latter Questions:

3c. The crystallization region corresponds to parameter values such that

$$\tanh \frac{\hbar(\alpha\varrho)^{1/2}}{2MkT} > \frac{3\hbar}{2(\alpha\varrho)^{1/2}} \frac{1}{\gamma^2} \left(\frac{\varrho}{M}\right)^{2/3},$$

or, numerically,

$$\tanh 20\frac{\sqrt{\varrho}}{T} > (10^{-16}\varrho)^{1/6}.$$

d. In the classical region, characterized by $\beta\hbar\omega \ll 1$, we can put the hyperbolic tangent equal to its argument and the melting curve is given by $\varrho = 10^{-12} T^3$. This approximation is valid when $20\sqrt{\varrho}/T \lesssim 0.5$, or $T \lesssim 5 \times 10^8$ K, $\varrho \lesssim 10^{14}$ kg m^{-3}.

e. In the limit as $\beta\hbar\omega = 20\sqrt{\varrho}/T \to \infty$, we have $\tanh \frac{1}{2}\beta\hbar\omega = 1$ which gives $(10^{-16}\varrho)^{1/6} = 1$, or $\varrho = 10^{16}$ kg m^{-3}, $T \to 0$. The corresponding point on the curve (Fig.16.19) lies on the ϱ-axis. Since the melting curve, which is known for small T and ϱ (classical limit), is continuous, it closes on itself, as indicated in Fig.16.19, which was obtained from a numerical calculation.

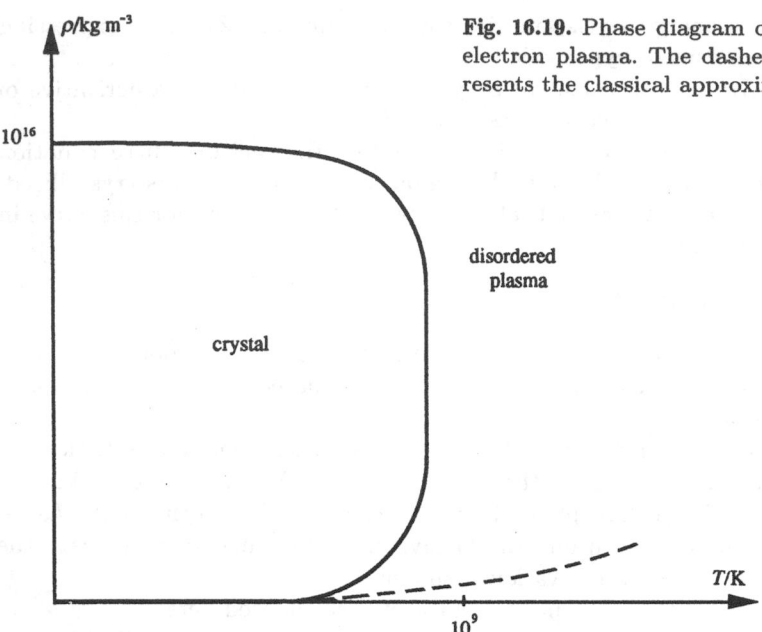

Fig. 16.19. Phase diagram of the ^{12}C-electron plasma. The dashed line represents the classical approximation

f. The star corresponds to the values $\varrho_0 = 10^{11}$ kg m^{-3}, $T_0 = 10^8$ K. Its matter can thus be described classically. As $\varrho_0 T_0^{-3} = 10^{-13}$ kg m^{-3} K^{-3} we are lying below the melting curve and the matter is thus a disordered dense classical plasma somewhat similar to a classical liquid. Nevertheless, we note that the uncertainties of the calculations and of the astrophysical data do not allow us to exclude the possibility of an error of a factor 10 in the (experimental or theoretical critical) value of $\varrho_0 T_0^{-3}$. Crystallization is thus not completely excluded.

g. The electron pressure can be calculated as if the temperature were zero, as $T_0/\Theta_F \ll 1$. It is given by the formula

$$\mathcal{P}_{\text{el}} = (9\pi^4)^{1/3} \frac{\hbar^2}{5m} \left(\frac{6\varrho}{M} \right)^{5/3}.$$

Therefore, $\mathcal{P}$ is an increasing function of ϱ and the melting curve in the $\mathcal{P}, T$-diagram shows essentially the same behaviour as in the ϱ, T-diagram. The maximum pressure along this curve is 1.5×10^{33} Pa as $T \to 0$.

h. We note here that the crystal can melt at low temperatures and very high pressures whereas an ordinary solid remains crystallized at high pressures; the orders of magnitude clearly have nothing in common. In an ordinary crystal the forces between the ions or the atoms are strongly repulsive at short distances apart due to the Pauli principle which prohibits atomic shells to interpenetrate, so that at high pressures the crystal is an ordered arrangement of hard spheres. Here there is no repulsion at short distances to prohibit the nuclei to vibrate. Their zero-point motion at zero temperature gives $\Delta = \sqrt{3\hbar/2m\omega}$ which varies as $d^{3/4}$ and exceeds d as $d \to 0$. It is thus the quantum zero-point motion which prevents the plasma to order as a crystal. In the region where $\varrho > \varrho_{\text{max}}$ the plasma becomes a Bose liquid of carbon ions embedded in a "degenerate" electron gas, that is, practically frozen in in its ground state. Increasing the pressure thus produces melting as the internuclear distances decrease more rapidly than the zero-point fluctuations. If ϱ continues to increase, the carbon nuclei touch one another and nuclear fusion reactions take place.

16.11 Landau Diamagnetism

The aim of this problem is to study the magnetism due to the motion of the electrons in a metal subject to a uniform magnetic field $\boldsymbol{B}$.

We make some simplifications, some of which are rather crude, in order to get a model where the calculations become simple.

We neglect the interactions of the electrons with the ions and with the other electrons, which we represent by a constant average potential. As this potential is equivalent to a change in the origin of the energy levels without any other consequences for the problem we can neglect it. Except in question 5, we neglect the effect of the magnetic field on the magnetic moment connected with the electron spin, retaining solely its effect on the orbital magnetic moment. Except in question 2, we treat the problem in two space dimensions.

These approximations produce a model where the Hamiltonian of a single electron in the field $\boldsymbol{B}$ directed along the z-axis is

$$\widehat{h} = \frac{1}{2m}\left[\widehat{p}_x^2 + (\widehat{p}_y + eB\widehat{x})^2\right];$$

the electron charge is $-e$, its mass m, and the vector potential has the components $(0, Bx)$. The metal is described as a square box with edge-length L.

We recall the solution of the relevant quantum mechanical problem. In the limit of large L the eigenvalues of $\widehat{h}$ have the form

$$\varepsilon_n = (2n+1)\mu_{\mathrm{B}}B,$$

where n is a quantum number which is a non-negative integer ($n = 0, 1, 2, \cdots$) and where $\mu_{\mathrm{B}} = e\hbar/2m$ is the Bohr magneton (elementary magnetic moment). Each level is degenerate: the degree of degeneracy g of the energy level ε_n is independent of n and equal to

$$g = 2\frac{eBL^2}{h},$$

where the factor 2 represents the spin degeneracy.

1. Consider a system of electrons in thermal equilibrium in the box and in the field $\boldsymbol{B}$. Treat these electrons as independent fermions to be studied

using the grand canonical formalism, with a temperature T and a chemical potential μ. We use the notation $\zeta = e^{\beta\mu}$ and $\beta = 1/kT$.

a. Find an expression for the grand potential $A(\mu, T, \boldsymbol{B}) = -kT \ln Z_{\mathrm{G}}$, without summing over n.

b. Write down the expression which gives the electron density per unit area, $\varrho = N/L^2$ as function of μ and T, without solving it.

c. Write an expression in the form of a simple integral for the grand potential $A_0(\mu, T)$ of an electron gas in two dimensions without a magnetic field. Check that it is the same as the limit as $B \to 0$ of the expression found in 1a.

2. Consider a very thin metallic film in three dimensions in a perpendicular magnetic field. In this case the single-electron Hamiltonian contains not only the term corresponding to the x- and y-directions, which was analyzed in the introduction to this problem, but also the kinetic energy $\hat{p}_z^2/2m$ in the z-direction. If l is the thickness of the layer, the wavefunctions must vanish for $z = 0$ and $z = l$.

Show that in a temperature region which you must find the grand potential of the layer reduces to the grand potential A just calculated, provided one modifies the chemical potential μ in a way to be specified. As a numerical application give this temperature range for $l = 50$ Å.

3a. Show that the magnetic susceptibility per unit area is given by the expression

$$\chi = -\frac{1}{L^2} \frac{\partial^2 A}{\partial B^2}\bigg|_{B=0,\mu,T}.$$

b. The parameter ζ is small at high temperatures, for fixed values of the electron density and the field B. Evaluate A retaining only the lowest order terms in ζ.

c. Find the magnetic susceptibility χ in this approximation. Give χ as function of ϱ and T. Show that the system is diamagnetic.

d. Calculate A to second order in B for arbitrary values of μ and T. You may use the Euler-Maclaurin formula from the end of this book.

e. Use the expression for A to find the susceptibility χ first as function of μ and T and then as function of ϱ and T. Compare the result with that found in c.

4. We shall show that, at sufficiently low temperatures, the magnetic moment of the metal as a function of the field B displays oscillations. This is the so-called *de Haas-van Alphen effect*. To simplify the situation assume the temperature to be zero and use a canonical ensemble taking the electron number N to be fixed. The system is then in a macro-state characterized by the fact that its energy is a minimum. Amongst the single-electron states which may be filled to produce this macro-state, those with lower energies $(n < \nu)$ are completely occupied and those with higher energies $(n > \nu)$ are completely empty; the single-particle states of energy ε_ν are, in general, partially occupied. Let φ be the average occupation number of those states.

a. Calculate φ and ν as functions of N and B; denote by $E(x)$ the integral part of a real positive number x. Sketch the behaviour of φ and ν as functions of $1/B$ for a given density ϱ.

b. Calculate the energy U of the macro-state as function of B for a given density and sketch its behaviour.

c. Show that at zero temperature the magnetization per unit area is given by

$$M = -\frac{1}{L^2}\frac{\partial U}{\partial B} \qquad (N \text{ fixed}).$$

Show that M oscillates as B changes and sketch these oscillations.

d. Indicate without calculations in what ranges of temperature and magnetic field one may expect to observe oscillations in M as important as at $T = 0$.

5. In order to include the effect of the coupling between the electron spin and the magnetic field, which we have so far neglected, we need to add to $\widehat{h}$ a term $\mu_B B\widehat{\sigma}$, where $\widehat{S}_z = \frac{1}{2}\hbar\widehat{\sigma}$ is the spin component along the magnetic field.

a. Find the new single-electron energies and their degree of degeneracy. Use them to find an expression for the grand potential.

b. Give as in 3d an expression for the grand potential to second order in B and find as in 3e the susceptibility as function of ϱ and T, now including spin effects.

c. Try to think of a kind of material in which the diamagnetism predicted in 3 would persist notwithstanding the spin paramagnetism.

Notes:

We obtain the eigenvalues of $\widehat{h}$ for an infinite two-dimensional layer by noting that the operators $\widehat{p}_x$ and $\widehat{x}' \equiv \widehat{x} + \widehat{p}_y/eB$ have the same commutation relations as $\widehat{p}_x$ and $\widehat{x}$, so that $\widehat{h}$ has the eigenvalues of a harmonic oscillator of frequency eB/m. In order to take care of the finite dimensions $L \times L$, it is convenient to impose periodic boundary conditions along the y-direction and to assume that a box potential along the x-direction forces the wavefunctions to vanish for $x = 0$ and $x = L$. This fixes the eigenvalues $p_y = qh/L$, where q is an integer. For given q, $\widehat{h}$ becomes a one-dimensional harmonic oscillator, centred at the point $x_q = -qh/LeB$ and confined to the box $0 \leq x \leq L$. The presence of this box has practically no effect on the eigenfunctions of the oscillator which are localized inside the box. The extension Δ_n of these functions is given by $\Delta_n^2 \equiv \langle (x - x_q)^2 \rangle = (n + \frac{1}{2})\hbar/eB$. When $0 \ll x_q - \Delta_n$, $x_q + \Delta_n \ll L$, so that the boundary conditions produce hardly any changes in the eigenvalues and eigenfunctions. Beyond this the eigenvalues are shifted upwards due to the introduction of the rigid walls; if q is such that the centre x_q of the oscillator leaves the box, the lowest energy levels are above the potential minimum, that is, the value of the harmonic potential $(x - x_q)^2 e^2 B^2/2m$ at $x = 0$ or $x = L$. To summarize, when $n\hbar \ll eBL^2$ we get practically the oscillator levels with a degree of degeneracy equal to the number of q-values such that $0 < x_q < L$, that is, to eBL^2/h, as we stated in the introduction of the problem. The effect of the upward shift of the levels introduces extra levels in intervals between the ε_n, but with a density which is small compared to the density of the levels that we have kept, as the interval involved for x_q by these extra levels is small compared to L; hence the corresponding number of q-values is small compared to g.

For question 3d we use the Euler-Maclaurin formula in the form

$$\sum_{n=0}^{\infty} f\left[\left(n+\frac{1}{2}\right)\varepsilon\right] \;=\; \frac{1}{\varepsilon}\int_0^{\infty} f(x)\,dx \;+\; \frac{\varepsilon}{24}f'(0) + \mathcal{O}(\varepsilon^3),$$

which provides $-A/kTg$ for the function $f(x) = \ln\left(1 + e^{\beta\mu - x}\right)$ and for $\varepsilon = 2\mu_{\mathrm{B}}B/kT$. This gives us

$$A \;=\; A_0 + \frac{(\mu_{\mathrm{B}}B)^2 L^2 m}{6\pi\hbar^2\left(1 + e^{-\mu/kT}\right)}.$$

Hence we find

$$\chi \;=\; -\frac{\mu_{\mathrm{B}}^2 m}{3\pi\hbar^2\left(1 + e^{-\mu/kT}\right)} \;=\; -\frac{e^2}{12\pi m\left(1 + e^{-\mu/kT}\right)}.$$

In zero field the electron density is

$$\varrho \;=\; \frac{2}{h^2}\int \frac{d^2 p}{e^{\beta(p^2/2m - \mu)} + 1} \;=\; \frac{mkT}{\pi\hbar^2}\ln\left(1 + e^{\mu/kT}\right),$$

which we can use to eliminate μ, whence

$$\chi \;=\; -\frac{\mu_{\mathrm{B}}^2 m}{3\pi\hbar^2}\left(1 - e^{-\pi\hbar^2\varrho/mkT}\right).$$

At low densities, $\varrho \ll mkT/\hbar^2$, we find a Curie-type diamagnetic relation, $\chi = -\mu_{\mathrm{B}}^2\varrho/3kT$, as in question 3c; however, in the opposite limit, which holds for a metal, the susceptibility tends to a constant.

The calculation is similar for question 5b. It is convenient to use here the Euler-Maclaurin formula

$$\frac{1}{2}f(0) + \sum_{n=1}^{\infty} f(n\varepsilon) \;=\; \frac{1}{\varepsilon}\int_0^{\infty} f(x)\,dx - \frac{\varepsilon}{12}f'(0) + \mathcal{O}(\varepsilon^3),$$

taking the new spectrum $2n\mu_{\mathrm{B}}B$ and its degree of degeneracy, $\frac{1}{2}g$ for $n = 0$ and g for $n \geq 1$, into account. From a simple comparison with the earlier result we then get

$$\chi \;=\; \frac{2\mu_{\mathrm{B}}^2 m}{3\pi\hbar^2}\left(1 - e^{-\pi\hbar^2\varrho/mkT}\right).$$

Pauli paramagnetism, which is three times as large as Landau diamagnetism, dominates and produces a total susceptibility which is positive. Note that to higher orders in B the orbital and spin effects do not add simply.

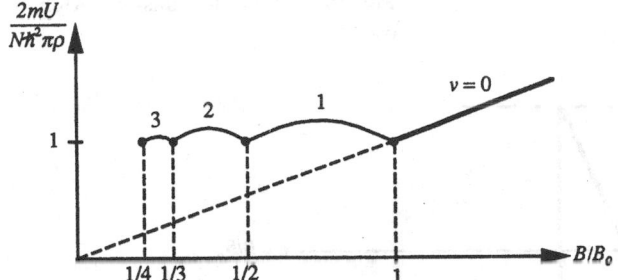

Fig. 16.20. The ground state energy of a paramagnetic film as function of the field

In a realistic material the potential due to the ions replaces the kinetic energy by the band energy. This means that the electron mass m is replaced by an effective mass m^* in the orbital part of $\widehat{h}$, but not in the coefficient μ_B in the coupling $\mu_B B \widehat{\sigma}$ with the spins, whence we find

$$\chi = -\frac{e^2}{12\pi m^* \left(1 + e^{-\mu/kT}\right)} \left[1 - 3\left(\frac{m^*}{m}\right)^2\right].$$

Conducting materials with a small effective mass, such as bismuth, are therefore diamagnetic.

The result of question 4b is

$$U = g\mu_B B \left[\sum_{n=0}^{\nu-1} (2n+1) + (2\nu+1)\varphi\right]$$

$$= \frac{N\pi\hbar^2 \varrho}{2m} \left[(2\nu+1)\frac{B}{B_0} - \nu(\nu+1)\frac{B^2}{B_0^2}\right]$$

$$= \frac{N\pi\hbar^2 \varrho}{2m} + \frac{e^2 B^2 L^2}{2\pi m}\varphi(1-\varphi),$$

where $B_0 = \varrho\pi\hbar/e$. This function is shown in Fig.16.20; it has singularities when B_0/B is an integer, since $\nu = E(B_0/B)$.

In question 4c we find

$$M = \varrho\mu_B \left[\frac{2B}{B_0}\nu(\nu+1) - (2\nu+1)\right], \qquad \nu = E(B_0/B),$$

which shows a saw-tooth shape with discontinuities when B_0/B is an integer (see Fig.16.21). These oscillations can be observed experimentally and are the so-called de Haas-van Alphen effect. Deviations from the model, extended to three dimensions, give information about the interactions between the electrons and the ions.

The oscillations are due to the Fermi level passing successively through the discrete energy levels ε_n. The phenomenon will continue without much change at non-zero temperatures as long as the Fermi factor has an abrupt change on the scale of the distance $2\mu_B B$ between successive levels ε_n, that is, as long as

$$kT \ll \mu_B B;$$

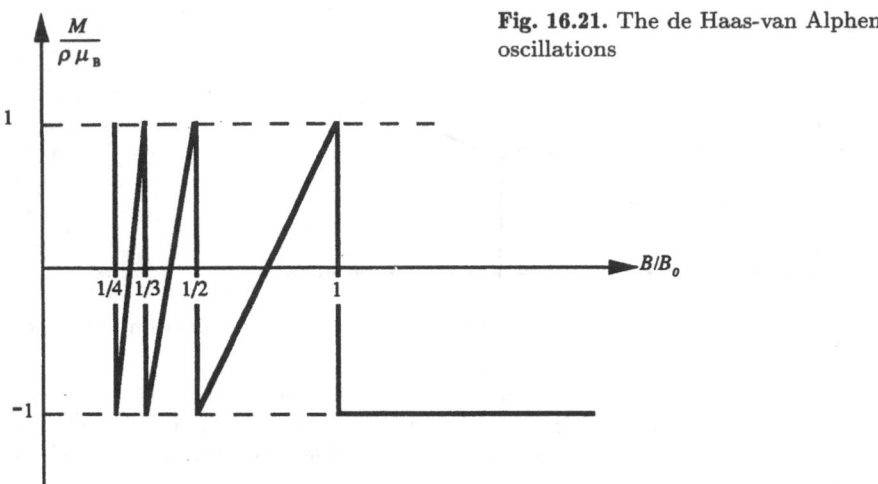

Fig. 16.21. The de Haas-van Alphen oscillations

the peaks in the $M(B)$-curve will just be rounded off over an interval $\Delta B \sim kT/\mu_B$. However, this range of temperatures and fields, $T \ll B$, if T is in kelvins and B in teslas, is difficult to reach, but faint oscillations can be observed even beyond it.

16.12 Electron-Induced Phase Transitions in Crystals

Certain metallic alloys with a chemical composition A_3B, for instance, the vanadium-silicon alloy V_3Si, can be in two crystalline phases which have a slightly different structure, depending on temperature. The first of these structures, which will be called "cubic", is the equilibrium state of the alloy *above* the temperature T_c where there is a phase transition between the two structures; in the case of V_3Si, $T_c = 21$ K. The other structure, called "tetragonal", is the equilibrium state of the alloy *below* T_c. The problem has the aim to explain the existence of this phase transition starting from the particular form of the electronic density of states of the A_3B alloys, which differs from that of the free electrons of an ordinary metal.

The cubic structure consists of cubic cells of edgelength $2a$ (Fig.16.22 shows two adjacent cells). The B atoms which do not play any explicit role in this problem are not shown; they are situated at the centre and at the vertices of the cubes. Each face of a cubic cell contains the centres of two A atoms. One can distinguish three kinds of positions for the A atoms, denoted, respectively, by A_X, A_Y, and A_Z. The A_X atoms form chains which are parallel to the X-axis. Similarly, the A_Y and A_Z atoms form, respectively, chains parallel to the Y- and Z-axes. In each chain two consecutive atoms are separated by a distance a.

The cell of the tetragonal structure (Fig.16.23) is obtained from that of the cubic structure by stretching the sides parallel to the X-axis and contracting

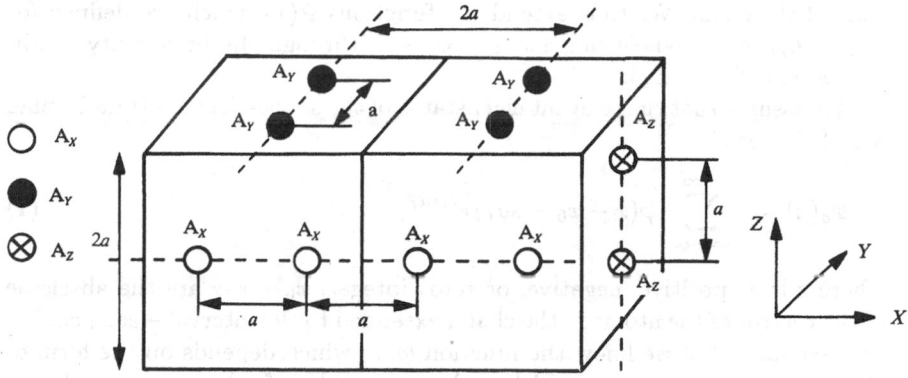

Fig. 16.22. The cubic structure of an A_3B alloy

those parallel to the Y- and Z-axes by equal amounts; the cell volume remains unchanged. Thus, the distances between the A_X atoms is now $a_X = a(1+\eta)$, whereas the distance between consecutive atoms in the chains parallel to the Y- and Z-axes is $a_Y = a_Z = a(1-\frac{1}{2}\eta)$. The positive parameter η measures the deformation of the tetragonal cell as compared to the cubic cell. In this problem we examine essentially situations where $\eta \ll 1$ (Fig.16.23 shows for the sake of clarity a very large deformation). We assume nevertheless that the above formulae giving a_X, a_Y, and a_Z as functions of η are exact, even if η is not infinitesimal.

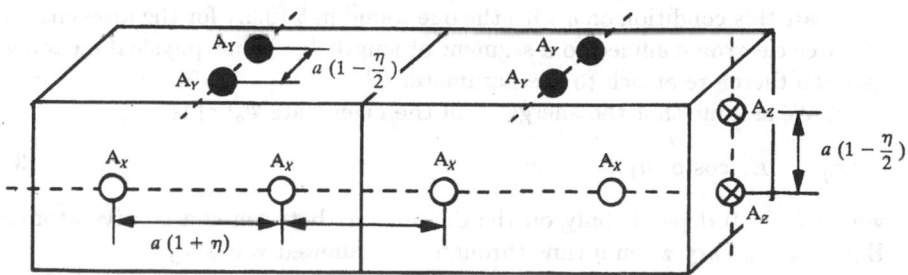

Fig. 16.23. The tetragonal structure of an A_3B alloy

Single-electron Density of States in the Two Phases of the Alloy

1. First of all we consider a single chain of A_I atoms ($I = X, Y$, or Z) containing n atoms. The length of the chain will be $L_I = na_I$, a_I being the distance between successive atoms. We treat this chain as a *one-dimensional* system of *non-interacting electrons*, with each electron being in a periodic potential $V(x)$ of period a_I. We are interested in the wavefunctions $\Psi(x)$ representing single-electron states in the chain. We impose on the functions $\Psi(x)$ the periodic boundary condition $\Psi(0) = \Psi(L_I)$. This allows us to simplify the treatment by getting rid of the boundary effects due to the

ends of the chain. We thus extend the functions $\Psi(x)$ which are defined for $x \in [0, L_I]$ to the whole range $x \in -\infty, +\infty$, through the periodicity condition $\Psi(x) = \Psi(x + L_I)$.

We assume that the relevant eigenstates of the single-electron Hamiltonian are of the form

$$\Psi_q(x) = \sum_{s=-\infty}^{+\infty} \varphi(x - x_0 - sa_I) e^{iqsa_I}, \tag{1}$$

where s is a (positive, negative, or zero) integer, $x_0 + sa_I$ are the abscissae of the centres of the atoms in the chain, extended to the interval $-\infty, +\infty$. We are assuming that we know the function $\varphi(x)$ which depends on the form of $V(x)$, and q is a parameter which enables us to identify the eigenstate $\Psi_q(x)$.

a. Show that when j is an integer the function (1) satisfies the relation

$$\Psi_q(x + ja_I) = e^{iqja_I} \Psi_q(x).$$

What can one say about the probability density $\mathcal{P}(x)$ for the presence of an electron at the abscissa x in the pure state Ψ_q?

b. Show that the periodic boundary conditions imposed on the functions $\Psi_q(x)$ determine n different functions $\Psi_q(x)$ corresponding to the following values of q:

$$q_j = \frac{2\pi j}{L_I}, \quad j = 0, 1, \cdots, n - 1. \tag{2}$$

Compare this condition on q with the one found in § 10.2.1 for the momentum of a free electron confined to a segment of length L_I. What physical meaning can you therefore attach to the parameter q?

c. We assume that the energy ε_q of the eigenstate $\Psi_q(x)$ is

$$\varepsilon_q = E_I \cos(q \cdot a_I), \tag{3}$$

where $E_I > 0$ depends only on the distance a_I between consecutive atoms. How does ε_q vary when q runs through its n allowed values q_j?

d. Show that in the limit of large lengths ($L_I \gg a_I$) the single-electron density of states $\mathcal{D}'_I$ in the chain is

$$\mathcal{D}'_I(\varepsilon) = \frac{2n}{\pi} \frac{\theta(E_I - |\varepsilon|)}{\sqrt{E_I^2 - \varepsilon^2}}, \tag{4}$$

where we have taken the spin into account and where $\theta(x)$ is the Heaviside function which is zero when $x \leq 0$ and 1 when $x > 0$.

2. The alloy has in each of the X-, Y-, and Z-directions m similar noninteracting chains of n atoms; we write $mn = \mathcal{N}$. An electron can occupy any single-electron state of the $3m$ chains and it can switch from one chain to another, in the same or in another direction. We assume that when it belongs to one chain it does not feel the potential from the other chains. We write

$\mathcal{D}_X(\varepsilon)$ for the single-electron density of states relating to the set of m chains which are parallel to the X-axis. Similarly, $\mathcal{D}_Y(\varepsilon)$ and $\mathcal{D}_Z(\varepsilon)$ are the densities of states relating to the m chains parallel to the Y- and Z-axes, respectively.

a. Compare the total number of single-electron states in the $3m$ chains with the total number of atoms in these chains. Give expressions for $\mathcal{D}_X, \mathcal{D}_Y$, and $\mathcal{D}_Z$ as functions of E_X, E_Y, and E_Z, which are the coefficients E_I occurring in Eq.(3) with $I = X, Y$, or Z.

b. Consider the cubic phase and put $E_X = E$. Determine the single-electron density of states $\mathcal{D}_c(\varepsilon)$ for the system of the $3m$ chains of the alloy.

c. Consider the tetragonal phase. What is the relation between the single-electron density of states $\mathcal{D}_t(\varepsilon)$ relating to the $3m$ chains and the densities of states $\mathcal{D}_X$ and $\mathcal{D}_Y$?

3. We assume that as a function of a_I we can express E_I in the form

$$E_I = E\left(1 - \alpha \frac{a_I - a}{a}\right), \tag{5}$$

where a is the distance between the A atoms in each of the chains in the cubic phase and where E was defined in 2b; α is a positive coefficient.

Write down the coefficients E_X and E_Y for the tetragonal phase as functions of α and the deformation η which was defined in the introduction (Fig.16.23). Draw graphs of $\mathcal{D}_c(\varepsilon)$ and $\mathcal{D}_t(\varepsilon)$ for given η.

Electronic Free Energy of the Alloy

In this section we assume that $3mn\gamma \equiv 3N\gamma$ non-interacting electrons in equilibrium at a temperature T occupy the single-electron states of the alloy given by (1). The number γ of electrons per atom is fixed, with $0 < \gamma < 2$. The other electrons occupy lower energy states; their only rôle is to ensure the electrical neutrality of the sample.

4. We assume that the structure of alloy is the cubic one.

a. Without performing the integrations, write down an expression for the grand potential $A_c(T, \mu)$, starting from $\mathcal{D}_c(\varepsilon)$ and E; μ is the chemical potential.

b. Write down the condition from which one can determine the value $\mu_c(T)$ of μ for a given filling γ, starting from $\mathcal{D}_c(\varepsilon)$, and E. You need not calculate $\mu_c(T)$ explicitly.

5. Assume now that the structure of the alloy is the tetragonal one.

a. Find, without performing the integrations, an expression for $A_t(T, \mu)$, starting from $\mathcal{D}_X(\varepsilon)$, $\mathcal{D}_Y(\varepsilon)$, E_X, and E_Y.

b. Write down the condition from which one can determine the value of $\mu_t(T)$ of μ for a given filling γ, starting from $\mathcal{D}_X(\varepsilon)$, $\mathcal{D}_Y(\varepsilon)$, E_X, and E_Y. You need not calculate $\mu_t(T)$ explicitly.

c. Show that

$$\mathcal{D}_I(\varepsilon) = \frac{E}{3E_I} \mathcal{D}_c\left(\frac{E}{E_I}\varepsilon\right); \qquad I = X \text{ or } Y.$$

Use this relation and the results from 3 to show that for a fixed value of the deformation η one can write the grand potential $A_t(T,\mu)$ in the form

$$A_t(T,\mu) = -\frac{1}{3\beta} \int_{-E}^{E} \left\{ \ln\left[1 + e^{\beta(\mu - \varepsilon + \varepsilon\alpha\eta)}\right] \right. $$
$$\left. + 2\ln\left[1 + e^{\beta(\mu - \varepsilon - \frac{1}{2}\varepsilon\alpha\eta)}\right] \right\} \mathcal{D}_c(\varepsilon)\, d\varepsilon. \tag{6}$$

6. We now assume that $\eta \ll 1$.

a. Show that to second order in η we can write for $A_t(T,\mu)$

$$A_t(T,\mu) = A_c(T,\mu) + \frac{1}{4}\alpha^2\eta^2 \int_{-E}^{E} \varepsilon^2 f' \mathcal{D}_c(\varepsilon)\, d\varepsilon, \tag{7}$$

where f' is the derivative with respect to ε of the electron Fermi factor $f(T,\mu,\varepsilon)$.

b. Use this result to show that for a given number of electrons, $3\mathcal{N}\gamma$, the difference $\mu_t(T) - \mu_c(T)$ is of second order in η.

c. Let F_c and F_t be the free energies of the $3\mathcal{N}\gamma$ electrons at a temperature T for the cubic and the tetragonal phases, respectively. Prove that

$$F_t = F_c + \frac{1}{4}\alpha^2\eta^2 \int_{-E}^{E} \varepsilon^2 f'_c \mathcal{D}_c(\varepsilon)\, d\varepsilon, \tag{8}$$

where $f'_c = f'[T,\mu_c(T),\varepsilon]$.

Stability of the Two Phases at Zero Temperature

In this section, we *put the temperature equal to zero*, and we retain the hypothesis that $0 < \gamma < 1$. Note that at zero temperature $f' = -\delta(\varepsilon - \mu)$.

7. Consider the cubic phase.

a. Show without calculations that the value of the chemical potential $\mu_c(T = 0)$, which we shall denote by μ_c^0, satisfies the relation

$$-E < \mu_c^0 < 0.$$

b. Calculate μ_c^0 explicitly as function of E and γ.

c. Calculate the energy U_c of the $3\mathcal{N}\gamma$ electrons as function of E and γ.

8. Consider the tetragonal phase with given η ($\eta \ll 1$), and write U_t for the energy of the $3\mathcal{N}\gamma$ electrons in this phase. Hence find from Eq.(8) for F_t the value of $U_t - U_c$ as function of η. What is the sign of $U_t - U_c$?

9. The total energy of the alloy contains, apart from the energy of the $3\mathcal{N}\gamma$ electrons, an energy U_d associated with the other degrees of freedom of the system. Assume that U_d, which is a minimum when $\eta = 0$, as a function of η has the form

$$U_d = \frac{3\mathcal{N}}{2} C\eta^2 \quad \text{with} \quad C > 0. \tag{9}$$

a. Compare the energy of the cubic phase with the energy of the tetragonal phase for a fixed value of η ($\eta \ll 1$). Discuss which of the phases is the ground state of the alloy, depending on the value of γ. Hence, find the condition on γ in order that the cubic phase cannot be the equilibrium state of the alloy at zero temperature. If that condition is satisfied we shall say that the cubic phase is *unstable*.

b. Find the numerical value of γ_0, that is, the maximum value of γ compatible with the instability of the cubic phase when $\alpha = 2$; $E = 1$ eV; $C = 20$ eV; these values correspond to the V_3Si alloy. Check that $\gamma_0 \ll 1$.

c. Compare the case $1 < \gamma < 2$ with the case $0 < \gamma < 1$ which was studied above, using the particle-hole symmetry.

10. Assume that the value of γ is fixed at

$$\gamma = 0.7\gamma_0, \tag{10}$$

and consider the tetragonal phase with fixed η ($\eta \ll 1$).

a. Calculate to first order in η the number of electrons contained in the m chains in the X-direction. Do the same for the m chains in the Y- or Z-directions. Show that the chains in the X-direction do not contain any electrons if η exceeds a value η_0 which you should calculate using the numerical data of 9b. What is the position of the chemical potential $\mu_t^0 = \mu_t(T = 0)$ with respect to the single-electron energy levels of the chains in the X-direction, when $\eta = \eta_0$?

b. Use the results of 9 to find out how the total energy of the alloy varies in the tetragonal phase when η increases, starting from zero. Show that, if η is left free to change, the equilibrium of the alloy corresponds to the chains in the X-direction being without electrons. Assume that the results from 9 retain their validity when $\eta \leq \eta_0$.

11. Consider now a deformation $\eta > \eta_0$, retaining the value of γ from 10.

a. Show that in this case μ_t^0 decreases linearly with η.

b. Calculate the energy U_t of the $3\mathcal{N}\gamma$ electrons in the tetragonal phase and show that $U_t + U_d$ has a minimum for a value η_1 of η which must be calculated numerically. Check that $\eta_1 > \eta_0$. What can one deduce about the equilibrium value of η at zero temperature?

c. Compare the values of $\mathcal{D}_c(\mu_c^0)$ and $\mathcal{D}_t(\mu_t^0)\big|_{\eta=\eta_1}$. Explain how one can use measurements of the magnetic susceptibility or of the specific heat to determine experimentally which of the two phases of the alloy is realized.

Stability of the Two Phases when $T \neq 0$

In this section we assume that the temperature T is non-zero and we fix the value of γ to be $\gamma = 0.7\gamma_0$. We assume that the deformation, η, of the structure does not contribute to the entropy of the alloy. Thus, the total free energy F of the alloy is $F = F_t + F_d$, where F_d reduces to the energy term U_d given in 9.

12. If $\eta \ll 1$, the total free energy F can be written as

$$F = F_c(T) + \tfrac{1}{2}K(T)\eta^2. \tag{11}$$

a. Express $K(T)$ in terms of C, α, E, and T, without performing the integrations.

b. In agreement with §4.2.2 we consider F as a *trial free energy* associated with the variational parameter η. Show that a necessary condition for the cubic phase to be the equilibrium state of the alloy at a temperature T is $K(T) > 0$.

In what follows we assume that this condition is sufficient and, on the other hand, that, if $K(T) < 0$, the tetragonal phase is the equilibrium state of the alloy. We shall discuss these two hypotheses in the comments below.

c. What is the equilibrium state of the alloy at $T = 0$?

d. What condition determines the temperature T_c of the phase transition between the cubic and the tetragonal phases, if it exists?

13a. Show that one can define a Fermi temperature, Θ_c for the electrons of the alloy in the cubic phase such that, if $T \ll \Theta_c$ the behaviour of the electrons is close to their behaviour at zero temperature in that phase.

b. Use the fixed value of γ and the numerical data from 9b to calculate the value of Θ_c. Explain why this value is much lower than that for free electrons in a metal.

c. Use the value of γ and the relation between C and γ_0, found in 9b, to express C as a function of α, E, and Θ_c.

14. Consider a temperature T_1, such that $k\Theta_c \ll kT_1 \ll E$.

a. Use the condition which determines the chemical potential $\mu_c(T_1)$ to show that

$$e^{(E + \mu_c(T_1))/kT_1} \;<\; \frac{2}{\sqrt{\pi}} \sqrt{\frac{\Theta_c}{T_1}}. \tag{12}$$

Write $u^2 = (\varepsilon + E)/kT_1$, and use the approximation

$$\int_0^{\sqrt{E/kT_1}} e^{-u^2}\,du \;\approx\; \int_0^{\infty} e^{-u^2}\,du \;=\; \frac{\sqrt{\pi}}{2}.$$

b. Use these results to show, by looking for a positive minorant of $K(T_1)$, that the cubic phase is the equilibrium state of the alloy at the temperature T_1.

c. Now find the order of magnitude of the temperature T_c for the transition between the tetragonal and the cubic phases. Compare this with the experimental value of T_c given in the introduction.

15. A computer calculation of $K(T)$ gives us the curve shown in Fig.16.24.

Examining for $T \ll \Theta_c$ the form of $K(T)$ determined in 12a, justify the existence of a minimum in the $K(T)$-curve. Remember that

$$\int g(x)\delta''(x - x_0)\,dx \;=\; g''(x_0).$$

Assume that $\mu_c(T) \approx \mu_c^0$ when $T \ll \Theta_c$.

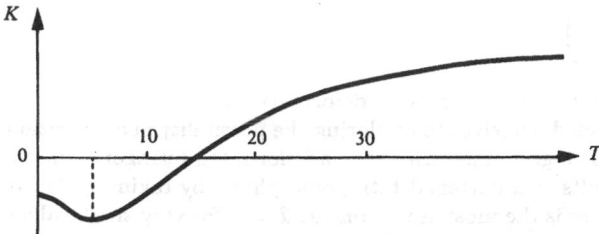

Fig. 16.24. The shear modulus of the cubic phase as function of the temperature

Comments:

The simplified model presented in this problem is inspired by the theory worked out by J.Labbé and J.Friedel[4] to explain the behaviour of a phase transition observed in solids such as V_3Si, Nb_3Sn, and V_3Ge. This transition from a cubic to a tetragonal structure is called a "martensite" transition – see Problem 19, where the extension of the crystal cell of steel in one of the three directions is caused by a completely different mechanism, namely, the migration of C atoms. In general, changes in structure of crystals are produced by forces between ions; the value of the free energy which is due to these forces depends on the temperature and the crystal structure; its minimum determines the latter, which thus depends on the temperature. Here we are in a remarkable situation where it is the contribution to the free energy not of the ions, but of a small fraction of the electrons which changes rapidly with temperature and with the lattice structure, even for small deformations, so that *the crystalline phase transition is induced by the electrons.*

The determination of the most stable phase was done by comparing the free energies for a fixed number of electrons. We could also have proceeded more directly by looking for the *minimum of the grand potential A*, which is the sum of (6) and (9), as function of η for fixed μ and T. A systematic study which can easily be carried out in the framework of the model used here leads thus to the phase diagram in the μ, T-plane, or in the γ, T-plane: the *absolute* minimum of A indicates which is the stable phase and gives the deformation η, if this is the tetragonal phase. A *local* minimum indicates a *metastable* phase. In particular, in question 12 the condition $K > 0$ only ensures the metastability of the cubic phase. To find a more conclusive result it is necessary to expand A (or F) up to fourth order in η in order to determine the presence of minima other than the one at $\eta = 0$ and to compare the value of A (or F) in these minima with their value at $\eta = 0$. In this way we get a transition temperature slightly higher than the one obtained in question 14, and the transition is a first-order one, with a sudden jump of the order parameter η at $T = T_c$.

The expansion up to fourth order in η makes it necessary to use a more realistic form for the parameter E_I than (5). Strong coupling band theory (§ 11.2.4) shows, in fact, that

---

[4] J.Physique (Paris) **27**(1966) 153, 303.

$$E_I = E \, \exp\left(-\alpha \, \frac{a_I - a}{a}\right),$$

of which (5) is an approximation valid for small deformations.

In the problem we restricted ourselves to exploring the possibility of a transition from a cubic phase, stable at high temperatures, to an elongated tetragonal phase. It is easy to extend the results to a flattened tetragonal phase, by taking $\eta < 0$. It then turns out that this phase is the most stable one at $T = 0$ for very small values of γ; when we let γ increase, it jumps to the elongated tetragonal phase, obtained in 11, and then to the cubic phase.

One can also consider orthorhombic structures which correspond to different values for the three parameters a_X, a_Y, a_Z. The extension of the calculations of the problem shows that those structures can only be metastable, as their grand potential is larger than that of a tetragonal structure. Other kinds of lattice deformations which are produced by changing the angles of the cell, with a_I being constant, are excluded since they do not affect the electron density of states.

The model could also be used to evaluate various thermodynamic quantities, such as the specific heat or the magnetic susceptibility, or mechanical quantities in each phase – for instance, $K = \partial^2 F/\partial\eta^2 = \partial^2 A/\partial\eta^2$ is the shear modulus in the cubic phase – or the latent heat for the transition from one structure to another. In particular, the agreement between experiments and predictions for the thermal properties of V_3Si has justified the above explanation of the martensite transition; it had been observed by crystal diffraction and electron microscopy without people knowing the reason for the change in the geometry of the cell.

16.13 Liquid-Solid Transition in Helium Three

Helium has two natural isotopes, helium four (^4He) with a nucleus composed of two protons and two neutrons, and helium three (^3He) with a nucleus composed of two protons and one neutron. We shall here study the liquid-solid equilibrium of *helium three*(^3He) at low temperatures, first when there is no magnetic field present, and then in the presence of a magnetic field which has the effect of polarizing the spins of different ^3He atoms in the same sense. In particular, we want to explain the difference between the two equilibrium curves represented in the temperature-pressure plane (Fig.16.25).

Preliminary Questions

1. Consider N atoms with average positions at the lattice sites of a cubic crystal lattice. Assume that each atom can oscillate, independent of the other atoms, around its average position, in the three perpendicular directions of the cubic lattice. The restoring force, which is proportional to the distance, is the same in the three directions. The characteristic frequency of the oscillations is ν.

a. Calculate the canonical partition function Z_1 of the N atoms, which we assume first to be spinless, in equilibrium at a temperature T. Hence find

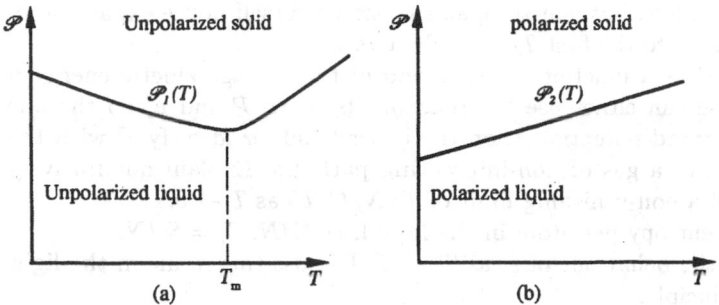

Fig. 16.25. The phase diagrams of unpolarized (a) and polarized (b) helium three

the internal energy U_1 and the entropy S_1 of the N atoms at the temperature T. What happens to U_1 and S_1 as $T \rightarrow 0$?

b. Now assume that each atom has a spin $\frac{1}{2}$, and that there is no interaction between the spins of the different atoms. Indicate the form of the partition function Z_2, the internal energy U_2, and the entropy S_2 of the N atoms which oscillate around the sites of the cubic lattice. Write $s_2 = S_2/N$ for the value of the entropy per atom. What is the limit of s_2 as $T \rightarrow 0$? Is this in agreement with Nernst's principle? Among the assumptions made, which is the one that may fail as $T \rightarrow 0$?

2. We recall the definition of the free enthalpy or Gibbs function,

$$ G = F + \mathcal{P}\Omega, $$

where $F = U - TS$ is the free energy, $\mathcal{P}$ the pressure, and Ω the volume.

a. Give expressions for the entropy S, the chemical potential μ, the volume Ω, and the internal energy U as function of $G(T, \mathcal{P}, N)$ and its partial derivatives with respect to the variables T, $\mathcal{P}$, and N.

b. Show that $G = \mu N$. You may, for example, use the extensivity of G.

Study of Unpolarized Liquid Helium Three

In the liquid state, helium three consists of interacting ^{3}He atoms. It is difficult to take these interactions into account when calculating the properties of the liquid. To simplify the study of liquid helium three we consider this system as a *gas of N non-interacting ^{3}He atoms*. At the same time, to simulate roughly the existence of interactions, we assume that the ^{3}He atoms in the gas have an effective mass m^* which differs from the real mass m of the ^{3}He atoms.

3. The total spin of the ^{3}He atom is $\frac{1}{2}$. How can one justify a half-odd-integral value for this spin? Which kind of statistics should one use to study a gas of N non-interacting ^{3}He atoms?

4a. Find the pressure $\mathcal{P}(N/\Omega, T)$ of liquid helium three at very low temperatures as a function of m^*, the density N/Ω, and the temperature T.

Assume that T is low, but non-zero, and restrict yourself in the expansion of $\mathcal{P}$ as function of T to the first T-dependent term.

b. Express $\mathcal{P}$ as a function of N/Ω and of the average kinetic energy u of an atom. You can either use the relations between $\mathcal{P}$ and u, on the one hand, and the grand potential A, on the other hand, or identify $\mathcal{P}$ with the kinetic pressure of a gas of non-interacting particles. Explain qualitatively the existence of a non-vanishing limit of $\mathcal{P}(N/\Omega, T)$ as $T \to 0$.

c. Find the entropy per atom in the liquid, $s_1(\Omega/N, T) = S_1/N$.

d. What is the behaviour of s_1 as $T \to 0$? Discuss this result in the light of the Pauli principle.

5a. Use the expression for $\mathcal{P}(N/\Omega, T)$ obtained in 4 to show that the Fermi temperature Θ_F of liquid helium three is of the form

$$\Theta_F(P) = a\mathcal{P}^{2/5},$$

where a is a constant. Give an expression for a.

b. Calculate the effective mass m^* of a helium atom as function of its real mass m by requiring that at atmospheric pressure $\mathcal{P}_0$ the Fermi temperature is $\Theta_F(\mathcal{P}_0) = 8.2$ K. Take for $\mathcal{P}_0$ the value 10^5 Pa.

c. Calculate the average volume $v_1(\mathcal{P})$ occupied by an atom in the liquid at a pressure $\mathcal{P}$ and for $T = 0$. What is the numerical value of $v_1(\mathcal{P}_0)$?

d. Give s_1 as function of Θ_F and T. Calculate the numerical value, in eV K^{-1}, of s_1 at a pressure $\mathcal{P}_0$ and at a temperature $T_0 = 10^{-2}$ K. What is the shape of the adiabats in the $T, \mathcal{P}$-plane?

6. Give for $T = 0$ the free enthalpy G_1 of the liquid as function of its Fermi temperature $\Theta_F(\mathcal{P})$. Hence find G_1 as function of $\Theta_F(\mathcal{P}_0)$ and $\mathcal{P}/\mathcal{P}_0$. Justify the fact that G_1 retains the same expression when $T \ll \Theta_F$, if one restricts oneself to an expansion of G_1 up to first order in T.

Study of Unpolarized Solid Helium Three

The ^{3}He atoms occupy in the solid the sites of a crystal lattice which we assume to be cubic (in actual fact it is body-centred cubic). The binding energy of an atom in the lattice is $-\varepsilon$ with $\varepsilon = 0.58 \times 10^{-3}$ eV. Moreover, each atom can independently of the others vibrate in the three directions of the lattice with a characteristic frequency ν, such that $h\nu/k = 20$ K. We assume the solid to be incompressible and unexpandable so that its volume is $\Omega = Nv_s$, where v_s is the unchangeable volume occupied by an atom and N the number of atoms of the solid.

7a. Show that the partition function Z_s of the solid is $Z_s = e^{N\beta\varepsilon} Z_2$, where Z_2 is the partition function calculated in 1b and where $\beta = 1/kT$. Find the average energy $u_s = U_s/N$ and the entropy $s_s = S_s/N$ per ^{3}He atom in the solid as functions of ε and ν.

b. Give the numerical value, in eV K^{-1}, of s_s at the temperature $T_0 = 10^{-2}$ K. Compare the entropy of the solid with that of the liquid at the temperature T_0. Why is the solid less ordered than the liquid?

Assume in the remainder of this problem that the model considered for the solid remains valid down to $T = 0$ so that if there is no external field the solid remains in the disordered state studied above.

8. Give the free enthalpy of the solid G_s as function of T, N, $\mathcal{P}$, and v_s (i) for $T = 0$; (ii) at very low temperatures. In the expansion of G_s as function of T we restrict ourselves to the term linear in T. With which physical characteristic of the solid is this linear term connected?

Study of the Liquid-Solid Equilibrium Curve in Unpolarized Helium Three

The liquid and solid phases of helium three are at a temperature T in equilibrium for a pressure $\mathcal{P}_1(T)$ (see Fig.16.25a).

9a. Show that on the equilibrium curve we have $G_l(T, \mathcal{P}_1(T), N) = G_s(T, \mathcal{P}_1(T), N)$, where G_l and G_s are, respectively, the free enthalpies of the liquid and the solid for N atoms.

b. Use the result of a to find the relation between the slope $d\mathcal{P}_1(T)/dT$ of the equilibrium curve, the change in entropy, $s_l - s_s$, and the change in the volume occupied per atom, $v_l - v_s$, at a temperature T and at the pressure $\mathcal{P}_1(T)$ (Clapeyron's law).

10. The value of the volume v_s is 3.94×10^{-29} m^3. We want to determine the equilibrium pressure $\mathcal{P}_1(0)$ at zero temperature.

a. From graphs of $G_s(T = 0, \mathcal{P}, N)$ and of $G_l(T = 0, \mathcal{P}, N)$ as functions of $\mathcal{P}$ show that there exist two possible values for the solid-liquid equilibrium pressure. In the remainder of the problem assume that only the smallest of these has a physical significance. The model ceases to be valid at too high a pressure.

b. Determine the numerical value of $\mathcal{P}_1(0)$.

c. Remember that for given temperature and pressure the stable phase of helium three is the one for which the free enthalpy is the lowest. Show that at $T = 0$ the solid is stable when $\mathcal{P} \geq \mathcal{P}_1(0)$ and the liquid is stable when $\mathcal{P} \leq \mathcal{P}_1(0)$.

11. Consider a temperature $T \neq 0$ such that $T \ll \Theta_F$, where Θ_F is the Fermi temperature of the liquid.

a. Use the results of 6 and 8 to show that the equilibrium pressure $\mathcal{P}_1(T)$ decreases when the temperature increases, starting from zero. Calculate the slope $d\mathcal{P}_1/dT|_{T=0}$. What is the physical origin of this negative slope?

b. Check that the value found for $d\mathcal{P}_1/dT|_{T=0}$ is compatible with Clapeyron's law found in 9b.

12. The $\mathcal{P}_1(T)$-curve shows, in fact, a minimum at $T = T_m$ (see Fig.16.25a).

a. Consider the variation of the entropies S_l and S_s along the equilibrium curve to explain why there is this minimum (neglect all terms of higher than first order in T). Estimate T_m and $\mathcal{P}_1(T_m)$.

In the remainder of this question we consider the effect of an *adiabatic compression*.

b. We start at a temperature $T_1 \leq T_m$ and a pressure $\mathcal{P} < \mathcal{P}_1(T_1)$ from the *liquid phase* and compress the system until the solid starts to appear at a temperature T_2. Compare T_2 with T_1 and discuss graphically the position of the point reached along the coexistence curve in this transformation; compare T_2 with T_m for different initial states T_1, $\mathcal{P}$.

c. Consider now the case when $T_2 < T_m$. If we continue to compress the system adiabatically we follow the coexistence curve, creating a certain amount of solid. What is the effect of this transformation on the temperature? Write down the equations which allow us to calculate as function of T the number of atoms produced in the solid phase – you get the total entropy of the mixture by adding the entropies of the liquid and of the solid. What is the final point of this process?

Study of Polarized Helium Three

We consider very low temperatures and we apply a magnetic field of sufficient strength to align *all* the spins of the $^3\mathrm{He}$ atoms in the same sense. We neglect the (constant) interaction energy between the spins and the magnetic field.

13a. What is, for fixed N/Ω, the relation between the single-particle densities of state $\mathcal{D}(u)$ and $\mathcal{D}'(u)$ which are associated, respectively, with the *unpolarized liquid* and with the *polarized liquid*, where u is the kinetic energy per atom?

b. Find, for fixed N/Ω, the Fermi temperature $\Theta'_F(\mathcal{P})$ of *polarized helium three* as a function of the Fermi temperature $\Theta_F(\mathcal{P})$ of the unpolarized liquid. Give the numerical value of $\Theta'_F(\mathcal{P}_0)$ at atmospheric pressure $\mathcal{P}_0$.

c. Show that for fixed density N/Ω polarizing the spins increases the pressure of the liquid, and give a physical explanation of this fact. Calculate this increase in pressure for $T = 0$.

14a. Write down expressions for the free enthalpies $G'_1(T = 0, \mathcal{P}, N)$ and $G'_s(T = 0, \mathcal{P}, N)$ of the polarized liquid and solid, for $T = 0$.

b. Hence find the value of the new equilibrium pressure $\mathcal{P}_2(0)$ between the polarized solid and liquid at $T = 0$.

15. Justify the positive slope of $\mathcal{P}_2(T)$ at low temperatures (see Fig.16.25b).

Comments:

The degeneracy 2^N of the ground state of unpolarized solid helium three, which is associated with the possibility that each spin takes on two values, entails a violation of Nernst's principle, at least for the model considered. In actual fact there are magnetic interactions between the $^3\mathrm{He}$ nuclei. They are very weak because the nuclear magneton is so small and they do not play any rôle, except at very low temperatures of the order of mK, whereas we consider here temperatures of the order of a fraction of one K. Nevertheless, they produce in the solid a phase transition at 1.03 mK below which the spins show anti-ferromagnetic ordering, that is, two neighbouring spins are oriented in opposite directions $\pm u$. The ground state remains degenerate, since the orientation of u is not fixed, but so little that the entropy per atom tends to zero for $T \ll 1$ mK.

In the liquid phase the Pauli principle ensures the vanishing of the entropy per atom at low temperatures. In the solid phase the sites are distinguishable; this allows each of them to have a spin which can take on two values.

When we compress the liquid its temperature grows as $\mathcal{P}^{2/5}$ until one reaches the $\mathcal{P}_1(T)$-curve. If the temperature is then below T_m, the compression along this curve lowers the temperature while ^{3}He gradually crystallizes (§ 12.2.3). In our model we could reach the absolute zero in this way. In practice, the existence of a spin ordering of the solid below 1 mK, as well as losses, limit the lowering of the temperature.

The spin polarization modifies profoundly the phase diagram of ^{3}He because it increases the Fermi level of the liquid and because it removes the spin entropy in the solid phase. We have restricted ourselves in the problem to two extreme situations where the magnetic field B is either zero, or sufficiently large to align the spins completely. It would be easy to extend the study to arbitrary magnetic fiels. The model thus allows us to explain the phase diagram of ^{3}He as function of the variables $\mathcal{P}$, T, and B.

16.14 Phonons and Rotons in Liquid Helium

The heavy, most abundant, isotope of helium (^{4}He) has two liquid phases which are traditionally denoted by I and II. The phase I is an ordinary liquid. The phase II has remarkable properties such as superfluidity which we want here to explain.

At atmospheric pressure $\mathcal{P}_0$ the phase II is stable below the temperature $T_\lambda = 2.18$ K (Fig.12.1). The value of its mass density, ϱ, which is practically independent of $\mathcal{P}$ and T, is 145 kg m^{-3}.

The Landau Model

The model of a gas of non-interacting bosons (§ 12.3.1) allows us to explain only a few of the properties of helium at low temperatures. In fact, the interactions between the atoms play an essential rôle in a liquid, especially for bosons (§ 12.3.2), and it is difficult to take them into account through a diagonalization of the Hamiltonian, even making very crude approximations. Landau's theory[5] of liquid helium II is based upon a semi-empirical description of the low-energy micro-states, which has been justified theoretically afterwards[6]. These micro-states can be approximately constructed in two stages:

a) We first determine the configuration of the atoms in the liquid which corresponds to the ground state. This state we shall call the *substrate*. The

---

[5] L.D.Landau, J.Phys.USSR **5** (1941) 71.

[6] R.P.Feynman, Progr.Low Temp.Phys. **1** (1955) 17.

wavefunction of the substrate is not simple, but we need not know it. For N atoms in a volume Ω, the energy E_s of the substrate is an extensive quantity.

b) Starting from the substrate we construct the first excited micro-states, the only ones which are useful for a study of the properties of helium II. More precisely, these micro-states describe the vibrations of the atoms in the substrate, considered as a fixed reference state. A vibrational micro-state will be regarded as a *superposition* of collective oscillations of the atoms. Each of these oscillations, which are called vibrational *modes*, is characterized by a wavevector q and a frequency $\omega(q)/2\pi$.

This model of helium II is analogous to the one introduced in Chap. 11 to describe the vibrations and the specific heat of a crystalline solid. In that case, the substrate consisted of a system of atoms at positions close to the crystal lattice sites, which deviated from these sites only through small fluctuations of their positions and their velocities, necessary to satisfy the Heisenberg principle in the ground state. The modes were the phonons, that is, the quantized collective vibrations of the atoms in the crystal. The structure of the substrate is here more complicated, since the atoms are so light that they fluctuate widely in the whole available volume, but the excitations still look like the phonons in a crystal. The distribution of the excited states of helium resembles also that of electromagnetic radiation in a cavity (Chap.13).

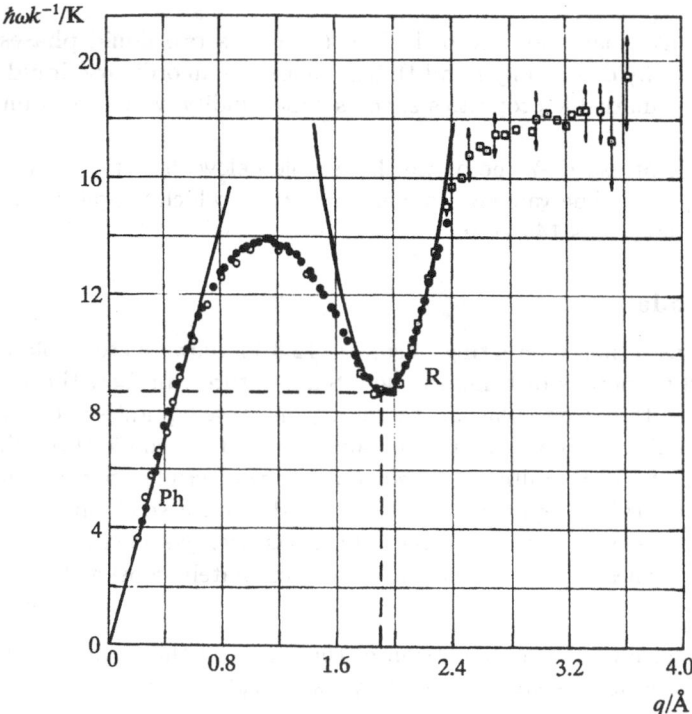

Fig. 16.26. The dispersion curve $\omega(q)$ of elementary excitations in helium II. The wavenumber q is measured in Å^{-1}, and the ordinate represents $\hbar\omega/k$ in kelvins

As in these earlier problems, we associate with *each mode* of helium II a *quantum harmonic oscillator*, the successive excited states of which lie at a distance of $\varepsilon(q) = \hbar\omega(q)$ from one another. An *elementary excitation* $[\varepsilon(q), q]$ is thus interpreted as adding to the system a *quasi-particle with an energy* $\varepsilon(q)$, a *momentum* $p = \hbar q$, and *zero spin*. The volume Ω of the system is taken care of through periodic boundary conditions.

In the case of helium II we can determine the function $\omega(q)$ either theoretically by taking the interactions between the atoms in the liquid into account, or experimentally by measuring the energy and momentum absorbed by the liquid when a single mode is excited by a low energy neutron. As the liquid is isotropic, the function $\omega(q)$ depends solely on the length $q = |q|$ of the wavevector. Figure 16.26 shows the so-called *dispersion curve*, that is the function $\omega(q)$, for a pressure equal to $\mathcal{P}_0$ and a temperature $T = 1.1$ K. In fact, this curve depends little on T and $\mathcal{P}$. As we shall see in what follows we can explain a number of properties of helium II, knowing this curve. In fact, a remarkable agreement with experiment is obtained up to 1.6 K.

For a treatment of some questions it will be necessary to know the *quasi-particle velocity*. In a medium in which vibrations can propagate this velocity is that of the centre of a wavepacket formed through a superposition of modes with wavevectors which are close to the wavevector q of the quasi-particle. The centre of the wavepacket, which resembles a classical particle, moves with a velocity

$$v(q) = \nabla_q(\omega) = \frac{d\omega(q)}{dq}\frac{q}{q}, \tag{1}$$

which can be found, once we know the dispersion curve $\omega(q)$.

Preliminary Questions

1. Consider a system of identical, non-interacting, zero-spin bosons contained in a box of volume Ω. Their single-particle quantum states are characterized by the value of the wavevector q. We write $\mathcal{D}(q)\,d^3q$ for the number of such states in the volume element $d^3q = dq_x\,dq_y\,dq_z$ centred around q. The energy of the state q is $\varepsilon(q)$, where ε is assumed to be a known function of q. Give $\mathcal{D}(q)$ as function of Ω. Use the expressions for the internal energy U and for the grand potential A of a grand canonical system in equilibrium as functions of the occupation factor f_q and of $\mathcal{D}(q)$ to derive expressions for U and A as functions of $\beta = 1/kT$, $\Omega, \varepsilon(q)$, and the chemical potential μ.

2. What is the most convenient statistical ensemble to use for studying the above system, if the number N of bosons is not conserved, because bosons can be created or absorbed by the walls of the vessel? How can you in that case use the expressions from 1 to find the free energy $F = U - TS$ as function of T, Ω, and ε? Give an expression for the pressure $\mathcal{P}$ of the boson gas.

3. Consider a discrete system of quantum harmonic oscillators with frequencies $\omega_i/2\pi (i = 1, 2, \cdots)$. Repeating the arguments from the main text,

show that the equilibrium of such a system can be described as the equilibrium of a set of non-interacting quasi-particles with possible energies $\hbar\omega_i$ which obey Bose-Einstein statistics. Justify in this way the equivalence, which we mentioned above when defining the model for helium II, between the modes $[\hbar\omega(\boldsymbol{q}), \boldsymbol{q}]$ and a gas of quasi-particles, provided the chemical potential has a value which should be given.

Specific Heat of Helium II

We now consider the model which represents helium II as consisting of a *substrate* of energy E_s and a system of *modes* $[\hbar\omega(q), \boldsymbol{q}]$, where $\omega(q)$ is given by Fig.16.26.

4. Show that the free energy F of a volume Ω of this fluid has the form

$$F = E_s + bT\Omega \int_0^\infty q^2 \ln\left(1 - e^{-\beta\hbar\omega}\right) dq, \tag{2}$$

where b is a constant which must be found and $q = |\boldsymbol{q}|$.

5. We shall, when evaluating F, neglect the exponential $e^{-\beta\hbar\omega}$ when $\beta\hbar\omega \geq 5$. Show that when $T < T_\lambda$ the thermodynamic properties of helium II depend only on two parts of the dispersion curve, defined by two intervals $[0, q_0]$ and $[q_1, q_2]$ for the values of the wavenumber. Give the order of magnitude of the quantities q_0, q_1, and q_2.

We shall systematically use in what follows the approximation which consists of neglecting $e^{-\beta\hbar\omega}$ when $\beta\hbar\omega \geq 5$.

6. We are interested in the contribution to F from quasi-particles with wavenumbers in the $[0, q_0]$ interval. The corresponding excitations are called *phonons* since they propagate at a well defined velocity c and, as in a crystalline solid, correspond to vibrations with *acoustic* frequencies. Show that to a good approximation the contribution F_{ph} of the phonons to F can be calculated by replacing q_0 by $+\infty$ and replacing the dispersion curve by a straight line $\omega = cq$. Give F_{ph} as a function of temperature, first algebraically and then numerically. Hence, find the contribution C_{ph} of the phonons to the molar specific heat at constant volume of helium II. Compare the numerical value of C_{ph} at a temperature of 1 K with typical values of specific heats of materials which you know of, such as gases, liquids, or solids, at room temperature.

7. The excitations corresponding to the $[q_1, q_2]$ interval are called *rotons*. Find an expression for the average number $\langle n(\boldsymbol{q}) \rangle$ of rotons in each mode $[\hbar\omega(q), \boldsymbol{q}]$. Show that this number remains always small when $T < T_\lambda$. Obtain a numerical upper bound for $\langle n(\boldsymbol{q}) \rangle$ in the roton region $[q_1, q_2]$ using Fig.16.26.

8. Show that the contribution F_R of the rotons to F has the form

$$F_R = -rT^{3/2}\Omega e^{-\Delta/T}, \tag{3}$$

where r and Δ are coefficients which you must find, both analytically and numerically as functions of fundamental constants and of characteristic features of the curve of Fig.16.26. To derive this result, replace $\omega(q)$ in the region considered by a parabola and show that it is a good approximation to replace after that $[q_1, q_2]$ by $[-\infty, +\infty]$.

9. Calculate the contribution C_R of the rotons to the molar specific heat at constant volume of helium II. Compare the numerical values of C_{ph} and C_R as functions of temperature and give a table showing for several values of the temperature the variation of the total molar specific heat of helium II, between $T = 0$ and $T = T_\lambda$.

Dynamical Properties of Helium II

We shall now use the above results to study some properties of uniformly flowing helium II. When it is set in motion this liquid can acquire momentum in two different ways which may be combined:

a) by moving the walls of a tube which surrounds the liquid parallel to themselves (Fig.16.27a). We assume that in this motion the *substrate is not dragged along*: the substrate velocity is $v_s = 0$. On the other hand, the walls can give up (or absorb) energy and momentum to (from) the fluid by creating (or absorbing) one (or more) quasi-particle(s) of momentum p and energy $\hbar\omega$. As a consequence, near the wall the quasi-particles are in equilibrium with the wall: they have the temperature of the wall and an average velocity u equal to the translational velocity v_w of the wall. For a narrow tube (for instance, one of the micro-channels of a porous medium) *all* the quasi-particles are thus in equilibrium with the wall.

b) by moving the substrate at a velocity v_s (for instance, using a piston which pushes the fluid) (Fig.16.27b). For a wide tube, far from the walls, the quasi-particles are in equilibrium with the substrate, as in the earlier parts of this problem, and their *relative* mean velocity u, *with respect to the substrate*, is zero.

The general situation corresponds to a combination of a and b where the walls influence the equilibrium of the quasi-particles while the substrate moves without interacting with the walls. If, for example, we consider a *narrow tube* with fixed walls, in which the substrate moves with a velocity $v_s = v$, the quasi-particles, in equilibrium with the walls, remain on average at rest in the laboratory frame so that their mean velocity u *with respect to the*

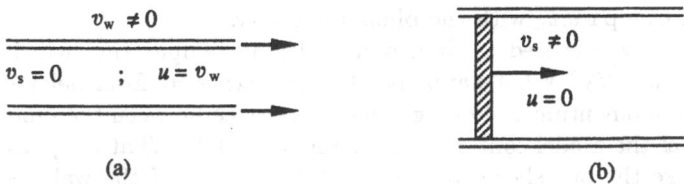

Fig. 16.27a,b. Two types of flow in helium II

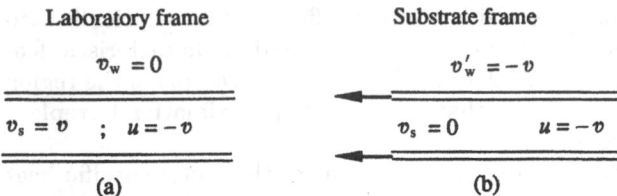

Laboratory frame   Substrate frame

$v_{\rm w} = 0$ $v'_{\rm w} = -v$

$v_s = v$; $u = -v$ $v_s = 0$ $u = -v$

(a)   (b)

Fig. 16.28a,b. Flow of superfluid helium in a narrow tube

substrate will be $u = -v$ (Fig.16.28a). In fact, to simplify the discussion we shall most often work in a reference frame fixed to the substrate (Fig.16.28b). In such a change of reference frame the *relative* velocity u is, of course, unchanged while $v_{\rm s} = 0$, but the walls are no longer fixed and move with a non-vanishing velocity $v'_{\rm w} = -v$.

10a. Justify briefly the fact that in the frame in which the substrate is at rest and where the quasi-particles are thus dragged along by the walls the density operator $\widehat{D}$ of the fluid is of the form

$$\widehat{D} = \frac{1}{Z}\, e^{-\beta\widehat{H}-(\boldsymbol{\Lambda}\cdot\widehat{\boldsymbol{P}})}, \qquad (4)$$

where $\widehat{\boldsymbol{P}}$ denotes the operator which is associated with the total momentum $\boldsymbol{P}$ of the quasi-particles, $\boldsymbol{P} = \sum \boldsymbol{p}_i = \sum \hbar\boldsymbol{q}_i$, and $\widehat{H}$ the Hamiltonian considered earlier.

b. Write down an expression for the average number $\langle n(\boldsymbol{q})\rangle$ of quasi-particles in the mode $\boldsymbol{q}$ as function of β and $\boldsymbol{\Lambda}$.

c. Give in the form of a triple integral over d^3q the total average momentum $\langle \boldsymbol{P}\rangle$ of the quasi-particles as function of β, Ω, and $\boldsymbol{\Lambda}$.

11. Show that the average $\boldsymbol{u} = \langle \boldsymbol{v}(\boldsymbol{q})\rangle$ of the quasi-particle velocities relative to the substrate equals $-\boldsymbol{\Lambda}/\beta$. To do this use fact that the integral

$$\int \nabla_q \left[\ln\left(1 - e^{-\beta\hbar\omega-(\boldsymbol{\Lambda}\cdot\hbar\boldsymbol{q})}\right)\right] d^3q \qquad (5)$$

vanishes. Use also Eq.(1) for the quasi-particle velocity as function of $\omega(q)$.

12a. Taking a frame of reference fixed to the substrate show that there exists a critical velocity u_c, the numerical value of which must be determined by using Fig.16.26, which has the following property: the number of quasi-particles remains negligible in the limit of very low temperatures $(T \to 0)$ as long as the relative velocity u of the quasi-particles relative to the substrate remains below u_c. Compare u_c with the phonon velocity.

b. Can the velocity u exceed u_c at a non-vanishing temperature which is, however, lower than T_λ? What happens when it reaches u_c from below? What exchanges of momentum and energy then take place between the fluid and the walls? Does the model considered here remain valid? What happens if in a situation like the one shown in Fig.16.27a the speed of the walls is increased sufficiently to exceed u_c?

13a. By staying in the frame fixed to the substrate, show that if the mean velocity $|\boldsymbol{u}|$ of the quasi-particles is *small compared to* u_c $(u \ll u_c)$, the total momentum $\boldsymbol{P}$ of the quasi-particles in a volume Ω of the fluid has an average value of the form

$$\langle \boldsymbol{P} \rangle = \varrho_n \Omega \boldsymbol{u}, \tag{6}$$

and express ϱ_n as an integral over q.

b. Write $F = -kT \ln Z$. Use the thermodynamic relation between $\langle \boldsymbol{P} \rangle$ and $\tilde{F}$ and the result of question 13a to show that $\tilde{F}$ differs from its value at rest $(u = 0)$, which is the free energy F considered earlier in this problem, as follows:

$$\Delta \tilde{F} = \tilde{F} - F = \tfrac{1}{2} \varrho_n \Omega u^2. \tag{7}$$

c. Show that one can interpret this situation as the dragging along by the walls of a fictitious fluid, with a velocity $\boldsymbol{u}$ and a mass density ϱ_n which changes with temperature.

14. In the case of a general flow of helium II, where the tube is not necessarily narrow, we consider a volume element Ω of the fluid which moves with the velocity $\boldsymbol{v}_s$ of the substrate. The quasi-particles which are contained in it move with a relative mean velocity $\boldsymbol{u}$, and hence with a mean velocity $\boldsymbol{v}_n \equiv \boldsymbol{u} + \boldsymbol{v}_s$ with respect to the laboratory frame of reference.

We define a mass density ϱ_s by the relation

$$\varrho = \varrho_n + \varrho_s, \tag{8}$$

where ϱ is the total mass density of helium II, and ϱ_n the mass density defined in Eq.(6). The densities ϱ_n and ϱ_s are called, respectively, the *normal* and *superfluid* mass densities.

a. Show that the mass current density $\boldsymbol{J} = \langle \boldsymbol{P}_{\text{lab}} \rangle / \Omega$ and the free energy of the moving fluid $\tilde{F}_{\text{lab}}$ are, respectively, equal to

$$\boldsymbol{J} = \varrho_n \boldsymbol{v}_n + \varrho_s \boldsymbol{v}_s, \tag{9}$$

$$\tilde{F}_{\text{lab}} = F + \tfrac{1}{2} \Omega \left(\varrho_n v_n^2 + \varrho_s v_s^2 \right), \tag{10}$$

where $\boldsymbol{P}_{\text{lab}}$ and $\tilde{F}_{\text{lab}}$ are relative to the laboratory frame of reference. We use the fact that, if a mass M of fluid has, in the frame of reference fixed to the substrate, an energy E and a momentum $\boldsymbol{P}$, these two quantities have in the laboratory frame the values

$$E_{\text{lab}} = E + (\boldsymbol{P} \cdot \boldsymbol{v}_s) + \tfrac{1}{2} M v_s^2, \tag{11}$$

and

$$\boldsymbol{P}_{\text{lab}} = \boldsymbol{P} + M \boldsymbol{v}_s, \tag{12}$$

while the entropy S remains unchanged (see § 14.4.4).

b. Deduce from this that helium II at low velocities behaves as a mixture of *two independent fluids* with a composition which depends solely on the temperature.

Note that the first fluid, the so-called *normal fluid*, is similar to an ordinary fluid, while the second, the so-called *superfluid*, exchanges neither momentum, nor energy with the walls of the tube.

15a. Show that the density ϱ_n of the normal fluid is the sum of two contributions, from the phonons and rotons, respectively.

b. Show that the contribution ϱ_n^{ph} of the phonons to ϱ_n has the form

$$\varrho_n^{ph} = AT^m, \tag{13}$$

and find the value of m.

c. Use the same approximations as in question 8 to evaluate algebraically and numerically the contribution ϱ_n^R of the rotons to ϱ_n.

Assuming that $\varrho_n^R \gg \varrho_n^{ph}$ when $T > 0.7$ K, show the behaviour of the $\varrho_n(T)$-curve.

d. Determine numerically the temperature at which the superfluid disappears. Interpret and discuss this result.

16. Two vessels A and B which are thermally insulated and contain helium II are initially at the same pressure and temperature and are connected by a *narrow tube*. We introduce a little amount of heat into A.

a. Due to this heat, the *superfluid* fraction decreases initially in one of the two vessels; which one? How is the system going to evolve?

b. After a short lapse of time we obtain in the system of the two vessels a stationary, near-equilibrium state in which there is a small temperature difference ΔT and a small pressure difference ΔP between A and B. Show that we have the relation

$$\frac{\Delta P}{\Delta T} = \frac{S}{\Omega}. \tag{14}$$

Assume that the chemical potentials μ_A and μ_B *of the helium atoms* in the vessels A and B are the same in the quasi-equilibrium state considered.

c. Using an order of magnitude estimate (for a temperature $T \simeq 1.5$ K) show that the above effect can lead to a significant difference between the fluid levels in the two vessels even when $T_A - T_B$ is very small, of the order of a few hundredths of a degree. This effect which can lead to a spouting of liquid helium is called the *fountain effect* (Fig.12.8).

d. Use thermodynamic arguments to justify the hypothesis $\mu_A = \mu_B$, made in 16b.

Solution

1. If $\mathcal{D}'(\varepsilon)$ is the density of states at energy ε, we have $\mathcal{D}(q)\,d^3q = \mathcal{D}'(\varepsilon)\,d\varepsilon$. The possible values of each component q_i of q are $2\pi n/L_i$, where $i = x, y,$ or z, where n is an integer, and where $\Omega = L_x L_y L_z$ is the volume, so that

$$\mathcal{D}(q) = \frac{\Omega}{8\pi^3}. \tag{1}$$

The internal energy U and the grand potential A of a quantum gas of non-interacting spinless bosons are given by the equations

$$U = \int \varepsilon f(\varepsilon) \, \mathcal{D}'(\varepsilon) \, d\varepsilon,$$
$$A = -\frac{1}{\beta} \int \ln\{1 + f(\varepsilon)\} \, \mathcal{D}'(\varepsilon) \, d\varepsilon. \tag{2}$$

Moreover, we can express $f(\varepsilon)$ as a function of q:

$$f\big(\varepsilon(q)\big) = \frac{1}{e^{\beta[\varepsilon(q)-\mu]} - 1}, \tag{3}$$

for non-interacting bosons. Hence, using (1), (2), and (3), we get

$$U = \int \varepsilon(q) \, f\big(\varepsilon(q)\big) \, \mathcal{D}(q) \, d^3q = \frac{\Omega}{8\pi^3} \int \frac{\varepsilon(q) \, d^3q}{e^{\beta[\varepsilon(q)-\mu]} - 1} \tag{4}$$

and

$$A = -\frac{1}{\beta} \int \ln\left[1 + f\big(\varepsilon(q)\big)\right] \mathcal{D}(q) \, d^3q$$

$$= \frac{\Omega}{8\pi^3\beta} \int \ln\left(1 - e^{-\beta[\varepsilon(q)-\mu]}\right) d^3q. \tag{5}$$

2. If N is not conserved, one can fix that quantity neither on average, nor exactly. It does not enter into the determination of thermodynamic equilibrium and the corresponding statistical ensemble has only the natural variables β, which is conjugated to $< U >$, and Ω, if we take $\ln Z$ as a thermodynamic potential. The expression for Z is the same as the grand partition function, which must be summed over all N-particle Hilbert spaces with N going from 0 to ∞, provided $\alpha = \beta\mu = 0$. The thermodynamic potential associated with the variables T and Ω is equivalently A or F, since $A = F - \mu N = F$ when $\mu = 0$. Hence the free energy F has the form (5) with $\mu = 0$:

$$F = \frac{kT\Omega}{8\pi^3} \int \ln\left(1 - e^{-\beta\varepsilon(q)}\right) d^3q. \tag{6}$$

The pressure of the boson gas is

$$P = -\frac{\partial F}{\partial \Omega} = -\frac{A}{\Omega} = -\frac{F}{\Omega} = -\frac{kT}{8\pi^3} \int \ln\left(1 - e^{-\beta\varepsilon(q)}\right) d^3q. \tag{7}$$

3. In the main text we have discussed phonons (§ 11.4) and photons in an enclosure (§ 13.1), as we did also in Exerc.4f: each quantum oscillator with frequency $\omega_i/2\pi$ makes a contribution to the total energy which is equal to $(n_i + \frac{1}{2})\hbar\omega_i$ where n_i is a positive integer or zero. A quantum micro-state of the whole system is characterized by giving the n_i of the different oscillators (each state n_i is non-

degenerate). If we change the energy origin to get rid of the $\frac{1}{2}\hbar\omega_i$, the micro-state $|n_1,\ldots,n_i,\ldots,\rangle$ has an energy

$$U = \sum n_i\,\hbar\omega_i = n_i\,\varepsilon_i \qquad \text{with} \qquad \varepsilon_i = \hbar\omega_i.$$

Let us describe the system as consisting of fictitious identical particles, the quasi-particles; the single-particle states are the set of the $\varepsilon_i = \hbar\omega_i$. We can then interpret the n_i as the occupation numbers of the different single-particle quantum states and they together determine the micro-state of the system. As $n_i = 0, 1, \cdots, \infty$, everything happens, as if the quasi-particles were bosons. Their total number $N = \sum n_i$ is clearly not conserved as the quasi-particles are at the same time energy quanta: even if the total energy is conserved, an energy quantum ε_i can disappear, being replaced by several other energy quanta. We thus cannot specify the number of particles; N is not a constant of the motion. In the model for helium II the above correspondence is implemented by moreover associating a quantum oscillator with a mode characterized by a wavevector q. We have therefore a quasi-continuous set of oscillators with a density of states $\mathcal{D}(q)$; each oscillator q has a frequency $\omega(q)/2\pi$, where $\omega(q)$ is given by the dispersion curve, and an energy $\varepsilon(q) = \hbar\omega(q)$. As in the discrete case of oscillators with ω_i, the quasi-particle number N is not conserved and the chemical potential of the system is zero (cf. question 2).

4. A micro-state $\{n_q\}$ of helium II has an energy

$$E = E_s + \sum n_q\,\hbar\omega_q, \tag{8}$$

where the summation is over all values of q which are allowed in the box of volume Ω.

The partition function can then be written in the form

$$Z = \sum_{\{n_q\}} e^{-\beta E_s}\, e^{-\beta\sum_q n_q\hbar\omega_q} = e^{-\beta E_s}\, Z_{\mathrm{qp}}, \tag{8'}$$

where Z_{qp} is associated with single quasi-particles; hence

$$F = -\frac{1}{\beta}\ln Z = E_s + F_{\mathrm{qp}},$$

where F_{qp} is the free energy (6) associated with a system of spinless non-interacting bosons in a box, whose number is not conserved. Replacing the integration over q_x, q_y, and q_z by integration over $q = |q|$ we get

$$F = E_s + \frac{kT\Omega}{2\pi^2}\int_0^\infty \ln\left(1 - e^{-\beta\hbar\omega(q)}\right) q^2\,dq. \tag{9}$$

5. If we agree to neglect $e^{-\beta\hbar\omega}$ for $x = \beta\hbar\omega \gtrsim 5$, that is, $e^{-x} \lesssim 7\times 10^{-3}$, this means that for given β we eliminate large values of $\hbar\omega$. The acceptable range $[0, \hbar\omega]$ is thus the narrower, the higher β, that is, the lower T. Therefore, $\hbar\omega_{\max}$ corresponds to the upper limit $kT = kT_\lambda$ of the stability range of helium II, and $\omega_{\max}$ is defined by

$$x = \frac{\hbar\omega_{\max}}{kT_\lambda} \lesssim 5, \qquad \text{or}$$

$$\frac{\hbar\omega_{max}}{k} \lesssim 11 \text{ K}. \tag{10}$$

As we have already indicated at the beginning of this problem, the dispersion curve, $\omega(q)$, does not depend strongly on T. Using that curve we see that condition (10) is satisfied in two q-value ranges, namely, $[0, q_0]$, with $q_0 \simeq 0.65 \text{ Å}^{-1}$ and $[q_1, q_2]$ around the point R, with $q_1 \simeq 1.65 \text{ Å}^{-1}$ and $q_2 \simeq 2.15 \text{ Å}^{-1}$. The abscissa q_R of the minimum R of the curve is 1.9 Å^{-1} and we thus have

$$|q_1 - q_R| \simeq |q_2 - q_R| \simeq 0.25 \text{ Å}^{-1}.$$

6. In the phonon region $[0, q_0]$, the dispersion curve is a straight line and we have $\omega = cq$, where c is the slope of the dispersion curve at the origin. From Fig.16.26 we find that $\hbar c/k = 18.3 \text{ Å K}$. To calculate F_{ph} we replace ω by cq in (9) in the region $[0, q_0]$. This gives

$$F_{ph} = \frac{kT\Omega}{2\pi^2} \int_0^{q0} \ln\left(1 - e^{-\beta\hbar cq}\right) q^2 \, dq. \tag{11}$$

Integration by parts and change of variable, $x = \beta\hbar cq$, give

$$F_{ph} \simeq -\frac{k^4 T^4 \Omega}{6\pi^2 \hbar^3 c^3} \int_0^5 \frac{x^3 \, dx}{e^x - 1}.$$

As

$$\int_5^\infty \frac{x^3 \, dx}{e^x - 1} \simeq \int_5^\infty x^3 e^{-x} \, dx \simeq 6 e^{-5} \simeq 0,$$

we can extend the integral over the interval $(0, +\infty)$. The resulting definite integral can be found in the table at the end of the book:

$$\int_0^\infty \frac{x^3 \, dx}{e^x - 1} = \Gamma(4)\zeta(4) = \frac{\pi^4}{15},$$

whence

$$F_{ph} = -\frac{\pi^2}{90} \frac{k^4 \Omega T^4}{\hbar^3 c^3} = -247 \, \Omega T^4 \text{ J}. \tag{12}$$

The entropy is $S = -\dfrac{\partial F}{\partial T}$ and the specific heat of a volume Ω of helium II is therefore

$$C_{ph} = T\frac{\partial S}{\partial T} = \frac{2\pi^2}{15} \frac{k^4 \Omega T^3}{\hbar^3 c^3} \simeq 2964 \, \Omega T^3 \text{ J K}^{-1}. \tag{13}$$

Using the values of the density, $\varrho_0 \simeq 145 \text{ kg m}^{-3}$, of helium II and the mass, 4×10^{-3} kg, of a mole of ^4He, we find for the molar specific heat

$$C_{ph} \simeq 0.082 \, T^3 \text{ J K}^{-1} \text{ mol}^{-1}.$$

For $T = 1$ K this value is small compared to those of condensed matter at room temperature: water, 75 J K^{-1} mol^{-1}, or solids (Dulong-Petit law), $3R = 25$ J K^{-1} mol^{-1}, or compared to those of gases $\frac{3}{2}R = 12.5$ J K^{-1} mol^{-1}.

7. The average number of rotons in a mode q equals the corresponding Bose-Einstein factor:

$$\langle n(q) \rangle = f_q = \frac{1}{e^{\beta \varepsilon(q)} - 1} = \frac{1}{e^{\hbar \omega(q)/kT} - 1}.$$

This number has an upper bound which is the value of f_q for $\omega_R = \omega(q_R)$ and $T = T_\lambda$, where

$$\frac{\hbar \omega_R}{kT_\lambda} \simeq 4.$$

We have thus

$$\langle n(q) \rangle_R \lesssim e^{-\hbar \omega_R/kT_\lambda} \simeq e^{-4} \simeq 0.02,$$

which is, in fact, a small number.

8. In the roton region $\omega(q)$ has a parabolic shape and we can therefore replace the dispersion relation by

$$\frac{\hbar(\omega - \omega_R)}{k} = a(q - q_R)^2, \tag{14}$$

where $a = 40.4$ K Å$^2 = 40.4 \times 10^{-20}$ K m^2. Moreover, as $e^{-\hbar \omega/kT} < e^{-\hbar \omega_R/kT_\lambda} \simeq 0.02 \ll 1$ we can expand the logarithm in Eq.(9) for the free energy as for a classical limit. We find

$$F_R \simeq -\frac{kT\Omega}{2\pi^2} \int_{q_1}^{q_2} q^2 \, dq \, e^{-\hbar \omega(q)/kT}, \tag{15}$$

with $\omega(q)$ given by (14). As indicated in the text, we assume that, if we extend the integral (15) to the interval $(-\infty, +\infty)$, the physical results remain essentially unchanged. In fact, if we make this approximation, the error is of the order of 1 % due to the fact that the integrand is so small outside the interval (q_1, q_2). Using that approximation we get, after changing the variable q to $x = \sqrt{a/T}(q - q_R)$:

$$F_R = -\frac{kT\Omega}{2\pi^2} e^{-\hbar \omega/kT} \left(\frac{T}{a}\right)^{3/2} \int_{-\infty}^{+\infty} \left(x + q_R \sqrt{\frac{a}{T}}\right)^2 e^{-x^2} \, dx. \tag{16}$$

The integral has two non-vanishing terms, the first one with x^2 and the second one with $q_R^2 a/T$. We need retain only the latter, as $q_R^2 a/T = 150/T$, and we get

$$F_R = -\frac{\Omega kT^{3/2} q_R^2}{2\pi^{3/2} a^{1/2}} e^{-\hbar \omega_R/kT},$$

which is of the expected form, $-rT^{3/2}\Omega e^{-\Delta/T}$, with

$$r = \frac{kq_R^2}{2\pi^{3/2} a^{1/2}} \quad \text{and} \quad \Delta = \frac{\hbar \omega_R}{k}.$$

Table 16.1. The phonon and roton contributions to the specific heat of He II, in $J \ K^{-1} \ mol^{-1}$.

T	0.2	0.4	0.6	0.9	1.2	1.5	1.8	2.0	2.18
C_{ph}	6.5×10^{-4}	0.0052	0.0175	0.059	0.14	0.27	0.47	0.65	0.84
C_R	–	2.4×10^{-6}	0.0018	0.127	0.97	3.05	6.30	8.92	11.46
C	6.5×10^{-4}	0.0052	0.019	0.186	1.11	3.32	6.77	9.57	12.30

The numerical values, equal to $r = 6.9 \times 10^5 \ J \ m^{-3} \ K^{-3/2}$ and $\Delta = 8.67$ K, are found by taking the values of a, $q_R \simeq 1.9 \ \mathring{A}^{-1}$, and ω_R from Fig.16.26.

9. The roton contribution to C, evaluated in the same way as the phonon contribution, is

$$C_R = -T \frac{\partial^2 F_R}{\partial T^2} = r\Omega T^{1/2} e^{-\Delta/T} \left(\frac{3}{4} + \frac{\Delta}{T} + \frac{\Delta^2}{T^2} \right). \tag{17}$$

Hence, taking for Ω the molar volume, we find

$$C_R \simeq 60 \left(\frac{T}{\Delta} \right)^{1/2} e^{-\Delta/T} \left(\frac{3}{4} + \frac{\Delta}{T} + \frac{\Delta^2}{T^2} \right) \ J \ K^{-1} \ mol^{-1}.$$

The temperature dependences of C_{ph} and C_R are very different. Because of the factor $e^{-\Delta/T}$ the roton contribution C_R is negligible compared to the phonon contribution at very low temperatures. Nevertheless, although the number of rotons per mode always stays small, the roton contribution cannot be neglected when the temperature increases, as the region $q \sim 0$ in wavevector space related to the phonons is much smaller than the neighbourhood of the sphere $q \sim q_R$ relating to the rotons.

Up to $T = 0.6$ K, the phonon contribution in T^3 dominates (Table 16.1). From $T = 1.4$ K onwards, the inverse is true. The curve obtained in this way is in good agreement with experiment, except at the highest temperatures, where measurements show that C diverges at $T = T_\lambda = 2.18$ K. In fact, the model is then no longer valid; it does not predict the phase transition to helium I, an ordinary liquid.

10a. The total average momentum is non-vanishing for the quasi-particles. It is imparted to them by the walls which play the role of a bath which determines the constant of motion $\boldsymbol{P}$. We can therefore fix $\langle \boldsymbol{P} \rangle$ by means of a Lagrangian multiplier $\boldsymbol{\Lambda}$. At equilibrium the Boltzmann-Gibbs density operator is thus

$$\widehat{D} = \frac{1}{Z} e^{-\beta \widehat{H} - (\boldsymbol{\Lambda} \cdot \widehat{\boldsymbol{P}})}. \tag{18}$$

b. As in 8 we have

$$E = E_s + \sum_q n(q) \hbar \omega(q). \tag{19}$$

On the other hand, we have

$$P = \sum_q n(q)\, p = \sum_q n(q)\, \hbar q. \tag{20}$$

Hence we find

$$Z = \sum_{\{n_q\}} \exp\left[-\beta E_{\mathrm{s}}\right] \exp\left[-\beta \sum_q n_q \hbar \omega(q) - \left(\boldsymbol{\Lambda} \cdot \sum_q n_q\, \hbar q\right)\right], \tag{21}$$

which is the same as Eq.(8') for the partition function, except that $\beta \omega(q)$ has been replaced by $\beta \omega(q) + (\boldsymbol{\Lambda} \cdot q)$. If we make the same substitution in the average occupation number $\langle n(q) \rangle$, we get

$$\langle n(q) \rangle = \frac{1}{e^{\beta \hbar \omega(q) + (\boldsymbol{\Lambda} \cdot \hbar q)} - 1}.$$

c. We find

$$\langle \boldsymbol{P} \rangle = \int \langle n(q) \rangle\, \hbar q\, \mathcal{D}(q)\, d^3 q = \frac{\Omega}{8\pi^3} \int \frac{\hbar q\, d^3 q}{e^{\beta \hbar \omega(q) + (\boldsymbol{\Lambda} \cdot \hbar q)} - 1}. \tag{22}$$

11. The integral in question is equal to

$$I = \int d^3 q\, \nabla_q \left[\ln\left(1 - e^{-\beta \varepsilon - (\boldsymbol{\Lambda} \cdot \boldsymbol{p})}\right)\right] = \int \frac{d^3 q\, \nabla_q \big(\beta \varepsilon + (\boldsymbol{\Lambda} \cdot \boldsymbol{p})\big)}{e^{\beta \varepsilon + (\boldsymbol{\Lambda} \cdot \boldsymbol{p})} - 1}.$$

Using the fact that $\boldsymbol{v}(q) = \nabla_q(\omega)$, $\boldsymbol{p} = \hbar q$, and $\varepsilon = \hbar \omega(q)$, we find

$$I = \int d^3 q \langle n(q) \rangle \big\{\beta \hbar \boldsymbol{v}(q) + \hbar \boldsymbol{\Lambda}\big\} = \frac{8\pi^3}{\Omega} N \big[\beta \hbar \boldsymbol{u} + \hbar \boldsymbol{\Lambda}\big].$$

Because the logarithm vanishes at infinity, the integral I vanishes, and we find therefore that $\beta \boldsymbol{u} + \boldsymbol{\Lambda} = 0$, that is, the required result:

$$\boldsymbol{u} = -\frac{\boldsymbol{\Lambda}}{\beta}. \tag{23}$$

12a. We write $\langle n(q) \rangle$ in the form

$$\langle n(q) \rangle = \frac{1}{e^{\beta \hbar [\omega(q) - (\boldsymbol{u} \cdot \boldsymbol{q})]} - 1}. \tag{24}$$

As long as $\omega(q) - (\boldsymbol{u} \cdot q) > \omega(q) - uq$ remains positive, this number tends to zero as $T \to 0$. Therefore, if

$$u < \frac{\omega(q)}{q}$$

for all values of q, the average number of quasi-particles per mode is negligible as $T \to 0$, exactly as for a zero velocity. The minimum of ω/q is the slope u_{c} of the tangent from the origin to the dispersion curve, which touches this curve in the roton region at q_{c}, ω_{c}; this minimum is slightly smaller than $\omega(q_{\mathrm{R}})/q_{\mathrm{R}}$ (Fig.16.26).

Hence, if $T \simeq 0$ and $u < u_c$, the liquid remains practically at rest in its ground state. Note that the integral I of question 11 is defined only when $u < u_c$ or $\Lambda < \Lambda_c$.

b. Let us consider a wavevector q of length q_c in the direction of u. The average number of rotons in the corresponding mode,

$$\langle n(q) \rangle = \frac{1}{e^{\hbar q_c(u_c - u)/kT} - 1},$$

tends to infinity as u increases and reaches u_c. Infinite amounts of energy and of momentum are thus imparted to the fluid through the creation of rotons due to the motion of the walls. This *catastrophe* implies that the model ceases to be valid for $u \geq u_c$. Therefore, if we let the velocity v_w of the walls increase, their motion starts to produce an anisotropic quasi-particle distribution (24) with $u = v_w$; when v_w reaches u_c, helium changes from helium II into normal helium I; after this phase change it is globally dragged along at the velocity v_w, just like any other fluid.

Figure 16.26 gives us the numerical value $\hbar u_c/k = 4.3 \times 10^{-10}$ m K, or $u_c = 56$ m s^{-1}, which is not in agreement with the experimental value of the critical velocity, being too large by two orders of magnitude. In fact, before the walls create a large number of rotons, other excitations appear which we have not discussed, the so-called *vortex lines* associated with gradients in the displacement velocity of the substrate, which we have here assumed to be rigid.

13a. The total quasi-particle momentum is

$$P = \frac{\Omega}{8\pi^3} \int d^3q \, \langle n(q) \rangle \, \hbar q.$$

Expanding $\langle n(q) \rangle$ up to first order in u, we get

$$\langle n(q) \rangle - \langle n(q) \rangle_{u=0} = \frac{\partial \langle n(q) \rangle}{\partial (u \times q)} (u \cdot q)$$

$$= \frac{e^{\hbar \omega(q)/kT}}{\left(e^{\hbar \omega(q)/kT} - 1\right)^2} \frac{\hbar}{kT} (u \cdot q).$$

The term $\langle n(q) \rangle_{u=0}$ corresponds to the rest state and contributes zero to $\langle P \rangle$. Therefore we find

$$\langle P \rangle = \frac{\Omega}{8\pi^3} \int \frac{e^{\hbar \omega(q)/kT}}{\left(e^{\hbar \omega(q)/kT} - 1\right)^2} \frac{\hbar^2}{kT} (u \cdot q) q \, d^3 q.$$

We take the z-axis along u and integrate over the angles. We find

$$\int (u \cdot q) q \sin \theta \, d\theta \, d\varphi = uq^2 \frac{u}{u} \int_0^\pi \int_0^{2\pi} \sin \theta \cos^2 \theta \, d\theta \, d\varphi = \frac{4\pi}{3} q^2 u,$$

whence

$$\langle P \rangle = \Omega \varrho_n u, \tag{25}$$

with

$$\varrho_n = \frac{\hbar^2}{6\pi^2 kT} \int_0^\infty \frac{e^{\hbar \omega/kT} q^4 \, dq}{\left(e^{\hbar \omega/kT} - 1\right)^2}. \tag{26}$$

b. From Eq.(18) we obtain

$$\langle P \rangle = -\frac{\partial \ln Z}{\partial \Lambda},$$

and hence, using the relations $\Lambda = -\beta u$ and $\widetilde{F} = -kT \ln Z$, we find that

$$\langle P_i \rangle = \frac{\partial \widetilde{F}}{\partial u_i} = \varrho_{\mathrm{n}} \Omega u_i.$$

Since $\widetilde{F}$ is the same as F when $\Lambda = 0$, we find by integration

$$\widetilde{F} - F = \tfrac{1}{2}\Omega \varrho_{\mathrm{n}} u^2. \tag{27}$$

c. When a fluid of mass density ϱ and volume Ω is set globally in motion with a velocity u, its momentum becomes $\Omega \varrho u$, its energy increases by $\tfrac{1}{2}\Omega \varrho u^2$, and its free energy increases by the same amount, while its entropy remains unchanged. Equations (25) and (27) can thus be interpreted as the total momentum and kinetic energy of a volume Ω of a fluid in motion, the density ϱ_{n} of which is a function of the temperature given by Eq.(26).

14a. In the frame fixed to the substrate $\langle P \rangle$ and $\widetilde{F}$ are given by Eqs.(25) and (27). Changing to the laboratory frame and using the formulæ for a change of frame which we gave, we find

$$P_{\mathrm{lab}} = \Omega \varrho_{\mathrm{n}} u + M v_{\mathrm{s}} = \varrho_{\mathrm{n}}\Omega\,(u + v_{\mathrm{s}}) + \varrho_{\mathrm{s}}\Omega v_{\mathrm{s}},$$

where $\varrho \equiv \varrho_{\mathrm{n}} + \varrho_{\mathrm{s}}$; hence we find, writing $v_{\mathrm{n}} = u + v_{\mathrm{s}}$,

$$J_{\mathrm{lab}} = \frac{P_{\mathrm{lab}}}{\Omega} = \varrho_{\mathrm{n}} v_{\mathrm{n}} + \varrho_{\mathrm{s}} v_{\mathrm{s}}.$$

On the other hand, as the entropy is the same in the two frames,

$$\widetilde{F}_{\mathrm{lab}} = \widetilde{F} + (E_{\mathrm{lab}} - E) = \widetilde{F} + P \cdot v_{\mathrm{s}} + \tfrac{1}{2}M v_{\mathrm{s}}^2.$$

If we now write $P = \Omega \varrho_{\mathrm{n}} u$, $M = (\varrho_{\mathrm{n}} + \varrho_{\mathrm{s}})\Omega,$ and $v_{\mathrm{n}} = v_{\mathrm{s}} + u$, we easily find that

$$\widetilde{F} = F + \tfrac{1}{2}\Omega\left(\varrho_{\mathrm{n}} v_{\mathrm{n}}^2 + \varrho_{\mathrm{s}} v_{\mathrm{s}}^2\right).$$

b. From a purely mechanical point of view we see that everything takes place, as if we had a mixture of two independent fluids with mass densities ϱ_{n} and ϱ_{s}, velocities v_{n} and v_{s}, fluxes $\varrho_{\mathrm{n}} v_{\mathrm{n}}$ and $\varrho_{\mathrm{s}} v_{\mathrm{s}}$, and kinetic energies per unit volume $\tfrac{1}{2}\varrho_{\mathrm{n}} v_{\mathrm{n}}^2 + \tfrac{1}{2}\varrho_{\mathrm{s}} v_{\mathrm{s}}^2$. The results referring to the frame fixed to the substrate show that only the normal fluid exchanges energy and momentum with the walls.

15a. If $\hbar\omega/kT$ is large in Eq.(26) for ϱ_{n}, the integrand is of order $e^{-\hbar\omega/kT} q^4$. We can thus again make the same approximations as before and neglect q-ranges corresponding to $\hbar\omega \gtrsim 5$, retaining only the phonon and roton regions. We shall calculate these two contributions.

b. We get the *phonon* contribution by putting $\omega = cq$. Writing $x = \beta\hbar cq$ and extending the domain of integration as in question 6, we find

$$\varrho_{\mathrm{n}}^{\mathrm{ph}} = \frac{\hbar^2}{6\pi^2 kT}\left(\frac{kT}{\hbar c}\right)^5 \int_0^\infty \frac{x^4 e^x\,dx}{(e^x - 1)^2},$$

or

$$\varrho_n^{ph} = AT^4.$$

The exact calculation shows that

$$A = \frac{2\pi^2}{45} \frac{k^4}{\hbar^3 c^5} \simeq 0.016 \text{ kg m}^{-3} \text{ K}^{-4}.$$

c. As in question 8, the *roton* contribution to the normal density is

$$\varrho_n^R = \frac{1}{2\pi^2} \int_0^{\infty} q^2 \, dq \, e^{-\hbar\omega_R/kT} \, e^{-a(q-q_R)^2/T} \frac{\hbar^2}{kT} \frac{1}{3} q^2$$

$$= \frac{\hbar^2}{6\pi^2 kT} q_R^4 \left(\frac{\pi T}{a}\right)^{1/2} e^{-\Delta/T} = 1800 \sqrt{\frac{\Delta}{T}} \, e^{-\Delta/T} \text{ kg m}^{-3},$$

with $\Delta \simeq 8.67$ K.

As $T \to 0$, the phonon contribution dominates, as ϱ_n^R is exponentially small. When $T \gtrsim 0.7$ K, we find that $\varrho_n \simeq \varrho_n^R$ increases fast. For instance, when $T \simeq 0.7$ K, we have $\varrho_n \simeq 2.6 \cdot 10^{-2}$ kg m^{-3}, and when $T = 2$ K, we have $\varrho_n \simeq 49$ kg m^{-3}.

d. Our model ceases to be valid, when ϱ_n exceeds the total density ϱ of the fluid. As $\varrho = \varrho_n + \varrho_s$, we see that ϱ_s is a decreasing function of T and that the superfluid disappears when $\varrho_n = \varrho$ ($\varrho_s = 0$). The characteristic temperature of this disappearance is given by

$$1800 \sqrt{\frac{\Delta}{T_0}} \, e^{-\Delta/T_0} = 145.$$

Hence we find that

$$T_0 \simeq 2.75 \text{ K}.$$

One expects that when ϱ_s vanishes the anomalous properties connected with an inviscid component also disappear: this vanishing should thus signal the phase transition of helium from its phase II to its phase I. Actually, T_0 is not very different from the experimentally observed transition temperature $T_\lambda = 2.18$ K. The agreement is remarkable, if we bear in mind that our model, which is based upon a simple description of low-energy excitations, becomes less and less adequate as the temperature rises. Morover, the model loses all its validity when ϱ_n, which is the effective density dragged along by the walls, exceeds the total density ϱ of the fluid. That part of the $\varrho_n(T)$-curve, where ϱ_n approaches ϱ, is thus rather unreliable.

16a. If we supply heat, the temperature in vessel A increases. We have seen in the previous question that $\varrho_s(T)$ is a decreasing function of T, so that the amount of superfluid will decrease in vessel A, before the system starts to approach equilibrium again. The narrow tube does not let normal fluid pass through, but lets superfluid pass through freely. There will thus be a tendency to make ϱ_s the same everywhere in the system through passing superfluid from B to A; as a consequence, the mass, and thus the pressure, will be higher in A.

b. The Gibbs-Duhem relation is

$$A = -\mathcal{P}\Omega = U - TS - \mu N,$$

or

$$\mu = \frac{F + \mathcal{P}\Omega}{N}.$$

We assume that $\mu_A = \mu_B$. Writing $\Delta x = x_A - x_B$, we find that

$$\mu_A - \mu_B = 0 = \Delta \frac{F}{N} + \mathcal{P}\Delta\frac{\Omega}{N} + \frac{\Omega}{N}\Delta\mathcal{P}, \tag{28}$$

where F/N and Ω/N are the free energy and the volume per atom.

On the other hand, we have the thermodynamic relation for an adiabatic transfer of matter from B to A:

$$\Delta\frac{F}{N} = -\frac{S}{N}\Delta T - \mathcal{P}\,\Delta\frac{\Omega}{N}. \tag{29}$$

Combining (28) and (29) we find

$$\frac{\Omega}{N}\,\Delta\mathcal{P} = \frac{S}{N}\,\Delta T,$$

which is the required formula:

$$\frac{\Delta\mathcal{P}}{\Delta T} = \frac{S}{\Omega}. \tag{30}$$

c. At $T \simeq 1.5$ K the entropy is dominated by the roton contribution (see question 8). We have therefore

$$\frac{S}{\Omega} = -\frac{1}{\Omega}\frac{\partial F_R}{\partial T} = r\frac{\partial}{\partial T}\left(T^{3/2}\,e^{-\Delta/T}\right) = r\,T^{1/2}\,e^{-\Delta/T}\left(\frac{3}{2} + \frac{\Delta}{T}\right),$$

or, at 1.5 K,

$$\frac{S}{\Omega} = 0.19 \times 10^5 \text{ Pa K}^{-1}.$$

Applying Eq.(30), we have $\Delta\mathcal{P} \simeq 2 \times 10^2$ Pa for $\Delta T \simeq 10^{-2}$ K. Using the hydrostatic formula for a fluid column, $\Delta\mathcal{P} = g\varrho\Delta z$, we finally get $\Delta z \simeq 14$ cm for $\Delta T \simeq 10^{-2}$ K.

d. In quasi-equilibrium the entropy is a maximum, taking into account the possibility of exchanges between the two vessels. These exchanges consist here merely in transfer of superfluid which does not transport any entropy. If dN atoms pass from A to B, or $-dN$ from B to A, the total entropy is thus stationary, and moreover S_A and S_B are separately unchanged, which leads to

$$dU_A = TdS_A + \mu_A dN - \mathcal{P}_A d\Omega_A = \mu_A dN - \mathcal{P}_A d\Omega_A,$$

$$dU_B = -\mu_B dN - \mathcal{P}_B d\Omega_B.$$

Moreover, the complete system only exchanges work with the outside,

$$dU_A + dU_B = -\mathcal{P}_A \, d\Omega_A - \mathcal{P}_B \, d\Omega_B.$$

Combining these equations, we find $(\mu_A - \mu_B) \, dN = 0$, or $\mu_A = \mu_B$.

16.15 Heat Losses Through Windows

We want to study how the use of composite windows can diminish the loss of heat from buildings through glazing. To do this we shall rely on rough numerical estimates.

1. Consider a window consisting of a simple pane of glass of area S. When the room is heated, a stationary regime is set up: we assume that the walls of the room are maintained at a uniform temperature T_i and that the outside temperature is T_e. The glass is sufficiently thin that the temperature hardly varies from one side to the other; we neglect the effects of conduction and air leaks near the window frame and we assume that the temperature T of the glass is uniform over the whole of its area S. The temperature of the inside air is close to T_i, except near the window where a thin boundary layer is established through which the temperature falls from T_i to T. A similar effect occurs on the outside, provided the air is calm. Nevertheless, when there is wind, the outside boundary layer cannot be established, and the temperature T of the window is equal to the outside temperature T_e.

Energy exchanges between the outside and the inside take place through radiation, through convection, and through conduction. The walls of the room absorb the infra-red. The glass is transparent to visible light, but *opaque* to infra-red, which it absorbs. We neglect the absorption of radiation by the air. We bear in mind that Stefan's constant is $\sigma = 5.67 \times 10^{-8}$ W m^{-2} K^{-4}.

The convection and conduction of heat through either of the boundary layers of air, inside or outside, are characterized by an empirical coefficient $\xi = 2$ W m^{-2} K^{-1}; the heat flux through a boundary layer with end temperatures T_1 and T_2 is $\xi(T_1 - T_2)$. We neglect the heat conduction across the glass and across the walls, which are assumed to be well insulated.

a. Evaluate the power W_1 lost through the window in the windy regime as function of the temperatures T_i and T_e.

b. Calculate the power W_1' lost through the window when the outside air is calm; compare this with W_1.

c. Calculate the annual cost corresponding to the energy losses through this window in the following conditions:

– window surface $S = 3$ m^2;
– price of 1 kWh for domestic heating is 0.2 Ffr;

- the room is heated 150 hours per year and for half of the time there is wind outside;
- the mean inside temperature is 20 °C, and the mean exterior temperature during heating periods is 5 °C.

2. We now consider a *double window*. In practice the two panes are mounted hot in a factory on a single tight frame; this makes it possible to eliminate water vapour between them. We neglect convection and conduction in the air which remains inside; in high-quality double windows the air is replaced by other gases to reduce these effects. In a steady regime the outside pane is at a temperature T_1 and the inside pane at a temperature T_2.

a. Evaluate the power W_2 lost in a windy regime, and compare it with the power W_1 lost through an ordinary window.

b. Same question, but for a calm regime (lost power W_2').

c. With the same hypotheses as in 1c and assuming that a double window costs 300 Ffr more than a simple window, find how many years it will be before the extra cost is recouped.

3. Consider a window consisting, more generally, of n parallel panes. Such windows are not yet available commercially, except with $n = 3$ in cold countries, but one may imagine a cheap construction where we have $n - 2$ transparent plastic sheets, which are light and cost very little, stretched on a frame between two pieces of glass to protect them. Let $T_1, T_2, \ldots, T_n$ be the temperatures of all these sheets.

a. Calculate the power W_3 lost in a windy regime, and compare it with the power W_1 lost through an ordinary window.

b. Repeat this for the case of calm outside air (power lost W_3').

c. Comparing W_3'/W_3 with W_1'/W_1 can you imagine an extra advantage of multiple windows?

4. Another possible type of windows, used in a number of modern buildings, makes use of glass with one of its sides coated by a *thin film* of semiconducting oxide, such as SnO_2 doped with F, typically 300 nm thick, or of metal, such as Ag, 10 nm thick. This coating is transparent to visible light and more or less *reflecting* in the infra-red, because of its good electrical conductivity. Let $R = 0.8$ be its infra-red reflection coefficient and $A = 1 - R$ its absorption coefficient. Remember that glass absorbs infra-red completely. We retain the hypotheses of 1, especially that, when there is wind, convection effects ensure that $T = T_e$. The temperature of the coating is the same as that of the glass.

a. Evaluate the power W_4 lost in a windy regime when the oxide layer is on the outside. Repeat the calculation for the case when the oxide layer is on the inside (power lost W_4'). Is it more advantageous to place the layer on the outside or on the inside?

b. Same questions for the case when the air is calm.

5. Double windows made out of one ordinary glass pane and one coated pane are also commercially available (Fig.16.29). The oxide or metal coating is applied to the face 3 (we number the glass faces from the outside inwards). We retain the same hypotheses as before.

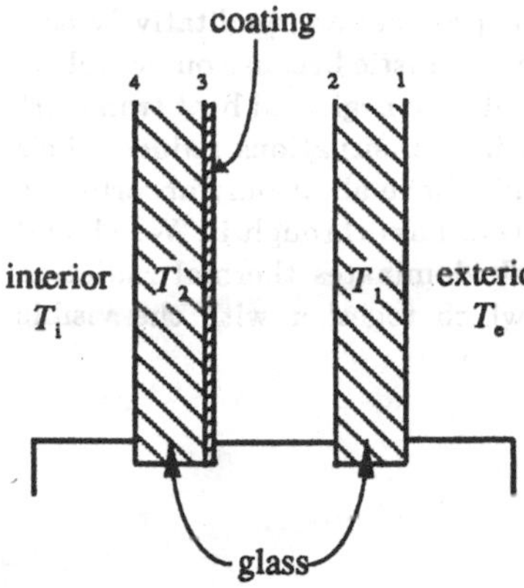

coating

Fig. 16.29. Double window with coated glass

4   3   2   1

interior T_2 T_1 exterior
T_i T_e

glass

a. Evaluate the power losses when there is wind and when there is no wind. Show that it is thermally more advantageous to coat either face 2 or face 3 than faces 1 or 4 (Fig.16.29) – this also protects the film against accidental scratching.

b. Would this conclusion have to be changed, if we take into account that glass is not perfectly absorptive for infra-red radiation, as we assumed so far, but has a reflectivity $R' = 0.1$?

c. One could imagine double windows with coatings on both faces 2 and 3, but for economic reasons these are not produced. Use a comparison of the efficiencies of double windows without coating, with one coating, and with two coatings to explain why this is the case.

6. In the preceding questions we have assumed that the radiation received from the outside was that of a black body at a temperature T_e; this hypothesis is only valid for the night. During the day one should add the solar radiation received by the window either directly, when the sun shines on the window, or indirectly after the radiation is scattered by the atmosphere, when the skies are dull or for a North facing window. The energy of this radiation is for 60 % in the visible spectrum and for 40 % in the infra-red. Its power Φ per unit area can be estimated to be $\Phi = 70$ W m^{-2} for a dull sky and to be $\Phi = 700$ W m^{-2} when the Sun is shining brightly. Study for these two alternatives the power transfer through the window for the various kinds of windows considered above: simple glass, double glass, glass with an oxide layer. Assume that glass is perfectly transparent for visible light while the oxide layer has a reflectivity $r = 0.10$ and an absorptivity $a = 0.15$. Estimate the temperatures attained within the room when there is no heating or cooling, for various values of the external temperature and of the flux Φ. Reach some conclusions about the energy economies for heating in winter and air-conditioning in summer. Explain why it is more advantageous for heating purposes to coat face 3 rather than face 2 in windows of the type of Fig.16.29.

Note. Even though the conclusions drawn from this problem are qualitatively correct, the numerical results obtained are often not very realistic because our simplifying hypotheses have been too all-embracing. Especially, we neglected heat transport by the gas enclosed inside double windows, which in real situations reduces their efficiency. Moreover, glass is partially reflecting; and, above all, it only absorbs the far infra-red and thus lets most of the solar radiation pass through it. We should at least have distinguished the far infra-red, which dominates thermal radiation at room temperatures, from the near infra-red, which together with the visible dominates solar radiation.

16.16 Incandescent Lamps

We want to study how incandescent lamps with a tungsten filament operate. A schematic picture of a light bulb is shown in Fig.16.30. The filament has the form of a helix; we shall assume that as far as its emission is concerned we can treat it as a rectilinear cylinder with diameter d and length l, behaving like a black body. The electrical current is brought to the filament by conductors of negligible electrical resistance. We also neglect their thermal conductivity and that of the gas which fills the bulb.

As this is a practical problem, we attach importance to numerical calculations; the results must be correct as to order of magnitude, but a relative accuracy within 30 % will most often be sufficient. The fundamental constants can be found at the end of the book. Tungsten has the following properties:

- Resistivity: $\varrho = 10^{-6}\ \Omega$ m, independent of temperature in the conditions under which the lamp is operating – bear in mind that the resistance of a wire of length l and cross-section s is $\varrho l/s$.
- Atomic mass: 184 g mol^{-1}.
- Mass density: 18 g cm^{-3}.
- Its solid-vapour equilibrium is characterized by the sublimation curve, $\mathcal{P}_s(T)$, empirically given by the equation

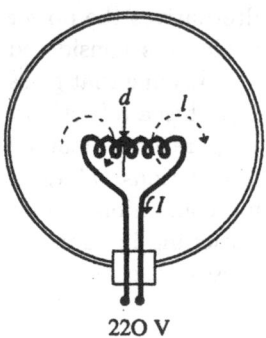

220 V

Fig. 16.30. Sketch of a light bulb. The filament has a diameter d and a length l, if we stretch it out. The glass bulb is modelled by a sphere of diameter δ

$$\log_{10} \mathcal{P}_s \ = \ -\frac{A}{T} + B, \tag{1}$$

where $\mathcal{P}_s$ is in pascal, T in kelvin, $A = 4.0 \times 10^4$ K, and $B = 11.3$.

General Properties

1. One wants to produce a lamp with an electric power consumption of $P = 100$ W for an effective alternating voltage equal to $\mathcal{V} = 220$ V.

a. Show that if one fixes the temperature T_f of the filament, its geometric dimensions (diameter d and length l) are determined. Give numerical values for $T_f = 2700$ K. How should d and l be modified to change the power P of the lamp for a given $\mathcal{V}$?

b. What are the advantages of a higher filament temperature? Estimate the luminous flux of a 100 W lamp operating at 2700 K.

Vacuum Lamp

2. We cannot increase the filament temperature arbitrarily because the tungsten will sublimate, which has two bad consequences: on the one hand, the tungsten vapour will condense on the walls of the bulb, producing an *opaque deposit*; on the other hand, when the mass loss of the filament becomes important, it becomes fragile and may *melt* at points where it has become thin and is heated more strongly. The melting temperature is 3400 °C. We assume that the filament stops operating when it has lost 3 % of its mass through sublimation.

We first of all consider a lamp which has been evacuated.

a. Evaluate the sublimation flux Φ, that is the number of tungsten atoms leaving the filament per unit area per unit time; they will stick to the walls of the lamp, which therefore remains practically a vacuum. To do this, assume that this flux depends solely on the temperature and remains unchanged when the filament becomes surrounded by tungsten vapour rather than by a vacuum. Evaluate, as in kinetic theory, the flux of atoms from this vapour hitting the metal surface; assume that these atoms have unit probability to be incorporated into the metal and write down the balance equation for the case where the metal is in equilibrium with its saturated vapour, eventually to find Φ as function of the temperature.

b. Calculate the period τ during which the lamp can operate for $T_f = 2700$ K.

c. We require that the lamp will have a lifetime of 1000 hrs for the same power consumption and the same voltage. At what temperature T_f' must we keep the filament? What should be its diameter d' and its length l' so that this temperature can be reached?

d. Evaluate the loss of lighting efficiency for a vacuum lamp operating at this latter temperature T_f' as compared to a lamp operating at $T_f = 2700$ K;

use as criterion for comparison the radiating luminous power around 0.6 μm, which is the centre of the visible spectrum.

Iodine Lamp

3. A recent technique to improve the efficiency and lifetime consists in filling the lamp with a gas, such as iodine, I. This forms with tungsten, W, a compound, C, which we shall assume to have the chemical formula WI and which has the following properties:

- The compound C is stable at the temperature of the gas which fills the lamp; this temperature is practically uniform and equal to the temperature T_B of the bulb. All tungsten atoms which sublimate thus combine with iodine atoms.
- The compound C is volatile so that one avoids its deposition on the inside of the bulb, provided the temperature T_B is sufficiently high. For practical and safety reasons one does not exceed $T_B = 600$ K. Under those conditions, if a solid deposit of the compound C were formed on the bulb, it would be in equilibrium with the C vapour in the lamp.
- Above 2000 K the compound is unstable; therefore, it decomposes when in contact with the filament, at $T_f = 2700$ K, so that the tungsten is reincorporated into the filament when a C molecule hits it.

This procedure therefore allows us to offset the tungsten losses from the filament by sublimation through a *permanent regeneration*, while at the same time *suppressing the deposit* on the bulb.

a. The high temperature T_B of the bulb is maintained because it absorbs part of the radiation from the filament, which is dissipated afterwards in the form of thermal radiation. Neglect thermal conductivity and convection effects in the gas. Show that the lamp must be rather small if one wishes the temperature T_B to be high. Calculate the diameter δ of the bulb, assumed to be a sphere, for $T_B = 600$ K. Assume that the material of the bulb – quartz – is transparent for wavelengths under 4 μm and opaque for infra-red wavelengths above 4 μm. Use the expansion

$$\frac{x}{e^x - 1} = 1 - \frac{x}{2} + \frac{x^2}{12} - \frac{x^4}{720} + \cdots,$$

and the result

$$\int_0^\infty \frac{x^3\, dx}{e^x - 1} = \frac{\pi^4}{15}.$$

b. The equilibrium vapour pressure of the compound C which may condense on the bulb is given by the empirical formula

$$\log_{10} \mathcal{P}'_s = -\frac{A'}{T} + B',$$

similar to Eq.(1) for tungsten, but with coefficients $A' \neq A$ and $B' \simeq B$. What approximate condition must be satisfied by A' so that the process is efficient? What is the property of crystalline C with which this condition is connected? The atomic mass of iodine is 127 g mol^{-1}.

c. Justify the hypotheses we have made about the stability of the compound C. Assume that the vapours of W and I are monatomic, that a molecule of $C = $ WI has a single bound state with a binding energy $u = 2$ eV, and that the bulb contains 10^{-4} g of iodine.

Argon Lamp

4. Usually lamps are filled with a neutral gas, like argon or krypton, which has a pressure of about 0.5 atm at room temperature. In fact, for safety reasons we do not want that pressure to exceed 1 atm when the lamp is lit.

a. Knowing that the radii of the argon and tungsten atoms are of the order of 2 Å and 1.5 Å, estimate the mean free path of a tungsten atom in an argon gas under the operating conditions.

b. What happens to the tungsten atoms which are emitted by the filament? Explain qualitatively why there is an improvement as compared to a vacuum lamp. Try to estimate the increase in lifetime.

Solution

1a. The electrical power consumed,

$$P = \frac{\mathcal{V}^2}{\mathcal{R}} = \frac{\mathcal{V}^2}{\varrho l[\pi(d/2)^2]^{-1}} = \frac{\pi d^2 \mathcal{V}^2}{4\varrho l},$$

which is transformed into heat through the Joule effect in the filament, is completely emitted as radiation. The Stefan-Boltzmann formula then gives

$$P = l\pi d\sigma T_f^4.$$

These two relations determine d and l. Eliminating l, we get, in SI units,

$$d^3 = \left(\frac{2P}{\pi\mathcal{V}}\right)^2 \frac{\varrho}{\sigma T_f^4} = \left(\frac{2 \times 100}{\pi \times 220}\right)^2 \frac{10^{-6}}{5.67 \times 10^{-8} \times (2700)^4},$$

or, $d = 0.03$ mm, and hence

$$l = \frac{P}{\pi d\sigma T_f^4} = 0.35 \text{ m} = 35 \text{ cm}.$$

The dimensions of the filament should increase with the power P as $d \propto P^{2/3}$ and $l \propto P^{1/3}$.

b. Given the power consumed, the yield, that is, the power emitted in the visible, is

$$\frac{\int_{\nu_1}^{\nu_2} u(\nu)\, d\nu}{\int_0^\infty u(\nu)\, d\nu} = \frac{\int_{x_1}^{x_2} (x^3\, dx)/(e^x - 1)}{\int_0^\infty (x^3\, dx)/(e^x - 1)} \simeq \frac{e^{-x_1}\left(x_1^3 + 3x_1^2\right)}{\frac{1}{15}\pi^4},$$

where ν_1 and ν_2 are the limits of the visible spectrum, and where

$$x_1 = \frac{h\nu_1}{kT_f} = \frac{hc}{kT_f \lambda_1}$$

corresponds to the infra-red limit ($\lambda_1 = 0.8$ μm). The approximation in the numerator is reasonable, as x_1 is sufficiently large ($x_1 = 6.7$ for 2700 K, $x_2 = 2x_1$). The yield increases rapidly with temperature; it is 8.5 % at 2700 K, and grows by a factor

$$\left(1 - \frac{3\Delta T}{T_f}\right) \exp\left(\frac{h\nu_1 \Delta T}{kT_f^2}\right) = \left(1 - \frac{3\Delta T}{T_f}\right) e^{6.7\Delta T/T_f} = 1 + \frac{3.7\Delta T}{T_f},$$

when the filament temperature is increased by ΔT, becoming, for instance, 10 % at 2800 K.

Moreover, as the visible range is in the tail of the spectrum emitted by the filament, the power emitted in the blue is much smaller than that emitted in the red. The ratio is approximately

$$\left(\frac{\nu_2}{\nu_1}\right)^3 e^{x_1 - x_2} = 8\, e^{-x_1} = 1\%$$

at 2700 K. This ratio improves significantly as the temperature is raised since it increases by a factor

$$\exp\left(\frac{h\nu_1}{k}\frac{\Delta T}{T_f^2}\right) = 1 + \frac{6.7\,\Delta T}{T_f}$$

when T_f increases by ΔT. One must therefore increase the filament temperature to get a whiter illumination.

These two effects, a larger fraction being emitted in the visible and a less red colour, can be described more accurately if we use photometric units (§ 13.3.4) which weight the power emitted by the sensitivity of the eye. The latter is characterized by a function $\varphi(\nu)$ which vanishes at both ends of the visible spectrum and which has a maximum, put equal to 1, corresponding to a wavelength of 5550 Å. The unit, the lumen, corresponds to a weighted power equal to $\frac{1}{683}$ W, which gives for the theoretical luminous flux of a P watt bulb

$$683\, P\, \frac{15k^4 T^4}{\pi^4 h^4} \int \frac{\varphi(\nu)\, \nu^3\, d\nu}{e^{h\nu/kT} - 1} = 105\, P \int \varphi\!\left(\frac{kTx}{h}\right) \frac{x^3\, dx}{e^x - 1}\, \text{lm},$$

where the losses due to the geometry of the filament and of the lamp have been neglected. Assuming a parabolic shape for φ one finds numerically a flux of 3300 lm for $T = 2300$ K and $P = 100$ W, which corresponds to a photometric efficiency of 5 %.

Apart from the difficulties mentioned earlier, increasing the temperature of the filament for a given consumed power P makes it necessary, according to question 1a, to use a thinner filament; this makes it more brittle and increases the risk of melting.

2a. The number of particles from the vapour hitting unit surface area in unit time is

$$n \int_{(p_x > 0)} \frac{p_x}{m} g(p) \, dp,$$

where n is the number of gas particles per unit volume. Using the Maxwell distribution and the equation of state of the gas we find for this flux

$$\frac{P}{kT} \frac{1}{m} \frac{1}{(2\pi mkT)^{1/2}} \int_{(p_x > 0)} p_x \, e^{-p_x^2/2mkT} \, dp_x = \frac{P}{(2\pi mkT)^{1/2}}.$$

When the vapour is saturated, that is, when the pressure equals $\mathcal{P}_s(T_f)$, it is in equilibrium with the solid at the temperature T_f so that the flux Φ lost by the latter is exactly compensated by the flux we have just evaluated of vapour atoms hitting the solid and being absorbed by it. We have therefore

$$\Phi = \frac{\mathcal{P}_s(T_f)}{(2\pi mkT_f)^{1/2}} = \frac{1}{(2\pi mkT_f)^{1/2}} 10^{-(A/T_f)+B}.$$

b. The total number of atoms lost by the filament during a period τ, which equals $\Phi \cdot l\pi d \cdot \tau$, should not exceed 3 % of its total number of atoms, $l\pi(d/2)^2 n_f$, where n_f is the number of atoms per unit volume of filament,

$$n_f = N_0 \frac{18\,000}{0.184} = 5.9 \times 10^{28} \text{ m}^{-3}.$$

The lifetime of the lamp is thus

$$\tau = \frac{3}{100} \frac{d}{4} n_f \frac{1}{\Phi}.$$

At 2700 K we have

$$\tau = 11\,600 \text{ s} \simeq 3 \text{ hours}.$$

c. Using the expressions for τ, Φ, and d (see question 1) for two different values of the temperature, T_f and T_f', we have

$$\frac{\tau'}{\tau} = \frac{d'}{d} \frac{\Phi}{\Phi'} = \left(\frac{T_f}{T_f'}\right)^{4/3} \left(\frac{T_f'}{T_f}\right)^{1/2} 10^{-(A/T_f)+(A/T_f')},$$

or

$$\frac{A}{T_f'} = \frac{A}{T_f} + \log_{10} \frac{\tau'}{\tau} - \frac{5}{6} \log_{10} \frac{T_f}{T_f'},$$

$$\frac{4 \times 10^4}{T_f'} = \frac{4 \times 10^4}{T_f} + \log_{10} \frac{3600 \times 10^3}{11600} - \frac{5}{6} \log_{10} \frac{T_f}{T_f'}.$$

The last term is expected to be small as compared to the preceding one, which equals 2.49. Neglecting it we find

$$T_f' = 2300 \text{ K}.$$

If we substitute this value into the neglected term, which then equals 0.06, we see that the result for T_f' is not changed. In order to operate at that temperature, a 100 W lamp should have a filament with a diameter $d' = 0.04$ mm and a length $l' = 50$ cm.

d. From Planck's law it follows that the loss in the efficiency of the lamp, measured by the power emitted at ν_0 and taking into account the change in the surface of the filament which is necessary to maintain the consumed power at a constant value, is equal to

$$\frac{l'd'}{ld} \frac{e^{x_0} - 1}{e^{x_0'} - 1}, \qquad \text{where}$$

$$x_0 \equiv \frac{h\nu_0}{kT_f} = \frac{hc}{kT_f\lambda_0} = 8.89,$$

$$x_0' \equiv \frac{h\nu_0}{kT_f'} = x_0 \frac{T_f}{T_f'}$$

are large. The loss in efficiency, if we use the results of question 1a, therefore reduces to

$$\left(\frac{T_f}{T_f'}\right)^4 e^{x_0 - x_0'} = 0.4.$$

We lose thus a factor of 2 to 3 in the illumination. More precisely, the luminous flux found for a 100 W lamp with a filament at 2300 K is 1400 lm, as compared to 3300 lm at 2700 K.

3a. The fraction of the power emitted by the filament which is absorbed by the bulb is, when we use the Stefan-Boltzmann and Planck laws,

$$P_1 = l\pi d\,\sigma T_f^4 \frac{\int_0^y (x^3\,dx)/(e^x - 1)}{\int_0^\infty (x^3\,dx)/(e^x - 1)},$$

where

$$y \equiv \frac{h\nu_c}{kT_f} = \frac{hc}{kT_f\lambda_c} = x_0 \frac{\lambda_0}{\lambda_c} = 1.33$$

is the dimensionless variable associated with the cut-off wavelength ($\lambda_c = 4$ μm) and with the filament temperature.

If we use Kirchhoff's law, we find that the bulb emits to the outside an infra-red power equal to

$$P_2 = \pi\delta^2\,\sigma T_B^4 \frac{\int_0^z (x^3\,dx)/(e^x - 1)}{\int_0^\infty (x^3\,dx)/(e^x - 1)},$$

where $\pi\delta^2$ is the surface of the sphere and where

$$z \equiv \frac{h\nu_c}{kT_B} = y\frac{T_f}{T_B} = 6$$

(the bulb emits as a black body for $\nu < \nu_c$ and does not emit anything for $\nu > \nu_c$). For a stationary regime we must write down the balance between the total power lost by the bulb and the total power absorbed by it; in the model considered by us these exchanges are exchanges of infra-red radiation. The total emitted power is $2P_2$, half to the outside and half to the inside. Nevertheless, the total energy P_2 emitted to the inside by one point of the bulb is again absorbed by another point of the bulb – we can neglect the surfaces of the filament and of the base as compared to that of the bulb. The total power absorbed by the bulb is thus $P_1 + P_2$ and the balance equation is $P_1 = P_2$, or

$$\delta^2 = ld \left(\frac{T_f}{T_B}\right)^4 \frac{\int_0^y (x^3 \, dx)/(e^x - 1)}{\int_0^z (x^3 \, dx)/(e^x - 1)}.$$

Even at the upper limit $y = 1.33$ of the integral in the numerator the expansion

$$\frac{x}{e^x - 1} = 1 - \frac{x}{2} + \frac{x^2}{12} - \frac{x^4}{720} + \cdots$$

converges well with successive terms decreasing rapidly. We have thus

$$\int_0^y \frac{x^3 \, dx}{e^x - 1} \simeq \int_0^y \left(x^2 - \frac{x^3}{2} + \frac{x^4}{12}\right) dx = \frac{y^3}{3} - \frac{y^4}{8} + \frac{y^5}{60} = 0.46.$$

On the other hand, the upper limit $z = 6$ of the integral in the denominator is large and we can write

$$\int_0^z \frac{x^3 \, dx}{e^x - 1} = \int_0^\infty \frac{x^3 \, dx}{e^x - 1} - \int_z^\infty \frac{x^3 \, dx}{e^x - 1}$$

$$\simeq \frac{\pi^4}{15} - \int_z^\infty x^3 e^{-x} \, dx$$

$$= \frac{\pi^4}{15} - e^{-z}\left(z^3 + 3z^2 + 6z + 1\right) = 5.6.$$

Altogether we thus have

$$\delta^2 = 0.35 \times 3 \times 10^{-5} \times \left(\frac{2700}{600}\right)^4 \times \frac{0.46}{5.6} \, \text{m}^2 = 3.5 \times 10^{-4} \, \text{m}^2,$$

or

$$\delta = 0.019 \, \text{m} = 2 \, \text{cm}.$$

b. In order that the process be efficient it is first of all necessary that the partial pressure $\mathcal{P}'$ of the compound $\mathcal{C}$ in the lamp is less than the pressure $\mathcal{P}'_s(T_B)$ of the saturated vapour at the bulb temperature. In fact, this condition not only makes it possible to avoid any solid deposit on the bulb during its operation, but also ensures that any possible deposit, for instance when the lamp cools off after it is switched off, will evaporate.

As far as the filament is concerned, the flux of tungsten atoms which leave it is the quantity Φ calculated in 2a. On the other hand there is a flux Φ' of molecules of the gaseous WI compound which will hit the filament and deposit on it a number Φ' of tungsten atoms per unit time and unit surface area. This flux, calculated at the start of 2a, equals

$$\Phi' = \frac{\mathcal{P}'}{(2\pi m' k T_B)^{1/2}},$$

where $\mathcal{P}'$ and T_B are the pressure and temperature of the compound $\mathcal{C}$ in the lamp while m' is the mass of one of its molecules. In a permanent regime no $\mathcal{C}$-molecule should be deposited on the bulb and the continuous regeneration of the filament means that $\Phi = \Phi'$. The partial pressure of the compound $\mathcal{C}$ in the lamp is thus given by the expression

$$\mathcal{P}' = \left(\frac{m' T_B}{m T_f}\right)^{1/2} \mathcal{P}_s(T_f).$$

This pressure must be less than $\mathcal{P}'_s(T_B)$, whence

$$\left(\frac{m' T_B}{m T_f}\right)^{1/2} 10^{-(A/T_f)+B} < 10^{-(A'/T_B)+B'},$$

or

$$A' < A\frac{T_B}{T_f} + (B' - B)T_B + \frac{1}{2} T_B \log_{10}\frac{m T_f}{m' T_B}.$$

With the given numerical data the last two terms are negligible and we find the condition

$$A' < A\frac{T_B}{T_f}, \qquad \text{or}$$

$$A' < 9 \times 10^3 \text{ K}.$$

The empirical form (1) of the sublimation function $\mathcal{P}_s(T)$ does not differ much from the form found in Probs.3 and 8 using simple theoretical models. In those two cases we saw the appearance of a factor $e^{-\varepsilon/kT}$ in $\mathcal{P}_s(T)$, where ε represented the energy needed to tear one of its atoms from the solid. The remaining factor varies more slowly with temperature than this exponential, and its replacement by a constant leads to Eq.(1), where $kA\ln 10$ can be identified with ε. The condition $A' < 9 \times 10^3$ K thus means that the cohesion energy per molecule $kA'\ln 10$ of a $\mathcal{C}$ crystal must be less than 1.7 eV.

c. It follows from §8.2.2 that the respective chemical potentials μ, μ', and μ'' in the W, $\mathcal{C}$, and I vapours must satisfy the relation $\mu' = \mu + \mu''$, in chemical equilibrium at a temperature T. Hence, it follows that

$$\frac{N(N_I - N')}{N'} = \Omega \left(\frac{m m'' kT}{2\pi \hbar^2 m'}\right)^{3/2} e^{-u/kT},$$

where N_I denotes the total number of iodine atoms, N that of free tungsten atoms, and N' the number of WI molecules. Provided $N' \ll N_I$, we find that $N/N' = 6 \times 10^{-8}$ at $T = T_B$ and $N'/N = 5 \times 10^{-8}$ at $T = T_f$. The complete determination of the densities of each of the chemical species in the bulb as functions of their distance from the filament would make it necessary to solve the transport equations, as in question 4b, including moreover a relation from chemical kinetics between the fluxes and the affinities.

4a. Consider a tungsten atom travelling a distance L; the volume swept through by a sphere centred on this atom and with a radius $a = 3.5$ Å, which is the sum of the radii of a tungsten and an argon atom, is

$$\pi a^2 L.$$

The mean free path l, that is, the mean distance traversed between two collisions, is obtained by dividing L by the number of centres of argon atoms in the above volume. If n_A is the number of argon atoms per unit volume we have, with $\mathcal{P}_A = 0.5 \times 10^5$ Pa for $T_A = 300$ K, or $\mathcal{P}_A = 10^5$ Pa for $T_A = 600$ K:

$$l = \frac{L}{\pi a^2 L n_A} = \frac{k T_A}{\pi a^2 \mathcal{P}_A} = 2 \times 10^{-7} \text{ m} = 0.2 \text{ μm}.$$

In the immediate vicinity of the filament the temperature is higher, but l does not exceed 1 μm, which corresponds to the temperature $T_f = \frac{2700}{600} T_A$ and the pressure 10^5 Pa.

b. The tungsten atoms which leave the filament collide with the argon atoms and thus change their velocities. There is thus some probability that they will hit the filament again and be incorporated in it. Qualitatively, that probability will be important if the first collisions occur before the particle is far from the filament at a distance exceeding the filament diameter. In fact, beyond that the trajectory of a W-atom has less chance to return to the filament than to end up at the bulb where it will condense. One condition for the operating period of the lamp to be improved is thus $l < d$ which is well satisfied by the preceding calculations of l.

The calculation of the increase in the operating period can be performed by analyzing the transport of tungsten atoms. We can obtain an order of magnitude estimate by studying a cylindrical geometry: bulb in the form of a cylinder of diameter δ, with on its axis a filament in the form of a rectilinear cylinder of diameter d. The tungsten atom flux Φ_g in the gas is conserved (div $\Phi_g = 0$). Following Chap.15 we introduce a diffusion coefficient D which relates this flux to the density gradient of the tungsten atoms,

$$\Phi_g = -D \nabla n,$$

and which is given by the expression

$$D = \lambda v_g l,$$

where λ is a numerical constant and v_g the mean square velocity of the tungsten atoms in the gas. The conservation of flux implies $\nabla^2 n = 0$, whence we have for $\frac{1}{2}d < r < \frac{1}{2}\delta$

$$n(r) = -\nu \ln r + \text{constant},$$

and $\Phi_g(r) = D\nu/r$. The constant ν will be determined from the flux emitted by the filament. Moreover, the calculations at the start of 2a show that the flux of atoms which hit the bulb is

$$\Phi_g(\tfrac{1}{2}\delta) = \frac{n(\tfrac{1}{2}\delta)\,v_g}{\sqrt{6\pi}} = \frac{2\lambda v_g l\nu}{\delta}.$$

This flux is deposited completely. Near the surface of the filament, the flux $\Phi_g(\tfrac{1}{2}d)$ is the difference between the sublimation flux Φ for a vacuum lamp calculated in question 2a and the return flux, or

$$\Phi_g(\tfrac{1}{2}d) = \Phi - \frac{n(\tfrac{1}{2}d)\,v_g}{\sqrt{6\pi}} = \frac{2\lambda v_g l\nu}{d}$$

(we neglect the variation of the temperature in the argon). Eliminating ν gives the net flux $\Phi_g(\tfrac{1}{2}d)$ lost by the filament and deposited on the bulb,

$$\left(\frac{d}{\delta}+1\right)\Phi_g(\tfrac{1}{2}d) = \Phi - \frac{1}{\sqrt{6\pi}}\ln\frac{\delta}{d}\frac{d}{2\lambda l}\Phi_g(\tfrac{1}{2}d),$$

whence we find that the lifetime of the lamp is multiplied by

$$\frac{d}{l}+1+\frac{1}{2\lambda\sqrt{6\pi}}\frac{d}{l}\ln\frac{\delta}{d},$$

which is of the order of d/l, as $l \ll d \ll \delta$.

In practice, the filament has the shape of a helix of diameter d'. The result depends on the pitch of the helix and we expect that we should replace d by a length which lies between d and d'. The lifetime, multiplied by a factor which is at least equal to 150, reaches thus a reasonable value of at least 500 hours.

16.17 Neutron Physics in Nuclear Reactors

The essential element of a nuclear reactor is the "fuel" which produces a fission chain reaction through a fissile nucleus producing two fragments with the emission of secondary neutrons, for instance,

$$^{235}_{92}\mathrm{U} + \mathrm{n} \rightarrow {}^{94}_{38}\mathrm{Sr} + {}^{140}_{54}\mathrm{Xe} + 2\mathrm{n}.$$

The average number of secondary neutrons which is emitted per primary neutron is ν. They can in turn be absorbed, which makes it possible that the reactions can continue, provided ν is sufficiently large. Each fission produces an energy W_f of the order of 200 MeV so that the secondary neutrons have a high kinetic energy. This energy is exchanged with various materials which constitute the core of the reactor, and is recovered in the form of heat thanks to circulating water which is either boiling or under pressure – molten sodium in the breeder Superphénix at Creys-Malville (France).

In a breeder the fuel, plutonium, produces a chain reaction even in the case of fast neutrons. Nevertheless, most reactors use as fuel uranium which is enriched with the fissile isotope ^{235}U. The fission reaction is then efficiently produced only if the primary neutron is "thermal", that is, slow with a kinetic energy comparable with the value $3kT/2$ which is characteristic of thermal equilibrium at the temperature T of the reactor. It is therefore important to slow the secondary neutrons down before they react again and this is the rôle of the "moderator", a material which surrounds the uranium bars and in which the secondary neutrons undergo collisions which thermalize them (§ 15.2.5). The moderator is often water which in that case acts both as moderator and as coolant; one also uses graphite or heavy water.

Apart from the fuel, the moderator, and the cooling fluid, the reactor core contains structural materials – cladding containing the fuel, the constituents of the heat exchangers – and movable control bars, made of boron or cadmium steel, meant to absorb neutrons so as to keep the flux at the wanted level. Finally, the core is surrounded by several barriers which play different rôles: to contain and to circulate the cooling fluid, to be a protection against radiation and against overpressures in the case of overheating, and to reflect the neutrons towards the core.

For the operation of the reactor neutron physics plays a prime rôle, especially the study of the neutron densities and fluxes. In fact, one needs to trace in detail what happens to the secondary neutrons, which can in the various elements of the reactor diffuse, start new fissions, get lost through capture or be absorbed by ^{238}U and produce Pu, leave the core, or be reflected. When the reactor operates normally, a "critical" regime is reached where the neutron density is stationary.

We shall first of all, in questions 1 to 3, establish an approximate equation governing the neutron distribution, where we assume them to be monoenergetic. We shall neglect their slowing down process, assuming it to be so efficient that the secondary neutrons reach almost immediately a "thermal" velocity. By considering very schematic models we shall then be able to treat simple questions and to understand some of the problems posed in the design of reactors. In the whole problem we assume that the reactor elements, except the fuel, are a single composite material which is uniformly distributed in space. The fuel which is the source for neutrons will be considered either (question 7) also to be distributed uniformly in the reactor, or (questions 4 to 6) to be located in a plane.

We denote by $f(\boldsymbol{r}, \boldsymbol{p}, t)$ the reduced single-particle density which describes the neutron distribution. As in Chap.15 we write down the neutron density:

$$\varrho(\boldsymbol{r}, t) = \int f(\boldsymbol{r}, \boldsymbol{p}, t) \, d^3\boldsymbol{p},$$

as well as the current density

$$\boldsymbol{J}(\boldsymbol{r}, t) = \int \frac{\boldsymbol{p}}{m} f(\boldsymbol{r}, \boldsymbol{p}, t) \, d^3\boldsymbol{p}.$$

We look for the balance equation for neutrons with a given momentum in a volume element $V\,d^3p$, where V is a finite volume in coordinate space, bounded by a surface S. In writing down the balance we take into account the flux of particles across S and the collisions which are of two kinds:

- reactions which make neutrons disappear from the volume element $V\,d^3p$ when they are captured. Each of those collisions is chacterized by the absorption cross-section σ_a. The absorbing centres involved in this process are uniformly distributed and have a density ϱ_a.
- collisions, changing the velocity. They are characterized by a differential cross-section $\sigma_{sc}(p \to p')$, where $\sigma_{sc}\,d^3p'$ describes the inelastic processes which lead to the volume element d^3p' in momentum space, and by the density ϱ_{sc} of the scatterers, assumed to be distributed uniformly. The total scattering cross-section is $\sigma_{sc}(p) = \int d^3p'\,\sigma_{sc}(p \to p')$.

Finally, we take into account the existence of neutron sources which are distributed in phase space with an intensity $Q(r, p, t)$ and we write

$$Q(r, t) = \int Q(r, p, t)\,d^3p,$$

for the number of particles emitted per unit volume and unit time due to fissions.

1a. Establish the evolution equation for the neutron distribution

$$\frac{\partial f}{\partial t} + (v \cdot \nabla_r)f + v\sigma_a\varrho_a f = Q(r, p, t) + K,$$

where $v = p/m$ and

$$
\begin{aligned}
K \equiv \varrho_{sc} \int d^3p' &\left[\frac{p'}{m} f(r, p', t)\sigma_{sc}(p' \to p) \right. \\
&\left. - \frac{p}{m} f(r, p, t)\sigma_{sc}(p \to p') \right].
\end{aligned}
\tag{1}
$$

b. Use this equation to find the evolution equation for the neutron density $\varrho(r, t)$.

2. We assume that the neutrons are monoenergetic so that $p = p_0\omega = mv_0\omega$, where ω is a unit vector along the direction of the momentum. We can then write

$$f(r, p, t) = \frac{1}{mp_0}\,g(r, \omega, t)\,\delta(\varepsilon - \varepsilon_0),$$

where $\varepsilon = \varepsilon(p) = p^2/m$ is the particle energy.

a. Show that

$$J(r, t) = \int \omega\,v_0 g(r, \omega, t)\,d^2\omega,$$

$$\varrho(r,t) \;=\; \int g(r,\omega,t)\,d^2\omega,$$

where $d^2\omega = \sin\theta\,d\theta\,d\varphi$ is an element of solid angle in momentum space.

b. We assume that the sources are monoenergetic and isotropic:

$$Q(r,p,t) \;=\; \frac{1}{mp_0}\,\delta(\varepsilon-\varepsilon_0)\,\frac{1}{4\pi}\,Q(r,t).$$

Moreover, we assume that the collisions do not modify the speed but change the direction of the velocity, and we introduce the elastic scattering cross-section $\sigma_{sc}(\omega' \to \omega)$ through the equation

$$\sigma_{sc}(p' \to p) \;=\; \frac{1}{p^2}\,\delta(p-p')\,\sigma_{sc}(\omega' \to \omega).$$

This assumption is justified for elastic collisions of neutrons with sufficiently heavy nuclei so that there is little recoil in a collision.

Write down the evolution equation for $g(r,\omega,t)$.

3. Expand g in powers of ω up to the term linear in ω:

$$g(r,\omega,t) \;=\; h(r,t) + \omega \cdot H(r,t);$$

this is justified provided the anisotropy in the velocity distribution is small.

a. Determine h and H as functions of $\varrho(r,t)$ and $J(r,t)$.

b. Use the evolution equation for g and the expansion of g just introduced to derive the evolution equation for $J(r,t)$:

$$\frac{1}{v_0}\,\frac{\partial J(r,t)}{\partial t} + (\sigma_a\varrho_a + \sigma_{sc}\varrho_{sc})\,J(r,t) \;=\; -\frac{v_0}{3}\,\nabla_r\varrho(r,t),$$

where we have assumed that the collisions which lead to the scattering cross-section $\sigma_{sc}(p' \to p)$ are isotropic.

c. Show that in the stationary case this equation is Fick's law and give an expression for the diffusion coefficient D. What is now the form of the equation for ϱ which we found in 1b, for the stationary case?

We now focus on a stationary regime and use the results of the earlier questions. In question 4 we find boundary conditions convenient for reactor calculations. In question 5 we study an equilibrium regime for the case of a planar source localized at the centre of the reactor. In question 6 we examine a technical device which allows us to limit the loss of neutrons.

In what follows we restrict ourselves to the case of a system which is uniform in the x- and y-directions; the problem thus depends only on a single variable, z. The current density J now has only a single component, $J(z,t)$ in the z-direction.

The boundary conditions along discontinuities of the reactor materials are the following ones:

- the density $\varrho(z)$ is continuous;
- the current $J(z)$ is continuous except when we pass through a surface which contains sources.

4a. Consider an infinite planar source along x and y in the $z = 0$ plane, which is characterized by its surface density $Q(z,t) = Q_0\,\delta(z)$. Show that

$$\lim_{z \to +0} [J(z) - J(-z)] = Q_0.$$

b. Split the total current $J(z)$ into two parts, $J_+(z)$ and $J_-(z)$, which are associated with the particles moving in the $z > 0$ and the $z < 0$ directions, respectively, and where $J(z) = J_+(z) - J_-(z)$. Show that

$$J_\pm(z) = \frac{1}{4}v_0\,\varrho(z) \mp \frac{1}{2}D\,\frac{d\varrho}{dz}.$$

c. Let $z = z_1 > 0$ be the boundary separating the active materials of the reactor, situated at $z < z_1$, from their surroundings, $z > z_1$, assumed to be a vacuum. How do $\varrho(z)$ and $J_\pm(z)$ vary for $z > z_1$? Considering the limits of $\varrho(z)$ and $d\varrho/dz$ as $z \to z_1$ with $z < z_1$ enables us to extrapolate by a straight line the curve $\varrho(z)$ for $z < z_1$ into the vacuum region. We thus define the "extrapolated" boundary of the reactor as the surface $z = z_0$ where $\varrho(z)$ would vanish, if it were linearly extended towards the outside of the actual material boundary, $z = z_1$, with its tangent equal to that at the latter boundary. Determine the distance $\lambda = z_0 - z_1$ as a function of the characteristics of the material in the region $z \lesssim z_1$.

In what follows we replace the boundary conditions along a natural boundary with the vacuum by requiring that ϱ vanishes along the extrapolated boundary, which is further away by a distance λ.

5. We model the reactor by the same planar source $Q_0\,\delta(z)$ as in question 4a, surrounded by a homogeneous material which fills the $-z_1 < z < +z_1$ region.

a. Give expressions for the neutron density $\varrho(z)$ and the neutron currents $J_\pm(z), J(z)$ as functions of the characteristics of this material, when the reactor extends to $z_1 = \pm\infty$.

b. Repeat these calculations for a finite-size reactor, $z_1 = \pm(z_0 - \lambda)$, where $\pm z_0$ is the position of the extrapolated boundary.

6. Study how one can reduce the neutron loss from a reactor. We use a model for a reactor where a first kind of material (material I with diffusion coefficient D_{I}) contains as above a source Q at its centre. This is surrounded by a second kind of material (material II with diffusion coefficient D_{II}) the effect of which we wish to study. To do this we shall determine for various geometries the ratio $J_+(z)/J_-(z)$ at the I–II interface.

a. The material II extends to infinity. Note that the solution $\varrho(z)$ in material II is of the same type as in question 5a.

b. Repeat the calculation for the case where the thickness of material II is e. In this case, a solution of the type found in 5b determines the fluxes in material II.

c. Comment qualitatively on the reflection effect of the cover provided by material II and its efficiency as function of its thickness. Estimate numerically the values of λ and e, the thickness of a cover which has an efficiency equal to 80 % of that of an infinite medium. Assume that in material II we have $\sigma_{sc}\varrho_{sc} = 1$ cm^{-1} and $\sigma_a\varrho_a = 0.2$ cm^{-1}.

7. We now investigate the non-stationary solutions. To do this we introduce new approximations in the evolution equations for $J(z,t)$ and $\varrho(z,t)$ which we found in 3. We treat the various elements of the reactor, both the fissile material and the scattering and absorbing materials, as a single homogeneous block.

a. Show that we can neglect the partial derivative with respect to the time in the evolution equation for $J(r,t)$, provided we make the following assumptions:

- the quantity $(\sigma_{sc}\varrho_{sc})^{-1}$ and also the characteristic distances over which the spatial density variations occur are of the order of centimeters $(\sigma_a\varrho_a \ll \sigma_{sc}\varrho_{sc})$.
- the temporal variations in $J(r,t)$ take place over times exceeding or equal to 10^{-3} s.
- the neutrons are thermal neutrons at a temperature $T = 300$ K.

b. Use the approximate evolution equation obtained in this way to derive the evolution equation for the neutron density $\varrho(r,t)$. In this case we cannot neglect the partial derivative of ϱ with respect to the time, if we take into account the presence of sources and the fact that $\sigma_a\varrho_a$ is small.

c. Consider the non-stationary case for a reactor extending from 0 to z_0. Take here the actual boundary the same as the extrapolated boundary. The source term for the reactor is given explicitly in the form

$$Q(z,t) = \nu v_0 \sigma_f \varrho_f \varrho(z,t),$$

where ν is the number of neutrons produced in a single fission reaction, ϱ_f is the number of fissile fuel nuclei, and σ_f the cross-section for the fission reaction. The factors ϱ_f and $\varrho(z,t)$ stem from the fact that a fission reaction requires a fissile nucleus and a neutron.

Determine the general form of the solution $\varrho(z,t)$, assuming that you know the initial distribution $\varrho(z,0)$. Start by looking for particular solutions of the form $e^{-\gamma t}\sin qz$ and give the possible values of q and γ. How does $\varrho(z,t)$ behave for large t?

d. One says that a reactor is operating *critically*, when it has reached a stationary regime involving energy production. Give the condition for critical operation. Show that there exists a minimum value for the density ϱ_f of the fissile product below which the reactions cannot reach the critical threshold. Show that a reactor has a *minimum size* z_0 for given values of the densities and the cross-sections; discuss the effect of enriching uranium with ^{235}U on the reactor size. What value must we give to $\sigma_f \varrho_f$, by adjusting ϱ_f, for a reactor of size $z_0 = 1$ m with the average values $\sigma_{sc} \varrho_{sc} = 1$ cm^{-1}, $\sigma_a \varrho_a = 0.2$ cm^{-1}, $\nu = 2.5$?

e. Estimate the power of a reactor of 1 m^3 in the critical regime taking in the x, y-direction a cross-section $S = 1$ m^2. One single fission produces on average an energy $W_f = 200$ MeV, which is sufficiently large that the capture energy can be neglected. The neutron density at the centre of the reactor is 10^7 cm^{-3}.

f. Assume that the absorption $\sigma_a \varrho_a$ is increased by a relative amount 10^{-4} during 0.01 s through a perturbation, and that it afterwards regains its initial value, corresponding to critical operation. What happens to the power of the reactor?

Note. The last question shows how sensitive the reactor is to perturbations and suggests that a reactor might easily stop operating or blow up. Luckily, on time scales larger than 0.01 s there are braking mechanisms in the nuclear reactions which we have neglected in our model and which stabilize the evolution of the neutron density $\varrho(t)$ by preventing it from decreasing or growing exponentially. That makes it possible to manage the reactor efficiently by modifying the absorption through shifting the control bars.

16.18 Electron Gas with a Variable Density

This problem concerns an electron gas the density of which can be made to vary. A structure which enables us to obtain such a gas is the electrical condenser shown in Fig.16.31. It has two electrodes of which one, A, is a perfect conductor (equipotential); the other, B, is a parallelepiped of dimensions

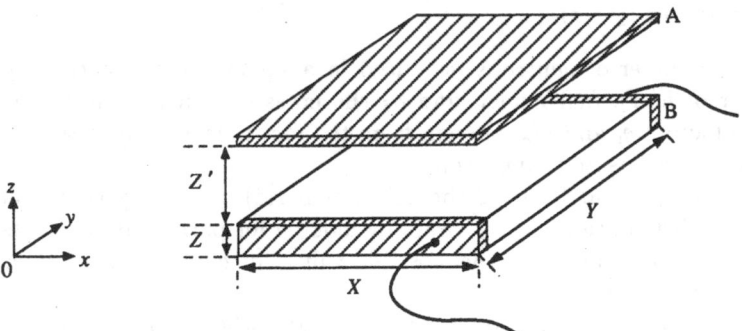

Fig. 16.31. A device allowing control of the electron density in the layer B

X, Y, Z $(X \gg Z, Y \gg Z)$ with its edges parallel to the x-, y-, and z-axes. The material of B does not contain free electrons when there is no charge on the condenser. If one introduces a charge $+Q$ on the electrode A by connecting A and B to a generator, it will be compensated by a charge $-Q$ on the electrode B, which consists of free electrons enclosed in a "box" of dimensions X, Y, Z. As energy zero we take the potential energy inside the box when the condenser is uncharged $(Q = 0)$. We assume that the potential outside the box is much larger than the thermal energy kT. The two electrodes are separated by an insulator of thickness Z'. We assume that the material of B and the insulator have the vacuum dielectric constant ε_0.

This model aims to give some idea about the operation of some of the semiconductor devices mentioned at the end of § 11.3.5. We shall see below that one can control the capacitance by acting on the potential applied between A and B. The so-called MIS (metal-insulator-semiconductor)*variable capacitance diodes* are precisely realized on these principles, by superimposing layers of silicon (electrode B), of SiO_2 (an insulating oxide), and of metal (electrode A). We shall also see that the potential between A and B controls the resistance of B along y, that is, the current flowing between two electrodes placed at $y = 0$ and $y = Y$ (Fig.16.31). In practice, the latter are replaced by n-doped pieces of semiconductor, while the bulk of B is p-doped, and this increases the effect. The so-called MOSFET *field effect transistors* operate along these lines.

We shall use the Poisson equation from electrostatics (§ 11.3.3), which is satisfied by the electrostatic potential ϕ:

$$\nabla^2 \phi = \frac{e}{\varepsilon_0} \varrho,$$

where ϱ is the electron density and ∇^2 the Laplace operator. The corresponding potential energy of an electron is $-e\phi$. If we use the fact that $X \gg Z, Y \gg Z$, we find that the solution of this equation leads to the following relations between the electric field in the insulator and the charge of the condenser:

$$E_z = -\frac{Q}{XY} \cdot \frac{1}{\varepsilon_0}, \qquad E_x = E_y = 0, \tag{1}$$

where E_z is the z-component of the electric field in the insulator and E_x and E_y are the transverse components.

One has the following technological restrictions:

- $Z > 10^{-8}$ m,
- $Z' > 10^{-8}$ m,
- $E_z < 10^8$ V/m, to prevent break-down in the insulator.

The necessary numerical values of the fundamental constants are given at the end of the book. Bear in mind that, if $f(x)$ is a function which has a zero at x_0, the δ-function has the following property:

$$\delta\big(f(x)\big) \;=\; \frac{\delta(x-x_0)}{|f'(x_0)|}.$$

Classical Gas

In the whole of this part we shall assume that the conditions are such that we can treat the electrons in the electrode B as a *classical gas* at a uniform temperature T.

We charge the condenser through an outside generator. We assume that the system is in equilibrium after the generator is disconnected and is characterized by an electrostatic charge Q on the electrode A.

1. Why is the chemical potential μ of the electrons the same in each point of the electrode B, but different from the chemical potential of A?

2. To start with we neglect both the electrostatic interaction with the electrode A and the electrostatic interaction between the electrons in the electrode B. Give the relation which connects the chemical potential μ with the electron density at a point of the electrode B. Give the relation connecting μ with Q.

3. We now introduce the electrostatic interaction with the electrode A. This amounts to introducing *in the material* B an electric field E_z which is uniform and connected with Q through Eq.(1).

   a. Find the electron density $\varrho(z)$ in the electrode B as function of z; calculate the ratio $\varrho(z)/\varrho(0)$.

   b. There exists a charge Q_1 such that when $Q \ll Q_1$ the density can be regarded as approximately independent of z. Find the order of magnitude of Q_1 as function of the thickness Z and the temperature T.

   c. Why does the presence of the electrical field E_z not imply the existence of an electrical current parallel to J_z?

   d. Is it technologically possible to build a device such that the condition $Q \ll Q_1$ is realized at room temperatures for the whole range of values of Q which are practically accessible?

4. We finally include the electrostatic interaction between the electrons in the electrode B. We make the assumption that in order to take it into account it is sufficient to assume that each electron sees the extra electrostatic potential created by a charge density corresponding to the mean electron density in each point. This approximation is justified in Chap.11.

   a. Write down the differential equation which for a given value of μ must be satisfied by the electrostatic potential $\phi(z)$.

   b. Show that the electrical field E_z vanishes in the $z = 0$ plane.

   c. Show that, if $Q \ll Q_1$, the z-dependence of $\varrho(z)$ is weak, as in the preceding question.

d. Evaluate $\phi(z)$ in this limit, considering $\varrho(z)$ to be a constant.

e. Find the functions $\varrho(z)$, $E_z(z)$, and $\phi(z)$ qualitatively in the case when $Q \gg Q_1$. Confirm these results quantitatively by solving the differential equation for ϕ or ϱ.

Capacitance of the Condenser

5. After the external generator has been switched off, its effect has been to produce between the electrodes A and B a difference in chemical potential:

$$\Delta\mu = -ev,$$

where v is the e.m.f. of the generator. The electrode A which is assumed to be a perfect metal is such that one can neglect the change in the chemical potential due to the extra charge $+Q$. The differential capacitance of the condenser is the ratio between a small change dQ in the charge and the corresponding change dv in the e.m.f..

a. Calculate this differential capacitance as a function of the charge Q on the condenser in the limit when $Q \ll Q_1$.

b. Under what conditions is it approximately equal to the capacitance one would obtain, if the electrode B were identical with the electrode A which is a perfect conductor?

c. Calculate for the cases when $Q' = Q_1/20$ and $Q'' = Q_1/10$ the difference between the e.m.f. corresponding, respectively, to Q'' and Q', at room temperature with $Z = 10^{-8}$ m and $Z' = 10^{-6}$ m.

Validity of the Classical Approximation

We shall now discuss the limits of the validity of the classical gas approximation for the electrons.

6. There exists a charge $Q_2(T)$ such that when $Q \gg Q_2(T)$ the electron density ϱ is so large that one should use the Fermi-Dirac statistics for free electrons. Discuss the relative magnitudes of Q_1 and Q_2 for possible values of the thickness Z for $T = 300$ K and for $T = 4$ K. To do that, evaluate $Q_2(T)$, assuming that $Q_2 \ll Q_1$, and discuss whether this assumption is consistent.

7. For small Z we cannot systematically replace the summation over the quantum number m_z, which is associated with the momentum p_z, by an integral.

a. Give an expression for the density of electronic states, $\mathcal{D}(\varepsilon)$, in the layer B as function of the energy for a value of Z of the order of 10^{-8} m.

b. What conditions must we impose on Q and T in order that we can treat the electron gas as a two-dimensional gas?

c. What conditions must we impose on Q and T in order that we can treat the electron gas as a *classical* two-dimensional gas?

Electrical Conduction

By connecting the layer B with two planar metallic contacts in the $y = 0$ and $y = Y$ planes (Fig.16.31), we can measure the electrical transport properties of the electron gas with a variable density. We assume that the electron densities remain those calculated above notwithstanding the perturbations introduced by the presence of the contacts and the current. At any rate we shall assume that the voltages applied to these contacts are always very small.

We assume that the electrons are scattered by uniformly distributed centres which have a density ϱ_{sc}. We shall use the Lorentz model to describe the interactions of the electrons with these centres. The scattering potential is taken as a hard sphere potential with a minimum distance apart equal to δ.

8. To start with we assume that we can use the classical approximation.

a. Give an expression for the conductivity of the electron gas as function of the charge Q, assuming a uniform electron density ($Q \ll Q_1$).

b. Give an expression for the electrical resistance of the layer B, measured between the two contacts, neglecting their own resistance.

9. We now increase the charge Q, still remaining within the framework of the classical approximation, into the $Q \gg Q_1$ domain.

a. Is the relation between the charge on the condenser and the resistance changed?

b. Calculate numerically the smallest resistance which one can reach, taking into consideration technological limitations, at room temperatures for $X = Y$, with $\varrho_{sc} = 10^{21}$ m^{-3} and $\delta = 10^{-8}$ m.

c. In general, we have no *a priori* information about the density and size of the scattering centres. What partial information about the scattering centres can we obtain from this kind of experiment?

10a. What qualitative deviation from the classical approximation can we see in the conduction when we increase the charge beyond the limit Q_2 defined in question 6?

b. The preceding questions assumed tacitly that the sizes X and Y were macroscopic. Technologically one can reduce Y to values of the order of 10^{-7} m. Discuss for what conditions one should question the calculation of the electrical resistance made in question 8.

c. Suggest properties other than the electrical conduction which one might study as function of the charge Q.

11. Assume that the thickness Z of the electrode B is as small as possible and that the charge Q is such that we can treat the electron gas as a classical two-dimensional gas. Treat the scattering centres in a two-dimensional Lorentz model by introducing a surface density ϱ'_{sc} of the centres, each of which is surrounded by a "hard circle" potential of radius δ' which has properties similar to the three-dimensional hard sphere potential.

a. Give for this case an expression for the collision term.

b. Evaluate the electrical resistance of the layer B.

Hints. The transition probability $\widetilde{W}(p, p')$, evaluated as in § 15.1.3, is no longer isotropic. Apply the Chapman-Enskog method and use symmetry considerations to evaluate the collision integral for the conditions when there flows a stationary electric current.

16.19 Snoek Effect and Martensitic Steels

Carbon is a classical "impurity" of iron. Up to concentrations c_0 (c_0 is the ratio of carbon atoms to iron atoms) of the order of 8×10^{-2} the Fe-C alloys are *steels*. Beyond that we enter the domain of *cast iron*.

We use an experiment thought up by Snoek to study the arrangements of small concentrations ($c_0 \lesssim 10^{-3}$) of carbon atoms in steels.

At room temperatures the iron atoms occupy all the sites of a body centred cubic lattice with cube edge 2λ. In such a lattice the vertices and the centres of the cubes are equivalent since a translation by a vector $(\lambda, \lambda, \lambda)$ interchanges centres and vertices. When their concentration is low, the carbon atoms may sit either at the middle of an edge (α) or in the centre of a face (β) of the elementary cube (Fig.16.32). Here also, a translation by the vector $(\lambda, \lambda, \lambda)$ interchanges the rôles of the edges and the faces; each possible site for a carbon atom is midway between two iron atoms at a distance 2λ from one another. We shall therefore distinguish *three families*, (x), (y), and (z) for these carbon sites according to the direction to the nearest iron neighbour: for instance, in Fig.16.32 α is an (x)-site, being between two iron atoms which are situated in the x-direction from α; similarly, β is a (y)-site.

If there are no constraints, these three families obviously play the same rôle. In equilibrium the corresponding concentrations are equal provided c_0 is sufficiently small:

$$c_x = c_y = c_z = \frac{c_0}{3}. \tag{1}$$

We note for what follows that there are 4 nearest neighbour carbon sites at a distance λ from a given carbon site, say, a (z)-site. Of these two are (x)-sites and two (y)-sites.

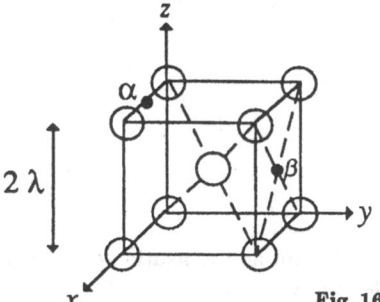

Fig. 16.32. Two possible sites for C in the iron lattice

We also note that a macroscopic volume Ω of the metal contains $\Omega/8\lambda^3$ elementary cubes, $N = 2 \times \Omega/8\lambda^3$ iron atoms, and $6 \times \Omega/8\lambda^3$ carbon sites.

Snoek's Experiment

We shall now try to break the equipartition expressed by (1). To do this we take a single-crystal sample of steel in the form of a prism of volume Ω with edgelengths a, a, and b along the x-, y-, and z-directions, respectively, which is initially without stresses and in equilibrium, $c_x = c_y = c_z = c_0/3$. At time $t = 0$ we submit it to a *uniaxial traction* σ in the z-direction, where σa^2 is the applied force. After that the stress remains constant. We measure the length of the sample in the z-direction as function of time and we denote the strain $\Delta b/b$ by ε.

At time $t = 0$ we observe the classical *elastic* strain

$$\varepsilon_1 = \frac{\Delta b}{b} = \frac{\sigma}{Y}, \tag{2}$$

where Y is the elastic Young modulus of pure iron. We then see an additional strain, the so-called *anelastic* strain: $\varepsilon_2(t)$ which increases with time and approaches a saturation value ε_2^∞ (Fig.16.33).

If we lift after a long time t' the applied stress, two consecutive contractions, $\varepsilon_1' = -\varepsilon_1$ and $\varepsilon_2' = -\varepsilon_2$ reproduce the initial shape of the sample: this so-called *Snoek effect* is thus reversible. Experiments show that:

(i) ε_2 varies exponentially with time. The equilibrium value ε_2^∞ is reached the faster, the higher the temperature T;
(ii) ε_2^∞ is proportional to the carbon concentration c_0;
(iii) ε_2^∞ is proportional to the stress σ and inversely proportional to the temperature T.

Statistical Thermodynamics Treatment

The idea is the following one. By increasing the distance between the iron atoms in the z-direction the stress σ tends to stabilize the carbon atoms

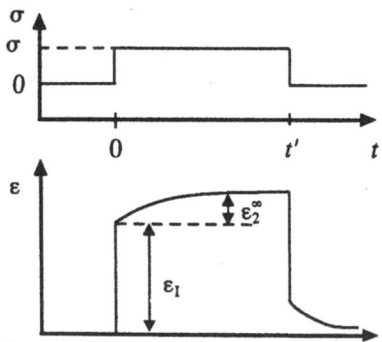

Fig. 16.33. Strain and relaxation of a steel bar under the traction σ

in the (z)-sites at the cost of the (x)- and (y)-sites. In fact, the latter are more strongly squeezed by their iron neighbours so that the energy of a carbon atom on a (z)-site becomes lower, by an amount $2w$, than if it were on an (x)- or (y)-site. This change in energy produces an asymmetry so that the populations c_x, c_y, c_z have no longer any cause to be equal. We shall characterize the deviation relative to the isotropic situation (1) by a parameter

$$\eta = c_z - \frac{c_0}{3}, \tag{3}$$

which is called the *polarization* of the system. In equilibrium, this parameter reaches the value η^∞.

We shall show in what follows (question 5) that the polarization η of the carbon atoms in the z-direction gives rise to a strain ε_2 which must be added to the elastic strain ε_1, defined in (2), and which is proportional to η:

$$\varepsilon_2 = B\,\eta, \tag{4}$$

and especially

$$\varepsilon_2^\infty = B\,\eta^\infty.$$

This will enable us to explain the Snoek effect.

We shall beforehand calculate the value η^∞ of η in equilibrium at a temperature T and for a given value of the total strain ε. To do this we evaluate the free energy $F(\varepsilon, \eta, T)$ of the *quasi-equilibrium* states of the sample in various stages of the variation of the polarization effect. These states are characterized not only by the temperature but also by giving the total strain $\varepsilon = \varepsilon_1 + \varepsilon_2$ and the polarization η, considered to be independent variables. The equilibrium polarization η^∞ is a function of ε which will be determined by expressing that in equilibrium the two systems consisting of the (z)-sites and of the (x)- and (y)-sites can, for fixed ε, exchange carbon atoms. We assume that ε_2 *can be neglected compared to* ε_1, a condition which will be satisfied, provided c_0 is sufficiently small.

We assume that the only important changes in the free energy come from the following contributions:

(i) *the elastic free energy*, which is associated with the interactions between the iron atoms and has the same expression as for pure iron:

$$\tfrac{1}{2}\Omega Y \varepsilon^2;$$

(ii) *the decrease by $2w$ in the internal energy*, which is produced by each jump (x) or $(y) \mapsto (z)$, when the carbon atom in (z) finds a more favourable position than if it were in (x) or (y). We take the energy $2w$ to be proportional to the deformation ε, which is practically constant and equal to ε_1 during the polarization since $\varepsilon_2 \ll \varepsilon_1$:

$$2w \ = \ A \ \varepsilon; \qquad\qquad\qquad (5)$$

(iii) *the change in the configuration entropy* S_m of the carbon atoms.

We denote by N the number of iron atoms in the sample, of volume $a^2 b$, and therefore by $N c_0$ the number of carbon atoms, and by ϱ the number of iron atoms per unit volume.

1. Determine the configuration entropy S_m corresponding to the concentrations c_x, c_y, and c_z which we assume to be arbitrary to begin with. Find the expression for the value S_m^0 of S_m for the case when there is equilibrium and there are no stresses ($c_x = c_y = c_z = c_0/3$).

2. What happens to this entropy when the carbon atoms become polarized? We assume here that σ is sufficiently small that we have $\eta \ll c_0$ and we remember that $c_0 \ll 1$. Calculate $S_m - S_m^0$ as function of η up to second order in η.

3. Use the hypotheses (ii) and (iii) mentioned earlier to determine the value of η^∞ at equilibrium as a function of ε. Express this quantity as a function of the parameters σ, c_0, and T which can be controlled experimentally.

4. Starting from the expression for $F(\varepsilon, \eta, T)$, which results from the hypotheses (i), (ii), and (iii) given earlier, give a relation between the stress σ and the total strain ε for a given polarization η.

5. Use this relation to prove that Eq.(4) is the consequence of these hypotheses and that the parameter B can be expressed as

$$B \ = \ \frac{A \varrho}{Y}$$

in terms of the quantities A and Y, which are defined through Eqs.(5) and (2).

6. Give an expression for the anelastic strain at equilibrium. Which of the experimental results (i) to (iii) given earlier can be explained by this model?

7. Evaluate w numerically in eV, taking $\sigma = 1000$ bar ($= 100$ MPa). Take for B the experimentally found value 0.8. The atomic mass of iron is 56 and its density 7.8 g cm^{-3}. Evaluate the anelastic strain ε_2^∞ at room temperature for a steel with a concentration $c_0 = 10^{-3}$.

Kinetic Treatment

We now want to study the way the directional ordering is established in time, that is, the evolution of the polarization η.

This needs carbon atoms to jump from (x)- or (y)-sites to (z)-sites. We assume that the carbon atoms jump only from one site to a nearest neighbour site. Each jump occurs along a diffusion path of length λ which joins two nearest neighbour sites of different kinds (Fig.16.34 shows an $(x) \mid \to \ (z)$ jump). We also assume that in the jump the carbon atom passes, in the $E(r)$ space, over an energy barrier at the middle of the two, departure and arrival, sites. Let ΔE be the height of this barrier and let ν be the characteristic frequency of vibration for a carbon atom on each of the sites.

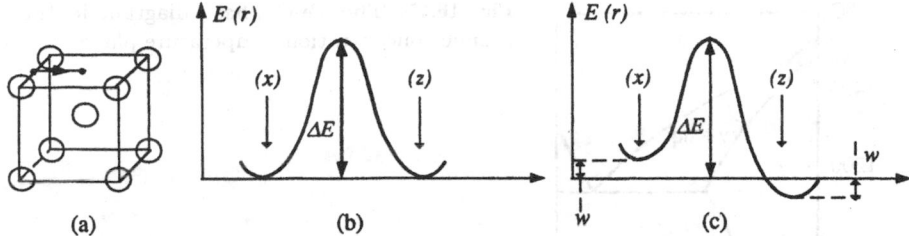

Fig. 16.34. (a) Jump of a carbon atom from an (x)- to a neighbouring (z)-site; (b) variation of the potential energy $E(r)$ along that path for $\varepsilon = 0$; (c) the same for $\varepsilon > 0$

A carbon atom jump from a *given* departure site to a *given* (nearest neighbour) arrival site is a random process characterized by a probability per unit time for it to occur that will depend on the temperature T and the height of the barrier which must be passed over. One can show that this probability per unit time equals $\nu \exp[-(E_M - E_0)/kT]$, where E_M is the energy at the top of the barrier and E_0 the energy at the departure site.

Take into account that the concentration c_0 is very small.

8. What is the mean number of jumps $\nu_{\rm j}$ of a given carbon atom per unit time? What is the mean time $\tau_{\rm j}$ which a carbon atom will spend at a given site at room temperature? Use the experimental values $\nu = 1.5 \times 10^{14}$ s^{-1} and $\Delta E = 0.87$ eV.

9. The stress in Snoek's experiment changes the energy profile of Fig. 16.34b to that of Fig. 16.34c: the (z)-sites have under this constraint an energy which is lower by $2w$ than that of the (x)- or (y)-sites (see Eq. (5) above). On the other hand, the energy of the top of the barrier, E_M, is not changed. We assume that the stress σ is sufficiently small that w will be small compared to both ΔE and kT. We neglect the change in w during the experiment, since $\varepsilon_2 \ll \varepsilon_1$. The carbon atoms on a (z)-site with a concentration c_z can jump to (x)- and to (y)-sites. Conversely, (x) and (y) atoms can jump to (z)-sites.

Write down a differential equation describing the balance for the change dc_z in c_z at time t during an interval dt much larger than $\tau_{\rm j}$.

10. Use that equation to find, starting from time $t = 0$, the general form of the functions $c_z(t)$ and $\varepsilon_2(t)$ in Fig. 16.33, to first order in w/kT, taking as the initial conditions $c_x = c_y = c_z = c_0/3$. Find especially the quantities c_z^{∞} and ε_2^{∞}, checking that the latter is the same as that found in 6, and also the characteristic time τ for the evolution of the anelastic strain. How does this time change with temperature?

Which of the experimental results (i) to (iii) of the introduction can be explained by this theory?

Use the numerical values calculated earlier to give the order of magnitude of τ. Check that the assumptions that $w \ll \Delta E$ and $w \ll kT$ were justified.

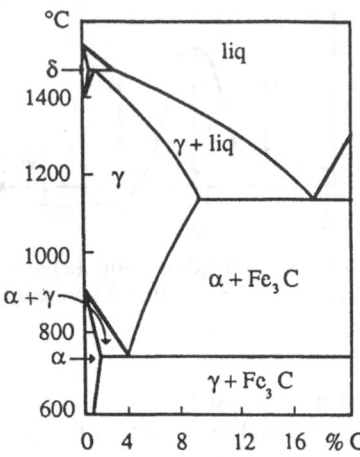

°C

δ
1400

1200 γ

1000

α + γ
800

α
600

0 4   8   12   16 % C

liq

γ + liq

α + Fe₃ C

γ + Fe₃ C

Fig. 16.35. The Fe-C phase diagram in the atomic concentration-temperature plane

11. The numerical values of the frequency ν and of the energy ΔE were given so far without justification (question 8). What experiments would you devise to measure them?

Martensitic Steels

The equilibrium Fe-C diagram is given in Fig.16.35. The α-phase, or *ferrite*, is the body-centred cubic phase studied above. The γ-phase, or *austenite*, is a face-centred cubic phase into which the carbon is more soluble than into the α-phase. The Fe_3C phase is called *cementite*.

Let us take a steel with $c_0 = 5 \times 10^{-2}$ at $1100°C$ and let us cool it down suddenly (temper or quench it) to room temperature.

12. What is the equilibrium structure of this steel at $1100°C$? What would normally be its equilibrium structure at room temperature or at $600°C$?

13. Experiments show that, in fact, neither the one nor the other of these two structures is the one seen after tempering. The tempered steel consists of a pseudo-ferrite similar to the phase studied earlier, but one in which locally *all* carbon atoms occupy *one* kind of site, say, (z)-sites. This structure is called *martensite* and is not in thermodynamic equilibrium: it is metastable. The formation of martensite, that is, the *total* polarization of the carbon atoms into (z) positions, is a dynamic process due to the tempering and which we shall not describe here.

What is the value of the polarization? What is, in your opinion, the shape of the elementary cell of martensite? What are its dimensions as compared to the edgelength 2λ of the ferrite cube of Fig.16.32? Can one use X-ray experimental measurements of these dimensions to find the value of the coefficient B in Eq.(4)?

14. Although metastable, martensite can exist a long time: one often finds martensite in prehistoric tools.

Can you explain qualitatively what leads to this longevity of martensite?

15. The polarization axis of martensite can be just as well in the x- or the y-direction as in the z-direction. Going from one region to another of a piece of tempered steel one finds small domains, of a few hundred Å to a few micron, of (x)- or (y)- or (z)-type martensite.

Use the answers to 13 to say something about the state of internal constraint in this piece. Can you find a connection between your answer and the great hardness of tempered steel? Do you understand why the hardness of such steels increases with the carbon concentration? We assume here without any further justification that the deformability of a crystal is connected with the facility with which certain crystal defects – dislocations – can move.

16. Remembering what we have seen so far how would you set out to decrease the hardness of a piece of martensitic steel?

In an actual examination students were given two nails which looked identical, but one of which was of tempered steel and the other had been annealed, that is, reheated during an hour at 500°C. They were then asked the following question: "How do you think the most ductile, that is, the most easily deformed, of the two enclosed nails has been treated?"

Solutions to the Last Three Questions:

14. The strain of the (z) polarization axis calculated in 13, which equals $2Bc_0/3 \simeq 3 \times 10^{-2}$, is much larger, by a factor of order 1000, than the anelastic strain calculated in 7. This strong distortion implies a large degree of stabilization of the carbon atoms in the (z) position thanks to the increase in w. In 7 we found a value $w \simeq 3 \times 10^{-3}$ eV for an elongation of 4×10^{-5}; the proportionality assumed in (5) would now give for w a value of a few eV, but that is incompatible with the value of ΔE and with the shape of Fig.16.34. It is therefore reasonable to estimate that w saturates for large deformations to a value of the order of ΔE, say, 0.5 eV. Under those conditions the characteristic time for $(z) \mapsto (x)$ or (y) jumps, which are necessary for a return to equilibrium, is increased considerably, as it contains a factor $e^{w/kT}$. For $w \simeq 0.5$ eV and at room temperatures it will be increased by $e^{20} \simeq 5 \times 10^8$ and changes from the value of 2.2 s, obtained in 8, to 10^9 s $\simeq 30$ years. For $w \simeq 0.6$ eV we would get 2000 years. The carbon atoms are thus frozen in for a very long period in the metastable (z) positions.

15. The (x), (y), and (z) domains exert considerable stresses upon one another, the strain of one of them being opposed by the different orientation of the polarization in the next ones. Interactions of an elastic nature between these internal stresses and the crystal defects which are responsible for deformations inhibit the mobility of these defects and the hardness is increased. The strength of the internal stresses, and thus the hardening effect, increases with the polarization of the domains which itself is proportional to c_0. The hardness of tempered steel therefore increases with carbon content.

16. To accelerate the (z) martensite $\mapsto$ ferrite transition we must make the $(z) \mapsto (x)$ or (y) jumps more favourable. It is sufficient for this to heat the sample, as this reduces the characteristic time through the factor $\exp[(\Delta E + w)/kT]$ which changes considerably. The hard nail is made of tempered martensitic steel; the other

one, which is easily bent, has been heated, at less than 700°C, so that it reached the equilibrium phase of ferrite + cementite which involves practically no internal constraints.

"Ni si haut, ni si bas! simple enfant de la terre,
Mon sort est un problème, et ma fin un mystère."

Lamartine, Méditations

Conclusion:
The Impact of Statistical Physics

Applications of Statistical Physics

The different examples which we have treated in the present book have shown the rôle played by statistical mechanics in *other branches of physics*: as a theory of matter in its various forms it serves as a bridge between macroscopic physics, which is more descriptive and experimental, and microscopic physics, which aims at finding the principles on which our understanding of Nature is based. This bridge can be crossed fruitfully in both directions, to explain or predict macroscopic properties from our knowledge of microphysics, and inversely to consolidate the microscopic principles by exploring their more or less distant macroscopic consequences which may be checked experimentally. The use of statistical methods for deductive purposes becomes unavoidable as soon as the systems or the effects studied are no longer elementary.

Statistical physics is a tool of common use in the *other natural sciences*. Astrophysics resorts to it for describing the various forms of matter existing in the Universe where densities are either too high or too low, and temperatures too high, to enable us to carry out the appropriate laboratory studies. Chemical kinetics as well as chemical equilibria depend in an essential way on it. We have also seen that this makes it possible to calculate the mechanical properties of fluids or of solids; these properties are postulated in the mechanics of continuous media. Even biophysics has recourse to it when it treats general properties or poorly organized systems.

If we turn to the domain of *practical applications* it is clear that the *technology of materials*, a science aiming to develop substances which have definite mechanical (metallurgy, plastics, lubricants), chemical (corrosion), thermal (conduction or insulation), electrical (resistance, superconductivity), or optical (display by liquid crystals) properties, has a great interest in statistical physics. For instance, ferromagnetism, a phenomenon which is of great importance as it has enabled the development of electrical engineering (alternators, transformers, motors), is the result of an order-disorder phase transition; we have sketched its theory in Chap.9. Here, observations and practical applications preceded understanding. However, we have shown in Chap.11 how the theory of another kind of substances, the semiconductors, has led to the construction of devices such as diodes or transistors and everybody knows their innumerable applications (micro-electronics, computers, energetics).

Through thermodynamics, which has become its extension, but also directly, statistical physics contributes to the development of various *energy techniques*, whether the extraction, transformation, transport, storage, saving, or efficient use of thermal, electrical, radiative, chemical, mechanical, or nuclear energy. In passing we have mentioned some applications to the quest for new energy sources, such as fusion or solar energy.

Methods and Concepts of Statistical Physics

Statistical physics has also played a major rôle in the realm of ideas. Like relativity and quantum mechanics, and maybe even more deeply, it has changed our world picture through its philosophical impact and its epistemological implications. As we have seen several times it has accustomed us, even outside the framework of quantum theory, to incorporate in science a partially subjective, or at least a *relative*, aspect inherent in the probability concept. We see now scientific progress as a devolution of more and more probable, more and more faithful, images of reality, but the belief in an absolute scientific Truth, which was so widespread amongst scientists and philosophers in the nineteenth century, can hardly be found nowadays: more or less great likelihood, yes, but no final certainty.

Also noteworthy are the contributions of statistical physics to the development of methods and concepts. It was one of the first disciplines where the *statistical method*, that is the art of making predictions when one has little knowledge, was worked out. A macroscopic substance is a typical example of a system which is too complicated to be described in all detail. The properties in which we are interested concern only a few degrees of freedom, or are average effects; the statistical treatment allows us to discard those aspects that are too complicated, not well known, or of no interest. This fruitful approach is met with in all applications of statistics. We have also illustrated another remarkable possibility opened up by statistics, to wit, the possibility to make nearly certain predictions even though the system studied has a random nature, provided it is large.

Another method, examples of which can be found in Chaps.7 or 15, dealing with transport, and 13, dealing with Kirchhoff's laws, consists in making up a *detailed balance* of exchanges which can take place between one part of the system and another. Conservation laws, such as those for energy, momentum, or particle numbers, play an important rôle in establishing this balance and this enables us to reduce the problem. These ideas can be transposed to economic sciences. We also saw in Chap.14 how fruitful is the systematic exploitation of *symmetry* and invariance properties.

Statistical physics is a field where the use of *models* is a nearly permanent feature. The ideal method, which would consist of starting from a known Hamiltonian to use a Boltzmann-Gibbs distribution for deriving the properties of a system, often turns out to be intractable. Therefore, as we did on many occasions in this book, one does not start from first principles, but one uses a model, that is, a simple, idealized structure, which is meant to depict consistently a more complicated reality. It even happens that one introduces contradictory models to describe different aspects of the same system. One knows that the social sciences, such as economics and sociology, make great use of models. Statistical physics can help us understand the significance of their use. In fact, in fields where theory is not so well developed, one is prepared to justify a model by its operational fertility and by more or less intuitive similarity arguments. In statistical physics, we can go further, as we saw at the end of Chap.1, by constructing a series of more and more refined models which are based partly upon more and more precise experiments, and partly upon known "first principles" from which approximations enable us to construct models. These nested models are successive steps in the capture of Nature.

Several *concepts* introduced in statistical physics have already passed into everyday usage. It suffices to mention *energy*, its transformations and the consequences of its conservation: the limitation of our energy resources, on the one hand, and the inevitability of "thermal pollution" if one uses energy, on the other hand, have become commonplace.

We have studied the relations between the concepts of *entropy, information*, and *disorder*. Energy degradation, and irreversibility, are manifestations of the general tendency towards the loss of order. We also saw that when we increase our knowledge about a system, we put order into it. These kinds of considerations by and large go beyond the framework of physics. Especially, information theory, the birth of which was partly due to statistical physics, enables us to improve the efficiency of transmission methods by using a suitable treatment of messages which avoids their being tangled up by noise. From this point of view, a sizeable part of biology appears as the study of mechanisms through which living organisms create order by using the information contained in their genetic code. Is even intellectual activity nothing but a struggle against chaos?

A set of remarkable effects, which may also be found outside statistical physics, is connected with the *establishing of order* in a system. We have been able to appreciate the *diversity* of the forms that order can take, from magnetism to the superfluidity of helium. The huge variety of crystal structures provide us with a spectacular example of this wealth of types of ordering. We saw also (§§ 11.1.2 and 12.2.3) that different forms of order may *compete* and that depending on the situation, one or other of them dominates. The *a priori* most surprising phenomenon is the sudden appearance of order: we have encountered several examples of phase transitions where a *qualitative* change is suddenly produced when the temperature is gradually lowered. Statistical physics enables us to understand the origin of such discontinuities which occur only thanks to the large size of the systems studied.

Finally, statistical physics clearly illustrates the concept of *hierarchical structures*, which is fashionable in human sciences. Depending on the scale in which we are interested, the laws of physics differ: a substance is on the microscopic scale a set of electrons and nuclei interacting through Coulomb forces; on the macroscopic scale, it is a gas or a solid with an order which we could not expect from the microscopic laws. However, the physicist believes in the existence of universal fundamental laws, or at least in the possibility of gradually unifying the various pieces of information available at a given time. Statistical physics helps us to resolve this contradiction between the unity of Nature and the change of her laws depending on the scale, by enabling us to derive a large number of macroscopic laws from the underlying microscopic laws. On the level of principles, the latter, which are simpler, appear to be more fundamental. On the practical level, the fact that this derivation is possible in relatively simple cases, all the same, does not transform macroscopic physics or chemistry into disciplines with the logical and consistent structure of mathematics: the constituent elements of reality are simple, but combining them leads to enormously rich complex phenomena.

Deduction in Science

This remark leads us to finish with a warning. Statistical mechanics stresses the deductive aspect of science, an aspect which is favoured in French teaching by the dominance of mathematics. In this book itself, at least starting from Chap.2, for the sake of pedagogical efficiency we have adopted a deductive approach. This should not lead one to develop erroneous ideas about the practice of science or to believe that the deductive method, of which statistical mechanics is one of the most beautiful examples, lies at the roots of our knowledge.

Certainly, most scientists adhere to *reductionism*, at least in their daily practice. According to this doctrine, sciences can be classified in a hierarchy with levels corresponding to increasing degrees of complexity of the objects studied. The deepest foundations are at the level of elementary particle physics. Atomic physics is based on it, and in its turn it is directly the basis of chemistry, and indirectly of the physics of materials, thanks to the techniques of statistical mechanics. The ladder extends from chemistry to molecular biology, then to cellular biology on which the study of multi-cellular organisms is based, and one may imagine that one day it will end up with psychology and sociology. According to reductionism, each level is completely governed by the lower level, without needing any new hypotheses, and its laws are, as a matter of principle, merely the consequences of the laws of the underlying level.

This idea of science, however, does not imply that deduction plays the dominant rôle in it. Most scientists think that biology can be reduced in successive steps to microscopic physics, which has as its fundamental laws quantum mechanics and electromagnetic interactions between electrons and nuclei. However, one would be completely wrong in believing that, conversely, one might some day solve biological problems, starting from these laws. Even though deduction is relatively easy for a composite system, consisting of a small number of elements, and even though, as we have seen, it is, thanks to statistical methods, still possible for materials which are rather poorly organized, usually it is not feasible. Using models often enables us to circumvent this difficulty, but at the risk of losing view of the reality which they have to take into account.

The relation between science and deduction has thus a paradoxical nature. Reductionism aims at constructing a consistent framework which, *in principle*, is deductive and based upon few fundamental laws. This attitude is fruitful, and everything so far leads us to believe that it enables us to acquire an increasingly adequate knowledge of reality. However, the approach itself which is followed to reduce one branch of science to another goes in the *opposite direction* to deduction: only rarely does one discover new effects when one attempts to construct them, starting from simple fundamental laws. On the contrary, most often one starts from a new effect, discovered at a complicated level through empirical kinds of research, to find elementary laws. For instance, phase transitions have been known a long time; however, it needed the whole arsenal of statistical mechanics to construct a theory accounting for them, and that, only many years after the laws of microphysics had been well established. One could never, starting from those laws, have imagined their existence, if one had not observed them previously at the macroscopic level. More generally, even though one knows that sciences such as thermodynamics or mechanics are based on microphysics, they are not in any way endangered – quite the contrary – and new effects continue to be discovered in those fields. At a different level, the duplication of DNA is a purely chemical effect; one would never have understood it without the reductionistic urge to look for its explanation merely in the structure of the molecules involved, but also one would never have invented it, just from knowing the laws of chemistry. In the progress of discovery, science thus goes from the complex to the elementary, checking *a posteriori* that the new observations are compatible with the known, or as yet unknown, simple fundamental laws.

Why is it so difficult to find the properties of complex systems, starting from the laws governing the behaviour of their elementary constituents? The example of phase transitions is significant. From a microscopic point of view, a phase transition, such as solidification, is a rather extravagant effect: nothing at the atomic scale indicates that there may appear a discontinuity in the macroscopic properties; moreover, the translational and rotational invariance of the microscopic laws is well reflected in the high-temperature, liquid, phase, but not in the low-temperature, crystalline, phase. Statistical mechanics explains these paradoxical differences in behaviour by appealing to approximation methods which violate certain properties of the microscopic laws, but which are eventually found to be justified by the enormous difference in scale between the substance and the atoms which are its constituents. Thus, once a certain degree of complexity is reached, a system consisting of a large number of constituents can behave in a manner which is radically different from that of its elements. There then appear *qualitatively new*, unpredictable properties; one of the essentials tasks of science is to uncover and to explain those. Altogether, the existence of a hierarchy between the sciences does not imply that it suffices to apply the most "fundamental" ones for building the others. Each level calls for a conceptual structure, laws, and even a methodology, all of its own. Unlike the physical sciences, biology or computer science use the so-called system approach which right from the start stresses the function, rather than the structure of objects.

It behooves us therefore to put the deductive method in its right place. A living science is characterized by a continuous interplay between fundamental laws or hypotheses and properties, considered to be their consequences. This two-way movement, using deductions every time – unfortunately rare at the start – when this is possible, is an essential source of the progress and the gradual unification of the sciences. In the last stages when a discipline has reached maturity and especially when one tries to apply it, deductive methods, like those of statistical physics, gain the upper hand. The steam engine preceded thermodynamics, but later thermodynamics played a significant rôle in improving it. Similarly, the extraordinary development of the techniques of electronic components rests upon an understanding in depth of the properties of substances, which is based upon a study of their microscopic structure and which cannot be attained from purely empirical research. Does this justify the strong accent we have placed on deduction?

Subject Index

Page numbers in upright type refer to Volume I, in slanted type to Volume II.

Units and Physical Constants

We use the international system of units, the so-called SI system, which is adopted by most official international organizations. Its fundamental units are the metre (m), the kilogram (kg), the second (s), the ampere (A), the kelvin (K), the mole (mol), and the candela (cd).

Derived SI units with special names are the radian (rad), the steradian (sr), the hertz (Hz $= s^{-1}$), the newton (N $=$ m kg s^{-2}), the pascal (Pa $=$ N m^{-2}), the joule (J $=$ N m), the watt (W $=$ J s^{-1}), the coulomb (C $=$ A s), the volt (V $=$ W A^{-1}), the farad (F $=$ C V^{-1}), the ohm ($\Omega =$ V A^{-1}), the siemens (S $=$ A V^{-1}), the weber (Wb $=$ V s), the tesla (T $=$ Wb m^{-2}), the henry (H $=$ Wb A^{-1}), the Celsius temperature ($^\circ$C), the lumen (lm $=$ cd sr), the lux (lx $=$ lm m^{-2}), the becquerel (Bq $= s^{-1}$), the gray (Gy $=$ J kg^{-1}), and the sievert (Sv $=$ J kg^{-1}).

Prefixes used with SI units to indicate powers of 10 as factors are: deca (da $= 10$); hecto (h $= 10^2$); kilo (k $= 10^3$); mega (M $= 10^6$); giga (G $= 10^9$); tera (T $= 10^{12}$); peta (P $= 10^{15}$); exa (E $= 10^{18}$); deci (d $= 10^{-1}$); centi (c $= 10^{-2}$); milli (m $= 10^{-3}$); micro ($\mu = 10^{-6}$); nano (n $= 10^{-9}$); pico (p $= 10^{-12}$); femto (f $= 10^{-15}$); atto (a $= 10^{-18}$).

Constants for electromagnetic units
$$\mu_0 = 4\pi \times 10^{-7}\,\mathrm{N\,A^{-2}} \text{ (definition of the ampere)}$$
$$\varepsilon_0 = \frac{1}{\mu_0 c^2}, \quad \frac{1}{4\pi\varepsilon_0} \simeq 9 \times 10^9\,\mathrm{N\ m^2\,C^{-2}}$$

velocity of light
$$c = 299\,792\,458\,\mathrm{m\ s^{-1}} \text{ (definition of the metre)}$$
$$c \simeq 3 \times 10^8\,\mathrm{m\,s^{-1}}$$

Planck's constant
$$h = 6.6260755 \times 10^{-34}\,\mathrm{J\,s}$$

Dirac's constant
$$\hbar = \frac{h}{2\pi} \simeq 1.055 \times 10^{-34}\,\mathrm{J\,s}$$

Avogadro's number
$$N_A \simeq 6.022 \times 10^{23}\,\mathrm{mol^{-1}} \text{ (by definition the mass of one mole of }^{12}\mathrm{C} \text{ is 12 g)}$$

(Unified) atomic mass unit
$$1\,\mathrm{u} = 1\,\mathrm{g}/N_A \simeq 1.66 \times 10^{-27}\,\mathrm{kg} \text{ (or dalton or amu)}$$

neutron and proton masses $m_\mathrm{n} \simeq 1.0014\,m_\mathrm{p} \simeq 1.008\,\mathrm{u}$

electron mass $m \simeq 1\,\mathrm{u}/1823 \simeq 9.11 \times 10^{-31}\,\mathrm{kg}$

Elementary charge $e \simeq 1.602 \times 10^{-19}\,\mathrm{C}$

Faraday's constant $N_A e \simeq 96\,485\,\mathrm{C\ mol^{-1}}$

Bohr magneton
$$\mu_\mathrm{B} = \frac{e\hbar}{2m} \simeq 9.27 \times 10^{-24}\,\mathrm{J\ T^{-1}}$$

nuclear magneton
$$\frac{e\hbar}{2m_\mathrm{p}} \simeq 5 \times 10^{-27}\,\mathrm{J\ T^{-1}}$$

Fine structure constant
$$\alpha = \frac{e^2}{4\pi\varepsilon_0 \hbar c} \simeq \frac{1}{137}$$

Hydrogen atom:

Bohr radius
$$a_0 = \frac{\hbar}{mc\alpha} = \frac{4\pi\varepsilon_0 \hbar^2}{me^2} \simeq 0.53\,\text{\AA}$$

binding energy
$$E_0 = \frac{\hbar^2}{2ma_0^2} = \frac{m}{2\hbar^2}\left(\frac{e^2}{4\pi\varepsilon_0}\right)^2 \simeq 13.6\,\mathrm{eV}$$

Rydberg constant
$$R_\infty = \frac{E_0}{hc} \simeq 109\,737\,\mathrm{cm^{-1}}$$

Boltzmann's constant	$k \simeq 1.381 \times 10^{-23}$ J K^{-1}
molar gas constant	$R = N_A k \simeq 8.32$ J K^{-1} mol^{-1}
Normal conditions: pressure	1 atm = 760 Torr = 1.01325×10^5 Pa
temperature	Triple point of water 273.16 K (definition of the kelvin)
	or 0.01°C (definition of the Celsius scale)
molar volume	22.4×10^{-3} m^3 mol^{-1}

Gravitational constant	$G \simeq 6.67 \times 10^{-11}$ m^3 kg^{-1} s^{-2}
gravitational acceleration	$g \simeq 9.81$ m s^{-2}

Stefan's constant	$\sigma = \dfrac{\pi^2 k^4}{60 \hbar^3 c^2} \simeq 5.67 \times 10^{-8}$ W m^{-2} K^{-4}
Definition of photometric units	A 1 W luminous power, emitted at a frequency of 540 THz, is equivalent to 683 lm

Energy units and equivalents	1 erg = 10^{-7} J (non SI)
	1 kWh = 3.6×10^6 J
electric potential	1 eV $\leftrightarrow$ 1.602×10^{-19} J $\leftrightarrow$ 11600 K
heat	1 cal = 4.184 J (non SI; specific heat of 1 g of water)
chemical binding	23 kcal mol^{-1} $\leftrightarrow$ 1 eV (non SI)
temperature (kT)	290 K $\leftrightarrow$ $\frac{1}{40}$ eV (room temperature)
mass (mc^2)	9.11×10^{-31} kg $\leftrightarrow$ 0.511 MeV (electron rest mass)
wavenumber (hc/λ)	109 700 cm^{-1} $\leftrightarrow$ 13.6 eV (Rydberg)
frequency $(h\nu)$	3.3×10^{15} Hz $\leftrightarrow$ 13.6 eV

It is useful to keep these equivalents handy for quickly finding orders of magnitude.

Various non SI units	1 angstrom (Å) = 10^{-10} m (atomic scale)
	1 fermi (fm) = 10^{-15} m (nuclear scale)
	1 barn (b) = 10^{-28} m^2
	1 bar = 10^5 Pa
	1 gauss (G) = 10^{-4} T
	1 nautical mile = 1852 m
	1 knot = 1 nautical mile per hour = 0.51 m s^{-1}
	1 astronomical unit (AU) $\simeq 1.5 \times 10^{11}$ m (Sun-Earth distance)
	1 parsec (pc) $\simeq 3.1 \times 10^{16}$ m (1 AU/arc sec)
	1 light year (ly) $\simeq 0.95 \times 10^{16}$ m

Solar data	Radius 7×10^8 m = 109 Earth radii
	Mass 2×10^{30} kg
	Average density 1.4 g cm^{-3}
	Luminosity 3.8×10^{26} W

A Few Useful Formulae

Normalization of a Gaussian function:

$$\int_{-\infty}^{+\infty} dx \, e^{-ax^2} = \sqrt{\frac{\pi}{a}};$$

differentiation of this formula with respect to a gives us the moments of the Gaussian distribution.

Euler's gamma-function:

$$\Gamma(t) \equiv \int_0^\infty x^{t-1} e^{-x} \, dx = (t-1)\Gamma(t-1),$$

$$\Gamma(t)\Gamma(1-t) = \frac{\pi}{\sin \pi t}, \qquad \Gamma(\tfrac{1}{2}) = \sqrt{\pi}.$$

Stirling's formula:

$$t! = \Gamma(t+1) \underset{t\to\infty}{\sim} t^t e^{-t} \sqrt{2\pi t}.$$

Binomial series:

$$(1+x)^t = \sum_{n=0}^{\infty} \frac{x^n}{n!} \frac{\Gamma(t+1)}{\Gamma(t+1-n)} = \sum_{n=0}^{\infty} \frac{(-x)^n}{n!} \frac{\Gamma(n-t)}{\Gamma(-t)}, \qquad |x| < 1.$$

Poisson's formula:

$$\sum_{n=-\infty}^{+\infty} f(n) = \sum_{l=-\infty}^{+\infty} \tilde{f}(2\pi l) \equiv \sum_{l=-\infty}^{+\infty} \int_{-\infty}^{+\infty} dx \, f(x) \, e^{2\pi i l x}.$$

Euler-Maclaurin formula:

$$\frac{1}{\varepsilon} \int_a^{a+\varepsilon} dx \, f(x) \approx \frac{1}{2}[f(a) + f(a+\varepsilon)] - \frac{\varepsilon}{12} f'(x)\Big|_a^{a+\varepsilon} + \frac{\varepsilon^3}{720} f'''(x)\Big|_a^{a+\varepsilon} + \dots$$

$$\approx f\left(a + \tfrac{1}{2}\varepsilon\right) + \frac{\varepsilon}{24} f'(x)\Big|_a^{a+\varepsilon} - \frac{7\varepsilon^3}{5760} f'''(x)\Big|_a^{a+\varepsilon} + \dots ;$$

this formula enables us to calculate the difference between an integral and a sum over n, when we put $a = n\varepsilon$.

Constants:

$$e \simeq 2.718, \qquad \pi \simeq 3.1416,$$

$$\gamma \equiv \lim \left(1 + \cdots + \tfrac{1}{n} - \ln n\right) \simeq 0.577 \qquad \text{Euler's constant.}$$

Riemann's zeta-function:

$$\zeta(t) \equiv \sum_{n=1}^{\infty} \frac{1}{n^t}, \qquad \int_0^\infty \frac{x^{t-1}\, dx}{e^x - 1} = \Gamma(t)\,\zeta(t),$$

$$\int_0^\infty \frac{x^{t-1}\, dx}{e^x + 1} = \left(1 - 2^{-t+1}\right) \Gamma(t)\,\zeta(t).$$

t	1.5	2	2.5	3	3.5	4	5
ζ	2.612	$\frac{1}{6}\pi^2$	1.341	1.202	1.127	$\frac{1}{90}\pi^4$	1.037

Dirac's δ-function:

$$\frac{1}{2\pi} \int_{-\infty}^{+\infty} dx\, e^{ixy/a} = \delta\left(\frac{y}{a}\right) = |a|\delta(y);$$

$$\frac{1}{|a|} \sum_{l=-\infty}^{+\infty} e^{2\pi i l y/a} = \delta(y \bmod a) \equiv \sum_{n=-\infty}^{+\infty} \delta(y - na);$$

$$\lim_{t\to\infty} \frac{\sin tx}{x} = \lim_{t\to\infty} \frac{1 - \cos tx}{tx^2} = \pi\,\delta(x);$$

$$f(x)\,\delta(x) = f(0)\,\delta(x), \qquad f(x)\,\delta'(x) = -f'(0)\,\delta(x) + f(0)\,\delta'(x).$$

If $f(x) = 0$ in the points $x = x_i$, we have

$$\delta[f(x)] = \sum_i \frac{1}{|f'(x_i)|}\,\delta(x - x_i).$$

9 783540 454786